Risto S. Laitinen, Raija Oilunkaniemi (Eds.)

Selenium and Tellurium Reagents

Also of interest

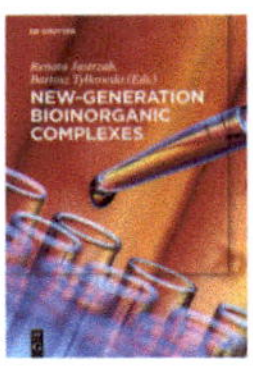

New-Generation Bioinorganic Complexes
Jastrzab, Tylkowski (Eds), 2016
ISBN 978-3-11-034880-4, e-ISBN 978-3-11-034890-3

Organophosphorus Chemistry
Novel Developments
Keglevich (Ed.), 2018
ISBN 978-3-11-053453-5, e-ISBN 978-3-11-053583-9

Iron-Sulfur Clusters in Chemistry and Biology
Volume 1: Characterization, Properties and Applications
Rouault (Ed.), 2017
ISBN 978-3-11-047850-1, e-ISBN 978-3-11-048043-6

Iron-Sulfur Clusters in Chemistry and Biology
Volume 2: Biochemistry, Biosynthesis and Human Diseases
Rouault (Ed.), 2017
ISBN 978-3-11-047939-3, e-ISBN 978-3-11-047985-0

Physical Sciences Reviews.
e-ISSN 2365-659X

Selenium and Tellurium Reagents

In Chemistry and Materials Science

Edited by
Risto S. Laitinen, Raija Oilunkaniemi

DE GRUYTER

Editors
Prof. Risto S. Laitinen
Laboratory of Inorganic Chemistry, Environmental and Chemical Engineering
University of Oulu
P.O. Box 3000
90014 Oulu, Finland
risto.laitinen@oulu.fi

Dr. Raija Oilunkaniemi
Laboratory of Inorganic Chemistry, Environmental and Chemical Engineering
University of Oulu
P.O. Box 3000
90014 Oulu, Finland
Raija.Oilunkaniemi@oulu.fi

ISBN 978-3-11-052794-0
e-ISBN (PDF) 978-3-11-052934-0
e-ISBN (EPUB) 978-3-11-052810-7

Library of Congress Control Number: 2018958583

Bibliographic information published by the Deutsche Nationalbibliothek
The Deutsche Nationalbibliothek lists this publication in the Deutsche Nationalbibliografie; detailed bibliographic data are available on the Internet at http://dnb.dnb.de.

Typesetting: Integra Software Services Pvt. Ltd.
Printing and binding: CPI books GmbH, Leck
Cover image: guvendemir/iStock/Getty Images Plus

www.degruyter.com

Preface

The chalcogen elements sulfur, selenium, and tellurium play a central role in both inorganic and organic chemistry. Sulfur chemistry, in particular, is very extensive and the element has a major impact on modern chemical industry. By contrast, selenium and tellurium research has seen slower progress possibly due to lack of suitable reagents. During recent decades, however, significant advances have been made also in the chemistry of these elements.

One of the objectives in fundamental research in chemistry is to understand the relationships between the molecular and electronic structures of the compounds and to correlate them with the properties observed in the bulk materials. This is also reflected in research on the preparation, structures, and chemical properties of compounds of heavy chalcogen elements selenium and tellurium, many of which currently find utility as versatile reagents both in inorganic and organic syntheses. They are also useful as electric conductors, semiconductors, or insulators and find applications, for instance, in photovoltaic cells and infrared sensors. Their use as catalysts, polymers, and thin films all play a significant role in the current main group chemistry research as well as in the modern technological society. The design of new synthetic strategies is instrumental to this progress.

Although several selenium and tellurium species are similar to the analogous sulfur compounds, there are often also very pronounced differences. The decreasing propensity of the elements to engage in multiple bonding and the increasing strength of the hypervalent and secondary bonding interactions, when going down Group 16, have a profound effect on the material properties. Understanding the trends in the electronic structures and their relationship with both molecular structures and the intermolecular interactions naturally has an impact in designing strategies for synthesis of new compounds and in the fabrication of novel materials. The current volume emphasizes the development of synthetic methods in selenium and tellurium chemistry. Structural, spectroscopic, and bonding properties of chalcogens have recently been reviewed several times and are treated elsewhere.

The field of synthetic chalcogen chemistry is vast and it is possible to cover only the tip of the ice berg in one volume. While some basic features in the chemical properties of the heavy chalcogen elements need to be described and reviewed, the emphasis of this treatise is on recent developments. Elemental selenium and tellurium, the alkali metal and some organyl selenides, tellurides, diselenides, and ditellurides, as well as selenium and tellurium dioxide, tetrachloride, -bromide, and -iodide are commercially available. Ultimately, both selenium and tellurium can be introduced into any reaction system starting from the elements. The main routes for their reactivity are considered in Chapter 1. Selenium and tellurium halogenides are also versatile reagents in various areas of both inorganic and organic chemistry and are discussed in Chapter 2. Selenium- and tellurium-oxygen species are considered in Chapter 3, chalcogen-nitrogen chemistry in Chapter 4, and

https://doi.org/10.1515/9783110529340-201

chalcogen-phosphorus chemistry in Chapter 5. Chapter 6 summarizes the synthetic methods in the preparation of cyclic heavy chalcogenoureas featuring heterocyclic cores and explores their diverse coordination chemistry with p- and d-block metals. The organic chemistry of chalcogen compounds with electrophiles is described in Chapter 7, and the development of the chemistry of species containing the chalcogen-hydrogen bonds in Chapter 8. The reduction of the Se-Se or Se-halogen bonds mediated by elemental zinc is particularly attractive for the simplicity and the efficiency of the protocols. This chemistry and its applications are discussed in Chapter 9. The use of reagents containing bonds between Group 14 elements and selenium or tellurium for the self-assembly of polynuclear metal-chalcogen compounds is covered in Chapter 10. The materials aspects of selenium and tellurium chemistry are discussed in Chapter 11, which comprises the synthesis of transition-, main group and f-block metal chalcogenolate and/or chalcogenide clusters as precursors for the metal chalcogenide thin films and nanostructures.

We would like to thank all authors of this book for their contributions. We are especially indebted to Prof. Tristram Chivers for helpful discussions. In addition to contributing to this volume, he provided us many insights for the form and content of this book. Several people in Walther de Gruyter also gave us invaluable help. Dr. Oleg Lebedev was the originator of the idea for this book. He, and Drs. Lena Stoll, Jana Habermann, Holger Kleessen, Ria Fritz, and Frauke Schafft supported us by providing advice for many technical problems in different phases of this project and have guided the book through the publication process.

October 2018, Oulu
Risto Laitinen and Raija Oilunkaniemi

Contents

Preface — V

List of contributing authors — XIII

Risto S. Laitinen, Raija Oilunkaniemi and Tristram Chivers

1 Introduction of selenium and tellurium into reaction systems — 1

1.1 Introduction — **1**
1.2 Elemental selenium — **2**
1.2.1 General — **2**
1.2.2 Anionic selenides and their derivatives — **3**
1.2.3 Selenium halogenides — **6**
1.2.4 Heterocyclic selenium sulfides — **6**
1.2.5 Homopolyatomic selenium cations — **8**
1.2.6 Triorganylphospine selenides — **8**
1.2.7 Grignard and organyllithium reagents — **8**
1.2.8 Selenium and heavy group 14 reagents — **9**
1.2.9 Selenium–boron chemistry — **10**
1.2.10 Transition metal selenides — **11**
1.3 Elemental tellurium — **12**
1.3.1 General — **12**
1.3.2 Anionic tellurides — **12**
1.3.3 Derivatives of anionic tellurides — **14**
1.3.4 Tellurium-containing chalcogen rings — **14**
1.3.5 Homopolyatomic tellurium cations — **15**
1.3.6 Triorganylphosphine tellurides — **15**
1.3.7 Grignard and organyllithium reagents — **16**
1.3.8 Silane- and germanetellones — **16**
1.3.9 Group 13 chemistry with tellurium — **16**
1.3.10 Titanocene chemistry — **18**
1.3.11 Solid-state and hydrothermal reactions — **19**
1.4 Summary — **19**
References — **20**

Tristram Chivers and Risto S. Laitinen

2 Selenium- and tellurium–halogen reagents — 27

2.1 Scope and introduction — **27**
2.2 Binary chalcogen–halogen reagents — **28**
2.2.1 Selenium tetrahalides — **28**
2.2.2 Selenium oxychloride — **29**
2.2.3 Selenium dihalides — **31**

2.2.4 Diselenium dihalides — 34
2.2.5 Mixtures of selenium halides — 36
2.2.6 Tellurium tetrahalides — 36
2.2.7 Tellurium dihalides and ditellurium dihalides — 41
2.3 Organochalcogen–halogen reagents — 44
2.3.1 Organoselenium and tellurium trihalides — 44
2.3.2 Diorgano-selenium and -tellurium dihalides — 46
2.3.3 Mixed-valent state compounds $RTeX_2TeR$ — 48
2.3.4 Organo-selenium and -tellurium monohalides — 48
2.4 Conclusions — 51
References — 52

Alexandra Pop, Cristian Silvestru and Anca Silvestru
3 Organoselenium and organotellurium compounds containing chalcogen-oxygen bonds in organic synthesis or related processes — 61
3.1 Introduction — 61
3.2 Organoselenium compounds — 64
3.2.1 Organoselenium(II) compounds — 64
3.2.2 Organoselenium(IV) compounds — 69
3.2.3 Organoselenium(VI) compounds — 84
3.3 Organotellurium compounds — 91
3.3.1 Organotellurium(II) compounds — 91
3.3.2 Organotellurium(IV) compounds — 92
3.3.3 Organotellurium(VI) compounds — 107
3.4 Conclusion and outlook — 109
Abbreviations — 110
References — 111

Tristram Chivers and Risto S. Laitinen
4 Selenium- and tellurium–nitrogen reagents — 123
4.1 Scope and introduction — 123
4.2 Binary selenium–nitrogen and tellurium–nitrogen reagents — 124
4.2.1 Neutral species — 124
4.2.2 Cations — 127
4.2.3 Anions — 127
4.3 Chalcogen–nitrogen halides — 127
4.3.1 Ternary selenium–nitrogen halides — 127
4.3.2 Imido chalcogen halides — 129
4.4 $E(NSO)_2$ reagents (E = Se, Te) — 130
4.5 Selenium/tellurium–nitrogen–silicon reagents — 131
4.6 Chalcogenylnitrosyls ArN=Se — 134

4.7 Chalcogen(IV) diimides — 135
4.7.1 Synthesis, structures and applications in organic synthesis — 135
4.7.2 Thermal decomposition — 136
4.7.3 Insertion and cycloaddition reactions — 137
4.7.4 Reactions with nucleophiles and electrophiles — 138
4.8 Tris(*tert*-butylimido)tellurite dianion — 138
4.9 Chalcogen(II) amides and diamides — 139
4.10 Radical chemistry — 141
4.11 Conclusions — 143
References — 143

Guoxiong Hua and J. Derek Woollins
5 Organophosphorus-selenium/tellurium reagents: from synthesis to applications — 151
5.1 Introduction — 151
5.2 Binary phosphorus-selenium/tellurium species — 152
5.2.1 Binary phosphorus-selenium species — 152
5.2.2 Binary phosphorus-tellurium species — 153
5.3 Organophosphorus(III)-selenium/tellurium compounds — 153
5.3.1 Organophosphorus(III)-selenium compounds — 153
5.3.2 Organophosphorus(III)-tellurium compounds — 155
5.4 Phosphorus (V)-selenides/tellurides — 156
5.4.1 Triorganylphosphine selenides — 156
5.4.2 Phosphinoselenoic chlorides and derivatives — 157
5.4.3 Phosphinoselenoic acids and derivatives — 162
5.4.4 *O,O,O*-Triorganorylphosphoroselenoates $(RO)_3P{=}Se$ — 163
5.4.5 Tertiary phosphine tellurides $R_3P{=}Te$ — 164
5.5 Diselenophosphinates/ditellurophosphinates — 166
5.5.1 Diselenophosphinates — 166
5.5.2 Tellurophosphinites and ditellurophosphinates — 172
5.6 Diselenaphosphetane diselenides and Woollins' reagent — 173
5.7 Phosphorus-selenium/tellurium amides and imides — 185
5.8 Outlook — 195
References — 196

Jamie S. Ritch
6 Synthesis and coordination chemistry of cyclic seleno- and telluroureas — 207
6.1 Introduction — 207
6.2 Structural variety of cyclic HCUs — 208
6.3 Synthesis of heavy chalcogenoureas — 209

6.3.1 Direct chalcogenation — **209**
6.3.2 Deprotonation-chalcogenation — **209**
6.3.3 One-pot reactions — **213**
6.3.4 Other methods — **214**
6.4 General properties of HCUs — **215**
6.5 ^{77}Se and ^{125}Te NMR spectroscopic properties — **217**
6.6 Coordination chemistry — **217**
6.6.1 p-block (groups 13–16) — **219**
6.6.2 Group 12 — **222**
6.6.3 Group 11 — **224**
6.6.4 Group 10 — **231**
6.6.5 Group 9 — **233**
6.6.6 Group 8 — **235**
6.6.7 Group 6 — **235**
6.7 Conclusions and outlook — **236**
Abbreviations — **237**
References — **237**

Fateh V. Singh and Thomas Wirth
7 Selenium and Tellurium Electrophiles in Organic Synthesis — 243
7.1 Introduction — **243**
7.2 Selenium electrophiles — **243**
7.2.1 Selenenylation reactions — **244**
7.2.2 Selenenylation of aliphatic C–H bonds — **256**
7.2.3 Selenenylation of aromatic C–H bonds — **258**
7.2.4 Cyclizations — **264**
7.2.5 Metal-catalyzed coupling reactions — **269**
7.2.6 Carbene insertion reaction — **273**
7.2.7 Rearrangements — **273**
7.2.8 Catalytic reactions — **275**
7.3 Chiral selenium electrophiles in stereoselective reactions — **281**
7.4 Tellurium electrophiles in organic synthesis — **284**
References — **289**

Peter C. Ho, Jin Wang and Ignacio Vargas-Baca
8 Reagents that Contain Se-H or Te-H Bonds — 301
8.1 Introduction — **301**
8.2 Chalcogen dihydrides (H_2Ch) — **301**
8.3 Sodium Hydrochalcogenides (NaHCh) — **303**
8.4 Lithium dihydroselenoxo aluminate $Li(AlH_2Se)$ — **305**
8.5 Chalcogenols RChH — **305**

8.6 Monochalcogenocarboxylic acids RC(O)ChH — 306
8.7 N,N-dimethyl Chalcogenocarbamic acids $Me_2NC(O)ChH$ — 308
References — 309

Claudio Santi, Lucia Capoccia and Bonifacio Monti
9 Zinc-Selenium reagents in organic synthesis — 315
9.1 Introduction — 315
9.2 Oxidative zinc insertion in Se-Se and Se-X bond — 316
9.2.1 Synthesis and reactivity of RSeZnSeR complexes — 316
9.2.2 A versatile and recyclable biphasic system — 320
9.2.3 Synthetic applications of TMEDA stabilized PhSeZnSePh — 322
9.2.4 PhSeZn-halides and the unexpected on-water reactivity — 323
References — 327

Alexander M. Polgar and John F. Corrigan
10 Recent advances in the self-assembly of polynuclear metal–selenium and –tellurium compounds from 14–16 reagents — 331
10.1 Introduction — 331
10.2 Silane elimination methods — 332
10.2.1 Bis(trimethylsilyl)chalcogenide reagents — 333
10.2.2 Organic trimethylsilylchalcogenides — 337
10.2.3 Metal chalcogenide/chalcogenolate clusters — 345
10.2.4 Metal trimethylsilylchalcogenolates — 351
10.3 Coordination of 14–16 ligands — 356
10.3.1 Chalcogenidotetrelates — 356
10.3.2 Organotetrel chalcogenides — 366
10.4 Summary — 370
Notes — 371
References — 371

Vimal K. Jain and G. Kedarnath
11 Applications of metal selenium/tellurium compounds in materials science — 383
11.1 Introduction — 383
11.2 Synthesis of chalcogenolate/chalcogenide complexes — 385
11.2.1 Reactions of nucleophilic chalcogen reagents with metal salts — 385
11.2.2 Oxidative addition of dichalcogenides — 386
11.2.3 Ligand-exchange reactions — 388
11.2.4 Chalcogenolysis method — 389
11.2.5 Insertion of chalcogen into M-C bond — 389
11.3 Transition metal chalcogenolate/chalcogenide clusters — 390

11.4 Transition metal chalcogenolate precursors for metal chalcogenides — **397**
11.4.1 Group 4 (Ti, Zr, Hf) chalcogenides — **397**
11.4.2 Group 5 (V, Nb, Ta) chalcogenides — **398**
11.4.3 Group 6 (Cr, Mo, W) chalcogenide materials — **399**
11.4.4 Group 7 (Mn, Tc, Re) chalcogenides — **399**
11.4.5 Group 8 (Fe, Ru, Os) chalcogenides — **400**
11.4.6 Group 9 (Co, Rh, Ir) chalcogenides — **401**
11.4.7 Group 10 (Ni, Pd, Pt) chalcogenides — **402**
11.4.8 Group 11 (Cu, Ag, Au) chalcogenides — **404**
11.4.9 Group 12 (Zn, Cd, Hg) chalcogenides — **407**
11.5 Main group chalcogenolates as precursors for metal chalcogenides — **411**
11.5.1 Group 13 chalcogenides — **411**
11.5.2 Group 14 chalcogenides — **413**
11.5.3 Group 15 chalcogenides — **416**
11.6 f-block metal chalcogenolates as precursors for metal chalcogenides — **420**
11.7 Ternary metal chalcogenides — **421**
Abbreviations — **427**
References — **427**

Index — 445

List of contributing authors

Tristram Chivers
Department of Chemistry
University of Calgrary
2500 University Drive NW
Calgary, Alberta
Canada T2N 1N4
Chapter 1, 2, 4

Risto S. Laitinen
Laboratory of Inorganic Chemistry,
Environmental and Chemical Engineering
University of Oulu
P.O. Box 3000
90014 Oulu, Finland
Chapter 1, 2, 4

Vimal K. Jain
UM-DAE Centre for Excellence in Basic Sciences
Health Centre Building
University of Mumbai
Kalina Campus
Santacruz (E), Mumbai-400098 (India)
E-mail: jainvk@cbs.ac.in
Chapter 11

G. Kedarnath
Chemistry Division
Bhabha Atomic Research Centre
Tromaby, Mumbai-400 085 (India)
and
Homi Bhabha National Insitute
Anushaktinagar, Mumbai-400 094 (India)
E-mail: kedar@barc.gov.in
Chapter 11

Raija Oilunkaniemi
Laboratory of Inorganic Chemistry,
Environmental and Chemical Engineering
University of Oulu
P.O. Box 3000
90014 Oulu, Finland
Chapter 1

Jamie S. Ritch
Department of Chemistry
The University of Winnipeg
515 Portage Avenue
Winnipeg, Manitoba, R3B 2E9, Canada
j.ritch@uwinnipeg.ca
Chapter 6

Claudio Santi
Department of Pharmaceutical Sciences
University of Perugia
Perugia, Italy
Chapter 9

Lucia Capoccia
Department of Pharmaceutical Sciences
University of Perugia
Perugia, Italy
Chapter 9

Bonifacio Monti
Department of Pharmaceutical Sciences
University of Perugia
Perugia, Italy
Chapter 9

https://doi.org/10.1515/9783110529340-202

Alexandra Pop
Department of Chemistry
Supramolecular Organic and Organometallic Chemistry Centre (SOOMCC)
Faculty of Chemistry and Chemical Engineering
Babeş-Bolyai University
11 Arany Janos
400028 Cluj-Napoca, ROMANIA
Chapter 3

Cristian Silvestru
Department of Chemistry
Supramolecular Organic and Organometallic Chemistry Centre (SOOMCC)
Faculty of Chemistry and Chemical Engineering
Babeş-Bolyai University
11 Arany Janos
400028 Cluj-Napoca, ROMANIA
Chapter 3

Anca Silvestru
Department of Chemistry
Supramolecular Organic and Organometallic Chemistry Centre (SOOMCC)
Faculty of Chemistry and Chemical Engineering
Babeş-Bolyai University
11 Arany Janos
400028 Cluj-Napoca, ROMANIA
ancas@chem.ubbcluj.ro
Chapter 3

Peter C. Ho
McMaster University
Department of Chemistry and Chemical Biology
1280 Main Street West
Hamilton, Ontario
Canada, L8S 4M1.
Chapter 8

Jin Wang
McMaster University
Department of Chemistry and Chemical Biology
1280 Main Street West
Hamilton, Ontario
Canada, L8S 4M1.
Chapter 8

Ignacio Vargas-Baca
McMaster University
Department of Chemistry and Chemical Biology
1280 Main Street West
Hamilton, Ontario
Canada, L8S 4M1.
vargas@chemistry.mcmaster.ca
Chapter 8

Fateh V. Singh
Cardiff University
School of Chemistry
Main Building
Park Place, Cardiff
CF10 3AT United Kingdom and
VIT University
Chennai Campus
Vandalur-Kelambakkam Road
Chennai-600127, Tamil Nadu, India
Chapter 7

Thomas Wirth
Cardiff University
School of Chemistry
Main Building
Park Place, Cardiff
CF10 3AT United Kingdom
wirth@cf.ac.uk
Chapter 7

Guoxiong Hua
EaStCHEM School of Chemistry
University of St Andrews
Fife, KY16 9ST, UK
Chapter 5

J. Derek Woollins
EaStCHEM School of Chemistry
University of St Andrews
Fife, KY16 9ST, UK
Chapter 5

Alexander M. Polgar
Department of Chemistry
Western University
1151 Richmont Street
London, Ontario
Canada N6A 5B7
Chapter 10

John F. Corrigan
Department of Chemistry
Western University
1151 Richmont Street
London, Ontario
Canada N6A 5B7
corrigan@uwo.ca
Chapter 10

Risto S. Laitinen, Raija Oilunkaniemi and Tristram Chivers

1 Introduction of selenium and tellurium into reaction systems

Abstract: The introduction of selenium and tellurium into both organic and inorganic compounds frequently begins with the elements. This chapter provides an overview of the main reactivity of the hexagonal allotropes of selenium and tellurium, which are the most stable form of the elements under ambient conditions. While the two elements have very similar chemical properties, there are also notable differences. Upon reduction, both elements form mono- and poly-chalcogenides, which are useful nucleophilic reagents in several reactions. The elements also react with many main group compounds as well as with transition metal complexes. They also form homopolyatomic cations upon oxidation. Both selenium and tellurium react with Grignard reagents and organyllithium compounds affording organylchalcogenolates, which upon oxidation form dichalcogenides that are themselves useful reagents in organic synthetic chemistry as well as in materials applications. This chapter provides a short introduction to the various topics that will be developed further in the subsequent chapters of this book.

Keywords: elemental selenium, elemental tellurium, selenides, tellurides, polyselenides, polytellurides, organyl selenolates, organyl tellurolates, homopolyselenium cations, homopolytellurium cations

1.1 Introduction

The use of selenium and tellurium compounds in diverse applications has grown significantly during the past decades. Consequently, chalcogen chemistry has been reviewed several times during the last 20 years, as exemplified by a selection of review articles [1–6] and books [3–12], which describe recent advances in synthetic inorganic and organic chalcogen chemistry, as well as in more applied areas such as medicinal chemistry and materials science. The current monograph is an overview of this progress and discusses the use of different classes of selenium and tellurium compounds as reagents.

The preparation of selenium and tellurium compounds generally involves simple molecular species such as the elements, inorganic or organic chalcogenides and polychalcogenides, selenium and tellurium halogenides, selenium and tellurium

This article has previously been published in the journal *Physical Sciences Reviews*. Please cite as: Laitinen, R. S., Oilunkaniemi, R., Chivers, T. Introduction of selenium and tellurium into reaction systems. *Physical Sciences Reviews* [Online] **2018** DOI: 10.1515/psr-2018-0059

https://doi.org/10.1515/9783110529340-001

oxides, as well as organyl phosphine selenides and tellurides. These classes are treated in detail in the subsequent chapters of this monograph. In this chapter, the introduction of commercially available selenium and tellurium into the reaction system is considered. The first step in many reactions of selenium and tellurium is the reduction of the elements to form anionic selenides and tellurides, which are the actual reagents.

Alkyl or aryl selenides, tellurides, diselenides, and ditellurides are also common starting materials in many organic syntheses. While many of them are commercially available, they can also be conveniently prepared from the elements. The dimer PhPSe $(\mu\text{-Se})_2P(Ph)Se$, which is known as Woollins' reagent, is also commercially available. This P-Se compound is proving to be an important source of selenium in a wide range of synthetic applications. Its chemistry is discussed in detail in Chapter 5.

1.2 Elemental selenium

1.2.1 General

Some reactions of elemental selenium are summarized in Figure 1.1. Many of the products are themselves useful reagents in further syntheses. Special examples have been discussed in the appropriate chapters of this monograph.

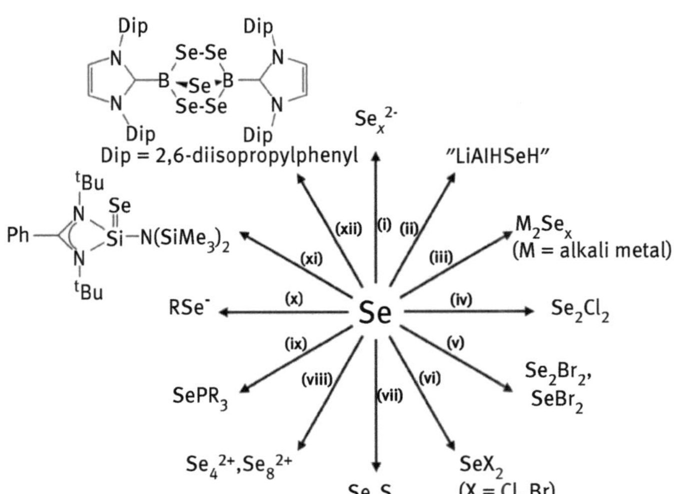

Figure 1.1: Some reactions of elemental selenium. (i) $M/NH_3(l)$ (M = Li, Na, K), $Na[BH_4]$, or $Li[AlEt_3H]$, (ii) $Li[AlH_4]$, (iii) M_2Se (M = alkali metal), (iv) $SeCl_4$, (v) Br_2, (vi) SO_2X_2/THF (X = Cl, Br), (vii) S_8, (viii) $AsF_5/SO_2(l)$ or $SbCl_5/SO_2(l)$, (ix) PR_3 (R = alkyl or aryl), (x) RMgBr/THF or RLi/THF (R = alkyl or aryl), (xi) $PhC(N^tBu)_2Si[N(SiMe_3)_2]$/THF, (xii) $\{[N(Dip)_2]_2(CH)_2CB\}_2/C_6H_6$.

1.2.2 Anionic selenides and their derivatives

Selenides and polyselenide anions are useful nucleophiles in synthetic applications. Alkali metal and alkaline earth metal chalcogenides are commonly prepared by mixing the elements at high temperatures, by the reaction of chalcogen elements and alkali-metal carbonates in solution, by solvothermal reactions, and by oxidation–reduction reactions involving Lewis acids [13, 14]. The alkali metal or alkaline earth metal selenides and tellurides thus formed can serve as reagents for the preparation of transition-metal or main-group element chalcogenides. A large organic cation or an encapsulating agent such as the crown-ether complex of an alkali-metal cation may stabilize the otherwise unstable anions.

Lithium, sodium, and potassium selenides and diselenides have traditionally been prepared by dissolving the chalcogen elements and the alkali metals in appropriate molar ratios in liquid ammonia. The early work by Bergström [15] and Klemm et al. [16], as well as that by Brandsma and Wijers [17], has been followed and developed by numerous studies, as exemplified by Müller et al. [18], Björgvinsson and Schrobilgen [19], and Colombara et al. [20]. Thiele et al. [21] have recently reviewed the preparation, structures, and properties of different classes of polyselenides, which have been stabilized by use of bulky cations.

In addition to alkali metals, polyselenides can be produced in liquid ammonia or in methylamine by other less-noble metals such as Mn and Fe to afford $[Mn(NH_3)_6]Se_3$, $[Mn(NH_3)_6]Se_4$, $[M(NH_3)_6]Se_6$ (M = Mn, Fe) and disordered phases $[Mn(NH_2Me)_6](S_5)_{0.15}(Se_{2.43}S_{1.57})_{0.85}$ and $[Mn(NH_3)_6](Se_{2.29}S_{2.71})$ [22].

Sandman et al. [23] have shown that alkali-metal selenides can also be produced upon reduction of selenium by elemental alkali metals in high-boiling polar aprotic solvents. This concept has been developed by Thompson and Boudjouk [24], who have demonstrated that lithium, sodium, and potassium selenides and diselenides can conveniently be also prepared in THF by use of ultrasound and in the presence of small amounts of a charge-transfer agent such as naphthalene. The electrochemical reduction of selenium powder in aprotic media such as acetonitrile or THF also provides an effective method to synthesize Se^{2-} and Se_2^{2-} selectively [25, 26].

Elemental selenium can also conveniently be reduced by using lithium tetrahydridoaluminate [27], sodium tetrahydridoborate [28], or lithium triethylhydridoborate [29]. The reaction with lithium tetrahydridoaluminate ("lithium aluminium hydride") has been reported to afford an intermediate product "LiAlHSeH", which can be used *in situ* to produce a wide variety of organic selenoethers and selenones [27]. This reagent is described in more detail in Chapter 8.

Since the distribution of the products in the reduction of elemental selenium with sodium tetrahydridoborate seemed to be sensitive to experimental conditions, Cusic and Dance [30] and Marques de Oliveira et al. [31] have studied the reaction in detail using ^{77}Se NMR (nuclear magentic resonance) spectroscopy. Since HSe^- and Se^{2-}

co-exist in fast equilibrium, there is no NMR evidence for the presence of Se^{2-}. Similarly, no resonance attributable to Se_2^{2-} has been observed. By contrast, the presence of polyselenide anions Se_x^{2-} (x = 3–6) is clearly evident in solution, as indicated in Figure 1.2 by the ^{77}Se NMR spectrum of a reaction mixture recorded in DMF (dimethylformamide) [30]. These observations are consistent with the detailed ^{77}Se NMR study of Björgvinsson and Schrobligen in liquid ammonia and ethylenediamine [19] and with the ESI (electrospray ionization) mass spectrometric study of the polyselenide speciation by Raymond et al. [32].

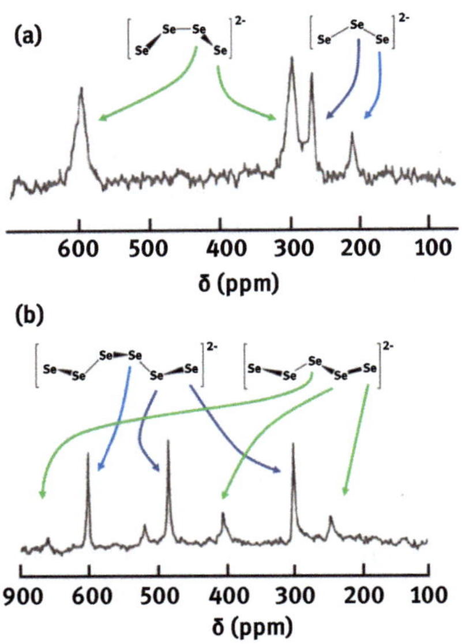

Figure 1.2: ^{77}Se NMR resonances of (a) Se_3^{2-} and Se_4^{2-} and (b) Se_5^{2-} and Se_6^{2-}. All spectra have been recorded in DMF [30]. Adapted from Cusick J, Dance I. The characterization of $[HSe]^-$ and $[Se_x]^{2-}$ ions by ^{77}Se NMR. Polyhedron 1991, 10, 2629–2640. Copyright 1991 Elsevier Ltd.

Klayman and Griffin [28] and Marques de Oliveira et al. [31] have shown that the treatment of Se_x^{2-} solutions with organyl halogenides with careful control of the reaction stoichiometry produces organyl selenides and diselenides in high purity. The solvothermal synthesis of $CuInSe_2$ also involves Se_x^{2-}, which has been produced by the reduction of selenium with $NaBH_4$ [33].

The reduction of selenium by lithium triethylhydridoborate ("superhydride") also results in the formation of a mixture of polyselenides Se_x^{2-} [29]. The distribution of polyselenide anions is dependent on the molar ratio of selenium and the reducing agent. The treatment of the anions with $[TiCp_2Cl_2]$ affords $[TiCp_2Se_5]$ [34–36] or dinuclear $[TiCp_2(\mu\text{-}Se_2)_2TiCp_2]$ [Cp = cyclopentadienyl ligand $C_5H_5^-$ or its alkyl-substituted derivatives] [37].

Both $[TiCp_2Se_5]$ and $[TiCp_2(\mu\text{-}Se_2)_2TiCp_2]$ are useful reagents for the preparation of homo- and heterocyclic chalcogen rings by the reaction with Se_2Cl_2 or $SeCl_2$ [38, 39] [see eqs. (1.1) and (1.2)]. These reactions are similar to those of

[$TiCp_2S_5$] and S_nCl_2, which afford a large number of homocyclic sulfur ring molecules S_{n+5} (for reviews, see Refs. [40–42]). The hybrid reactions of [$TiCp_2S_5$] or [$TiCp_2(\mu\text{-}S_2)_nTiCp_2$] ($n$ = 2, 3) with Se_2Cl_2 [37] and those of [$TiCp_2Se_5$] or [$TiCp_2(\mu\text{-}Se_2)_2TiCp_2$] with S_nCl_2 (n = 1, 2) [37, 43, 44] afford heterocyclic selenium sulfides, as do the reactions of [$TiCp_2Se_nS_{5\text{-}n}$] mixtures with sulfur or selenium chlorides [44, 45].

$$[TiCp_2Se_5] + Se_nCl_2 \rightarrow Se_{5+n} + [TiCp_2Cl_2] \tag{1.1}$$

$$[TiCp_2((\mu - Se_2)_2TiCp_2] + 2\ Se_nCl_2 \rightarrow Se_{4+2n} + 2\ [TiCp_2Cl_2]\ (n = 1,\ 2) \tag{1.2}$$

An intriguing application of the reduction of elemental selenium by lithium triethylhydridoborate involves the reaction of the reduction product with acyl chlorides or carboxylic acids followed by treatment with sugar azides [46] (see Figure 1.3). A new amide bond is formed, and elemental selenium is recovered for further use.

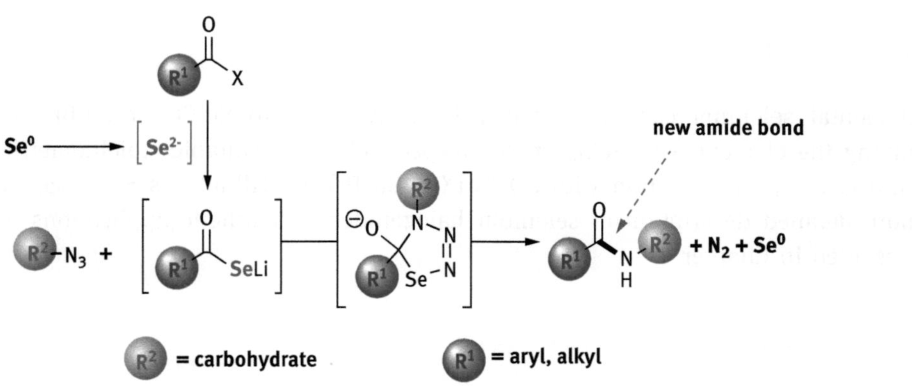

Figure 1.3: The preparation of sugar-derived amides by reduction of elemental selenium with lithium triethylhydridoaluminate followed by a reaction with an acyl chloride [46]. Reproduced with permission from Silva L, Affeldt RF, Lüdtke DS. Synthesis of glycosyl amides using selenocarboxylates as traceless reagents for amide bond formation.J Org Chem 2016, 81, 5464–5473. Copyright 2016 American Chemical Society.

Elemental selenium can also be reduced to Se^{2-} in aqueous alkaline solution [47], though it was reported that the yield of the Se^{2-} ion is only moderate owing to disproportionation [see eq. (1.3)]. When tin(II) chloride was added to the basic media, the yield was significantly improved. It has been deduced that tin(II) oxide was *de facto* the reducing agent. However, when selenium powder was irradiated in the microwave oven in an NaOH solution and subsequently treated with $CuSO_4 \cdot 5H_2O$, bulk Cu_5Se_4 could be precipitated from the solution [48].

$$3\ X + 6\ MOH \rightarrow 2\ M_2X + M_2XO_3 + 3\ H_2O \quad (1.3)$$

A direct *in situ* reduction of selenium using aqueous DMF or sodium formaldehyde-sulfoxylate followed by the reaction with lead acetate afforded PbSe powder [49]. $NiSe_2$ has been prepared by mixing metallic nickel and elemental selenium in mercaptoethanol at room temperature for a prolonged period, resulting in a rust-coloured solution, which upon annealing at 350 °C under a nitrogen atmosphere produced the phase-pure product [50]. The material is an efficient hydrogen evolution reaction catalyst. The reaction of selenium or tellurium and iron(II) bromide in octane-9-enylamine ("oleylamine") at 170 °C resulted in the formation of FeE_2 (E = Se, Te) nanocrystals [51]. The amine acted as a solvent and as a reducing agent for the chalcogen. The iron dichalcogenides have attractive electronic properties and can have utility in different device applications.

Hydrazine can also be utilized in the reduction of selenium and tellurium [47, 52]. The concurrent treatment of the solution with silver nitrate yielded Ag_2Se nanowires [52].

1.2.3 Selenium halogenides

Elemental selenium can conveniently be converted into Se_2Cl_2 or Se_2Br_2 by mixing the element with $SeCl_4$ or Br_2, respectively, in a suitable stoichiometric ratio [53, 54]. The reaction with SO_2X_2 (X = Cl, Br) in THF affords SeX_2 [55]. A more detailed description of selenium halogenides in synthetic applications is presented in Chapter 2.

1.2.4 Heterocyclic selenium sulfides

Crystalline solid solutions of selenium sulfides Se_nS_{8-n} can be produced by melting the elements in *vacuo*, quenching the molten mixtures, and extracting the quenched melt with CS_2 followed by crystallization [41]. The composition of the products has been established by ^{77}Se NMR spectroscopy [56] (see Figure 1.4).

It was deduced that the main components in the CS_2 solutions of the crystalline phases extracted from quenched molten mixtures of sulfur and selenium are sulfur-rich SeS_7 and 1,2-Se_2S_6 ring molecules even in mixtures of high initial selenium content [56]. Other identified main species are 1,2,3-Se_3S_5, 1,2,3,4-Se_4S_4, 1,2,3,4,5-Se_5S_3, 1,2,3,4,5,6-Se_6S_2, and Se_8, in which the number of homonuclear bonds is maximized.

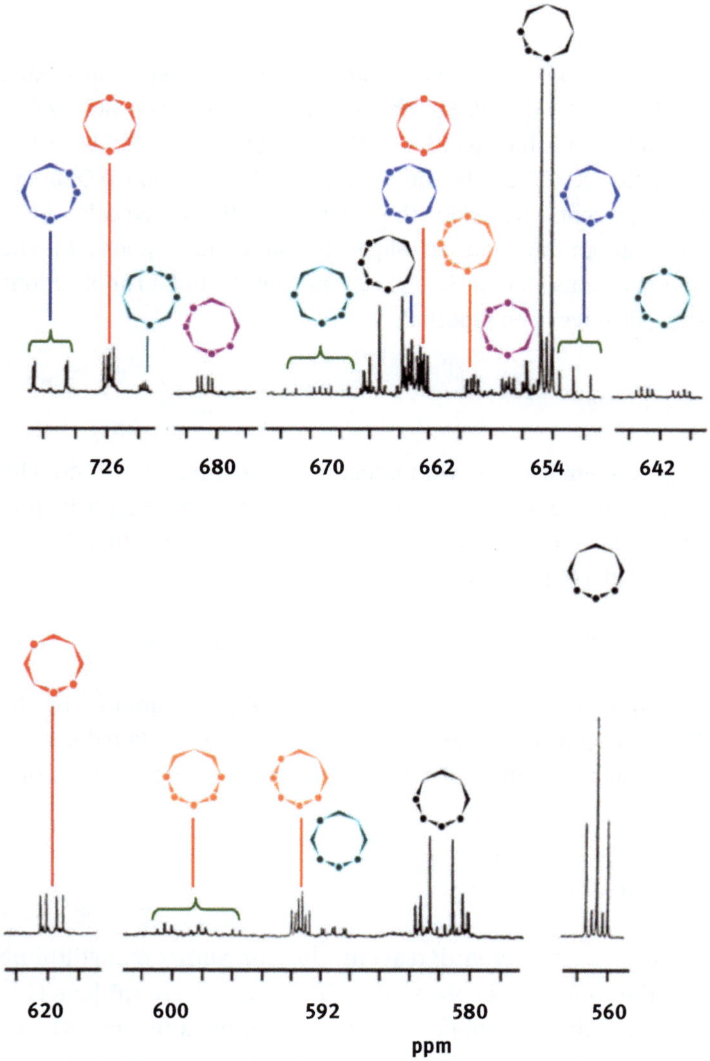

Figure 1.4: ^{77}Se NMR spectrum of the CS_2 solution of quenched sulfur–selenium melt involving ^{77}Se-enriched selenium (enrichment 92 %) [56, 57]. The selenium content in the molten mixture of sulfur and selenium is 30 mol% of selenium. The resonances from different individual Se_nS_{8-n} species have been shown in different colours. The closed circles indicate selenium atoms and the empty corners sulfur atoms. Reproduced with permission from Laitinen RS, Oilunkaniemi R. Catenated compounds group 16 (Se, Te). In Poeppelmeier, K, Reedijk J. Eds. Comprehensive Inorganic Chemistry II, 2nd Ed., Elsevier, Amsterdam 2013, pp. 197–231. Copyright 2013 Elsevier SA.

1.2.5 Homopolyatomic selenium cations

Elemental selenium can be oxidized to homopolyatomic cations in super acidic media, in acidic melts, and in liquid SO_2 [1, 58]. Suitable oxidizing agents are $AlCl_3$, AsF_5, $SbCl_5$, $BiCl_5$, and transition metal halogenides OsF_6, IrF_6, PtF_6, VCl_6, $ZrCl_6$, $HfCl_6$, $NbCl_6$, $TaCl_6$, and WX_6, ReX_6 (X = F, Cl). In addition, $VOCl_3$, $NbOCl_3$, and $WOCl_4$ can also be used as oxidizing agents. Recently, the selenium cations have also been prepared using chemical vapour transport techniques [59] and ionic liquids [58]. The known homopolyselenium cations include Se_4^{2+}, Se_8^{2+}, and Se_{10}^{2+}, but no applications of these species in synthesis have been reported.

1.2.6 Triorganylphospine selenides

Selenium and organic polyselenides react with phosphines yielding phosphine selenides R_3PSe. The reaction with phosphines is often used for abstracting selenium atoms from a polyselenium chain, as exemplified by the formation of $[TiCp_2(\mu\text{-}Se_2)_2TiCp_2]$ from $[TiCp_2Se_5]$ [37] [eq. (1.4)].

$$2\ [TiCp_2Se_5] + 6\ Ph_3P \rightarrow [TiCp_2(\mu - Se_2)_2TiCp_2] + 6\ Ph_3PSe \qquad (1.4)$$

The interaction of elemental selenium with phosphines [60] is exemplified by the one-pot synthesis of diselenophosphinic selenoesters [61]. A more detailed discussion of the preparation and application of phosphorus–selenium reagents is presented in Chapter 5.

1.2.7 Grignard and organyllithium reagents

Elemental selenium reacts with Grignard reagents [62] or with organyllithium reagents [63] in dry THF to afford selenolates RSe^- (R = organyl group) [eqs (1.5) and (1.6)]. Upon oxidation, the selenolates can be converted into organyl diselenides, which are themselves convenient reagents in synthetic chemistry (see Chapters 5–11) [62–64]. There are several methods to carry out the oxidation. The best yields have been observed by use of potassium hexacyanidoferrate(III) [63, 64] [eq. (1.7)].

$$2\ Se + 2\ RMgBr \rightarrow 2\ RSe^- + 2\ Mg^{2+} + 2\ Br^- \qquad (1.5)$$

$$Se + RLi \rightarrow RSe^- + Li^+ \qquad (1.6)$$

$$2\ RSe^- + 2\ [Fe(CN)_6]^{3-} \rightarrow RSeSeR + 2\ [Fe(CN)_6]^{4-} \qquad (1.7)$$

Organyl monoselenides can be prepared from lithium selenolate by treatment with an organyl halogenide [eq. (1.8)]. By suitable selection of the organic substituent, a number of different selenides can be produced, as exemplified in Refs. [65, 66].

$$RSe^-Li^+ + R'X \rightarrow RSeR' + Li^+Cl^- \tag{1.8}$$

The cyclopentadienyl rings in $[V(C_5H_5)(C_7H_7)]$ [67] or $[Fe(C_5H_5)_2]$ [68] can be lithiated followed by the insertion of selenium or tellurium [67, 68]. The treatment of the former complex affords dichalcogenides. In the case of the latter, a chain of three chalcogen atoms links the two cyclopentadienyl rings in the same complex (see Figure 1.5).

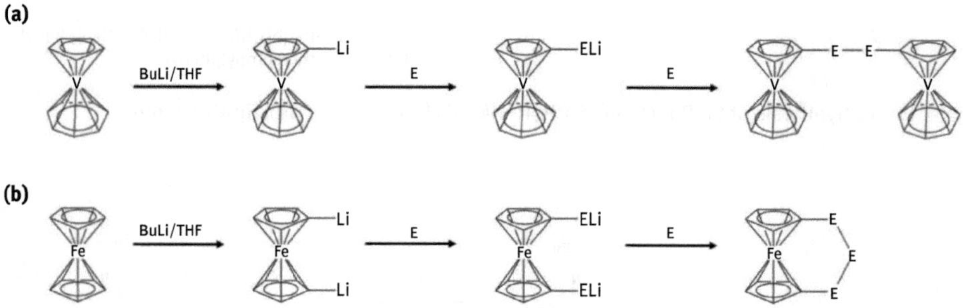

Figure 1.5: Elemental chalcogens as soft oxidants (E = S, Se, Te). Preparation of (a) $[\{V(C_7H_7)(C_5H_4E)\}_2]$ [67] and (b) $[Fe(C_5H_4E)_2E]$ [68].

1.2.8 Selenium and heavy group 14 reagents

Stable silanechalcogenones containing a terminal Si=E (E = Se, Te) functionality can be formed by the reaction of the amidinate-stabilized silicon(II)bis(trimethylsilyl) amide and selenium or tellurium [69] (see Figure 1.6(a)) and germaneselones by the abstraction of selenium from RR'$GeSe_4$ with triphenylphosphine [70]. 1,2,3,4,5-Tetraselenagermolane can be prepared by lithiation of RR'$GeBr_2$ followed by the treatment with elemental selenium (see Figure 1.6(b)). The formation of RR'GeSe in reasonable yields requires very bulky organic substituents.

Another application of germylenes in the formation of a Ge=Se double bond involves the oxidative addition of elemental selenium to a Ge(II) centre [71, 72] (see Figure 1.7). The actual reactant, aminotroponiminato(trimethylsilathio)germylene was obtained by the reaction of the germylene monochloride with $Li^+(SSiMe_3^-)$ [71]. When treated with selenium, the oxidative addition was followed by a virtually quantitative condensation step (see Figure 1.8). By contrast, when $Li^+(SSiMe_3^-)$ was substituted by $Li^+(OSiR_3^-)$ (R = Me, Ph), only a clean single-stage oxidative addition took place [72].

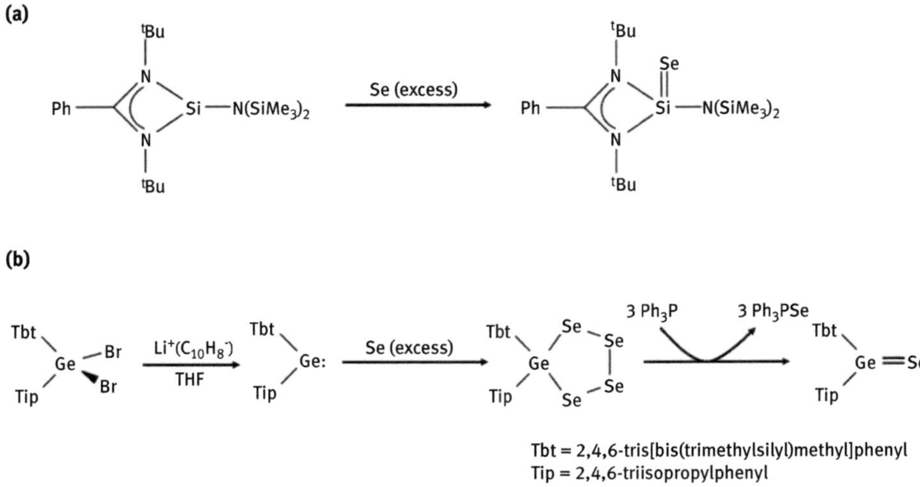

Figure 1.6: Synthesis of (a) $PhC(N^tBu)_2Si(Se)N(SiMe_3)_2$ [68] and (b) (Tbt)(Tip)GeSe [69].

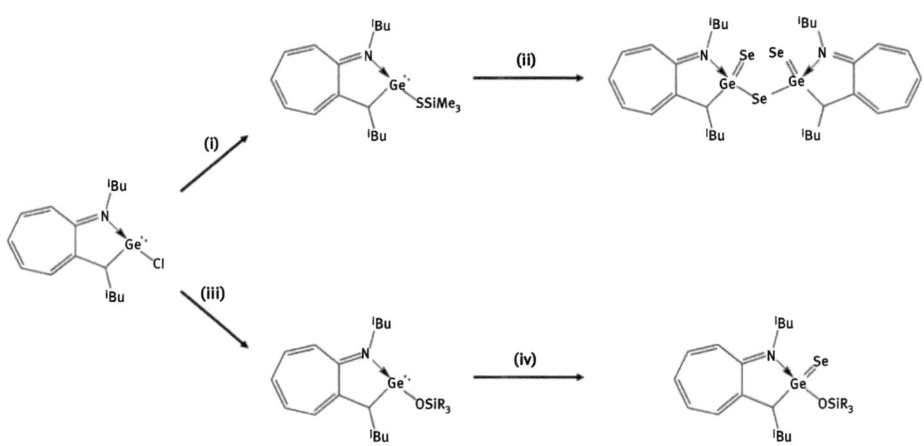

Figure 1.7: Oxidative addition of selenium to the Ge(II) centre [71, 72]. (i) 2 $Li^+(SSiMe_3^-)$·THF/hexane 0 °C, (ii) Se/THF r.t., (iii) $Li^+(OSiR_3^-)$/toluene r.t. (R = Me, Ph), (iv) Se/THF r.t.

1.2.9 Selenium–boron chemistry

Organic diborylenes RB=BR can be stabilized by bulky substituents such as Tbt {2,3,6-tris[bis(trimethylsilyl)methyl]phenyl} [73]. Selenium can be inserted into the boron–boron double bond under the influence of uv-radiation. The formal reactant is $TbtB(SeMe)_2$, but it has been postulated that upon irradiation the

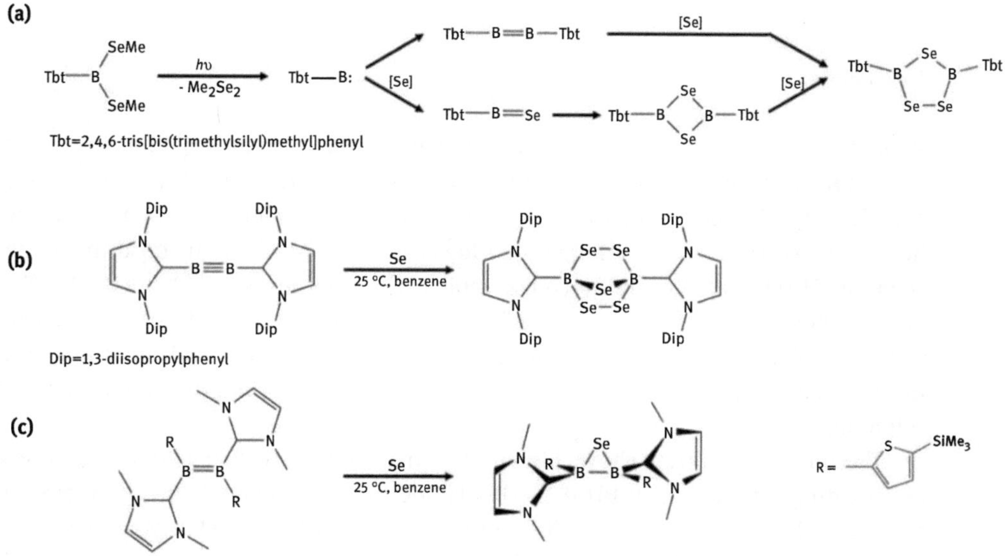

Figure 1.8: Reactions of boron–boron multiple bonds with elemental selenium. (a) Ref. [73]. (b) Ref. [74]. (c) Ref. [75].

reactive component is the borylene Tbt-B:, which is formed upon elimination of Me_2Se_2 from $TbtB(SeMe)_2$. The monomer Tbt-B: dimerizes forming Tbt-B=B-Tbt, which then reacts with elemental selenium [73]. It has very recently been shown that that the diboryne RB≡BR (R = 1,2-bis(2,6-diisopropylphenyl)imidazol-2-ylene) reacts with an excess of elemental selenium to form $RB(\mu\text{-}Se_2)_2(\mu\text{-}Se)BR$ [74]. By choice of suitable organic groups, it was shown that the diborylene RR'B=BRR' (R = 1,3-dimethylimidazol-2-ylidene, R' = 5-trimethylsilyl-thien-2-yl) reacts with elemental selenium yielding diboraselenirane [75]. All three reactions are compared in Figure 1.8.

1.2.10 Transition metal selenides

Transition metal and main group polyselenides can conveniently be prepared and crystallized by utilizing the molten salt technique [76–78], in which alkali metal polychalcogenides are used as molten fluxes. The reactions are conducted in the temperature range 200–450 °C to ensure the stability of the produced materials. Transition metal polyselenides generally have extended 1D or 2D structures. There is a recent review on the synthesis of lanthanoid-containing heterometallic chalcogenides, which have been prepared by utilizing alkali metal polychalcogenide flux [79]. Polynuclear selenium- and tellurium-containing complexes are discussed in more detail in Chapter 10.

1.3 Elemental tellurium

1.3.1 General

Though the chemistry of tellurium and its compounds is mainly similar to that of selenium, there are also notable differences [2]. The general bonding features are significantly different compared to those of the lighter chalcogen congeners. The homonuclear Te–Te bond energies are lower (149 kJ mol^{-1}) than those involving selenium (192 kJ mol^{-1}) or sulfur (266 kJ mol^{-1}). The heteronuclear E-X bonds (E = Te, Se, S) follow the same trend (see Ref. [80] for the compilation of bond energies). Consequently, tellurium compounds are generally more labile and more air- and moisture-sensitive. Tellurium also has a weaker tendency for multiple bonding than selenium and sulfur.

By contrast, tellurium shows higher propensity to hypervalency [81] and secondary bonding interactions (SBI) [82]. This is due to the decreased energy difference between σ(Te-X) and σ*(Te-X) orbitals and stronger n(X)→ σ*(Te-X) interactions. This accounts for stronger intermolecular interactions in tellurium compounds compared to those in related selenium and sulfur species. Electrostatic effects further enhance the strength of the secondary bonding interactions due to the existence of σ-holes in tellurium (see Ref. [2] and references cited therein).

An overview of the use of elemental tellurium as a reagent is presented in Figure 1.9 with some illustrative examples.

1.3.2 Anionic tellurides

As in the case of selenium, the reduction of tellurium to telluride anions is the most convenient method for initial introduction of tellurium into reaction systems. The reduction has been carried out in a similar fashion to selenium, i.e. with alkali metals in liquid ammonia [20, 21, 23] or in dipolar aprotic solutions [23], sodium tetrahydridoborate [83–87], and lithium triethylhydridoaluminate [88, 89]. As the effects due to hypervalence and secondary bonding interactions are more significant in the case of tellurium than with the lighter chalcogen congeners [81, 82], tellurium shows more diverse polyanion chemistry than sulfur and selenium. Polytellurides can therefore exhibit charges that deviate from −2.

In addition to selenide and polyselenide anions, Björgvinsson and Schrobilgen [19] have also explored the product distribution of tellurium-containing chalcogenide and polychalcogenide anions in the liquid ammonia or in the ethylenediamine solution using ^{125}Te and ^{77}Se NMR spectroscopy. 2,2,2-crypt was added into the solutions to complex with the alkali metal cation and ensure narrow NMR resonances. As the stoichiometry of the starting materials was varied [see eq. (1.9)], the presence of Te_n^{2-} (n = 1–4) and HTe^- was inferred. The synthesis of stable salts of HE^- (E = Se, Te) in ionic

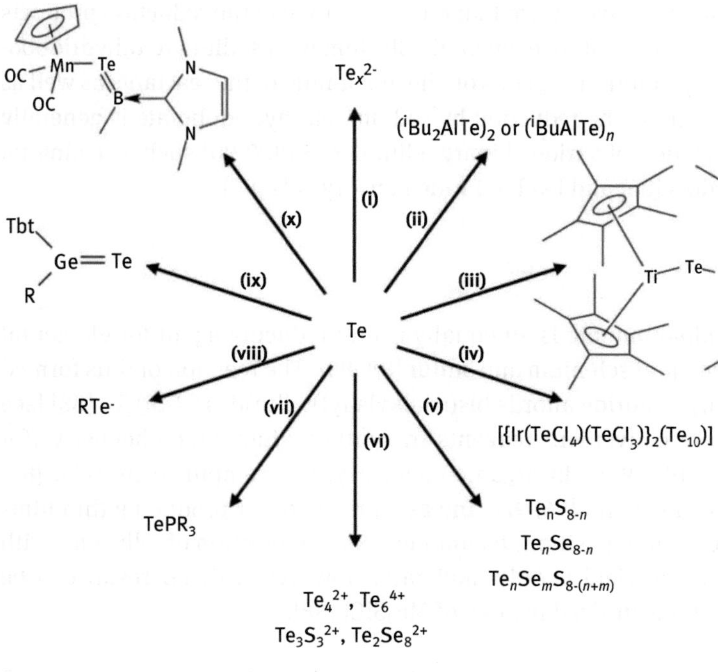

Figure 1.9: Illustrative examples on the reactivity of elemental tellurium. (i) $M/NH_3(l)$ (M = Li, Na, K), $NaBH_4$/THF or Li[$AlEt_3H$], (ii) $^iBu_2AlH$/toluene, (iii) [$Ti(C_5Me_5)_2H$]/toluene, (iv) Te, Ir, and $TeCl_4$/250 °C, (v) molten mixtures of tellurium, selenium, and sulfur, (vi) $SbF_5/SO_2(l)$, (vii) PR_3 (R = alkyl), (viii) RMgX/THF (X = Cl, Br) or RLi/THF, (ix) Ge(Tbt)RGe(Ph_2C_2)/benzene, (x) [MnCp(CO)$_2${EB(^{t}Bu)(C[NMe]$_2$C$_2$H$_2$)}].

liquids using large counter-cations has also recently been reported [90]. Their use in syntheses has been discussed in Chapter 8.

$$2\ \mathrm{K} + n\mathrm{Te} \rightarrow \mathrm{K_2Te}_n \tag{1.9}$$

Polytelluride and polyselenide anions could also be prepared from the stoichiometric amounts of K_2E and E (E = Te, Se) [eq. (1.10)]. In the case of mixtures of selenium and tellurium, the reaction affords also a series of tellurium selenide anions. The formation of open-chain potassium-cryptated salts of $TeSe_2^{2-}$ and $Te_xSe_{4-x}^{2-}$, as well as pyramidal $TeSe_3^{2-}$, has been reported [91].

$$\mathrm{K_2E} + n\mathrm{E} \rightarrow \mathrm{K_2E}_{n+1} \tag{1.10}$$

The mixing of Te_n^{2-} and Se_n^{2-}, which have been prepared upon reduction by metallic lithium in liquid ammonia, yielded in addition to open-chain $Te_2Se_2^{2-}$ and pyramidal $TeSe_3^{2-}$ anions also the spirocyclic anion $TeSe_{10}^{2-}$ and extended anionic networks $(Te_3Se_6)_n^{2n-}$ and $(Te_3Se_7)^{2n-}$ [92].

Basmadijan et al. [83] have shown that the distribution of the reduction products HTe^-, Te^{2-}, and Te_2^{2-} upon treating elemental tellurium with sodium tetrahydridoborate $NaBH_4$ in aqueous solution depends on the molar ratio of the reactants as well as on the pH of the solution. The reduction by sodium tetrahydridoborate is generally applied in the production of various binary tellurides [84][88-90] such as cadmium telluride quantum dots [86] and lead telluride nanocrystals [84].

1.3.3 Derivatives of anionic tellurides

Lithium ethylhydridoaluminate is an equally useful reducing agent for elemental tellurium as for elemental selenium and sulfur [88, 89]. The reaction of thus formed Te_x^{2-} with trialkylsilyl chloride affords bis(trialkylsilyl)tellurides $(R_3Si)_2Te$ [88] [see eq. (1.11)], which are convenient reagents in synthetic inorganic chemistry (for examples, see Refs. [93, 94]), in organic chemistry, as exemplified in Refs. [88, 95], in coordination chemistry [96, 97], and as a precursor for producing thin films by the ALD (atomic layer deposition) technique [98]. The reaction of tellurides with Me_3SiCl is slow and the yield is only moderate. However, a better result can be obtained if $^tBuMe_2SiCl$ is utilized instead of Me_3SiCl [88].

$$1/xTe_x^{2-} + 2\ R_3SiCl \rightarrow (R_3Si)_2Te + 2\ Cl^- \quad (1.11)$$

The reduction of tellurium by lithium triethylhydridoaluminate in THF followed by treatment with TeX_4 (X = Cl, Br) [eq. (1.12)] affords Te_2X_2 [99]. Ditellurium dichloride is a yellow liquid, which, though unstable, can be stored for hours in an inert atmosphere. Ditellurium dibromide is orange-red liquid, which is more stable than the corresponding chloride. These tellurium(I) reagents have been used to prepare the heteroatomic rings 1,2-Te_2S_5 and 1,2-Te_2Se_5 by reactions with $[Ti(C_5H_5)_2E_5]$ (E = S, Se) [eq. (1.13)].

$$Te_x^{2-} + xTeX_4 \rightarrow xTe_2X_2 + 2xX^- (X = Cl, Br) \quad (1.12)$$

$$[Ti(C_5H_5)_2E_5] + Te_2X_2 \rightarrow 1,2-Te_2E_5 + [Ti(C_5H_5)_2X_2] \quad (1.13)$$

Na_2Te or NaTeH can also be utilized in the syntheses of tellurium macrocycles, as reported by Panda [100].

1.3.4 Tellurium-containing chalcogen rings

Tellurium forms both binary and tertiary heterocyclic molecules with selenium and sulfur, though the systems involving tellurium are much simpler than the binary system of selenium and sulfur (see Section 1.2.4) [41, 57]. As in the case of the Se_nS_{8-n} molecules, the eight-membered rings TeS_7 and 1,2-, 1,3-, and 1,4-Te_2S_6 are formed in the binary melt [101], while the ternary species 1,2-, 1,3-, 1,4-, and 1,5-$TeSeS_6$ are observed in sulfur-rich, sulfur–selenium–tellurium melts, as indicated by the ^{125}Te NMR spectrum of a molten mixture of sulfur (97 mol%), selenium (1,5 mol%), and tellurium (1.5 mol%) shown in Figure 1.10.

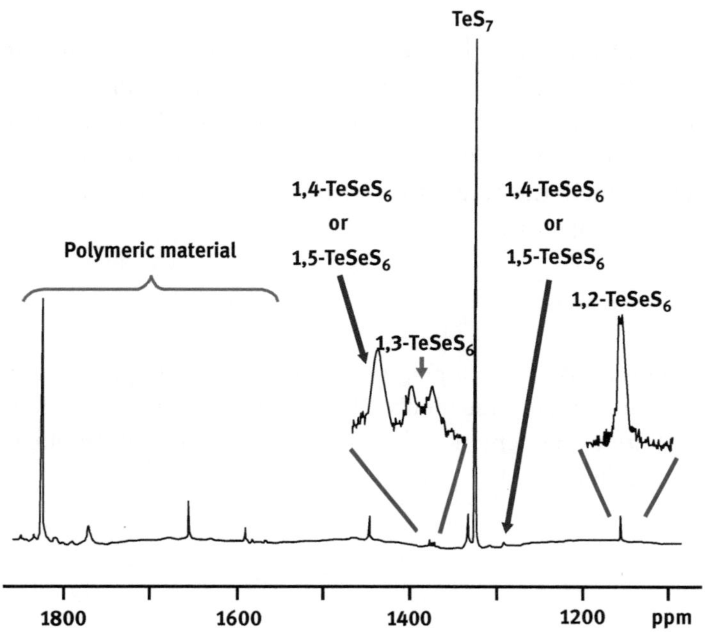

Figure 1.10: ^{125}Te NMR spectrum of ternary sulfur–selenium–tellurium melt containing ^{77}Se-enriched selenium (1.5 mol%) and ^{125}Te-enriched tellurium (1.5 mol%) (enrichment in both cases 92 %). The spectrum has been recorded at 145 °C [101]. Adapted with permission from Chivers T, Laitinen RS, Schmidt KJ, Taavitsainen, J. Inorg Chem 1993, 32, 337–340. Copyright 1993 American Chemical Society.

1.3.5 Homopolyatomic tellurium cations

Tellurium can be oxidized to several different homopolyatomic cations in the same fashion as selenium [1][62,63], e.g. Te_4^{2+}, Te_6^{2+}, Te_6^{4+}, Te_8^{2+}, Te_8^{4+}, etc. However, applications of these cations for the synthesis of other tellurium compounds have not been reported.

1.3.6 Triorganylphosphine tellurides

Triorganylphosphine tellurides can be prepared from the corresponding phosphine and elemental tellurium [102, 103]. Trialkylphosphine tellurides are sufficiently stable for subsequent synthetic work and have been widely used in the synthesis of semi-conducting metal tellurides [104–109]. More detailed discussion on the phosphorus–tellurium chemistry is presented in Chapter 5.

1.3.7 Grignard and organyllithium reagents

As in the case of selenium, both Grignard reagents and organyllithium reagents are efficient in producing organyltellurolates [110], which can easily be oxidized to the corresponding ditellurides that are themselves useful reagents in synthetic inorganic and organic chemistry (for some examples in different areas of chemistry, see Refs. [111]-[118]). The reactions of ditellurides are discussed in Chapters 5–11 and the ligand chemistry of tellurolates in Chapter 11.

1.3.8 Silane- and germanetellones

Recent years have seen progress in developing synthetic strategies to prepare silicon–tellurium and germanium-tellurium double bonds. A stable dialkylsilylene reacts with elemental tellurium or selenium to form a silanetellone and silaneselone, respectively [112], as shown in Figure 1.11.

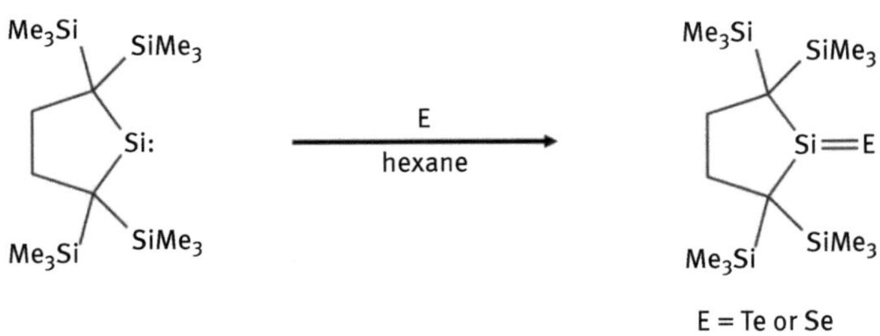

Figure 1.11: Preparation of silanechalcogenones [112].

Tokitoh et al. [113] carried out a direct telluration of germylene Tbt(R)Ge: containing very bulky substituents on germanium (Tbt = 2,4,6-tri[tris(trimethylsilyl)methyl]phenyl; R = 2,4,6-triisopropylphenyl or bis(trimethylsilyl)methyl) and obtained thermally stable germanetellones (see Figure 1.12), which were moisture-sensitive. The presence of a Ge=Te double bond was verified by the X-ray structure determination. The treatment of the germanetellone with mesitonitrile oxide or 2,3-dimethyl-1,3-butadiene resulted in [3 + 2] or [4 + 2] cycloaddition products, respectively.

1.3.9 Group 13 chemistry with tellurium

Tellurium is the heaviest element known to form a double bond with boron [114]. The reaction involves the insertion of tellurium into the Mn=B bond in the base-stabilized

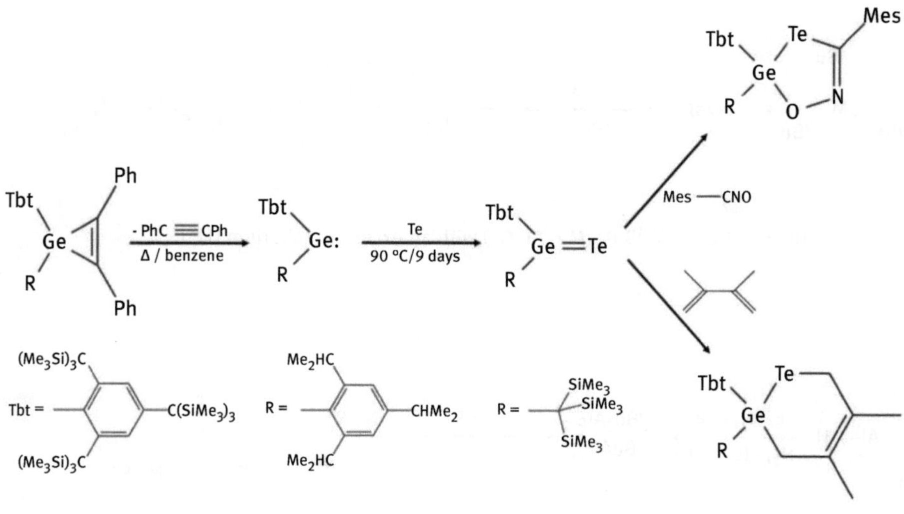

Figure 1.12: Formation of germanetellone and its cycloaddition reactions [113].

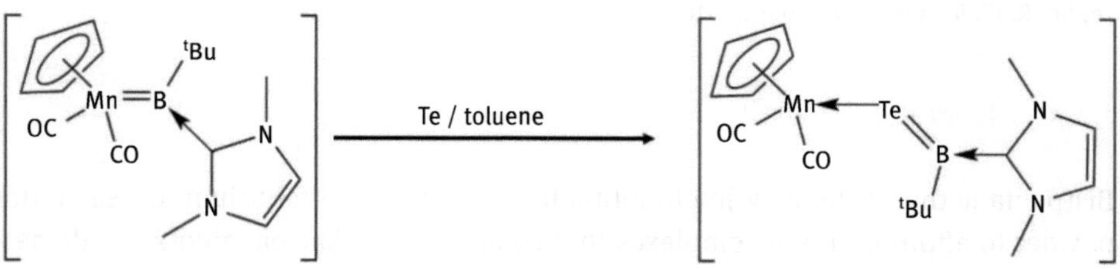

Figure 1.13: The formation of a B=Te double bond [114].

complex [Mn{=B(^{t}Bu)C[(MeN)$_2$(CH$_2$)]$_2$}(C$_5$H$_5$)(CO)$_2$], as shown in Figure 1.13. Lighter chalcogens react in a similar fashion.

An excess of elemental tellurium in refluxing toluene reacts with M(^{t}Bu)$_3$ (M = Al, Ga) to form the cubane [M(^{t}Bu)(μ_3-Te)]$_4$ and ^{t}BuTeTe^{t}Bu [115, 116] (see Figure 1.14). The structure of the tetrameric cage molecule was deduced by comparison of the spectroscopic properties with those of [M(^{t}Bu)(μ_3-Se)]$_4$ and [M(^{t}Bu)(μ_3-S)]$_4$. The crystal structure of [Ga(^{t}Bu)(μ_3-S)]$_4$ is known [117].

Elemental tellurium and selenium also react with diisobutylaluminium hydride ^{i}Bu$_2$AlH in toluene at 120–130 °C [118]. Mass spectrometric and NMR spectroscopic information indicated the formation a mixture of the dimer (^{i}Bu$_2$AlE)$_2$ and the oligomer (^{i}Bu$_2$AlE)$_n$ together with an unspecified product containing carbon, chalcogen, and aluminium. These *in situ* reagents react with amides to give chalcogenamides (see Figure 1.15). While selenoamides are relatively stable, telluroamides easily decompose.

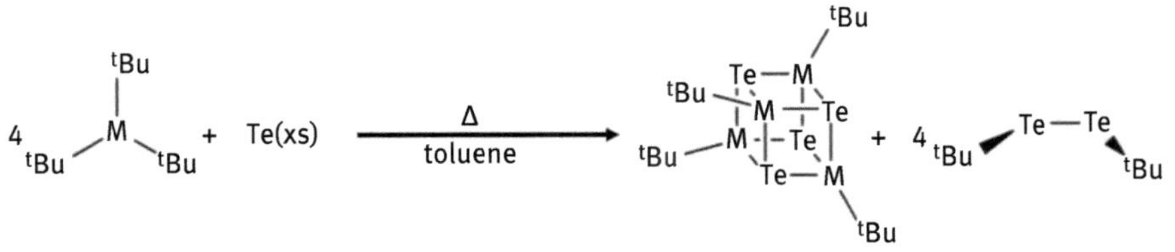

Figure 1.14: The reaction of M(^{t}Bu)$_3$ (M = Al, Ga) with an excess of tellurium [115].

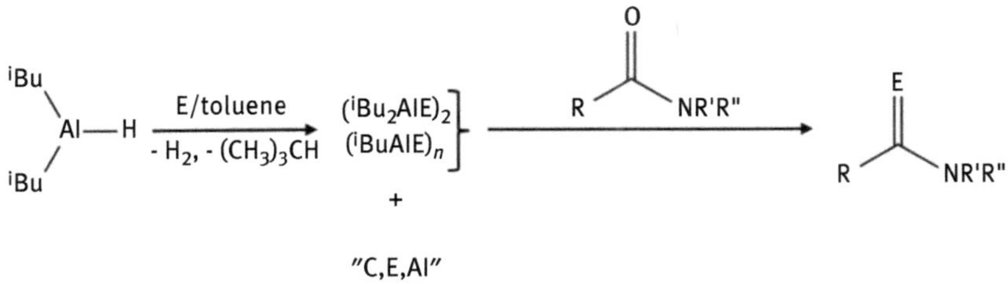

Figure 1.15: The reaction of tellurium and selenium with diisobutylaluminium hydride and amides (E = Te, Se; R, R', R" = organyl groups) [118].

1.3.10 Titanocene chemistry

Bis(pentamethylpentadienyl)hydridotitanium reacts with tellurium or selenium powder to afford titanium complexes that contain monochalcogenido(2-) or dichalcogenido(2-) ligands (see Figure 1.16) [119, 120].

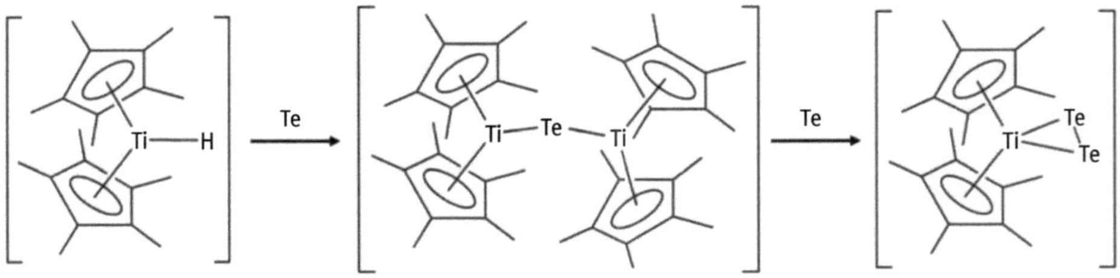

Figure 1.16: The reaction of [Ti(C$_5$Me$_5$)$_2$H] with tellurium [119].

While both tellurium and selenium react with [Ti(C$_5$Me$_5$)$_2$H], the reaction proceeds more smoothly in the case of tellurium, if tributylphosphine telluride is used as a tellurium source [119, 120]. Interestingly, [Ti(C$_5$Me$_5$)$_2$Te$_2$] is mononuclear, while the related complex, which involves the methylcyclopentadienyl ligand coordinated to titanium, is dinuclear [Ti(C$_5$H$_4$Me)$_2$(μ-Te$_2$)$_2$Ti(C$_5$H$_4$Me)$_2$] [92]. This appears to be due to steric effects.

1.3.11 Solid-state and hydrothermal reactions

Intriguingly, solid-state (hydrothermal) reactions of tellurium with transition metals and tellurium(IV) tetrahalides can produce materials that incorporate unusual homoatomic tellurium rings. For example, prolonged heating of a mixture of elemental tellurium with a late transition metal and tellurium tetrachloride in an evacuated ampoule at 250–300 °C produces $[\{M(TeX_4)(TeX_3)\}_2(Te_{10})]$ and $[\{M(TeI_4)(TeI_2)\}_2(TeI_4)(Te_2I_2)(Te_{10})]$ (M = Rh, Ir; X = Cl, Br), as well as $[\{Ru(TeI_4)(TeI_2)\}_2(Te_{10})]$, which all contain a neutral tellurium cage molecule Te_{10} [121, 122]. The same methodology involving tellurium, suitable metals, and tellurium tetrahalogenides yields $[Ru(Te_9)][InCl_4]_2$ and $[Ru(Te_8)]Cl_2$ [123], $[M(Te_6)]X_3$ (M = Rh, Ir; X = Cl, Br, I), and $[Ru_2(Te_6)(TeBr_3)_4(TeBr_2)_2]$ [124], all of which are one-dimensional coordination polymers containing an uncharged homonuclear tellurium ring.

The hydrothermal reaction of tellurium or selenium with silver and hydrogen iodide at 220 °C afforded $[(AgI)_2(Te_6)]$ or the isomorphous selenium analogue $[(AgI)_2(Se_6)]$ [125].

1.4 Summary

Progress in the synthetic chemistry of selenium and tellurium is largely due to the development of a variety of reagents that can be applied in both inorganic and organic chemistry, as well as in materials science. Many of these useful reagents are available commercially, for instance, elemental selenium and tellurium, selenium and tellurium dioxide, selenium and tellurium tetrahalogenides, some inorganic and organic selenides and tellurides, as well as diselenides and ditellurides. They can be used to synthesize more complex species, which can themselves act as precursors for further applications.

This chapter is an overview of the different reactions involving elemental selenium and tellurium. The synthetic details and reactions of their derivatives will be discussed in greater detail in chapters 2–11. The first step in synthetic strategies often involves the reduction of the elements to chalcogenide or polychalcogenide ions, since they are stronger nucleophiles than the elements themselves. The most convenient reducing agents are alkali metals in liquid ammonia, sodium tetrahydridoborate, and lithium triethylhydridoaluminate ("superhydride"). Alkali metal and alkaline earth metal chalcogenides can also be prepared by direct mixing of the elements at high temperatures, by the reaction of chalcogen elements and alkali metal carbonates in solution, by solvothermal reactions, or by oxidation–reduction reactions involving Lewis acids.

Unlike commercially available SeX_4 and TeX_4 (X = Cl, Br, I), the less-stable reagents, such as $SeCl_2$ or $SeBr_2$, need to be prepared and used *in situ*. This can conveniently be accomplished by the treatment of elemental selenium with sulfuryl chloride or bromide in THF. Both dihalides are stabilized in coordinating solvents. Tellurium dihalogenides cannot be prepared by the same route. Elemental selenium and tellurium also react with the Grignard reagents or organyllithium compounds affording organyl chalcogenolates that can easily be oxidized into organyl dichalcogenides. The latter play a role among the key reagents in organic and organometallic syntheses.

There are several methods to oxidize elemental selenium and tellurium to homo- and heteropolychalcogen cations. Their most significant feature is that they have been a source of new understanding of chemical bonding in main group compounds and have provided information of the relationships between the electronic structures and molecular structures.

References

[1] Krossing I. Structure and bonding of the neutral chalcogens and their polyatomic cations. In: Devillanova FA, Du Mont W-W, editor(s). Handbook of chalcogen chemistry: new perspectives in sulfur, selenium and tellurium, 2nd ed. Vol. 1. Cambridge, UK: RSC Publishing. ch. 7.1 2013:349–81.

[2] Chivers T, Laitinen RS. Tellurium: a maverick among the chalcogens. Chem Soc Rev. 2015;44:1725–39.

[3] Back TG, editor. Organoselenium chemistry – a practical approach. Oxford, UK: Oxford University Press, 1999.

[4] Wirth T, editor. Organoselenium chemistry – modern developments in organic synthesis. Topics in current chemistry, Vol. 208. Berlin, Germany: Springer, 2000.

[5] Chivers T. A guide to chalcogen-nitrogen chemistry. New Jersey: World Scientific Publishing, 2005.

[6] Devillanova FA, editor. Handbook of chalcogen chemistry: new perspectives in sulfur, selenium and tellurium. Cambridge UK: RSC Publishing, 2007.

[7] Woollins JD, Laitinen RS, editors. Selenium and tellurium chemistry; from small molecules to biomolecules and materials. New York USA: Springer, 2011.

[8] Wirth T, editor. Organoselenium chemistry. Weinheim Germany: Wiley-VCH GmbH & Co. KGaA, 2012.

[9] Patai S, Rappoport Z, editors. The chemistry of organic selenium and tellurium compounds, Vol. 3, Part 2. Chichester UK: John Wiley & Sons Ltd, 2012.

[10] Devillanova FA, Du Mont W-W, editors. Handbook of chalcogen chemistry, new perspectives in sulfur, selenium and tellurium, 2nd ed. Cambridge UK: RSC Publishing, 2013.

[11] Jain VK, Priyadarsini KI, editors. Organoselenium compounds in biology and medicine. Cambridge UK: RSC Publishing, 2018.

[12] Poeppelmeier K, Reedijk J, editors. Comprehensive inorganic chemistry II, 2nd ed., Vol. 1. Amsterdam: Elsevier, 2013.

[13] Kanatzidis MG, Huang SP. Coordination chemistry of heavy polychalcogen ligands. Coord Chem Rev. 1994;130:509–621.

[14] Kanatzidis MG, Das BK. Superheated solvent media for organometallic (poly)chalcogenide cluster synthesis. Comm Inorg Chem. 1999;21:29–51.

[15] Bergström FW. The polysulfides and polyselenides of lithium, sodium and potassium. J Am Chem Soc. 1926;48:146–51.

[16] Klemm W, Sodomann H, Langmesser P. Beiträge Zur Kenntnis der Alkalimetallchalkogenide. Z Anorg Allg Chem. 1939;241:281–304.

[17] Brandsma L, Wijers H. A simple laboratory scale preparation method for dialkyl sulfides, selenides, tellurides, thiols, and selenols. Rec Trav Chim Pays-Bas. 1963;82:68–74.

[18] Müller V, Frenzen G, Dehnicke K. Synthese, FIR-Spektren und Kristallstrukturen der Pentaselenide. Z Naturforsch. 1992;47b:205–10.

[19] Björgvinsson M, Schrobilgen GJ. Homo- and heteropolychalcogenide anions Ch^{2-}, HCh^{-}, Ch_2^{2-}, Ch_3^{2-}, and Ch_4^{2-} (Ch = Se and/or Te): solution ^{1}H, ^{77}Se, ^{123}Te, and ^{125}Te NMR study. Inorg Chem. 1991;30:2540–7.

[20] Colombara D, Goncalves A-M, Etcheberry A. Synthesis of K_2Se solar cell dopant in liquid NH_3 by solvated electron transfer to elemental selenium. Electrochem Commun. 2018;93:44–8.

[21] Thiele G, Vondung L, Donsbach C, Pulz S, Dehnen S. Organic cation and complex cation-stabilized (poly-)selenides, $[cation]_x(Se_y)_z$: diversity in structures and properties. Z Anorg Allg Chem. 2014;640:2684–700.

[22] Kysliak O, Beck J. Zn, Mn, and Fe hexammine and hexakis(methylamine) polychalcogenides – ionic compounds by solvothermal synthesis. Inorg Chem Commun. 2013;38:146–51.

[23] Sandman DJ, Stark JC, Acampßora LA, Gagne P. A direct broadly applicable approach to the synthesis of aromatic molecular and supramolecular selenium and tellurium compounds. Organometallics. 1983;2:549–51.

[24] Thompson DP, Boudjouk P. A convenient synthesis of alkali metal selenides and diselenides in tetrahydrofuran and the reactivity differences exhibited by these salts toward organic bromides. Effect Ultrasound J Org Chem. 1988;53:2109–12.

[25] Gautheron B, Degrand C. Electroreduction of selenium at a mercury pool cathode in aprotic media. Application to the electrosynthesis of dibenzyl selenide and diselenide. J Electroanal Chem. 1984;163:415–8.

[26] Gautheron B, Tainturier G, Degrand C. Ultrasound-induced electrochemical synthesis of the anions Se_2^{2-}, Se^{2-}, Te_2^{2-}, and Te^{2-}. J Am Chem Soc. 1985;107:5579–81.

[27] Ishihara H, Koketsu M, Fukuta Y, Nada F. Reaction of lithium aluminum hydride with elemental selenium: its application as a selenating reagent into organic molecules. J Am Chem Soc. 2001;123:8408–9.

[28] Klayman DL, Griffin TS. Reaction of selenium with sodium borohydride in protic solvents. A facile method for the introduction of selenium into organic molecules. J Am Chem Soc. 1973;95: 197–9, and references cited therein.

[29] Gladysz JA, Hornby JL, Garbe JE. Convenient one-flask synthesis of dialkyl selenides and diselenides via lithium triethylborohydride reduction of Se_x. J Org Chem. 1978;43:1204–8.

[30] Cusick J, Dance I. The characterization of $[HSe]^{-}$ and $[Se_x]^{2-}$ ions by ^{77}Se NMR. Polyhedron. 1991;10:2629–40.

[31] Marques de Oliveira AR, Piovan L, Simonelli F, Barison A, de Fatima de Santos M, Marcondes de Mello MB. A ^{77}Se NMR study of elemental selenium reduction using $NaBH_4$. J Organomet Chem. 2016;806:54–9.

[32] Raymond CC, Dick DL, Dorhout PK. Speciation of main-group metal ions by electrospray mass spectrometry. 1. Investigation of aqueous polyselenide species and effects of cations and pH. Inorg Chem. 1997;36:2678–81.

[33] Zhang J, Zhang S, Zhang H, Zhang Y, Zheng Z, Xiang Y. Activated selenium for promoted formation of metal selenide nanocrystals in solvothermal synthesis. Mater Lett. 2014;122:306–8.

[34] Fenske D, Adel J, Dehnicke K. The crystal structure of bis(η5-cyclopentadienyl)titana(TiIV) pentaselenacyclohexane, Cp_2TiSe_5. Z Naturforsch. 1987;42b:931–3.

[35] Albrecht N, Weiss J. Bis(η^5-cyclopentadienyl)(pentaselenido)metal complexes, Cp_2MSe_5, of titanium, zirconium, and hafnium and (μ_2-O)(μ_2-Se_4)($Cp_2Hf)_2$, a binuclear compound with tetraselenide and oxygen bridges. J Organomet Chem. 1988;355:89–98.

[36] Pekonen P, Hiltunen Y, Laitinen RS. The selenium-selenium coupling in bis(cyclopentadienyl) titanium pentaselenide $[Ti(C_5H_5)_2Se_5]$. Acta Chem Scand. 1989;43:914–6.

[37] Giolando DM, Papavassiliou M, Pickardt J, Rauchfuss TB, Steudel R. Synthesis and structure of 1,4-$[(RCp)_2Ti]_2Se_4$ and its application to the chalcogenospecific synthesis of 1,2,5,6-Se_4S_4. Inorg Chem. 1988;27:2596–600.

[38] Steudel R, Papavassiliou M, Strauss E-M, Laitinen R. Sulfur compounds 95. A series of selenium rich chalcogen ring (Se_5S, Se_5S_2, Se_7) from titanocene pentaselenide. Angew Chem Int Ed Engl. 1986;25:99–101.

[39] Pekonen P, Hiltunen Y, Laitinen RS, Pakkanen TA. Chalcogen ring interconversion pathways. ^{77}Se NMR spectroscopic study of the decomposition of 1,2,3,4,5-Se_5S_2 to 1,2,3,4,5,6-Se_6S_2 and 1,2,3,4-Se_4S_2. Inorg Chem. 1991;30:3679–82.

[40] Steudel R. The synthesis of sulfur- and selenium-containing organic and inorganic rings from titanocene precursors. Studies Inorg Chem. 1992;14:233–53.

[41] Pekonen P, Suontamo R, Laitinen RS. Homo- and heteroatomic chalcogen rings. Coord Chem Rev. 1994;130: 1–62, and references cited therein.

[42] Steudel R, Eckert B. Solid sulfur allotropes. Top Curr Chem. 2003;230:1–79.

[43] Steudel R, Jensen D, Papavassiliou M. Sulfur and selenium compounds 123. Dinuclear titanocene chalcogenides as sulfur and selenium transfer reagents in ring synthesis. Phosphorus Sulfur Silicon Relat Chem. 1989;41:349–53.

[44] Papavassiliou M, Pickardt J, Steudel R. Sulfur compounds 153. Synthesis, structure and reactions of $Cp'_2TiSe_3S_2$ and Cp'_2TiSe_4S ($Cp' = \eta^5$-$C_5H_4CH_3$). Phosphorus Sulfur Silicon Relat Elem. 1992;65:161–4.

[45] Pekonen P, Hiltunen Y, Laitinen RS, Valkonen J. ^{77}Se NMR spectroscopic and X-ray crystallographic characterization of bis(cyclopentadienyl)titanium selenide sulfide mixtures, $[Ti(C_5H_5)_2Se_xS_{5-x}]$. Inorg Chem. 1991;30:1874–8.

[46] Silva L, Affeldt RF, Lüdtke DS. Synthesis of glycosyl amides using selenocarboxylates as traceless reagents for amide bond formation. J Org Chem. 2016;81:5464–73.

[47] Potapov VA, Amosova SV. New methods for preparation of organoselenium and organotellurium compounds from elemental chalcogens. Russ J Org Chem. 2003;39: 1373–80, and references cited therein.

[48] Giribabu K, Suresh R, Manigandan R, Thirumal E, Stephen A, Narayanan V. Aqueous based synthesis of Cu_5Se_4 nanosheets. J Mater Sci. 2013;24:1888–94.

[49] Khanna PK, Subbarao VVVS, Wagh M, Jadhav P, Patil KR. Synthesis of fine PbE (E=S, Se) powder from direct in situ reduction of sulphur or selenium. Mater Chem Phys. 2005;93:91–4.

[50] McCarthy CL, Downes CA, Brutchey RL. Room temperature dissolution of bulk elemental Ni and Se for solution deposition of $NiSe_2$ HER electrocatalyst. Inorg Chem. 2017;56:10143–6.

[51] Bastola E, Bhandari KP, Matthews AJ, Shrestha N, Ellingson RJ. Elemental anion thermal injection synthesis of nanocrystalline marcasite iron dichalcogenide $FeSe_2$ and $FeTe_2$. RSC Adv. 2016;6:69708–14.

[52] Tan L, Fu J, Liu S. Growth of photoluminescent Ag_2Se nanowires from simple precursor solution. CrystEngComm. 2014;16:10534–8.

[53] Lamoureux M, Milne J. The disproportionation of diselenium dichloride Se_2Cl_2 and diselenium dibromide Se_2Br_2. Can J Chem. 1989;67:1936–41.

[54] Milne J, Williams AJ. Exchange process in diselenium and selenium-sulfur dihalides, Se_2X_2 and $SeSX_2$ (X = Br, Cl). Inorg Chem. 1992;31:4534–8.

[55] Maaninen A, Chivers T, Parvez M, Pietikäinen J, Laitinen RS. Syntheses of THF solutions of SeX_2 (X = Cl, Br) and a new route to selenium sulfides Se_nS_{8-n} (n = 1-5); crystal structures of $SeCl_2(tht)_2$ and $SeCl_2$•tmtu (tmtu = tetramethylthiourea). Inorg Chem. 1999;38:4093–7.
[56] Laitinen RS, Pakkanen TA. ^{77}Se NMR spectroscopic characterization of selenium sulfide ring molecules Se_nS_{8-n}. Inorg Chem. 1987;26:2598–603.
[57] Laitinen RS, Oilunkaniemi R. Catenated compounds: group 16 (Se, Te). In: Poeppelmeier K, Reedijk J, editors. Comprehensive inorganic chemistry II, 2nd ed. Amsterdam: Elsevier, 2013: 197–231.
[58] Ahmed E, Ruck M. Homo- and heteroatomic polycations of groups 15 and 16. Recent advances in synthesis and isolation using room temperature ionic liquids. Coord Chem Rev. 2011;255: 2892–903, and references cited therein.
[59] Beck J. Rings, cages and chains – the rich structural chemistry of the polycations of the chalcogens. Coord Chem Rev. 1997;163: 55–70, and references cited therein.
[60] Artem'ev AV, Chernysheva NA, Yas'ko SV, Gusarova NK, Bagryanskaya IY, Trofimov BA. Straightforward solvent-free synthesis of tertiary phosphine chalcogenides from secondary phosphines, electron-rich alkenes, and elemental sulfur or selenium. Heteroatom Chem. 2016;27:48–53
[61] Artem'ev AV, Chernysheva NA, Gusarova NK, Yas'ko SV, Liao J-H, Liu CW, et al. Reaction of vinyl selenides with secondary phosphines and elemental selenium: one-pot selective synthesis of a new family of diselenophosphinic Se-esters. Heteroatom Chem. 2014;25:135–9
[62] Kuwajima I, Shimizu M, Urabe H. Oxidation of alcohols with *tert*-butyl hydroperoxide and diaryl diselenide. J Org Chem. 1982;47: 837–42, and references cited therein.
[63] Jen K-Y, Cava MP. Improved synthesis of aromatic diselenides. J Org Chem. 1983;48:1449–51. and references cited therein.
[64] Oilunkaniemi R, Laitinen RS, Ahlgrén M. The solid state conformation of diaryl ditellurides and diselenides: the crystal and molecular structures of $(C_4H_3E)_2E'_2$ (E = O, S; E' = Te, Se). Z Naturforsch. 2000;55b:361–8.
[65] Ilie A, Crespo O, Gimeno MC, Holthausen MC, Laguna A, Diefenbach M, et al. (N,Se) and (Se,N,Se) ligands based on carborane and pyridine fragments – reactivity of 2,6-[(1′-Me-2′,3′-*closo*-$C_2B_{10}H_{10}$)$SeCH_2$]$_2C_5H_3N$ towards copper and silver. Eur J Inorg Chem. 2017;2643–52
[66] Gujral G, Gulati SD, Bhasin KK, Potapov VA, Amosova SV. Synthesis and characterization of unsymmetric 4-picolyl selenides. Phosphorus Sulfur Silicon. 2016;191:55–8.
[67] Lu F. Elemental sulfur as soft oxidant for facile one-pot preparation of air-sensitive organometallic di[5]trovacenyldichalcogenides. Inorg Chem Commun. 2013;37:148–50.
[68] Karjalainen MM, Sanchez-Perez C, Rautiainen JM, Oilunkaniemi R, Laitinen RS. Chalcogen-chalcogen secondary bonding interactions in trichalcogenaferrocenophanes. CrystEngComm. 2016;18: 4538–45, and references cited therein.
[69] Chan Y-C, Li Y, Gangly R, So C-W. Acyclic amido-containing silanechalcogenones. Eur J Inorg Chem. 2015;3821–4
[70] Matsumoto T, Tokitoh N, Okazaki R. The first kinetically stabilized germanethiones and germaneselenones: syntheses, structures, and reactivities. J Am Chem Soc. 1999;121:8811–24.
[71] Karwasara S, Yadav D, Jha CK, Rajaraman G, Nagendran S. Single-step conversion of silathiogermylene to germaacid anhydrides; unusual reactivity. Chem Commun. 2015;51:4310–3.
[72] Karwasara S, Siwatch RK, Jha CK, Nagendran S. Aminotroponiminatosilathio- and siloxygermylenes: reactivity comparison. Organometallics. 2015;34:3246–54.
[73] Ito M, Tokitoh N, Kawashima T, Okazaki R. Formation of borylene by photolysis of an overcrowded bis(methylseleno)borane. Tetrahedron Lett. 1999;40:5557–60.

[74] Braunschweig H, Dellermann T, Ewing WC, Kramer T, Schneider C, Ullrich S. Reductive insertion of elemental chalcogens into boron-boron multiple bonds. Angew Chem Int Ed. 2015;54:10271–5.
[75] Braunschweig H, Constantinidis P, Dellermann T, Ewing WC, Fischer I, Hess M, et al. Highly strained heterocycles constructed from boron-boron multiple bonds and heavy chalcogens. Angew Chem Int Ed. 2016;55:5606–9
[76] Kanatzidis MG. Molten alkali-metal polychalcogenides as reagents and solvents for the synthesis of new chalcogenide materials. Chem Mater. 1990;2:353–63.
[77] (a) Kanatzidis MG, Sutorik AC. The application of polychalcogenide salts to the exploratory synthesis of solid state multinary chalcogenides at intermediate temperatures. Prog Inorg Chem. 1995;43:151–265; (b) Sheldrick WS, Wachhold M. Solventothermal synthesis of solid-state chalcogenidometalates. Angew Chem Int Ed Engl. 1997;36:206–24.
[78] Sheldrick WS. Polychalcogenides. In: Devillanova FA, Du Mont W-W, editors. Handbook of chalcogen chemistry – new perspectives in sulfur, selenium and tellurium, 2nd ed., Vol. 1. Cambridge: RSC Publishing, 2013: 514–45.
[79] Zhou J. Synthesis of heterometallic chalcogenides containing lanthanide and group 13–15 metal elements. Coord Chem Rev. 2016;315:112–34.
[80] Housecroft CE, Sharpe AG. Inorganic chemistry, 5th ed. Harlow, UK: Pearson Education Ltd., 2018.
[81] Hayashi S, Nakanishi W Hypervalent compounds. In: Devillanova FA, Du Mont W-W, editors. Handbook of chalcogen chemistry: new perspectives in sulfur, selenium and tellurium, 2nd ed., Vol. 2. Cambridge: RSC Publishing, 2013: 335–72.
[82] Cozzolino AF, Elder PJW, Vargas-Baca I. A survey of tellurium-centered secondary-bonding supramolecular synthons. Coord Chem Rev. 2011;255:1426–38.
[83] Basmadijan GP, Parker GR, Magarian RA, Ice RD, Mills SL, Kirschner AS. Synthesis of tellurium-123 m-labeled nucleophiles: potential use in the synthesis of new tellurium-123 m-labeled radiopharmaceuticals. Int J Appl Rad Isot. 1980;31:517–20.
[84] Ramasamy K, Nejo AO, Ziqubu N, Rajasekhar PVSR, Nejo AA, Revaprasadu N, et al. A new route to lead chalcogenide nanocrystals. Eur J Inorg Chem. 2011;5196–201
[85] Yao D, Liu Y, Zhao W, Wei H, Luo X, Wu Z, et al. A totally phosphine-free synthesis of metal telluride nanocrystals by employing alkylamides to replace alkylphosphines for preparing highly reactive tellurium precursors. Nanoscale. 2013;5:9593–7
[86] Kumar J, Mahesh HM. Concentration-dependent optical properties of TGA stabilized CdTe quantum dots synthesized via the single injection hydrothermal method in the ambient environment. Superlatt Microstruct. 2017;104:118–27.
[87] Kumar A, Dittmer DC. A catalytic process for the transposition of allylic hydroxyl groups and carbon-carbon double bonds. Tetrahedron Lett. 1994;35:5583–6.
[88] Detty MR, Seidler MD. Bis(trialkylsilyl) chalcogenides. 1. Preparation and reduction of group 6A oxides. J Org Chem. 1982;47:1354–6.
[89] Taavitsainen J, Laitinen RS. Tellurium-containing chalcogen rings. Main Group Chem. 1999;3:59–67.
[90] Finger LH, Sundermeyer J. Halide-free synthesis of hydrochalcogenide ionic liquids of the type [cation][HE] (E = S, Se, Te). Chem Eur J. 2016;22:4218–30.
[91] Björgvinsson M, Sawyer JF, Schrobilgen GJ. Crystal structures of potassium cryptated salts of the TeSe22-, pyramidal TeSe32-, and mixed compounds containing pyramidal TeSe32- and chain Te*x*Se4-*x*2- anions and 77Se and 125Te solution NMR studies of the pyramidal selenothiotellurite anions TeS*m*Se3-*m*2- (*m* = 0-3). Inorg Chem. 1991;30:4238–4245
[92] Sekar P, Ibers JA. Syntheses and characterization of some mixed Te/Se polychalcogenide anions $[Te_mSe_n]^{2-}$. Inorg Chem. 2004;43:5436–41.

[93] Fenske D, Grissinger A. Zur Reaktion von [Cp'_2TiCl_2] und [$Cp'TiCl_3$] mit $E(SiMe_3)_2$ (E = Se, Te; Cp' = $C_5H_4CH_3$). Die Kristalstrukturen von [$Cp'_4Ti_2Te_4$], [$Cp'_5Ti_5Se_6$] und [$Cp'_5Ti_5Te_6$]. Z Naturforsch. 1990;34b:1309–13.
[94] Haas A, Kutsch HJ. Preparation and characterization of hexakis(trifluoromethyl)digermachalcogenides. Chem Ber. 1988;121:803–4.
[95] Sewing D, Du Mont -W-W, Pohl S, Saak W, Lenoir D. Diacyltelluride: synthese durch Reaktionen von Acylchloriden mit Bis(trialkylsilyl)telluriden; Strukturbestimmungen an Di(1-adamantoyl) tellurid und Adamantancarbonsäureanhydrid. Z Anorg Allg Chem. 1998;624:1363–8.and references cited therein
[96] Borecki A, Corrigan JF. New copper and silver trimethylsilylchalcogenolates. Inorg Chem. 2007;46: 2478–84, and references cited therein.
[97] Merzweiler K, Kraus H. New organometallic tin-tellurium compounds. Syntheses and crystal structures of [$\{Cp(CO)_2Fe\}_4Sn_2Te_2$] and [$\{Cp(CO)_3Mo\}_4Sn_4Te_6$]. Z Naturforsch. 1994;49b:621–6.
[98] Pore V, Hatanpää T, Ritala M, Leskelä M. Atomic layer deposition of metal tellurides and selenides using alkylsilyl compounds of tellurium and selenium. J Am Chem Soc. 2009;131:3478–80.
[99] Pietikäinen J, Laitinen RS. Novel tellurium halides Te_2Cl_2 and Te_2Br_2. Chem Commun. 1998;2381–2.
[100] Panda A. Developments in tellurium containing macrocycles. Coord Chem Rev. 2009;253:1946–65.
[101] Chivers T, Laitinen RS, Schmidt KJ, Taavitsainen J. NMR spectroscopic evidence for the existence of tellurium-containing chalcogen rings and polymers in sulfur-tellurium and sulfur-selenium-tellurium melts. Inorg Chem. 1993;32:337–40.
[102] Zingaro RA. Tributylphosphine telluride. J Organomet Chem. 1963;1:200.
[103] Zingaro RA, Steeves BH, Irgolick K. Phosphine tellurides. J Organomet Chem. 1965;4:320–3.
[104] Nordheider A, Woollins JD, Chivers T. Organophosphorus-tellurium chemistry: from fundamentals to applications. Chem Rev. 2015;115:10378–406.
[105] Brennan JG, Siegrist T, Stuczynski SM, Steigerwald ML. Cluster intermediates in an organometallic synthesis of palladium telluride PdTe. J Am Chem Soc. 1990;112:9233–6.
[106] Mastai Y, Hodes G. Size quantization in electrodeposited CdTe nanocrystalline films. J Phys Chem B. 1997;101:2685–90.
[107] Daniliuc C, Druckenbrodt C, Hrib CG, Ruthe F, Blaschette A, Jones PG, et al. The first trialkylphosphane telluride complexes of Ag(I): molecular, ionic and supramolecular structural alternatives. Chem Commun. 2007;2060–2.
[108] Mallek R, Sanhoury MAK, Ben Dhia MT, Khaddar MR. Cadmium(II) complexes with phosphine tellurides: synthesis and multinuclear (^{31}P, ^{125}Te, and ^{113}Cd) NMR characterization in solution. J Coord Chem. 2014;67:1541–9.
[109] Mallek R, Sanhoury MAK, Bahri L, Khaddar MR, Ben Dhia MT. Synthesis, characterization, and solution behavior of mercury(II) chloride complexes with phosphine tellurides. J Coord Chem. 2016;69:726–34.
[110] Engman L, Cava MP. Organotellurium compounds 6. Synthesis and reactions of some heterocyclic lithium tellurolates. Organometallics. 1982;1:470–3.
[111] Bhasin KK, Arora V, Klapötke TM, Crawford M-J. One-pot synthesis of pyridyltellurium derivatives from a reaction with isopropylmagnesium chloride and X-ray crystal structures of various pyridyl ditellurides. Eur J Inorg Chem. 2004;4781–8
[112] Iwamoto T, Sato K, Ishida S, Kabuto C, Kira M. Synthesis, properties and reactions of a series of stable dialkyl-substituted silicon-chalcogen doubly bonded compounds. J Am Chem Soc. 2006;128:16914–20.

[113] Tokitoh N, Matsumoto T, Okazaki R. First stable germanetellones: syntheses and crystal structures of the heaviest germanium-chalcogen double-bond compound. J Am Chem Soc. 1997;119:2337–8.
[114] Liu S, Légaré M-A, Auerhammer D, Hofmann A, Braunschweig H. The first boron-tellurium double bond: direct insertion of heavy chalcogens into a Mn=B double bond. Angew Chem Int Ed. 2017;56:15760–3.
[115] Cowley AH, Jones PR, Harris PR, Atwood DA, Contreras L, Burek CJ. Tetrameric gallium and aluminium chalcogenides $[Me_3CMX]_4$ (M = Al, Ga; X = S, Se, Te). A new class of heterocubanes. Angew Chem Int Ed Engl. 1991;30:1143–6.
[116] Power MB, Ziller JW, Tyler AN, Barron AR. Interaction of tri-*tert*-butylgallium with elemental sulfur, selenium, and tellurium. Organometallics. 1992;11:1055–63.
[117] Power MB, Barron AR. Isolation of the first gallium hydrosulfido complex and its facile conversion to Ga_4S_4 cubane: X-ray structure of [*tert*-$Bu_2Ga(\mu\text{-}SH)]_2$ and [*tert*-$BuGaS]_4$. J Chem Soc Chem Commun. 1991;1315–7.
[118] Li GM, Zingaro RA. Reaction of diisobutylaluminium hydride with selenium and tellurium: new reagents for the synthesis of seleno- and telluro-amides. J Chem Soc, Perkin Trans I. 1998;647–9.
[119] Fischer JM, Piers WE, Ziegler T, MacGillivray LR, Zaworotko MJ. Permethyltitanocene derivatives with naked chalcogen ligands: synthesis of $[(Cp^*{}_2Ti)\ {}_2(\mu\text{-}E)]$ and $Cp^*{}_2Ti(\eta^2\text{-}E_2)]$ and the role of the terminal chalcogenides $[Cp^*{}_2Ti(E)]$ in their interconversion (E = Se, Te). Chem Eur J. 1996;2:1221–9.
[120] Fischer JM, Piers WE, MacGillivray LR, Zaworotko MJ. Permethyltitanocene tellurides and ditellurides. X-ray structures of $[(C_5Me_5)_2Ti]_2(\mu\text{-}Te)$ and $(C_5Me_5)_2Ti(\eta^2\text{-}Te_2)$. Inorg Chem. 1995;34:2499–500.
[121] Günther A, Heise M, Wagner FR, Ruck M. An iridium-stabilized formally uncharged Te10 molecule with 3-center-4-electron bonding. Angew Chem Int Ed. 2011;50:9987–90.
[122] Günther A, Isaeva A, Ruck M. Stabilization of decatellurium molecules in isolated and concatenated clusters. Z Anorg Allg Chem. 2013;639:254–60.
[123] Günther A, Isaeva A, Baranov AI, Ruck M. Neutral tellurium rings in the coordination polymers $[Ru(Te_9)](InCl_4)_2$, $[Ru(Te_8)]Cl_2$, and $[Rh(Te_6)]Cl_3$. Chem Eur J. 2011;17:6382–8.
[124] Günther A, Ruck M. Hexatellurium rings in coordination polymers and molecular clusters: synthesis and crystal structures of $[M(Te_6)]X_3$ (M = Rh, Ir; X = Cl, Br, I) and $[Ru_2(Te_6)](TeBr_3)_4$ $((TeBr_2)_2$. Z Anorg Allg Chem. 2012;638:317–23.
[125] Deiseroth H-J, Wagener M, Neumann E. $(AgI)_2Te_6$ and $(AgI)_2Se_6$: new composite materials with cyclic Te_6 and Se_6 molecules stabilized in the "solid solvent" AgI. Eur J Inorg Chem. 2004;4755–8.

Tristram Chivers and Risto S. Laitinen

2 Selenium– and tellurium–halogen reagents

Abstract: Selenium and tellurium form binary halides in which the chalcogen can be in formal oxidation states (IV), (II) or (I). They are versatile reagents for the preparation of a wide range of inorganic and organic selenium and tellurium compounds taking advantage of the reactivity of the chalcogen–halogen bond. With the exception of the tetrafluorides, the tetrahalides are either commercially available or readily prepared. On the other hand, the low-valent species, EX_2 (E = Se, Te; X = Cl, Br) and E_2X_2 (E = Se, Te; X = Cl, Br) are unstable with respect to disproportionation and must be used as *in situ* reagents. Organoselenium and tellurium halides are well-known in oxidation states (IV) and (II), as exemplified by REX_3, R_2EX_2 and REX (R = alkyl, aryl; E = Se, Te; X = F, Cl, Br, I); mixed-valent (IV/II) compounds of the type $RTeX_2TeR$ are also known. This chapter surveys the availability and/or preparative methods for these widely used reagents followed by examples of their applications in synthetic inorganic and organic selenium and tellurium chemistry. For both the binary halides and their organic derivatives, the discussion is subdivided according to the formal oxidation state of the chalcogen.

Keywords: binary selenium halogenides, organyl selenium halogenides, binary tellurium halogenides, organyl tellurium halogenides, mixed-valent organic tellurium halogenides

2.1 Scope and introduction

The diverse applications of halides of selenium and tellurium as reagents for the synthesis of organic and inorganic compounds stem from a combination of the reactivity of the E–X bond (E = Se, Te; X = halogen) and the availability of these reagents in a variety of oxidation states, namely EX_4, OEX_2, EX_2 and E_2X_2. This chapter will begin with a survey of the reactions of tetrahalides $E(IV)X_4$. The latter are generally more easily handled than the halides in which the chalcogen is in a lower oxidation state, $E(II)X_2$ and $E(I)_2X_2$, which are discussed subsequently. In keeping with this sequence, the reactions of organochalcogen halides will be treated in the order, $RE(IV)X_3$, $R_2E(IV)X_2$, mixed-valent $RTe(IV)X_2Te(II)R$ compounds and, finally, RTe(II)X (R = alkyl, aryl) reagents. Triorganochalcogen monohalides, which exist as chalcogenonium salts $[R_3E(IV)]X$ in the solid state and in solution, are not discussed because they do not incorporate covalent E–X bonds [1]. In each section,

This article has previously been published in the journal *Physical Sciences Reviews*. Please cite as: Chivers, T., Laitinen, R. S. Selenium- and Tellurium-Halogen Reagents. *Physical Sciences Reviews* [Online] **2018** DOI: 10.1515/psr-2018-0060

https://doi.org/10.1515/9783110529340-002

the discussion will commence with a review of the available procedures for the preparation of each of the chalcogen halides together with some comments on their stability. Where appropriate, cross-references will be made to accompanying chapters in this book in which selenium or tellurium halides are used to generate other functional groups, e. g. E-O (Chapter 3) and E-N (Chapter 4). For information on the structural chemistry and spectroscopic data of heavy chalcogen halides, the reader is referred to a recent book chapter [2].

2.2 Binary chalcogen–halogen reagents

2.2.1 Selenium tetrahalides

Selenium tetrachloride and tetrabromide are commercially available white solids, which have been used widely in the preparation of inorganic and organic selenium compounds. Although mildly moisture sensitive, they are thermally stable and easily manipulated under a dry atmosphere. By contrast, the use of teflon vessels is recommended for the use of highly reactive SeF_4, which is sometimes preferred over the heavier tetrahalides SeX_4 (X = Cl, Br), e. g. in the low-temperature (–50°C) synthesis of the binary selenium azide $Se(N_3)_4$ in liquid SO_2 [eq. (2.1)] [3]. The formation of the very strong Si–F bond in volatile Me_3SiF provides a powerful driving force in reactions of binary and organochalcogen fluorides with trimethylsilyl reagents (*vide infra*).

$$SeF_4 + 4Me_3SiN_3 \rightarrow Se(N_3)_4 + 4Me_3SiF \quad (2.1)$$

The reactions of organolithium reagents with selenium tetrahalides are not a fruitful source of selenenuranes R_4Se. Consequently, recent investigations of the preparation and stability of $(C_6H_5)_4Se$ have used $(C_6H_5)_2SeF_2$ as the starting material (Section 2.3.2) [4]. Selenium tetrachloride is, however, a convenient source of selenium(IV) for the preparation of selenium(IV) diimides RN=Se=NR by reaction with the appropriate primary amine RNH_2 (R = ^tBu [5] Ad [6]a) or the monolithiated primary amine Mes*N(H)Li [6]b]; the applications of RN=Se=NR reagents are discussed in Chapter 4, Section 7.1. However, the analogous reaction of $SeCl_4$ with MesNHLi does effect reduction to afford the selenium(II) diamide $Se[N(H)Mes]_2$ [6]b. A series of tetraalkoxyselenuranes $Se(OR)_4$ (R = Me, Et, ^iPr, neo-pentyl, CF_3CH_2) has been prepared by the reaction of the appropriate alcohol with $SeCl_4$ in the presence of trimethylamine; however, the R = Me or Et derivatives are thermally unstable [7]. By contrast, ethylene glycol, pinacol and neopentyl glycol give selenuranes that can be isolated as white crystalline solids [7]. Solutions of $SeCl_4$ in liquid ammonia act as an *in situ* source of binary selenium–nitrogen anions, e.g. $[Se_2N_2]^{2-}$, for the preparation of metal complexes [8] (see Chapter 4, Section 4.2.3).

Selenium tetrahalides also exhibit a propensity for reduction in reactions with Lewis bases (two-electron ligands) as illustrated by the examples depicted in Figure 2.1 [9]. An

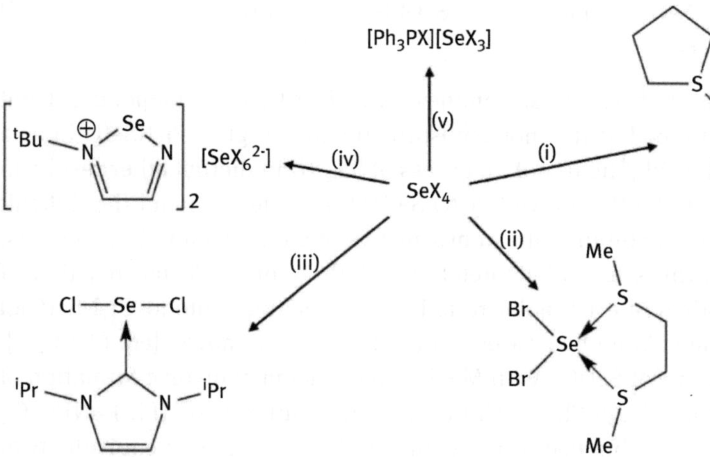

Figure 2.1: Reactions of SeX_4 (X = Cl, Br) with Lewis bases: (i) tetrahydrothiophene; (ii) MeS $(CH_2)_2SMe$; X = Br; (iii) 2,5-diisopropylimidazole-2-ylidene; X = Cl; (iv) 1,4-di-*tert*-butyl-1,3-diazabutadiene; X = Cl, Br; (v) PPh_3; X = Cl, Br.

early example of this behaviour is the formation of base-stabilized $SeCl_2$•tmtu from treatment of $SeCl_4$ with tetramethylthiourea (tmtu) [10]. More, recently, it has been shown that thioether ligands also engender the formation of adducts of Se(II)X_2 either as the monomer *trans*-[SeX_2(tht)$_2$] (X = Cl, Br; tht = tetrahydrothiophene) or as halo-bridged polymers [SeX_2(SMe_2)] (X = Cl, Br), whereas bidentate sulfur-centred ligands form *cis*-[$SeBr_2$(MeS(CH_2)$_2$SMe)] [11]. A 1:1 adduct of $SeCl_2$ is also produced from the reaction of $SeCl_4$ with the carbene 2,5-diisopropylimidazole-2-ylidene in tetrahydrofuran (THF) [12]a. Treatment of SeX_4 (X = Cl, Br) with the bidentate nitrogen-centred ligand 1,4-di-*tert*-butyl-1,3-diazabutadiene also results in a two-electron-redox process accompanied, in this case, by reductive elimination of X_2 to give 1,2,5-selenadiazolium cations as $[SeX_6]^{2-}$ salts [12]b. Triphenylphosphine reacts with SeX_4 (X = Cl, Br) to form the $[Ph_3PX]^+$ cation accompanied by the Se(II)-centred anion $[SeX_3]^-$ [12]a.

Although reactions of selenium tetrahalides with alkenes and alkynes have been studied extensively in early work, recent attention has been focused on the applications of selenium dihalides for the synthesis of organoselenium compounds (see Section 2.2.3) [13].

2.2.2 Selenium oxychloride

Selenium oxychloride $SeOCl_2$ is an alternative to selenium tetrahalides for applications in both inorganic and organic chemistry. This selenium(IV) reagent is readily obtained as a moisture-sensitive orange liquid in 84 % yield from the reaction of selenium dioxide and chlorotrimethylsilane [eq. (2.2)] [14].

$$SeO_2 + 2Me_3SiCl \rightarrow SeOCl_2 + (Me_3Si)_2O \quad (2.2)$$

Reactions of $SeOCl_2$ with secondary amines, e. g. $^iPr_2NH$, in a nonpolar solvent (pentane) can be stopped at the monosubstituted stage to give $^iPr_2NSe(O)Cl$ as a white solid in 65 % yield. The use of an excess of $^iPr_2NH$ in diethyl ether results in reduction of Se(IV) to Se(II) to give $(^iPr_2N)_2Se$ [14]. In order to avoid the difficult separation of dialkylammonium salts from sensitive selenium-containing products, trimethylsilyl derivatives are often preferable to the amines. Thus, reaction of $SeOCl_2$ with $^iPr_2NSiMe_3$ in a 1:1 molar ratio is an alternative route to $^iPr_2NSe(O)Cl$ (68 % yield) and the volatile by-product Me_3SiCl is easily removed [eq. (2.3)] [14]. However, the reactions of $SeOCl_2$ with Me_3Si-N reagents may involve formation of $(Me_3Si)_2O$ rather than Me_3SiCl, as shown by the synthesis of $(Ph_3P{=}N)_2SeCl_2$ [eq. (2.4)], which is also obtained from treatment of $SeCl_4$ with two equivalents of $Ph_3P{=}NSiMe_3$ [15].

$$^iPr_2NSiMe_3 + SeOCl_2 \rightarrow {}^iPr_2NSe(O)Cl + Me_3SiCl \quad (2.3)$$

$$2Ph_3P{=}NSiMe_3 + SeOCl_2 \rightarrow (Ph_3P{=}N)_2SeCl_2 + (Me_3Si)_2O \quad (2.4)$$

The reaction of the primary amine $^tBuNH_2$ with $SeOCl_2$ in a 3:1 molar ratio in diethyl ether at −10 °C produced the first selenenylamine ^tBuNSeO as a brown-orange oil (90 % yield) [5], which can be recrystallized from THF at −20°C to reveal a dimeric structure [6]. In the case of diamines, e. g. 4,5-dimethyl-1,2-phenyldiamine, the reaction with $SeOCl_2$ in boiling toluene results in cyclocondensation to give the corresponding 2,1,3-benzoselenadiazole (61 % yield) (Figure 2.2) [15]. Selenium oxychloride has also been used as a “selenium transfer reagent” for the synthesis of cyano-substituted selenophenes [16].

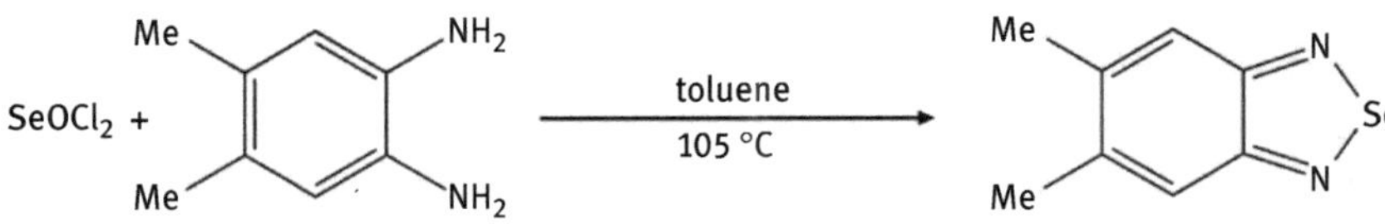

Figure 2.2: Synthesis of a selenophene using $SeOCl_2$.

The potentially useful reagent $Me_3SiNSeNSiMe_3$ is generated by the reaction of $SeOCl_2$ with $LiN(SiMe_3)_2$ in a 1:2 molar ratio in diethyl ether. However, this thermally unstable selenium(IV) diimide must be used *in situ*, and applications have been limited to the preparation of the six-membered P_2N_3Se heterocycle (Chapter 4, Section 5) [17].

2.2.3 Selenium dihalides

The strong tendency of selenium dihalides SeX_2 (X = Cl, Br) to disproportionate in solution according to eq. (2.5) and (2.6) necessitates the use of these important selenium sources as *in situ* reagents [18].

$$3SeCl_2 \leftrightarrow Se_2Cl_2 + SeCl_4 \tag{2.5}$$

$$2SeBr_2 \leftrightarrow Se_2Br_2 + Br_2 \tag{2.6}$$

The reduction of $SeCl_4$ with an equimolar amount $SbPh_3$ has been employed successfully to generate $SeCl_2$ *in situ* for the synthesis of Se–N heterocycles [19], but the separation of Ph_3SbCl_2 from the desired product can be difficult. The use of $PhSO_2SeCl$ as a synthetic equivalent of $SeCl_2$ was demonstrated in reactions with arylaminines to give ephemeral selenonitrosoarenes ArN=Se [20] (see Chapter 4, Section 6). Subsequently, however, it was shown that pure $SeCl_2$ is readily obtained by treatment of elemental selenium with an equimolar amount of SO_2Cl_2 [eq. (2.7)] in THF; such solutions are stable towards disproportionation for 1 day at 23°C as determined by ^{77}Se NMR spectroscopy [21]. Similarly, the facile preparation of solutions of pure selenium dibromide is achieved by treatment of grey selenium with bromine (Br_2) in a chlorocarbon solvent such as $CHCl_3$ or CCl_4 [eq. (2.8)] [22, 23]

$$Se + SO_2Cl_2 \rightarrow SeCl_2 + SO_2 \tag{2.7}$$

$$Se + Br_2 \rightarrow SeBr_2 \tag{2.8}$$

THF solutions of $SeCl_2$ prepared according to eq. (2.7) have been widely used for the preparation of inorganic selenium–nitrogen heterocycles. The reactions of $SeCl_2$ with primary amines in various molar ratios in THF have been investigated in particular detail [24]. In the case of *tert*-butylamine, these studies have revealed the successive formation of a series of acyclic imidoselenium dichlorides $ClSe[N(^tBu)Se]_nCl$ (n = 1–3) via the initial intermediate $^tBuN(H)SeCl$, which serves as a source of ^tBuNSe units in the build-up of longer chains (Figure 2.3) [25–27]. Cyclocondensation reactions or reduction of the bifunctional dichlorides $ClSe[N(^tBu)Se]_nCl$ with an excess of $^tBuNH_2$ account for the formation of the variety of cyclic selenium imides that are the ultimate products of this reaction [24].

The cyclocondensation reaction with primary amines is a lot cleaner if a bis-silylated derivative is employed. Thus, the treatment of $SeCl_2$ with an equimolar amount of $(Me_3Si)_2NMe$ at −10°C affords the eight-membered ring 1,3,5,7-$Se_4(NMe)_4$ exclusively in 90 % yield [eq. (2.9)] [28]. The reagent $SeCl_2$ also plays a key role in the stepwise generation of the cage molecule 1,5-$Se_2S_2N_4$. In the first step, the reaction of

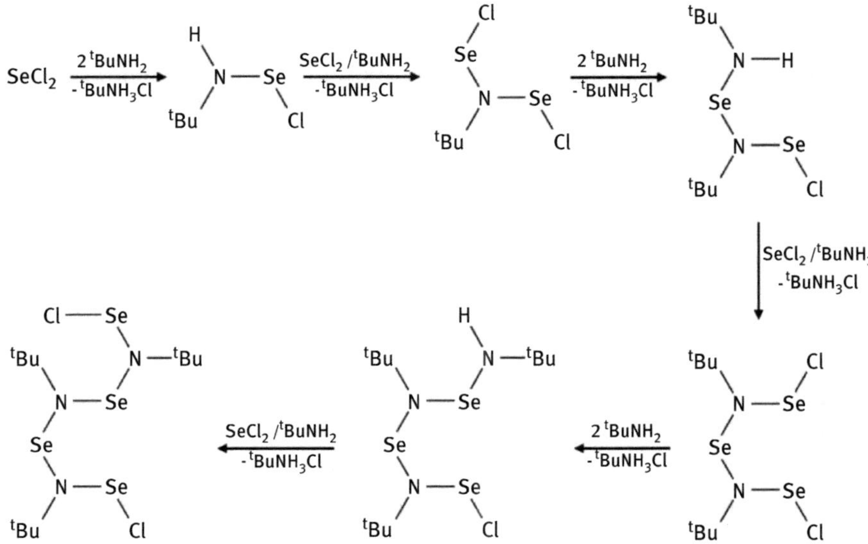

Figure 2.3: Formation of imidoselenium(II) dichlorides $^{t}ClSe[N(^{t}Bu)Se]_{n}Cl$ (n = 1–3) from reactions of $SeCl_2$ with $^{t}BuNH_2$.

$SeCl_2$ with two molar equivalents of the sulfur(IV) diimide $Me_3SiNSNSiMe_3$ produces yellow crystals of $(Me_3SiNSN)_2Se$ (65 % yield), which undergoes cyclocondensation with $SeCl_2$ to dark-red 1,5-$Se_2S_2N_4$ (73 %) [eq. (2.10)] [29].

$$(Me_3Si)_2NMe + SeCl_2 \rightarrow 1/4\ Se_4(NMe)_4 + 2Me_3SiCl \qquad (2.9)$$

$$(Me_3SiNSN)_2Se + SeCl_2 \rightarrow 1{,}5\text{-}Se_2S_2N_4 + 2Me_3SiCl \qquad (2.10)$$

In the area of organometallic chemistry, the $SeCl_2$ electrophile has been employed to introduce a μ_4-Se bridge in double butterfly cluster complexes (Figure 2.4) [30].

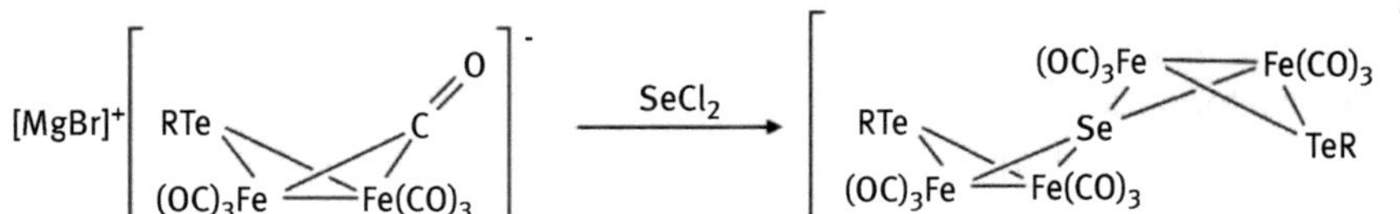

Figure 2.4: Insertion of selenium into a metal cluster using $SeCl_2$.

In the past 15 years, solutions of $SeCl_2$ or $SeBr_2$ (eqs (2.7) and (2.8) respectively) have been shown to have wide applications for the synthesis of organoselenium compounds, especially selenium-containing heterocycles [31]. Some representative examples are

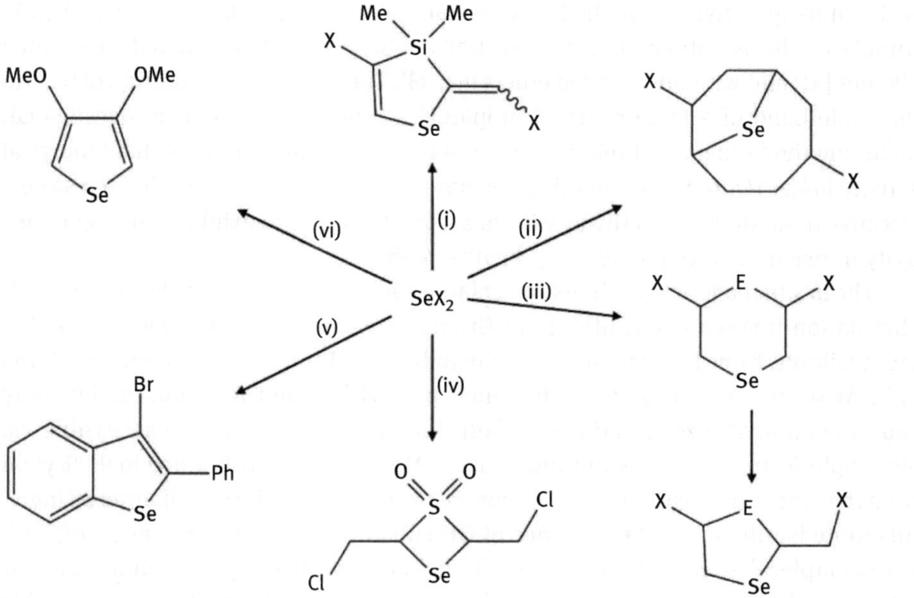

Figure 2.5: Preparation of selenium-containing heterocycles using SeX_2 reagents (X = Cl, Br): (i) $(HC{\equiv}C)_2SiMe_2$, (ii) *cis,cis*-1,5-cyclooctadiene, (iii) $E(CH{=}CH_2)_2$ (E = S, Se), (iv) $O_2S(CH{=}CH_2)_2$ (X = Cl), (v) PhC≡CPh (X = Br) and (vi) 2,3-dimethoxy-1,3-butadiene.

illustrated in Figure 2.5. An early report revealed the formation of selenasilafulvenes from the reactions of SeX_2 (X = Cl, Br) with diorganyl diethynylsilanes [32], and this methodology has been extended to germanium-containing systems [22]. These reactions proceed with chemo-, regio- and stereoselectivity in chloroform at room temperature [22, 33]. The transannular addition of SeX_2 to *cis,cis*-1,5-cyclooctadiene takes place chemo- and regioselectively to give 2,6-dihalo-9-selenobicyclo[3.3.1]nonanes in high yields [23]. The reactions of SeX_2 with divinyl sulfide or selenide initially produces the six-membered 2,6-dihalo-1,4-chalcogenaselenanes, which readily undergo isomerization to the thermodynamically preferred five-membered rings, 1,3-chalcogenaselenolanes [34, 35]. The involvement of a seleniranium cation in this ring-contraction process is supported by quantum chemical calculations [34]. In the case of divinyl sulfone $O_2S(CH{=}CH_2)_2$, the presence of the strongly electron-withdrawing group results in the predominant formation of a four-membered ring in 85 % yield [36]. The use of $SeCl_2$ to generate selenophene monomers that undergo polymerization to give materials with potentially useful electrochromic and photothermal properties is a promising development [37–39]. For example, the reaction of $SeCl_2$ with 2,3-dimethoxy-1,3-butadiene gives 3,4-dimethoxyselenophene in 41 % yield [37].

Annulation reactions are observed in the reactions of SeX_2 (X = Cl, Br) with substrates that incorporate phenyl substituents. A simple example is the formation of 3-bromo-2-phenylbenzo[*b*]selenophene in 92 % yield from the reaction of $SeBr_2$

with diphenylacetylene in diethyl ether at room temperature (Figure 2.5) [40]. Annulation is also observed in the reactions of SeX_2 with thiophenenyl-substituted alkynes [41] and with unsaturated ethers [42, 43]. The electrophilic addition of SeX_2 to the triple bond of a large range of propargyl alcohols has also been investigated, including the formation of selenium-containing spiroketals with potential biological activity [44]a, [44]b. Selenium dihalides have also been shown to undergo a stereo-selective (*anti*-addition) pathway with hex-2-yne [45]a and to exhibit high regioselectivity in reactions with methyl vinyl ketone [45]b.

The *in situ* reagent $SeCl_2$ is also a versatile source of diaryl selenides Ar_2E via salt elimination involving aryl-lithium or Grignard reagents [46–48]. Derivatives that are available from this methodology include Ar = Ph, *o*-tolyl, 2-methylnaphthyl [46], Ar = bromo- or dibromo-substituted pyridyl [47] and Ar = intramolecularly coordinated azobenzene derivatives [48]. Diaryl selenides are also accessible via electrophilic aromatic substitution. Thus, $(4\text{-}MeOC_6H_4)_2Se$ is obtained in 91 % yield from the reaction of $SeCl_2$ with 4-methoxybenzene in chloroform [49]. Interestingly, this route is tolerant to the presence of OH substituents on the aromatic ring [50]. For example, the reaction of $SeCl_2$ in diethyl ether with 2-naphthol in a 1:2 molar ratio, in the presence of a catalytic amount of $AlCl_3$, yields the selenide $(2\text{-}HOC_{10}H_6)_2Se$ in 56 % yield [50]. A promising application of $SeCl_2$ in electrophilic aromatic substitution involves the formation of tetrameric macrocycles in good yields from 3,5-dialkoxythiophenes in diethyl ether at room temperature as illustrated in Figure 2.6 [50].

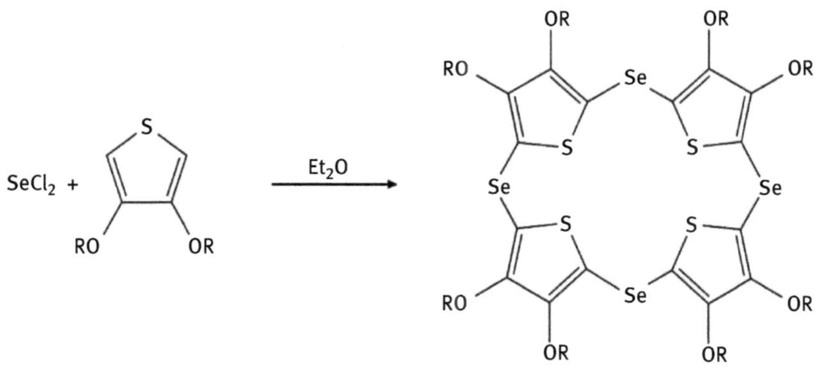

Figure 2.6: Synthesis of selenacalix[4]thiophenes from $SeCl_2$ with 3,4-dialkoxythiophenes (R = methyl, hexyl).

2.2.4 Diselenium dihalides

Diselenium dihalides Se_2X_2 are yellow-brown (X = Cl) or blood-red (X = Br) moisture-sensitive liquids. With careful control of the stoichiometry, these selenium(I) reagents can be prepared by the reactions shown in eqs (2.11) and (2.12), respectively, carried below 0°C in the case of X = Br [51]. On the basis of Raman and ^{77}Se NMR spectra, diselenium dihalides Se_2X_2 exist in equilibrium with small amounts selenium(II) dihalides SeX_2 and Se_nX_2 ($n = 3, 4$) [52]. This observation is significant when considering the outcome of reactions of these selenium(I) halides. For example, the reaction of C_6F_5Li with Se_2Cl_2 affords the monoselenide $(C_6F_5)_2Se$ in 60 % yield [53]. For comparison, $SeCl_4$ is reduced by C_6F_5Li to give $(C_6F_5)_2Se$ in only 20 % yield [53].

$$3Se + SeCl_4 \rightarrow 2Se_2Cl_2 \tag{2.11}$$

$$2Se + Br_2 \rightarrow Se_2Br_2 \tag{2.12}$$

Diselenium dichloride also reacts effectively as $SeCl_2$ for the synthesis of the monoselenides $Se[N(SiMe_3)R]_2$ (R = $SiMe_3$ [54], ^tBu [55]) in 75 and 65 % yields, respectively, in hexane with concomitant formation of red selenium [eq. (2.13)]. By contrast, the reaction of the amine $(Me_3Si)_2NH$ with Se_2Cl_2 in dichloromethane at 0°C produces a mixture of polyselenides $Se_n[N(SiMe_3)_2]_2$ (n = 2,3,4) from which the predominant triselenide (n = 3) cannot be separated in pure form [56].

$$2Li[N(SiMe_3)R] + Se_2Cl_2 \rightarrow Se[N(SiMe_3)R]_2 + Se \tag{2.13}$$

The outcome of the reaction with $Me_3SiNPPh_3$ is another example of Se_2Cl_2 behaving as $SeCl_2$. In acetonitrile solution, the ion-separated salt $[SeCl(NPPh_3)_2]^+SeCl_3^-$, in which selenium(II) is present in both the cation and anion, is formed together with elemental selenium [eq. (2.14)] [57].

$$3Se_2Cl_2 + 2Me_3SiNPPh_3 \rightarrow \left[SeCl(NPPh_3)_2\right]^+ SeCl_3^- + 4Se + 2ClSiMe_3 \tag{2.14}$$

The reaction of Se_2Cl_2 with trimethylsilyl azide in dichloromethane produces an insoluble black powder with the composition $Se_3N_2Cl_2$, which incorporates the five-membered $Se_3N_2Cl^+$ cation with one Se(IV) and two Se(II) centres (a better synthesis of $Se_3N_2Cl_2$ is described in Section 2.2.5) [58, 59]. The dimeric selenium–nitrogen chloride $(Se_3N_2Cl)_2$ is isolated as an explosive dark brown powder in 75 % yield from the reaction of equimolar amounts Se_2Cl_2 with the cage molecule Se_4N_4 [59].

Diselenium dichloride has also been utilized to generate cyclic selenium allotropes. Thus, the reaction of titanocene pentaselenide with Se_2Cl_2 in CS_2 at 20 °C produces dark-red *cyclo*-Se_7 in 75% yield [eq. (2.15)] [60]. Unpredictably, heating $RhCl_3{\bullet}4H_2O$ with an excess of Se_2Cl_2 yields the complex $Rh_2Se_9Cl_6$ in which the nine-membered cyclo-Se_9 is sandwiched between two Rh(III) centres [61].

$$\left[(C_5H_5)_2\right]TiSe_5 + Se_2Cl_2 \rightarrow Se_7 + \left[(C_5H_5)_2\right]TiCl_2 \tag{2.15}$$

The reactions of gallium metal with Se_2X_2 (X = Cl, Br) are the most convenient preparation of ternary gallium(III) selenide halides GaSeX, which are soluble in pyridine to give the six-membered rings [GaSeX•Py]$_3$ [62].

Se_2Cl_2 has been utilized to generate compounds with discrete Se=N bonds by reaction with dichloramines according to eq. (2.16) [63, 64].

$$\begin{array}{l} XNCl_2 + Se_2Cl_2 \rightarrow XN{=}SeCl_2 + SeCl_4 \\ (X = SF_5, CF_3, C_2F_5) \end{array} \quad (2.16)$$

The passage of acetylene gas through a solution of Se_2Cl_2 in CCl_4 under pressure (10–12 atm., 30–40 °C) produces *E,E*-bis(2-chlorovinyl)selenide in 98 % yield together with elemental selenium [eq.(17)] [65].

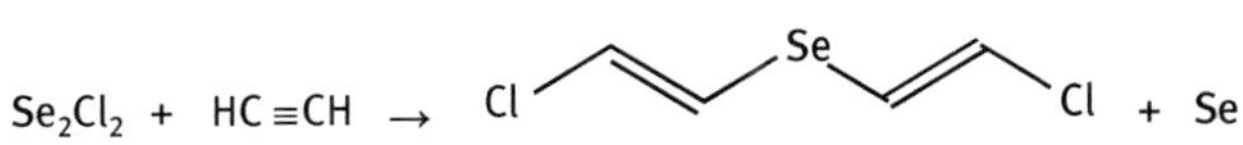

(2.17)

2.2.5 Mixtures of selenium halides

A mixture of selenium halides has been used effectively for the synthesis of selenium compounds with (a) different functional groups on selenium, (b) selenium centres in formal oxidation states II and IV or (c) selenium in formal oxidation state III as illustrated by the examples in Figure 2.7. In reactions with *tert*-butylamine, a mixture of $SeOCl_2$ with $SeCl_4$ produces the four-membered ring $^tBuN{=}Se(\mu\text{-}N^tBu)_2Se{=}O$ with terminal Se=NR and Se=O functionalities [26], whereas the use of $SeCl_2$ in combination with $SeOCl_2$ generates the acyclic compound ClSeN(^{t}Bu)Se(O)Cl with Se(II) and Se(IV) centres [66]. The reaction of $LiN(SiMe_3)_2$ with a mixture of Se_2Cl_2 and $SeCl_4$ in a 1:4 molar ratio (designed to mimic "$SeCl_3$") produces pure Se_4N_4 in 66 % yield [67], whereas the "$SeCl_3$" reagent reacts with the monoselenide $Se[N(SiMe_3)_2]_2$ to give the five-membered ring $[Se_3N_2Cl]Cl$ (95 % yield) [58, 59].

2.2.6 Tellurium tetrahalides

Tetrahalides of tellurium TeX_4 are white (X = Cl, Br) or yellow (X = I) solids that are available from commercial sources and have been used for the preparation of a wide range of inorganic and organic tellurium compounds. Tellurium tetrafluoride is a very moisture-sensitive, colourless solid that can be obtained in 78 % yield by heating a mixture of TeO_2 with sulfur tetrafluoride at 130 °C for 48 h in a stainless steel or Monel (a nickel–copper alloy) autoclave [68, 69]. As in the case of selenium tetrahalides, TeF_4 is preferred over $TeCl_4$ for the preparation of the extremely explosive binary azide, Te $(N_3)_4$, by reaction with trimethylsilyl azide in $CFCl_3$ at 0 °C [cf. eq. (2.1)]; the use of $TeCl_4$

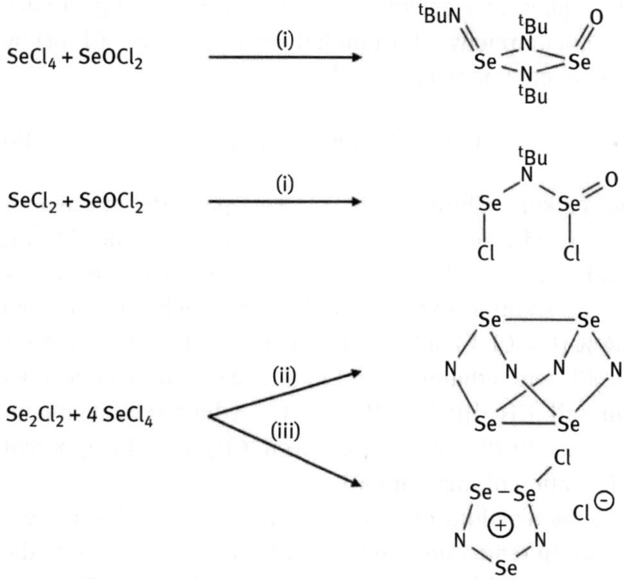

Figure 2.7: Preparation of selenium–nitrogen compounds from mixtures of selenium halides: (i) $^tBuNH_2$, (ii) $LiN(SiMe_3)_2$ and (iii) $Se[N(SiMe_3)_2]_2$.

results in only partial metathesis to give $TeCl_{4-n}(N_3)_n$ (n = 1,2) [70]a. The reaction of the hexafluoride TeF_6 with trimethylsilyl azide in MeCN also produces $Te(N_3)_4$ [70]b. Tellurium tetrafluoride has also been used to prepare $Te(CN)_4$ as solvates with THF or MeCN by reactions with trimethylsilyl cyanide in those solvents [71].

Tetraaryl-tellurium(IV) compounds can be prepared from aryl-lithium or Grignard reagents; however, yields may be low owing to concomitant reductive elimination to give diaryl-tellurium(II) compounds and biphenyls. Tetraphenyl-tellurium(IV) and tetra(p-tolyl)-tellurium(IV) have been prepared from ArLi and $TeCl_4$ by salt metathesis in 70 % and 52 % yields, respectively, and it was shown that phenyl radicals are not involved in the formation of biphenyls [72]. The perfluoro derivative $(C_6F_5)_4Te$ is obtained in 75 % yield from the reaction of the Grignard reagent C_6F_5MgBr and $TeCl_4$ at 30 °C in diethyl ether or in "good yields" from AgC_6F_5 and $TeCl_4$ in EtCN at −78 °C [73]. Tetraalkyl-tellurium(IV) derivatives R_4Te (R = Me, ^{n}Bu, Me_3SiCH_2, $CH_2{=}CH$) are available from the reactions of the appropriate lithium or Grignard reagent in ether solvents at low temperature; however they exhibit a strong tendency to decompose to the corresponding dialkyl-tellurium(II) species [74, 75].

The reactions of aryl-lithium reagents with $TeCl_4$ have also been used to generate hypervalent tellurium compounds. Thus, the one-pot reaction of $TeCl_4$ with LiAr (Ar = 4-$CF_3C_6H_4$) in a 1:4 molar ratio in diethyl ether at room temperature produces the hexaaryl compound $TeAr_6$ in 17 % yield, accompanied by Te(IV) and Te(II) derivatives [eq. (2.18)] [76, 77]. Changing the molar ratio to 1:6 did not improve the yield of $TeAr_6$;

however, the reaction of five equivalents of PhLi with TeX_4 (X = Cl, Br) generates Ph_5TeLi, which is converted to pentaphenyltellurium halides Ph_5TeX (X = Cl, Br) by treatment with SO_2Cl_2 or Br_2 in a one-pot reaction [78].

$$TeCl_4 + 4ArLi \rightarrow TeAr_6 + Ar_3TeCl + TeAr_2 \tag{2.18}$$

In an early report, several tetraalkoxytelluranes $Te(OR)_4$ were prepared in unspecified yields by reactions of $TeCl_4$ with the corresponding alcohol (R = Me, Et, ^iPr, neopentyl) or sodium alkoxide (R = CF_3CH_2, $(CF_3)_2CH$) in a 1:4 molar ratio [7]. A subsequent investigation of the sodium alkoxide–$TeCl_4$ system achieved isolated yields of 58 % (R = Me) or 64 % (R = CF_3CH_2) [79]. An unexpected transformation is observed in the reaction of $TeCl_4$ with ethane-1,2-diol in a 1:1 molar ratio *in acetonitrile* at reflux. The ammonium salt $[NH_4]^+[(OCH_2CH_2O)TeCl_3]^-$, which exhibits immunomodulatory activity, is obtained in 79 % yield; the solvent CH_3CN is the apparent source of nitrogen in the NH_4^+ cation of this salt [80].

As depicted in Figure 2.8, the reactions of tellurium tetrahalides with nitrogen-centred nucleophiles may result in reduction to tellurium(II), as exemplified by the formation of the tellurides $Te[N(SiMe_3)R]_2$ (50 %, R = $SiMe_3$ [54]; 35 %, ^tBu [55]) from salt metathesis of $TeCl_4$ with the corresponding lithium reagents. Similarly, the polymeric tellurium(II) complex $[Te(NMe_2)_2]_\infty$ is isolated as a yellow, moisture-sensitive solid (74 % yield) from the reaction of $LiNMe_2$ with $TeCl_4$ in a 4:1 molar ratio in THF–Et_2O solutions [81]. This propensity for reduction has been used in the synthesis of a wide range of dialkynyl tellurides in moderate-to-high yields from the reactions of $TeCl_4$ with lithium amides, e. g. LiN^iPr_2, followed by treatment with terminal alkynes [82]. However, dialkynyl tellurides $Te(C{\equiv}CR)_2$ (R = Me, Et, ^nPr, ^tBu, Me_3Si, Ph) can also be prepared in 42–69 % yields by the reactions of $TeCl_4$ with four equivalents of the corresponding alkynyl-lithium reagent [83].

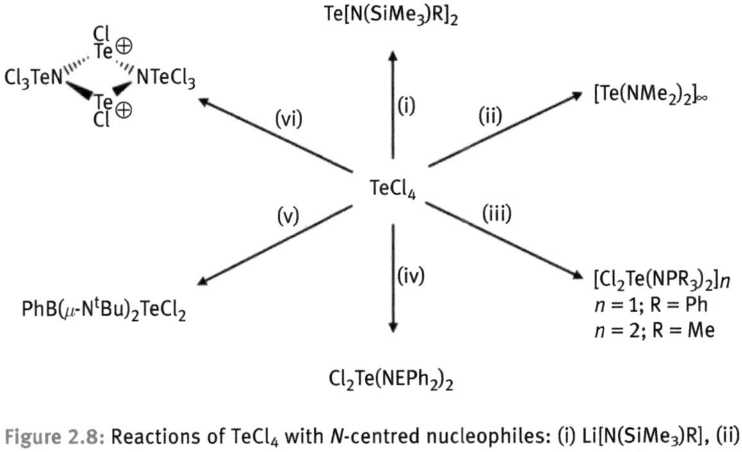

Figure 2.8: Reactions of $TeCl_4$ with *N*-centred nucleophiles: (i) $Li[N(SiMe_3)R]$, (ii) $LiNMe_2$, (iii) Me_3SiNPR_3 (R = Me, Ph), (iv) $Me_3SiNEPh_2$ (E = C, S), (v) $Li_2[PhB(N^tBu)_2]$ and (vi) $(Me_3Si)_3N$.

The use of Me_3Si-N derivatives and/or a chelating ligand reduces the tendency towards reduction as exemplified by the syntheses of the tellurium(IV) derivatives $[TeCl_2(NPR_3)_2]_n$ (n = 1: R = Ph [84]; n = 2: R = Me [85]), $[TeCl_2(NEPh_2)_2]$ (E = C, S) [86], $PhC(NSiMe_3)_2TeCl_3$ [87] and $PhB(\mu\text{-}N^tBu)_2TeCl_2$ [88] (Figure 2.8). Moreover, the reaction of $TeCl_4$ with one equivalent of dilithium *meso*-tetra-*p*-tolylporphyrin, Li_2(ttp), in hexane at reflux yields the dark green tellurium(IV) porphyrin complex $[Te(ttp)Cl_2]$ in which only three nitrogen atoms of the porphyrin ligand are coordinated to tellurium [89]. However, the reaction of $TeCl_4$ with the multifunctional reagent $N(SiMe_3)_3$ in a 2:1 molar ratio produces a dimer of the hypothetical tellurium(IV) imide $[Cl_3Te\text{-}N{=}TeCl]^+$ [90] (Figure 2.8). The tellurium(IV) imide $^tBuN{=}TeCl_2$ is obtained as a trimer from the redistribution reaction of $TeCl_4$ with the tellurium(IV) dimide dimer $[^tBuNTe(\mu\text{-}N^tBu)_2TeN^tBu]_2$ [91].

Tellurium tetrahalides are key reagents for the synthesis of 1,2,5-telluradiazoles and benzo-2,1,3-telluradiazoles, which exhibit supramolecular chromotropism. The latter are generally prepared by cyclocondensation of $TeCl_4$ with the appropriate phenylenediamine in pyridine followed by the addition of triethylamine. Yields of pure benzo-2,1,3-telluradiazoles, after removal $[Et_3NH]Cl$, range from 43 % for the 3,6-dibromo derivative [92] to 84 % for the 4,5,6,7-tetrafluoro derivative [93]. The reaction of 2,3-diaminomaleonitrile with TeX_4 (X = Cl, Br) in the presence of pyridine and triethylamine gives a product that is contaminated with $[Et_3NH]Cl$ and pyridine. Pure dicyano-1,2,5-telluradiazole is obtained in 57 % yield after treatment of this product with iced water [94].

The reactions of tellurium tetrahalides with a wide range of N, P, O, S and Se-centred Lewis bases have been investigated and reduction is sometimes observed, especially for the heavier halogens (Figure 2.9) [9]. For example, tellurium(II) diodide is trapped by the chelating nitrogen ligand in the reaction of TeI_4 with a 1,2-bis(arylimino)acenaphthene [95]a; however, similar reactions with TeX_4 (X = Cl, Br) result in C-H activation of an isopropyl group in the ligand to give a *N,N,C*-bound TeX_3 complex [95]a,b. The reaction of TeI_4 with a carbene ligand, 2,3-dihydro-1,3-diisopropyl-4,5-dimethylimidazol-2-ylidene, in THF at reflux also produces a 1:1 adduct with TeI_2 (66 % yield) via reductive elimination of I_2 from the initially formed TeI_4 adduct [96]. The 1:1 adduct of TeI_2 with the *Se,Se'*-chelating ligand $SePPh_2CH_2Ph_2PSe$ has been prepared in 85 % yield from a combination of elemental tellurium and I_2 with an equimolar amount of the ligand in CH_2Cl_2 for 4 days at room temperature [97].

In contrast to the behaviour of SeX_4 (Figure 2.1), the tellurium tetrahalides TeX_4 (X = Cl, Br) form 1:1 adducts with 1,4-di-*tert*-butyl-1,3-diazabutadiene (Figure 2.9) and TeI_4 does not react at room temperature with this *N,N'*-chelating ligand [11]. The reactions of TeX_4 (X = Cl, Br) with 2,2-bipyridine (bipy) also produce 1:1 adducts in the ionic form $[Te(bipy)X_3]X$, which for X = Cl can be reduced by $SbPh_3$ in dichloromethane to give the tellurium(II) complex $[Te(bipy)Cl_2]$ (Section 2.2.7) [98].

In an early work, a 2:1 *trans*-octahedral complex was isolated in 34 % yield from the reaction of tetramethylthiourea (tmtu) with $TeCl_4$ [99]. Recently, the reactions of TeX_4 (X = F, Cl, Br, I) with chalcogen-centered ligands, including phosphine

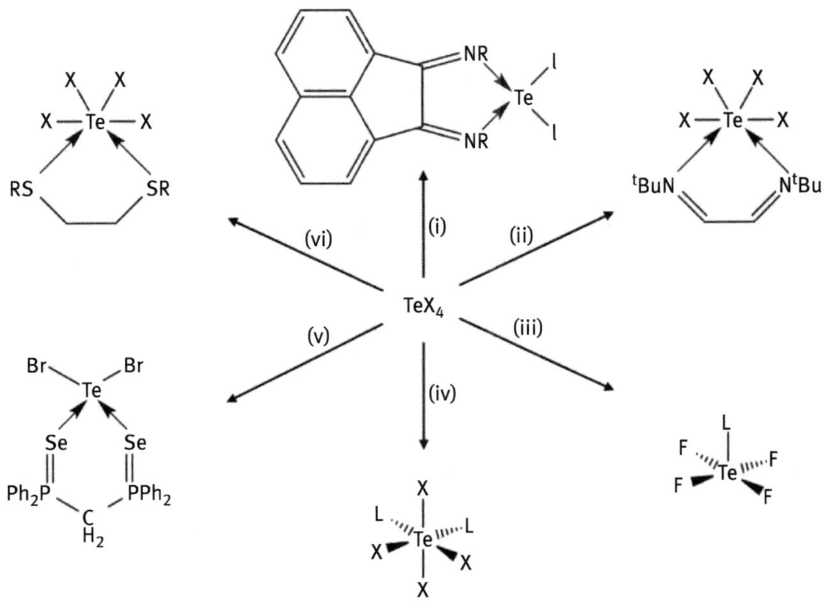

Figure 2.9: Reactions of TeX_4 (X = F, Cl, Br, I) with Lewis bases: (i) bis(arylimino)acenaphthene; R = 2,6-diisopropylphenyl, mesityl; (ii) 1,4-di-*tert*-butyl-1,3-diazabutadiene, X = Cl, Br; (iii) OPR_3 (R = Me, Ph), X = F; (iv) L = OPR_3 (R = Me, Ph), X = Cl, Br; (v) $Ph_2P(Se)CH_2P(Se)PPh_2$ and (vi) $RSCH_2CH_2SR$ (R = Me, ^iPr), X = Cl, Br.

chalcogenides and thio- or seleno-ethers, have been investigated extensively [100, 101]. In the case of phosphine oxides OPR_3 (R = Me, Ph), the adduct obtained is dependent upon the nature of the halogen. Thus, TeF_4 forms the square-pyramidal 1:1 complexes $[TeF_4(OPR_3)]$ [100], whereas the heavier tetrahalides give distorted *cis*-octahedral 1:2 complexes $[TeX_4(OPR_3)_2]$ (X = Cl, Br; R = Me, Ph) (Figure 2.9) [101]a. Chelating thioether ligands also give *cis*-octahedral 1:2 complexes $[TeX_4(RS(CH_2)_2SR]$ (X = Cl, Br; R = Me, ^iPr), whereas the monodentate thioether dimethyl sulfide generates 1:1 complexes, which exist as halo-bridged dimers $[\{X_3(Me_2S)Te\}_2(\mu\text{-}X)_2]$ [101]b. Soft selenoether ligands form 2:1 *trans*-octahedral complexes of the type $[TeX_4(SeMe_2)]$ (X = F, Cl. Br) and chelating selenium ligands, e. g. o-$C_6H_4(SeMe)_2$, produce *cis*-octahedral 1:1 complexes [101]a. However, reduction of the tellurium centre does occur in the reaction of $TeBr_4$ with a chelating phosphine selenide to give $[TeBr_2(Ph_2P(Se)CH_2P(Se)Ph_2]$ (Figure 2.9) [100].

In contrast to the reactions of selenium tetrahalides with triphenylphosphine Ph_3P, which involve oxidative addition to the phosphorus(III) centre (Figure 2.1), the corresponding reaction with $TeCl_4$ results in the reduction of one-half of the tellurium to tellurium metal [eq. (2.19)] [12].

$$2Ph_3P + 2TeCl_4 \rightarrow [Ph_3PCl]_2[TeCl_6] + Te \tag{2.19}$$

Surprisingly, tellurium tetrachloride also acts as a Lewis acid upon treatment with the nucleophile $[Mn(CO)_5]^-$ in THF at −15 °C to give the six-coordinate adduct $[Mn(CO)_5(TeCl_4)]^-$ as a PPN^+ salt [102]. Alternatively, tellurium tetrahalides may act as sources of the TeX_3 ligand in oxidative–addition reactions with Rh(I), Ru(I) or Ir(I) complexes as exemplified by the formation of the ion-separated rhodium(III) complex $[RhCl(TeF_3)(CO)(PEt_3)_2][TeF_5]$ [103] and the octahedral iridium(III) complex $[IrCl_2(CO)(PPh_3)_2(TeCl_3)]$ [104]. However, $TeCl_4$ may also behave as a mild chlorinating agent rather than an electrophile with other metal carbonyl complexes, e. g. the formation of $[FeCl_2(CO)_3L_2]$ from reactions with $[Fe(CO)_3L_2]$ (L = CO, PPh_3) [104].

There have been extensive investigations of the addition of $TeCl_4$ to unsaturated organic compounds, which are covered in the earlier review literature. A recent example involves reactions of TeX_4 (X = Cl, Br) with 3-hexyne and 4-octyne which give either monoaddition products with *E*-stereochemistry or bis-addition products (Figure 2.10) [105]. These addition products can be considered as examples of $RTeCl_3$ (Section 2.3.1) and R_2TeCl_2 derivatives, respectively, and they are readily reduced to either ditellurides or monotellurides (Figure 2.10) in high yields using either aqueous sodium metabisulfite in CCl_4 or aqueous $NaBH_4$ sodium borohydride in THF [105].

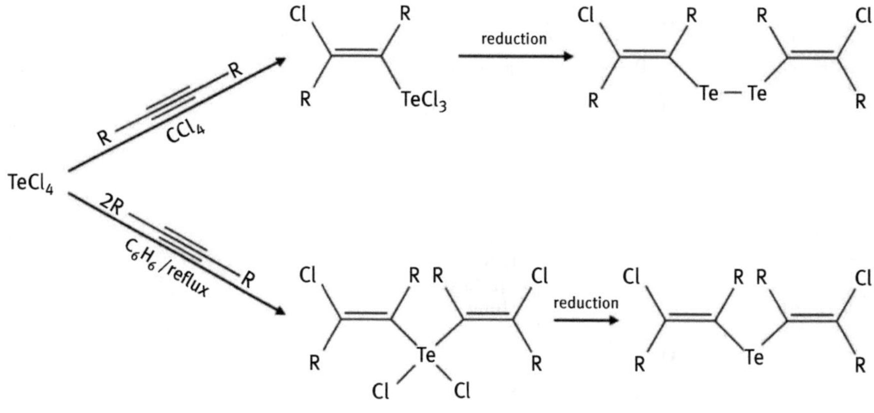

Figure 2.10: Addition of $TeCl_4$ to alkynes RC≡CR (R = Et, Pr).

2.2.7 Tellurium dihalides and ditellurium dihalides

In an early work, tellurium dihalides TeX_2 (X = Cl, Br) were prepared by the reaction of molten tellurium with CF_2Cl_2 or CF_3Br, respectively [106]a, [106]b. These dihalides disproportionate readily in ether solvents to give TeX_4 and elemental tellurium. However, the reaction of $TeCl_2$ with liquid bromine was reported to give $TeCl_2Br_2$ as a yellow solid [106]a. The isolation of $TeCl_2$ in 60 % yield from the reaction of $TeCl_4$ with an equimolar quantity of hexamethyldisilane at reflux has been claimed and Me_3SiCl was identified as a by-product [107]. This source of *in situ* $TeCl_2$ has been used to prepare the silane or germane tellones [(Tbt)(Dip)M=Te] (M = Si, Ge) by reaction with [(Tbt)(Dip)MLi_2] (Tbt = 2,4,6-tris[bis(trimethylsilyl)methyl]phenyl, Dip = 2,6-diisopropylphenyl) [108]. The same reagent has also been used in a ring-closure reaction with 2,3-dimethoxy-1,3-butadiene for the efficient synthesis of 3,4-dimethoxytellurophene [Figure 2.11(a)] [109]. Alternatively, a Lewis base adduct of $TeCl_2$ (*vide infra*) may be used to generate a tellurophene ring via Zr/Te transmetallation [Figure 2.11(b)] [110].

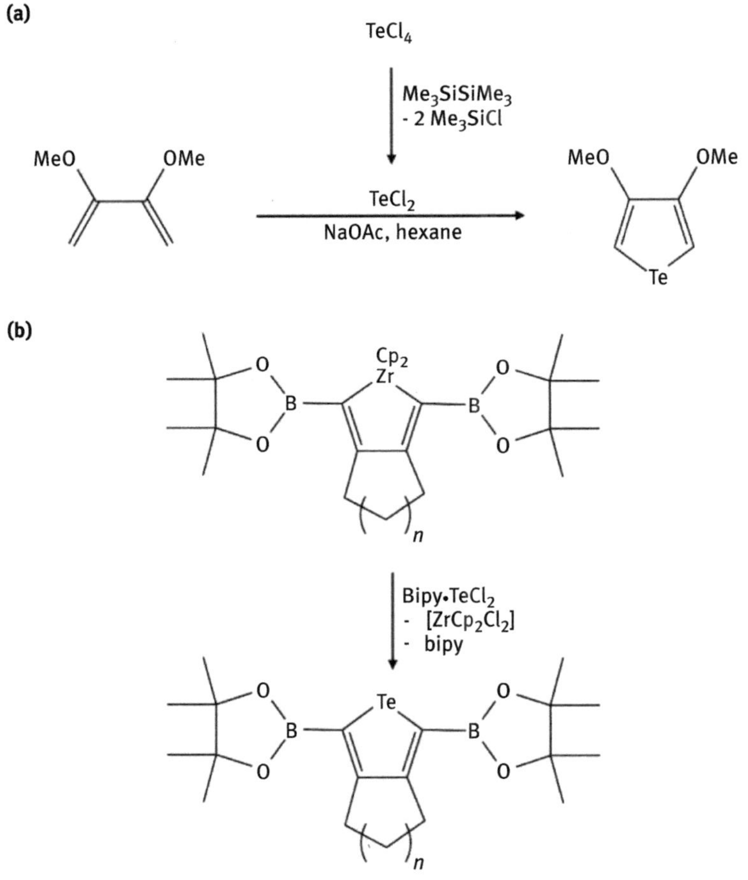

Figure 2.11: Synthesis of tellurophenes using (a) *in situ* $TeCl_2$ and (b) the Lewis base adduct (bipy) $TeCl_2$.

In fact, Lewis base adducts of TeX_2 (X = Cl, Br, I) are used more frequently than the *in situ* reagent for the synthesis of inorganic or organic tellurium compounds [9]. Examples of these sources of TeX_2 include the 1:1 adduct with tmtu, which has been known for many years [111] and the more recent discoveries of [Te(bipy)Cl_2]

(Section 2.2.6) [97] and [Te(PEt$_3$)X$_2$] (X = Cl, Br, I) [112]. In addition to the example shown in Figure 2.11(b), some other illustrations of the applications of these adducts in the synthesis of tellurium compounds are depicted in Figure 2.12. The Te(bipy)Cl$_2$ complex provides a source of diaryl tellurides via reactions with Grignard reagents [9]. The bidentate bipy ligand in this complex can be replaced by two N-heterocylic carbenes to give the 2:1 adduct Te(^{i}Pr$_2$IM)$_2$Cl$_2$ [9]. Treatment of [(Dipp$_2$BIAN)TeI$_2$] with 2.5 equivalents of silver triflate in CH$_2$Cl$_2$ rapidly generates the corresponding bis-triflate, which acts as a source of a tellurium-centred dication upon reaction with two equivalents of ^{i}Pr$_2$IM [113]. The metathetical reaction of (tmtu)TeCl$_2$ with 2.5 equivalents of Li[HC(PPh$_2$Se)$_2$] results in disproportionation to give elemental tellurium and an unusual tellurium(IV) complex (Figure 2.12) [114].

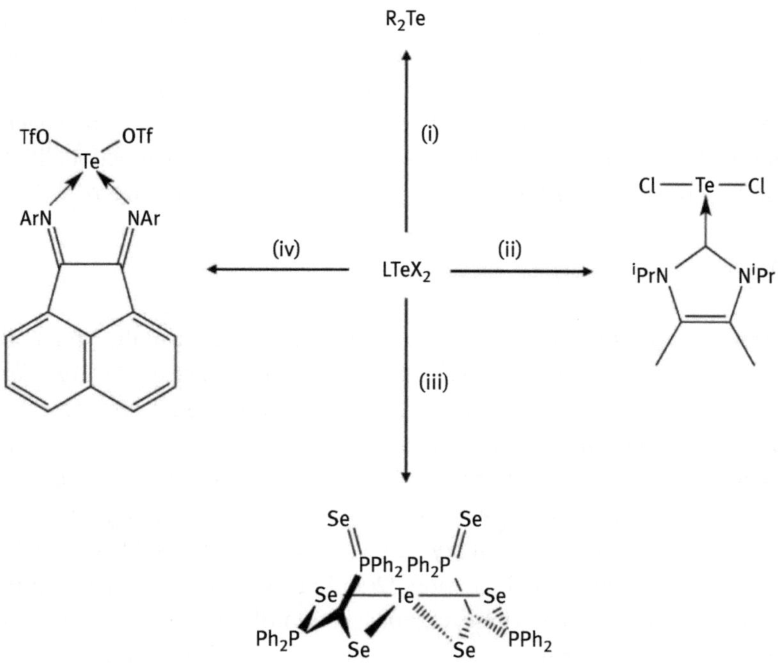

Figure 2.12: Synthesis of tellurium compounds from Lewis base adducts of TeX$_2$: (a) RMgCl (R = Ph, CH$_2$Ph), L = bipy, X = Cl, (ii) ^{i}Pr$_2$IM, L = bipy, X = Cl, (iii) Li[HC(PPh$_2$Se)$_2$], L = tmtu, X = Cl and (iv) 2.5AgOTf, L = Dipp$_2$BIAN, X = I.

Ditellurium dihalides Te$_2$X$_2$ (X = Cl, Br) are prepared by reducing elemental tellurium with the superhydride Li[BHEt$_3$] and treating the resulting telluride with the appropriate tellurium tetrahalide TeX$_4$ [115]. These tellurium(I) halide reagents are obtained as yellow and orange liquids, respectively, but they have limited stability. However, they have been used to generate the seven-membered, heteroatomic chalcogen rings 1,2-Te_2E_5 (E = S, Se) via metathesis with titanocene pentachalcogenides [eq. (2.20)] [115].

$$Cp_2TiE_5 + Te_2Cl_2 \rightarrow 1,2\text{-}Te_2E_5 + Cp_2TiCl_2 \qquad (2.20)$$

2.3 Organochalcogen–halogen reagents

A wide variety of organochalcogen(IV)–halogen reagents of the type REX_3 and R_2EX_2 (E = Se, Te; X = F, Cl, Br, I) are known and their applications in organic chemistry have been surveyed [116]. Organochalcogen(II) compounds, especially PhSe(II) halides and their derivatives, are widely used as electrophilic reagents in the functionalization of carbon–carbon multiple bonds [117]. In addition, understanding the structural features of organochalcogen halides, especially the tellurium derivatives, has been a major goal of recent investigations and this aspect is addressed in several surveys [118]. After a brief introduction to the most common representatives of each compound class, the following sections will focus on the synthesis and reactions of derivatives that incorporate bulky R groups on the chalcogen or intramolecular heteroatom–chalcogen interactions, since these strategies enable access to derivatives with functionalities that are not stable with simple alkyl or aryl groups attached to the chalcogen.

2.3.1 Organoselenium and tellurium trihalides

Arylselenium trihalides have been known since the 1930s. They are usually prepared by halogenation of the corresponding arylselenocyanate or diaryldiselenide [119]. For example, the series 4-$XC_6H_4SeBr_3$ (X = H, Br, Cl, CH_3) was obtained by treating the appropriate diaryldiselenide with three equivalents of a solution of bromine in chloroform [119]. These solid tribromides are moisture-sensitive and decompose upon prolonged storage or when exposed to a laser beam or heat with loss of Br_2. Aryl or alkylselenium trichlorides are prepared similarly, e. g. by chlorination of PhSeSePh or MeSeSeMe with Cl_2 gas in methylene dichloride [120]a,b; $PhSeCl_3$ is isolated as hygroscopic, pale yellow crystals [120]a. These organoselenium(IV) trichlorides form 1:1 adducts with $SbCl_5$, which are formulated as ionic salts $[RSeCl_2^+][SbCl_6^-]$ [120]a.

In an early work, a series of arylselenium trifluorides $ArSeF_3$ (Ar = C_6H_5, 4-$CH_3C_6H_4$, 4-FC_6H_4, 2-$C_2H_5C_6H_4$, 2-$NO_2C_6H_4$) were prepared by oxidative fluorination of the corresponding diaryl diselenide with silver difluoride; yields were in the range 24–77 % [121]a. As a representative example, $PhSeF_3$ was obtained from the reaction of PhSeSePh with an excess of AgF_2 in boiling CF_2ClCCl_2F (bp. is 47 °C) in a *polypropylene* vessel (to avoid etching of glass). The very moisture-sensitive arylselenium trifluorides are readily hydrolysed to seleninic acids $ArSeO_2H$ upon exposure to moisture [121]a. More recently, the use of XeF_2 in CH_2Cl_2 at 0 °C as fluorinating agent in perfluoroalkoxy-copolymer vessels has extended the range of known $RSeF_3$ derivatives to alkyl (R = Me, ^iPr) or bulky aryl derivatives [R = 2,4,6-$R'_3C_6H_2$

(R' = Me, ^{i}Pr, ^{t}Bu)], as well as the R = 2-$Me_2NCH_2C_6H_4$ compound [eq. (2.21)] [121]b. Because of the limited thermal stability above 0 °C, reactions of these selenium trifluorides with trimethylsilyl azide occur rapidly in CH_2Cl_2 at −50 °C to give the corresponding triazides, which have even lower thermal stability [121]b.

$$RSeSeR + 3XeF_2 \rightarrow 2ArSeF_3 + 3Xe \qquad (2.21)$$

In contrast to the formation of $TeCl_2$ from the reaction of $TeCl_4$ with hot $Me_3SiSiMe_3$ (Section 2.2.7), when combined in CS_2 these reagents generate a polymeric methyltellurium trichloride $(MeTeCl_3)_\infty$ as very moisture-sensitive, brownish crystals in 43 % yield [122]. The ethyl derivative $EtTeCl_3$ is obtained in 75 % yield by chlorination of EtTeTeEt with Cl_2 gas in cold (−20 °C) methylene dichloride [120]. The bis-substituted derivative, $CH_2(TeCl_3)_2$, is a an interesting example of an organotellurium trichloride, which is obtained as a white crystalline solid from the reaction of acetic anhydride with $TeCl_4$ in a 1:3 molar ratio at reflux for ca. 1 day [123]. Reduction of $CH_2(TeCl_3)_2$ with aqueous sodium metabisulfite ($Na_2S_2O_5$) produces an almost quantitative yield of the grey polymer $(CH_2TeTe)_n$ [123].

Halogenation of ditellurides with an excess of Cl_2 (or SO_2Cl_2), Br_2 or I_2 is a common route to aryl derivatives $ArTeX_3$, which are also accessible from reactions of TeX_4 with arylmercury halides or via Friedel–Crafts reactions of $TeCl_4$ with arenes [124]. Supermesityltellurium trichloride Mes*$TeCl_3$ (Mes* = 2,4,6-$^{t}Bu_3C_6H_2$) is obtained as bright yellow crystals in 96 % yield by chlorination of Mes*TeTeMes* with SO_2Cl_2 in a molar ratio of *ca*. 1:2.8 in diethyl ether [125]. However, the installation of the very bulky aryl group on tellurium engenders some unusual reactivity. As depicted in Figure 2.13, intramolecular HCl elimination involving an ortho *tert*-butyl group with the formation of a Te-C bond occurs in the solid state at room temperature. On the other hand, an equilibrium exists between Mes*$TeCl_3$ and the tellurium (II) species Mes*TeCl and chlorine gas in solution (Section 2.3.4) [125]. On exposure to moist air, partial hydrolysis occurs to give Mes*$Te(OH)Cl_2$ [125].

The stabilizing influence of intramolecular heteroatom interactions on organotellurium(IV) trihalides is exemplified by 2-dimethylaminomethylphenyl derivatives $ArTeX_3$ (Ar = 2-$Me_2NCH_2C_6H_4$), which are prepared by halogenation of the corresponding ditelluride ArTeTeAr with SO_2Cl_2 (X = Cl) or Br_2 (X = Br) [118]a, [126]. More recently, fluorination of this ditelluride with XeF_2 in dichloromethane at room temperature was shown to produce almost quantitative yields of the colourless trifluoride (2-$Me_2NCH_2C_6H_4$)TeF_3, which reacts with trimethylsilyl azide to give the thermally unstable triazide (2-$Me_2NCH_2C_6H_4$)$Te(N_3)_3$ [127].

Similar to the behaviour of $TeCl_4$ (Section 2.2.6), reactions of aryltellurium trichlorides with Me_3Si-N derivatives take place without reduction of the tellurium centre. Examples of aryltellurium(IV) compounds prepared in this manner include Ph_3P=NTe(Ar)Cl_2, $Me_2S(O)NTe(Ar)Cl_2$, Me_3SiN=S=NTe(Ar)Cl_2 (Ar=4-$MeOC_6H_4$) [84], and Ar(R'$_2$N)$TeCl_2$ (Ar = C_6H_5, 4-$MeOC_6H_4$, 4-$C_6H_5OC_6H_4$; R'$_2$ = Et_2, C_4H_8) [128].

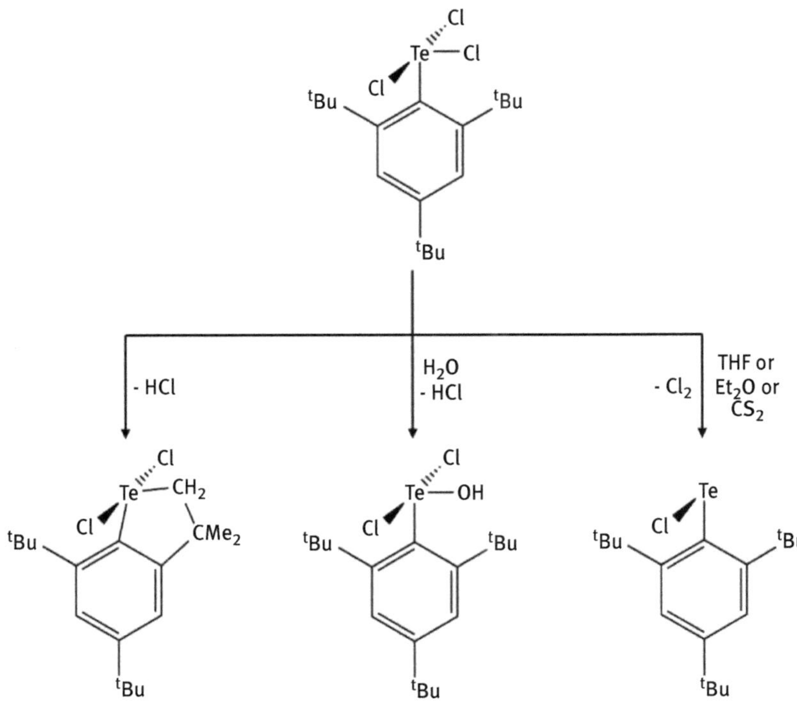

Figure 2.13: Decomposition pathways of $ArTeCl_3$ (Ar = 2,4,6-$^tBu_3C_6H_2$).

The base hydrolysis of aryltellurium trichlorides provides a general route to aryltellurinic acids $[ArTe(O)OH]_n$, which are usually polymeric, high-melting solids with low solubility in organic solvents [129]. However, the installation of a highly bulky group on tellurium enables the isolation of a partial hydrolysis product, e. g. Mes*Te(OH)Cl_2 Figure 2.13) [125]. In the case of $ArTeCl_3$ (Ar = 2,6-$Mes_2C_6H_3$), the controlled base hydrolysis proceeds in two steps to give the aryltellurinic chloride $[ArTe(O)Cl]_2$ and, subsequently, the aryltellurinic acid $[ArTe(O)OH]_2$, both of which are dimeric (Figure 2.14) [130]. The hydrolysis of the intramolecularly coordinated $ArTeCl_3$ (Ar = 8-$Me_2NC_{10}H_6$) with an excess of aqueous NaOH generates a series of polynuclear telluroxanes as crystalline materials [131].

2.3.2 Diorgano-selenium and -tellurium dihalides

Simple dialkyl and diaryl-selenium dihalides R_2SeX_2 (R = Me, Ph; X = Cl, Br) are obtained as monomeric solids from halogenation of the corresponding selenide R_2Se with an equivalent amount of Cl_2 or Br_2 [132]. By contrast, the difluorides are liquids (R = Me, Et, ^{n}Pr, ^{i}Pr) or low-melting solids (R = $(CH_2)_4$, Ph) that are generated by treatment of R_2Se with AgF_2 [133]. Alternatively, XeF_2 in CH_2Cl_2 can be used as the

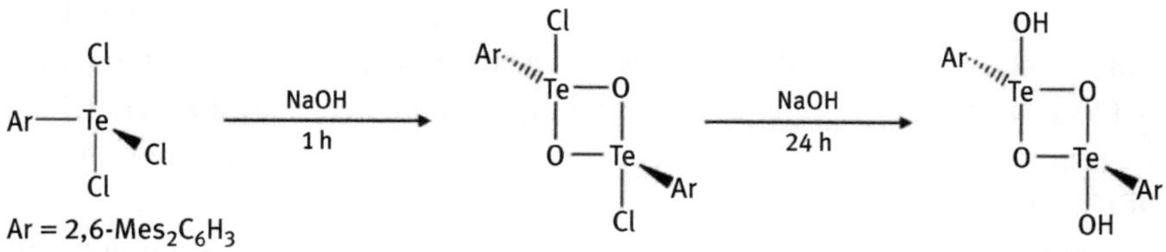

Figure 2.14: Hydrolysis of $ArTeCl_3$ (Ar = 2,6-$Mes_2C_6H_3$).

fluorinating agent to prepare R_2SeF_2 (R = Me, Et, ^iPr, Ph, Mes, Mes*, 2,4,6-$^iPr_3C_6H_2$, 2-$Me_2NCH_2C_6H_4$) [127]. Ph_2SeF_2 is obtained as a dark purple crystalline solid, which reacts with two equivalents of PhLi in diethyl ether to produce the explosive tetraphenylselenurane Ph_4Se (Section 2.2.1) [4].

Diorganotellurium dihalides are accessible in a number of ways: (a) condensation reactions with aromatic compounds such as ROC_6H_5 or $R_2NC_6H_5$, (b) arylation of $TeCl_4$ with ArHgCl, (c) halogenation of tellurides R_2Te with Cl_2 (or SO_2Cl_2), Br_2, I_2 and (d) reactions of elemental tellurium with alkyl halides [124]. As a recent example of method (c), the diorganotellurium diiodides $TeI_2(CH_2SiMe_3)_2$, $TeI_2Th(CH_2SiMe_3)$, $TeI_2Ph(CH_2SiMe_3)$ and TeI_2Th_2 (Th = 2-thienyl) are obtained as air-stable orange or red crystals in >90 % yields by treatment of the appropriate telluride with I_2 in THF [134]. The oxidative-addition of alkyl bromides or iodides to elemental tellurium [method (d)] is a well-established procedure that has been applied to the preparation of $(PhCOCH_2)_2TeX_2$ as colourless (X = Br) or red (X = I) crystals, in 50 % or 65 % yields, respectively [135].

In the case of 1,5-ditelluracyclooctane, reaction with Cl_2 or I_2 [method (c)] generates dihalogenoditelluranes with a transannular Te–Te bond in almost quantitative yields (Figure 2.15) [136]a. The thermolysis of the dichloroditellurane in boiling dimethylformamide (DMF) (160 °C) for 5 min engenders the ring-expansion product Cl_6[12]aneTe_3, which exhibits redox behaviour upon treatment with six equivalents of PhSH in dimethyl sulfoxide (DMSO) at 50 °C to give a cyclic tritelluride in 90 % yield (Figure 2.15) [136]b.

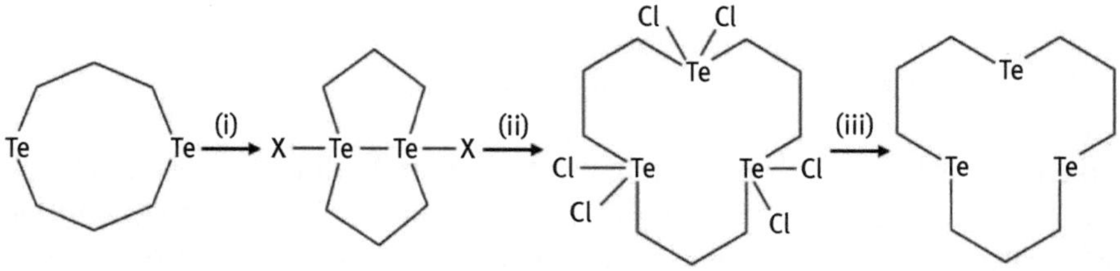

Figure 2.15: Halogenation of 1,5-ditelluracyclooctane and subsequent ring expansion upon thermolysis: (i) Cl_2 or I_2 in benzene at RT, (ii) DMF, 160 °C, X = Cl, (iii) 6PhSH, NEt_3, DMSO, 50°C.

Diaryltellurium difluorides $(4\text{-}ROC_6H_4)_2TeF_2$ (R = Me, Et) are obtained as colourless crystals in low yields by fluorination of the corresponding dichlorides with AgF_2 in boiling toluene [124]b. Similar to the preparation of aryltellurium trifluorides (*vide supra*), the reaction of XeF_2 in dichloromethane at room temperature with the mono-telluride $(2\text{-}Me_2NCH_2C_6H_4)_2Te$ produces the colourless difluoride $(2\text{-}Me_2NCH_2C_6H_4)_2TeF_2$, which reacts with trimethylsilyl azide to give the thermally unstable diazide $(2\text{-}Me_2NCH_2C_6H_4)_2Te(N_3)_2$ [127].

The hydrolysis of diaryltellurium dihalides with an excess of base in aqueous ethanol is an important route for the preparation of the ditellurium oxides (telluroxanes), $(R_2TeO)_n$, as crystalline solids [129]. Incomplete hydrolysis using water gives rise to either diaryltellurium hydroxyhalides $R_2Te(OH)X$ or their condensation products $XR_2TeOTeR_2X$. A better route to the latter class of ditelluroxane is the redistribution reaction between R_2TeO and R_2TeX_2, e.g. R = 4-$MeOC_6H_4$, X = I [137].

The bifunctionality of diaryltellurium dihalides provides an opportunity to generate tellurium-containing macrocycles through cyclocondensation reactions with carboxylic or phosphinic acids. Thus, the reactions of Ar_2TeCl_2 (Ar = 4-$MeOC_6H_4$) with 1,1´-ferrocenedicarboxylic acid in the presence of NEt_3 in benzene generates an almost quantitative yield of a 16-membered macrocycle [138], while a similar reaction with 1,1,2,3,3-pentamethyltrimethylenephosphinic acid produces a 12-membered $Te_4P_2O_6$ ring [139].

2.3.3 Mixed-valent state compounds $RTeX_2TeR$

Tellurium can form an unusual class of mixed-valent [Te(IV)/Te(II)] compounds if a deficiency of the halogen reagent is used in the halogenation of ditellurides. Thus, the reaction of PhTeTePh with only one molar equivalent of Br_2 produces red-brown $PhTeBr_2TePh$ in almost quantitative yield [140]. Derivatives with a bulky aryl group $ArTeX_2TeAr$ (Ar = 2,6-$Mes_2C_6H_3$; X = Cl, Br) are obtained in a similar manner from ArTeTeAr and SO_2Cl_2 or Br_2 [140] and a mixed-valent difluoride $BbtTeF_2TeBbt$ is prepared as air-stable, red-brown crystals in 64 % yield by treatment of the ditelluride with XeF_2 in dichloromethane at −78 °C (Figure 2.16) [141].

The applications of ditelluride dihalides $RTeX_2TeR$ for the generation of aryltellurenyl cations are discussed at the end of the next section.

2.3.4 Organo-selenium and -tellurium monohalides

In marked contrast to the thermal instability of selenium(II) dihalides (Section 2.2.3), phenyl selenium(II) halides PhSeX (X = Cl, Br) are commercially available solids that can be readily prepared by treatment of PhSeSePh with one equivalent of SO_2Cl_2 or Cl_2 in hexane or Br_2 in THF. By contrast, the treatment of PhSeSePh with I_2 produces the 1:1 complex $Ph_2Se{\bullet}I_2$ [142]. However, the presence of a bulky aryl group causes Se–Se bond cleavage to give the black selenenyl iodide ArSeI (Ar = Mes*) in almost quantitative yield [143]. A bulky alkyl group also stabilizes the Se-I functionality in $(Me_3Si)_3CSeI$ [144]. The complex $Ph_2Se{\bullet}I_2$ is commercially available and behaves as

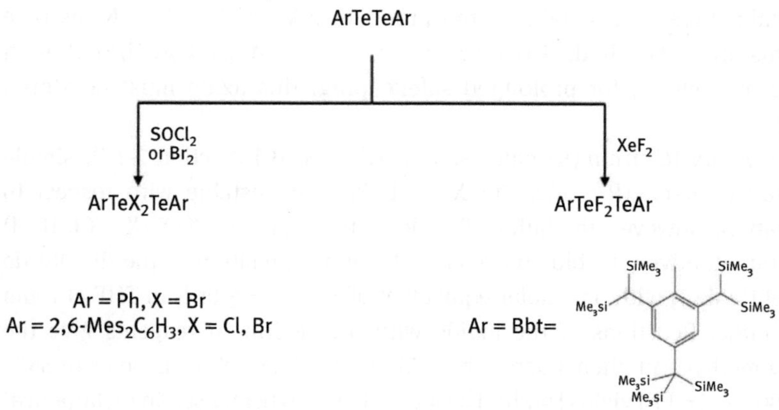

Figure 2.16: Synthesis of a mixed-valent aryltellurenyl halides $ArX_2TeTeAr$.

the reagent "PhSeI" in the iodoselenation of 2-hexyne [142]. Arylselenenyl fluorides ArSeF in which a substituent in the *ortho* position provides steric or electronic protection are obtained by treatment of diselenides ArSeSeAr or aryl selenotrimethylsilanes $ArSeSiMe_3$ with XeF_2 in dichloromethane at low temperatures in perfluoroethylene-perfluorovinyl ether copolymer (PFA) vessels [145]. They were identified by low-temperature ^{19}F and ^{77}Se NMR spectroscopy but, at ambient temperature, they readily undergo disproportionation.

The applications of phenylselenenyl reagents in organic chemistry, primarily for the functionalization of carbon–carbon multiple bonds, are discussed in Chapter 7. The halide ion in PhSeX is often replaced *in situ* in order to avoid the incorporation of halide nucleophiles during selenenylation reactions [117]. For example, the following $PhSe^+X^-$ equivalents have been used for addition to alkynes: $Ph_2Se_2/XeF_2/2SbF_5$, $Ph_2Se_2/Br_2/AgSbF_6$, $Ph_2Se_2/2NOSbF_6$ and $PhSeCl/SbCl_5$ [146]. The latter reagent reacts with alkynes RC≡ CR in dichloromethane at −40 °C to give selenirenium cations as $SbCl_6^-$ salts which, in the case of R = ^{t}Bu, Ad, can be isolated as stable crystalline solids (Figure 2.17) [146].

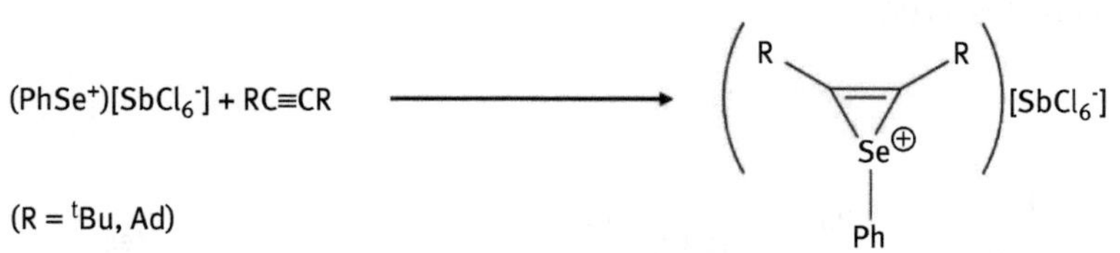

Figure 2.17: Preparation of selenirenium ions from $PhSeCl/SbCl_5$ and alkynes.

The reactions of arylselenium chlorides ArSeCl with trimethylsilyl azide produce the corresponding azides $ArSeN_3$, which decompose with loss of N_2 to give diselenides ArSeSeAr even for very bulky aryl derivatives. However, the strategy of intramolecular heteroatom stabilization produces 2-$Me_2NCH_2C_6H_4SeN_3$ as pale yellow crystals in 88 % yields from the reaction of 2-$Me_2NCH_2C_6H_4SeCl$ with sodium azide in acetone; for prolonged safekeeping, this azide must be stored at 0 °C [147].

Similar to binary tellurium(II) halides, TeX_2 (X = Cl, Br) (Section 2.2.7), simple organic derivatives RTeX (R = Me, Ph; X = Cl, Br) are unstable with respect to disproportionation. However, the bulky alkyl derivatives $(Me_3Si)_3CTeX$ (X = Cl, Br, I) are obtained quantitatively as blue-black crystals by halogenation of the ditelluride $(Me_3Si)_3CTeTeC(SiMe_3)_3$ with one molar equivalent of SO_2Cl_2, Br_2 or I_2 in THF at room temperature [148]a. Reactions of the iodide with organolithium reagents give the corresponding methyl and phenyl derivatives $(Me_3Si)_3CTeR$ as yellow needles in 85 % (R = Me) and 80 % (R = Ph) yields [148]b. The iodide has also been used in metathetical reactions with silver or potassium salts of pseudo-halides or pseudo-chalcogenides to generate $(Me_3Si)_3CTeX$ (X = CN, SCN, SeCN, NCO, N_3) and $(Me_3Si)_3CTeYTeC(SiMe_3)_3$ (Y = NCN, NSN), respectively, in yields ranging from 72 to 92 % [148]b.

Bulky substituents or intramolecular heteroatom–tellurium coordination can also serve to stabilize aryltellurenyl halides, as illustrated by the monomeric species 2,4,6-$R_3C_6H_2TeX$ (R = ^{i}Pr, ^{t}Bu; X = Br, I), which are prepared by reactions of the corresponding ditellurides with equimolar amounts of Br_2 or I_2; the iodides are thermally more stable than the bromides [149]. Stabilization can also be achieved by intramolecular heteroatom–tellurium interaction as in 2-$Me_2NCH_2C_6H_4TeI$ [150]. The kinetic or donor stabilization afforded by bulky aryl groups or the presence of coordinating heteroatoms extends to the corresponding azides, e.g. 2,4,6-$R_3C_6H_2TeN_3$ (R = ^{i}Pr, ^{t}Bu) and 2-$Me_2NCH_2C_6H_4TeN_3$, which are prepared by reactions of the iodides with silver azide in acetonitrile [150].

Aryltellurenyl iodides behave as iodine-like acceptors in reactions with iodide ion, phosphine selenides or $TeMes_2$ (Figure 2.18). For example, the green iodide 2,4,6-$^i Pr_3C_6H_2TeI$ forms the stable, red adduct $[Et_4N][2,4,6\text{-}^iPr_3C_6H_2TeI_2]$ upon treatment with $[Et_4N]I$ [149]. The bifunctional phosphine selenides $SePPh_2(CH_2)_nPh_2PSe$ (n = 1,2) react with MesTeI in dichloromethane to give 1:2 complexes in *ca.* 75 % yields [97]. In related behaviour, attempts to recrystallize MesTeI always produced the 1:1 adduct $Mes(I)Te(TeMes_2)$ [151]

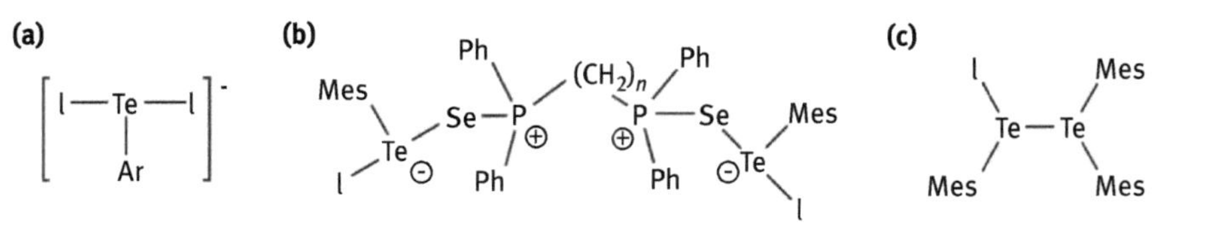

Figure 2.18: Adducts of ArTeI with (a) iodide ion (Ar = 2,4,6-$^i Pr_3C_6H_2$), (b) $SePPh_2(CH_2)_nPh_2PSe$ (n = 1,2) (Ar = Mes) and (c) $TeMes_2$ (Ar = Mes).

Halogenation reactions of a ditelluride with bulky aryl substituents RTeTeR (R = Bbt = 2,4,6-[$(Me_3Si)_2CH]_3C_6H_2$) have also been investigated using different molar ratios of the reagents [152]. With one molar equivalent of SO_2Cl_2, Br_2 or I_2 in CH_2Cl_2, the thermally stable monohalides BbtTeX were isolated in 82–95 % yields as red-purple, blue and green crystals, respectively. Changing the molar ratio to 3:1 produced the trihalides $BbtTeX_3$ (X = Cl, yellow crystals, 93 %; X = Br, red crystals, 99 %), but only the monoiodide BbtTeI when three equivalents of I_2 were used [152].

The heteroatom (*ortho*-methoxy) substituents in the ditelluride RTeTeR (R = 2,6-dimethoxyphenyl) have a notable influence on the outcome of the reaction with I_2 [153]. When this reaction is carried out in a 1:1 molar ratio in dichloromethane at −10 °C the mixed-valent compound $RTeTeI_2R$ is isolated in very high yields. At room temperature the tellurenyl iodide is formed, but it reacts further with an additional molar equivalent of I_2 to give the adduct $RTe(I)I_2$ [153].

The reaction of PhTeTePh with one equivalent of I_2 in diethyl ether produces black crystals of stoichiometry "PhTeI" which, in contrast to the other monohalides PhTeX (X = Cl, Br) (*vide supra*), is stable in the solid state. In the absence of bulky substituents on the aryl group, this species exists as the tetramer $Ph_4Te_4I_4$ with a four-membered Te_4 ring [154]. However, the tetrameric structure is easily disrupted upon addition of PPh_3 to give the 1:1 adduct $Ph_3PTe(Ph)I$ [154]. Although this monomeric complex does not act as a source of $PhTe^+$, aryltellurenyl cations stabilized by two-electron donors, e. g. PPh_3 or NHC, can be generated from mixed-valent ditellurium dihalides ArX_2TeAr (Section 2.3.3) as illustrated in Figure 2.19 [155, 156].

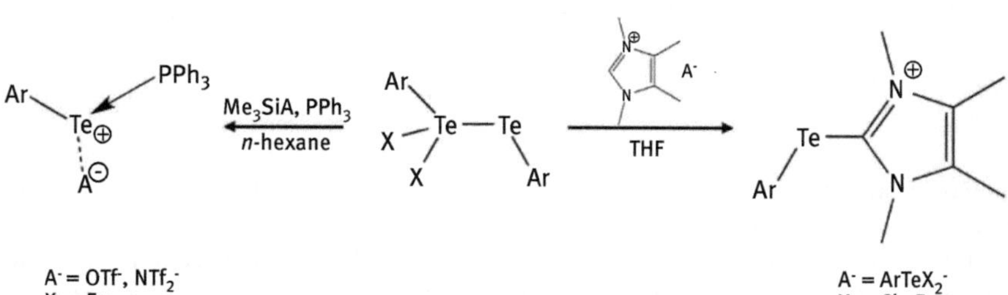

Figure 2.19: Synthesis of phosphine and carbene adducts of $PhTe^+$: Ar = Bbt (left-hand reaction); 2,6-$Mes_2C_6H_3$ (right-hand reaction).

2.4 Conclusions

The reactivity, variety and, in some cases, ready access to Se/Te halides facilitates the wide use of these reagents in the preparation of both inorganic and organic compounds. In particular, the commercial availability of the easily manipulated tetrahalides $E(IV)X_4$ (E = Se, Te; X = Cl, Br) enables their applications in inorganic and organic synthesis. Although lower-valent dihalides $E(IV)X_2$ (E = Se, Te; X = Cl, Br) have a propensity for disproportionation, methods have been developed to allow their use as *in situ* reagents. Specifically, an extensive chemistry of selenium dihalides especially in organic synthesis, but also for the preparation of inorganic compounds, has emerged. To a lesser extent, tellurium dihalides are also effective reagents in the form of adducts with Lewis bases, provided that the electron donors released in these reactions do not interfere with the desired transformation. In the case of selenium, the existence of halides in a variety of formal oxidation states, e.g. $SeCl_4$, $SeOCl_2$, $SeCl_2$ and Se_2Cl_2, has permitted the creative use of mixtures of these reagents for the synthesis of unusual selenium compounds.

In contrast to the binary systems, organoselenium(II) halides PhSe(II)X (X = Cl, Br) are commercially available, thermally and reasonably air-stable reagents that have wide applications in organic synthesis, notably for additions to carbon–carbon multiple bonds. On the other hand, the tellurium analogues PhTe(II)X, like their inorganic counterparts, are unstable towards disproportionation. However, this tendency can be suppressed by invoking the strategies of kinetic stabilization using either very bulky substituents on the tellurium atom or groups that provide intramolecular heteroatom–tellurium coordination. These approaches have also been applied to the preparation of reasonably stable organotellurium(IV) fluorides. The heavier organotellurium(IV) halides, as exemplified by REX_3 and R_2TeX_2 (R = alkyl, aryl; X = Cl, Br), are readily prepared and easily handled in the case of aryl derivatives. The multifunctionality of these reagents presages wide applications in synthesis, which have only been partially realized. A unique feature of organotellurium halides is the existence of mixed-valent species of the type $RTe(IV)X_2Te(II)R$. The synthetic potential of these reagents has only recently been exploited, e.g. in the generation of RTe^+ cations.

References

[1] Klapötke TM, Krumm B, Mayer P, Piotrowski H, Schwab I, Vogt M. Synthesis and structures of triorganotelluronium pseudohalides. Eur J Inorg Chem. 2002;2701–9.

[2] Wang J, Xu Z. Recent developments in binary halogen-chalcogen compounds, polyanions and polycations. In: Devillanova FA, editor. Handbook of chalcogen-nitrogen chemistry: new perspectives in sulfur, selenium and tellurium. Vol. 1. 2nd ed. Cambridge UK: RSC Publishing, 2013. p. 425–7.

[3] Klapötke T, Krumm BM, Scherr M, Haiges R, Christe KO. The binary selenium(IV) azides $Se(N_3)_4$, $[Se(N_3)_5]^-$ and $[Se(N_3)_6]^{2-}$. Angew Chem Int Ed. 2007;46:8686–90.

[4] Bienfait AM, Kubella P, Mueller B, Seppelt K. Isolation of tetraphenylselenurane. Heteroatom Chem. 2011;22:576–8.

[5] Herberhold M, Jellen W. New selenium-nitrogen compounds *tert*-butyl-seleninylamine and di (*tert*-butyl)selenium diimide $Se(N^tBu)_2$. Z Naturforsch. 1986;41b:144–8.

[6] (a) Maaninen T, Laitinen R, Chivers T. A monomeric selenium(IV) diimide and a dimeric seleninylamine. Chem Commun. 2002;1812–3.(b) Maaninen T, Tuononen HM, Kosunen K, Oilunkaniemi R, Hiitola J, Laitinen R, Chivers T. Formation, structural characterization, and calculated NMR chemical shifts of selenium-nitrogen compounds from SeCl4 and ArNHLi (Ar = supermesityl, mesityl). Z. Anorg. Allg. Chem. 2004, 630, 1947–54.

[7] Denney DB, Denney DZ, Hammond PJ, Hsu YF. Preparation and NMR studies of tetraalkoxyselenuranes and tetraalkoxytelluranes. J Am Chem Soc. 1981;103:2340–7.
[8] Parkin IP, Woollins JD. Preparation and characterization of $[Pt(SeSN_2)(PR_3)_2]$, $[Pt(Se_2N_2)(PR_3)_2]$, $[Pt(SeSN_2H)(PR_3)_2][BF_4]$, and $[Pt(Se_2N_2H)(PR_3)_2][BF_4]$. J Chem Soc Dalton Trans. 1990;925–30.
[9] Dutton JL, Ragogna PJ. Recent developments in the Lewis acidic chemistry of selenium and tellurium halides and *pseudo*-halides. In: Woollins JD, Laitinen RS, editors. Selenium and tellurium chemistry; from small molecules to biomolecules and materials, Ch. 8. New York: Springer, 2011:179–99.
[10] Wynne KJ, Pearson PS, Newton MG, Golen J. Preparation and properties of dichloro- and dibromo(tetramethylthiourea)selenium(II). Crystal and molecular structure of $SeBr_2$•tmtu. Inorg Chem. 1972;11:1192–6.
[11] Jolleys A, Levason W, Reid G. Thioether coordination to divalent selenium halide acceptors – synthesis, properties and structures. Dalton Trans. 2013;42:2963–72.
[12] (a). Dutton JL, Sutrisno A, Schurko RW, Ragogna PJ. Synthesis and characterization of cationic selenium-nitrogen heterocycles from *tert*-butyl-DAB (DAB = 1,4-di-*tert*-butyl-1,3-diazabutadiene) and SeX_4 via reductive elimination of X_2 (X = Cl, Br): a distinct contrast with tellurium. Dalton Trans. 2008:3470–7. (b) Dutton JL, Tabeshi R, Jennings MC, Lough AJ, Ragogna PJ. Redox reactions between phosphines (R_3P; R = ^{n}Bu, Ph) or carbene (^{i}Pr$_2$IM) and chalcogen tetrahalides ChX_4 (^{i}Pr$_2$IM = 2,5-diisopropylimidazole-2-ylidene; Ch = Se. Te; X = Cl, Br). Inorg. Chem. 2007, 46, 8594–602.
[13] Potapov VA, Musalov MV, Musalova MV, Amosova SV. Recent advances in organochalcogen synthesis based on reactions of chalcogen halides with alkynes and alkenes. Curr Org Chem. 2016;20: 136–45.
[14] King RB, Sangokoya SA. (Dialkylamido)selenium derivatives. Inorg Chem. 1987;26:2727–30.
[15] Roesky HW, Weber K-L, Seseke U, Pinkert W, Noltemeyer M, Clegg W, et al. Structural and NMR studies of short selenium-nitrogen bonds. J Chem Soc Dalton Trans. 1985;565–71.
[16] Amaresh R, Lakshmikantaham MV, Baldwin JW, Cava MP, Metzger RM, Rogers RD. Condensed thiophenes and selenophenes: thionyl chloride and selenium oxychloride as sulfur or selenium transfer agents. J Org Chem. 2002;67:2453–8.
[17] Bestari K, Cordes AW, Oakley RT, Young KM. Heterocyclic 1,2,4,6-thia- and 1,2,4,6-selenatriazinyl radicals; spin distributions and modes of association. J Am Chem Soc. 1990;112:2249–55.
[18] Lamoureux M, Milne J. Selenium chloride and bromide equilibria in aprotic solvents: A ^{77}Se NMR study. Polyhedron. 1990;9:589–95.
[19] Belluz PDB, Cordes AW, Kristof PV, Liblong SW, Oakley RT. 1,2,3,5-Diselenadiazolyls as building blocks for molecular metals. J Am Chem Soc. 1989;111:9276–8.
[20] Bryce MR, Chesney A. Phenylsulfinylselenyl chloride ($PhSO_2SeCl$): A new reagent for the formation of C-Se and N-Se bonds. Generation and *in situ* Diels-Alder trapping of selenonitrosoarene intermediate (Ar-N=Se). J Chem Soc Chem Commun. 1995;195–6.
[21] Maaninen A, Chivers T, Parvez M, Pietikäinen J, Laitinen RS. Syntheses of THF solutions of SeX_2 (X = Cl, Br) and a new route to selenium sulfides Se_nS_{8-n} (n = 1-5); crystal structures of $SeCl_2$ $(tht)_2$ and $SeCl_2$•tmtu (tmtu = tetramethylthiourea). Inorg Chem. 1999;38:4093–7.
[22] Amosova SV, Penzik MV, Martynov AV, Makhaeva NA, Yarish NO, Voronkov MG. Unsaturated five-membered selenium-germanium containing heterocycles based on the reactions of selenium di- and tetrahalides with diorganyl diethynyl germanes. J Organomet Chem. 2008;693:3346–50.
[23] Potapov VA, Amosova SV, Abramova EV, Musalov MV, Lyssenko KA, Finn MG. 2,6-Dihalo-9-selenabicyclo[3.3.1]nonanes and their complexes with selenium dihalides; synthesis and structural characterisation. New J Chem. 2015;9:8055–9.
[24] Chivers T, Laitinen RS. Insights into the formation of inorganic heterocycles via cyclocondensation of primary amines with group 15 and 16 halides. Dalton Trans. 2017;46: 1357–67.

[25] Maaninen T, Chivers T, Laitinen R, Wegelius E. Acyclic imidoselenium(I) halides; synthesis and X-ray structures of $ClSe[N(^tBu)Se]_nCl$ (n = 1-3). Chem Commun. 2000;759–60.

[26] Maaninen T, Chivers T, Laitinen R, Schatte G, Nissinen M. Selenium imides: ^{77}Se NMR investigations of the $SeCl_2$-$^tBuNH_2$ reaction and X-ray structures of $Se_3(N^tBu)_3$, $^tBuNSe(\mu{-}n^tBu)_2SO_2$ and $^tBuNSe(\mu{-}n^tBu)_2SeO$. Inorg Chem. 2000;39:5341–7.

[27] Karhu AJ, Pakkanen OJ, Rautiainen JM, Oilunkaniemi R, Chivers T, Laitinen RS. The role of imidoselenium(II) chlorides in the formation of cyclic selenium imides via cyclocondensation. Dalton Trans. 2016;45:6210–21.

[28] Karhu AJ, Pakkanen OJ, Rautiainen JM, Oilunkaniemi R, Chivers T, Laitinen RS. Experimental and computational ^{77}Se NMR investigations of the cyclic eight-membered selenium imides 1,3,5,7-$Se_4(NR)_4$ (R = Me, ^tBu) and 1,5-$Se_6(NMe)_2$. Inorg Chem. 2015;54:4990–7.

[29] Konu J, Maaninen A, Paananen K, Ingman P, Laitinen RS, Chivers T, et al. Preparation and structural characterization of $(Me_3SiNSN)_2Se$, a new synthon for sulfur-selenium nitrides. Inorg Chem. 2002;41:1430–5.

[30] Song LC, Hu QM, Fan HT, Tang MY, Yang ZY, Lu GL. Reactions of complex anions of the type $[(\mu\text{-RTe})(\mu\text{-CO})Fe_2(CO)_6]^-$ with electrophiles. A new route to μ_4-Se-containing double-butterfly clusters. Organometallics. 2002;21:2468–72.

[31] Musalov MV, Potapov VA. Selenium dihalides: new possibilities for the synthesis of selenium-containing heterocycles. Chem Heterocycl Compd. 2017;53: 150–2.

[32] Potapov VA, Amosova SV, Belozerova OV, Albanov AI, Yarosh OG, Voronkov MG. Synthesis of 3,6-dihalo-4,4-dimethyl-1,4-selenasilafulvenes. Chem Heterocycl Compd. 2003;39:549–50.

[33] Amosova SV, Martynov AV, Makhaeva NA, Belozerova OV, Penzik MV, Albanov AI, et al. Unsaturated five-membered selenium-silicon containing heterocycles based on the reactions of selenium di- and tetrahalides with diorganyl diethynylsilanes. J Organomet Chem. 2007;692:946–52.

[34] Potapov VA, Shagun VA, Penzik MV, Amosova SV. Quantum chemical studies of the reaction of selenium dichloride with divinyl sulfide and comparison with experimental results. J Organomet Chem. 2010;695: 1603–8.

[35] Potapov VA, Amosova SV, Volkova KA, Penzik MV, Albanov AI. Reactions of selenium dichloride and dibromide with divinyl selenide: synthesis of novel selenium heterocycles and rearrangement of 2,6-dihalo-1,4-diselenanes. Tetrahedron Lett. 2010;51: 89–92.

[36] Potapov VA, Kurkutov EO, Musalov MV, Amosova SV. Reaction of selenium dichloride and dibromide with divinyl sulfone; synthesis of novel four- and five-membered selenium heterocycles. Tetrahedron Lett. 2010;51:5258–361.

[37] Patra A, Wijsboom YH, Zade SS, Li M, Sheynin Y, Leitus G, et al. Poly(3,4-ethylenedioxyselenophene). J Am Chem Soc. 2008;130:6734–6.

[38] Kim B, Kim J, Kim E. Visible to near-IR electrochromism and photothermal effect of poly(3,4-propylenedioxyselenophene)s. Macromolecules. 2011;44:8791–7.

[39] Das S, Zade SS. Poly(cyclopenta[*c*]selenophene): a new polyselenophene. Chem Commun. 2010;46:1168–70.

[40] Potapov VA, Khuriganova OI, Amosova SV. Efficient procedure of preparation of 3-bromo-2-phenylbenzo[*b*]selenophene from selenium dibromide and diphenylacetylene. Russ J Org Chem. 2010;46:1421–2.

[41] Arsenyan P. A simple method for the preparation of selenopheno[3,2-*b*] and [2,3-*b*]thiophenes. Tetrahedron Lett. 2014;55:2527–9.

[42] Musalov MV, Potapov VA, Amosova SV. Reaction of selenium dichloride with allyl phenyl ether. Russ J Org Chem. 2011;47:948–9.

[43] Potapov VA, Musalov MV, Amosova SV. Reaction of selenium dichloride and dibromide with unsaturated ethers. Annulation of 2,3-dihydro-1,4-oxaselenine to the benzene ring. Tetrahedron Lett. 2011;52:4606–12.

[44] (a). Braverman S, Cherkinsky M, Jana R, Kalendar Y, Sprecher M. Reaction of selenium and tellurium halides with propargyl alcohols. The regio- and stereoselectivity of addition to the triple bond. J Phys Org Chem. 2010;23:1114–20. (b) Braverman S, Cherkinsky M, Jana R, Kalendar Y, Gottlieb HE, Mats EM, Gruzman A, Goldberg I, Sprecher M. One-pot three-component preparation of novel selenium-containing spiroketals. J. Phys. Org. Chem. 2013, 26, 102–8.

[45] (a). Potapov VA, Musalov MV, Khuriganova OI, Musalova MV, Amosova SV. Stereoselective reaction of selenium dihalide with hex-3-yne. Russ J Org Chem. 2013;49:1836–8. (b) Potapov VA, Kurkutov EO, Musalov MV, Amosova SV. Regioselective reaction of selenium dihalides with methyl vinyl ketone. Russ. J. Org. Chem. 2013, 49, 1831–3.

[46] Zade SS, Panda S, Singh HB, Wolmershäuser G. Synthesis of diaryl selenides using the in situ reagent $SeCl_2$. Tetrahedron Lett. 2005;46:665–9.

[47] Bhasin KK, Singh RS, Kumar H, Mehta SK. A one-flask synthesis and characterization of novel symmetrical pyridyl monoselenides and X-ray crystal structure of bis(5-bromo-2-pyridyl)selenide and bis(2-bromo-5-pyridyl)selenide. J Organomet Chem. 2010;695:648–52.

[48] Srivastava K, Chakraborty T, Singh HB, Butcher RJ. Intramolecularly coordinated azobenzene selenium derivatives. Effect of the Se•••N intramolecular interaction on luminescence. Dalton Trans. 2011;40:4489–96.

[49] Potapov VA, Khuriganova OI, Amosova SV. First example of aromatic substitution using selenium(II) chloride. Russ J Org Chem. 2009;45:1581–2.

[50] Kumar P, Kashid VS, Mague JT, Balakrishna MS An efficient approach for the synthesis of functionalized selenoethers and selencalix[4]thiophenes. Tetrahedron Lett. 2014;55:5232–5.

[51] Milne J, Williams AJ. Exchange processes in diselenium and selenium-sulfur dihalides, Se_2X_2 and $SeSX_2$ (X = Br, Cl). A ^{77}Se 2D-EXSY study. Inorg Chem. 1992;31:4534–8.

[52] Lamoureux M, Milne J. The disproportionation of diselenium dichloride, Se_2Cl_2, and diselenium dibromide, Se_2Br_2. Can J Chem. 1989;67:1936–41.

[53] Cohen SC, Reddy MLN, Massey AG. Perfluoroaromatic derivatives of sulphur, selenium and tellurium. J Organomet Chem. 1968;11:563–8.

[54] Björgvinsson M, Roesky HW, Payer F, Stalke D, Sheldrick GM. Preparation and structural characterization of bis[bis(trimethylsilyl)amido]chalcogenides of selenium and tellurium. Inorg Chem. 1990;29:5140–3.

[55] Björgvinsson M, Roesky HW, Payer F, Stalke D, Sheldrick GM. Preparation and structure characterization of bis[*tert*-butyl(trimethylsilyl)amino]chalcogenides of selenium and tellurium. Eur J Solid State Inorg Chem. 1992;29:759–76.

[56] Siivari J, Maaninen A, Haapaniemi E, Laitinen RS, Chivers T. Formation and identification of bis [bis(trimethylsilyl)amino]tri- and tetrachalcogenides. Z Naturforsch. 1995;50b:157–82.

[57] Folkerts H, Dehnicke K, Magull J, Goesmann, Fenske D. Phosphorane iminato-trichloroselenate (II): syntheses and crystal structures of $[SeCl(NPPh_3)_2]^+SeCl_3^-$ and $[Me_3SiN(H)PMe_3^+]_2[Se_2Cl_{62}^-]$. Z Anorg Allg Chem. 1994;620:1301–6.

[58] Siivari J, Chivers T, Laitinen R. $Se_3N_2Cl_2$, A novel selenium-nitrogen chloride: reinvestigations of "Se_4N_2". Angew Chem Int Ed Engl. 1992;31:1518–9.

[59] Siivari J, Chivers T, Laitinen RS. Synthesis and characterization of selenium-nitrogen chlorides; force-field calculations for the $Se_3N_2Cl^+$ cation. Inorg Chem. 1993;32:4391–5.

[60] Steudel R, Papavassiliou M, Strauss E-M, Laitinen R. Selenium-rich chalcogen rings Se_5S, Se_5S_2 and Se_7 from titanocene pentaselenide. Angew Chem Int Ed Engl. 1986;25:99–101.

[61] Demchneko PY, Gladyshevskii RE, Volkov SV, Yanko OG, Kharkova LB, Fokinan ZA, et al. The first nonaselenium ring. Chem Commun. 2010;46:4510–22.

[62] Nogai SD, Schier A, Schmidbaur H. Soluble pyridine complexes of the ternary gallium(III) chalcogenide halides $(GaEX)_3$ with E = S, Se and X = Cl, Br. Z Naturforsch. 2001;56b:711–8.
[63] Thrasher JS, Bauknight CW, DesMarteau DD. (Perfluoroalkylimino)selenenyl chlorides. Inorg Chem. 1985;24:1598–9.
[64] Thrasher JS, Seppelt K. Reactions of pentafluorosulfanyldichloramine. Z Anorg Allg Chem. 1983;507:7–11.
[65] Musalov MV, Potapov VA, Amosova SV. Reaction of selenium dichloride with acetylene. Russ J Org Chem. 2011;47:1115–6.
[66] Karhu AJ, Jämsä J, Rautiainen JM, Oilunkaniemi R, Chivers T, Laitinen RS. A selenium-nitrogen chain with selenium in different oxidation states. Z Anorg Allg Chem. 2017;643:495–500.
[67] Siivari J, Chivers T, Laitinen RS. A simple, efficient synthesis of Se_4N_4. Inorg Chem. 1993;32:1519–20.
[68] Lentz D, Pritzkow H, Seppelt K. Novel tellurium fluorides; *cis*- and *trans*-$F_4Te(OTeF_5)_2$, FTe $(OTeF_5)_5$ and $Te(OTeF_5)_6$. Inorg Chem. 1978;17:1926–31.
[69] Hector AL, Jolleys A, Levason W, Reid G. TeX_4 (X = F, Cl, Br) as Lewis acids – complexes with soft thio-and seleno-ether ligands. Dalton Trans. 2012;41:10988–99.
[70] (a). Klapötke TM, Krumm B, Mayer P, Schwab I. Binary tellurium(IV) azides $Te(N_3)_4$ and $[Te(N_3)_5]^-$. Angew Chem Int Ed. 2003;42:5843–6. (b) Haiges R, Boatz JA, Vij A, Gerken M, Schneider S, Schroer T, Christe KO. Polyazide chemistry: preparation and characterization of $Te(N_3)_4$ and $[P(C_6H_5)_4]_2[Te(N_3)_6]$ and evidence for $[N(CH_3)_4][Te(N_3)_5]$. Angew. Chem. Int. Ed. 2003, 42, 5847–51.
[71] Lentz D, Szwak MD. Synthesis and structure determination of tellurium tetracyanide solvates: pseudopolymorphism of $Te(CN)_4$ and TeF_4. Angew Chem Int Ed. 2005;44:5079–82.
[72] Glover SA. Mechanism of thermal decomposition of tetra-aryltellurium species. JCS Perkin I. 1980;1338–44.
[73] Naumann D, Tyrra W, Herrmann R, Pantenburg I, Wickleder MS Syntheses and properties of tetrakis(pentafluorophenyl)tellurium, $Te(C_6F_5)_4$ and related compounds. Z Anorg Allg Chem. 2001;628:833–42.
[74] Gedridge RW, Harris DC, Higa KT, Nissan RA. Isolation and characterization of tetramethyltellurium(IV). Organometallics. 1989;8:2817–10.
[75] Gedridge RW, Higa KT, Nissan RA. Synthesis and mechanistic studies of symmetric tetraorganyl-tellurium(IV) (R_4Te) and diorganyltellurium(II) (R'_2Te) compounds (R = R' = Me, ^nBu, Me_3SiCH_2, and CH_2=CH; R = ^tBu and allyl). Organometallics. 1991;10:286–91.
[76] Minoura M, Sagami T, Akiba K, Modrakowski C, Sudau A, Seppelt K, et al. Hexaaryltellurium, the first neutral compounds comprising hexaarylated elements. Angew Chem Int Ed Engl. 1996;35:2660–1.
[77] Minoura M, Sagami T, Miyasato M, Akiba K. Synthesis and characterization of hexaaryltellurium ($TeAr_6$), the first neutral hexaarylated element compound. Tetrahedron. 1997;53:12195–202.
[78] Minoura M, Sagami T, Akiba K. Convenient one-pot synthesis and structures of pentaphenyltellurium monohalides. Organometallics. 2001;20:2437–9.
[79] Betz R, Stapel M, Pfister M, Roessner FW, Reichvilser MM, Klüfers P. Tetramethoxy-λ^4-tellane and tetrakis(2,2,2-trifluoroethoxy)-λ^4-tellane. Z Anorg Allg Chem. 2008;634:2391–6.
[80] Sundberg MR, Uggla R, Laitalainen T, Bergman J. Influence of secondary bonding on the intradimer distance of trichloro(ethane-1,2-diolato-*O,O'*)tellurate(IV). J Chem Soc Dalton Trans. 1994;3279–83.
[81] Allan RE, Gornitzka H, Kärcher J, Paver MA, Rennie M-A, Russell CA, et al. Structure and reactivity of $[Te(NMe_2)_2]_\infty$: application to the preparation of metalloorganic tellurium(II) compounds. J Chem Soc Dalton Trans. 1996;1727–30.

[82] Murai T, Imaeda K, Kajita S, Ishihara H, Kato S. Reductive dialkynylation of tellurium tetrachloride with lithium amides and terminal alkynes. J Chem Soc Chem Commun. 1991;832–3.
[83] Gedridge RW, Brandsma L, Nissan RA, Verkruijesse HD, Harder S, De Jong RLP, et al. Preparation of symmetrical di-1-alkynyl tellurides and diynes by reduction of $TeCl_4$ with 1-alkynyllithium reagents. Organometallics. 1992;11:418–22.
[84] Roesky HW, Münzenberg J, Bohra R, Noltemeyer M. Syntheses and crystal structures of compounds containing short Te-N bonds. J Organomet Chem. 1991;418:339–48.
[85] Folkerts H, Dehnicke K, Magull J. Synthesis and crystal structure of $[TeCl_2(NPMe_3)]_2$. Z Naturforsch. 1995;50b:573–6.
[86] Münzenberg J, Roesky HW, Noltemeyer M, Besser S, Herbst-Irmer R. Syntheses and structural comparison of some tellurium(IV) iminates. Z Naturforsch. 1993;48b:199–208.
[87] Hey E, Ergezinger C, Dehnicke K. Synthesis and crystal structure of the amidinato complex $PhC(NSiMe_3)_2TeCl_3$. Z Naturforsch. 1989;44b:205–7.
[88] Chivers T, Fedorchuk C, Schatte G, Brask JK. Syntheses and X-ray structures of boraamidinate complexes of lithium, phosphorus, and tellurium. Can J Chem. 2002;80:821–31.
[89] Grubisha DS, Guzei IA, Al-Salim N, Boyd PDW, Brothers PJ, Woo LK. Novel coordination in the first tellurium porphyrin complex: synthesis and crystal structure of $[Te(ttp)Cl_2]$. Angew Chem Int Ed. 2001;40:4743–5.
[90] Passmore J, Schatte G, Cameron TS. Preparation and structure of $Te_4N_2Cl_8[AsF_6]_2{\bullet}2SO_2$ containing $Te_4N_2Cl_8^{2+}$, the first tellurium-nitrogen-halogen cation. J Chem Soc Chem Commun. 1995;2311–2.
[91] Chivers T, Enright GD, Sandblom N, Schatte G, Parvez M. Synthesis and reactions of *tert*-butylimidotellurium dihalides: X-ray structures of $[Cl_2Te(\mu\text{-}N^tBu)_2TeCl_2]_3$ and $[(^tBuO)_2Te(\mu\text{-}N^tBu)_2Te(^tBuO)_2]_3$. Inorg Chem. 1999;38:5431–6.
[92] Cozzolino AF, Britten JF, Vargas-Baca I. The effect of steric hindrance on the association of telluradiazoles through Te—N secondary bonding. Cryst Growth Des. 2006;6:181–6.
[93] Cozzolino AF, Whitfield PS, Vargas-Baca I. Supramolecular chromotropism of the crystalline phase of 4,5,6,7-tetrafluorobenzo-2,1,3-telluradiazole. J Am Chem Soc. 2010;132:17265–70.
[94] Semenov NA, Pushkarevsky NA, Beckmann J, Finke P, Lork E, Mews R, et al. Tellurium-nitrogen heterocyclic chemistry – synthesis, structure, and reactivity towards halides and pyridine of 3,4-dicyano-1,2,5-telluradiazole. Eur J Inorg Chem. 2012;3693–403.
[95] (a). Reske G, Cowley AH. Direct reactions of tellurium halides with chelating nitrogen ligands. Trapping of TeI_2 by a 1,2-bis(arylimino)acenaphthene ligand and C-H activation of an α,α'-diiminopyridine ligand. Chem Commun. 2006:4856–8. (b) Martin CD, Ragogna PJ. Reactions of diiminopyridine ligands with chalcogen halides. Inorg. Chem. 2012, 51, 2947–53.
[96] Kuhn N, Abu-Rayyan A, Piludu C, Steimann M. Reaction of tellurium tetraiodide with 2,3-dihydro-1,3-diisopropyl-4,5-dimethylimidazol-2-ylidene. Heteroatom Chem. 2005;16:316–9.
[97] Hrib CG, Jones PG, Du Mont W-W, Lippolis V, Devillanova FA. Complexes of bidentate phosphane selenide ligands with mesityltellurenyl iodide and tellurium diiodide. Eur J Inorg Chem. 2006;1294–302.
[98] Dutton JL, Farrar GJ, Sgro MJ, Battista TL, Ragogna PJ. Lewis base sequestered chalcogen dihalides: synthetic sources of ChX_2 (Ch = Se, Te; X = Cl, Br). Chem Eur J. 2009;15:10263–71.
[99] Foss O, Johannessen W. Complexes of tellurium tetrachloride and tetrabromide with tetramethyl thiourea. Acta Chem Scand. 1961;15:1939–48.
[100] Levason W, Reid G, Victor M, Zhang W. Tellurium(II) and tellurium(IV) complexes of phosphine chalcogenide ligands, synthesis and X-ray structures. Polyhedron. 2009;28:4010–6.
[101] (a). Hector AL, Jolleys A, Levason W, Reid G. TeX_4 (X = F, Cl, Br) as Lewis acids – complexes with soft thio- and seleno-ether ligands. Dalton Trans. 2012;41:10988–99. (b) Gurnani C, Jura M,

Levason W, Ratnani R, Reid G, Webster M. Preparation and structures of tellurium(IV) halide complexes with thioether coordination. Dalton Trans. 2009, 4122–8.
[102] Liaw W-F, Chiou S-J, Lee G-H, Peng S-M. A discrete chlorotellurate $[Cl_4Te\text{-}Mn(CO)_5]^-$: coordinative addition of the metalloanion $[Mn(CO)_5]^-$ to $TeCl_4$. Inorg Chem. 1998;37:1131–4.
[103] Ebsworth EAV, Holloway JH, Watson PG. Five-coordinate rhodium(III) complexes containing the novel TeF_3 ligand. J Chem Soc Chem Commun. 1991;1443–4.
[104] Dyson PJ, Hill AF, Hulkes AG, White AJP, Williams DJ. Tetravalent tellurium ligands. Angew Chem Int Ed. 1999;38:512–4.
[105] Potapov VA, Musalov MV, Musalova MV, Rusakov YY, Khabibulina AG, Rusakova IL, Amosova SV., et al. Stereoselective synthesis of E-2-halovinyl tellanes, ditellanes and selenides based on tellurium tetrahalides, selenium dihalides and internal alkynes. J Organomet Chem. 2018;867:300–5 and references cited therein.
[106] (a). Aynsley EE. The preparation and properties of tellurium dichloride. J Chem Soc. 1953:3016–9. (b) Aynsley EE, Watson RH. The preparation and properties of tellurium dibromide. J. Chem. Soc. 1955, 2603–6.
[107] Paul RC, Arneja A, Narula SP. Reactions between hexamethyldisilane and some inorganic halides. Inorg Nucl Chem Lett. 1969;5:1013–5.
[108] Tokitoh N, Sadahiro T, Hatano K, Sasaki T, Takeda N, Okazaki R. Syntheses of kinetically stabilized silaneselone and silanetellone. Chem Lett. 2002;31:34–5.
[109] Patra A, Wijsboom YH, Leitus G, Bendikov M. Synthesis, structure and electropolymerization of 3,4-tellurophene; comparison with selenium analogue. Org Lett. 2009;11:1487–90.
[110] He G, Delgado T, Schatz DJ, Merten C, Mohammadpour A, Mayr L, et al. Coaxing solid-state phosphorescence from tellurophenes. Angew Chem Int Ed. 2014;53:4587–91.
[111] Foss O, Johannessen W. Complexes of tellurium dichloride and dibromide with tetramethylthiourea. Acta Chem Scand. 1961;15:1940–1.
[112] Konu J, Chivers T. Synthesis, spectroscopic and structural characterization of tertiary phosphine tellurium dihalides Et_3PTeX_2 (X = Cl, Br, I). Dalton Trans. 2006;3941–6.
[113] (a). Dutton JL, Tuononen HM, Ragogna PJ. Tellurium(II)-centered dications from the pseudohalide "$Te(OTf)_2$". Angew Chem Int Ed. 2009;48:4409–13. (b) Dutton JL, Ragogna PJ. Dicationic tellurium analogues of the classic *N*-heterocyclic carbene. Chem. Eur. J. 2010, 16, 12454–61.
[114] Konu J, Chivers T. Novel carbon-centred reactivity of $[HC(PPh_2Se)_2]^-$ in the formation of structurally diverse Sn(IV), Te(IV) and Hg(II) complexes of the triseleno ligand $[SeC(PPh_2Se)_2]^{2-}$. Chem Commun. 2010;46:1431–3.
[115] Pietikäinen J, Laitinen RS. Novel tellurium halides Te_2Cl_2 and Te_2Br_2. Chem Commun. 1998;2381–2.
[116] Petragnani N, Stefani HA. Advances in organic tellurium chemistry. Tetrahedron. 2005;61:1613–78.
[117] Freudendahl DM, Wirth T. New selenium electrophiles and their reactivity. In: Woollins JD, Laitinen RS, editors. Selenium and tellurium chemistry: from small molecules to biomolecules and materials, New York, Vol. Ch. 2. Springer, 2011:41–55.
[118] (a). McWhinnie WR. Intra-molecular coordination – a route to novel organotellurium compounds. Phosphorus Sulfur Silicon Relat Elem. 1991;67:107–23. (b) Zukerman-Schpector J, Haiduc I. Diorganotellurium(IV) dihalides and secondary bonding; revisiting the coordination polyhedra. Phosphorus, Sulfur, and Silicon and Related Elements 2001, 171, 73–112.
[119] Clark ER, Al-Turaihi MA. On the structure of arylselenium tribromides. J Organomet Chem. 1977;124:391–7.
[120] (a). Wynne KJ, Pearson PS. Adducts of organoselenium and -tellurium trichlorides with antimony pentachloride. Inorg Chem. 1971;10:1871–4. (b) Wynne KJ, George JW. Lewis-base behavior of selenium(IV) halides. II. Methylselenium trichloride and trimethylselenium chloride. J Am Chem Soc. 1969;91:1649–52.

[121] (a). Maxwell WM, Wynne KJ. Arylselenium trifluorides. Inorg Chem. 1981;20:1701–12. (b) Klapötke TM, Krumm B, Scherr, M. Studies on the properties of organoselenium(IV) fluorides and azides. Inorg Chem. 2008;47:4712–22.
[122] Pietkäinen J, Laitinen RS, Konu J, Valkonen J. Formation and crystal structure of polymeric $(MeTeCl_3)_\infty$. Z Naturforsch. 2001;56b:1369–72.
[123] Dirk CW, Nalewajek D, Blanchet GB, Schaffer H, Moraes F, Boysel RM, et al. Poly(methylene ditelluride). J Am Chem Soc. 1985;107:675–7.
[124] (a) Laitinen RS, Oilunkaniemi R. Tellurium: organotellurium chemistry. In: King RB, editor(s). Encyclopedia of Inorganic Chemistry, 2nd ed. Vol. IX. Chichester, UK: John Wiley & Sons Ltd. 2005:5539–56. (b) Berry FJ, Kustan EH, Roshani M, Smith BC. 1H NMR spectra of some aryltellurium compounds. J Organomet Chem. 1975;99:115–7.
[125] Beckmann J, Heitz S, Hesse M. Four distinctively different decomposition pathways of metastable supermesityltellurium(IV) trichloride. Inorg. Chem. 2007;46:3275–82.
[126] Singh HB, Sudha N, Weat AA, Hamor TA. Orthotellurated derivatives of *N,N*-dimethylbenzylamine: crystal and molecular structure of [2-(dimethylaminomethyl)phenyl]tellurium(IV) tribromide and [2-(butyldichlorotelluro)benzyl]dimethylammonium chloride. J Chem Soc Dalton Trans. 1990;907–13.
[127] Hammerl A, Klapötke TM, Krumm B, Scherr M. Tellurium(IV) fluorides and azides containing the nitrogen donor substituent R = $2\text{-}Me_2NCH_2C_6H_4$: crystal structure of $RTeF_3$ and of an unusual tellurium(VI) fluoride salt. Z Anorg Allg Chem. 2007;633:1618–26.
[128] Miller JM, Chadha RK. Reactions of tellurium(IV) compounds with trimethyl(dialkylamino) silanes. J Organomet Chem. 1981;216:177–84.
[129] Beckmann J, Finke P. Organotelluroxanes. In: Woollins JD, Laitinen RS, editors. Selenium and tellurium chemistry; from small molecules to biomolecules and materials, Ch.7. New York: Springer, 2011:165–71.
[130] Beckmann J, Finke P, Hesse M, Wettig B. Well-defined stibonic and tellurinic acids. Angew Chem Int Ed. 2008;47:9982–4.
[131] Beckmann J, Bolsinger J, Duthie A. Intramolecularly coordinated telluroxane clusters and polymers. Chem Eur J. 2011;17:930–40.
[132] (a) Wynne KJ, George JW. Adducts of dimethylselenium dichloride and trimethylamine. Inorg Chem. 1965;4:256–7. (b) Wynne KJ, George JW. Dimethylselenium dihalides and their adducts with boron trihalides. J Am Chem Soc. 1965;87:4750–4.
[133] Wynne KJ. The preparation and properties of diorganoselenium difluorides. Inorg Chem. 1970;9:299–301.
[134] Poropudas MJ, Vigo L, Oilunkaniemi R, Laitinen RS. Structural trends in TeI2RR': crystal structures of $TeI_2(CH_2SiMe_{32}$, $TeI_2Th(CH_2SiMe_3)$, $TeI_2Ph(CH_2SiMe_3)$, and TeI_2Th_2 (Th = 2-thienyl, C_4H_3S). Heteroatom Chem. 2011;22:348–57.
[135] Chauhan AKS, Kumar A, Srivastava RC, Butcher RJ. Synthesis and characterization of monomeric diorganotelluirum dihalides: crystal and molecular structures of diphenylacyltellurium dibromide and diiodide. J Organomet Chem. 2002;658:169–75.
[136] (a). Fujihara H, Takaguchi Y, Ninoi T, Erata T, Furukawa N. Preparation and characterization of a four-coordinated ditellurane stabilized by a transannular tellurium-tellurium bond from 1,5-ditelluracyclooctane: a new multicenter hypervalent species. J Chem Soc Perkin Trans 1. 1992:2583–4. (b) Takaguchi Y, Horn E, Furukawa N. Preparation and X-ray structure analysis of 1,1,5,5,9,9-hexachloro-1,5,9-tritelluracyclododecane (Cl_6[12]aneTe_3) and its redox behavior. Organometallics 1996, 15, 5112–5.
[137] Beckmann J, Bolsinger J, Spandl J. The structural diversity of Te-I interactions within tetraorganoditelluroxane diiodides and related compounds. J Organomet Chem. 2008;693:957–64.

[138] Chandrasekhar V, Thirumoorthi R. 1,1′-Ferrocenedicarboxylate-bridged redox-active organotin and tellurium-containing 16-membered macrocycles: synthesis, structure, and electrochemistry. Organometallics. 2007;26:5415–22.
[139] Chandrasekhar V, Thirumoorthi R. Halide-capped tellurium-containing macrocycles. Inorg Chem. 2009;48:10330–7.
[140] Beckmann J, Hesse M, Poleschner H, Seppelt K. Formation of mixed-valent aryltellurenyl halides RX_2TeTeR. Angew Chem Int Ed. 2007;46:8277–80.
[141] Sugamata K, Sasamori T, Tokitoh N. Fluorination reaction of a ditelluride bearing bulky aryl substituents: formation of mixed-valent Te(IV)-Te(II) ditelluride difluoride. Chem Asian J. 2011;6:2301–3.
[142] Kubiniok S, Du Mont W-W, Pohl S, Saak W. The reagent diphenyldiselane/iodine: no phenylselenenyl iodide but a charge transfer complex with cyclic moieties. Angew Chem Int Ed. 1988;27:431–3.
[143] Du Mont W-W, Kubiniok S, Peters K, Von Schnering H-G. Synthesis and structure of a stable iodoselenide. Angew Chem Int Ed. 1987;26:780–1.
[144] Du Mont W-W, Wagner I. Iodo[tris(trimethylsilyl)methyl]selane: synthesis of the first alkyl selenenyl iodide. Chem Ber. 1988;121:2109–10.
[145] Poleschner H, Seppelt K. First detection of a selenenyl fluoride ArSeF by NMR spectroscopy; the nature of the Ar_2Se_2/XeF_2 and $ArSeSiMe_3/XeF_2$ reagents. Chem Eur J. 2004;10:6565–74.
[146] Poleschner H, Seppelt K. Selenirenium and tellurirenium ions. Angew Chem Int Ed. 2008;47: 6461–4.
[147] Klapötke TM, Krumm B, Polborn K. Isolation of a stable selenium azide $RSeN_3$. J Am Chem Soc. 2004;126:710–1.
[148] (a). Giselbrecht K, Bildstein B, Sladky F. Tris(trimethylsilyl)methanetellurenyl halides $(Me_3Si)_3CTeX$ (X = Cl, Br, I): synthesis of stable alkanetellurenyl halides. Chem Ber. 1989;122:1255–6. (b) Fimml W, Sladky F. Synthesis of stable alkanetellurenyl pseudohalides and pseudochalcogenides: $(Me_3Si)_3CTeX$ (X = CN, SCN, SeCN, NCO, N_3) and $(Me_3Si)_3CTe$-Y-TeC$(SiMe_3)_3$ (Y = NCN, NSN). Chem Ber. 1991;124:1131–3.
[149] Du Mont -W-W, Meyer H-U, Kubiniok S, Pohl S, Saak W. Cleavage of bulky diaryl ditellurides with bromine and iodine: crystal structure of Et4N+ 2,4,6-(i-$C_3H_{73}C_6H_2TeI_2$)-. Chem Ber. 1992;125:761–6.
[150] Klapötke TM, Krumm B, Nöth H, Galvez-Ruiz JC, Polborn K. Kinetic and donor stabilization of organotellurenyl iodides and azides. Inorg Chem. 2005;44:5254–65.
[151] Ledesma GN, Lang ES, Vázquez-López EM, Abram U. Synthesis and characterization of the first aryltellurium(II) halide complex stabilized by a Te-Te bond from a tellurium ether. Inorg Chem Commun. 2004;7:478–80.
[152] Sasamori T, Sugamata K, Tokitoh N. Halogenation reactions of a ditelluride having bulky aryl groups leading to the formation of organotellurium halides. Heteroatom Chem. 2011;22:405–11.
[153] Manzoni De Oliveira G, Faoro E, Lang ES. New aryltellurenyl iodides with uncommon valences: synthetic and structural characteristics of $[RTeTeI_2R]$, $[R_2TeTeR_2][Te_4I_{14}]$, and $[RTe(I)I_2]$ (R = 2,6-dimethoxyphenyl). Inorg Chem. 2009;48:4607–9.
[154] Boyle PD, Cross WI, Godfrey SM, McAuliffe CA, Pritchard RG, Sarwar S, et al. Synthesis and characterization of $Ph_4Te_4I_4$, containing a Te_4 square, and $Ph_3PTe(Ph)I$. Angew Chem Int Ed. 2000;39:1796–8.
[155] Beckmann J, Finke P, Heitz S, Hesse M. Aryltellurenyl cation $[RTe(CR'_2)]^+$ stabilized by an N-heterocyclic carbene. Eur J Inorg Chem. 2008;1921–5.
[156] Sugamata K, Sasamori T, Tokitoh N. Generation of an organotellurium(II) cation. Eur J Inorg Chem. 2012;775–8.

Alexandra Pop, Cristian Silvestru and Anca Silvestru

3 Organoselenium and organotellurium compounds containing chalcogen-oxygen bonds in organic synthesis or related processes

Abstract: This chapter emphasizes aspects related to the role of organochalcogen (Se, Te) compounds with single E–O and/or double E=O (E=Se, Te) bonds in organic synthesis, as reagents, intermediates, or catalysts, and it gives a larger extent mainly to data reported in the field during the last ten years. For each of these two heavier chalcogens the material is structured according to the oxidation state of the chalcogen and, for the same oxidation state, in sections dedicated to a particular type of compounds. Functionalization or cyclization reactions in which the organochalcogen compounds take part as nucleophiles, electrophiles or radicals, employed in various synthetic transformations, are discussed and, where available, the mechanistic aspects are outlined. New chiral species and new strategies were developed during last years in order to increase the yield, the reaction rate and the stereoselectivity in specific organic transformations, i.e. addition, oxidation, elimination, cyclization or rearrangement reactions. A notably attention was devoted to easily accessible and environmental friendly catalysts, re-usable and "green" solvents, as well as waste-free procedures.

Keywords: organoselenium, organotellurium, oxidation, organic synthesis, catalysis

3.1 Introduction

Organochalcogen compounds attracted a considerable increased attention during last decades due to their importance in different fields, including organic synthesis, materials science, medicine and biology or coordination chemistry. Organoselenium species proved to be valuable reagents in chemo-, regio- stereoselective or stereospecific reactions in various synthetic transformations such as selenenylation, selenocyclization, selenoxide elimination and 2,3-sigmatropic rearrangement processes, as well as in asymmetric catalysis. At a lower extent, organotellurium analogues were employed in organic synthesis as well. Plenty of books or book chapters [1] as well as review articles [2] were dedicated along the last decades to the organoselenium chemistry, in general, and to the role of organoselenium species in organic synthesis,

This article has previously been published in the journal *Physical Sciences Reviews*. Please cite as: Silvestru, A., Silvestru, C., Pop, A. Organoselenium and Organotellurium Compounds Containing Chalcogen-Oxygen Bonds in Organic Synthesis or Related Processes. *Physical Sciences Reviews* [Online] **2018** DOI: 10.1515/psr-2018-0061

https://doi.org/10.1515/9783110529340-003

in particular, while less contributions were summarizing aspects regarding a similar role of organotellurium compounds [1]a–d, [1]g, [1]h, [2]e, [3]. This chapter is dedicated to the use of organoselenium and -tellurium compounds with chalcogen-oxygen bonds in organic synthesis and related processes and is based on a full search of the original literature performed using the SciFinder database. However, since various reviews and book chapters were dedicated over years to topics related to the subject of this survey, the literature older than 2008 was cited only when required to emphasize some important aspects.

Already at the beginning of the twentieth-century elemental selenium was used to obtain aromatic compounds by dehydrogenation of saturated cyclic and polycyclic hydrocarbons [4], while selenium dioxide was firstly mentioned in 1929 in a patent from I.G. Farbenindustrie AG as oxidant in synthetic organic chemistry [5]. Furthermore, SeO_2 was mentioned as reagent in allylic hydroxylations and in oxidation of ketones to α-diketones [6]a–g, and more recently the I_2/SeO_2 system was reported as an efficient catalyst for the selective synthesis of mono- and bis-sulfenylindoles, the selectivity being controlled by the SeO_2 amount [6]h. The role of tellurium and TeO_2 in reactions with various organic substrates is mentioned only later, about 1980th, by Bergman and Engman [7]. The chemistry of organoselenium compounds goes back to 1847 when the first derivative, ethylselenol, was synthesized [8], but the use of organoselenium species in organic synthesis was marked only in 1970 by the formation of alkenes by decomposition of organoselenoxides under mild conditions [9]. From structural point of view the organoselenium compounds are closely related to their sulfur and tellurium analogues, but the chemical behavior of organoselenium reagents is quite different in comparison with those of the lighter sulfur or the heavier tellurium. In several cases, organoselenium compounds give reactions which are not possible with sulfur or tellurium in similar conditions. In comparison with sulfur, much milder conditions are required for the reactions employing organoselenium reagents, i. e. *syn*-elimination of selenoxides is about three orders of magnitude faster than the elimination of the corresponding less polar and less basic sulfoxides, sigmatropic rearrangement reactions proceed at markedly lower temperatures, selenuranes are more easily obtained than the sulfur analogues, selenides and selenolate anions are less basic and more nucleophilic than the corresponding sulfur derivatives, etc [1].

Different functional groups can be introduced into organic substrates by processes based on organoselenium and -tellurium compounds, when the chalcogen behaves as an electrophile, as a nucleophile or as a radical towards the organic substrate. The higher polarity of the Te=O or the Te–O bonds results in a considerably different reactivity of the tellurium containing species, compared with that one of the lighter group 16 congeners.

Among the large diversity of heavy organochalcogen derivatives, we highlight in this chapter the role of organoselenium and -tellurium compounds with chalcogen-oxygen bonds, giving a special attention to the achievements in this area during the last ten years. Besides the types of organic transformations to which such species take part, several criteria might be used for their classification: the oxidation state of the chalcogen (+ 2, + 4 or + 6), the number of E–O or/and E=O bonds, the presence of O→E secondary interactions, their role as starting material, intermediate or catalyst in the described process. The types of organochalcogen compounds with chalcogen–oxygen bonds which were already employed in organic synthesis are summarized in Figure 3.1.

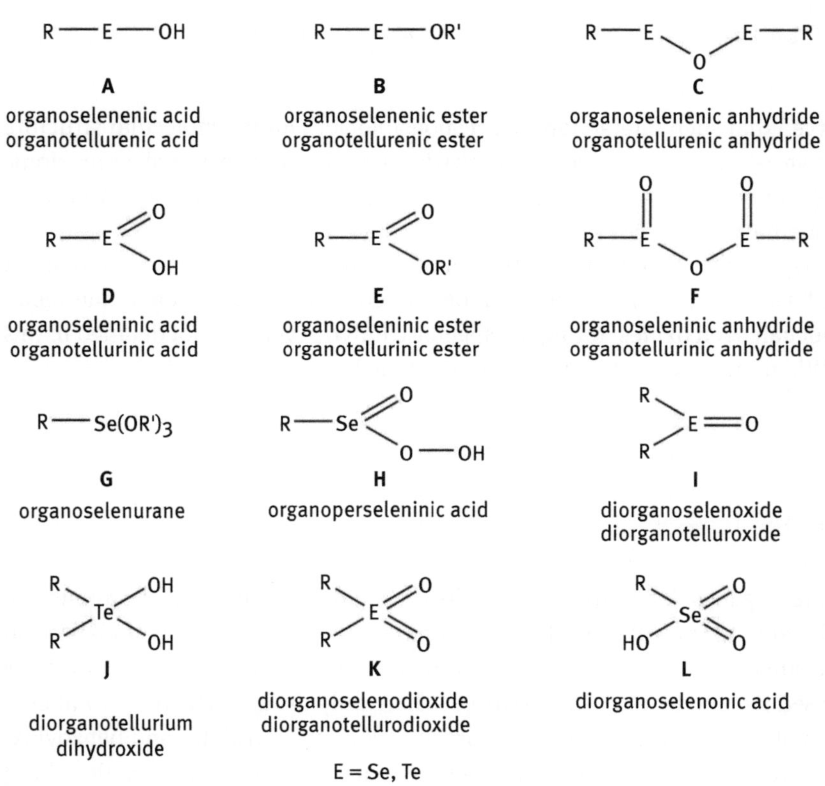

Figure 3.1: Organochalcogen compounds with E–O and/or E=O bonds (E=Se, Te).

The presence of a heteroatom capable to establish an intramolecular interaction with the chalcogen usually results in heterocyclic rigid conformations which are responsible for an increased stereoselectivity of these species in catalytic processes [10]. Chiral organoselenium species with O→E intramolecular interactions (Figure 3.2) were employed in functionalization (i.e. methoxyselenylation) reactions [11].

In such species the strength of the intramolecular interaction increases with decreasing the size of the R group, better selectivities being obtained for R=H. It

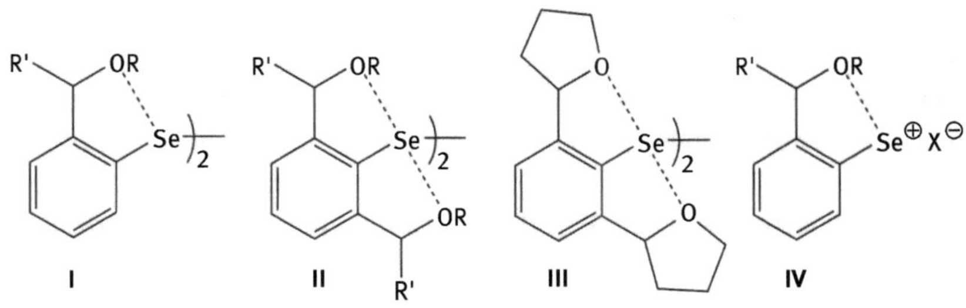

Figure 3.2: Organoselenium species with O→Se intramolecular coordination.

was observed that such interactions are responsible both for the molecular structure of the active reagent and for the asymmetric induction to the stereocenter. Ionic compounds of type **IV** were successfully used in the stereoselective synthesis of isoquinoline-based alkaloids [12].

We have to mention that the glutathione peroxidase (GPx)-like antioxidant processes based on diorganodichalcogenides which undergo oxidation to intermediary organochalcogen oxides during the specific catalytic cycle are not detailed in this chapter. Where necessary, such processes are only briefly mentioned.

3.2 Organoselenium compounds

3.2.1 Organoselenium(II) compounds

Most of the organoselenium(II) species involved in organic transformations are organoselenenic acids, ArSeOH (**1**), which are eliminated during functionalization of organic substrates by using organoselenium(IV) compounds and such processes are discussed in subchapter 3.2.2. We discuss here only the catalytic role of PhSeOH (**2**) in the synthesis of nitriles, as well as organoselenium(II) triflates and benzoyloxyselenides involved in selenenylation reactions, thus resulting in Se-functionalized organic products. Such products are versatile reagents for other organic transformations in which the advantage of the selenium heteroatom is exploited.

3.2.1.1 Organoselenenic acids

Arylselenenic acids, ArSeOH (**1**), generated *in situ* from either diaryl diselenides, Ar_2Se_2, or diorganoselenides of type Ph(R)Se (R=^{i}Pr, Et, *c*-C_6H_{11}) and H_2O_2, proved to be effective and reusable catalysts for the synthesis of organonitriles by dehydration of aldoximes under mild conditions (Figure 3.3) [13]. The best results were obtained by using (3-$FC_6H_4)_2Se_2$ or Ph_2Se_2 as pre-catalysts and it was observed that the yields in nitriles are influenced by the Ar_2Se_2/H_2O_2 molar ratio and the organic

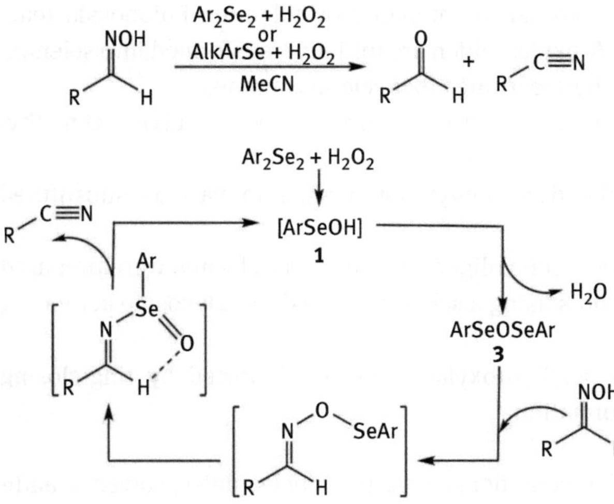

Figure 3.3: ArSeOH as catalyst in aldoxime dehydration reaction.

group in the starting aldoxime. The proposed reaction mechanism in the dehydration process involves the self-condensation of the *in situ* generated unstable areneselenenic acids **1** and the reaction of the intermediary formed arylselenenic anhydrides, $(ArSe)_2O$ (**3**), with the corresponding aldoxime.

3.2.1.2 Organoselenium(II) triflates and sulfates

Organoselenium(II) triflates, ArSeOTf (**4**), were used in organic synthesis as strong electrophilic reagents. The phenylselenenyl triflate, PhSeOTf (**5**), was employed in various reactions which occur with high regio- and stereoselectivity and processes as those mentioned below were previously reviewed [14]:

- addition to alkenes bearing terminal hydroxy or carboxy groups results in cyclic compounds, e.g., *γ*-lactones are obtained by addition of PhSeOTf (**5**) to *γ,δ*-unsaturated carboxylic acids, while hydroxyalkenes can be easily transformed into tetrahydrofuran or -pyran derivatives;
- lactamization of olefinic amides;
- intramolecular oxyselenenylation resulted in fused bicyclic systems, e.g., dioxanes and oxazines;
- *β*-phenylselenoazide was prepared by the stereospecific *trans* addition of NaN_3 to phenyl- or PhSe-substituted alkenes;
- addition to alkynes results initially in alkenyl selenirenium ions, which can be easily converted to either *trans*-substituted alkenes or a mixture of regioisomers in the presence of an external nucleophile (e.g., $Et_3N{\cdot}3HF$), depending on the type of the initial alkyne. Further hydrolysis of the PhSe-substituted alkenes led easily to ketones;

- secondary and tertiary amines can be prepared by selenium Polonovski reactions, starting from amine *N*-oxides which are initially transformed into selenoxyammonium triflates and then rearrange to *α*-selenoxyamines;
- 5,6-dihydro-1,3-oxazines were obtained by cyclization of allylic *O*-methyl isoureas;
- intramolecular nucleophilic deselenenylation results in various substituted heterocycles;
- *α*-cyclodextrine derivatives, cyclic oligozacharides and glycans were prepared by *O*-glycosylation of alcohols using a selenium-based thioglycoside activating agent;
- carbohydrate addition to polyhydroxylated alkenes followed by ring-closing proceeds with glycoside formation.

The influence of the reaction conditions, including nucleophiles, solvents, additives, and counterions on the phenylselenofunctionalization reactions were systematically investigated [15] and the regio- and stereoselectivity of such processes were documented at theoretical level as well [16].

Recently the electrophilic, chiral arylselenium triflate [2-{(*S*)-Me(MeS)CH}C_6H_4] SeOTf (**6**) was employed as promoting reagent in the highly stereoselective selenocyclization of several 2-methoxycarbonylalk-3-enols by using dynamic kinetic resolution [17]. It was observed previously that aromatic groups with pendant arms bearing sulfur atoms capable for intramolecular interactions with the electrophilic selenium have higher selectivity than the similar reagents bearing oxygen or nitrogen donor atoms. By reacting the selenenylating agent with the racemic starting alkenol in a 1:2 molar ratio, a mixture of stereoselectively enriched tetrahydrofuran and the corresponding enantiomerically enriched alkenol was obtained with a moderate to good enantioresolution (40–80 % ee) (Figure 3.4).

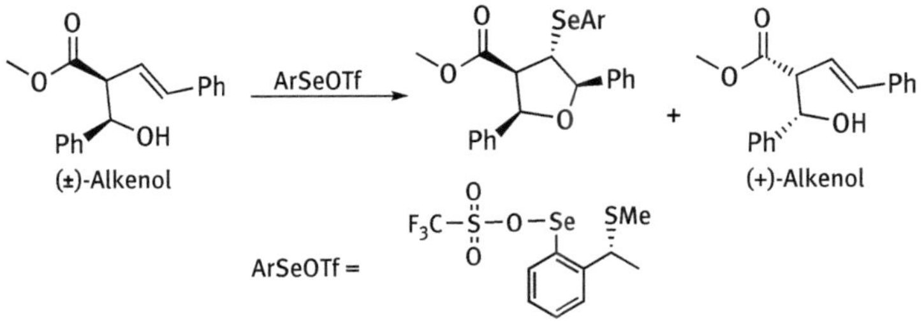

Figure 3.4: ArSeOTf as selenenylating agent of alkenols.

The organoselenium(II) triflates **7–9**, with aromatic substituents bearing one or two pendant arms with oxygen donor atoms, were used for the asymmetric selenenylation

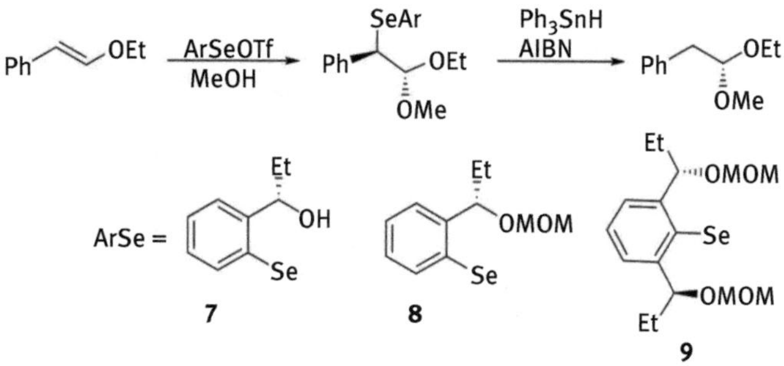

Figure 3.5: Hypercoordinated ArSeOTf in the asymmetric selenenylation of (*E*)-ethoxystyrene.

of (*E*)-ethoxystyrene, which was transformed in the corresponding chiral acetal with an enantiomeric excess ranging from 46 % to 86 % (Figure 3.5) [18].

Several terpenyl-based organoselenium(II) triflates were used for the methoxyselenenylation of styrene and it was observed that the species containing phenyl- or naphtyl-substituted terpenyl groups showed a lower selectivity compared with the unsubstituted species [19].

An efficient glycosylation method was developed by *Kim* and *co-workers* based on glycosyl pentenoates as donors and PhSeOTf (**5**) as promoter [20]. The triflate **5** proved to be more efficient than iodonium dicollidine perchlorate, N-iodosuccinimide-trifluoromethanesulfonic acid mixture or 1,3-dithian-2-yl tetrafluoroborate for such processes.

The one-pot conversion of 1,4-disubstituted cyclohexene into the corresponding 1,2-diols can be promoted by the electrophile $PhSeOSO_3H$ (**10**), generated *in situ* by oxidation of a catalytic amount of Ph_2Se_2 with an excess of ammonium persulfate in $MeCN/H_2O$ mixture, following the mechanism proposed in Figure 3.6, **path a** [21]. The use of the chiral diselenide $[2\text{-}\{(S)\text{-Me(MeS)CH}\}C_6H_4]_2Se_2$ in a similar reaction generated *in situ* the intermediate $[2\text{-}\{(S)\text{-Me(MeS)CH}\}C_6H_4]SeOSO_3H$ (**11**) and stereospecifically converted 1-phenylcyclohexene into a *cis*-1,2-diol with excellent facial selectivity [21]. The oxidation of alkynes to 1,2-unprotected dicarbonyl derivatives was also achieved by using as catalyst the electrophile **10**, produced *in situ* from a mixture of $(NH_4)_2S_2O_8$ and Ph_2Se_2 (Figure 3.6, **path b**) [22].

3.2.1.3 Organoselenium(II) carboxylates

The mixed anhydrides between an organoselenenic acid and a carboxylic acid, i. e. ArSeOC(O)R (**12**), are key intermediates in selenation of both internal, $R–C≡C–R^1$, and terminal, R–C≡C–H, alkynes. While the reaction between internal alkynes and a binary system of benzoyl peroxide and Ph_2Se_2, which generates *in situ* the efficient PhSeOC(O)Ph (**13**) reagent, proceeds *via* addition to the triple bond and the formation of *β*-(benzoyloxy)alkenyl selenides, the same binary system and a terminal alkyne affords a C(*sp*)–H substitution reaction with formation of the corresponding alkynyl

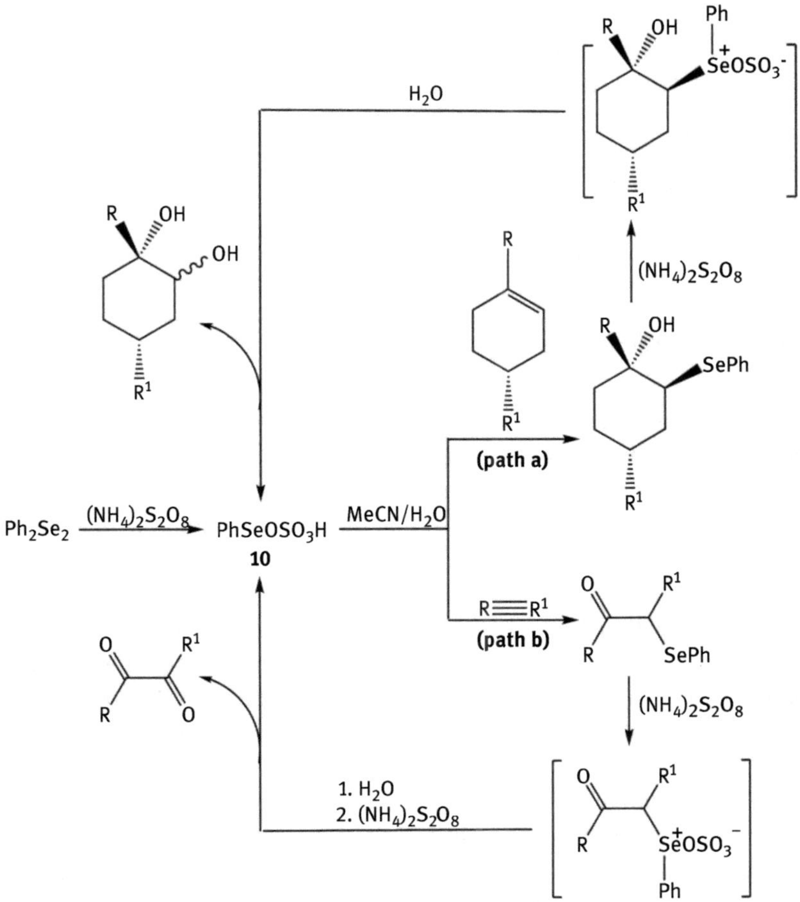

Figure 3.6: $PhSeOSO_3H$ in the synthesis of 1,2-diols (**path a**) and 1,2-dicarbonyl derivatives (**path b**).

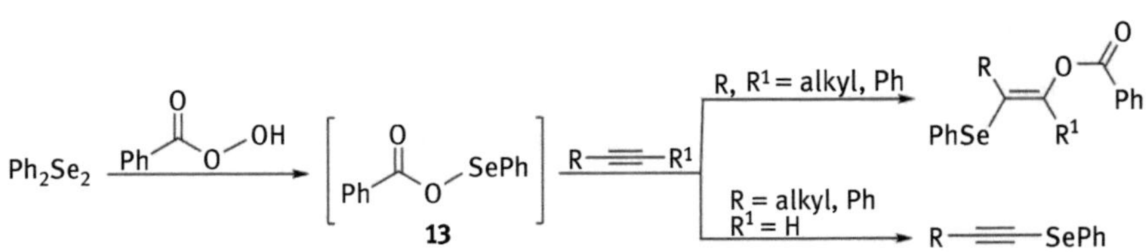

Figure 3.7: PhSeOC(O)Ph as intermediate in selenation of alkynes.

selenide, as depicted in Figure 3.7 [23]. Similar benzoyloxyselenation reactions based on mixtures of benzoyl peroxide and other aromatic or aliphatic diorgano diselenides, R_2Se_2 (R=4-$MeOC_6H_4$, 4-ClC_6H_4, 4-$CF_3C_6H_4$, ^{n}Bu) also produced acceptable yields of the expected addition product [23].

3.2.2 Organoselenium(IV) compounds

3.2.2.1 Organoseleninic acids and their derivatives

Alkylseleninic acids were successfully employed in the synthesis of various arylselenylated products and alkyl selenosulfides with relevant therapeutically phosphatase inhibitory activity [24].

PhSe(O)OH (**14**) and, at lower extent, other arylseleninic acids containing aromatic rings with electron-withdrawing substituents were used as pre-catalysts in various oxidation reactions of different organic substrates with H_2O_2 as oxidizing agent (e.g., alcohols to aldehydes or ketones, phenols to *ortho*- or *para*-quinones, aldehydes to carboxylic acids, alkenes to epoxides, sulfides to sulfones, amines to ketones, etc.). PhSe(O)OH (**14**) was employed in Baeyer–Villiger oxidation reactions, dehydrogenation of steroidal derivatives and to regenerate ketones from their hydrazones, oximes, or semicarbazones, as well [25]. Actually, in most processes the corresponding anhydride ArSe(O)OSe(O)Ar (**15**) or the perseleninic acid, ArSe(O)OOH (**16**), are generated *in situ* by self-condensation or oxidation with hydrogen peroxide, respectively, and these latter species act as effective catalysts in the mentioned processes.

Several recent contributions during last years revealed new perspectives and applications of the seleninic acids as pre-catalysts in organic transformations, as followings:

Dehydration of oximes

PhSe(O)OH (**14**) proved to be a suitable pre-catalyst for the dehydration of aldoximes to nitriles in open atmosphere and MeCN solution [26, 27], as the selenium(IV) compound is more stable and milder than the Ar_2Se_2/H_2O_2 systems and more tolerant to different functional organic groups than the active species discussed in subchapter 2.1.1. Such a catalytic process was applied to aldoximes RCH=NOH bearing different organic groups, e.g., Ph, mono- and trisubstituted phenyl, naphtyl, ^{t}Bu, thiophene or pyridyl groups. Almost quantitative yields were obtained for the dehydration of aldoximes bearing 4-$MeOC_6H_4$ (98 %) and 1-$C_{10}H_7$ (99 %) groups. The process can be described by a mechanism similar to that one based on the Ar_2Se_2/H_2O_2 system (Figure 3.3). In the initial stage of such a process the pre-catalyst PhSe(O)OH (**14**) is reduced, in two steps, to diphenyl diselenide by the aldoxime used as an organic substrate and the Ph_2Se_2 is then continuously re-oxidized by air to the catalyst PhSeOH (**2**). Stronger oxidants, as pure oxygen, were used to re-generate the organoselenium(IV) pre-catalyst **14** (Figure 3.8).

A complex study of the dehydration of anisaldoxime revealed that the yield in anisonitrile is influenced by the solvent, the organic group in the selenium catalyst and the catalyst loading. PhSe(O)OH (**14**) and electron-deficient species, e.g., 4-FC_6H_4Se(O)OH (**17**), showed good catalytic activity and a low content of the aldehyde by-product, while electron-enriched catalysts, e.g., 4-MeC_6H_4Se(O)OH (**18**), showed a poor catalytic activity. Based on the cheaper PhSe(O)OH pre-catalyst, a solvent free

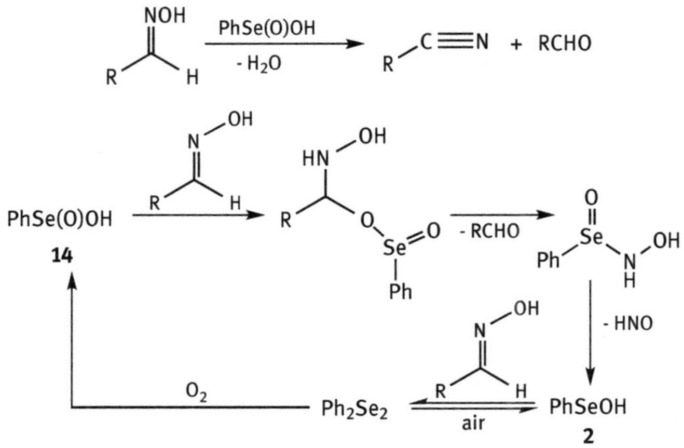

Figure 3.8: PhSe(O)OH as pre-catalyst for dehydration of aldoximes.

and waste-free dehydration method was developed to produce anisonitrile in almost quantitative yield, at large scale, with total recovery of the fully active catalyst [27].

By using appropriate conditions, mainly in the presence of H_2O_2 (5 % mol) as an oxidizing agent, the formation of the carbonylic product (aldehyde or ketone, depending on the type of oxime) can be favored in the deoximation process [28].

Oxidation reactions

A green procedure was developed for the regio- and diastereo-selective continuous synthesis of hydroxy lactones by using a selenium-mediated catalytic process in flow conditions [29]. By such a procedure several alkanoic acids were transformed into the corresponding hydroxy lactones, *via* the intermediary oxidation to epoxides, similarly with the classic protocol reported previously (Figure 3.9) [30]. In both pocedures PhSe(O)OH (**14**) was used as pre-catalyst for the active PhSe(O)OOH (**19**) species. The advantages of the flow protocol reside in the use of green solvents (H_2O/acetone and EtOAc), the lack of wastes and the possibilities to recover and reuse starting materials and solvents, as well as the short time (about 1 h) for a similar or even a higher conversion compared to the classic procedure.

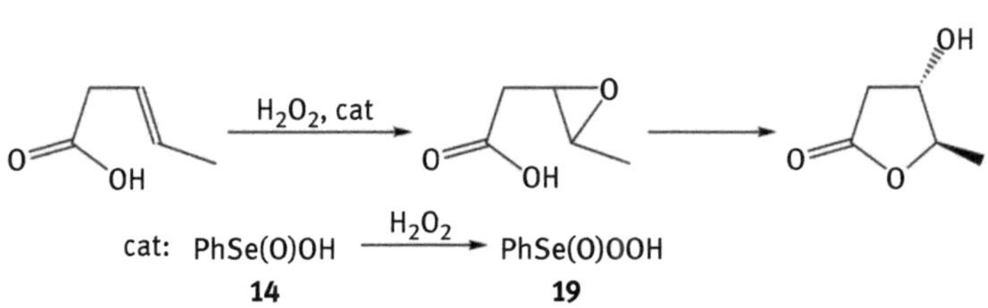

Figure 3.9: PhSe(O)OH as pre-catalyst in the synthesis of hydroxy lactones.

2-Trifluoromethansulfonylbenzeneseleninic acid, $[2\text{-}(TfO)C_6H_4]Se(O)OH$ (**20**), was employed as catalyst in Baeyer-Villiger oxidation of various organic substrates [31]. The oxidation of 4-formylpyrazole (Figure 3.10) was used as an intermediary step in the synthesis of various 4-allyloxy-1-substituted pyrazoles, further employed in the divergent synthesis of withasomnine alkaloids [31]c.

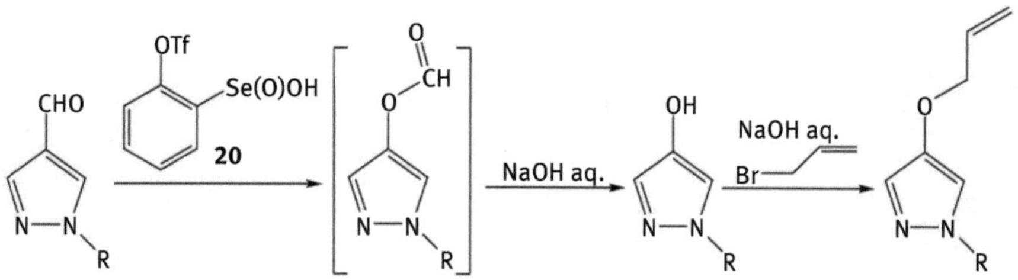

Figure 3.10: Oxidation of 4-formylpyrazole with $[2\text{-}(TfO)C_6H_4]Se(O)OH$ (**20**).

Dehydrogenation reactions

PhSe(O)Cl (**21**) was used as electrophilic reagent for the dehydrogenation of the nitroalkenes (Figure 3.11) employed in the asymmetric synthesis of the alkaloid daphnilactone B [32].

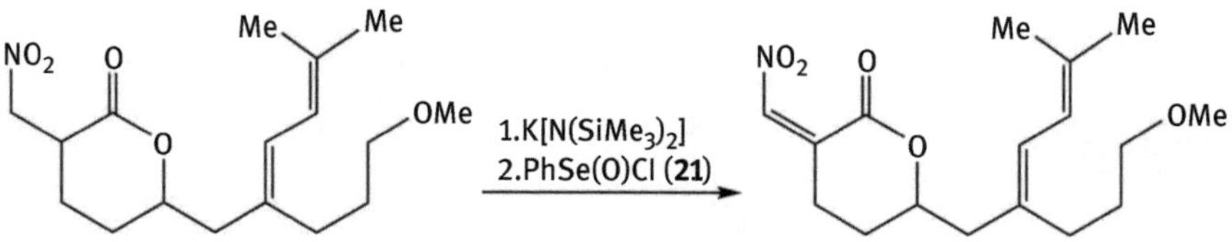

Figure 3.11: PhSe(O)Cl in dehydrogenation of nitroalkenes.

3.2.2.2 Heterocyclic seleninate esters and related compounds

The heterocyclic seleninate esters **22**–**24** (Figure 3.12) exhibited an efficient catalytic activity in various oxidation processes with H_2O_2, including GPx mimetics based on thiols oxidation to disulfides on the valuable expend of hydroperoxide reduction [33, 34], oxidation of sulfides to sulfoxides, alkenes to epoxides, enamines to α-hydroxyketones [35], disulfides to thiosulfinates [36] and secondary alcohols to ketones [35].

The catalyst **23f** displayed the best performance, as the undesirable further oxidation of sulfoxides to sulfones, as well as the hydrolysis of epoxides to diols or of the formed enamines to their parent ketones were strongly reduced (Figure 3.13). Compared with the benzeneseleninic acid previously used in such transformations,

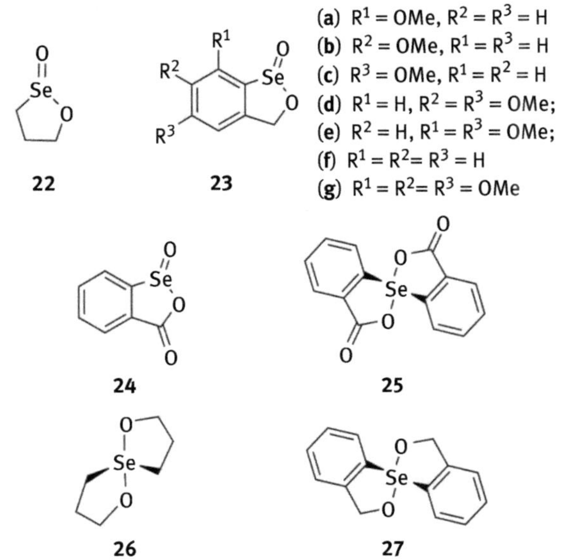

Figure 3.12: Heterocyclic seleninate esters and spirodioxyselenuranes used as catalysts in H_2O_2-based oxidation processes.

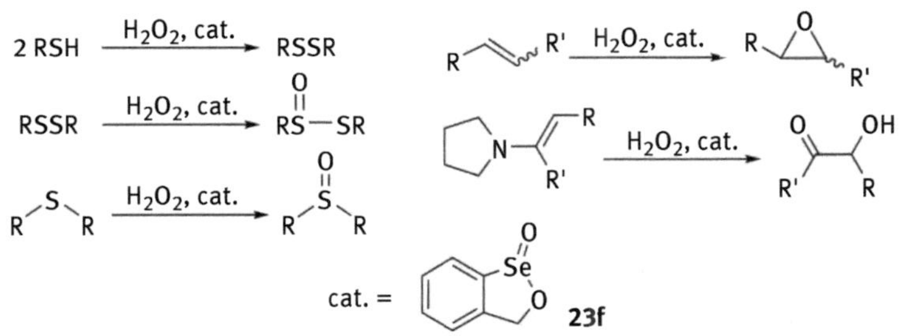

Figure 3.13: Oxidation reactions catalyzed by the heterocyclic compound **23 f.**

23f has the advantages of higher rates and enhanced selectivity in the desired sulfoxide and epoxide products [35, 37].

Back and *co-workers* noticed that the alkylic seleninate esters (e.g., **22** in Figure 3.12), as well as the spirodioxyselenurane derivatives (e.g., **26** in Figure 3.12) exhibit better catalytic activity compared to the aromatic related species (e.g., **27** in Figure 3.12) [38] as GPx mimetics in the reduction of *tert*-butyl hydroperoxide or hydrogen peroxide, but the catalytic cycle is different from that one of GPx. Compared with the well-known catalytic cycle of GPx, the advantage of the mechanisms described in Figure 3.14 is the absence of any selenol or selenolate intermediate capable to further generate oxygen active species.

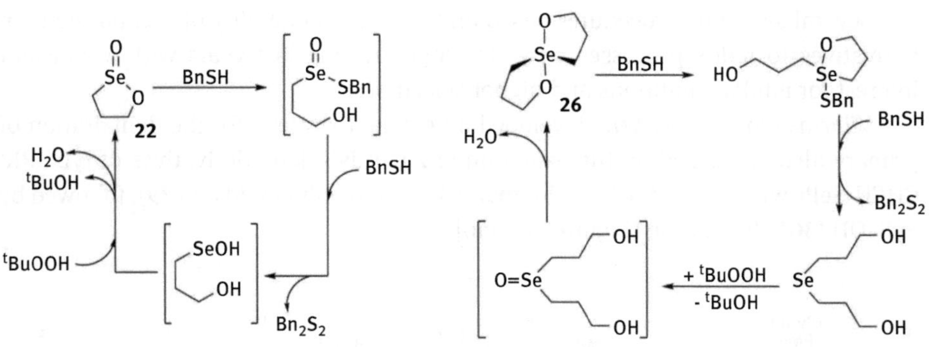

Figure 3.14: Catalytic mechanism in the oxidation of thiols to disulfides with compounds **22** or **26** as catalysts.

It was noticed that 4-MeO-substituted phenyl groups enhanced the catalytic activity, while MeO substituents in positions 3 or 2 of the aromatic ring decreased or totally suppressed it, respectively. Moreover, di- or three substituted species showed no increased catalytic activity in comparison with the active 4-MeO derivative, despite the expected increase through mesomeric electron donation in 2- and 3-substituted esters, the steric effects by O→Se intramolecular coordination of substituents in *ortho* position or a reduced electron-withdrawing inductive effect of 2- and 3-MeO groups determined by the additional 3-substituent [34].

For certain seleninate esters overoxidation of disulfides to thiolsulfinates was observed during GPx mimetic experiments. This observation was used to develop a synthetic procedure for thiolsulfinates by oxidation of disulfides with hydrogen peroxide in the presence of benzo-1,2-oxaselenolane *Se*-oxide in mild conditions. Overoxidation to thiolsulfonates could be suppressed by using a strict stoichiometric amount of H_2O_2. The oxidation of unsymmetrical substituted disulfides afforded a mixture of regioisomers and in aryl alkyl derivatives the alkyl-substituted sulfur was preferentially oxidized [34].

On another hand, it was observed that such heterocyclic species can catalyze undesired oxidation of disulfides, either formed in GPx-like processes or existing in spectator disulfide-containing peptides [34].

3.2.2.3 Diorganoselenium(IV) oxides

3.2.2.3.1 Intermediates in organic synthesis

Organoselenium derivatives containing selenium-oxygen bonds are valuable intermediates in different organic transformations and a rich chemistry was developed based both on aliphatic and aromatic organoselenium oxides. Such species were employed in various organic transformations, including *β*-selenoxide elimination or 2,3-sigmatropic rearrangement reactions [1], [15]c, [39].

Several synthetic procedures based on the high adaptability of organoselenium derivatives to redox processes were developed during last years with a constant interest for milder conditions and higher selectivity.

Vilarrasa and *co-workers* developed an efficient strategy for the dehydration of primary alcohols based on the oxidation of 2-pyridylseleno derivatives of type RR′CHCH$_2$SePy with Dess–Martin periodinane (i.e. intermediates **28** and **29**), followed by PySeOH (**30**) elimination (Figure 3.15) [40].

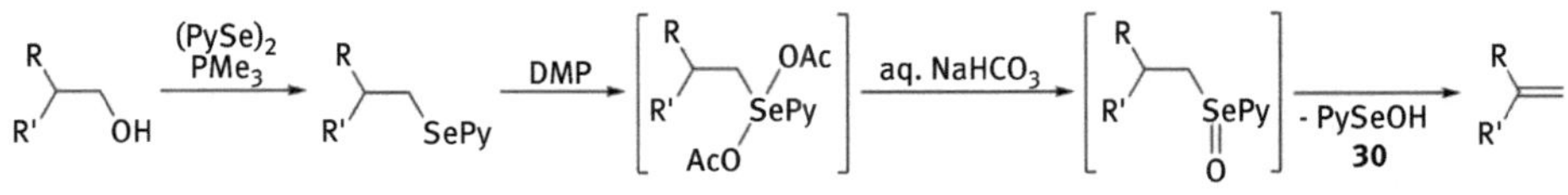

Figure 3.15: Dehydration of primary alcohols to alkenes, based on the oxidation of 2-pyridylseleno derivatives.

Franck and Outurquin investigated the oxidation of several selenenylated dihydropyrans with H_2O_2 in a two-steps process, in mild conditions (H_2O_2 35 % in water and CH_2Cl_2 as solvent) [41]. The selenoxides **31** and **32** formed in the first step were observed to undergo subsequently either (*i*) a ring contraction, thus resulting in tetrahydrofuran-2-ones or other furan-based derivatives (Figure 3.16), or (*ii*) an oxidative cleavage of the C=C double bond, thus leading to a dicarboxylic acid (Figure 3.17).

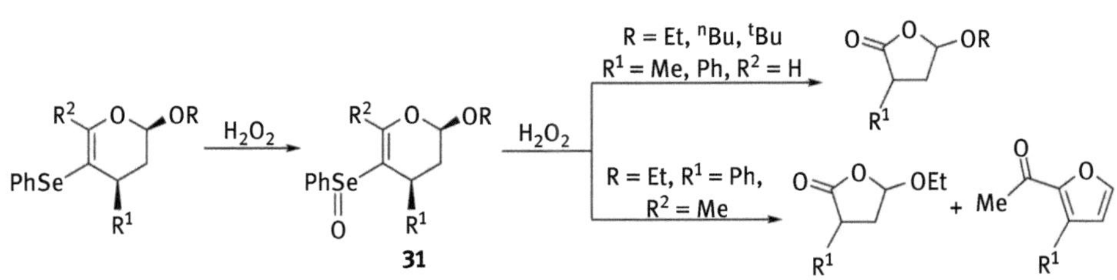

Figure 3.16: Oxidation of selenenylated dihydropyrans *via* ring contraction.

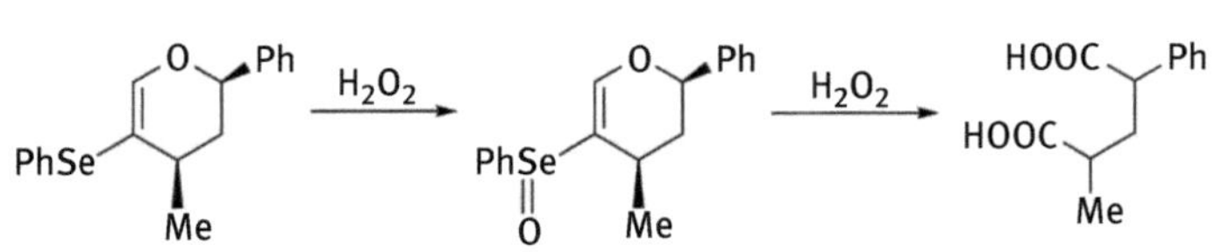

Figure 3.17: Oxidation of selenenylated dihydropyrans *via* C=C bond cleavage.

Different reaction mechanisms, depending on the nature of the ring substituents, were proposed for these oxidative processes, all of them based on the intermediary epoxidation of the enolic double bond. For alkoxy-substituted derivatives (see Figure 3.16), the R^2 substituent makes the difference between mechanisms **A** and **B** (Figure 3.18). For R^2=H (**31a**) (mechanism **A**), PhSeOH (**2**) formed by hydrolysis is subsequently oxidized to phenylperseleninic acid, PhSe(O)OOH (**19**), capable to oxidize the keto-semiacetal to the corresponding lactone-semiacetal in a Baeyer–Villiger oxidation reaction. Further rearrangement, elimination of formic acid and ring closure reactions resulted in 5-alkoxy-tetrahydrofuran-2-one. The diastereoselectivity of the rearrangement process was observed to depend on the substituents as well.

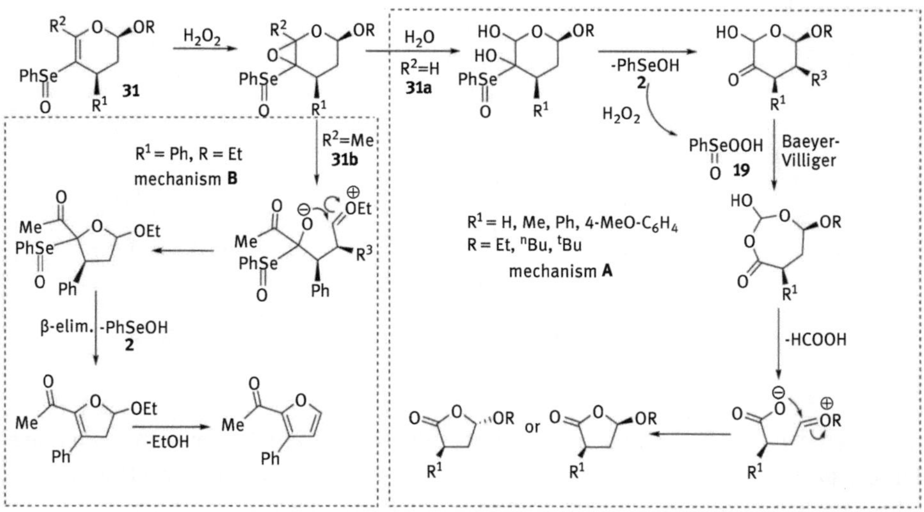

Figure 3.18: Proposed mechanisms based on ring contraction for the oxidation of selenenylated dihydropyrans.

For R^2=Me, R=Et (**31b**), mechanism **B** was proposed. Due to the steric effect of the methyl group, the attack of water is much slower in the first stage and a different rearrangement process followed by further β-elimination of PhSeOH (**2**) took place. Finally, EtOH elimination generated the corresponding furan.

When the alkoxy group was replaced by a phenyl group (see **32** in Figure 3.17), the low electron donating ability of the latter did not allow the ring opening accompanied by the formation of an oxonium cation (as described in mechanism **A**, Figure 3.18) but only the cleavage of the double bond, thus resulting in the corresponding dicarboxylic acid (Figure 3.19).

Arylseleno derivatives of condensed polycyclic systems (Figure 3.20) are valuable intermediates in the synthesis of several analogues of natural products, e.g., the

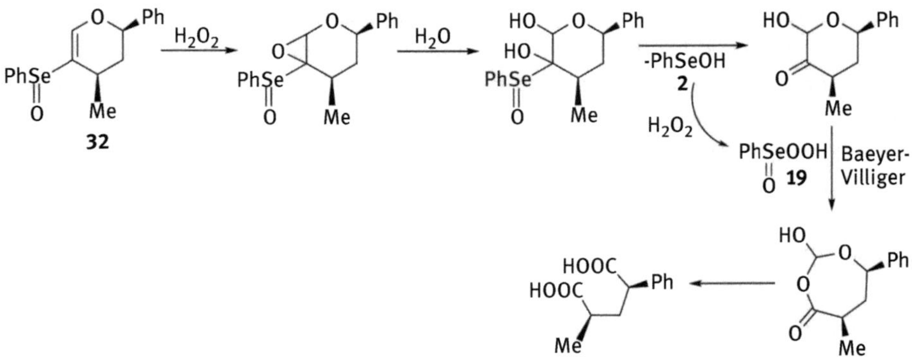

Figure 3.19: Proposed mechanism based on C=C bond cleavage in the oxidation of selenenylated dihydropyrans.

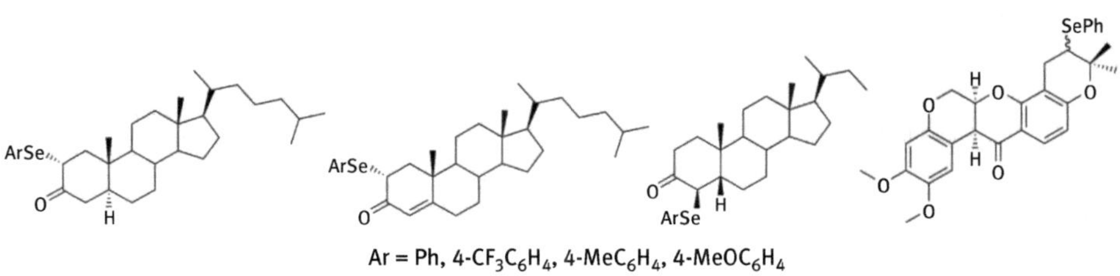

Figure 3.20: Arylselenium(II)-substituted derivatives of condensed polycyclic systems.

flavonoid *deguelin* [42], the diterpene glycoside *eleutherobin* [43] or steroidal cholestenone-based therapeutic agents [42], as they are easily oxidized to arylselenoxides which undergo further elimination of the arylselenenic acid.

The mechanism of such an oxidative elimination process involves a cyclisation transition step [44], as appears in the model depicted in Figure 3.21.

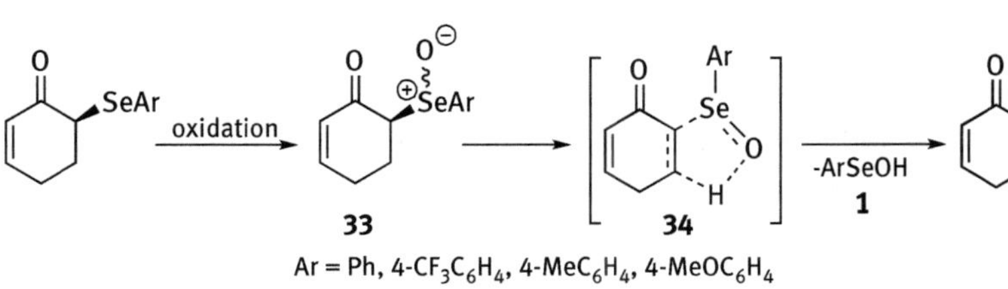

Figure 3.21: Mechanism of oxidative elimination of arylselenenic acid.

Experimental and theoretical investigations regarding the influence of the aryl group attached to selenium upon the oxidation and the elimination reactions showed that electron-withdrawing substituents (4-$CF_3C_6H_4$) stabilize aryl selenides towards oxidation and electron-donating groups (4-$MeOC_6H_4$) accelerate the oxidation process. Moreover, the elimination step is not influenced by the nature of the aryl substituents on selenium and oxidation is the rate-determining step in this type of oxidative elimination processes [44].

Slow oxidation of [pyrenyl(CH_2)$_n$](2-$NO_2C_6H_4$)Se with H_2O_2 led to fluorescent formyl- (n = 1) and vinyl- (n = 2) pyrene, respectively, by successive formation of [pyrenyl(CH_2)$_n$]($C_6H_4NO_2$-2)Se=O (**35**) and (2-$O_2NC_6H_4$)SeOH (**36**) elimination [45].

Electrophilic selenium compounds are easily added to alkenes in the presence of different external nucleophiles (Figure 3.22) [19, 46]. Further oxidation and *syn*-elimination of the selenium moiety from the intermediary formed diorganoselenoxide resulted in new optically active species with carbon–carbon, carbon–nitrogen or carbon–oxygen bonds, depending on the organic substrate [9, 47].

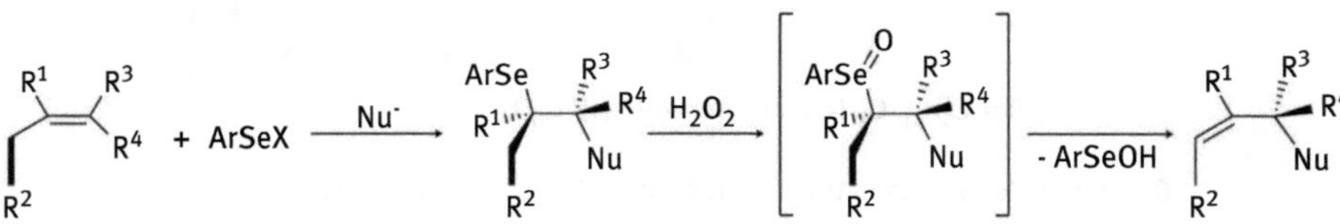

37

Figure 3.22: Arylselenium moiety oxidation and *syn*-elimination reactions resulting in optically active species.

Hard nucleophiles such as piperidine and phenol were added to the electron-poor double bond in α-(phenylseleninyl)sulfinyl and -sulfonyl alkenes [48]. The nucleophile addition of piperidine to a racemic mixture of α-(phenylseleninyl)sulfinyl alkenes resulted in diastereomeric mixture of unstable α-(phenylseleninyl)sulfinyl alkanes which spontaneously underwent a stereospecific *syn*-elimination of PhSeOH (**2**), thus resulting in a single stereoisomer, namely (*E*)-2-butyl-1-(2-(phenylsulfinyl) vinylpiperidine. It seems that the stereogenic selenoxide group is remarkably effective in inducing asymmetry and outweighing the influence of the sulfinyl group.

In contrast with the large number of synthetic methods based on *syn*-elimination reactions, only few examples are known in which *anti*-elimination occurs after oxidation [49, 50]. Recently, *Nishide* and *co-workers* evidenced, both at experimental and theoretical level, such an *anti*-elimination process in the synthesis of alkynyl selenoxides, favored by an intramolecular O→Se interaction (Figure 3.23) [50]. (*Z*)-1,2-bis[4-(trimethylsilyl)phenylseleno]-1-alkenes were oxidized with *m*CPBA to the corresponding 1,2-bis(selenoxides), which undergo *anti*-elimination by heating. A similar protocol applied to the sulfur analog showed no elimination and this behavior was attributed to the absence of any intramolecular O→S non-bonding interaction.

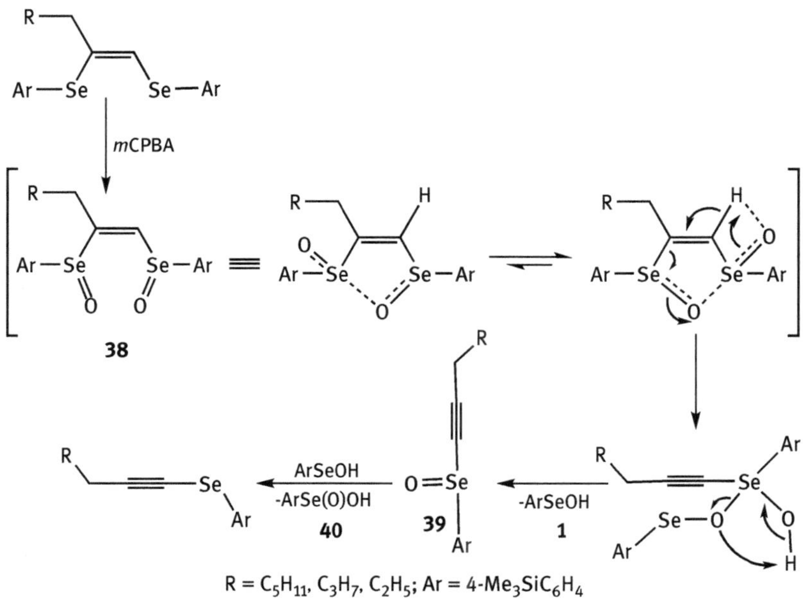

Figure 3.23: Diorganoselenoxides in the synthesis of (alkynyl)(aryl)selenides.

3.2.2.3.2 Catalysts in organic synthesis

Hydroxylation reactions

Organoselenium compounds were largely used as Lewis base catalysts in various transformations [51]. Most of the reported studies are devoted to the functionalization of alkenes, either by halogenation or sulfenylation, when the organoselenium catalysts activate the electrophilic reagent in a selenium-captured ionic complex, the latter being further involved in the substrate functionalization.

Recently *Zhao* and *co-workers* [52] developed a very efficient method for the synthesis of trifluoromethylthiolated organic compounds by difunctionalization of multisubstituted alkenes in mild conditions. Various trifluoromethylthiolated tertiary alcohols were obtained by a vicinal CF_3S hydroxylation of 1,1-multisubstitued alkenes (Figure 3.24).

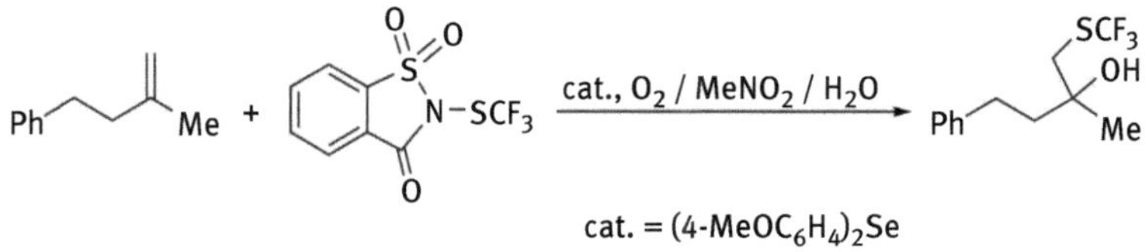

Figure 3.24: Catalytic procedure for the synthesis of trifluoromethylthiolated tertiary alcohols by hydroxylation of alkenes.

These types of transformations were performed with moderate to very high yields in catalytic conditions based on a diaryl selenide/O_2/$MeNO_2$ system which is involved in a reversible redox cycle between Ar_2Se and $Ar_2Se{=}O$ (**41**). The role of the intermediary formed diarylselenoxide is to enable the formation of triflic acid (TfOH), the properly active species in further functionalization of the organic substrate in a Lewis base selenium catalysis (Figure 3.25) [52].

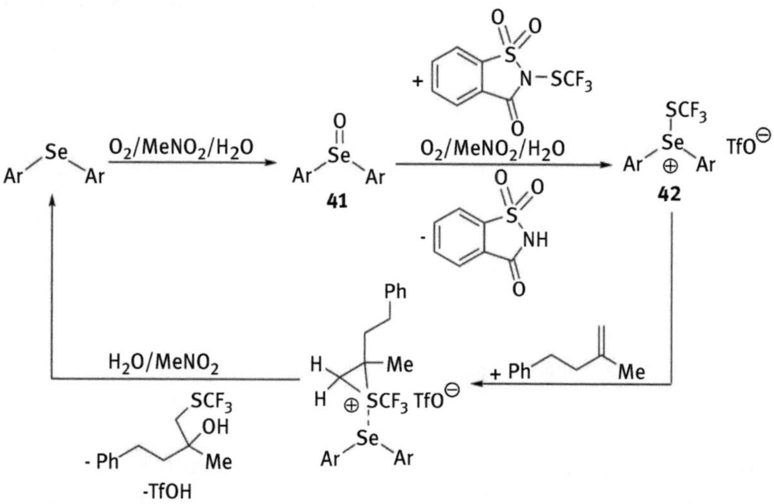

Figure 3.25: Proposed mechanism for the catalytic synthesis of trifluoromethylthiolated tertiary alcohols.

Intramolecular cyclization reactions

Following a similar procedure, alkenes tethered by nucleophile carboxylic, ester, hydroxy or sulfamide groups underwent intramolecular cyclization reactions (Figure 3.26) [52].

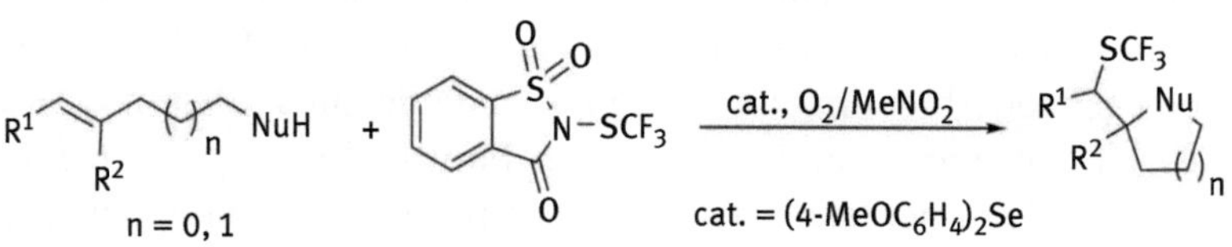

$n = 0$; $R^1 = H$; $R^2 = Ph$; NuH = COOH, COOMe
$n = 0$; $R^1 = CH_2CH_2Ph$; $R^2 = H$; NuH = COOH
$n = 1$; $R^1 = H$, $R^2 = Ph$; NuH = OH, NHTs
$n = 1$; $R^1 = CH_2CH_2Ph$; $R^2 = H$; NuH = COOH

Figure 3.26: Intramolecular cyclization reactions.

Hydroformylation reactions

Recently diarylselenoxides were used by *Börner* and *co-workers* as additives in the homogeneous rhodium-catalyzed hydroformylation of olefins (Figure 3.27) in order to improve the catalytic properties of the commonly used $[Rh(COD)Cl]_2$ complex [53]. It was observed that $Ph_2Se{=}O$ (**43**) behaves as an oxo ligand towards the metal centre (see complex **44** in Figure 3.27) and the yield in the transformation of *n*-octene to the *n*-nonanal increased, independently from the concentration of the co-catalyst. The yield was higher compared with the catalytic process without selenium additives.

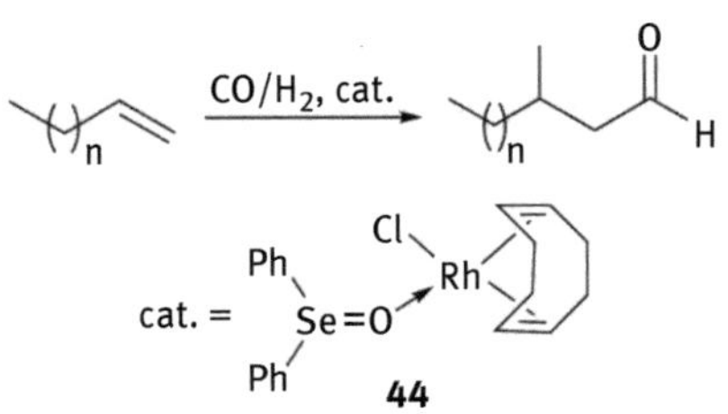

44

Figure 3.27: Ph_2SeO as ligand in a rhodium catalyst for hydroformylation of olefins.

A similar behavior in the catalytic process was observed for additives with bulky substituted aryl groups, while the selectivities were almost the same (approx. 28 %). A slight increase of selectivity was observed when diphenylselenodioxide, Ph_2SeO_2 (**45**), was used as additive, but in this case a dramatic decrease of the yield was observed with the increased content of the additive.

Noteworthy, under a syngas atmosphere, the diorganoselenium monoxide was reduced to the corresponding diorganoselenide which determined an increased yield of hydroformilation as well, but at higher concentrations it was observed to inhibit the catalytic process.

Bromination reactions

Detty and *co-workers* have shown that dendrimeric diorganoselenoxides are very efficient to activate H_2O_2 for oxidation of bromide to HOBr and Br_2 [54]. They reported the homogeneous bromination of various organic substrates in a two phase system (CH_2Cl_2/ phosphate buffer pH 6) with NaBr and H_2O_2, i.e. 4-pentenoic acid and 2,2-diphenyl-4-pentenoic acid to the corresponding 4-bromomethyl *γ*-lactones, 1,3,5-trimethoxybenzene to 2-bromo-1,3,5-trimethoxybenzene, *N,N*-dimethylaniline to a 2:1 mixture of 4-bromo-*N,N*-dimethylaniline and 2-bromo-*N,N*-dimethylaniline, *N*-phenylmorpholine to *N*-(4-bromophenyl)morpholine [55]. The activation mechanism involves the formation of a (hydroxy)(perhydroxy)selenane of type RR′Se(OH)(OOH) (**47**) which further acts as a powerful oxidizing agent towards Br^- anions (Figure 3.28).

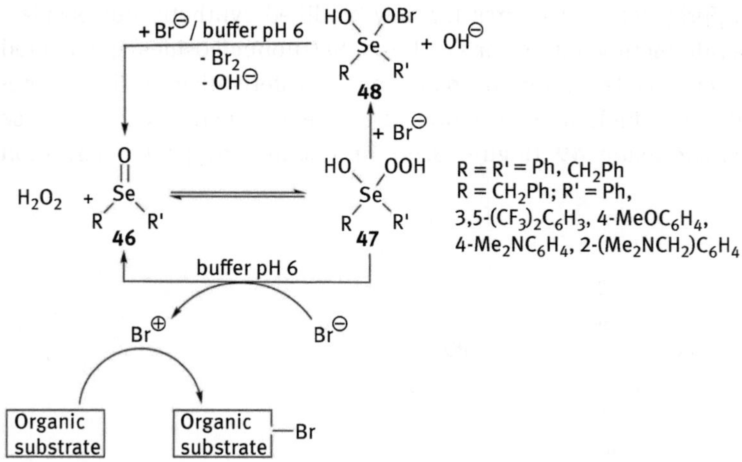

Figure 3.28: Diorganoselenoxides in oxidative bromination of organic substrates.

Various selenoxides of type RR′Se=O (**46**), having no *β*-hydrogen in their structure and thus showing no tendency to undergo selenoxide elimination, were employed as catalysts. It was observed that the catalytic activity increased by increasing the Lewis basicity of the selenoxide oxygen and the stability of the intermediary formed $[RR'SeOH]^+$ species, so that the species with aromatic groups functionalized with electron-donating substituents have shown increased reaction rates. The most active species was $[2\text{-}(Me_2NCH_2)C_6H_4](PhCH_2)Se{=}O$ (**46a**) with an enhanced donating ability due to the intramolecular N→Se interaction [55].

Recently, *Detty* and *co-workers* reported the bromination of the same substrates in “green” conditions, avoiding the use of organic solvents and using a heterogenised catalyst [56]. [4-(Hydroxymethyl)phenyl](benzyl)selenoxide (**46b**) was sequestered within an APTES/TEOS (10/90 [mol/mol] xerogel. This procedure had several advantages: an enhanced reactivity and a much faster reaction on the large porous surface offered by the xerogel support (23-fold faster than the solution process), a “green” approach for oxidation and bromination, as well as a valuable possibility to separate, recover and recycle the catalyst. The mechanism of bromination is similar to that one observed in solution, and involves the addition of H_2O_2 to the diorganoselenium oxide with formation of the corresponding (hydroxy)(perhydroxy)selenane (**47**), capable to oxidize Br^- to HOBr, which finally brominates the organic substrate (Figure 3.28).

3.2.2.4 Heterocyclic selenoxides

Condensation reactions

The current interest in nucleic acids-based therapies, as well as the applications of modified oligonucleotides in antisense and RNA interference, prompted the interest to find appropriate procedures for new modified synthetic nucleosides, despite the elaborate methods used for synthesis. Selenonucleosides, considered third-generation nucleosides, after the oxo- and thio derivatives [57], are prepared by several steps, including oxidation of a 4-selenosugar and further use of the 4-selenosugar oxide **49** (Figure 3.29) in Pummerer-type condensation reactions [58].

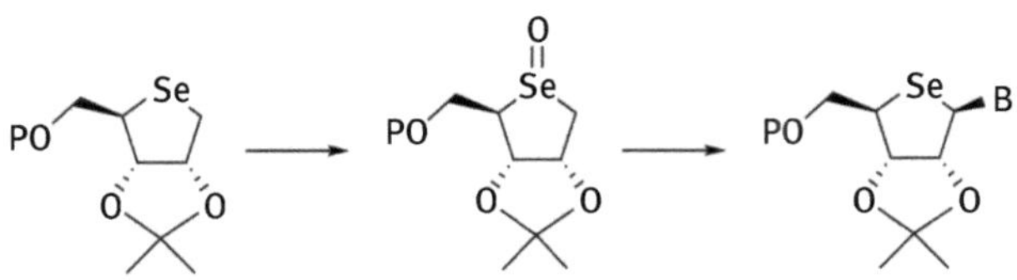

49

P = Protecting group; B = Nucleoside base

Figure 3.29: Heterocyclic selenoxides in the synthesis of selenonucleosides.

It was noted that the selenoxide resulted either as a mixture of two inseparable diastereomers or only one of them, depending on the sugar type and the protecting group. *Pinto* and *co-workers* reported that the oxidation of 1,4-anhydro-5-*tert*-butyldimethylsilyl-2,3-*O*-isopropylidene-4-seleno-D-ribitol with *m*CPBA resulted in a single selenoxide diastereomer with an excellent yield. Further Pummerer-like condensation reactions with the appropriate nucleobase and deprotection with aqueous acetic acid lead to 4′-selenonucleosides (seleno-thymidine, -cytidine, -adenosine and –uridine, respectively) with high *β*-stereoselectivity [58]a.

The selenosugar **50** led to a mixture of *R* and *S* diastereomers of the selenoxide **51** in a 3:2 molar ratio, regardless which oxidizing reagent was used (*m*CPBA, ozone or $NaIO_4$) and the two diastereomers could not be separated. Further treatment of this mixture with trimethylsilyl trifluoromethanesulfonate and Et_3N in the presence of silylated uracil led to the desired 4′-selenouridine **52** in low yield (36 %). This behavior was explained by the instability of the selenoxides which underwent a partial deoxygenation process before the Pummerer-like reaction, thus resulting in a mixture of 4′-selenouridine, diselenide and the starting selenosugar (Figure 3.30) [58]b.

The same group reported a deoxygenation process of selenoxides **51** and **53**, which occurs much slower for **51** than for the latter (i. e. about one week *vs* one day) in $CDCl_3$ solution [58]c. The different behavior was explained in terms of different conformations of the two selenosugars, with the silyloxymethyl substituent at 4-position located pseudoequatorial in **51**, at a larger distance from hydrogens, and pseudoaxial in **53**, in closer proximity to the hydrogen atoms located at 5-position, thus favouring *syn*-elimination of oxygen from the latter compound (Figure 3.31) [58]c.

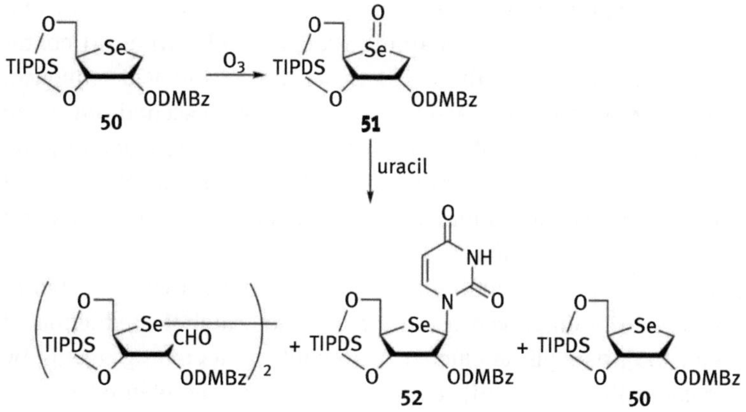

Figure 3.30: Heterocyclic selenoxides in the synthesis of selenouridine.

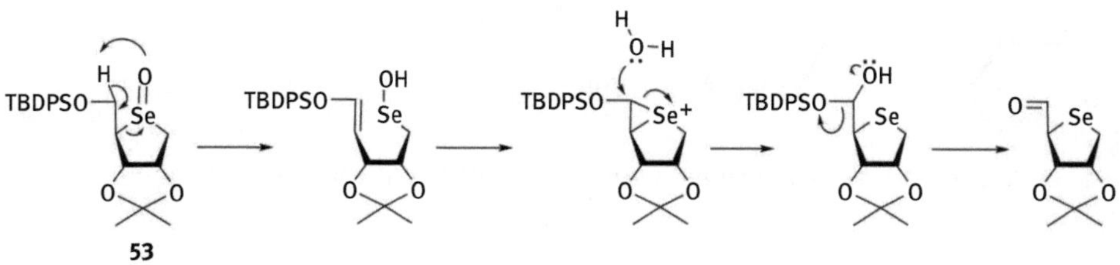

Figure 3.31: Heterocyclic selenoxides in the synthesis of the corresponding heterocyclic aldehydes.

Oxidative S–S bond formation

The water soluble *trans*-3,4-dihydroxyselenolane oxide **54** was employed as an oxidative folding reagent in studies regarding the disulfide bond formation and the redox behavior of biological systems [59]. It was demonstrated that the S–S bond is formed during a two-steps process (Figure 3.32), involving a thioselenurane intermediate **55** formation in a rate determining bimolecular reaction, followed by a fast intramolecular rearrangement which results in disulfide formation concomitant with elimination of the reduced species **56** and water.

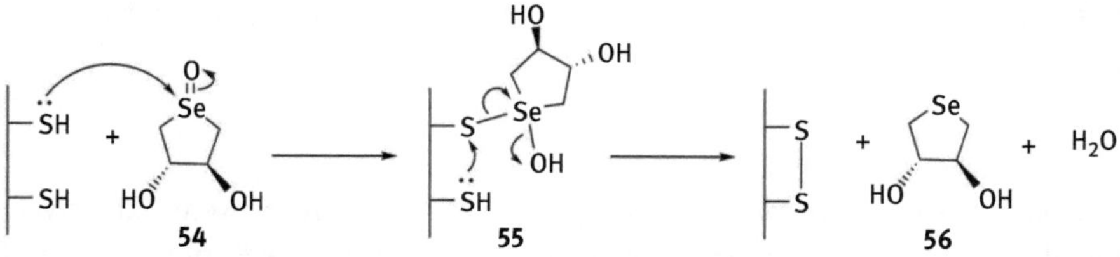

Figure 3.32: Redox behavior of the dihydroxyselenolane oxide **54** in the oxidation of thiols to disulfides.

The selenolane oxide **54** proved several advantages: (*i*) it behaves as a strong and selective oxidant for various S–S bond-containing proteins under wide pH conditions, not depending on the number of the S–S bonds or the amino acid sequence, (*ii*) the S–S bond formation proceeds in a stochastic manner, with second-order rate constants proportional to the number of free thiol groups along the reactant polypeptide, if there are no steric congestions or other electronic factors affecting the SH groups, and (*iii*) such a behavior offers valuable information about the reactivity and the structure of the folding intermediates.

Both reduced species **56** and the amino functionalized derivatives depicted in Figure 3.33 were successfully employed as GPx mimics in catalytic reduction of hydroperoxides [59]b, [60], as well as efficient peroxynitrite degrading agents by isomerization of peroxynitrite to NO_3^- [61] *via* formation of a stable selenoxide.

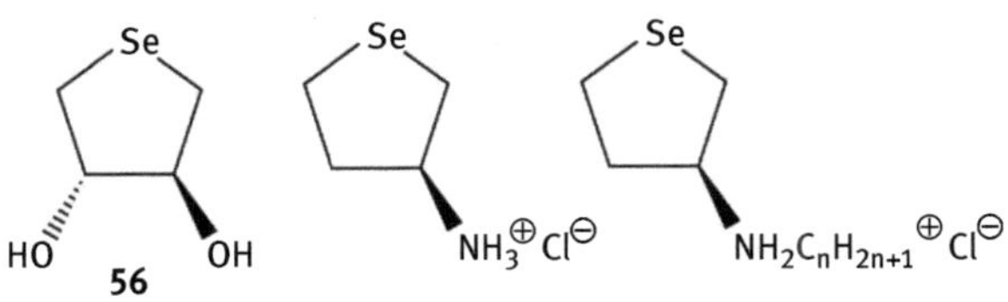

Figure 3.33: Heterocyclic functionalized selenides evaluated as GPx mimics.

3.2.3 Organoselenium(VI) compounds

By far the organoselenium(VI) compounds most used in organic synthesis are the readily available (aryl)(vinyl)selenones, which were employed in Michael addition reactions of various organic substrates. The selenone group activates the carbon-carbon double bond towards the addition and, at the same time, it acts as a good leaving group in subsequent functionalization or cyclization processes.

Several strategies based on catalytic Michael addition reactions were developed during last years for the synthesis of valuable organic species, most of them analogues of natural products or compounds with relevant biological activity. In some procedures organic catalysts based on ureido or thioureido skeletons, capable for a high stereoselectivity, high yields and environmental friendly conditions of synthesis were employed, as alternatives for the previously used inorganic metal-based catalysts.

Michael addition and functionalization

The excellent leaving properties of the phenylselenonyl group were extensively exploited in functionalization reactions under mild conditions with preservation of the enantiomeric purity (Figure 3.34) [62].

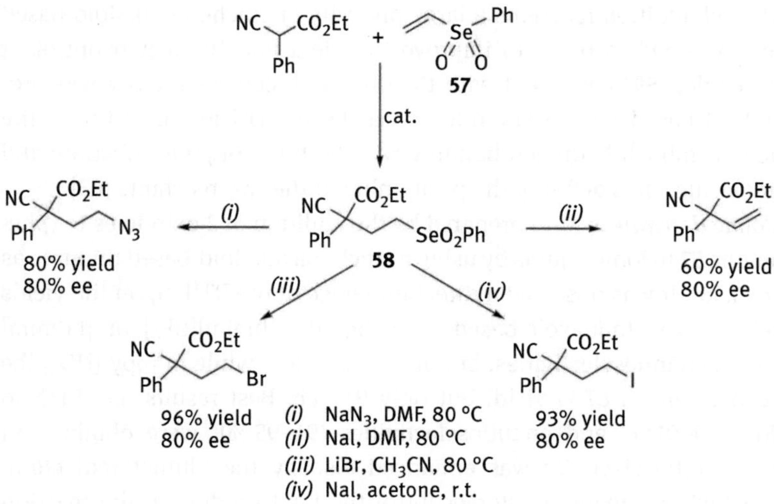

Figure 3.34: Diorganoselenodioxide **58** in functionalization reactions.

Michael adducts between methyl *α*-aryl-*α*-isocyanoacetates and (vinyl)(aryl)selenones were employed in the synthesis of the enantio-enriched *α*-aryl-*α*-(2′-phenylselenonylethyl)-*α*-isocyanoacetates (**60**) (Figure 3.35), which were subsequently converted into various functionalized *α*-aryl-*α*-(2′-FG-alkyl)-*α*-amino acids (FG = functional group) or heterocyclic oxindoles and pyrrolidinones [63, 64].

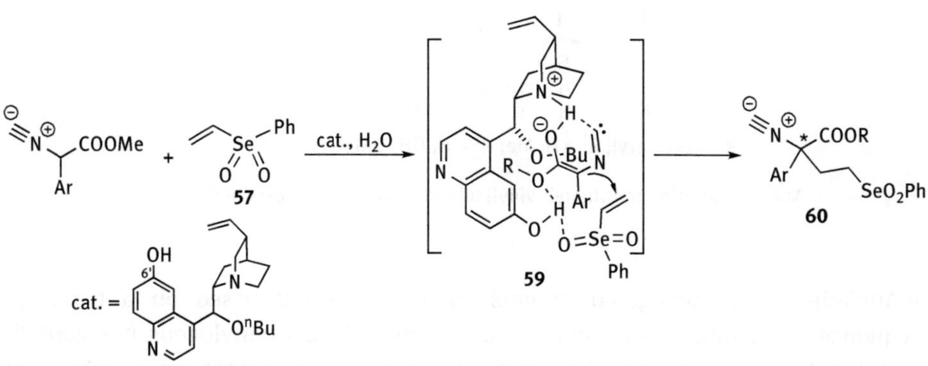

Figure 3.35: (Phenyl)(vinyl)selenone in the synthesis of **60**.

Michael adducts based on (aryl)(vinyl)selenone and *α*-alkyl-*α*-nitroacetates were used for the synthesis of *α*,*α*-disubstituted *α*-nitroacetates and quaternary *α*-amino acids [65], while those based on methyl 2-isocyano-2-(4-methoxy-2-nitrophenyl)acetate were employed in the synthesis of (±)-Trigonoliimine A and (±)-Trigonoliimine B [63, 64].

For the Michael addition reactions a large diversity of cinchona-alkaloid-based catalysts were developed in order to improve the yield and the stereocontrol. A transition state model (**59**) consistent with the observed stereoselectivity was proposed (Figure 3.35) based on the key role played by the OH group at C6′ in the catalyst, prone to establish hydrogen bonding with both the organic substrate and the selenone molecule, thus defining the positioning of the two reactants [63].

Pyrroloindoline derivatives were prepared by the addition of 2-oxindoles to (phenyl)(vinyl)selenone **57** in ionic liquids by using a cinchona alkaloid-based thiourea as catalyst (Figure 3.36). It was observed a direct influence of the RTIL upon the yields and enantioselectivities. Imidazole-based ionic liquids, [bmim][BF_4] or [bdmim][BF_4], offered good enantioselectivities, but only low yields, while in [bpy][NO_3] the addition proceeded with a 87 % yield, but only 9 % *ee*. Best results, i.e. high to excellent yields (80–91 %) and enantioselectivities (91–95 %), were obtained in [bpy][BF_4]. The enantioselectivity was enhanced also by the bifunctional chiral organocatalyst which is capable to stereocontrol the final product in the reaction with the activated *α,β*-unsaturated selenone [66].

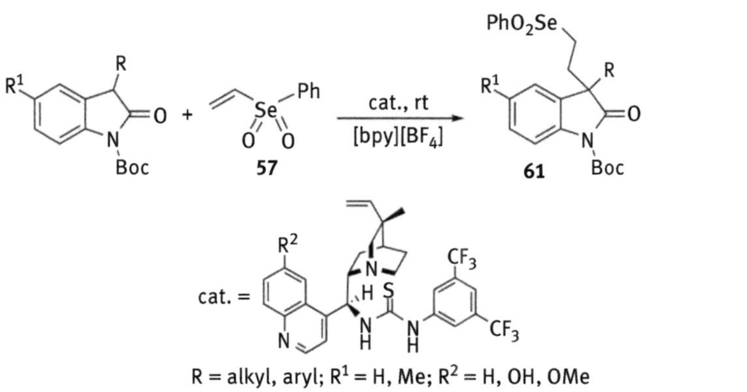

Figure 3.36: Michael addition to (phenyl)(vinyl)selenone in the synthesis of **61**.

γ-Allylation of deconjugated butenolides was reported based on a two-steps sequence consisting of a diastereo- and enantioselective vinylogous nucleophilic addition to (phenyl)(vinyl)selenone **57** catalyzed by a tertiary amino (thio)urea bifunctional catalyst, followed by a Julia-Kocienski olefination (Figure 3.37) [67].

(Vinyl)selenone-modified furanosides combining the (vinyl)selenone moiety with the furanose and pyranose skeletons of *d*-fructose act as densely functionalized Michael acceptors comprising a masked aldehyde, an electron-deficient double bond and an excellent leaving selenone group. These properties were successfully exploited in diversity oriented synthesis of a large variety of organic compounds, in reactions with various nucleophiles, thus resulting in enantiopure species: trisubstituted densely

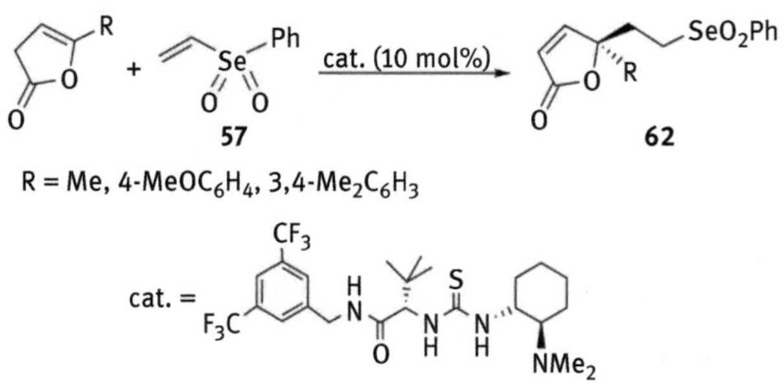

Figure 3.37: Michael addition to (phenyl)(vinyl)selenone in the synthesis of **62**.

functionalized triazoles [68], bicyclic azasugars, cyclopropanated carbohydrates, dihydrofuran- and dihydroisoxazole-substituted furanosides or isonucleosides [69], aziridines, thiiranes, cyclopropanes, dibenzyloxymethylfuran etc [70].

Michael addition, functionalization and ring closure reactions

During last years MIRC reactions were employed in one-pot synthesis of various cyclic compounds, among which several types are described below.

Three membered rings

Catalytic methods were developed for the stereoselective synthesis of substituted cyclopropanes [62, 71, 72], which are key synthons in the architecture of various natural or synthetic biologically active complex molecules, e.g., amino acids, peptides, enzyme inhibitors, therapeutic agents with antifungal, antibacterial, antiviral, antitumor or antipsychotic activities.

Nitrile-substituted cyclopropanes were prepared as single *Z*-isomers starting from *α*-substituted cyanoacetates in an one-pot procedure. The Michael addition was efficiently catalyzed by bifunctional ureidic catalysts and resulted in enantioenriched selenones containing an all-carbon quaternary stereocenter (**64**). The subsequent cyclization proceeds by an intramolecular alkylation induced by de-ethoxycarbonylation in KCN/DMF (Figure 3.38) [62], LiCl/HMPA or EtONa/EtOH solution [71].

It was observed that aryl substituents on the starting cyanoacetates favor the de-ethoxycarbonylation, while *α*-allylcyanoacetates gave no cyclopropane. Both aryl- and alkyl-substituted (vinyl)selenones were used as Michael acceptors. Even if the alkyl-substituted (vinyl)selenones showed a higher rate of addition, the selectivity in the cyclization step was only moderate in a LiCl/HMPA solution and it did not proceed in the environmentally benign EtONa/EtOH mixture [71].

In another experiment, the Michael addition of chiral non-racemic di-(−)-bornyl and di-(−)-menthyl malonate to (aryl)(vinyl)selenones resulted in an approx. 1:1 mixture of diastereoisomers. Further cyclization afforded a diastereomeric mixture of cyclopropanes in similar yields. From such mixtures highly enantiomerically

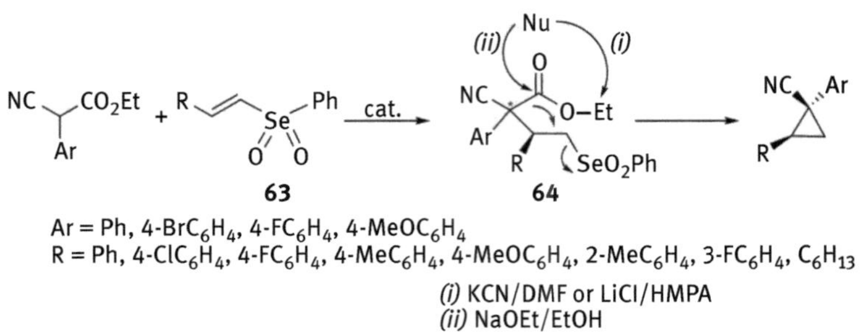

Figure 3.38: Synthesis of functionalized cyclopropane *via* Michael addition to (phenyl)(2-substituted-vinyl)selenone.

enriched cyclopropanes could be obtained after medium pressure column chromatography and removal of the bornyl and menthyl groups [72].

(Phenyl)(2-substituted-vinyl)selenones (**63**) were successfully employed in the preparation of various aziridines *via* a tandem Michael addition - intramolecular nucleophilic substitution process, starting from primary amines, aminoalcohols or diamines. High yields were obtained both in toluene and water, but a significant increased rate of the process was observed in aqueous conditions, thus suggesting that water participates to the process as well (Figure 3.39) [73].

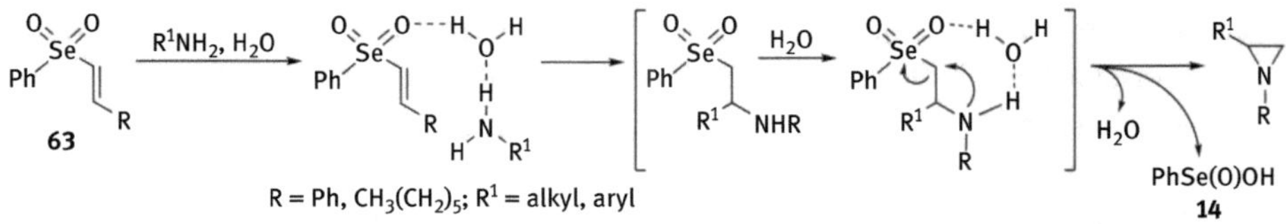

Figure 3.39: Synthesis of aziridines *via* Michael addition to (phenyl)(2-substituted-vinyl)selenones.

Six- and seven-heterocyclic rings

MIRC processes applied to enantiopure 1,2-diols, N-protected-1,2-aminoalcohols, thiols and diamines resulted in substituted enantiopure 1,4-dioxanes, morpholines, and piperazines [74]. The addition of the organic substrate to the selenone **63** proceeds in the presence of NaH, through the intermediary formation of a carbanion (**65**) which subsequently undergoes a proton transfer and cyclization by PhSe(O)OH (**14**) elimination, as exemplified in Figure 3.40 for the addition of 1,2-diols.

An integrated one-pot procedure based on a Brønsted base-catalyzed Michael addition of α-substituted α-isocyanoacetates to (phenyl)(vinyl)selenone and subsequent Brønsted acid-catalyzed domino oxidative cyclization afforded 1,3-oxazinan-2-ones in a procedure in which the (phenyl)(vinyl)selenone (**57**) had a triple role, behaving successively as an activator, a leaving group and a latent oxidant (Figure 3.41) [75, 76]

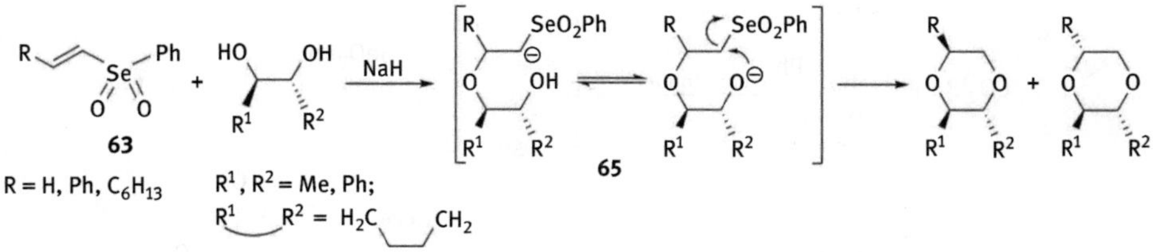

Figure 3.40: Reaction of (phenyl)(2-substituted-vinyl)selenones and 1,2-diols.

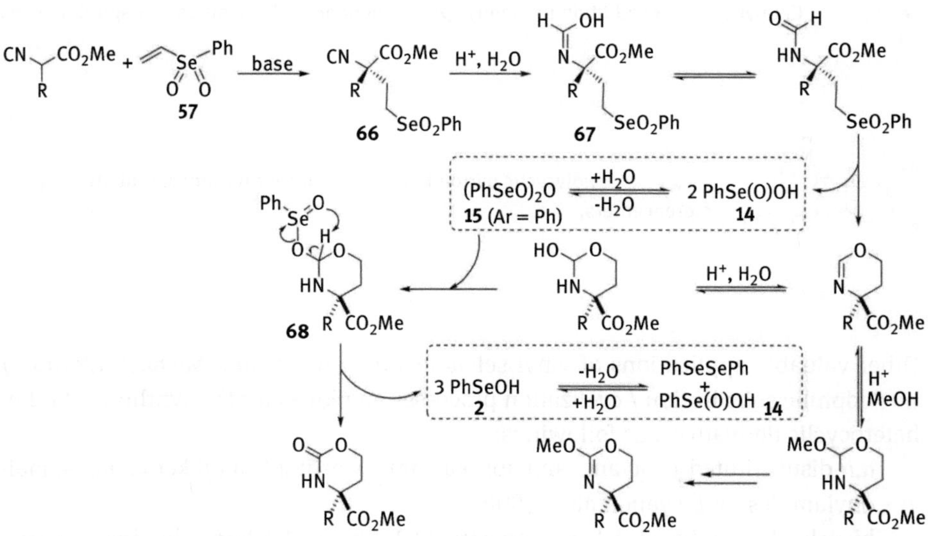

Figure 3.41: Proposed mechanism for the synthesis of 1,3-oxazinan-2-ones.

Spirolactones and polycyclic pyrrolidines

The reaction between *tert*-butyl ketoester and (phenyl)(vinyl)selenone (**57**) proceeded with higher efficiency and stereoselectivity (*ee* values in the range 90–98 %, yield 80–99 %) when cinchona-based catalysts were used compared with the similar reactions catalyzed by Na_2CO_3 (yield 71 %) (Figure 3.42) [77].

Indanone-based *tert*-butyl *β*-ketoesters with various substituents afforded the desired products in shorter reaction times, with high yields and enantioselectivities, while *β*-ketoesters with a cyclohexanone core could not be successfully used in such transformations. The ^{t}Bu group seems to be essential for the process as it is easily cleaved by silica during the cyclization / elimination step and the attempts to use ethyl substituted *β*-ketoesters afforded the spirolactone in a very poor yield.

A similar procedure was employed as a key step in the cascade synthesis of highly enantioenriched (up to 98 % ee) polycyclic pyrrolidines bearing an *α*,*β*-aminoester motif, with adjacent tertiary and quaternary stereocenters (Figure 3.43) [78, 79].

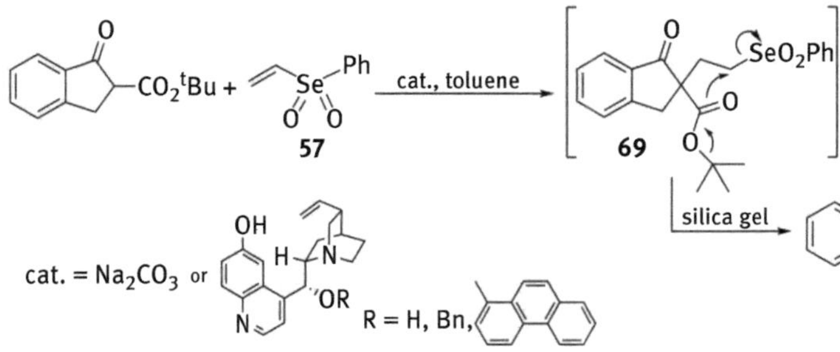

Figure 3.42: Catalytic Michael addition to (phenyl)(vinyl)selenone in the synthesis of spirolactones.

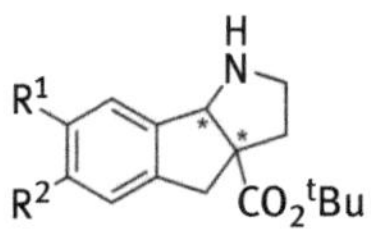

Figure 3.43: Polycyclic pyrrolidines with adjacent tertiary and quaternary stereocenters.

Other heterocyclic species

Other valuable applications of vinyl selenones refer to domino Michael addition / nucleophilic substitution / cyclization processes employed for the synthesis of other heterocyclic derivatives, as followings:

- *α,α*-disubstituted *γ*-lactams and spirolactams, starting from *β*-ketoamides, malonylamides, or *β*-cyanoamides [80];
- bicyclic benzo fused 1,4-heterocycles (2,3-dihydro-1,4-benzodioxines, benzodithiines and 3,4-dihydro-2H-1,4-benzoxazines), starting from benzo 1,2-diols, 1,2-thiols and 2-(benzylamino)phenols [81];
- spirocyclopropyl oxindoles, starting from variously substituted vinyl selenones and enolizable oxindoles, in aqueous NaOH medium. Due to the intermediary oxindole enolate formed by proton transfer, which enable the free rotation around the single C–C bond, a stereocontrol of the product based on the configuration of the starting substrate is not possible, but the two isomers can be easily separated by chromatography [82]. The method offers the advantage of diverse substitution, e.g., at the cyclopropane ring, the aromatic ring and at the nitrogen atom;
- *cis*-fused perhydrofuro[2,3-*b*]furans were obtained from 3,4,6-tri-*O*-acetyl-d-glucal and 2-(phenylselenyl)ethanol in a stereoselective procedure based on the PhSe(O)OH (**14**) elimination from the intermediary formed selenone glycoside [83];
- recently, *Temperini* and *co-workers* reported a mild, stereoselective and general approach to the preparation of substituted tetrahydropyrans and isochromans, from phenylseleno alcohols via oxidation to diorganoselenodioxide (**70**) and an efficient 6-*exo*-tet ring-closure (Figure 3.44) [84].

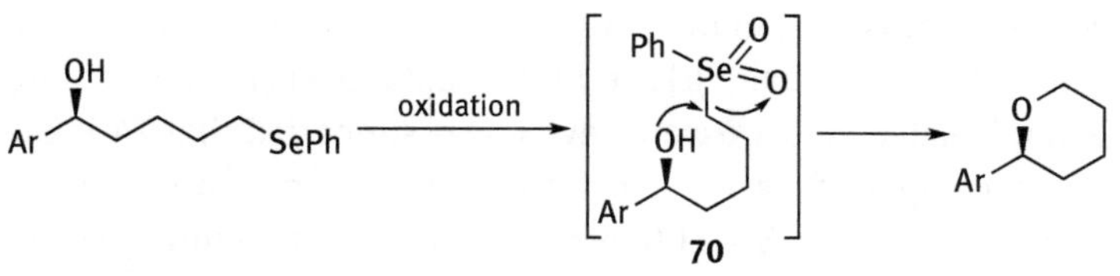

Figure 3.44: Diorganoselenodioxide **70** as intermediate in the synthesis of 2-substituted tetrahydropyrans.

Based on the excellent departing- and electron-withdrawing abilities of the selenonyl group, *Beng* and *co-workers* implemented a successful strategy for the synthesis of *α,β*-carbo-difunctionalized piperidine and azepane derivatives by a regioselective functionalization of *α*-selenonyl enecarbamates and eneformamides [85].

3.3 Organotellurium compounds

The use of tellurium compounds in organic synthesis was collected over years in books or books chapters as well as review articles by *Petragnani* and *co-workers* [3] a–c. Some important, mainly recent, achievements in this field are highlighted in the following sections.

3.3.1 Organotellurium(II) compounds

There are quite few studies related to the direct use of organotellurium(II) compounds containing a Te–O bond in organic synthesis and related processes [3]a, e.g., PhTeOAc (**71**) is an intermediate in the diacetoxylation of olefins by the anhydride $[PhTe(O)]_2O$ (**72**) in boiling acetic acid [86], while $PhTeOSO_2C_6H_4NO_2$-4 (**73**), obtained *in situ* from Ph_2Te_2 and 4-nitrobenzenesulfonyl peroxide, was reported to be an efficient electrophilic reagent for the tellurolactonization of unsaturated carboxylic acids or tellurocyclization of unsaturated alcohols (Figure 3.45) [87].

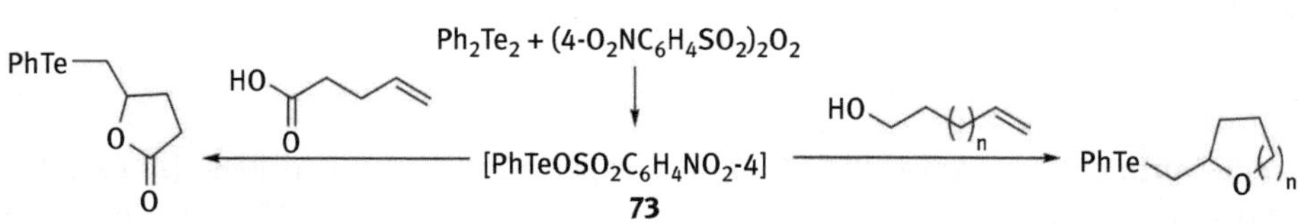

Figure 3.45: Tellurocyclization of unsaturated carboxylic acids or alcohols by *in situ* prepared $PhTeOSO_2C_6H_4NO_2$-4 (**73**).

The phenyltellurium(II) triflate, PhTeOTf (**74**), was reported as an efficient tellurophilic activator for various organotellurium derivatives which, in the presence of aromatic compounds, afford the corresponding Friedel–Crafts reaction products [88]. Compound **74** was generated *in situ* from PhTeTe(Ph)Br_2 and AgOTf (1:2 molar ratio) and, although it was not isolated, spectroscopic evidences supported that **74** is the activating species. This compound was proved to convert a mixture of 1,3,5-trimethoxybenzene and acyl telluride MeTeC(O)Ph into the corresponding Friedel–Crafts product (91 % yield at r.t.) (Figure 3.46), the effect being considerably higher than those obtained with PhSeOTf (**5**), PhSOTf or other conventional activators, under similar reaction conditions.

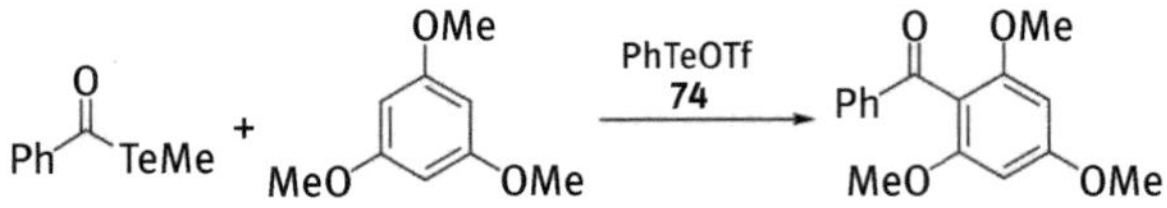

Figure 3.46: Conversion of 1,3,5-trimethoxybenzene / MeTeC(O)Ph mixture into the corresponding Friedel–Crafts product using PhTeOTf (**74**) as tellurophilic activator.

The same effect was exhibited by compound **74** when used with other alkyltellurium(II) substrates (e.g., MeTeCHMePh, ^{t}BuTePh) and nucleophiles (e.g., $MeOC_6H_5$, 1,3-$Me_2C_6H_4$ or thiophene) and the observed regioselectivities in the Friedel–Crafts reaction are consistent with the formation of carbocationic species from organotellurium compounds. The polymer-end organotellurium compound, obtained by living radical polymerization of styrene with the polymer-end mimetic initiator MeTeCHMePh [89], was also activated by **74** to provide the corresponding end-functionalized polymer when treated with 1,3,5-trimethoxybenzene as nucleophile [88].

3.3.2 Organotellurium(IV) compounds

Several aspects related to the synthesis and structure of telluroxanes as well as their applications in organic synthesis were recently reviewed by *Panda, Singh* and *co-workers* [3]d].

3.3.2.1 Organotellurinic acids and their anhydrides

Aryltellurinic acids, ArTe(O)OH (**75**), and especially their anhydrides, $[ArTe(O)]_2O$ (**76**), or mixed anhydrides, ArTe(O)X [**77**, e.g., X=OAc, $O(O)CCF_3$, OTf], were described to exhibit oxidizing properties, in many aspects similar to those of the selenium analogous (*vide supra*) or diaryltelluroxides (*vide infra*). The studies were extended to potential applications in other organic transformations, the results being periodically summarized in detail by several review articles and books very useful to a reader interested in this topic [3]c, [3]d, [90–92].

3.3.2.1.1 Reagents in organic synthesis

Oxidation reactions

Aryltellurinic anhydrides, $[ArTe(O)]_2O$ (**76**), and mixed anhydrides, ArTe(O)X (**77**), are mild and selective oxidizing agents [3]a, [3]d, [90]. Thus, in an earlier work, *Barton* and *co-workers* [93] reported on the oxidizing properties of several aryltellurinic anhydrides of the type $[(4\text{-}ROC_6H_4)Te(O)]_2O$ [R=Me (**78**), ^{n}Bu (**79**), Ph (**80**)] and $[(2\text{-}naphthyl)Te(O)]_2O$ (**81**). It was shown that, while the phenols as $2,5\text{-}^tBu_2C_6H_3OH$ and $3,5\text{-}^tBu_2C_6H_3OH$ were not oxidized in refluxing acetic acid, hydroquinones and thiophenols were readily oxidized to the corresponding quinones (e.g., $2,5\text{-}^tBu_2$-1,4-benzoquinone or 1,4-naphthoquinone) and disulfides [e.g., Ph_2S_2 or $(4\text{-}^tBu\text{-}2\text{-}MeC_6H_3)_2S_2$]. Pyrrolidine, indoline, phthaloylhydrazide and 4-phenylurazole were not sensitive to these aryltellurinic anhydrides [93]. In an independent work, *Ogura* and *co-workers* [94] described the $[(4\text{-}MeOC_6H_4)Te(O)]_2O$ (**78**) as a mild oxidizing agent towards thiol, phosphine, thioamide, thiourea, thionoester, benzylic alcohol, etc. (products in 60–95 % yields) (Figure 3.47). More recently, a comparative study on oxidizing properties of two other tellurinic acid anhydrides, i. e. $[(4\text{-}HOC_6H_4)Te(O)]_2O$ (**82**) and $[(3\text{-}Me\text{-}4\text{-}HOC_6H_4)Te(O)]_2O$ (**83**), was reported. The research was focused on the conversion of benzoin to benzil, PhC(O)C(O)Ph, 4-substituted benzyl alcohols to corresponding benzaldehydes, PhSH to Ph_2S_2, triphenylphosphine to $Ph_3P{=}O$ and hydroquinone to *p*-benzoquinone, and the obtained results indicated **83** as a better oxidizing agent (products in 85–95 % yields) than **82** (products in 72–85 % yields) [95].

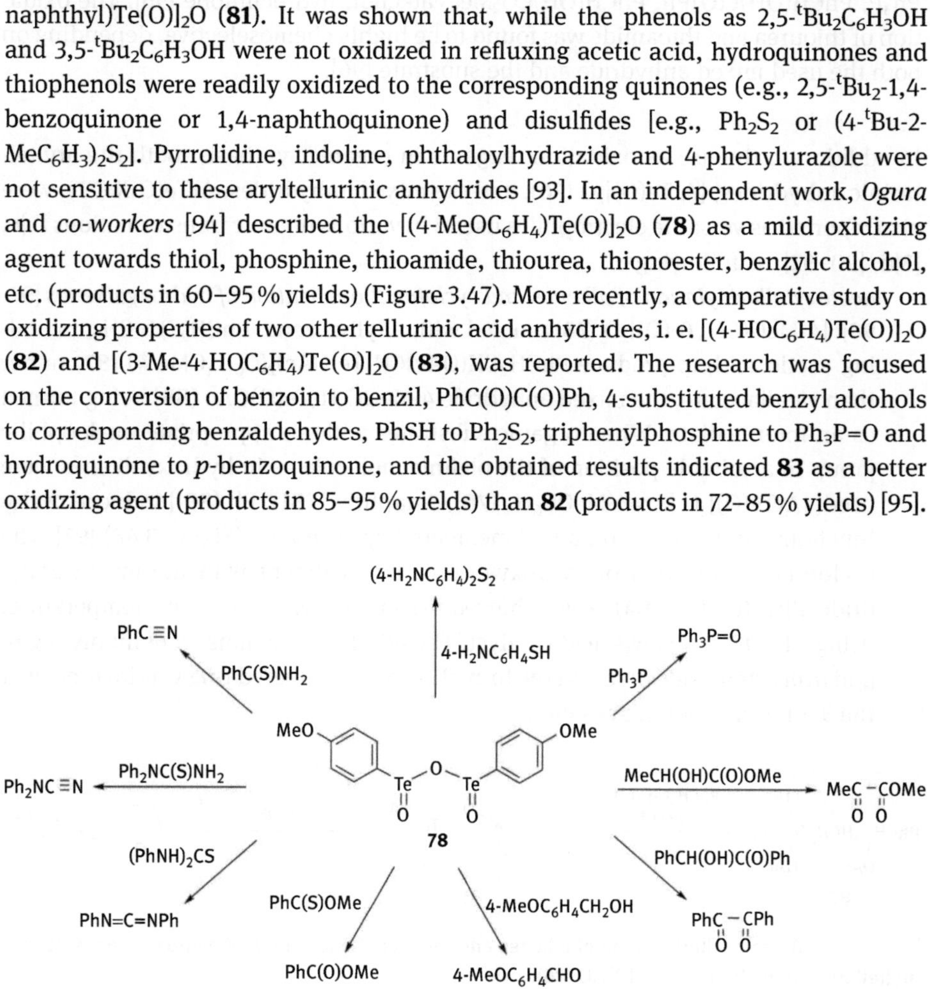

Figure 3.47: Oxidation reactions using the aryltellurinic anhydride $[(4\text{-}MeOC_6H_4)Te(O)]_2O$ (**78**) as stoichiometric oxidizing agent.

Although $[(4\text{-}MeOC_6H_4)Te(O)]_2O$ (**78**) does not generally react with the internal alkynes, RC≡CR′, the 1,2-diphenylethyne, PhC≡CPh, was oxidized to benzil, PhC(O)C(O)Ph, by equimolar amount of **78** and catalytic amount of H_2SO_4 [94].

Olefins were reported to be oxidized by $[PhTe(O)]_2O$ (**72**) in AcOH, under the catalysis of H_2SO_4, to give vicinal *syn*-diacetates and the benzenetellurenic acid derivatives formed *in situ* were suggested as the active species in the process [86].

As reported for aryltellurinic anhydrides, $[ArTe(O)]_2O$ (**76**) (*vide supra*), the related mixed anhydrides PhTe(O)X [X=OAc (**84**), $O(O)CCF_3$ (**85**), OTf (**86**)] were reported to act as mild oxidizing agents towards similar substrates, e.g., PhSH, Ph_3P, PhCH(OH)C(O)Ph, $PhCH(OH)CO_2Me$, catechol, hydrochinone [96]. The oxidation of thiourea and thioamide was found to be highly chemoselective, depending on both the used mixed anhydride and the substrate [96].

Electrophiles in organic transformations

In addition to the use as oxidation reagents in organic synthesis, both the aryltellurinic anhydrides, $[ArTe(O)]_2O$ (**76**), and the mixed anhydrides, ArTe(O)X (**77**), were also reported as versatile electrophiles useful in various organic transformations [3]c, [3]d, [90–92], as followings:

- *acetoxytellurinylation of alkenes and cyclofunctionalization of hydroxyalkenes* [97, 98]. $[4\text{-}MeOC_6H_4Te(O)]_2O$ (**78**) reacted with allylbenzene, $PhCH_2CH{=}CH_2$, in boiling acetic acid, to produce $PhCH_2CH(OAc)CH_2Te(OAc)_2(C_6H_4OMe\text{-}4)$ (**87**) rather than *vic*-diacetate, the mixed anhydride $4\text{-}MeOC_6H_4Te(O)OAc$ (**88**) being formed as an intermediate. When organic substrates with an appropriate nucleophilic group in a suitable position of the olefinic molecule as in hydroxyalkenes were used, $[ArTe(O)]_2O$ [Ar=4-$MeOC_6H_4$ (**78**), 2-naphthyl (**81**), Ph (**89**)] induced cyclofunctionalization to 5-, 6-, and 7-membered cyclic ethers (Figure 3.48) [97]. The cyclofunctionalization of hydroxyalkenes by treatment with the mixed anhydride PhTe(O)OAc (**84**) was achieved faster, in $CHCl_3$, at room temperature, using $BF_3{\cdot}OEt_2$ as Lewis acid catalyst. The addition reactions were highly regio- and *trans*-stereoselective and the formation of 5-membered ring was favored over the 3- or 4-membered ones [98].

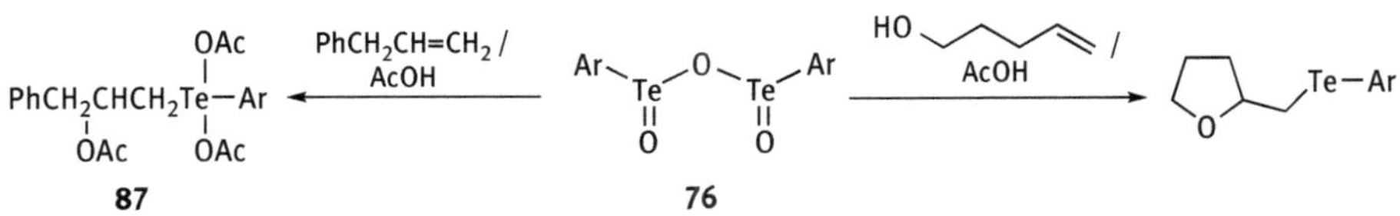

Figure 3.48: Acetoxytellurinylation of alkenes and cyclofunctionalization of hydroxyalkenes using aryltellurinic anhydrides, $[ArTe(O)]_2O$ (**76**).

- *aminotellurinylation of alkenes, cyclofunctionalization of alkenic carbamates, and one-pot synthesis of 2-oxazolidinones from alkenes* [99, 100]. A variety of olefins underwent regio- and stereoselective aminotellurinylation when treated with mixed anhydrides PhTe(O)X [X=OAc (**84**), $O(O)CCF_3$ (**85**), OTf (**86**)], in CH_2Cl_2 or $CHCl_3$, in presence of excess of ethyl carbamate, H_2NCO_2Et, as a nucleophile. The use of **84** or **85** required additional $BF_3{\cdot}OEt_2$, while the reaction with **86** worked in the absence of the Lewis acid. The tellurinylated reaction products were isolated as the corresponding phenyltellurenylated derivatives after reduction with hydrazine hydrate (Figure 3.49, a). The aminotellurinylation reaction was extended to intramolecular cyclofunctionalization of olefinic carbamates into nitrogen heterocycles (products in 73–96 % yields) (e.g., in Figure 3.49, b) [100]. Moreover, when the aminotellurinylation process of olefines was carried out at higher temperature, e.g., in refluxing 1,2-dichloroethane, 2-oxazolidinone derivatives were obtained directly in a high yield, this reaction thus providing a simple one-pot preparation method for compounds of this class directly from alkenes or cycloalkenes (e.g., in Figure 3.49, c) [99, 100].

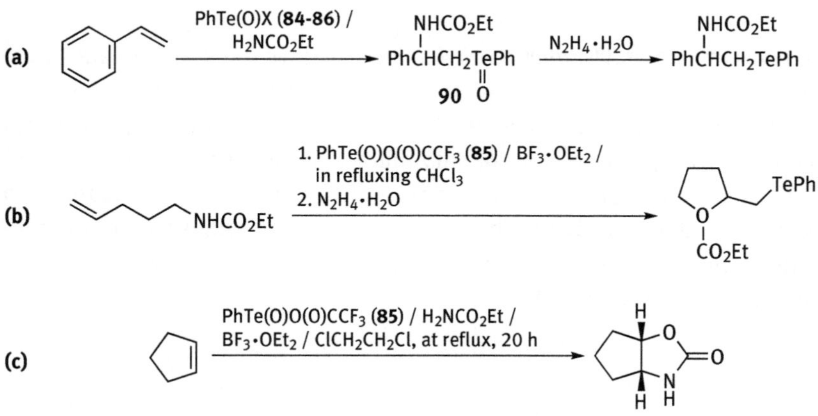

Figure 3.49: Aminotellurinylation of alkenes (a), cyclofunctionalization of alkenic carbamates (b) and conversion of alkenes to 2-oxazolidinones (c) using aryltellurinic mixed anhydrides **84–86**.

– *amidotellurinylation of alkenes and alkynes, and one-pot synthesis of 2-oxazolines from alkenes and oxazoles from alkynes* [101–104]. The mixed anhydride PhTe(O)O(O)CCF$_3$ (**85**), in the presence of BF_3·OEt_2, afforded at room temperature the amidotellurinylation of alkenes in MeCN, acting both as solvent and as nucleophile (Figure 3.50, a). The propionitrile and benzonitrile worked also as nucleophiles in such reactions. The resulting amidotellurinylation products could be converted at higher temperatures, following an *in situ* intramolecular substitution, into 2-oxazolines in high yields. These reactions are highly regio- and stereo-selective [101, 102]. Similarly, PhTe(O)OTf (**86**) undergoes (*E*)-stereoselective amidotellurinylation of alkynes in MeCN in the presence of a Brønsted acid, i. e. H_2SO_4. When terminal alkynes were used, the resulting addition products (*E*-**91** isomers) were found to isomerize thermally to the corresponding [(*Z*)-*β*-acetamidovinyl] phenyltellurium(IV) oxides (*Z*-**91** isomers), isolated as the corresponding tellurides. By contrast, the products obtained from internal alkynes were partially transformed into oxazoles *via* a spontaneous intramolecular cyclization at higher temperature (Figure 3.50, b) [103, 104].

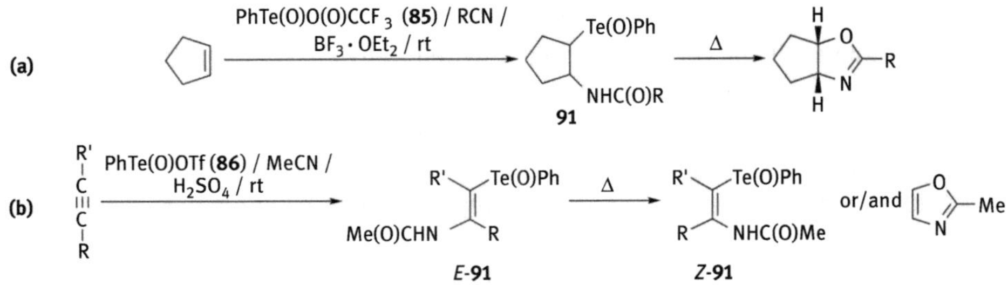

Figure 3.50: Conversion of alkenes to 2-oxazolines (a) and of alkynes to oxazole (b), respectively, *via* amidotellurinylation with aryltellurinic mixed anhydrides **85** and **86**.

– *one-pot conversion of allylsilanes to allylamines* [105]. Different allylsilanes were treated with $PhTe(O)O(O)CCF_3$ (**85**) in presence of $BF_3{\cdot}OEt_2$, in 1,2-dichloroethane at room or lower temperature, to give (allyl)(phenyl)telluroxides (**92**). Further *in situ* treatment with primary or secondary amines at higher temperature (65 °C) afforded the isolation of the corresponding allylamines. This transformation showed a high regioselectivity. Thus, γ-substituted allylsilanes gave γ-substituted allylamines with an arylamine, but generally α-substituted allylamines with an alkylamine (Figure 3.51).

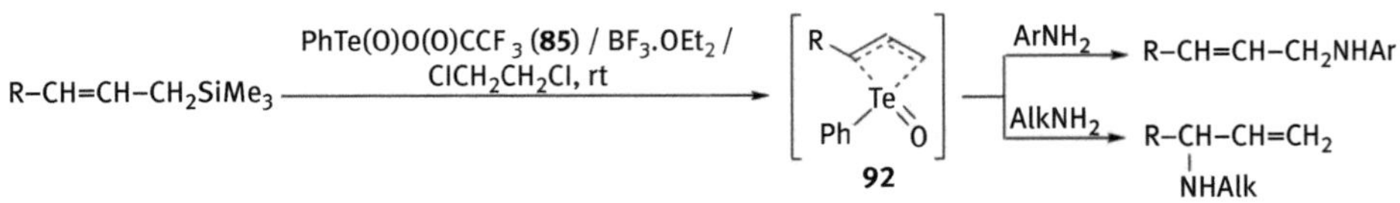

Figure 3.51: Conversion of allylsilanes to allylamines *via* phenyltellurinylation.

3.3.2.1.2 Catalysts in organic synthesis

In an earlier work it was reported the catalytic activity of a divinylbenzene-styrene copolymer-supported tellurinic acid (**93**) on the selective epoxidation of olefins with hydrogen peroxide (Figure 3.52) [106]. The activity of the Te-containing catalyst strongly depends on the degree of cross-linking in the divinylbenzene-styrene copolymer used for its preparation, the higher cross-linked polystyrene leading to higher conversions. The oxidation rate is olefin structure-dependent (e.g., increased alkyl substitution accelerates the rate) and the oxidation process is stereospecific, with retention of the *cis-trans* geometry in the epoxide product. It should be noted here that the polymeric structure of the Te-containing catalyst **93** is mandatory, with free tellurinic acids, i. e. $C_6H_5CH_2CH_2Te(O)OH$ (**94**) and $4\text{-}MeOC_6H_4Te(O)OH$ (**95**), no epoxidation being observed [106].

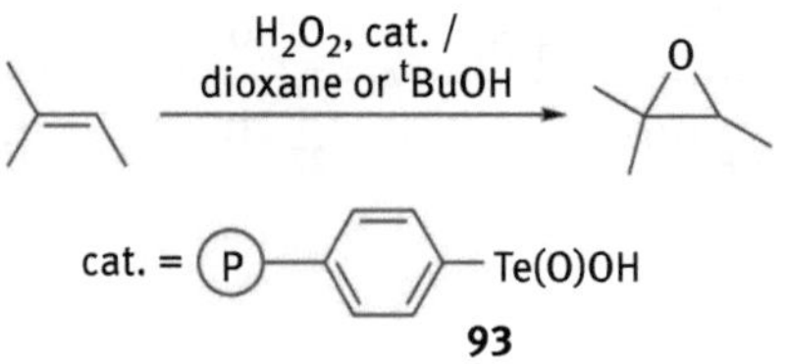

Figure 3.52: Copolymer-supported tellurinic acid (**93**) as catalyst for selective epoxidation of olefins with H_2O_2.

Recently, several diaryl ditellurides (Figure 3.53), in catalytic amounts, were reported to accelerate the oxidation of NaBr with H_2O_2 to electrophilic bromine species ($HOBr/Br_2$), in buffered aqueous solution. The resulting "Br^+" species prepared *in situ* were trapped by a series of alkenoic acids and activated aryl compounds [107].

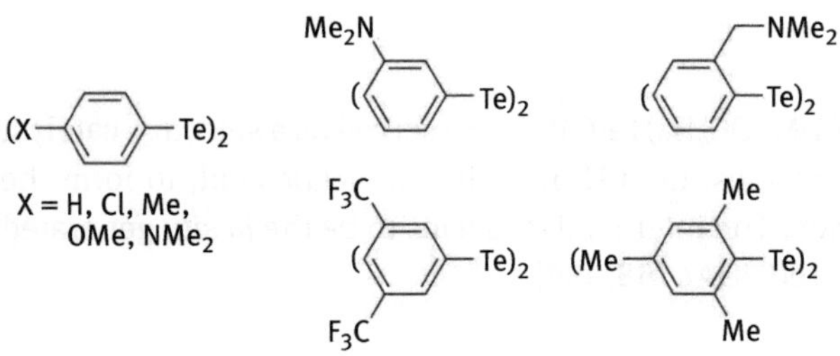

Figure 3.53: Diaryl ditellurides used as additives in catalytic amounts for activation of H_2O_2.

The active catalysts were proved to be in fact the corresponding aryltellurinic acids, ArTe(O)OH (**75**), generated *in situ* by the reaction of Ar_2Te_2 with H_2O_2 (Figure 3.54) [107]. The unsubstituted phenyltellurinic acid, PhTe(O)OH (**96**), was found to be the best catalyst (e.g., addition of Ph_2Te_2 in 0.20 mol % relative to substrate promoted a 240-fold increase in the rate of oxidation of NaBr with H_2O_2, as measured by the bromination of 4-pentenoic acid) and its use afforded bromolactonization of a series of alkenoic acids (products in 78–84 % isolated yields) and monobromination of various activated aromatic substrates (products in 92–95 % isolated yields). Both electron-donating and electron-withdrawing substituents on the phenyl ring of the catalyst diminished the rate of reactions. The stereoelectronic effects can affect the rate of bromination by influencing (*i*) the rate of oxidation of Ar_2Te_2 to the tellurinic acid, ArTe(O)OH (**75**), (*ii*) the rate of reaction of H_2O_2 and NaBr with the tellurinic acid, and (*iii*) the rate of oxidation of the tellurinic acid **75** to the telluronic acid, $ArTe(O)_2(OH)$ (**97**), the latter acting as a catalytic termination step since they were shown to be not capable to activate the peroxide. The balancing act also includes relative rates for conversion of the pertellurinic acid, ArTe(O)OOH (**98**) to the telluronic acid **97** or for reaction of the pertellurinic acid **98** with bromide (Figure 3.54) [107].

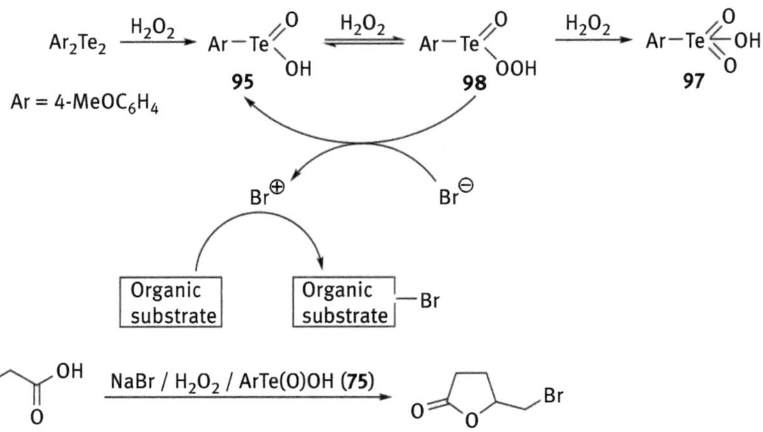

Figure 3.54: Proposed mechanism for the diarylditelluride/aryltellurinic acid-catalyzed oxidation of halides salts with H_2O_2.

The tellurinic anhydride $[(4\text{-MeOC}_6H_4)Te(O)]_2O$ (**78**) served as a selective catalyst for the hydration of terminal alkynes, RC≡CH, in refluxing acetic acid, to form the corresponding ketones, RC(O)Me. The intermediate seems to be the *in situ* generated mixed anhydride $(4\text{-MeOC}_6H_4)Te(O)OAc$ (**88**) [94].

3.3.2.2 Heterocyclic tellurinate esters

The heterocyclic tellurinate ester **99** (Figure 3.55) was evaluated as mimetic of the selenoenzyme glutathione peroxidase. Its superior catalytic properties compared to the selenium(IV) analog was proved in a model system based on the reduction of *tert*-butyl hydroperoxide or hydrogen peroxide with benzyl thiol, the rate of the reaction being measured by monitoring the formation of dibenzyl disulfide [37]d.

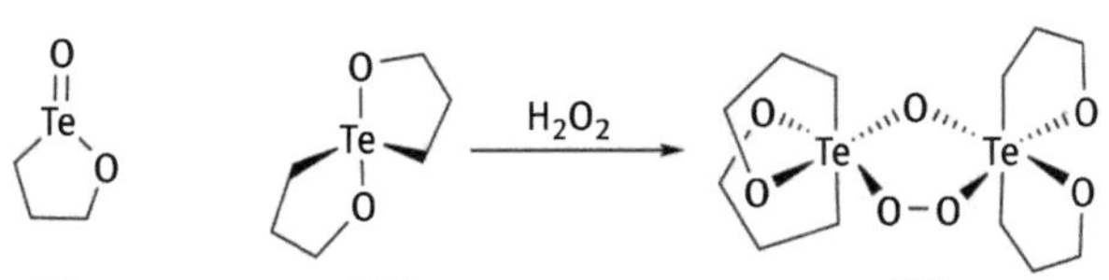

99 100 101

Figure 3.55: Heterocyclic organotellurium(IV) compounds evaluated as GPx mimetics.

3.3.2.3 Diorganotelluroxides

The reactivity of diorganotelluroxides, R_2TeO, differs considerably from that of the lighter congeners R_2SeO due to a higher polarity of the Te=O double bond compared to the corresponding Se=O double bond.

3.3.2.3.1 Reagents in organic synthesis

Diaryltelluroxides, Ar_2TeO (**102**) (e.g., Ar=Ph, 4-$MeOC_6H_4$, 4-$Me_2NC_6H_4$, Mes, Tip, etc.), were reported as mild and selective oxidants for various compounds, including organic substrates [3]c, [108, 109]. Thus, early works of *Barton* and *co-workers* described the oxidative properties of $(4\text{-}MeOC_6H_4)_2TeO$ (**103**) [93, 109, 110], a compound which is easily regenerated from the parent telluride produced during the oxidation reactions. It was reported that the telluroxide **103** can oxidize thio- and selenocarbonyl compounds into their oxo analogues. Thus, **103** reacted with derivatives like thioketones, thio- or selenoesters, dithioesters, xanthates, thiocarbonates, thioamides, thioureas, catechol, arylhydrazines, phosphines, thiols, etc. to give their corresponding oxo derivatives or disulfides, respectively (Figure 3.56) (for a review of these reactions, see Ref. [3]c). However, **103** was inert to a wide range of functional groups including ketones, phenols, amines, oximes, isonitriles etc. suggesting its potential use for selective oxidation of compounds. Exceptions are also reported, e.g., telluroxide **103** react with the bulky 2,6-di-tert-butylphenol to afford, following a carbon-carbon coupling reaction, the unexpected 3,3′,5,5′-tetra-tert-butyl-[1,1′-bi (cyclohexylidene)]-2,2′,5,5′-tetraene-4,4′-dione in 93 % yield, the same quinone being obtained by oxidation of the corresponding diphenohydroquinone (94 % yield) [111]. On the other hand, the telluroxide **103** was also able to produce the cleavage of the C–C bond in PhCH(OH)–CH(OH)Ph which afforded benzaldehyde in 75 % yield [112].

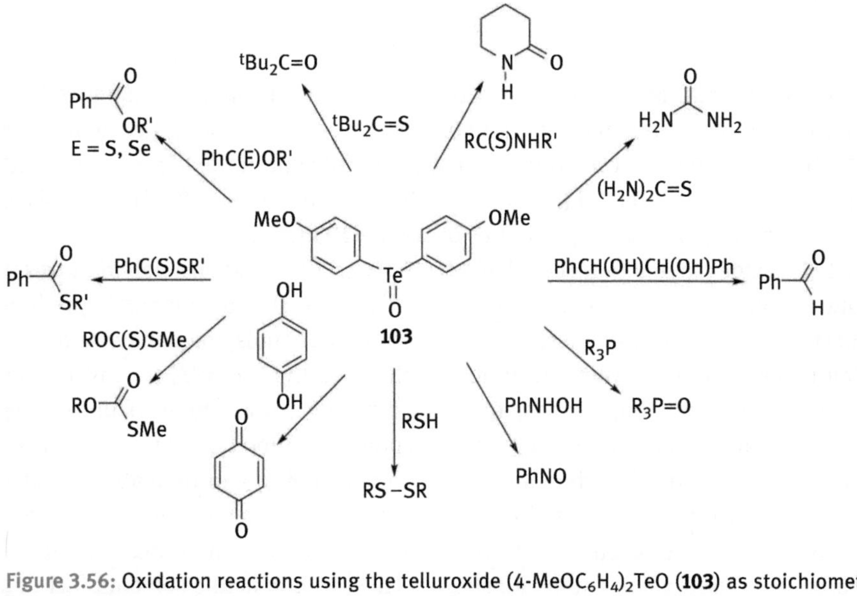

Figure 3.56: Oxidation reactions using the telluroxide $(4\text{-}MeOC_6H_4)_2TeO$ (**103**) as stoichiometric oxidizing agent.

The telluroxide **103** was reported either to give no reaction with simple alcohols [110] or to oxidize 4-bromobenzyl alcohol to 4-bromobenzaldehyde in a very low yield (12 %) after 3 h in refluxing xylene [113]. In similar reaction conditions the same alcohol was oxidized to the corresponding aldehyde in only 16 % yield by Ph_2TeO (**104**) but in 94 and 96 % yields by Mes_2TeO (**105**) and Tip_2TeO (**106**), respectively. It should be mentioned that benzylic alcohol, several of its 4-substituted derivatives with electron-donating and -withdrawing groups, as well as allylic alcohols were also efficiently oxidized to the corresponding aldehydes in good to excellent yields (70–100 %) by **104** and **106** in refluxing xylene. However, non-activated alcohols could not be oxidized, even at a higher temperature. It was suggested that the enhanced oxidizing ability of these two bulkier telluroxides might be corelated to the fact that they can exist in a monomeric form [113]. The mechanism proposed for the oxidation of alcohols by diaryltelluroxides **102** involved the formation of an adduct **107** which undergoes intramolecular dehydration (experimentally, the azeotropic removal of water was essential for an increased yield of conversion) to give the corresponding carbonyl compound (Figure 3.57).

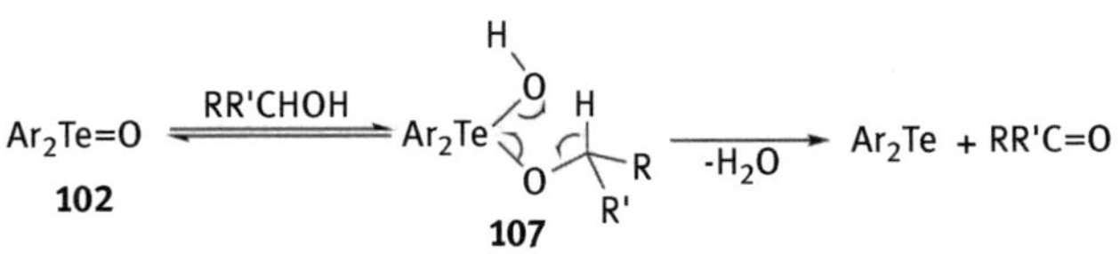

Figure 3.57: Proposed mechanism for oxidation of alcohols to carbonyl compounds using diaryltelluroxides **102**.

The bulky telluroxide Mes_2TeO (**105**) converted 77 % of triphenylphosphine into triphenylphosphine oxide after 2 h, in $CHCl_3$ at room temperature, while for $(4\text{-}MeOC_6H_4)_2TeO$ (**103**) only 13 % yield of triphenylphosphine oxide was detected in the reaction mixture after 2 h (80 % conversion required about 48 h [109]). This is indicative of the better oxidant properties of **105** *vs.* **103** [113].

The telluroxide reagent $(4\text{-}MeOC_6H_4)_2TeO$ (**103**) was anchored on a polymeric resin substrate to give polystyrene-bound diaryl telluroxide (**108**). This polymeric product was inocent towards simple amines, amides, alcohols or phenols, but behave as a mild oxidizing reagent towards other substrates. Similarly to the free **103**; it converted at room temperature, with high yields, thiols to disulfides, phosphine to phosphine oxides, hydroquinone and catechol to the corresponding quinones, as well as thioketones, thioesters and trithiocarbonates to the corresponding oxo derivatives, using CH_2Cl_2, $CHCl_3$ or acetic acid as reaction solvent. In addition, solvent-dependent reactions were reported to occur when thioamides, 1,3-diphenylthiourea or 1-phenylthiourea were used as oxidation substrates. Thus, in non-acidic solvents (CH_2Cl_2, $CHCl_3$) thioamides and 1,3-diphenylthiourea were desulfurated to give nitriles or the corresponding urea. When acetic acid was used as acidic solvent, thioamides were converted into 3,5-diorgano-1,2,4-thiadiazoles, and 1-phenylthiourea into 5-imino-N,4-diphenyl-4,5-dihydro-1,2,4-thiadiazol-3-amine (Figure 3.58) [114].

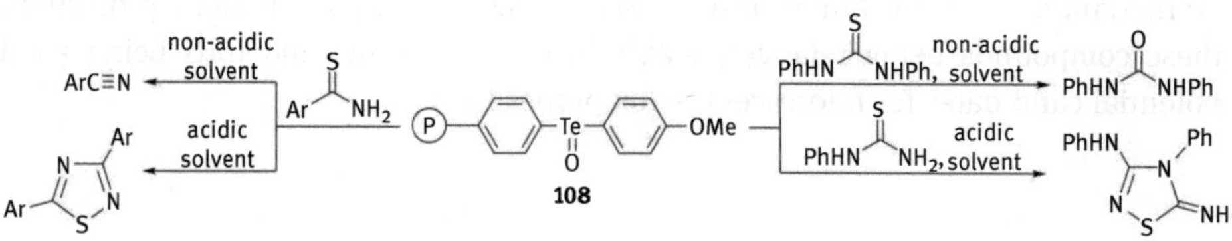

Figure 3.58: Solvent-dependent transformations of thioamides, 1,3-diphenylthiourea or 1-phenylthiourea using polystyrene-bound $(4\text{-}MeOC_6H_4)_2TeO$ (**108**).

The polymeric product **108** show important advantages over the isolated **103** species, e.g., easier product work-up, the reduced functionalized polymer being recovered just by filtration, and multiple recyclic use without decrease in activity. In addition, **108** requires shorter reaction times and milder experimental conditions compared to the related polymer-supported selenoxide [114].

A recent report described the unusual reaction of the intramolecularly coordinated $[1\text{-}(8\text{-}Me_2NC_{10}H_6)]_2TeO$ (**109**) species with acetonitrile in which the oxygen atom of the telluroxide molecule was transfered to a carbon atom to afford the novel zwitterionic diaryltelluronium(IV) acetimidate $[1\text{-}(8\text{-}Me_2NC_{10}H_6)]_2TeNC(O)Me$ (**110**) in moderate (57 %) yield. Hydrolysis of this acetimidate derivative with a HCl solution gave acetamide and the diarylhydroxytelluronium(IV) chloride $[\{1\text{-}(8\text{-}Me_2NC_{10}H_6)\}_2Te(OH)]Cl$ (**111**) (Figure 3.59) [115]. Preliminary results revealed that the reactions of hypercoordinated **109** with propionitrile and benzonitrile followed the same pathway to afford analogous acetamide products, the hydrolysis of which gave the corresponding propioamide and benzamide. The lack of intramolecular N→Te interactions in other Ar_2TeO (**102**) (Ar=Ph, $4\text{-}MeOC_6H_4$, $4\text{-}Me_2NC_6H_4$, Mes) seems to prevent similar reactions of these diaryltelluroxides with nitriles [115].

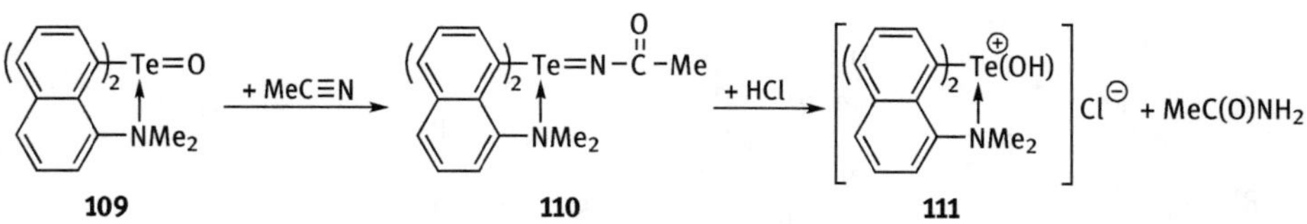

109 **110** **111**

Figure 3.59: Conversion of acetonitrile into acetamide following its reaction with the hypervalent telluroxide $[1\text{-}(8\text{-}Me_2NC_{10}H_6)]_2TeO$ (**109**).

The telluride/telluroxide pairs based on the *Te*-rhodamine core are mild oxidizing agents for the conversion of thiols to disulfides [116–118]. Such compounds were investigated as fluorescent redox probes for redox cycles in living cells [116, 117]. *Detty* and *co-workers* reported that the fluorescent tellurorosamine telluroxides **112** and **113** (Figure 3.60) are much more stable at physiological pH than the related species with phenyl or 2-methylphenyl substituents on the position 9 of the *Te*-rhodamine core. The bulky mesityl group in **112** and **113** provide steric protection, these compounds being relatively stable in protic solvents and thus being good potential candidates for fluorescent redox probes [117].

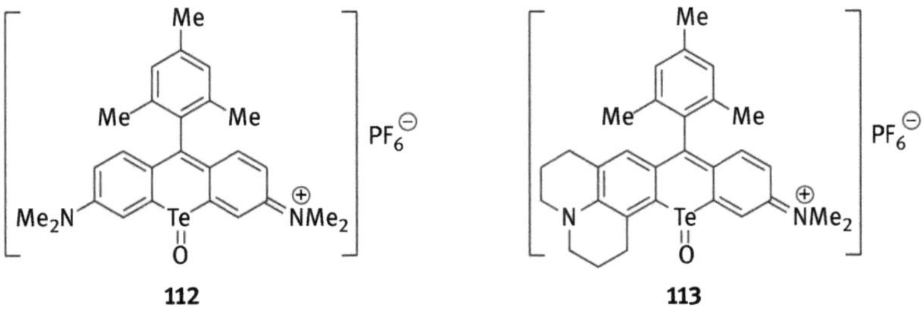

Figure 3.60: Examples of fluorescent tellurorosamine telluroxides.

When treated with acetylenic dienophiles the isotellurazole *Te*-oxides (**114**) can be easily converted, under mild reaction conditions, in low to moderate yields (22–77 %), into polysubstituted pyridines in a high regioselective manner *via* hetero-Diels–Alder reactions and the subsequent tellurium extrusion from the intermediary cycloadducts (**115**) (Figure 3.61) [119]. The reactions with dimethyl acetylenedicarboxylate occurred in CH_2Cl_2 solution even at room temperature. When unsymmetrical acetylenic dienophiles bearing an electron-withdrawing group (EWG) (R^3=EWG in Figure 3.61) were used the [4 + 2] reactions afforded, in a highly regioselective manner, polysubstituted pyridines bearing the EWG at the C-3 position.

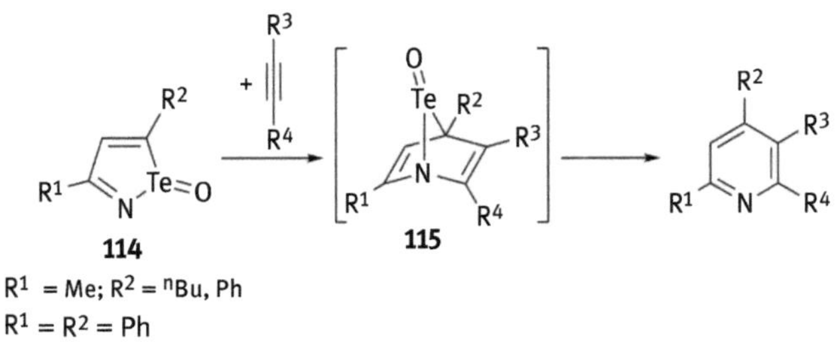

Figure 3.61: Synthesis of polysubstituted pyridines from isotellurazole *Te*-oxides (**114**) and acetylenic dienophiles.

3.3.2.3.2 Catalysts in organic synthesis

Diaryltelluroxides were reported as efficient catalytic species for various reactions. Earlier studies reported the oxidation of thiocarbonyl compounds to their oxo analogs using $(4\text{-MeOC}_6H_4)_2TeO$ (**103**) in a catalytic system with 1,2-dibromotetrachloroethane, in a two-phase system, at room temperature [109, 120]. The organic halide acted as a brominating agent for diaryltellurium(II) species, the resulting Ar_2TeBr_2 being then hydrolyzed to the telluroxide in the presence of aqueous base, thus closing the catalytic cycle. The telluroxide **103**, generated electrochemically from $(4\text{-MeOC}_6H_4)_2Te$ in the presence of water, reacts with thiobenzamide to afford the corresponding nitrile, PhCN [121]. It should be mentioned that when the electrolysis was performed in anhydrous conditions, $(4\text{-MeOC}_6H_4)_2Te$ reacted with the supporting electrolytes, i. e. $[Bu_4N]OAc$ or $[Et_4N]OTs$, to give $(4\text{-MeOC}_6H_4)_2Te(OAc)_2$ (**116**) and $(4\text{-MeOC}_6H_4)_2Te(OTs)_2$ (**117**) which behave different towards the same organic substrate, i. e. affording benzonitrile or 1,2,4-thiadiazole, respectively [121].

Diaryltelluroxides with bulky aromatic substituents, Ar_2TeO (**102**) (Ar=Ph, 4-$MeOC_6H_4$, Mes, Tip), are efficient oxygen carrier in aerobic photooxidation of phosphite esters to phosphates [122] and of triorganosilanes to triorganosilanols [123]. In these cases diaryltellurides were used as pre-catalysts being oxidized in photosensitized conditions by molecular oxygen which played the role of terminal oxidant (Figure 3.62). The reactivity of the phosphite esters was dependent on the phosphorus substituents. It was established that the efficiency of the catalyst was increased and completion was reached when the oxidation reaction of phosphite esters was carrying out at 40–45 °C in acetonitrile [122]. Best results for the oxidation of silanes to silanols were achieved using the system Mes_2Te / hematoporphyrin (as photosensitizer) in pyridine as solvent, at 50–70 °C. An optically active silane was stereoselectively oxidized to the corresponding silanol with retention of the configuration. It should be noted the excellent selectivity in favor of silanols (90 % to quantitative yields) and the absence of the corresponding disiloxanes (less than 1%). The use of (4-$MeOC_6H_4)_2Te$ and Ph_2Te as a catalyst failed to afford acceptable yields of silanols [123].

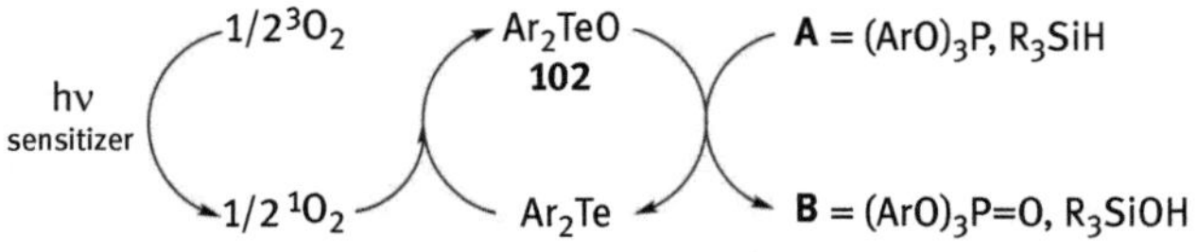

Figure 3.62: Proposed catalytic cycle for aerobic Ar_2Te-catalyzed oxidation of phosphite esters to phosphates and of triorganosilanes to triorganosilanols, respectively.

The activation of H_2O_2 in various oxidation processes of organic substrates, including the oxidation of X^- (X = halogen) to X_2 or HOX [124], and thiols to disulfides [38]d, [118, 125], was performed with diorganotellurides as pre-catalysts. The oxidation reactions with hydrogen peroxide typically proceed with the intermediary formation of the corresponding diorganotelluroxide or -dihydroxytellurane, followed by reductive elimination.

Recently, *Mugesh* and *co-workers* reported the synthesis of stable spirodiazatellurane by the oxidation of 2,2′-tellurobis(benzanilide) with H_2O_2/^tBuOOH, *via* the intermediary formation of diorganotelluroxide [125]d. The antioxidant activity of the new spirodiazatellurane (**118**) was evaluated in the transformation of $PhCH_2SH$ to $(PhCH_2)_2S_2$ and it was observed a higher catalytic activity as GPx mimic in comparison with the spirodiazaselenurane analogue, the latter exhibiting a similar activity as ebselen. The increased efficiency of the spirodiazatellurane as GPx mimic was attributed to an efficient redox cycle which involved the telluride **119** and the telluroxide **120** intermediary formed in the catalytic process (Figure 3.63). Moreover, it was noticed that at lower thiol concentrations the telluroxide **120** undergoes a reversible spirocyclization, which may protect the tellurium moiety from over oxidation to tellurium(VI), while at a large excess of thiol (> 10 equiv.) this reversible process is suppressed.

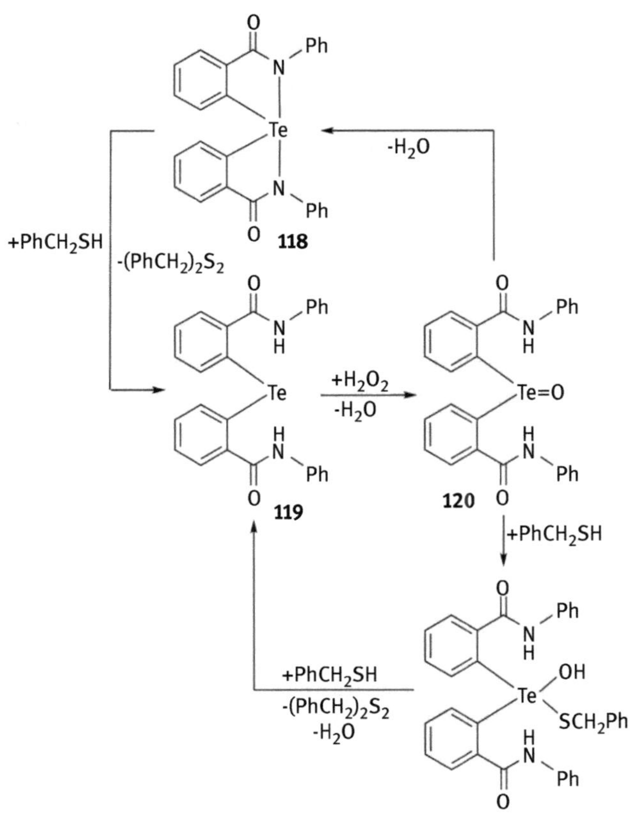

Figure 3.63: Proposed catalytic mechanism for the reduction of H_2O_2 by benzyl thiol using the spirodiazatellurane **118** as catalyst.

Thiols were efficiently oxidized to disulfides also by singlet oxygen photocatalytically produced under visible-light irradiation, in an aerobic process using as pre-catalyst either $(4\text{-}MeOC_6H_4)_2Te$ [126] or the 9-mesityl-3,6-bis(dimethylamino)telluroxanthylium hexafluorophosphate, a *Te*-rhodamine-based chromophore [118]. In the first case the active species was the telluroxide $(4\text{-}MeOC_6H_4)_2TeO$ (**103**) in aprotic solvents, while in organic solvents the oxidation occurred both through Ar_2TeO and Ar_2TeO_2 species (*vide infra* subchapter 3.3) [126]. The bulky tellurorhodamine oxide **113** was the effective species in the latter process [118].

The telluroxide $(4\text{-}MeOC_6H_4)_2TeO$ (**103**) was reported to act as a catalyst for aldol condensation [108]. In an earlier work *Engman* and *Cava* reported the condensation of aromatic aldehydes with active methylene compounds under mild aprotic conditions using **103** as a selective base catalyst. However, aldol condensations with very acidic methylene compounds gave only moderate yields [127]. The telluroxide **103** was also reported as catalyst for the preparation in moderate to excellent yields of (*i*) *α,β*-unsaturated ketones and 2,4-alkadienones through Claisen-Schmidt condensation [128]a, (*ii*) *α,β*-unsaturated nitriles through condensation of benzaldehydes and activated nitriles [128]b, and (*iii*) *α*-cyano-*α,β*-unsaturated sulfones by condensation of phenylsulfonylacetonitrile with aromatic aldehydes [128]c. In addition to **103**, several other diorganotelluroxides, i. e. Ar_2TeO (Ar=Ph, Mes, $3\text{-}MeOC_6H_4$, $4\text{-}Me_2NC_6H_4$) as well as the mixed-substituent species $(4\text{-}MeOC_6H_4)(4\text{-}Me_2NC_6H_4)TeO$ and MePhTeO, were also investigated as catalysts for aldol condensation [129]. Based on the condensation results obtained using the benzaldehyde-tetralone system, it was suggested that the catalytic efficiency of diorganotelluroxides is dependent on the substituent grafted on the aryl groups attached to tellurium. Best results were obtained with electron donating substituents, like methoxy or dimethylamino groups, which enhanced the basicity of the telluroxide oxygen *via* a resonance effect through the tellurium atom [129].

In recent studies the telluroxides $(4\text{-}HOC_6H_4)_2TeO$ (**121**) and $(3\text{-}Me\text{-}4\text{-}HOC_6H_4)_2TeO$ (**122**) were also investigated as catalysts in the synthesis of chalcones by aldol condensation of (*i*) benzaldehyde with acetophenone and 4-bromoacetophenone, and (*ii*) 4-anisaldehyde with acetophenone, 4-methylacetophenone, 4-chloroacetophenone and 4-bromoacetophenone, respectively. The higher catalytic activity of **122** compared to **121** was explained by the higher basicity of the former [130]. Similarly, 1-(4-bromophenyl)-3-(3,4-dimethoxyphenyl)-2-propene-1-one was obtained by the catalytic aldol condensation between 3,4-dimethoxybenzaldehyde and 4-bromo-acetophenone in the presence of $(4\text{-}MeOC_6H_4)_2TeO$ (**103**) as catalyst [131].

3.3.2.4 Diorganotellurium dihydroxides

Diorganotellurium(IV) dihydroxides oxidize in mild conditions thiols to disulfides [125]a, [132, 133] as well as halides to halogens and hydrohalous acids [124]c which are able to further oxidize various organic substrates. Generally, active $RR'Te(OH)_2$

species are generated *in situ* from diorganotellurides and hydrogen peroxide, in a two steps process involving the formation of the telluroxide which is subsequently hydrolyzed to the corresponding diorganotellurium(IV) dihydroxides.

In earlier works *Detty* and *co-workers* showed that tellurapyrylium dyes act as catalysts for the oxidation with hydrogen peroxide of thiophenol and leucodyes (Figure 3.64) [134, 135]. Similarly with the diaryltellurides, the tellurapyrylium dyes are easily oxidized by H_2O_2 to the corresponding telluroxides, the latter forming by hydrolysis heterocyclic dihydroxytelluranes (**123**), which are the active species as long as the unstable diorganotellurium(IV) dihydroxides undergo a thermal reductive elimination of hydrogen peroxide, thus sustaining the continuous oxidation of the organic substrate.

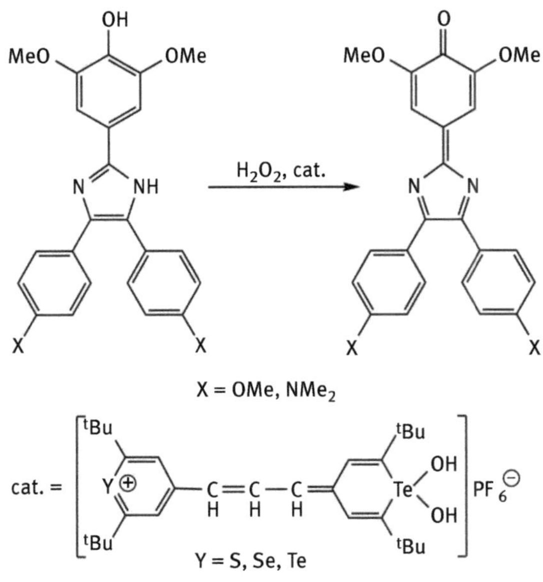

123

Figure 3.64: Oxidation of leucodyes with H_2O_2 using a dihydroxytellurane derivative as catalyst.

Recently, the same group employed a protocol for the oxidation of inorganic halides to oxidizing hypohalous acids and halogens, based on a dialkyltellurium(II) pro-catalyst, namely 3-(n-hexyltelluro)-1-propanol, sequestered on an APTES/TEOS xerogel [136]. The procedure was developed with the aim to prepare an active antifouling protective film containing the RR'Te pro-catalyst. The reported mechanism (Figure 3.65) might be exploited for various oxidation processes in organic synthesis also, as the tellurium-sequestered reagents were observed to display a similar reactivity as that one manifested by other previously reported diorganotellurides in solution.

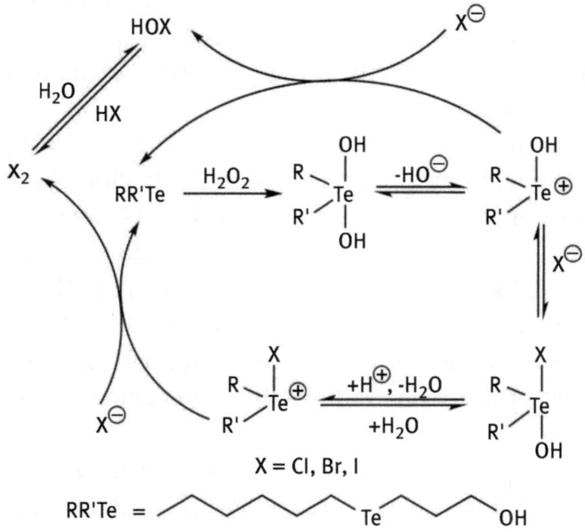

Figure 3.65: Proposed catalytic cycle for diorganotelluride-catalyzed oxidation of halide salts with H_2O_2.

3.3.3 Organotellurium(VI) compounds

Bis(4-methoxyphenyl)tellurone, $(4\text{-}MeOC_6H_4)_2TeO_2$ (**124**), was firstly used as a mild oxidizing reagent by *Cava* and *Engman* to convert benzenethiol to diphenyl disulfide (97 % yield), hydroquinone to p-benzoquinone (39 % yield) and benzoin to benzil (89 % yield) [112]. In contrast to the related telluroxide $(4\text{-}MeOC_6H_4)_2TeO$ (**103**), which was reported to be inert towards alcohols [110], several benzylic alcohols were oxidized by tellurone **124** to the corresponding aldehydes (70–80 % yield) [112]. Surprisingly, oxidation of hydrobenzoin with tellurone **124** resulted in cleavage of a C–C bond to produce benzaldehyde (79 % yield) [112]. Recently, *Ando* and *co-workers* obtained a more effective mild tellurone oxidant, namely the monomeric bis(2,4,6-triisopropylphenyl)tellurone, $(2,4,6\text{-}^iPr_3C_6H_2)_2TeO_2$ (**125**) [137]. This species was successfully employed in the oxidation of various primary alcohols, including benzyl- and benzyl-substituted alcohols, *trans*-$PhCH{=}CHCH_2OH$, *trans*-2-dodecenol, $PhCH_2CH_2CH_2OH$ and 1-dodecanol, when the corresponding aldehydes were obtained with excellent to quantitative yields. While non-activated secondary alcohols, like 2-dodecanol, were resistant to oxidation, other secondary alcohols, e.g., PhCH(OH)Me and $PhCH(OH)CO_2Me$, were converted with good yields (87 and 77 %, respectively) in the corresponding ketones by using the same tellurone oxidant. An enhanced reactivity of the diorganotellurones in comparison with the diorganotelluroxides was generally observed. Thus, it should be noted that the telluroxides $(2,4,6\text{-}^iPr_3OC_6H_4)_2TeO$ (**126**) could not oxidize non-activated alcohols, even at high temperature [137]. Under similar conditions, the intramolecularly coordinated tellurone $[1\text{-}(8\text{-}Me_2NC_{10}H_6)]_2TeO_2$ (**127**) has shown oxidation properties towards alcohols [138], this difference in reactivity being tentatively attributed to the intramolecular N→Te interactions in the latter.

The tellurone **124** oxidized quantitatively Ph_3P to Ph_3PO in refluxing $CDCl_3$, with recovery of the corresponding telluride [137]. The oxidant behavior of the 1:1 adduct formed in the reaction of the aryltellurinic anhydride $[(ArTeO)_2O]_n$ and the diaryltellurone Ar_2TeO_2, i. e. $[\{(ArTeO)_2O\}_2(Ar_2TeO_2)_2]$ (**128**, Ar=2,6-diethylphenyl), was also evidenced by converting Ph_3P to Ph_3PO, in $CDCl_3$ at room temperature, along with quantitative formation of Ar_2Te_2 and Ar_2Te, respectively [139]. It should be noted that due to the bulkiness and high stability of the adduct **128** the completeness of this oxidation reaction was achieved only after 6 days.

The oxidation of diaryltellurides to either Ar_2TeO (100 % product for Ar=Ph, $4\text{-}MeOC_6H_4$, Mes) or Ar_2TeO / Ar_2TeO_2 mixtures (Ar=$2,4,6\text{-}{}^iPr_3C_6H_2$, $2,6\text{-}Et_2C_6H_3$, $2,6\text{-}{}^iPr_2C_6H_3$, $2,6\text{-}Ph_2C_6H_3$) can be achieved with singlet oxygen, formed under photosensitized conditions, by using tetraphenylporphyrin, rose bengal or hematoporphyrin as photosensitizer [113, 122, 123, 126, 140]. It was observed that bulky aryl groups favour the formation of diaryltellurones *via* a nucleophilic pertelluroxide intermediate, $[Ar_2Te^+OO^-]$ [140]. The aerobic oxidation of thiols to disulfides was successfully realized by using various diaryltellurides, Ar_2Te (Ar=Ph, $4\text{-}MeOC_6H_4$, Mes, $2,4,6\text{-}{}^iPr_3C_6H_2$) as catalysts, under photosensitized conditions (77–100 % yields) [126]. The tellurides with bulkier aryl groups retarded the oxidation reaction rate and, in addition, induced the formation of some unidentified by-product. Under similar conditions, oxidation of aliphatic dithiols to cyclic disulfides was achieved in moderate yields (45–75 %) using $(4\text{-}MeOC_6H_4)_2Te$ and TPP or rose Bengal as sensitizers (Figure 3.66).

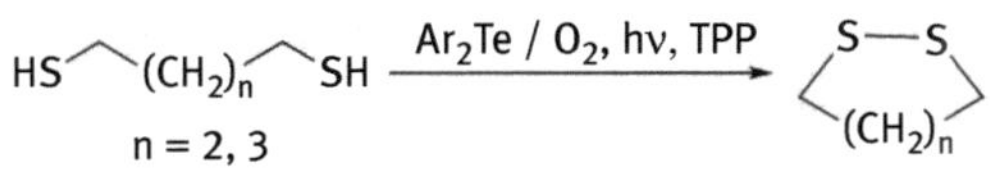

Figure 3.66: Aerobic oxidation of aliphatic dithiols to cyclic disulfides.

Detailed experimental studies based on $(4\text{-}MeOC_6H_4)_2Te$ suggested that in aprotic solvents the telluride is oxidized under photosensitized conditions to tellurone $(4\text{-}MeOC_6H_4)_2TeO_2$ (**124**), which is the effective oxidant in a two steps process (the corresponding diaryltelluroxide species acting as intermediates). By contrast, in protic medium the oxidation was realized only by the telluroxide $(4\text{-}MeOC_6H_4)_2TeO$ (**103**) (Figure 3.67).

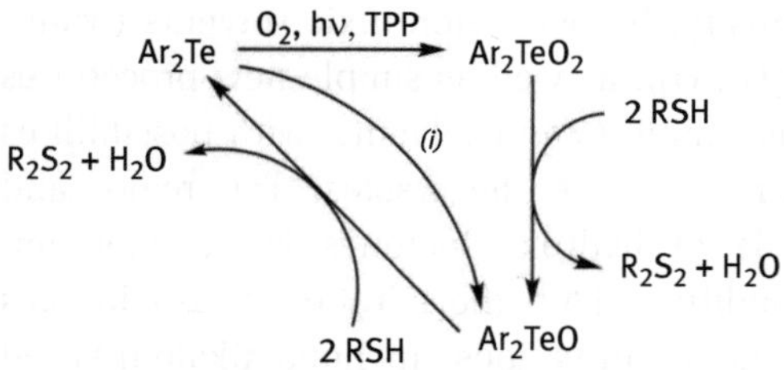

Figure 3.67: Proposed catalytic cycle for aerobic Ar_2Te-catalyzed oxidation of thiols to disulfides.

3.4 Conclusion and outlook

The last decades showed a tremendous interest in using organochalcogen (Se, Te) derivatives in organic synthesis, as starting materials, intermediates or catalysts. A pletora of organic transformations based on heavy organochalcogen compounds, mainly organoselenium derivatives, was developed, based on their flexible behavior in connection with a suitable organic substrate, as they can act as electrophilic, nucleophilic or radical reagents, generally in mild conditions, with easily tunable chemo-, regio- and stereoselectivity. The derivatives with E–O, E=O (E=Se, Te) or both types of bonds have a special place in organic synthesis as various processes occur via oxidation and ArEOH, RR′EO or ArE(O)OH elimination, even if compounds without oxygen in their structure are used as starting reagents, e.g., diorganochalcogenides or -dichalcogenides. Moreover, organochalcogen(IV) and -chalcogen(VI) derivatives are successfully employed in oxidation reactions of organic substrates. A continuous search for best synthesis conditions applied in order to obtain higher yields and a better control on the stereoselectivity in organic transformations led not only to new chiral organochalcogen species, but to new combinations of solvents, catalysts and oxidizing reagents. *m*CPBA, benzoyl peroxide, ammonium persulfate, ^tBuOOH, H_2O_2 or the $O_2/MeNO_2$ system were successfully used as *in situ* oxidizing agents for diorganodiselenides or diorganoselenides, the choice being correlated with their oxidizing strength, the additional reaction conditions and the target compounds. An increased tendency to use milder and green conditions led to improved methods based on hydrogen peroxide as oxidizing agent and moreover, to replace toxic organic solvents (toluene, methanol) with water or ionic liquids, where possible.

Taking into account the last achievements in the field, it clearly appears that the future concern in organic synthesis based on organoselenium and -tellurium derivatives is focused not only on developing new reagents, catalysts and strategies for higher yields and better stereoselectivities, but also on the tendency towards a green and sustainable chemistry, by using non-toxic solvents (water, ionic liquids) and oxidizing agents (H_2O_2, O_2), as well as simple new procedures based on recoverable starting materials, catalysts and solvents, with possibilities to recirculate them, and to significantly reduce the wastes. The regio- and diastereoselective continuous synthesis of hydroxy lactones by a selenium-mediated catalytic process in flow conditions [29], the addition of 2-oxindoles to vinyl selenone in ionic liquids [66], the use of various cinchona-alkaloid-based catalysts in Michael addition reactions to (aryl)(vinyl)selenones [63] or of hetero-genised catalysts in halogenation reactions [56, 136] are several examples of the recent achievements in organic synthesis with organochalcogens. The use of organochalcogen species in organic synthesis responds also to the increased interest to replace the expensive transition metal catalysts in certain processes [141], e.g., the recently reported activation of the carbon–nitrogen triple bond in nitriles by the telluroxide $(8\text{-}Me_2NC_{10}H_6)_2TeO$ [115] resembles the catalytic activity of the rhodium and platinum complexes in the hydrolysis of acetonitrile to acetamide [142].

Anyway, besides the organoselenium chemistry, which continues the increasing developing trend from the previous decades, the applications of organotellurium compounds in organic synthesis, despite several notable differences in their reactivity and the persistent characteristic smell, are expected to reach a more accelerated development in the future.

Abbreviations

Ac	acetyl
AcO	acetate
Alk	alkyl
Ar	aryl
AcOH	acetic acid
APTES/TEOS	3-aminopropyltriethoxysilane/tetraethoxysilane
bdmim	1-butyl-2,3-dimethylimidazolium
bmim	1-butyl-3-methylimidazolium
Bn	benzyl
Boc	*tert*-butyloxycarbonyl
bpy	4-methyl-N-butylpyridinium
^{n}Bu	*normal*-butyl
^{t}Bu	*tert*-butyl
cat.	catalyst
*m*CPBA	*meta*-chloroperoxybenzoic acid
DHS^{ox}	dihydroxyselenolane oxide
DHS^{red}	reduced dihydroxyselenolane
DMBz -	dimethoxybenzoyl
DMF	dimethylformamide

DMP	Dess–Martin periodinane
DOS	diversity oriented synthesis
FG	functional group
HMPA	hexamethylphosphoramide
Me	methyl
Mes	mesityl
MIRC	Michael initiated ring closure
MOM	methoxymethyl
Nu	nucleophile
TfO	triflate (trifluoromethanesulfonate)
TfOH	triflic acid
Ph	phenyl
^{i}Pr	*iso*-propyl
Py	pyridyl
RTILs	Room Temperature Ionic Liquids
TBDPS	*tert*-butyldiphenylsilyl
TIP	2,4,6-triisopropylphenyl
TIPDS	1,1,3,3-tetraisopropyldisiloxane
TMS	trimethylsilyl
OTs	tosylat

References

[1] See, for example: (a) Patai S, Rappoport Z, editors. Patai's chemistry of functional groups: the chemistry of organic selenium and tellurium compounds. Vol. 1. New York, NY, USA: John Wiley & Sons, Inc., 1986; (b) Patai S, editor. Patai's chemistry of functional groups: the chemistry of organic selenium and tellurium compounds. Vol. 2. New York, NY, USA: John Wiley & Sons, Inc., 1987; (c) Rappoport Z, editor. Patai's chemistry of functional groups: the chemistry of organic selenium and tellurium compounds. Vol. 3. Chichester, UK: John Wiley & Sons, Ltd., 2012; (d) Rappoport Z, editor. The chemistry of organic selenium and tellurium compounds. Vol. 4. Chichester, UK: John Wiley & Sons, Ltd., 2013; (e) Paulmier C. Selenium reagents and intermediates in organic chemistry. Tetrahedron Organic Chemistry. Vol. 4. (Baldwin JE, editor), Oxford, UK: Pergamon Press, 1986; (f) Wirth T, editor. Organoselenium chemistry: modern developments in organic synthesis. Topics in Current Chemistry. Vol. 208. Berlin, Germany: Springer Verlag, 2000; (g) Ogawa A. Selenium and tellurium in organic synthesis. In: Yamamoto H, Oshima K, editors. Main group metals in organic synthesis. Vol. 2. Weinheim, Germany: Wiley-VCH, 2004:813–66; (h) Woollins JD, Laitinen R, editors. Selenium and tellurium chemistry: from small molecules to biomolecules and materials. Berlin, Heidelberg, Germany: Springer-Verlag, 2011; (i) Wirth T, editor. Organoselenium chemistry – synthesis and reactions. Weinheim, Germany: Wiley-VCH, 2012; (j) Santi C, editor. Organoselenium chemistry: between synthesis and biochemistry. Saif, Sharjah, United Arab Emirates: Bentham Science Publications, 2014.

[2] See, for example: (a) Clive DLJ. Modern organoselenium chemistry. Tetrahedron. 1978;34:1049–132; (b) Mugesh G, du Mont W-W, Sies H. Chemistry of biologically important synthetic organoselenium compounds. Chem Rev. 2001;101:2125–80; (c) Freudendahl DM, Santoro S, Shahzad SA, Santi C, Wirth T. Green chemistry with selenium reagents: development of efficient catalytic reactions. Angew Chem Int Ed. 2009;48:8409–11; (d) Freudendahl DM, Shahzad SA, Wirth T. Recent advances in organoselenium chemistry. Eur J Org Chem. 2009:1649–64; (e) Godoi B, Schumacher RF, Zeni G. Synthesis of heterocycles via electrophilic

cyclization of alkynes containing heteroatom. Chem Rev. 2011:111:2937–80; (f) Sancineto L, Palomba M, Bagnoli L, Marini F, Santi C. Advances in electrophilic organochalcogen reagents. Curr Org Chem. 2016;20:122–35; (g) Santi C, Jacob RG, Monti B, Bagnoli L, Sancineto L, Lenardão EJ. Water and aqueous mixtures as convenient alternative media for organoselenium chemistry. Molecules. 2016;21:1482–98.

[3] See, for example: (a) Petragnani N, Stefani HA. Tellurium in organic synthesis: second, updated and enlarged edition. Oxford, UK: Academic Press, Elsevier Ltd, 2007; (b) Petragnani N. Tellurium. In: Abel EW, Stone FGA, Wilkinson G, editors. Comprehensive organometallic chemistry II. Vol. 11. Main-group metal organometallics in organic synthesis. Oxford, UK: Elsevier, 1995:571–601; (c) Petragnani N, Stefani HA. Advances in organic tellurium chemistry. Tetrahedron 2005;61:1613–79; (d) Srivastava K, Panda A, Sharma S, Singh HB. Telluroxanes: synthesis, structure and applications. J Organomet Chem. 2018;861:174–206.

[4] (a) Diels O, Karstens A. Über Dehydrierungen mit Selen (II. Mitteilung). Ber Dtsch Chem Ges B. 1927;60:2323–25; (b) Diels O, Gädke W, Körding P. Über die Dehydrierung des Cholesterins. (III. Mitteilung). Justus Liebigs Ann Chem. 1927;459:1–26; (c) Clemo GR, Dickenson HG. The action of selenium on compounds containing angular methyl groups. J Chem Soc. 1935;735–8; (d) House WT, Orchin M. A study of the selenium dehydrogenation of guaiol and related compounds. Selenium as a hydrogen transfer agent. J Am Chem Soc. 1960;82:639–42; (e) Silverwood HA, Orchin M. Aromatization reactions with selenium and aryl diselenides. J Org Chem. 1962;27:3401–4.

[5] Kacer F. (to I.G. Farbeindustrie AG) DE 557249, August 23, 1932.

[6] For selected reviews, see: (a) Waitkins GR, Clark CW. Selenium dioxide: preparation, properties, and use as oxidizing agent. Chem Rev. 1945;36:235–89; (b) Rabjohn N. Selenium dioxide oxidation. Org React. 1949;5:331–86; (c) Rabjohn N. Selenium dioxide oxidation. Org React. 1976;24:261–415; (d) Trachtenberg EN. Oxidation reactions with selenium dioxide. In: Augustine RL, editor. Oxidation: techniques and application in organic synthesis. Vol. 1, Ch. 3. New York: M. Dekker, 1969; (e) Fieser LF, Fieser M. Reagents for organic synthesis. Vol. 1. New York: Wiley, 1967: 992–1000; (f) Reich HJ. Organoselenium oxidations. In: Trahanovsky WS, editor. Oxidation in organic chemistry. Vol. 5C, Ch. 1. New York: Academic Press, 1978: 1–130; (g) Nicolaou KC, Petasis NA. Selenium in natural products synthesis, Ch. 2 and Ch. 3. Philadelphia, PA, USA: CIS Ins., 1984; (h) Thurow S, Penteado F, Perin G, Alves D, Santi C, Monti B, et al. Selenium dioxide-promoted selective synthesis of mono- and bissulfenylindoles. Org Chem Front. 2018;5:1983–91.

[7] (a) Bergman J, Engman L. Acetoxymethylation of benzene, toluene and related compounds with TeO_2/HOAc. Tetrahedron Lett. 1978;19:3279–82; (b) Bergman J, Engman L. Novel synthesis of 3-halogenobenzo[b]tellurophene-derivatives. Tetrahedron Lett. 1979;20:1509–10; (c) Bergman J, Engman L. Tellurium in organic synthesis, IX. Oxidation of some nitrogenous compounds with TeO_2. Z Naturforsch. 1980;35b:882–4.

[8] Siemens C. Ueber das selenmercaptan. Justus Liebigs Ann Chem. 1847;61:360–2.

[9] Jones DN, Mundy D, Whitehouse RD. Steroidal selenoxides diastereoisomeric at selenium; *syn*-elimination, absolute configuration, and optical rotatory dispersion characteristics. J Chem Soc D: Chem Commun. 1970;86–7.

[10] Mukherjee AJ, Zade SS, Singh HB, Sunoj RB. Organoselenium chemistry: role of intramolecular interactions. Chem Rev. 2010;110:4357–416.

[11] (a) Déziel R, Goulet S, Grenier L, Bordeleau J, Bernier J. Asymmetric selenomethoxylation of olefins involving a chiral C_2 symmetrical electrophilic organoselenium reagent. J Org Chem. 1993;58:3619–21; (b) Wirth T. Asymmetric reaction of arylalkenes with diselenides. Angew Chem Int Ed Engl. 1995;34:1726–8; (c) Déziel R, Malenfant E, Bélanger G. Practical synthesis of (*R,R*)- and (*S,S*)-bis[2,6-bis(1-ethoxyethyl)phenyl] diselenide. J Org Chem. 1996;61:1875–6;

(d) Déziel R, Malenfant E, Thibault C, Fréchette S, Gravel M. 2,6-Bis[(2*S*)-tetrahydrofuran-2-yl] phenyl diselenide: an effective reagent for asymmetric electrophilic addition reactions to olefins. Tetrahedron Lett. 1997;38:4753–6; (e) Wirth T, Fragale G. Asymmetric addition reactions with optimized selenium electrophiles. Chem Eur J. 1997;3:1894–902; (f) Bürgler FW, Fragale G, Wirth T. Stereoselective addition reactions with chalcogen electrophiles. ARKIVOC 2007;X:21–8.

[12] Wirth T, Fragale G. Stereoselective isoquinoline alkaloid synthesis with new diselenides. Synthesis. 1998;162–6.

[13] Yu L, Li H, Zhang X, Ye J, Liu J, Xu Q, et al. Organoselenium-catalyzed mild dehydration of aldoximes: an unexpected practical method for organonitrile synthesis. Org Lett. 2014;16:1346–9.

[14] Iwaoka M, Tomoda S, Kerns RJ, Wei P, Ramírez JS. Benzeneselenenyl trifluoromethanesulfonate. In: Fuchs P, Bode J, Charette A, Rovis T, editor(s). e-EROS – Encyclopedia of Reagents for Organic Synthesis. Hoboken, NJ, USA: John Wiley & Sons, 2005. DOI: 10.1002/047084289X.rb015.pub2.

[15] (a) Frederickson M, Grigg R. Electrophile mediated heteroatom cyclizations onto C-C π-bonds. Part 1: Halogen and chalcogen mediated cyclization. Org Prep Proced Int. 1997;29:33–67. (b) Robin S, Rousseau G. Electrophilic cyclization of unsaturated amides. Tetrahedron. 1998;54:13681–736; (c) Paulmier C. Inter and intramolecular nucleophilic substitutions of activated phenylselanyl groups. Phosphorus Sulfur Silicon. 2001;172:25–54; (d) Petragnani N, Stefani HA, Valduga CJ. Recent advances in selenocyclofunctionalization reactions. Tetrahedron. 2001;57:1411–48; (e) Robin S, Rousseau G. Formation of four-membered heterocycles through electrophilic heteroatom cyclization. Eur J Org Chem. 2002:3099–114; (f) Ranganathan S, Muraleedharan KM, Vaisha NK, Jayaraman N. Halo- and selenolactonisation: the two major strategies for cyclofunctionalisation. Tetrahedron. 2004;60:5273–308; (g) Khokhar SS, Wirth T. Selenocyclizations: control by coordination and by the counterion. Angew Chem Int Ed. 2004;43:631–3.

[16] Markovic Z, Konstantinovic S, Juranic I, Dosen-Micovic L. Molecular orbital study of the rearrangement of seleniranium ions. Gazz Chim Ital. 1997;127:429–34.

[17] Tomassini C, Di Sarra F, Monti B, Sancineto L, Bagnoli L, Marini F, et al. Kinetic resolution of 2-methoxycarbonylalk-3-enols through a stereoselective cyclofunctionalization promoted by an enantiomerically pure electrophilic selenium reagent. Arkivoc. 2017;II:303–12

[18] Uehlin L, Wirth T. Synthesis of chiral acetals by asymmetric selenenylations. Phosphorus Sulfur Silicon Relat Elem. 2009;184:1374–85.

[19] Ścianowski J, Szumera J, Kleman P, Pacula AJ. Synthesis and reactions of terpenyl diselenides functionalized with phenyl and naphthyl groups. Tetrahedron: Asymmetry. 2016;27:238–45.

[20] Choi TJ, Baek JY, Jeon HB, Kim KS. A new efficient glycosylation method employing glycosyl pentenoates and PhSeOTf. Tetrahedron Lett. 2006;47:9191–4.

[21] Santi C, Tiecco M, Testaferri L, Tomassini C, Santoro S, Bizzoca G. Diastereo and enantioselective synthesis of 1,2-diols promoted by electrophilic selenium reagents. Phosphorus Sulfur Silicon. 2008;183:956–60.

[22] Santoro S, Battistelli B, Gjoka B, Si CWS, Testaferri L, Tiecco M, et al. Oxidation of alkynes in aqueous media catalyzed by diphenyl diselenide. Synlett. 2010;1402–6.

[23] Kodama S, Saeki T, Mihara K, Higashimae S, Kawaguchi S, Sonoda M, et al. A benzoyl peroxide/diphenyl diselenide binary system for functionalization of alkynes leading to alkenyl and alkynyl selenides. J Org Chem. 2017;82:12477–84.

[24] Abdo M, Zhang Y, Schramm VL, Knapp S. Electrophilic aromatic selenylation: new OPRT inhibitors. Org Lett. 2010;12:2982–5.

[25] Back TG. Benzeneseleninic acid and benzeneseleninic anhydride. In: Fuchs P, Bode J, Charette A, Rovis T, editors. e-EROS - Encyclopedia of Reagents for Organic Synthesis. Hoboken, NJ, USA: John Wiley & Sons, 2007. DOI: 10.1002/9780470842898.rb016.pub2.
[26] Zhang X, Sun J, Ding Y, Yu L. Dehydration of aldoximes using PhSe(O)OH as the *pre*-catalyst in air. Org Lett. 2015;17:5840–2.
[27] Jing X, Wang T, Ding Y, Yu L. A scalable production of anisonitrile through organoselenium-catalyzed dehydration of anisaldoxime under solventless conditions. Appl Catal, A. 2017;541:107–11.
[28] Jing X, Yuan D, Yu L. Green and practical oxidative deoximation of oximes to ketones or aldehydes with hydrogen peroxide/air by organoselenium catalysis. Adv Synth Catal. 2017;359:1194–201.
[29] Cerra B, Mangiavacchi F, Santi C, Lozza AM, Gioiello A. Selective continuous flow synthesis of hydroxy lactones from alkenoic acids. React Chem Eng. 2017;2:467–71.
[30] Sancineto L, Mangiavacchi F, Tidei C, Bagnoli L, Marini F, Gioiello A, et al. Selenium catalyzed oxacyclization of alkenoic acids and alkenols. Asian J Org Chem. 2017;6:988–92.
[31] (a) Ichikawa H, Usami Y, Arimoto M. Synthesis of novel organoselenium as catalyst for Baeyer–Villiger oxidation with 30 % H_2O_2. Tetrahedron Lett. 2005;46:8665–8; (b) Ichikawa H, Watanabe R, Fujino Y, Usami Y. Divergent synthesis of withasomnines via synthesis of 4-hydroxy-1*H*-pyrazoles and Claisen rearrangement of their 4-*O*-allylethers. Tetrahedron Lett. 2011;52:4448–51; (c) Usami Y, Watanabe R, Fujino Y, Shibano M, Ishida C, Yoneyama H, et al. Divergent synthesis and evaluation of inhibitory activities against cyclooxygenases-1 and –2 of natural withasomnines and analogues. Chem Pharm Bull. 2012;60:1550–60.
[32] Denmark SE, Baiazitov RY, Nguyen ST. Tandem double intramolecular [4+2]/[3+2] cycloadditions of nitroalkenes: construction of the pentacyclic core structure of daphnilactone B. Tetrahedron. 2009;65:6535–48.
[33] Press DJ, Mercier EA, Kuzma D, Back TG. Substituent effects upon the catalytic activity of aromatic cyclic seleninate esters and spirodioxyselenuranes that act as glutathione peroxidase mimetics. J Org Chem. 2008;73:4252–5.
[34] Press DJ, McNeil NMR, Hambrook M, Back TG. Effects of methoxy substituents on the glutathione peroxidase-like activity of cyclic seleninate esters. J Org Chem. 2014;79: 9394–401.
[35] Mercier EA, Smith CD, Parvez M, Back TG. Cyclic seleninate esters as catalysts for the oxidation of sulfides to sulfoxides, epoxidation of alkenes, and conversion of enamines to α-hydroxyketones. J Org Chem. 2012;77:3508–17.
[36] McNeil NMR, McDonnell C, Hambrook M, Back TG. Oxidation of disulfides to thiolsulfinates with hydrogen peroxide and a cyclic seleninate ester catalyst. Molecules. 2015;20:10748–62.
[37] Reich HJ, Chow F, Peake SL. Seleninic acids as catalysts for oxidations of olefins and sulfides using hydrogen peroxide. Synthesis. 1978;299–301.
[38] (a) Back TG, Moussa Z. Remarkable activity of a novel cyclic seleninate ester as a glutathione peroxidase mimetic and its facile in situ generation from allyl 3-hydroxypropyl selenide. J Am Chem Soc. 2002;124:12104–5; (b) Back TG, Moussa Z. Diselenides and allyl selenides as glutathione peroxidase mimetics. Remarkable activity of cyclic seleninates produced in situ by the oxidation of allyl ω-hydroxyalkyl selenides. J Am Chem Soc. 2003;125:13455–60; (c) Back TG, Moussa Z, Parvez M. The exceptional glutathione peroxidase-like activity of di(3-hydroxypropyl) selenide and the unexpected role of a novel spirodioxaselenanonane intermediate in the catalytic cycle. Angew Chem Int Ed. 2004;43:1268–70; (d) Back TG, Kuzma D, Parvez M. Aromatic derivatives and tellurium analogues of cyclic seleninate esters and spirodioxyselenuranes that act as glutathione peroxidase mimetics. J Org Chem. 2005;70:9230–6; (e) Kuzma D, Parvez M, Back TG. Formation of a spirodiazaselenurane and its corresponding

azaselenonium derivatives from the oxidation of 2,2′-selenobis(benzamide). Structure, properties and glutathione peroxidase activity. Org Biomol Chem. 2007;5:3213–7.
[39] (a) Miniejew C, Outurquin F, Pannecoucke X. New phenylselanyl group activation: synthesis of aziridines and oxazolidin-2-ones. Org Biomol Chem. 2004;2:1575–6; (b) Redon S, Berthe-Berkaoui A-L, Pannecoucke X, Outurquin F. Selenylated dienes: synthesis, stereochemical studies by ^{77}Se NMR, and transformation into functionalized allenes. Tetrahedron. 2007;63:3707–17.
[40] Andreou T, Burés J, Vilarrasa J. Reaction of Dess–Martin periodinane with 2-(alkylselenyl) pyridines. Dehydration of primary alcohols under extraordinarily mild conditions. Tetrahedron Lett. 2010;51:1863–6.
[41] Redon S, Pannecoucke X, Franck X, Outurquin F. Synthesis and oxidative rearrangement of selenenylated dihydropyrans. Org Biomol Chem. 2008;6:1260–7.
[42] Russell DA, Freudenreich JJ, Ciardiello JJ, Sore HF, Spring DR. Stereocontrolled semi-syntheses of deguelin and tephrosin. Org Biomol Chem. 2017;15:1593–6.
[43] Wong LSM, Turner KA, White JM, Holmes AB, Ryan JH. Asymmetric synthesis of a hydroxylated nine-membered lactone from tartaric acid using the Claisen rearrangement. Aust J Chem. 2010;63:529–32.
[44] Macdougall PE, Smith NA, Schiesser CH. Substituent effects in selenoxide elimination chemistry. Tetrahedron. 2008;64:2824–31.
[45] Chen W, Bay WP, Wong MW, Huang D. Selenium Blue-α and -β: turning on the fluorescence of a pyrenyl fluorophore via oxidative cleavage of the Se–C bond by reactive oxygen species. Tetrahedron Lett. 2012;53:3843–6.
[46] See JY, Yang H, Zhao Y, Wong MW, Ke Z, Yeung Y-Y. Desymmetrizing enantio- and diastereoselective selenoetherification through supramolecular catalysis. ACS Catal. 2018;8:850–8.
[47] (a) Santi C, Santoro S. Electrophilic selenium. In: Wirth T, editor. Organoselenium chemistry – synthesis and reactions. Weinheim, Germany: Wiley-VCH, 2012:1–51; (b) Ścianowski J, Rafiński Z. Electrophilic selenium reagents: addition reactions to double bonds and selenocyclizations. In: Santi C, editor. Organoselenium chemistry: between synthesis and biochemistry. Saif, Sharjah, United Arab Emirates: Bentham Science Publications, 2014: 8–60; (c) Sancineto L, Palomba M, Bagnoli L, Marini F, Santi C. Advances in electrophilic organochalcogen reagents. Curr Org Chem. 2015;20:122–35.
[48] Aversa MC, Barattucci A, Bonaccorsi P, Temperini A. Regio- and stereocontrolled synthesis of (Z)-α-(phenylseleno)sulfinyl and -sulfonyl alkenes via sulfenic acids, and a study of their reactivity. Eur J Org Chem. 2011;5668–73.
[49] Ogawa A, Sekiguchi M, Shibuya H, Kuniyasu H, Takami N, Ryu I, et al. Oxidation of 1,2-bis (phenylselebo)-1-alkenes. A novel example of selenoxide anti-elimination. Chem Lett. 1991;20:1805–6.
[50] Tarao A, Tabuchi Y, Sugimoto E, Ikeda M, Uchimoto H, Arimitsu K, et al. A mechanistic investigation of *anti*-elimination in (*Z*)-1,2-bis(arylseleno)-1-alkenes and their sulfur analogs. Org Biomol Chem. 2015;13:5964–71.
[51] See, for example: (a) Balkrishna SJ, Prasad CD, Panini P, Detty MR, Chopra D, Kumar S. Isoselenazolones as catalysts for the activation of bromine: bromolactonization of alkenoic acids and oxidation of alcohols. J Org Chem. 2012;77:9541–52; (b) Chen F, Tan CK, Yeung Y-Y. C_2-Symmetric cyclic selenium-catalyzed enantioselective bromoaminocyclization. J Am Chem Soc. 2013;135:1232–5; (c) Luo J, Zhu Z, Liu Y, Zhao X. Diaryl selenide catalyzed vicinal trifluoromethylthioamination of alkenes. Org Lett. 2015;17:3620–3; (d) Verma A, Jana S, Prasad CD, Yadav A, Kumar S. Organoselenium and DMAP co-catalysis: regioselective synthesis of medium-sized halolactones and bromooxepanes from unactivated alkenes. Chem Commun.

2016;52:4179–82; (e) Wu J-J, Xu J, Zhao X. Selenide-catalyzed stereoselective construction of tetrasubstituted trifluoromethylthiolated alkenes with alkynes. Chem Eur J. 2016;22:15265–9; (f) Luo J, Liu Y, Zhao X. Chiral selenide-catalyzed enantioselective construction of saturated trifluoromethylthiolated azaheterocycles. Org Lett. 2017;19:3434–7; (g) Liu X, Liang Y, Ji J, Luo J, Zhao X. Chiral selenide-catalyzed enantioselective allylic reaction and intermolecular difunctionalization of alkenes: efficient construction of C-SCF_3 stereogenic molecules. J Am Chem Soc. 2018;140:4782–6.

[52] Zhu Z, Luo J, Zhao X. Combination of Lewis basic selenium catalysis and redox selenium chemistry: synthesis of trifluoromethylthiolated tertiary alcohols with alkenes. Org Lett. 2017;19:4940–3.

[53] Weilbeer C, Selent D, Dyballa KM, Franke R, Spannenberg A, Börner A. Evaluation of organoselenium based compounds as co-catalysts in rhodium-catalyzed hydroformylation. Chemistry Select. 2016;1:5421–9.

[54] Drake MD, Bright FV, Detty MR. Dendrimeric organochalcogen catalysts for the activation of hydrogen peroxide: origins of the “dendrimer effect” with catalysts terminating in phenylseleno groups. J Am Chem Soc. 2003;125:12558–66.

[55] Goodman MA, Detty MR. Selenoxides as catalysts for the activation of hydrogen peroxide. Bromination of organic substrates with sodium bromide and hydrogen peroxide. Organometallics. 2004;23:3016–20.

[56] Bennett SM, Tang Y, McMaster D, Bright FV, Detty MR. A xerogel-sequestered selenoxide catalyst for brominations with hydrogen peroxide and sodium bromide in an aqueous environment. J Org Chem. 2008;73:6849–52.

[57] Jeong LS, Tosh DK, Kim HO, Wang T, Hou X, Yun HS, et al. First synthesis of 4′-selenonucleosides showing unusual southern conformation. Org Lett. 2008;10:209–12.

[58] (a) Jayakanthan K, Johnston BD, Pinto BM. Stereoselective synthesis of 4′-selenonucleosides using the Pummerer glycosylation reaction. Carbohydr Res. 2008;343:1790–800. (b) Taniike H, Inagaki Y, Matsuda A, Minakawa N. Practical synthesis of 4′-selenopyrimidine nucleosides using hypervalent iodine. Tetrahedron. 2011;67:7977–82; (c) Ishii K, Saito-Tarashima N, Ota M, Yamamoto S, Okamoto Y, Tanaka Y, et al. Practical synthesis of 4′-selenopurine nucleosides by combining chlorinated purines and ‘armed’ 4-selenosugar. Tetrahedron. 2016;72:6589–94.

[59] (a) Iwaoka M, Kumakura F, Yoneda M, Nakahara T, Henmi K, Aonuma H, et al. Direct observation of conformational folding coupled with disulphide rearrangement by using a water-soluble selenoxide reagent – A case of oxidative regeneration of ribonuclease A under weakly basic conditions. J Biochem. 2008;144:121–30. (b) Kumakura F, Mishra B, Priyadarsini KI, Iwaoka M. A water-soluble cyclic selenide with enhanced glutathione peroxidase-like catalytic activities. Eur J Org Chem 2010, 440-5; (c) Arai K, Dedachi K, Iwaoka M. Rapid and quantitative disulfide bond formation for a polypeptide chain using a cyclic selenoxide reagent in an aqueous medium. Chem Eur J 2011, 17, 481–5.

[60] Arai K, Moriai K, Ogawa A, Iwaoka M. An amphiphilic selenide catalyst behaves like a hybrid mimic of protein disulfide isomerase and glutathione peroxidase 7. Chem Asian J. 2014;9:3464–71.

[61] Kumar PV, Singh BG, Kunwar A, Iwaoka M, Priyadarsini KI. Degradation of peroxynitrite by simple, recyclable selenolanes. Bull Chem Soc Jpn. 2016;89:490–7.

[62] Marini F, Sternativo S, Del Verme F, Testaferri L, Tiecco M. Enantioselective organocatalytic Michael addition of α-substituted cyanoacetates to α,β-unsaturated selenones. Adv Synth Catal. 2009;351:103–6.

[63] Buyck T, Wang Q, Zhu J. Catalytic enantioselective Michael addition of α-aryl-α-isocyanoacetates to vinyl selenone: synthesis of α,α-disubstituted α-amino acids and (+)- and (-)-trigonoliimine A. Angew Chem Int Ed. 2013;52:12714–8.

[64] Buyck T, Wang Q, Zhu J. From racemic to enantioselective total synthesis of trigonoliimines *via* development of an organocatalytic enantioselective Michael addition of α-aryl-α-isocyanoacetate to vinyl phenyl selenone. Chimia. 2014;68:211–4.
[65] Clemenceau A, Wang Q, Zhu J. Enantioselective synthesis of quaternary α-amino acids enabled by the versatility of the phenylselenonyl group. Chem Eur J. 2016;22:18368–72.
[66] Zhang T, Cheng L, Hameed S, Liu L, Wang D, Chen Y-J. Highly enantioselective Michael addition of 2-oxindoles to vinyl selenone in RTILs catalyzed by a *Cinchona* alkaloid-based thiourea. Chem Commun. 2011;47:6644–6.
[67] Simlandy AK, Mukherjee S. Catalytic asymmetric formal γ-allylation of deconjugated butenolides. Org Biomol Chem. 2016;14:5659–64.
[68] Bhaumik A, Samanta S, Pathak T. Enantiopure 1,4,5-trisubstituted 1,2,3-triazoles from carbohydrates: applications of organoselenium chemistry. J Org Chem. 2014;79:6895–904.
[69] Bhaumik A, Pathak T. Methyl-α-D-2-selenonyl pent-2-enofuranoside: a reactive selenosugar for the diversity oriented synthesis of enantiomerically pure heterocycles, carbocycles, and isonucleosides. J Org Chem. 2015;80:11057–64.
[70] Bhaumik A, Das A, Pathak T. Vinyl selenones derived from D-fructose: a new platform for fructochemistry. Asian J Org Chem. 2016;5:1048–62.
[71] Marini F, Sternativo S, Del Verme F, Testaferri L, Tiecco M. A new stereoselective synthesis of cyclopropanes containing quaternary stereocentres *via* organocatalytic Michael addition to vinyl selenones. Adv Synth Catal. 2009;351:1801–6.
[72] Bagnoli L, Scarponi C, Testaferri L, Tiecco M. Preparation of both enantiomers of cyclopropane derivatives from the reaction of vinyl selenones with di-(-)-bornyl malonate. Tetrahedron: Asymmetry. 2009;20:1506–14.
[73] Sternativo S, Marini F, Del Verme F, Calandriello A, Testaferri L, Tiecco M. One-pot synthesis of aziridines from vinyl selenones and variously functionalized primary amines. Tetrahedron. 2010;66:6851–7.
[74] Bagnoli L, Scarponi C, Rossi MG, Testaferri L, Tiecco M. Synthesis of enantiopure 1,4-dioxanes, morpholines, and piperazines from the reaction of chiral 1,2-diols, amino alcohols, and diamines with vinyl selenones. Chem Eur J. 2011;17:993–9.
[75] Buyck T, Wang Q, Zhu J. Triple role of phenylselenonyl group enabled a one-pot synthesis of 1,3-oxazinan-2-ones from *α*-isocyanoacetates, phenyl vinyl selenones, and water. J Am Chem Soc. 2014;136:11524–8.
[76] Buyck T, Wang Q, Zhu J. Integrated one-pot synthesis of 1,3-oxazinan-2-ones from isocyanoacetates and phenyl vinyl selenones. Chimia. 2015;69:199–202.
[77] Sternativo S, Calandriello A, Costantino F, Testaferri L, Tiecco M, Marini F. A highly enantioselective one-pot synthesis of spirolactones by an organocatalyzed Michael addition/cyclization sequence. Angew Chem Int Ed. 2011;50:9382–5.
[78] Sternativo S, Walczak O, Battistelli B, Testaferri L, Marini F. Organocatalytic Michael addition of indanone carboxylates to vinyl selenone for the asymmetric synthesis of polycyclic pyrrolidines. Tetrahedron. 2012;68:10536–41.
[79] Sardella R, Ianni F, Lisanti A, Scorzoni S, Marini F, Sternativo S, et al. Direct chromatographic enantioresolution of fully constrained β-amino acids: exploring the use of high-molecular weight chiral selectors. Amino Acids. 2014;46:1235–42.
[80] Sternativo S, Battistelli B, Bagnoli L, Santi C, Testaferri L, Marini F. Synthesis of *γ*-lactams *via* a domino Michael addition/cyclization reaction of vinyl selenone with substituted amides. Tetrahedron Lett. 2013;54:6755–7.
[81] Bagnoli L, Casini S, Marini F, Santi C, Testaferri L. Vinyl selenones: annulation agents for the synthesis of six-membered benzo-1,4-heterocyclic compounds. Tetrahedron. 2013;69:481–6.

[82] Palomba M, Rossi L, Sancineto L, Tramontano E, Corona A, Bagnoli L, et al. A new vinyl selenone-based domino approach to spirocyclopropyl oxindoles endowed with anti-HIV RT activity. Org Biomol Chem. 2016;14:2015–24.
[83] Sridhar PR, Reddy GM, Seshadri K. Stereoselective synthesis of *cis*-fused perhydrofuro [2,3-*b*]furan derivatives from sugar-derived allyl vinyl ethers. Eur J Org Chem. 2012; 6228–35.
[84] Temperini A, Barattucci A, Bonaccorsi PM, Rosati O, Minuti L. Stereoselective synthesis of substituted tetrahydropyrans and isochromans by cyclization of phenylseleno alcohols. J Org Chem. 2015;80:8102–12.
[85] Beng TK, Silaire AWV, Alwali A, Bassler DP. One-shot access to α,β-difunctionalized azepenes and dehydropiperidines by reductive cross-coupling of α-selenonyl-β-selenyl enamides with organic bromides. Org Biomol Chem. 2015;13:7915–9.
[86] Kambe N, Tsukamoto T, Miyoshi N, Murai S, Sonoda N. Oxidation of olefins with benzenetellurinic anhydride. Chem Lett. 1987;16:269–72.
[87] Yoshida M, Suzuki T, Kamigata N. Novel preparation of highly electrophilic species for benzenetellurenylation or benzenesulfenylation by nitrobenzenesulfonyl peroxide in combination with ditelluride or disulfide. Application to intramolecular ring closures. J Org Chem. 1992;57:383–6.
[88] Yamada T, Mishima E, Ueki K, Yamago S. Phenyltellanyl triflate (PhTeOTf) as a powerful tellurophilic activator in the Friedel–crafts reaction. Chem Lett. 2008;37:650–1.
[89] Yamago S, Iida K, Yoshida J. Organotellurium compounds as novel initiators for controlled/ living radical polymerizations. Synthesis of functionalized polystyrenes and end-group modifications. J Am Chem Soc. 2002;124:2874–5.
[90] Ogura F, Otsubo T. Benzenetellurinic anhydride. In: Paquette LA, editor(s). e-EROS – Encyclopedia of Reagents for Organic Synthesis. Hoboken, NJ, USA: John Wiley & Sons, 2001, and references cited therein. DOI: 10.1002/047084289X.rb024.
[91] Hu NX, Aso Y, Otsubo T, Ogura F. New synthetic reactions using arenetellurinic anhydrides. Phosphorus Sulfur Relat Elem. 1988;38:177–89.
[92] Ogura F, Otsubo T, Aso Y. New synthetic reactions using organotelluriums. Phosphorus Sulfur Silicon Relat Elem. 1992;67:223–38.
[93] Barton DHR, Finet J-P, Thomas M. Organotellurinic acid anhydrides as selective oxidants in organic synthesis. Tetrahedron. 1986;42:2319–24.
[94] Hu NX, Aso Y, Otsubo T, Ogura F. Novel oxidizing properties of p-methoxybenzenetellurinic acid anhydride. Tetrahedron Lett. 1986;27:6099–102.
[95] Narwal JK, Garg S, Verma KK. Studies on the oxidizing properties of some tellurinic acid anhydrides. Chem Sci Trans. 2013;2:1491–5.
[96] Fukumoto T, Matsuki T, Hu NX, Aso Y, Otsubo T, Ogura F. Benzenetellurinic mixed anhydrides as mild oxidizing agents. Chem Lett. 1990;19:2269–72.
[97] Hu NX, Aso Y, Otsubo T, Ogura F. Cyclofunctionalization of hydroxyolefins induced by arenetellurinic anhydride. Tetrahedron Lett. 1987;28:1281–4.
[98] Hu NX, Aso Y, Otsubo T, Ogura F. Cyclofunctionalization of hydroxyolefins induced by arenetellurinyl acetate. J Org Chem. 1989;54:4391–7.
[99] Hu NX, Aso Y, Otsubo T, Ogura F. Organotellurium-mediated synthesis of oxazolidin-2-ones from alkenes. J Chem Soc Chem Commun. 1987;1447–8.
[100] Hu NX, Aso Y, Otsubo T, Ogura F. Aminotellurinylation of olefins and its utilization for synthesis of 2-oxazolidinones. J Org Chem. 1989;54:4398–404.
[101] Hu NX, Aso Y, Otsubo T, Ogura F. Tellurium-based organic synthesis: a novel one-pot formation of 2-oxazolines from alkenes induced by amidotellurinylation. Tetrahedron Lett. 1988;29:1049–52.

[102] Hu NX, Aso Y, Otsubo T, Ogura F. Organotelluriums. Part 21. Amidotellurinylation of olefins and a novel one-pot synthesis of 4,5-dihydro-oxazoles from olefins. J Chem Soc Perkin Trans I. 1989;1775–80.
[103] Fukumoto T, Aso Y, Otsubo T, Ogura F. Stereoselective addition reactions of alkynes with benzenetellurinyl trifluoromethanesulfonate in acetonitrile: organotellurium-mediated one-pot synthesis of oxazoles from internal alkynes. J Chem Soc Chem Commun. 1992; 1070–2.
[104] Fukumoto T, Aso Y, Otsubo T, Ogura F. Syntheses of β-amidovinyltellurides and oxazoles by addition reactions of alkynes with benzenetellurinyl trifluoromethanesulfonate in acetonitrile. Heteroatom Chem. 1993;4:511–6.
[105] Hu NX, Aso Y, Otsubo T, Ogura F. Transformation of allylsilanes into allylamines via phenyltellurinylation. Tetrahedron Lett. 1988;29:4949–52.
[106] Brill WF. A site isolated tellurium oxidation catalyst having no soluble analog. J Org Chem. 1986;51:1149–50.
[107] Alberto EE, Muller LM, Detty MR. Rate accelerations of bromination reactions with NaBr and H_2O_2 via the addition of catalytic quantities of diaryl ditellurides. Organometallics. 2014;33:5571–81.
[108] Finet J-P. Dianisyl telluroxide. In: Paquette LA, editor(s). e-EROS – Encyclopedia of Reagents for Organic Synthesis. Hoboken, NJ, USA: John Wiley & Sons, 2001, and references cited therein. DOI: 10.1002/047084289X.rd009m.
[109] Ley SV, Meerholz CA, Barton DHR. Diaryl telluroxides as new mild oxidising reagents. Tetrahedron. 1981;37 (Suppl. 1):213–23.
[110] Barton DHR, Ley SV, Meerholz CA. Bis(*p*-methoxyphenyl) telluroxide: a new, mild oxidising agent. Chem Commun. 1979;755–6.
[111] Barton DHR, Finet J-P, Giannotti C, Thomas M. Telluramine derivatives as selective oxidants. Tetrahedron Lett. 1988;29:2671–4.
[112] Engman L, Cava MP. Synthesis and reactions of bis(p-methoxyphenyl)tellurone. J Chem Soc Chem Commun. 1982;164–5.
[113] Oba M, Endo M, Nishiyama K, Ouchi A, Ando W. Photosensitized oxygenation of diaryl tellurides to telluroxides and their oxidizing properties. Chem Commun. 2004;1672–3.
[114] Hu NX, Aso Y, Otsubo T, Ogura F. Polymer-supported diaryl selenoxide and telluroxide as mild and selective oxidizing agents. Bull Chem Soc Jpn. 1986;59:879–84.
[115] Mallow O, Bolsinger J, Finke P, Hesse M, Chen Y-S, Duthie A, et al. Oxygen transfer from an intramolecularly coordinated diaryltellurium oxide to acetonitrile. Formation and combined AIM and ELI-D analysis of a novel diaryltellurium acetimidate. J Am Chem Soc. 2014;136: 10870–3.
[116] Koide Y, Kawaguchi M, Urano Y, Hanaoka K, Komatsu T, Abo M, et al. A reversible near-infrared fluorescence probe for reactive oxygen species based on Te–rhodamine. Chem Commun. 2012;48:3091–3.
[117] Kryman MW, Schamerhorn GA, Yung K, Sathyamoorthy B, Sukumaran DK, Ohulchanskyy TY, et al. Organotellurium fluorescence probes for redox reactions: 9-aryl-3,6-diaminotelluroxanthylium dyes and their telluroxides. Organometallics. 2013;32:4321–33.
[118] Lutkus LV, Irving HE, Davies KS, Hill JE, Lohman JE, Eskew MW, et al. Photocatalytic aerobic thiol oxidation with a self-sensitized tellurorhodamine chromophore. Organometallics. 2017;36:2588–96.
[119] Shimada K, Takata Y, Osaki Y, Moro-oka A, Kogawa H, Sakuraba M, et al. Regioselective synthesis of polysubstituted pyridines via hetero-Diels–Alder reaction of isotellurazoles with acetylenic dienophiles. Tetrahedron Lett. 2009;50:6651–3.

[120] Ley SV, Meerholz CA, Barton DHR. Catalytic oxidation of thiocarbonyl compounds involving the use of 1,2-dibromotetrachloroethane as a brominating reagent for diaryl Te^{II} species. Tetrahedron Lett. 1980;21:1785–8.
[121] Matsuki T, Hu NX, Aso Y, Otsubo T, Ogura F. Indirect electrolytic oxidation of thioamides using organotellurium as a mediator. Bull Chem Soc Jpn. 1988;61:2117–21.
[122] Oba M, Okada Y, Nishiyama K, Ando W. Aerobic photooxidation of phosphite esters using diorganotelluride catalysts. Org Lett. 2009;11:1879–81.
[123] Okada Y, Oba M, Arai A, Tanaka K, Nishiyama K, Ando W. Diorganotelluride-catalyzed oxidation of silanes to silanols under atmospheric oxygen. Inorg Chem. 2010;49:383–5.
[124] See, for example: (a) Detty MR, Zhou F, Friedman AE. Positive halogens from halides and hydrogen peroxide with organotellurium catalysts. J Am Chem Soc. 1996;118: 313–8; (b) Francavilla C, Drake MD, Bright FV, Detty MR. Dendrimeric organochalcogen catalysts for the activation of hydrogen peroxide: improved catalytic activity through statistical effects and cooperativity in successive generations. J Am Chem Soc. 2001;123:57–67; (c) Higgs DE, Nelen MI, Detty MR. Iodination of organic substrates with halide salts and H_2O_2 using an organotelluride catalyst. Org Lett. 2001;3:349–52; (d) Abe M, You Y, Detty MR. 21-Telluraporphyrins. 2. Catalysts for bromination reactions with hydrogen peroxide and sodium bromide. Organometallics. 2002;21:4546–51.
[125] See, for example: (a) You Y, Ahsan K, Detty MR. Mechanistic studies of the tellurium(II)/tellurium(IV) redox cycle in thiol peroxidase-like reactions of diorganotellurides in methanol. J Am Chem Soc. 2003;125:4918–27; (b) Ahsan K, Drake MD, Higgs DE, Wojciechowski AL, Tse BN, Bateman MA, et al. Dendrimeric organotelluride catalysts for the activation of hydrogen peroxide. Improved catalytic activity through statistical and stereoelectronic effects. Organometallics. 2003;22:2883–90; (c) Braga AL, Alberto EE, Soares LC, Rocha JBT, Sudati JH, Roos DH. Synthesis of telluroamino acid derivatives with remarkable GPx like activity. Org Biomol Chem. 2009;7:43–5; (d) Sarma BK, Manna D, Minoura M, Mugesh G. Synthesis, structure, spirocyclization mechanism, and glutathione peroxidase-like antioxidant activity of stable spirodiazaselenurane and spirodiazatellurane. J Am Chem Soc. 2010;132:5364–74; (e) Singh VP, Poon J, Engman L. Turning pyridoxine into a catalytic chain-breaking and hydroperoxide-decomposing antioxidant. J Org Chem. 2013;78:1478–87.
[126] Oba M, Tanaka K, Nishiyama K, Ando W. Aerobic oxidation of thiols to disulfides catalyzed by diaryl tellurides under photosensitized conditions. J Org Chem. 2011;76:4173–7.
[127] Engman L, Cava MP. Bis(p-methoxyphenyl)telluroxide, a novel organotellurium aldol catalyst. Tetrahedron Lett. 1981;22:5251–2.
[128] (a) Zhong Q, Lu R. Catalytic synthesis of α,β-unsaturated ketones and 2,4-alkadienones by organotellurium oxide. Yingyong Huaxue. 1990;7(3):89–92. (Chem Abstr. 1991;114:61639); (b) Zhong Q, Shao J, Liu C, Lu R. Catalytic synthesis of α,β-unsaturated nitriles, cyanoesters and cyanoamides by organotellurium oxide. Yingyong Huaxue. 1991;8(5):17–20 (Chem Abstr. 1992;116:83329); (c) Zhong Q, Lu Y, Liu C, Tao W, Zou Y. Catalytic synthesis of α-cyano-α,β-unsaturated sulfones in the presence of organotellurium oxide. Chin Chem Lett. 1991;2:683–4 (Chem Abstr. 1992;116:173723).
[129] Akiba M, Lakshmikantham MV, Jen K-Y, Cava MP. Organotellurium chemistry. 9. Structural parameters in the telluroxide-catalyzed aldol condensation. J Org Chem. 1984;49:4819–21.
[130] Verma KK, Garg S, Narwal JK. Study on the applications of some bis (hydroxyaryl) telluroxides as aldol catalyst. Pharma Chem. 2011;3:218–25.
[131] Eliseeva EV, Red'kin NA, Gar'kin VP, Pytskii IS, Buryak AK. Chromatography–mass spectrometry study of aldol condensation reaction with an organotelluric catalyst. Prot Met Phys Chem Surf. 2015;51:1050–7.

[132] Engman L, Stern D, Pelcman M, Andersson CM. Thiol peroxidase activity of diorganyl tellurides. J Org Chem. 1994;59:1973–9.
[133] Kanda T, Engman L, Cotgreave IA, Powis G. Novel water-soluble diorganyl tellurides with thiol peroxidase and antioxidant activity. J Org Chem. 1999;64:8161–9.
[134] Detty MR. Reaction pathways of telluroxide equivalents. Reductive elimination of hydrogen peroxide from dihydroxytelluranes and oxidation of carbon via intramolecular transfer of oxygen. Organometallics. 1991;10:702–12.
[135] Detty MR, Gibson SL. Tellurapyrylium dyes as catalysts for oxidations with hydrogen peroxide and as scavengers of singlet oxygen. Dihydroxytelluranes as mild oxidizing agents. Organometallics. 1992;11:2147–56.
[136] McMaster DM, Bennett SM, Tang Y, Finlay JA, Kowalke GL, Nedved B, et al. Antifouling character of 'active' hybrid xerogel coatings with sequestered catalysts for the activation of hydrogen peroxide. Biofouling. 2009;25:21–33.
[137] Oba M, Okada Y, Nishiyama K, Shimada S, Ando W. Synthesis, characterization and oxidizing properties of a diorgano tellurone carrying bulky aromatic substituents. Chem Commun. 2008;5378–80.
[138] Beckmann J, Bolsinger J, Duthie A, Finke P. New series of intramolecularly coordinated diaryltellurium compounds. Rational synthesis of the diarylhydroxytelluronium triflate $[(8\text{-}Me_2NC_{10}H_6)_2Te(OH)](O_3SCF_3)$. Organometallics. 2012;31:238–45.
[139] Oba M, Nishiyama K, Koguchi S, Shimada S, Ando W. Synthesis and properties of tellurinic anhydride–tellurone adducts. Organometallics. 2013;32:6620–3.
[140] Oba M, Okada Y, Endo M, Tanaka K, Nishiyama K, Shimada S, et al. Formation of diaryl telluroxides and tellurones by photosensitized oxygenation of diaryl tellurides. Inorg Chem. 2010;49:10680–6.
[141] Power PP. Main-group elements as transition metals. Nature. 2010;463:171–7.
[142] Kukushkin VYu, Pombeiro AJL. Metal-mediated and metal-catalyzed hydrolysis of nitriles. Inorg Chim Acta. 2005;358:1–21.

Tristram Chivers and Risto S. Laitinen

4 Selenium- and tellurium–nitrogen reagents

Abstract: The reactivity of the chalcogen–nitrogen bond toward main-group element or transition-metal halides, as well as electrophilic and nucleophilic reagents, is the source of a variety of applications of Se–N and Te–N compounds in both inorganic or organic chemistry. The thermal lability of Se–N compounds also engenders useful transformations including the formation of radicals via homolytic Se–N bond cleavage. These aspects of Se–N and Te–N chemistry will be illustrated with examples from the reactions of the binary selenium nitride Se_4N_4, selenium–nitrogen halides $[N(SeCl_n)_2]^+$ (n = 1, 2), the synthons $E(NSO)_2$ (E = Se, Te), chalcogen–nitrogen–silicon reagents, chalcogen(IV) diimides RN=E=NR, the triimidotellurite dianion $[Te(N^tBu)_3]^{2-}$, chalcogen(II) amides and diamides $E(NR_2)_2$ (E = Se, Te; R = alkyl, $SiMe_3$), and heterocyclic systems.

Keywords: selenium nitride, selenium–nitrogen halogenides, bis(sulfinylamino) chalcogenanes, chalcogen diimides, triimidotellurite dianion, chalcogen amides, chalcogen diamides

4.1 Scope and introduction

The development of the chemistry of selenium–nitrogen (Se–N) and tellurium–nitrogen (Te–N) chemistry has lagged behind the remarkable advances in sulfur–nitrogen (S–N) chemistry that have occurred in the past 40 years [1]. In part, this slow progress can be attributed to the lack of simple, easily handled Se–N and Te–N reagents. Nevertheless, a number of such reagents are now available and the goal of this chapter is to provide an overview of their applications in inorganic and organic chemistry.

The chemistry of Se–N and Te–N compounds has been discussed in various reviews and book chapters over the past 30 years [2–11]. While several of these expositions cover the topic in general [2, 3, 6a, 7, 8, 10], other contributions address specific aspects, e.g. metal–Se–N compounds [4], binary Se–N species [5], heterocyclic Se–N and Te–N compounds [6b, 11] and imido-selenium- and –tellurium compounds [9]. A book [1] and three recent book chapters include a discussion of S–N as well as Se–N and Te–N chemistry [8a, 8b, 11].

In keeping with the focus of this treatise on synthetic chemistry, the emphasis of this chapter will be on the preparative methods for the most important reagents

This article has previously been published in the journal *Physical Sciences Reviews*. Please cite as: Chivers, T., Laitinen, R. S. Selenium- and Tellurium-Nitrogen Reagents. *Physical Sciences Reviews* [Online] **2018** DOI: 10.1515/psr-2017-0125

https://doi.org/10.1515/9783110529340-004

containing Se–N and Te–N bonds, as well as their reactions. Comparison will be made with related S–N reagents wherever appropriate. The structural chemistry of Se–N and Te–N compounds will not be discussed, since this aspect is covered well in many of the aforementioned book chapters and reviews.

This overview will start with a discussion of the applications of inorganic selenium–nitrogen reagents commencing with binary selenium nitrides, notably Se_4N_4, followed by sections on selenium- and tellurium-nitrogen halides, $E(NSO)_2$ synthons (E = Se, Te) and chalcogen-nitrogen-silicon (Se–N–Si and Te–N–Si) reagents. The second half of the chapter will deal with organic Se–N and Te–N reagents, including the ephemeral selenonitrosyls ArN=Se (Ar = aryl), chalcogen diimides RN=E=NR (E = Se, Te: R = alkyl, aryl), triimidochalcogenite dianions $[E(NR)_3]^{2-}$ (E = Se, Te) and chalcogen diamides $E(NR_2)_2$ (E = Se, Te; R = alkyl, $SiMe_3$).

4.2 Binary selenium–nitrogen and tellurium–nitrogen reagents

4.2.1 Neutral species

The quintessential sulfur–nitrogen cage molecule S_4N_4 is a rich source of novel sulfur–nitrogen chemistry [1, 8]. The selenium analogue Se_4N_4 may be prepared in several ways. One of these involves the reaction of $(CH_3CH_2O)_2SeO$ with gaseous ammonia in benzene; this procedure has been adapted for the preparation of ^{15}N-enriched $Se_4{}^{15}N_4$ by using stoichiometric amounts of $^{15}NH_3$ [12]. A second method employs the treatment of selenium tetrahalides SeX_4 (X = Cl or Br) with ammonia at elevated temperatures [13, 14]. A more convenient procedure involves the reaction of $(Me_3Si)_2NLi$ with a mixture of selenium chlorides [eq. (4.1)]. Pure Se_4N_4 is isolated in 66% yield after washing with 10% aqueous KCN solution to remove red selenium, selenium halides and selenium oxides and with water to remove LiCl [15].

$$12(Me_3Si)_2NLi + 2Se_2Cl_2 + 8SeCl_4 \rightarrow 2Se_4N_4 + 24Me_3SiCl + 12LiCl \qquad (4.1)$$

Caution: Dry Se_4N_4 is an extremely dangerous material, which can explode at the slightest provocation, e.g. when touched with a metal spatula. It is essential, therefore, to store and handle this reagent under an inert solvent, e.g. a hydrocarbon, to limit reactions to small amounts (<500 mg) of Se_4N_4, and to wear appropriate protective clothing [14, 15].

Despite its hazardous nature Se_4N_4 has been used as the source of a number of important selenium–nitrogen compounds either via oxidation or adduct formation (Figure 4.1). The first example involved reaction of Se_4N_4 with WCl_6 in boiling dichloromethane, which generates a complex of the monomeric chloroselenonitrene (NSeCl) ligand, *viz.* dimeric $[Cl_4W(NSeCl)]_2$ [16]; the analogous molybdenum complex is obtained from treatment of Se_4N_4 with $MoCl_5$ [17]. The oxidation of Se_4N_4 with $[Se_4][AsF_6]_2$ produces $[Se_3N_2]_2[AsF_6]_2$ in high yield as an orange solid that can be

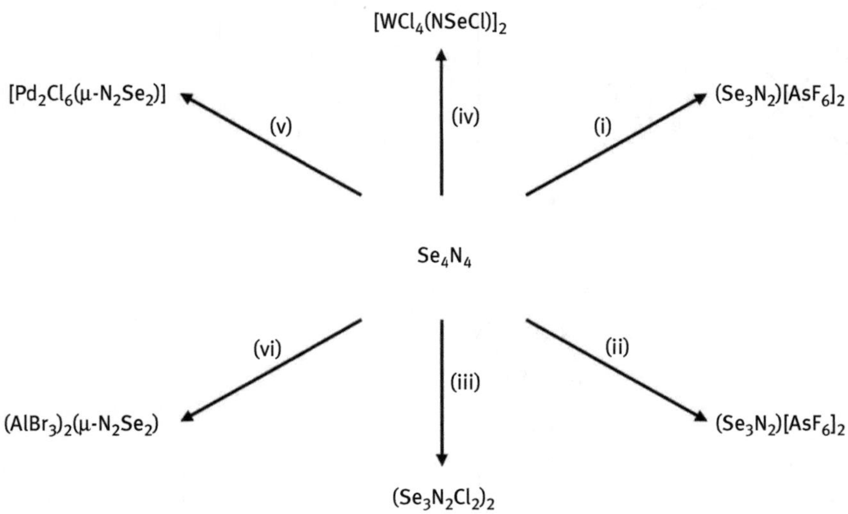

Figure 4.1: Preparation of Se–N compounds from Se_4N_4: (i) $[Se_4][AsF_6]_2$, (ii) AsF_5, (iii) Se_2Cl_2, (iv) $AlBr_3$, (v) $[PPh_4]_2[Pd_2Cl_6]$ and (vi) WCl_6.

manipulated with a metal spatula and is stable to heat [18]; the binary Se–N cation in this salt is a dimer of the radical cation $[Se_3N_2]^{+\bullet}$. The corresponding dication $[Se_3N_2]^{2+}$ is produced upon treatment of Se_4N_4 with AsF_5. The chlorinated dimer $[Se_3N_2Cl]_2$ is formed as an explosive, insoluble dark brown powder from the reaction of Se_2Cl_2 with Se_4N_4 in dichloromethane [19].

A major goal in investigations of selenium–nitrogen compounds has been the generation of polymeric $(SeN)_x$ [20], the selenium analogue of the well-known polysulfur nitride $(SN)_x$. The latter is a conducting polymer with metallic properties, which is generated by the polymerization of S_2N_2 formed by heating S_4N_4 [1]. By contrast, the pyrolysis of Se_4N_4 under vacuum at temperatures up to 220 °C results in decomposition to elemental selenium and N_2 gas [20]. Although the binary selenium nitride Se_2N_2 has not been isolated, the reaction of Se_4N_4 with $AlBr_3$ in dibromomethane at room temperature generates the adduct $[(AlBr_3)_2(\mu\text{-}Se_2N_2)]$, incorporating a bridging Se_2N_2 ligand, as an air-sensitive yellow solid [21] (Figure 4.1). Transition-metal complexes of Se_2N_2 are found in the dianion $[Pd_2Cl_6(\mu\text{-}Se_2N_2)]^{2-}$ formed in the high-temperature reaction of Se_4N_4 with $[Pd_2Cl_6]^{2-}$ salts in dichloromethane [22].

The reactions of Se_4N_4 with transition-metal reagents have also been employed to generate complexes of the chelating binary selenium–nitrogen anions, e.g. Se_3N^-, $Se_2N_2^{2-}$ and its protonated derivative $Se_2N_2H^-$, via oxidative-addition reactions (Figure 4.2) [4]. For example, the combination of Se_4N_4 with $[PtCl_2(PMe_2Ph)]_2$ in boiling chloroform produces a mixture of $[PtCl(Se_3N)(PMe_2Ph)]$ and $[PtCl(Se_2N_2H)(PMe_2Ph)]Cl$, which

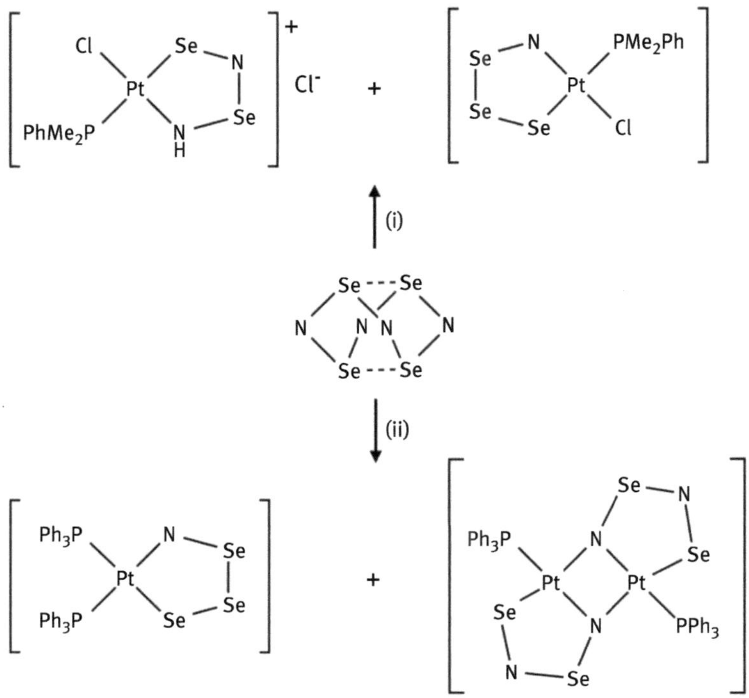

Figure 4.2: Oxidative-addition reactions of Se_4N_4 with (i) $[PtCl_2(PMe_2Ph)]_2$ in boiling CH_2Cl_2 and (ii) $Pt(PPh_3)_3$ in CH_2Cl_2 at 23 °C.

were separated (in low yields) by chromatography [23, 24] (Figure 4.2). Complexes of the N,Se chelating $Se_2N_2^{2-}$ dianion are also formed via the oxidative-addition of Se_4N_4 to the zerovalent platinum complex $Pt(PPh_3)_3$ in dichloromethane. The monomeric complex $[Pt(Se_2N_2)(PPh_3)_2]$ was characterized in solution by ^{31}P NMR spectroscopy while the solid-state structure of the dimer $[Pt(Se_2N_2)(PPh_3)]_2.CH_2Cl_2$, which was isolated in 30% yield, was determined by X-ray crystallography [25].

The binary sulfur nitride S_4N_2 is a six-membered ring with a sulfur diimide –N=S=N– functionality. It is a low melting red solid (Mp 23 °C), which must be stored below –20 °C to avoid decomposition [26]. The preparation of the selenium analogue Se_4N_2 as a black powder stable at room temperature "in almost quantitative yield" from the reaction of Se_2Cl_2 and trimethylsilyl azide has been claimed [27]. However, a reinvestigation of this reaction showed that the black powder is the selenium–nitrogen chloride $Se_3N_2Cl_2$ [28]. The facile thermal decomposition of acyclic selenium(IV) diimides RN=Se=NR (R = alkyl, $SiMe_3$) (Section 4.7) above 0 °C implies that cyclic Se_4N_2 is likely to have low thermal stability due to the presence of the –N=Se=N– functionality in the ring.

Unlike Se_4N_4 (and S_4N_4) the only known tellurium nitride has the composition Te_3N_4, as expected for a tellurium(IV) nitride. A highly explosive yellow powder identified as "Te_3N_4" is obtained from the treatment of potassium triimidotellurite

$K_2[Te(NH)_3]$ with an excess of ammonium nitrate in liquid ammonia [29]. The structure of "Te_3N_4" may involve a μ_3-nitrido bridging three tellurium(IV) centers of a Te_3N_3 ring [30]. Support for this suggestion comes from the identification of this structural motif in the tetra-adduct $Te_6N_8(TeCl_4)_4$ in which the central feature is a dimer of Te_3N_4 [31]. The highly explosive nature of "Te_3N_4" precludes synthetic applications.

4.2.2 Cations

The simple binary cations NS^+ and NS_2^+ are important reagents in sulfur–nitrogen chemistry [1]. The selenium analogues NSe^+ and NSe_2^+ are unknown as isolated species. However, the latter may be generated *in situ* from the acyclic cation $N(SeCl)_2^+$ upon reduction with $SnCl_2$ (see Section 4.3).

4.2.3 Anions

The acyclic sulfur–nitrogen anions $SSNS^-$ and $SNSNH^-$ may be used as isolated or *in situ* reagents, respectively, for the generation of metal complexes [32]. Although the selenium analogues have not been characterized in solution or in the solid state, solutions of Se_4N_4 in liquid ammonia behave as an *in situ* source of such anions. Thus, Se_4N_4 dissolves in liquid ammonia at high pressure (*ca.* 50 atm.) and reacts with $PtCl_2(PMe_2Ph)_2$ to give $Pt(Se_2N_2)(PMe_2Ph)_2$ [33]. It was observed, however, that the ^{15}N-labeled material $Se_4{}^{15}N_4$ can be recovered unchanged from such solutions implying that Se_4N_4 does not react with liquid ammonia. By contrast, liquid ammonia solutions of S_4N_4 have been shown to contain binary S–N anions such as cyclic $S_3N_3^-$ and $S_4N_5^-$ by ^{14}N NMR spectroscopy [34].

Solutions of $SeCl_4$ in liquid ammonia provide a safer route to metal complexes of the $Se_2N_2^{2-}$ dianion than the reactions of Se_4N_4 described in Section 4.1, as illustrated by the example shown in eq. (4.2); the orange platinum complex $[Pt(Se_2N_2)(DPPE)]$ (DPPE = $Ph_2PCH_2CH_2PPh_2$) is isolated in 74% yield [35]. Presumably the Cl_2 by-product formed in this reaction is purged under the stream of nitrogen that is used to remove the excess of ammonia, since it is known that complexes of the type $[Pt(Se_2N_2)(PR_3)_2]$ react with halogens to regenerate Se_4N_4 [36].

$$[PtCl_2(DPPE)] + 2SeCl_4 + 8NH_3 \rightarrow [Pt(Se_2N_2)(DPPE)] + 2Cl_2 + 6NH_4Cl \qquad (4.2)$$

4.3 Chalcogen–nitrogen halides

4.3.1 Ternary selenium–nitrogen halides

As a source of the –SeNSe– unit the cationic selenium–nitrogen halides $[Cl_2SeNSeCl_2]^+$ and $[ClSeNSeCl]^+$ are potential building blocks in Se–N chemistry; they can be prepared in gram-scale quantities from readily available reagents. The former species was obtained in an attempt to generate the binary cation NSe^+ from the reaction of $[SeCl_3][AsF_6]$ with $N(SiMe_3)_3$. When this reaction is carried out in $CFCl_3$ solution at 0 °C the orange solid $[(SeCl_2)_2N][AsF_6]$ is produced in 83% yield [eq. (4.3)] [37]. The use of the hexachloroantimonate $[SeCl_3][SbCl_6]$ rather than the hexafluoroarsenate salt results in loss of Cl_2 from the initial product to generate the $[ClSeNSeCl]^+$ cation as the orange $SbCl_6^-$ salt in 89% yield [38] [eq. (4.4)].

$$6[SeCl_3][AsF_6] + 5N(SiMe_3)_3 \rightarrow 3[(SeCl_2)_2N][AsF_6] + 9Me_3SiF + 6Me_3SiCl + 3AsF_3 + N_2 \quad (4.3)$$

$$2[SeCl_3][SbCl_6] + N(SiMe_3)_3 \rightarrow [(SeCl)_2N][SbCl_6] + SbCl_5 + Cl_2 + 3Me_3SiCl \quad (4.4)$$

The reaction of $[(SeCl)_2N][SbCl_6]$ with $CF_3C{\equiv}CCF_3$ and $SnCl_2$ in liquid SO_2 produces the five-membered cationic ring $CF_3CSeNSeCCF_3^+$ as the colorless $SbCl_6^-$ salt [39, 40] (Figure 4.3) This transformation probably occurs via the intermediate formation of NSe_2^+ (Section 4.2.2), which undergoes cycloaddition with the alkyne. The reduction of the cyclic $CF_3CSeNSeCCF_3^+$ cation with sodium dithionite in liquid SO_2 occurs with the loss of nitrogen to produce the cyclic diselenide $CF_3CSeSeCCF_3$ as a red liquid in 40% yield, which slowly dimerizes to an eight-membered ring (Figure 4.3). For comparison, the corresponding reaction with the sulfur analogue generates the neutral 7π-electron radical $CF_3CSNSCCF_3$ as a black-green liquid via a one-electron reduction [41].

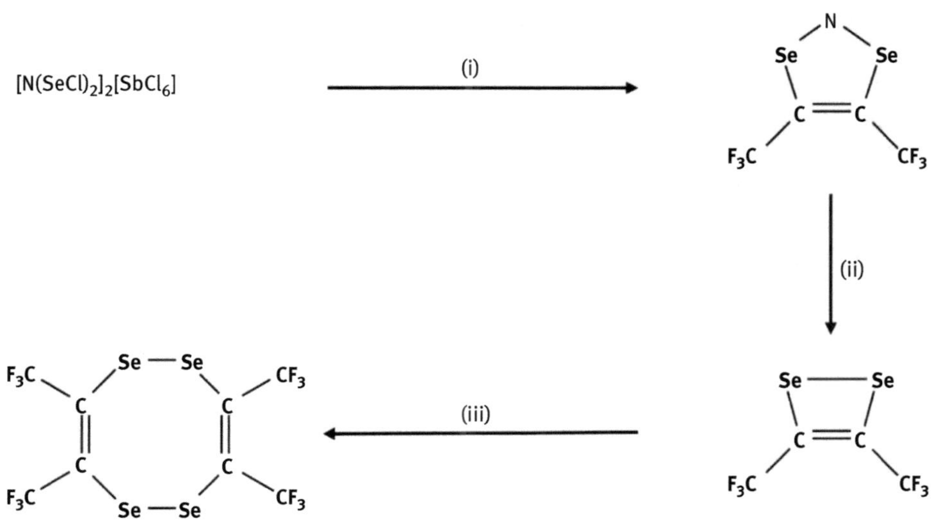

Figure 4.3: Oxidative-addition of *in situ*-generated "NSe_2^{+}" to an alkyne: (i) excess $SnCl_2$, $CF_3C{\equiv}CCF_3$, (ii) $Na_2S_2O_4$ and (iii) 0 °C, 7 d.

Red crystals of the selenium–nitrogen chloride Se_2NCl_3 are isolated in 57% yield from the reaction of tris(trimethylsilyl)amine with selenium tetrachloride in a 1:2 molar ratio in *boiling* dichloromethane [42]. If the same reaction is carried out *at 0–20* °C in CH_2Cl_2 the chlorine-rich product Se_2NCl_5 is obtained as a pale pink powder in 94% yield [43]. Under similar conditions the reaction of $SeBr_4$ with $N(SiMe_3)_3$ generates the explosive bromo derivative Se_2NBr_3 [43].

Reaction of Se_2NCl_3 with the chloride-ion acceptor $GaCl_3$ produces the acyclic $[ClSeNSeCl]^+$ cation (Figure 4.4(a)). The cyclic cation $[Se_3N_2Cl]^+$ is formed as the $SbCl_6^-$ salt upon treatment of Se_2NCl_5 with antimony(III) chloride in CH_2Cl_2 (Figure 4.4(b)); this cation is also obtained, as the $GaCl_4^-$ salt in 50% yield, upon reduction of $[ClSeNSeCl]^+$ with triphenylantimony in dichloromethane [44, 45].

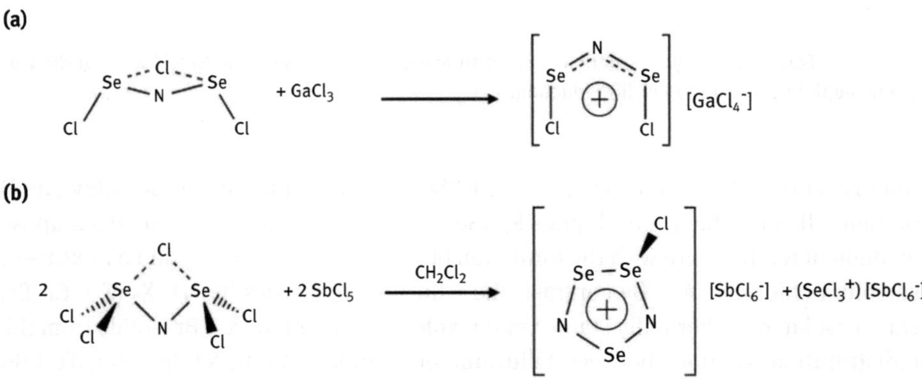

Figure 4.4: Formation of (a) acyclic $[ClSeNSeCl]^+$ and (b) cyclic $[Se_3N_2Cl]^+$ cations from selenium–nitrogen chlorides and Lewis acids.

There are no tellurium analogues of the selenium–nitrogen halides Se_2NCl_x (x = 3, 5). Instead the reaction of $N(SiMe_3)_3$ with tellurium tetrachloride in a 1:2 molar ratio in CH_2Cl_2 followed by treatment of the white precipitate with AsF_5 produces the $Te_4N_2Cl_8^{2+}$ dication as the AsF_6^- salt [45]. This dication is formally a dimer of the hypothetical tellurium(IV) imide $[Cl_3Te{-}N = TeCl]^+$, however no reactions of this species have been reported.

4.3.2 Imido chalcogen halides

A homologous series of moisture-sensitive, acyclic *tert*-butylimido selenium(II) chlorides $ClSe[N(^tBu)Se]_nCl$ (n = 1–3) have been isolated from the cyclocondensation reactions of *tert*-butylamine with $SeCl_2$ in various stoichiometries [46]. These intermediates react with *tert*-butylamine to form cyclic selenium imides via either nucleophilic substitution or reduction; the latter process produces rings with an Se–Se linkage

(Figure 4.5). The bifunctional reagents $ClSe[N(^tBu)Se]_nCl$ (n = 1–3) are potential synthons for the incorporation of other main-group elements into Se–N rings.

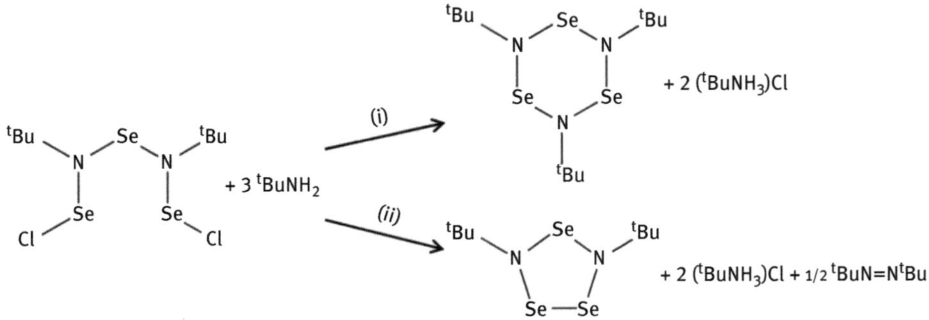

Figure 4.5: Formation of cyclic selenium imides from reactions of $ClSe[N(^tBu)Se]_2Cl$ with $^tBuNH_2$ *via* (i) nucleophilic substitution or (ii) reduction.

Imidoselenium(IV) dihalides of the type $RNSeCl_2$ are either unknown (R = alkyl, aryl) or thermally unstable yellow liquids R_FNSeCl_2 ($R_F = CF_3$ or C_2F_5); the latter decompose at ambient temperature with the formation of the diazene $R_FN = NR_F$ and a mixture of selenium chlorides [47]. By contrast, the tellurium analogues $^tBuNTeX_2$ (X = Cl, Br) can be isolated as thermally stable yellow-gold (X = Cl) or red (X = Br) solids from the redistribution reactions between tellurium tetrahalides and $^tBuNTe(\mu\text{-}N^tBu)_2TeN^tBu$ in THF at 23 °C [eq. (4.5)] [48]. The halide exchange between $[Cl_2Te(\mu\text{-}N^tBu)_2TeCl_2]_2$ and trimethylsilyl bromide provides a cleaner route to the corresponding bromide.

$$^t\mathrm{BuNTe}(\mu-\mathrm{N}^t\mathrm{Bu})_2\mathrm{TeN}^t\mathrm{Bu} + 2\mathrm{TeX}_4 \rightarrow 4/n[\mathrm{X_2Te}(\mu-\mathrm{N}^t\mathrm{Bu})_2\mathrm{TeX}_2]_2\ (\mathrm{X} = \mathrm{Cl,\ Br}) \tag{4.5}$$

The potential of the imidotellurium(IV) dichloride as a reagent in Te–N chemistry is indicated by the reaction with potassium *tert*-butoxide in THF, which proceeds at 23 °C to give the metathetical product in 71% yield [eq. (4.6)] [49].

$$1/2[\mathrm{Cl_2Te}(\mu-\mathrm{N}^t\mathrm{Bu})_2\mathrm{TeCl}_2]_2 + 4\mathrm{KO}^t\mathrm{Bu} \rightarrow (^t\mathrm{BuO})_2\mathrm{Te}(\mu-\mathrm{N}^t\mathrm{Bu})_2\mathrm{Te}(\mathrm{O}^t\mathrm{Bu})_2 + 4\mathrm{KCl} \tag{4.6}$$

4.4 $E(NSO)_2$ reagents (E = Se, Te)

Bis(sulfinylamino)chalcogenanes $E(NSO)_2$ (E = Se Te) reagents are readily prepared by the methods shown in eqs (4.7) and (4.8). The yield of yellow crystals of $Se(NSO)_2$

is 69% when the reaction is conducted in CH_2Cl_2 [49, 50]. Although $SeCl_2$ is thermally unstable, it is possible that this reagent could be used (to avoid loss of selenium) if the reaction is carried out in THF at 0 °C [51].

$$2Me_3SiNSO + Se_2Cl_2 \rightarrow Se(NSO)_2 + 1/8Se_8 + 2Me_3SiCl \quad (4.7)$$

There are a number of synthetic routes to the tellurium analogue $Te(NSO)_2$. The recommended procedure involves the reaction of $Te(SCF_3)_2$ with $Hg(NSO)_2$ in CS_2 at 50 °C for 5 days in the absence of light [eq. (4.8)] [52]. The sparingly soluble solid is isolated in 60% yield after sublimation. The unusual use of photolytically sensitive $Te(SCF_3)_2$ as a source of tellurium(II) in this synthesis probably reflects the instability of tellurium(II) dihalides TeX_2 (X = Cl, Br).

$$Te(SCF_3)_2 + Hg(NSO)_2 \rightarrow Te(NSO)_2 + Hg(SCF_3)_2 \quad (4.8)$$

Bis(sulfinylamino)chalcogenanes $E(NSO)_2$ (E = Se, Te) are versatile reagents in chalcogen–nitrogen chemistry [53, 54]. A common reaction of thionylimines RNSO is the thermal or base-promoted elimination of SO_2 to give the corresponding acyclic sulfur(IV) diimide RN = S = NR [1]. In the case of $Se(NSO)_2$, this process gives rise to heterocyclic compounds containing different chalcogens. For example, the reaction of $Se(NSO)_2$ with $TiCl_4$ produces the highly insoluble yellow adduct $SeSN_2{\bullet}TiCl_4$ (Figure 4.6), which is assumed to have a polymeric structure with bridging four-membered $SeSN_2$ rings [50]. By contrast, reactions with the oxidizing Lewis acids MF_5 (M = As, Sb) produce the dimeric cation $[Se_2SN_2]_2^{2+}$ as $[MF_6]^-$ salts [50]. A different type of behavior is observed with $SeCl_4$, which yields the cations $[ClSe_2SN_2]^+$ as the chloride salt [50]. The corresponding reactions with tellurium tetrahalides TeX_4 (X = Cl, Br) generate $X_2TeSeSN_2$, a five-membered ring that contains all three heavy chalcogens (Figure 4.6) [55].

In contrast to the behavior of $Se(NSO)_2$, the oxidation of the tellurium analogue $Te(NSO)_2$ with Cl_2 in CS_2 produces high yields of the tellurium(IV) derivative $Cl_2Te(NSO)_2$, which can also be obtained quantitatively by treatment of elemental tellurium with ClNSO in CS_2 at 0 °C for 48 h [52]. $Cl_2Te(NSO)_2$ undergoes SO_2 elimination upon reaction with Cl_2 to form the dinuclear complex $Cl_2Te(\mu\text{-}Cl)(\mu\text{-}Cl)(\mu\text{-}NSN)TeCl_2$ or when heated at 95 °C (Figure 4.7) [52]

4.5 Selenium/tellurium–nitrogen–silicon reagents

The susceptibility of Si–N bonds to cleavage by main-group element halides accompanied by facile elimination of volatile trimethylsilyl halides render selenium–nitrogen–silicon reagents versatile sources of other selenium–nitrogen compounds. However, the thermal instability of the selenium(IV) diimide $Me_3SiNSeNSiMe_3$ (Section 4.7) has limited the application of this reagent. It can be prepared by the reaction of $(Me_3Si)_2NLi$ with

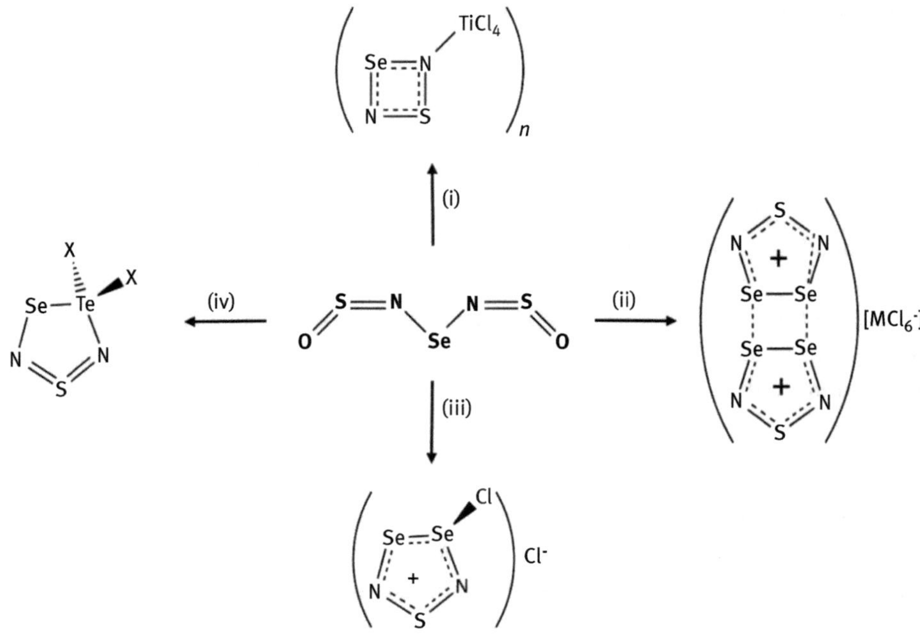

Figure 4.6: Formation of mixed-chalcogen heterocycles from $Se(NSO)_2$ and Lewis acids: (i) $TiCl_4$, (ii) MF_5 (M = As, Sb), (iii) $SeCl_4$ and (iv) TeX_4 (X = Cl, Br).

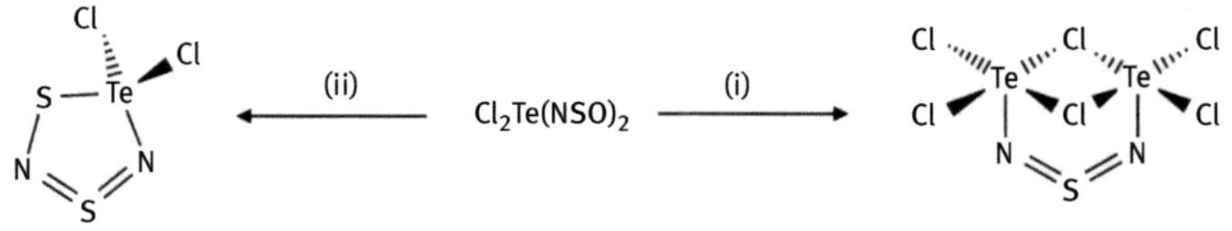

Figure 4.7: SO_2 elimination reactions of $Cl_2Te(NSO)_2$: (i) Excess Cl_2, CS_2 and (ii) 95 °C, 12 h.

$SeOCl_2$ in pentane [56] or diethyl ether [57]. Reaction of *in situ*-generated $Me_3SiNSeNSiMe_3$ with acyclic $[(NPPh_2Cl)_2]Cl$ produces the six-membered ring $(Ph_2PN)_2(NSeCl)$, which can be isolated as pale yellow crystals in 40% yield, suggesting that wider use of this reagent should be possible (Figure 4.8) [57].

By contrast to the selenium(IV) analogue (*vide supra*), the selenium(II) derivative $(Me_3Si)_2NSeN(SiMe_3)_2$ is thermally more stable and, hence, represents a more useful reagent in Se–N chemistry. The reaction of $LiN(SiMe_3)_2$ with Se_2Cl_2 in a 2:1 molar ratio in *n*-hexane at −78 °C produces this monoselenide as yellow crystals in 75% yield [eq. (4.9)] [58]. In a subsequent investigation of the same reaction, the diselenide $(Me_3Si)_2NSeSeN(SiMe_3)_2$ was isolated as a yellow oil in 19% yield, in addition to the monoselenide $(Me_3Si)_2NSeN(SiMe_3)_2$ (64%) [59]. The triselenide $(Me_3Si)_2NSeSeSeN(SiMe_3)_2$ is the major

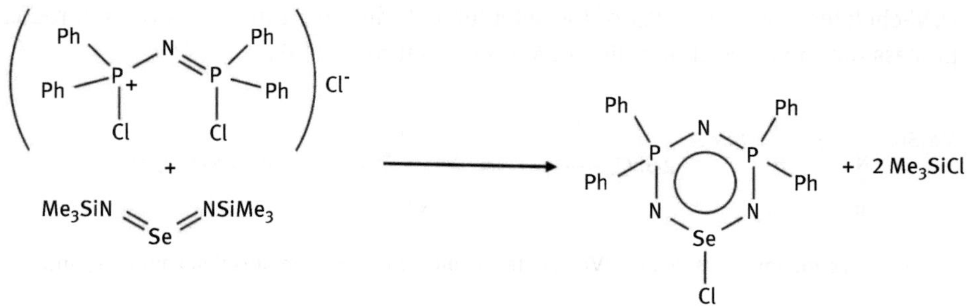

Figure 4.8: Generation of a P_2N_3Se heterocycle from $Me_3SiNSeNSiMe_3$.

product of the reaction of $(Me_3Si)_2NH$ with Se_2Cl_2 in a 2:1 molar ratio in CH_2Cl_2 at 0 °C; however, the triselenide could not be separated in pure form from the co-formed di- and tetra-selenides [59]

$$2LiN(SiMe_3)_2 + Se_2Cl_2 \rightarrow Se[N(SiMe_3)_2]_2 + 1/8Se_8 + 2LiCl \qquad (4.9)$$

The easily handled reagent $(Me_3Si)_2NSeN(SiMe_3)_2$ reacts with $SeCl_4$ to produce pure Se_4N_4 in 71% yield [15], but the method depicted in eq. (4.1) is the preferred route to this explosive material. An extension of this methodology to the reaction of $(Me_3Si)_2NSeN(SiMe_3)_2$ with a mixture of SCl_2 and SO_2Cl_2 in a 1:1:1 molar ratio produced good yields of the mixed-chalcogen cage 1,5-$Se_2S_2N_4$ as an insoluble red-brown powder accompanied by small amounts of red selenium [60]. By contrast, a mixture of six-membered rings, predominantly SeS_3N_2, is produced in the reaction of $(Me_3Si)_2NSN(SiMe_3)_2$ with a 3:1:1 mixture of S_2Cl_2, Se_2Cl_2 and $SeCl_4$ [61].

The explosive black selenium–nitrogen chloride $Se_3N_2Cl_2$ is obtained in 95% or 89% yields, respectively, by reaction of $(Me_3Si)_2NSeN(SiMe_3)_2$ with (a) a mixture of $SeCl_4$ and Se_2Cl_2 in a 4:1 molar ratio [eq. (4.10)] or (b) $SeOCl_2$ [19]; these reactions are carried out in CH_2Cl_2 at −78 °C. The related dimeric selenium–nitrogen chloride $(Se_3N_2Cl)_2$ is isolated as an insoluble dark brown powder in essentially quantitative yield by adjusting the stoichiometry of the reactants [eq. (4.11)] [19].

$$Se[N(SiMe_3)_2]_2 + 4SeCl_4 + Se_2Cl_2 \rightarrow Se_3N_2Cl_2 + 12Me_3SiCl \qquad (4.10)$$

$$2Se[N(SiMe_3)_2]_2 + 2SeCl_4 + Se_2Cl_2 \rightarrow (Se_3N_2Cl)_2 + 8Me_3SiCl \qquad (4.11)$$

The unsymmetrically substituted monoselenide $(Me_3Si)^tBuNSeN^tBu(SiMe_3)$ with only one Si–N bond per amido group is obtained in 75% yield as a yellow solid from the reaction of the corresponding lithium amide with Se_2Cl_2 in *n*-hexane [62]. Treatment of this monoselenide with $SnCl_4$ in a 1:2 molar ratio in dichloromethane produces the

N,N'-chelated $SnCl_4$ complex of the selenium(IV) diimide $^tBuN=Se=N^tBu$ in a redox process that involves elimination of Me_3SiCl (Figure 4.9) [63].

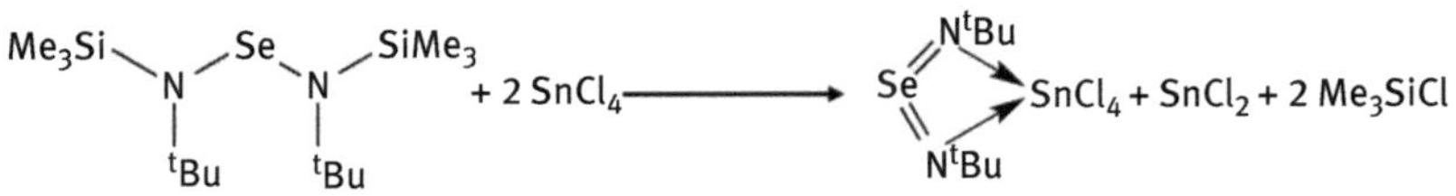

Figure 4.9: Formation of a selenium(IV) diimide complex from a bis(amido)selenium(II) reagent.

The formation of long chains typified by the polymer $(SN)_x$ is a prominent feature of sulfur–nitrogen chemistry [1]. In that context the preparation of the yellow monoselenide $(Me_3SiNSN)_2Se$ in 65% yield by the route shown in eq. (4.12) is significant [64]. The treatment of this monoselenide with $SeCl_2$ in CH_2Cl_2/THF at −78 °C, however, results in cyclization to give the cage molecule 1,5-$Se_2S_2N_4$ as a dark red precipitate in 73% yield [eq. (4.13)] [64].

$$2(Me_3SiN)_2S + SeCl_2 \rightarrow (Me_3SiNSN)_2Se + 2Me_3SiCl \tag{4.12}$$

$$(Me_3SiNSN)_2Se + SeCl_2 \rightarrow 1,5-Se_2S_2N_4 + 2Me_3SiCl \tag{4.13}$$

4.6 Chalcogenylnitrosyls ArN=Se

Chalcogenylnitrosyls RN=Se are formed as transient species when R = aryl. The selenium(II) synthon $PhSO_2SeCl$ has been used as a source of selenium in reactions with arylamines $ArNH_2$ in the presence of trimethylamine. The selenonitrosoarene ArN=Se can be trapped as a Diels–Alder adduct with dimethylbutadiene (Figure 4.10) [65]. The 1,2-selenazines formed in this way have limited stability (2–3 h) at room temperature when Ar = 4-XC_6H_4 (X = Br, Me), but this is improved to 3 days when Ar = 2-$MeSC_6H_4$. Dimers of RNSe have not been observed, however larger cyclic oligomers $(RNSe)_n$ (R = alkyl, n = 3, 4) have been isolated as stable crystalline solids from cyclocondensation reactions of $SeCl_2$ with primary amines [46] or the thermal decomposition of selenium(IV) diimides (Section 4.7).

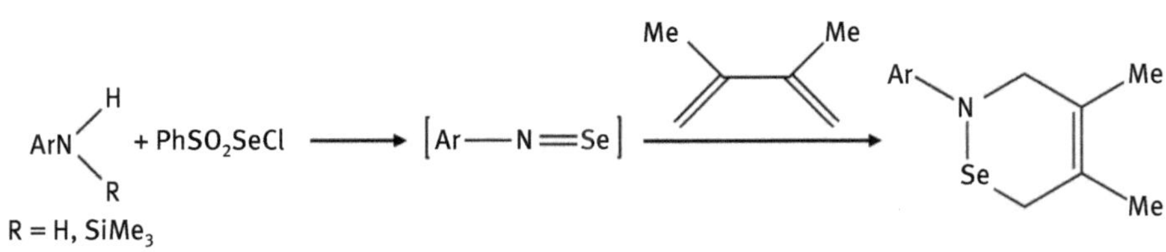

Figure 4.10: Generation and trapping of ArN=Se with dimethylbutadiene.

4.7 Chalcogen(IV) diimides

4.7.1 Synthesis, structures and applications in organic synthesis

The selenium(IV) diimides RN=Se=NR (R = ^{t}Bu, Ts) (Ts = *p*-toluenesulfonyl) were first reported and used as *in situ* reagents for the allylic amination of olefins or 1,3-dienes more than 40 years ago [66, 67]. In early experiments, these imidoselenium reagents were prepared by the reaction of $SeCl_4$ with two equivalents of the corresponding amine (*tert*-butylamine or *p*-toluenesulfonamide) in CH_2Cl_2 in the presence of four equivalents of an amine base. A more reactive aminating reagent is formed when anhydrous chloramine-T (TsNClNa) is stirred with elemental selenium in CH_2Cl_2 [66, 67]. Owing to the insolubility of the reactants in this synthesis, a subsequent modification involved the use of the more soluble 2-nitrobenzenesulfonamide (NsNClNa) for the generation of NsN=Se=NNs [68]. This *in situ* reagent reacts in a manner similar to selenium dioxide in the allylic amination of olefins and 1,2-diamination of 1,3-dienes (Figure 4.11) [68].

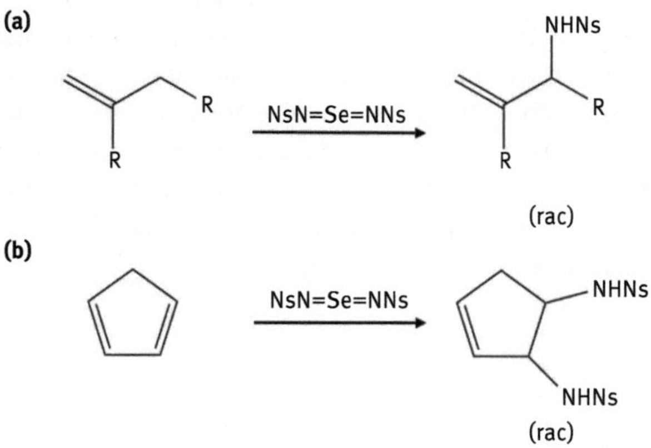

Figure 4.11: (a) Allylic amination and (b) 1,2-diamination with a diimidoselenium(IV) reagent (Ns = 2-nitrobenzenesulfonyl).

The thermally unstable dialkyl selenium(IV) diimides RN=Se=NR (R = ^{t}Bu, Ad) (Ad = adamantyl) have been isolated as an oil or yellow crystals, respectively, from the reaction of the appropriate primary alkylamine with $SeCl_4$ [69, 70]; the introduction of supermesityl substituents (R = Mes* = 2,4,6-^{t}Bu$_3$C$_6$H$_2$) enhances the thermal stability of the selenium(IV) diimide [71]. The X-ray analyses of RN=Se=NR (R = Ad, Mes*) reveal monomeric structures in the solid state [70, 71]. By contrast, the tellurium(IV) diimide ^{t}BuNTe(μ-N^{t}Bu)$_2$TeN^{t}Bu, which is obtained in good yields as a thermally stable orange solid from the reaction of LiNH^{t}Bu with $TeCl_4$ in THF, is dimeric [72]. The formation of the tellurium(IV) diimide is accompanied by smaller amounts of the cyclic tellurium(II) imide $(TeN^tBu)_3$ when this reaction is conducted in toluene [73]. The calculated dimerization energy for the [2 + 2] cycloaddition of two $E(NR)_2$ (E = Se, Te; R = ^tBu, $SiMe_3$) molecules is strongly exothermic for tellurium(IV) diimides, but approximately thermoneutral for selenium(IV) diimides, consistent with the experimental observations [74, 75]. However, the dimerization of selenium(IV) diimides is promoted by group 12 metal dihalides MCl_2 (M = Cd, Hg) to give *N,N′*-chelated complexes of $^tBuNSe(\mu\text{-}N^tBu)_2SeN^tBu$ [76], which are structurally analogous to the $HgCl_2$ complex of the tellurium(IV) diimide dimer $^tBuNTe(\mu\text{-}N^tBu)_2TeN^tBu$ [77].

4.7.2 Thermal decomposition

The thermal decomposition of RN=Se=NR (R = ^tBu, Ad) provides a source of several cyclic selenium imides (Figure 4.12) [75, 78]. Thus, a solution of $^tBuN{=}Se{=}N^tBu$ in toluene at 20 °C produces the 6-membered ring $Se_3(N^tBu)_3$, the 5-membered ring $Se_3(N^tBu)_2$ and its higher homologue the 15-membered ring $Se_9(N^tBu)_6$ [78]. Similarly, the thermal decomposition of AdN=Se=NAd in THF gives rise to $Se_3(NAd)_3$ and $Se_3(NAd)_2$; the five-membered ring structure of the latter was confirmed by X-ray analysis [75]. Cyclodimerization of acyclic RN=Se=NR to give a four-membered ring (Section 4.7.1) may be the first step in the formation of these cyclic selenium imides. However, the generation of selenium-rich systems must involve elemental selenium, which is formed during the thermal decomposition as evinced by the detection of $^tBuN{=}N^tBu$ as a by-product [78]. A more versatile route to the generation of cyclic selenium imides is provided by cyclocondensation reactions of $SeCl_2$ with $^tBuNH_2$, since the stoichiometry can be controlled (Section 4.3) [46].

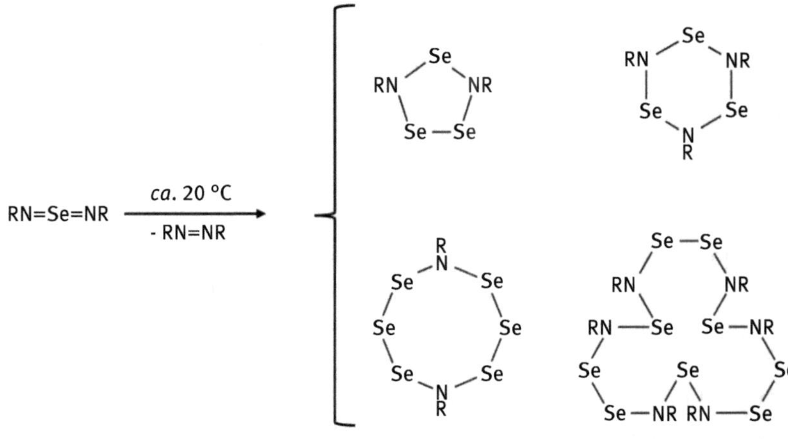

Figure 4.12: Formation of cyclic selenium imides from thermal decomposition of RN=Se=NR (R = ^tBu, Ad).

4.7.3 Insertion and cycloaddition reactions

Selenium(IV) diimides readily undergo insertion reactions into the Si–Si bond of hexachlorodisilane [eq. (4.14)] [79]. Insertion into the C–B bond of triethylborane occurs at low temperatures, but the initially formed cyclic product $^tBuN(\mu\text{-}SeEt)(\mu\text{-}BEt_2)N^tBu$ decomposes at –50 °C [80].

$$RN = Se = NR + Cl_3Si - SiCl_3 \rightarrow R(Cl_3Si)N - Se - N(SiCl_3)R \quad (4.14)$$

A different type of behavior is observed between selenium diimides and bis(amino) stannylenes [81]. In a 1:1 molar ratio these reactants produce a red spirocylic compound, which formally results from the oxidative-addition of the selenium diimide to the Sn(II) center. However, the use of an excess of the stannylene results in the insertion of a Sn(II) center into an Sn–N bond of the initially formed spirocyclic compound to give a yellow tricyclic compound with a Sn–Sn bond (Figure 4.13) [81].

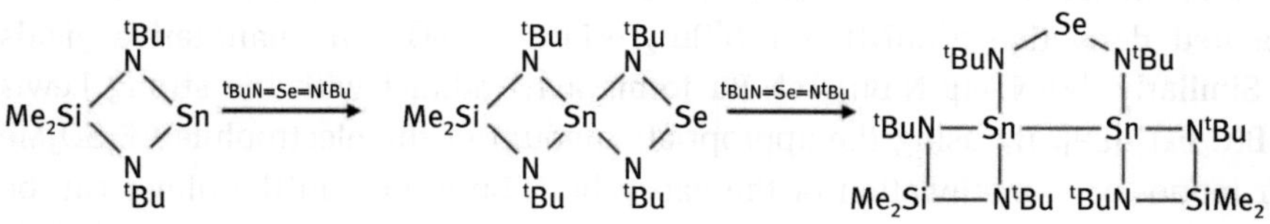

Figure 4.13: Reactions of selenium diimides with a bis(amino)stannylene.

A double cycloaddition occurs in the reaction of the tellurium(IV) diimide dimer $^tBuNTe(\mu\text{-}N^tBu)_2TeN^tBu$ with an excess of *tert*-butyl isocyanate (Figure 4.14) [82]. The initial step generates an *N,N′*-ureato tellurium imide, which can be viewed as the cycloaddition product of ^tBuNCO and monomeric tellurium(IV) imide. With additional ^tBuNCO, this intermediate is converted to the corresponding telluroxide, which is isolated as a dimer.

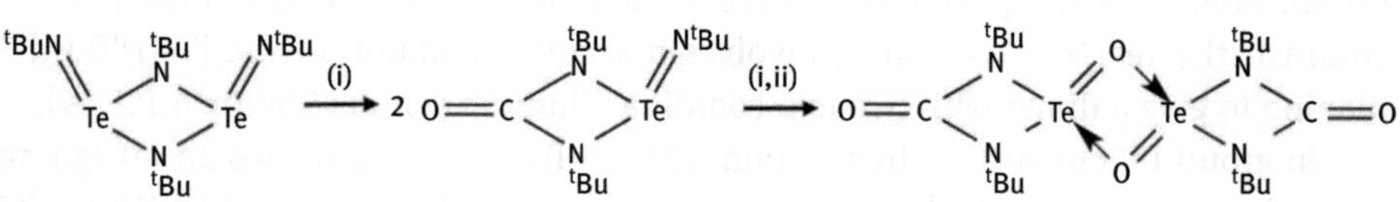

Figure 4.14: Cycloaddition reactions of a tellurium diimide dimer with *tert*-butyl isocyanate. (i) + 2^tBuNCO and (ii) – $2^tBuN{=}C{=}N^tBu$

4.7.4 Reactions with nucleophiles and electrophiles

The sensitivity of the polar chalcogen–nitrogen bonds in chalcogen(IV) diimides to nucleophilic attack at the chalcogen center is signified by the ready conversion of the >E=NR (E = Se, Te) functionality to the corresponding >E=O group upon exposure to moisture. The formation of the partial hydrolysis products AdNSe(μ-NAd)$_2$SeO and OSe(μ-NAd)$_2$SeO during the decomposition of AdN=Se=NAd in THF is a cogent example of this tendency [75]. In the case the tellurium(IV) diimide ^{t}BuNTe(μ-N^{t}Bu)$_2$TeN^{t}Bu deliberate hydrolysis using stoichiometric amounts of $(C_6F_5)_3B\bullet H_2O$ results in the successive replacement of the terminal imido groups by oxido ligands [83].

Both monomeric selenium(IV) diimides and dimeric tellurium(IV) diimides are susceptible to nucleophilic attack at the chalcogen center by reagents such as LiNH^{t}Bu or KO^{t}Bu to give the tripodal dianions $[E(N^tBu)_3]^{2-}$ [84, 85] and $[Te(O^tBu)(N^tBu)_2]^-$ [86], respectively. Reactions of the tris(*tert*-butylimido)tellurite dianion $Te(N^tBu)_3]^{2-}$ are discussed in Section 4.8 [87].

The terminal N^{t}Bu groups in the dimer ^{t}BuNTe(μ-N^{t}Bu)$_2$TeN^{t}Bu are readily protonated by Brønsted acids. For example, reaction with HCF_3SO_3 produces the monoprotonated derivative [(^{t}BuNH)Te(μ-N^{t}Bu)$_2$TeN^{t}Bu][CF_3SO_3] in quantitative yields [88]. Similarly, ^{t}BuNTe(μ-N^{t}Bu)$_2$TeN^{t}Bu forms a 1:1 adduct with the strong Lewis acid $B(C_6F_5)_3$ [89]. By using the appropriate amount of the electrophile CF_3SO_3Me either mono- or di-methylation of the exocyclic N^{t}Bu groups in the dimer can be achieved [90].

4.8 Tris(*tert*-butylimido)tellurite dianion

Tris(*tert*-butylimido)chalcogenite dianions $[E(N^tBu)_3]^{2-}$ (E = Se, Te) are potentially versatile reagents in chalcogen–nitrogen chemistry. In the case of the selenium derivative (E = Se) applications are limited due to facile air oxidation to give radical species [91]. However, detailed investigations of the reaction chemistry of the tellurium derivative have revealed a rich coordination chemistry that, in some cases, is accompanied by redox behavior (Figure 4.15) [87]. For example, in group 13 chemistry the reaction with $PhBCl_2$ produces an *N,N′*-chelated complex with a terminal N^{t}Bu group, which is an unique representative of a monomeric tellurium(IV) imide [84]. By contrast, the reaction with $InCl_3$ involves a second chelation of the $[Te(N^tBu)_3]^{2-}$ dianion to give a dimer with two five-coordinate indium centers (Figure 4.15) [92].

In group 14, chemistry, the reaction with tin(II) chloride provides an intriguing case of redox behavior [93]. The product incorporates the pyramidal $[Te(N^tBu)_3]^{2-}$ ligand bridging two tin atoms in different oxidation states, one of which is attached to a terminal telluride. The structure is completed by an N^{t}Bu bridge between the two tin centers resulting from Te–N^{t}Bu bond cleavage in the redox process. Redox behavior

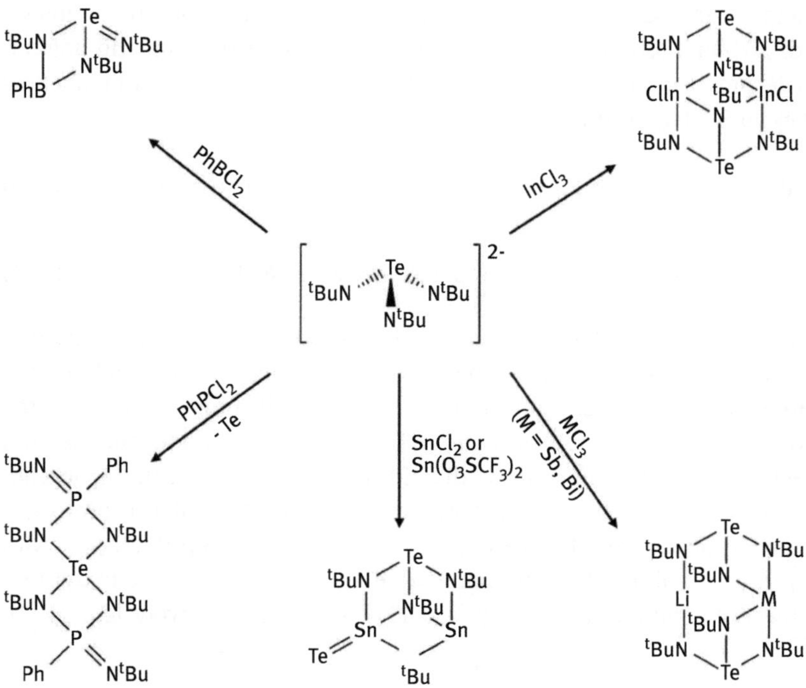

Figure 4.15: Reactions of the tris(*tert*-butylimido)tellurite dianion $[Te(N^tBu)_3]^{2-}$ with main-group element halides.

is also observed in the reaction of $[Te(N^tBu)_3]^{2-}$ with $PhPCl_2$ [87]. In this transformation, half of the tellurium in the dianion is reduced to tellurium metal while the two phosphorus centers are oxidized from the +3 to +5 oxidation states to give a spirocyclic complex of tellurium(IV) (Figure 4.15). By contrast, the reaction of $Li_2[Te(N^tBu)_3]$ with the heavier group 15 trihalides ECl_3 (E = Sb, Bi) generates the homoleptic monoanionic complexes $E[Te(N^tBu)_3]_2^-$ as their monolithium derivatives [94].

4.9 Chalcogen(II) amides and diamides

The selenium(II) derivative $(^iPr_2N)_2Se$ has been isolated as a white solid in 61% yield from the reaction of diisopropylamine with $SeOCl_2$ in diethyl ether at 0 °C; $(Me_2N)_2Se$ was obtained as an oil in 30% yield from the decomposition of $(Me_2N)_2SeO$, which was prepared by reaction an excess of dimethylamine with $SeOCl_2$ in hexane [95]. In contrast to the poorly characterized $Se(NR_2)_2$ (R = Me, Et) derivatives [95], polyselanes with terminal piperidino or morpholino groups can be prepared by heating black selenium powder with morpholine or piperidine in the presence of red lead. In the case of piperidine, this procedure provides the tetraselane as red crystals in an optimum yield of 26% based on selenium [eq. (4.15)], whereas morpholine gives a mixture of di-, tri- and tetra-selanes [96]. An intriguing property of piperidinotetraselane is its decomposition in CS_2 at room temperature to produce a new monoclinic form of red *cyclo*-Se_8 [eq. (4.16)] [97].

$$19Se + 8C_5H_{10}NH + Pb_3O_4 \rightarrow 4Se_4(NC_5H_{10})_2 + 3PbSe + 4H_2O \tag{4.15}$$

$$Se_4(NC_5H_{10})_2 + 2CS_2 \rightarrow 3/8Se_8 + Se(S_2CNC_5H_{10})_2 \tag{4.16}$$

The applications of the silylated chalcogen(II) diamides $E[N(SiMe_3)_2]_2$ resulting from the facile cleavage of the Si–N bond by main-group element halides are discussed in Section 4.5. A fascinating feature of the tellurium derivative is the formation of the monomeric radical cation $[Te[N(SiMe_3)_2]_2]^{+\bullet}$ upon one-electron oxidation with AsF_5 [98]. Tellurium(II) diamides are useful reagents in organotellurium chemistry owing to the susceptibility of the Te–N bonds to protolysis by weakly acidic substrates. For example, the reaction of $Te[N(SiMe_3)_2]_2$ with phenylacetylene in boiling THF produces a mixture of bis(phenylethynyl)telluride and 1,4-diphenylbutadiene in 38 and 12% yields, respectively [eq. (4.17)] [99, 100].

$$\begin{aligned}(Me_3Si)_2NTeN(SiMe_3)_2 + 4PhC\equiv CH \rightarrow\ & PhC\equiv CTeC\equiv CPh + PhC\equiv C-C\equiv CPh \\ & + 4(Me_3Si)_2NH\end{aligned} \tag{4.17}$$

In a similar manner, alkynyl tellurides RC≡CPh may be obtained in good yields (R = Ph, 72%; R = ^nBu, 85 %) by treatment of arenetellurenamides with terminal acetylenes in THF at 20 °C eq. (4.18); the reagent $PhTeN^iPr_2$ is generated *in situ* by treatment of benzenetellurenyl iodide with LiN^iPr_2 in THF [101]. The transformation shown in eq. (4.18) is specific for tellurium, since it does not occur when pure $PhSeN^iPr_2$ is used [101].

$$PhTeN^iPr_2 + RC\equiv CH \rightarrow RC\equiv CTePh + HN^iPr_2 \tag{4.18}$$

Although the dimethylamido derivative $Te(NMe_2)_2$ reacts readily with a range of amines, phosphines and thiols, the only stable products that have been isolated from such reactions are those containing bulky thiolate ligands, e. g. $Te(SCPh_3)_2$, which is obtained as air- and moisture-stable orange crystals from the reaction in toluene at −78 °C [eq. (4.19)] [102]. Polymeric $[Te(NMe_2)_2]_\infty$ is isolated in 74% yield as yellow needles from the reduction of $TeCl_4$ with an excess of $LiNMe_2$ in THF-diethyl ether; this moisture-sensitive reagent deposits black metallic tellurium on exposure to air [102].

$$1/\infty\left[\{Te(NMe_2)_2\}_\infty\right] + 2Ph_3CSH \rightarrow Te(SCPh_3)_2 + 2NHMe_2 \tag{4.19}$$

The susceptibility of the selenium–nitrogen bond to protonolysis is the key to the function of the cyclic selenenamide Ebselen, an anti-inflammatory drug that acts as a glutathione peroxidase mimic. Mechanistic studies show that the Se–N bond in Ebselen reacts readily with PhSH to form a selenenyl sulfide, which is subsequently converted to the corresponding selenol and selenenic acid [103, 104] (Figure 4.16). The potentially broad range of pharmacological applications of Ebselen are under active investigation [105].

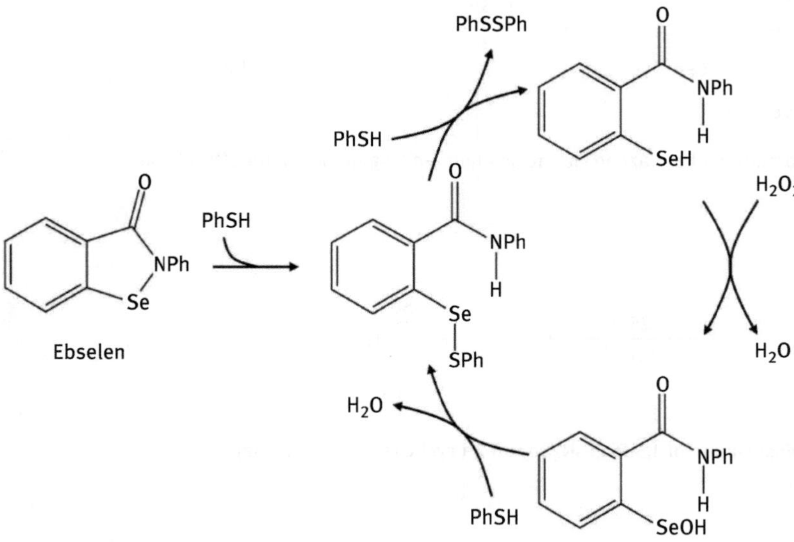

Figure 4.16: Catalytic mechanism of action of Ebselen.

4.10 Radical chemistry

The thermal lability of the Se–N linkage may engender unusual transformations resulting from homolytic cleavage and the concomitant formation of radicals. As a cogent example, the metathesis of $PhC(NSiMe_3)[N(SiMe_3)_2)]$ with three molar equivalents of PhSeCl gives rise to the resonance-stabilized radical $[PhC(NSePh)_2]^{\bullet}$ (detected by in situ EPR spectroscopy) via Se–N bond scission. This radical undergoes a rapid transformation to an intramolecularly Se–N stabilized diazene [106] (Figure 4.17).

Radical formation is also observed for P,N,Se heterocyles. A pertinent example is the dissociation of the eight-membered ring 1,5-$Ph_4P_2N_4Se_2$ into the four-membered cyclic radical $[Ph_2PNSe]^{\bullet}$ in CH_2Cl_2 solution at 25 °C [107] (Figure 4.18). In marked contrast, the sulfur analogue 1,5-$Ph_4P_2N_4S_2$ is thermally stable up to 150 °C [108].

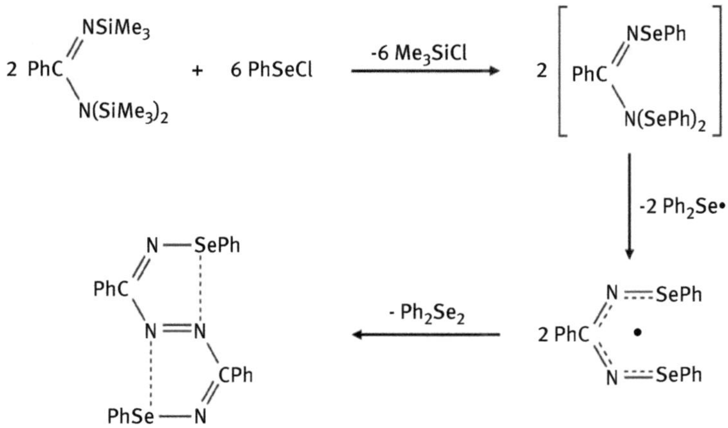

Figure 4.17: Formation of a diazene via the selenium–nitrogen radical $[PhC(NSePh)_2]^{\bullet}$.

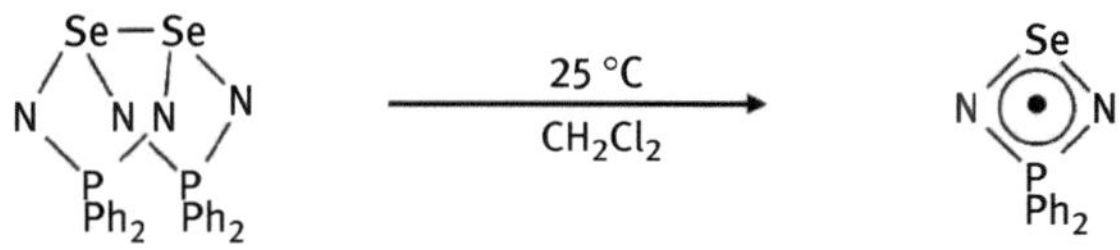

Figure 4.18: Dissociation of 1,5-$Ph_4P_2N_4Se_2$ into a cyclic radical $[Ph_2PNSe]^{\bullet}$.

Ring contraction (dissociation) or ring expansion is a common feature of inorganic heterocycles, but isomerization is rare. In this context, the slow *solid-state isomerization* of 1,5-$Ph_4P_2N_4Se_2R_2$ (R = alkyl) into the 1,3-isomer in essentially quantitative yields *at room temperature* is remarkable [109] [Figure 4.19]. The mechanism of this transformation is not known, but it represents another manifestation of novel chemistry generated by the lability of Se–N linkages.

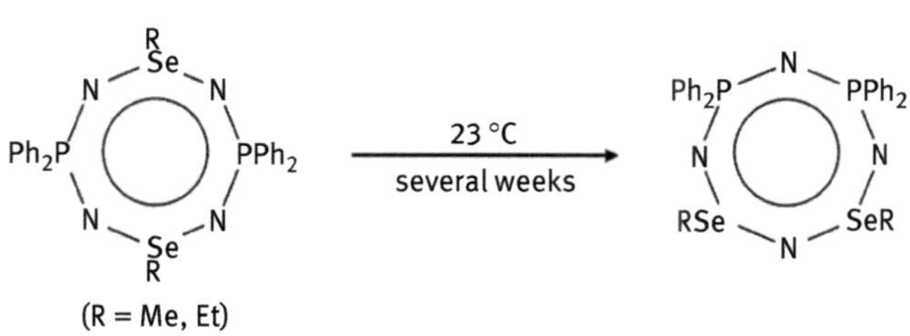

Figure 4.19: Solid-state transformation of 1,5-$Ph_4P_2N_4Se_2R_2$ into the 1,3-isomer.

4.11 Conclusions

Synthetic applications of selenium–nitrogen or tellurium–nitrogen reagents in inorganic or organic chemistry are less widespread than those of their sulfur–nitrogen analogues, in part, owing to the difficulty in handling highly reactive, unstable compounds. For example, although reactions of Se_4N_4 may generate cyclic Se–N cations or metal complexes of Se–N anions, via oxidation or reduction, respectively, the explosive nature of this binary selenium nitride has limited its applications (Section 4.2.1). Thus, all attempts to generate the selenium analogue of the conducting polymer $(SN)_x$, which is produced via S_2N_2 generated by the thermolysis of S_4N_4, have failed. Similarly, although the binary sulfur–nitrogen cations NS_x^+ (x = 1, 2) are readily prepared and exhibit a variety of insertion or cycloaddition reactions [1], the application of the selenium analogue NSe_2^+ is limited to a single example of its use as *situ* reagent (Section 4.3.1).

The use of selenium diimides RN=Se=NR in organic synthesis is well-established and aminations of unsaturated substrates by these reagents follow a similar pathway to reactions of the isoelectronic reagent selenium dioxide SeO_2 (Section 4.7.1). Selenium (IV) diimides are, however, thermally unstable and this lability has limited the applications of the silicon–nitrogen–selenium reagent $Me_3SiN=Se=NSiMe_3$ in reactions with non-metal or metal halides to a single example (Section 4.5); nonetheless, further examples of the use of this *in situ* reagent can be readily envisaged. On the other hand, the thermal decomposition of selenium(IV) diimides results in chemistry that is not observed for the sulfur analogues, notably the formation of a wide variety of thermally stable, cyclic selenium imides (Section 4.7.2). Tellurium(IV) diimides exhibit much higher thermal stability than their selenium analogues as a consequence of dimerization without further ring transformations. Consequently, the reactions of these tellurium–nitrogen compounds with electrophiles and nucleophiles have been studied in some detail (Sections 4.7.3 and 4.7.4). For example, the dianion $[Te(N^tBu)_3]^{2-}$, isolectronic with tellurite $[TeO_3]^{2-}$, is readily obtained and has been shown to have an extensive coordination chemistry with p-block elements (Section 4.8).

The susceptibility of the tellurium–nitrogen bond in tellurium(II) amides to protonolysis has been exploited in the preparation tellurium acetylides. However, the expansion of this application of tellurium diamides to the formation of other Te bonds (e. g., E = P, S) is so far limited to a single example (Section 4.9). This approach to the synthesis of tellurium compounds, as well as further applications of *in situ* reagents such as NSe_2^+ and $Me_3SiN=Se=NSiMe_3$, certainly merit further investigations.

References

[1] Chivers T. A guide to chalcogen-nitrogen chemistry. Singapore: World Scientific, 2005.

[2] Kirsch G. Compounds with Se-N and Te-N bonds. In: Patai S, editor. The chemistry of organic selenium and tellurium compounds. Vol. 2. John Wiley & Sons Ltd, 1987:421–61.

[3] Bjorgvinsson M, Roesky HW. The structures of compounds containing selenium-nitrogen and tellurium-nitrogen bonds. Polyhedron. 1991;10:353–70.
[4] Kelly PF, Slawin AMZ, Williams DJ, Woollins JD. Metal-selenium-nitrogen complexes. Chem Soc Rev. 1992:245–52.
[5] Klapötke T. Binary selenium-nitrogen species. In: Steudel. R, editor. The chemistry of inorganic ring systems. Elsevier Science Publishers. Ch. 20, 1992: 409–27.
[6] (a) Chivers T. Selenium-nitrogen chemistry. Main Group Chem News. 1993;1:6–14. (b) Chivers T, Doxsee DD. Heterocyclic selenium- and tellurium-nitrogen compounds. Comments Inorg Chem 1993;15:109–35.
[7] Haas A, Kasprowski J, Pryka M. Tellurium- and selenium-nitrogen compounds; preparation, characterization and properties. Coord Chem Rev. 1994;130:301–53.
[8] (a) Chivers T, Laitinen RS. Chalcogen-nitrogen chemistry. In: Devillanova FA, editor. Handbook of chalcogen-nitrogen chemistry: new perspectives in sulfur, selenium and tellurium. Cambridge UK: RSC Publishing, 2007:223–85. (b) Chivers T, Laitinen RS. Recent developments in chalcogen-nitrogen chemistry. In: Devillanova FA. editor. Handbook of chalcogen-nitrogen chemistry: new perspectives in sulfur, selenium and tellurium. Vol. 2. Cambridge, UK: RSC Publishing, 2013: 191–237.
[9] Laitinen RS, Oilunkaniemi R, Chivers T. Synthesis, structures, bonding, and reactions of imido-selenium and –tellurium compounds. In: Woollins JD, Laitinen RS, editor(s). Selenium and tellurium chemistry. Berlin: Springer-Verlag, 2011:103–22.
[10] Klapötke TM, Krumm B, Moll R. Compounds of selenium and tellurium with nitrogen bonds. In: Rappoport Z, editor. The chemistry of organic selenium and tellurium compounds. Vol. 3, Part 2. Chichester UK: John Wiley & Sons Ltd, 2012: 863–89.
[11] Vasiljeva J, Arsenyan P. Synthesis of Se-N and Te-N bond-containing heterocycles. Chem Heterocycl Compd. 2017;53:1061–7.
[12] Ginn VC, Kelly PF, Woollins JD. Facile introduction of ^{15}N into chalcogen-nitrogen systems using ^{15}N-labelled ammonia. J Chem Soc Dalton Trans. 1992;2129–30.
[13] Gowik PK, Klapötke TM. The Raman spectrum of tetraselenium tetranitride Se_4N_4. Spectrochim Acta. 1990;46A:1371–3.
[14] Awere EG, Passmore J, White PS. Preparation and characterization of thermally stable $[Se_3N_2]_n[AsF_6]_2$ containing the "electron-rich aromatic" 6π $[Se_3N_2]^{2+}$ (n = 1) and 7π $[Se_3N_2]^{+\bullet}$ (n = 2). J Chem Soc Dalton Trans. 1993;299–310.
[15] Siivari J, Chivers T, Laitinen RS. A simple, efficient synthesis of Se_4N_4. Inorg Chem. 1993;32:1519–20.
[16] Adel J, El-Kholi A, Willing W, Muller U, Dehnicke K. Se_4N_4 as a reagent for synthesis of chloroselenonitrene-tungsten complexes. Chimia. 1988;42:70–1.
[17] Adel J, Dehnicke K. Synthesis of the chloroselenonitrene complex with molybdenum $[Cl_4Mo(NSeCl)]_2$. Chimia. 1988;42:413–?
[18] Awere EG, Passmore J, White PS, Klapötke TM. The preparation, characterization in solution of the 7π radical 1,2,4-triseleno-3,5-diazolium and the 6π (1,2,4-triseleno-3,5-diazolium)$^{2+}$ cations, and the X-ray crystal structures of $[Se_3N_2]_2[AsF_6]_2$ and $[Se_3N_2][AsF_6]_2$ containing the first stable binary selenium-nitrogen species. J Chem Soc Chem Commun. 1989;1415–7.
[19] Siivari J, Chivers T, Laitinen RS. Synthesis and characterization of selenium-nitrogen chlorides: force-field calculation for the $Se_3N_2Cl^+$ cation. Inorg Chem. 1993;32:4391–5.
[20] Wolmershäuser G, Brulet CR, Street GB. Mixed sulfur-nitrogen-selenium compounds. Inorg Chem. 1978;17:3586–9.
[21] Kelly PF, Slawin AMZ. Preparation and crystal structure of $[(AlBr_3)_2(Se_2N_2)]$, the first example of a main-group element adduct of diselenium dinitride. J Chem Soc Dalton Trans. 1996;4029–30.

[22] Kelly PF, Slawin AMZ. The preparation of salts of $[Pd_2(\mu\text{-}Se_2N_2)Cl_6]^{2-}$, the first adducts of diselenium dinitride. Angew Chem Int Ed Engl. 1995;34:1758–9.

[23] Kelly PF, Slawin AMZ, Williams DJ, Woollins JD. The preparation and X-ray structure of $[Pt(Se_3N)Cl(PMe_2Ph)]$. J Chem Soc Chem Commun. 1989;408–9

[24] Kelly PF, Parkin IP, Slawin AMZ, Williams DJ, Woollins JD. Platinum complexes of the anions $Se_2N_2^{2-}$ and Se_2N_2H-; X-ray structure analysis of $[Pt(Se_2N_2H)(PMe_2Ph)]Cl$. Angew Chem Int Ed Engl. 1989;28:1047–9.

[25] Kelly PF, Slawin AMZ, Williams DJ, Woollins JD. Reaction of Se_4N_4 with $Pt(PPh_3)_3$: the preparation and X-ray structure of $[Pt(Se_2N_2)(PPh_3)_2].CH_2Cl_2$. Polyhedron. 1990;9:1567–91.

[26] Chivers T, Codding PW, Laidlaw WG, Liblong SW, Oakley RT, Trsic M. The crystal, molecular and electronic structures of tetrasulfur tetranitride. J Am Chem Soc. 1983;105:1186–92.

[27] Dehnicke K, Schmock F, Köhler KF, Frenking G. Se_4N_2, A novel selenium nitride. Angew Chem Int Ed Engl. 1991;30:577–8.

[28] Siivari J, Chivers T, Laitinen R. $Se_3N_2Cl_2$, A novel selenium-nitrogen chloride: reinvestigation of "Se_4N_2". Angew Chem Int Ed Engl. 1992;31:577–8.

[29] Schmitz-Dumont O, Ross B. Formation of selenanthrene from diphenyl selenide and of dipotassium triimidotellurite from diphenyl telluride. Angew Chem Int Ed Engl. 1967;6:1071–2.

[30] Chivers T. Tellurium compounds of main group elements; progress and prospects. J Chem Soc Dalton Trans. 1996;1185–94.

[31] Mosa W, Lau C, Möhlen M, Neumüller B, Dehnicke K. $[Te_6N_8(TeCl_4)_4]$ – tellurium nitride stabilized by tellurium tetrachloride. Angew Chem Int Ed Engl. 1998;37:2840–2.

[32] Chivers T, Laitinen RS. Fundamental chemistry of binary S,N and ternary S,N,O anions: analogues of sulfur oxides and N,O anions. Chem Soc Rev. 2017;46:5182–92.

[33] Kelly PF, Woollins JD. The reactivity of Se_4N_4 in liquid ammonia. Polyhedron. 1993;12:1129–33.

[34] Belton PS, Parkin IP, Williams DJ, Woollins JD. The reactions of sulphur-nitrogen species in liquid ammonia. J Chem Soc Chem Commun. 1988;1479–80.

[35] Parkin IP, Woollins JD. Preparation and characterization of $[Pt(SeSN_2)(PR_3)_2]$, $[Pt(Se_2N_2)(PR_3)_2]$, $[Pt(SeSN_2H)(PR_3)_2][BF_4]$, and $[Pt(Se_2N_2H)(PR_3)_2][BF_4]$. J Chem Soc Dalton Trans. 1990;925–30.

[36] Parkin IP, Slawin AMZ, Williams DJ, Woollins JD. Reaction of $[Pt(Se_2N_2)(DPPE)]$ with halogens – a new route to Se_4N_4. Phosphorus Sulfur Silicon Relat Elem. 1991;57:273–7.

[37] Broschag M, Klapötke TM, Tornieporth-Oetting IC, White PS. The preparation and structure of $[(SeCl_2)_2N][AsF_6]$•MeCN containing the first ternary selenium, nitrogen, chlorine cation. J Chem Soc Chem Commun. 1992;1390–1.

[38] Broschag M, Klapötke TM, Schulz A, White PS. The $ClSeNSeCl^+$ cation. An unusual structurally very versatile ion adopting different solid-state structures as deduced by X-ray and ab initio methods. Inorg Chem. 1993;32:5734–8.

[39] Broschag M, Klapötke TM. The intriguing diversity of neutral and cationic selenium-nitrogen heterocycles. Phosphorus Sulfur Silicon Relat Elem. 1994;93–94:181–4.

[40] Borisenko KB, Broschag M, Hargittai I, Klapötke TM, Schröder D, Schulz A, et al. Preparation of $N(SeCl)_2{}^{+}X^{-}$ (X = $SbCl_6$ or $FeCl_4$), $F_3CCSeNSeCCF_3{}^{+}SbCl_6{}^{-}$, $F_3CCSeNSeCCF_3$, $F_3CCSeSeCCF_3$ and $F_3CCSeSeC(CF_3)C\,(CF_3)SeSeCCF_3$. Electron diffraction study of $F_3CCSeSeCCF_3$ and crystal structure of the eight-membered heterocycle $F_3CCSeSeC(CF_3)C\,(CF_3)SeSeCCF_3$. J Chem Soc Dalton Trans. 1994;2705–12

[41] Awere EG, Burford N, Mailer C, Passmore J, Schriver MJ, White PS, et al. The high yield preparation, characterization, and gas phase structure of the thermally stable $[F_3CCSNSCCF_3]^{\bullet}$. J Chem Soc Chem Commun. 1987;66–7.

[42] Wollert R, Höllwarth A, Frenking G, Fenske D, Goesmann H, Dehnicke K. Se_2NCl_3 and $[Se_2NCl_2][GaCl_4]$, chloride nitrides of trivalent selenium. Angew Chem Int Ed Engl. 1992;31:1251–3.

[43] Lau C, Neumüller B, Hiller W, Herker M, Vyboishchikov SF, Frenking G, Dehnicke K. Se_2NBr_3, Se_2NCl_5, $Se_2NCl_6^-$: new nitride halides of selenium(III) and selenium(IV). Chem Eur J. 1996;2:1373–8.
[44] Wollert R, Neumüller B, Dehnicke K. Synthesis and crystal structure of $[Se_3N_2Cl][GaCl_4]$. Z Anorg Allg Chem. 1992;616:191–4.
[45] Passmore J, Schatte G, Cameron TS. Preparation and structure of $Te_4N_2Cl_8(AsF_6)_2{\bullet}2SO_2$ containing $Te_4N_2Cl_8^{2+}$, the first tellurium-nitrogen-halogen cation. J Chem Soc Chem Commun. 1995;2311–2.
[46] Chivers T, Laitinen RS. Insights into the formation of inorganic heterocycles *via* cyclocondensation of primary amines with group 15 and 6 halides. Dalton Trans. 2017;46:1357–67.
[47] Thrasher JS, Bauknight Jr. CW, DesMarteau DD (Perfluoroalkylimino)selenyl chlorides. Inorg Chem. 1985;24:1598–9.
[48] Chivers T, Enright G, Sandblom N, Schatte G, Parvez M. Synthesis and reactions of *tert*-butylimidotellurium dihalides: X-ray structures of $[Cl_2Te(\mu\text{-}N^tBu)_2TeCl_2]_2$ and $(^tBuO)_2Te(\mu\text{-}N^tBu)_2Te(O^tBu)_2$. Inorg Chem. 1999;38:5431–6.
[49] Haas A, Kasprowski J. Synthesis of bis(sulfinylamido)selane $Se(NSO)_2$. Chimia. 1987;41:340.
[50] Haas A, Kasprowski J, Angermund K, Betz P, Krüger C, Tsay Y-H, et al. Synthesis, structures, and properties of cyclothiaselenazenium cations $[Se_2N_2S]_2^{2+}$, $[XSe_2N_2S]_2^{2+}$, $[Se_2N_2S]^+$, $[S_3SeN_5]^+$ as well as $Cl_2Se_2N_2S$ and $SeSN_2{\bullet}TiCl_4$. Chem Ber. 1991;124:1895–906.
[51] Maaninen A, Chivers T, Parvez M, Pietikäinen J, Laitinen RS. Syntheses of SeX_2 (X = Cl, Br) in THF solution and a new route to selenium sulfides Se_nS_{8-n} (n = 1-5): X-ray crystal structure of *cis*-$SeCl_2(tht)_2$ and $SeCl_2{\bullet}$tmtu. Inorg Chem. 1999;38:4093–7.
[52] Boese R, Dworak J, Hass A, Pryka M. Synthesis, structures and properties of CF_3S-substituted tellurium compounds. Chem Ber. 1995;128:477–80.
[53] Haas A, Pryka M. New pathways to tellurium-chalcogen-nitrogen chemistry: preparations, structures, and properties of telluraheterocycles. Chem Ber. 1995;128:11–22.
[54] Haas A. Acyclic and heterocyclic tellurathianitrogen compounds: a review on recent publications. J Organomet Chem. 2002;646:80–93.
[55] Haas A, Kasprowski J, Pryka M. Synthesis, structures, and properties of Cl_2TeSeN_2S and $[Se_2N_2S]_2^{2+}$ $[XSO_3^-]_2$ (X = F, CF_3). Chem Ber. 1992;125:789–92.
[56] Fockenberg F, Haas A. Advances in perhalogeno thiocarbonyl, selenocarbonyl and selenodiimide chemistry. Z Naturforsch. 1986;41b:413–22.
[57] Bestari K, Cordes AW, Oakley RT, Young KM. Heterocyclic 1,2,4,6-thia- and 1,2,4,6-selenatriazinyl radicals; spin distributions and modes of association. J Am Chem Soc. 1990;112:2249–55.
[58] Björgvinsson M, Roesky HW, Pauer F, Stalke D, Sheldrick GM. Preparation and structural characterization of the bis[bis(trimethylsilyl)amido]chalcogenides of selenium and tellurium. Inorg Chem. 1990;29:5140–3.
[59] Siivari J, Maaninen A, Haapaniemi E, Laitinen RS. Formation and identification of bis[bis(trimethylsilyl)amino]tri- and tetrachalcogenides. Z Naturforsch. 1995;50b:1575–82.
[60] Maaninen A, Laitinen RS, Chivers T, Pakkanen TA. Preparation, X-ray structure, and spectroscopic characterization of 1,5-$Se_2S_2N_4$. Inorg Chem. 1999;38:3450–4.
[61] Maaninen A, Siivari J, Suontamo RJ, Konu J, Laitinen RS, Chivers T. Theoretical and experimental studies of six-membered selenium-sulfur nitrides $Se_xS_{4-x}N_2$ (x = 0-4). Preparation of S_4N_2 and SeS_3N_2 by the reaction of bis[bis(trimethylsilyl)amino]sulfane with chalcogen chlorides. Inorg Chem. 1997;36:2170–7.
[62] Björgvinsson M, Roesky HW, Pauer F, Stalke D, Sheldrick GM. Preparation and structure characterization of the bis[*tert*-butyl(trimethylsilyl)amino]chalcogenides of selenium and tellurium. Eur J Solid State Chem. 1991;29:759–76.

[63] Gindl J, Björgvinsson M, Roesky HW, Freire-Erdbrügger C, Sheldrick GM. Synthesis and structure of a stable selenodiimide complex. Dalton Trans. 1993;811–2.

[64] Konu J, Maaninen A, Paananen K, Ingman P, Laitinen RS, Chivers T, et al. Preparation and structural characterization of $(Me_3SiNSN)_2Se$, a new synthon for sulfur-selenium nitrides. Inorg Chem. 2002;41:1430–5.

[65] Bryce MR, Chesney A. Phenylsulfinylselenyl chloride ($PhSO_2SeCl$): a new reagent for the formation of C-Se and N-Se bonds. Generation and *in situ* Diels-Alder trapping of selenonitrosoarene intermediate (Ar-N=Se). J Chem Soc Chem Commun. 1995;195–6.

[66] Sharpless KB, Hori T, Truesdale LK, Dietrich CO. Allylic amination of olefins and acetylenes by imido selenium compounds. J Am Chem Soc. 1976;98:269–71.

[67] Sharpless KB, Singer SP. 1,2-Diamination of 1,3-dienes by imido selenium compounds. J Org Chem. 1976;41:2504–5.

[68] Bruncko M, Khong T-AV, Sharpless KB. Allylic amination with a modified diimidoselenium reagent. Angew Chem Int Ed Engl. 1996;35:454–6.

[69] Herberhold M, Jellen W. New selenium-nitrogen compounds: *tert*-butyl-seleninylamine, ^tBuNSeO, and di(*tert*-butyl)selenium diimide. Z Naturforsch. 1986;41b:144–8.

[70] Maaninen T, Laitinen R, Chivers T. A monomeric selenium(IV) diimide and a dimeric seleninylamine. Chem Commun. 1812–1813;2002.

[71] Maaninen T, Tuononen HM, Kosunen K, Oilunkaniemi R, Hiitola J, Laitinen R, et al. Formation, structural characterization, and calculated NMR chemical shifts of selenium-nitrogen compounds from $SeCl_4$ and ArNHLi (Ar = supermesityl, mesityl). Z Anorg Allg Chem. 2004;630:1947–54.

[72] Chivers T, Sandblom N, Schatte G. Tellurium-nitrogen compounds: bis(*tert*-butyl)tellurium diimide dimer and bis{lithium[tris-*tert*-butylimido)]}tellurite. Inorg Synth. 2001;34:42–8.

[73] Chivers T, Gao X, Parvez M. Tellurium-nitrogen double bonds and a novel Te_3N_3 ring: the formation and structures of $[(^tBuNH)(^tBuN)_3Te_2]Cl$, $[^tBuNTe(\mu\text{-}N^tBu)]_2$ and $[^tBuNTe]_3$. J Am Chem Soc. 1995;117:2359–60.

[74] Sandblom N, Ziegler T, Chivers T. Chalcogen diimides; relative stabilities of monomeric and dimeric structures $[E(NMe)_2]_n$ (E = S, Se, Te; n = 1, 2). Inorg Chem. 1998;37:354–9.

[75] Maaninen T, Tuononen HM, Schatte G, Suontamo R, Valkonen J, Laitinen R, et al. Experimental and theoretical investigations of structural trends for selenium(IV) diimides and oxides: X-ray structure of $Se_3(NAd)_2$. Inorg Chem. 2004;43:2097–104.

[76] Karhu AJ, Risto M, Rautiainen JM, Oilunkaniemi R, Chivers T, Laitinen RS. Mercury- and cadmium-assisted [2 +2] cycloaddition of *tert*-butylselenium diimide. Inorg Chem. 2015;54:9499–508.

[77] Chivers T, Schatte G. Cadmium complexes of the tripodal $[Te(N^txBu)_3]_2$- dianion and the $HgCl_2$ adduct of a tellurium diimide dimer. Can J Chem. 2003;81:1307–14

[78] Maaninen T, Chivers T, Laitinen R, Schatte G, Nissinen M. Selenium imides: ^{77}Se investigations of the $SeCl_2$-$^tBuNH_2$ reaction and X-ray structures of $Se_3(N^tBu)_3$, $^tBuN(\mu\text{-}N^tBu)_2SO_2$, and $^tBuN(\mu\text{-}N^tBu)_2SeO$. Inorg Chem. 2000;39:5341–7.

[79] Wrackmeyer B, Distler B, Herberhold M. ^{15}N and ^{77}Se nuclear magnetic resonance study of selenium diimides and amino selenanes. Z Naturforsch. 1993;48b:1307–14.

[80] Köster R, Schüssler W, Boese R, Herberhold M, Gerstmann S, Wrackmeyer B. Ethyloboration of selenium dioxide and selenium bis(*tert*-butylimide) – molecular structure of an organo-substituted eight-membered $[\text{-BOSeO-}]_2$ heterocycle. Chem Ber. 1996;129:503–7.

[81] Wrackmeyer B, Köhler C, Milius W, Herberhold M. Reaction of N,N'-dialkyl selenium diimides with bis(amino)stannylenes – formation of a tin-tin bond. Z Anorg Allg Chem. 1995;621:1625–31.

[82] Schatte G, Chivers T, Jaska C, Sandblom N. Cycloaddition reaction of *tert*-butyl isocyanate and a tellurium diimide dimer: extended helical structure of the ureato telluroxide $\{[OC(\mu\text{-}N^tBu)_2TeO]_2$. Chem Commun. 2000;1657–8.

[83] Schatte G, Chivers T, Tuononen HM, Suontamo R, Laitinen R, Valkonen J. Experimental and theoretical investigations of tellurium(IV) diimides and imidotelluroxanes; X-ray structures of $B(C_6F_5)_3$ adducts of OTe(μ-N^{t}Bu)$_2$TeN^{t}Bu and [OTe(μ-N^{t}Bu)$_2$Te(μ-O)]$_2$ and ^{t}BuNH$_2$. Inorg Chem. 2005;44:443–51.
[84] Chivers T, Gao X, Parvez M. A $Te_2N_6Li_4$ cage containing the tris(*tert*-butylimido)tellurite dianion. Angew Chem Int Ed Engl. 1995;34:2549–51.
[85] Chivers T, Parvez M, Schatte G. Preparation and structure of [Li$_2$Se(N^{t}Bu)$_3$]$_2$, containing the novel Se(N^{t}Bu)$_3$$^{2-}$ anion. Inorg Chem. 1996;35:4094–5.
[86] Chivers T, Gao X, Parvez M. Pyramidal tellurite ions Te(N^{t}Bu)$_2$(E^{t}Bu)$^-$ (E = O, NH); syntheses and structures of [K(THF)Te(N^{t}Bu)$_2$(O^{t}Bu)]$_2$ and {[LiTe(N^{t}Bu)$_2$(NH^{t}Bu)]$_2$LiCl}$_2$. Inorg Chem. 1996;35:553–4.
[87] Chivers T. The imido ligand in main group element chemistry. Can J Chem. 2001;79:1841–50.
[88] Konu J, Chivers T, Schatte G, Parvez M, Laitinen RS. Formation of N-I charge-transfer bonds and ion pairs in polyiodides with imidotellurium cations. Inorg Chem. 2005;44:2973–82.
[89] Chivers T, Schatte G. Cooperative ring-opening by $B(C_6F_5)_3$ and a tellurium diimide dimer. Eur J Inorg Chem. 2003;3314–7.
[90] Chivers T, Parvez M, Schatte G. Reactions of *cis*-[^{t}BuNTe(μ-N^{t}Bu)]$_2$ and M[O$_3$SCF$_3$] (M = Ag, Cu): chelation, cis → trans Isomerization, and the spirocyclic ligand [^{t}BuNTe(μ-N^{t}Bu)]$_2$Te(μ-O)]$_2$. Inorg Chem. 1999;38:5171–7.
[91] Brask JK, Chivers T, McGarvey B, Schatte G, Sung R, Boeré RT. ESR investigations of the radicals {Li$_3$[E(N^{t}Bu)$_3$]$_2$}$^\bullet$ (E = S, Se) and the radical anions SO$_x$(N^{t}Bu)$_{3-x}$$^{\bullet-}$ (x = 1, 2). Inorg Chem. 1998;37:4633–6.
[92] Chivers T, Schatte G. A new coordination mode for the tripodal [Te(N^{t}Bu)$_3$]$^{2-}$ dianion: preparation and structures of {ClIn[Te(NtBu)$_3$]}$_2$ and {LiIn[Te(N^{t}Bu)$_3$]$_2$}•LiCl. Eur J Inorg Chem. 2002;2266–9.
[93] Chivers T, Schatte G. Synthesis and characterization of [^{t}BuNSn(μ-N^{t}Bu)$_2$TeN^{t}Bu]μ^3-SnTe, a stannatellurone with four-coordinate tin. Chem Commun. 2001;2264–5.
[94] Chivers T, Parvez M, Schatte G, Yap GPA. Homoleptic bismuth and antimony complexes of the tripodal [Te(N^{t}Bu)$_3$]$^{2-}$ anion. Inorg Chem. 1999;38:1380–1.
[95] King RB, Sangokoya SA. (Dialkylamido)selenium derivatives. Inorg Chem. 1987;26:2727–30.
[96] Foss O, Janickis V. Synthesis of dimorpholino-di-, tri-, and tetra-selane, and dipiperidonotetraselane. Crystal and molecular structures of the two tetraselanes. JCS Dalton. 1980;620–3.
[97] Foss O, Janickis V. X-ray crystal structure of a new red, monoclinic form of *cyclo*-octaselenium, Se_8. JCS Chem Comm. 1977;834–5.
[98] Björgvinsson M, Heinze T, Roesky HW, Pauer F, Stalke D, Sheldrick GM. Synthesis and structure of the first tellurium(III) radical cation. Angew Chem Int Ed Engl. 1991;30:1677–9.
[99] Murai T, Imaeda K, Kajita S, Kimura K, Ishihara H, Kato S. Tellurium amides as a new tellurating agent. Phosphorus Sulfur Rel Elem. 1992;67:239–42.
[100] Murai T, Imaeda K, Kajita S, Ishihara H, Kato S. Reductive dialkynylation of tellurium tetrachloride with lithium amides and terminal alkynes. J Chem Soc Chem Commun. 1991;832–4.
[101] Murai T, Nomura K, Kimura K, Kato S. Organotellurium compounds involving tellurium(II)-nitrogen bonds. Synthesis of arenetellurenamides and their reactions with acetylenes. Organometallics. 1991;10:1095–8.
[102] Allan RE, Gornitza H, Kärcher J, Paver MA, Rennie M-A, Russell CA, et al. Structure and reactivity of [{Te(NMe$_2$)$_2$}$_\infty$]; application to the preparation of metalloorganic tellurium(II) compounds. J Chem Soc Dalton Trans. 1996:1727–30.
[103] Sarma BK, Mugesh G. Antioxidant activity of anti-inflammatory compound Ebselen: a reversible cyclization pathway via selenenic acid and seleninic acid intermediates. Chem Eur J. 2008;14:10603–14.

[104] Selvakumar K, Shah P, Singh HB, Butcher RJ. Synthesis, structure, and glutathione peroxidase-like activity of amino acid containing Ebselen analogues and diaryl diselenides. Chem Eur J. 2011;17:12741–55.
[105] Azad GK, Tomar RS. Ebselen, a promising antioxidant drug: mechanisms of action and targets of biological pathways. Mol Biol Rep. 2014;41:4865–79.
[106] Chandrasekhar V, Chivers T, Kumaravel SS, Parvez M, Rao MNS. Reactions of N,N,N'-tris (trimethylsilyl)benzamidine with organochalcogen halides: formation of diazenes via the resonance-stabilized radicals $[PhC(NEPh)_2]^\bullet$. Inorg Chem. 1991;30:4125–30.
[107] Chivers T, Doxsee DD, Parvez M. Preparation and spectroscopic characterization of 1,5- and 1,3-diphosphadiselenatetrazocines. Inorg Chem. 1993;32:2238–42.
[108] Chivers T, Hilts RW, Jin P, Chen Z, Lu X. Synthesis, properties, and bishomoaromaticity of the first tetrahalogenated derivative of a 1,5- diphosphadithiatetrazocine; a combined experimental and computational investigation. Inorg Chem. 2010;49: 3810–5 and references cited therein.
[109] Chivers T, Doxsee DD, Fait JF, Parvez M. Preparation and solid-state isomerization of *Se,Se'*-dialkyldiphosphadiselenatetrazocines. Inorg Chem. 1993;32:2243–8.

Guoxiong Hua and J. Derek Woollins

5 Organophosphorus-selenium/tellurium reagents: from synthesis to applications

Abstract: Organic selenium- and tellurium-phosphorus compounds have found wide application as reagents in synthetic inorganic and organic chemistry, such as oxygen/chalcogen exchange, oxidation/reduction, nucleophilic/electrophilic substitution, nucleophilic addition, free radical addition, Diels–Alder reaction, cycloadditions, coordination, and so on. This chapter covers the main classes of phosphorus-selenium/tellurium reagents, including binary phosphorus-selenium/tellurium species, organophosphorus(III)-selenium/tellurium compounds, phosphorus(V)-selenides/tellurides, diselenophosphinates/ditellurophopshinates, diselenaphosphetane diselenides, Woollins' reagent, phosphorus-selenium/tellurium amides, and imides. Given the huge amount of literature up to mid-2017, this overview is inevitably selective and will focus particularly on their synthesis, reactivity, and applications in synthetic and coordination chemistry.

Keywords: organophosphorus compounds, phosphorus-selenium reagents, phosphorus-tellurium reagents, Woollins reagent, phosphorus-chalcogen heterocycles

5.1 Introduction

The chemistry of compounds incorporating selenium or tellurium has been rapidly expanding in both inorganic and organic chemistry for three decades as a variety of selenium- and tellurium-based reagents have enabled chemists to explore new synthetic transformations [1–4]. Research attention for organophosphorus-selenium/tellurium chemistry dates back over two centuries and covers a wide range of compounds from simple phosphorus selenides or tellurides due to the wide variety of the compounds and rich bioactivities, the potential applications in pharmaceutical industry, agrochemical industry, material science areas, and their usefulness as versatile reagents in both organic and inorganic synthesis [2]. There are currently over 7500 compounds containing phosphorus-selenium bonds and just less than 1500 compounds containing phosphorus-tellurium bonds so far. The high reactivity of phosphorus-selenium/tellurium-containing compounds make them interesting reagents/building precursors giving rise to complicated and

This article has previously been published in the journal *Physical Sciences Reviews*. Please cite as: Hua, G., Woollins, J.D. Organophosphorus-Selenium/Tellurium Reagents: From Synthesis to Applications. *Physical Sciences Reviews* [Online] **2018** DOI: 10.1515/psr-2017-0132

https://doi.org/10.1515/9783110529340-005

diverse chalcogen- or non-chalcogen-containing molecules, expanding the horizon of organic phosphorus-selenium/tellurium chemistry ever further.

The aim of this chapter seeks to provide an overview of the main classes of phosphorus-selenium/tellurium-containing reagents, namely, binary phosphorus-selenium/tellurium species, organophosphorus(III)-selenium/tellurium compounds, phosphorus(V)-selenides/tellurides, diselenophosphinates/ditellurophosphinates, diselenaphosphetane diselenides and Woollins' reagent, and phosphorus-selenium/tellurium amides and imides. Given the huge amount of published materials, this chapter is inevitably selective and will focus particularly on their synthesis, reactivity, and applications in synthetic chemistry, coordination chemistry, and so on. More comprehensive reviews focusing solely or predominately on phosphorus-selenium and phosphorus-tellurium chemistry are available elsewhere [3–9]. The literature is covered up to mid-2017.

5.2 Binary phosphorus-selenium/tellurium species

5.2.1 Binary phosphorus-selenium species

A structural feature of binary phosphorus-chalcogen species is the existence of an extensive series of cage molecules of P_4E_n (E = S, Se), in which a P_4 tetrahedron is incorporated with various numbers of bridging and /or terminal chalcogen atoms [10]. The P_4S_n series (n = 3–10) is known and the first binary phosphorus-sulfur species P_4S_3 was applied in the heads of "strike-anywhere" matches [11]. In the case of selenium, there are some new forms such as P_2Se_5, P_2Se, $P_{14}Se$, P_4Se, P_4Se_2, P_2Se_{10}, P_4Se_4, and the polymeric catena-$[P_4Se_4]_x$ present apart from the binary species P_4Se_n (n = 3–7 and 10) which are similar with their sulfur analogues. P_4Se_3, derived from heating white phosphorus and selenium in tetralin [12], has the same structural features as the low temperature form of P_4S_3. P_4Se_5, prepared by reacting P_4Se_3 with bromine in CS_2, has the same structure as P_4S_5 [13]. The P_2Se_5 molecule adopts a norbornane-like structure with two phosphorus atoms with oxidation state +3 bridged by two diselenide units (–Se–Se–) and one selenide unit (–Se–) [14].

The potassium salts of diselenophosphates $(RO)_2P(Se)SeK$ **2** can be prepared from the reaction of phosphorus pentaselenide (P_2Se_5) with alcohols, followed by immediate addition of an ethanolic KOH solution (Figure 5.1) [15]. The formation of $(RO)_2P(Se)SeH$ **1** must be the primary reaction because it was found necessary to add an ethanolic KOH solution to give the salts. The potassium salts of diselenophosphates **2** are rather unstable and turn red upon short storage. The pure and colorless powders of $[NH_4][Se_2P(OR)_2]$ **3** can be readily obtained by bubbling anhydrous NH_3(g) into the solution of $(RO)_2P(Se)SeH$ **1** at 0 °C [16]. It has been observed that the ammonium salts of phosphorodiselenoic acid are much easier to handle than the potassium salts.

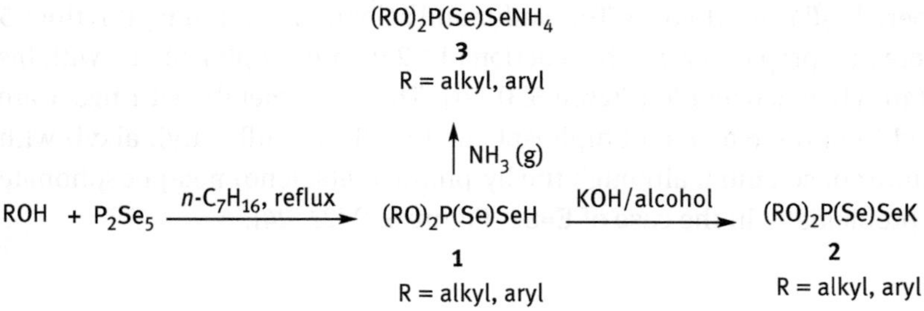

Figure 5.1: Reactivity of P_2Se_5 toward alcohols.

5.2.2 Binary phosphorus-tellurium species

Binary phosphorus-tellurium species have not reported yet, presumably because of the weakness of P-Te bonds. However, the reaction of the elemental sulfur, tellurium, and phosphorus at 700 K, followed by extraction with CS_2 led to a mixture of P_4S_3, P_4S_2Te, and P_4STe in the approximate ratio of 160:4:1 [17]. Binary P-Te anions can however be stabilized in solid-state materials, such as the ternary telluride BaP_4Te_2 and $Na_2P_4Te_2$. BaP_4Te_2 can be prepared by heating a stoichiometric mixture of Ba, Te, and P at 475 ºC [18]. The crystal structure of BaP_4Te_2 reveals a chain of $[P_4Te_2]^{2-}$ dianions comprising of condensed rings in the chair form, with tellurium atoms in the 1,4-positions and two covalent P-Te bonds. $Na_2P_4Te_2$ (**4**) can be generated in solution in 98 % yield by the reaction of white phosphorus with Na_2Te_2 in *N*-methylimidazole at ambient temperature (Figure 5.2) [19]. The ^{31}P and ^{125}Te NMR spectra of **4** reveal a bicyclo[1.1.0]tetraphosphane (butterfly) conformation with *exo* tellurium atoms, deriving from cleavage of one P-P bond of the P_4 tetrahedron *via* an oxidative addition of $[Te\text{-}Te]^{2-}$.

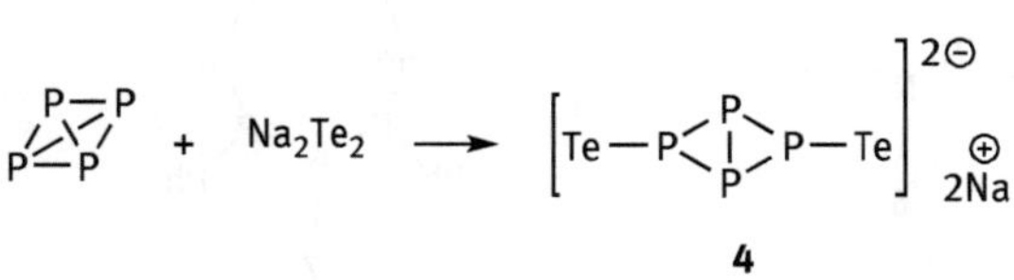

Figure 5.2: Synthesis of $Na_2P_4Te_2$**4** from P_4 and Na_2Te_2.

5.3 Organophosphorus(III)-selenium/tellurium compounds

5.3.1 Organophosphorus(III)-selenium compounds

P(III)/chalcogen heterocycles have received attention as ligands for the coordination of transition metals [20]. The simplest phosphorus-chalcogen heterocycles are three-membered P_2E rings (E=S, Se Te). Sulfur and selenium-containing P_2E rings **5** (E=S, Se) were first prepared from the reaction of 1,2-dichlorodiphosphane with bis (trimethylstannyl)chalcogenide (Figure 5.3) [21]. The three-membered rings **5** are also obtained from the reaction of diphosphene RP=PR (R=bulky aryl, alkyl) with elemental sulfur or selenium although the by-product di(seleno)metaphosphonate $RP(=Se)_2$ is problematic in the case of E=Se (Figure 5.3) [22–24].

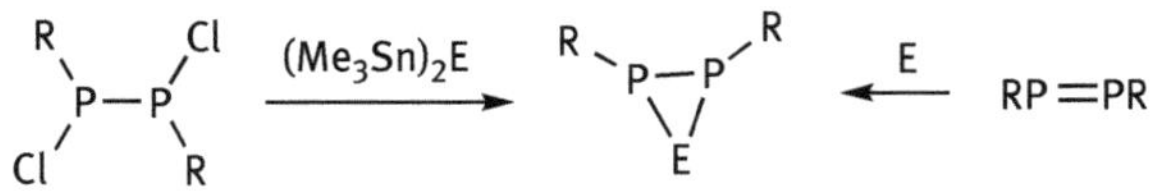

5
(E = S, Se)

Figure 5.3: Synthesis of sulfur- and selenium-containing P_2E rings **5**.

The reaction of $(Me_3Si)_2NMe$ with $PhPCl_2$ under solvent-free conditions gave an intermediate **6**, which undergoes cyclo-condensation with Na_2Se resulting in the formation of eight-membered macrocycles **7** in good yield (74 %) (Figure 5.4) [25]. The $^{31}P\{^1H\}$ NMR spectrum of **7** shows a singlet at $\delta = 115.2$ ppm with an indefinable set of satellites that can be attributed to the isotopologues containing ^{77}Se isotope (spin ½, natural isotopic abundance 7.6 %) [26]. The ^{77}Se NMR spectrum of **7** has a pseudo-quintet at $\delta = 199.8$ ppm, caused by virtual coupling [27], along with a ^1J(Se,P) coupling constant of 125 Hz of typical endocyclic P-Se-P fragments [28].

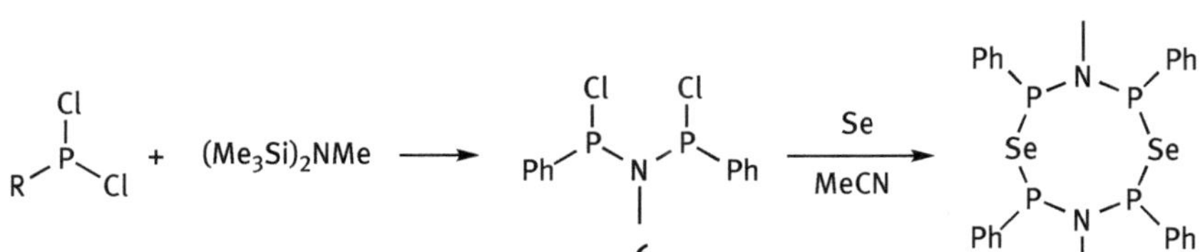

6

7

Figure 5.4: Formation of eight-membered macrocycle **7**.

Reaction of macrocycle **7** in toluene solution with elemental selenium at 111 ºC for 12 h via an oxidative ring contraction yielded a four-membered ring **8** in 82 % yield (Figure 5.5) [25]. The ^{31}P NMR spectrum of **8** has a singlet at δ_P = 43.3 ppm. Meanwhile, compared to compound **7**, the ^{77}Se NMR spectrum of **8** shows a down-field shifted triplet at δ_{Se} = 715.9 ppm for the endocyclic selenium atom due to the coupling to the two adjacent phosphorus atoms [^1J(Se,P) = 255 Hz] and a double doublets resonance at δ_P = -16.1 ppm [^1J(Se,P) = 850 Hz, ^3J(Se,P) = 7 Hz] for the exocyclic selenium atoms [29]. Polynuclear Ag(I) complexes can be used as versatile precursors for the synthesis of a variety of metal complexes via trans-metalation reactions [30, 31]. Macrocycle **7** can also be used as a small bite chelate for the synthesis of multinuclear metal complexes. Reacting macrocycle **7** with AgOTf in different ratios in CH_2Cl_2 led to the s mononuclear or dinuclear or trinuclear sandwich complexes **9** (Figure 5.5) [25].

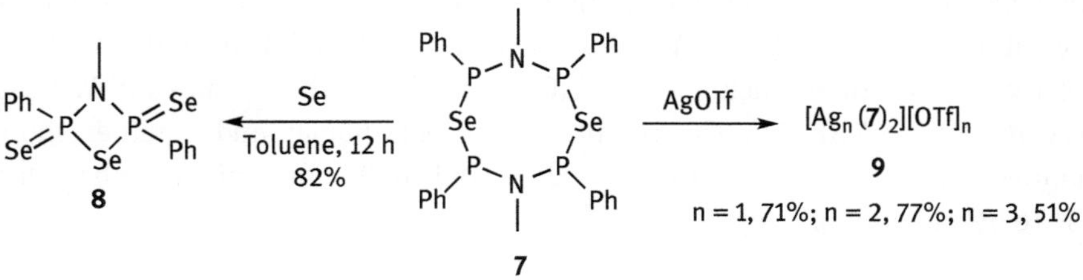

Figure 5.5: Synthesis and reactivity of macrocycle **7**.

5.3.2 Organophosphorus(III)-tellurium compounds

Acyclic organophosphorus(III)-tellurium compounds with either P^{III}-Te-C or P^{III}-Te-P^{III} linkage are potential reagents for synthetic chemistry. The organophosphorus (III)-tellurium compounds R_2PTeR' (tellurophosphinites) **10–12** were prepared by the slow redistribution of R_2P-PR_2 and $R'TeTeR'$ (Figure 5.6) [32–34]. The photolysis of compound **10** in a THF solution of $Cr(CO)_6$ led to the formation of the chromium complexes $Cr(CO)_5(CF_3)_2PTeMe$ (**13**).

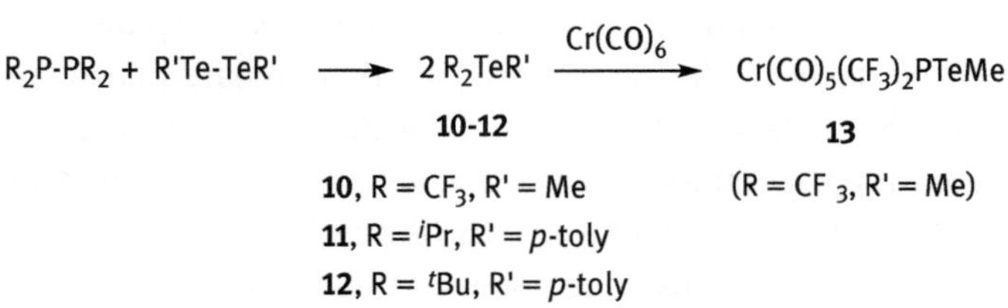

Figure 5.6: Synthesis of R_2PTeR'**10–12**.

The analogous telluradiphosphirane **14** can be synthesized from substitution and redox reaction of sodium telluride and tert-butyldichlorophosphine in pentane [35], or from the reaction of $Te(SiMe_3)_2$ with $(R_2N)PCl_2$ (Figure 5.7) [36]. Three-membered chalcogenadiphosphiranes $(RP)_2Te$ tend to dimerize and subsequently eliminate to give phosphorus-tellurium heterocycles of varying ring size products $(RP)_nTe_m$ (n = 2, 3, 4; m = 0, 1, 2, 3).

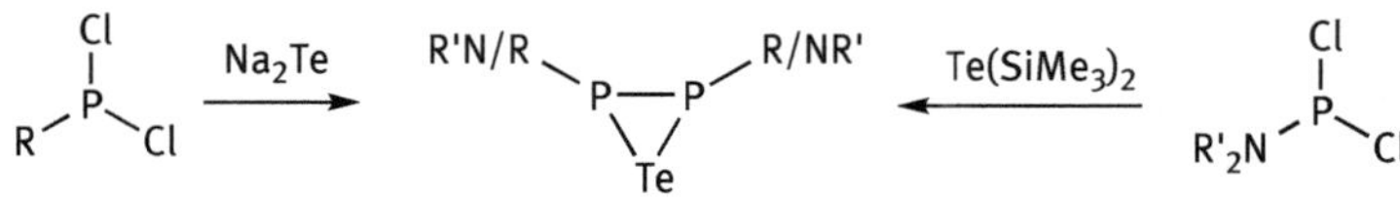

14

Figure 5.7: Synthesis of three-membered telluradiphosphiranes $(RP)_2Te$ **14**.

The 1,2,4-tellurophosphole **15** incorporating a C-P-Te functionality was obtained as a yellow oil by the reaction of $[Li(TMEDA)_2][SbP_2C_2{}^tBu]$ with $Te(S_2CNEt_2)_2$ in DME [37]. The heterocycle **15** further reacting with $[W(CO)_5(THF)]$ gave the P-bonded η^1-$W(CO)_5$ complex **16** (Figure 5.8) [38]. The complex **16** reveals planar and a rather short phosphorus-tellurium bond [d(P-Te) = 2.378(2) Å] contributing from higher s-character in the P-Te hybrid orbital.

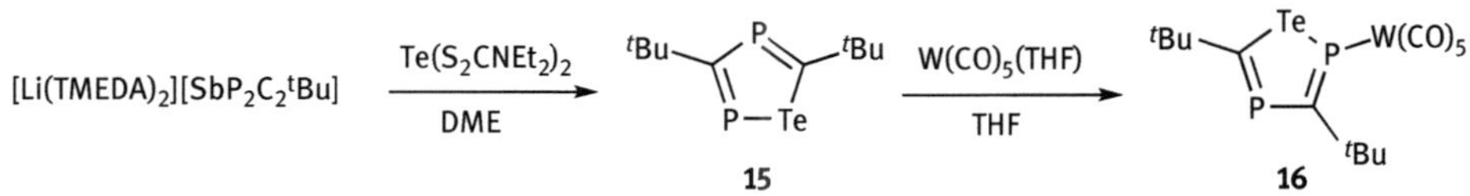

15 **16**

Figure 5.8: Synthesis of 1,2,4-tellurophosphole **15** and its tungsten complex **16**.

5.4 Phosphorus (V)-selenides/tellurides

5.4.1 Triorganylphosphine selenides

Preparation of a solid phosphine selenide from direct treatment of the appropriate tertiary phosphine with elemental selenium often requires numerous and time-consuming extractions to obtain the pure phosphine selenides. For example, tertiary phosphine selenides can be prepared by direct treatment of the appropriate tertiary phosphine either with elemental selenium or with KSeCN or with organic diselenides [39–42]. The reaction of triarylphosphine with potassium selenocyanate rapidly produced quite pure triarylphosphine selenides **17** in almost quantitative yields (Figure 5.9). Compared with the usual preparative routes for tertiary phosphine selenides, the reaction of a tertiary phosphine with potassium selenocyanate in acetonitrile is quite simple.

$$Ar_3P + KSeCN \xrightarrow{CH_3CN} Ar_3P(Se)$$

17

$Ar_3 = Ph_3$, 96%; $(4\text{-}CH_3C_6H_4)_3$, 95%; $(3\text{-}CH_3C_6H_4)_3$, 94%; $(Me_2NCH_2CH_2CH_2)Ph_2$, 87%; $Ph_2(CH_2CH{=}CH_2)$, 89%

Figure 5.9: Synthesis of triarylphosphine selenides **17**.

Triphenylphosphine selenide and its polymer-supported counterpart have been well known as an efficient selenium transfer reagent for the conversion of *H*-phosphonate diesters into the corresponding phosphoroselenoate [43]. Dinucleoside *H*-phosphonate **18** reacting with TPPSe (1.1 equiv.) in pyridine in the presence of trimethylsilyl chloride (TMS-Cl, 6 equiv.) led to the silylation agent. The reaction is very straightforward and completes within 15 min giving the silylated phosphoroselenoate (δ_P = 56.9 and 56.6 ppm) as the sole nucleotidic product. Similarly, selenation of diethyl *H*-phosphonate with 1.1 equiv. of TPPSE takes 5 min in pyridine in the presence of DBU (5 equiv.) to produce dinucleoside phosphoroselenoate **19** in *ca.* 90 % yield (Figure 5.10), which was identical to that prepared in another method [44]. Polymer-supported phosphine selenide TPPSe-2 was obtained by shaking in THF polystyryl diphenylphosphine resin with KSeCN (3 equiv.) for 4 h. The ^{31}P NMR spectrum of the suspended polymer beads showed a complete disappearance of the broad signal at −5.4 ppm of phosphine and a new broad signal at 35.5 ppm of phosphine selenide.

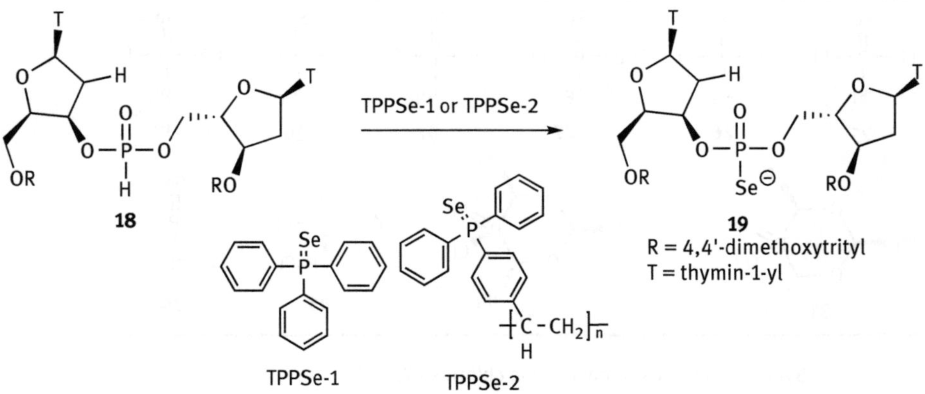

Figure 5.10: Synthesis of phosphoroselenoate **19**.

5.4.2 Phosphinoselenoic chlorides and derivatives

In contrast to phosphine selenides RP(=Se)(R^1)(R^2), much less attention has been paid to the selenium analogues of phosphinic acids and their derivatives **20–24** (Figure 5.11), wherein two alkyl or aryl groups and chlorine or chalcogen atom are attached to the phosphorus atom of the P=Se group.

Diethylphosphinoselenoic chloride **25** was the first reported example of phosphinoselenoic chlorides by the reaction of chlorophosphine with elemental selenium without any solvents and the purification by decantation and distillation [44, 45]. The synthesis of alkylphosphinoselenoic chlorides **26–29**, diphenyl phosphinoselenoic chloride **30**, cyclic phosphinoselenoic chlorides **31** and **32**, and two *P*-chiral chlorides

33 and **34** were achieved via the selenation of the corresponding chlorophosphines by a similar method (Figure 5.12) [46–49].

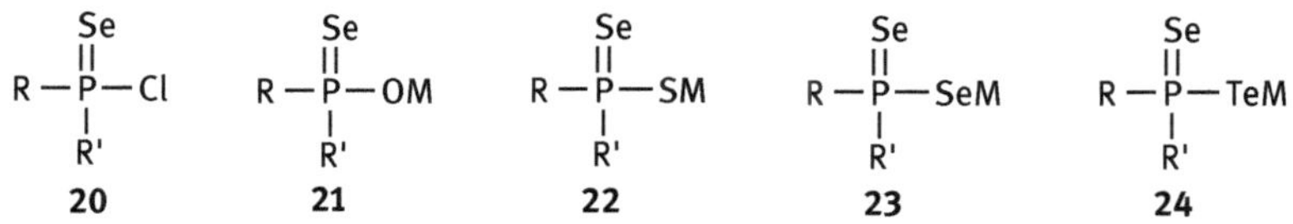

Figure 5.11: Selenium isologs of phosphinic acids and their derivatives **20–24**.

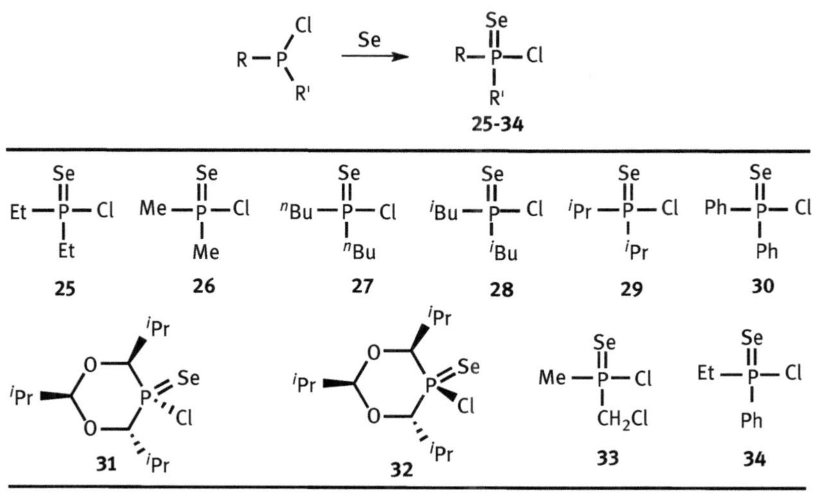

Figure 5.12: Synthesis of phosphinoselenoic chlorides **25–34**.

Another efficient method for the synthesis of *P*-chiral phosphinoselenoic chlorides without isolating intermediate trivalent chlorophenylphosphine has been developed [50, 51]. Elemental selenium and carbon nucleophiles were added to dichlorophenylphosphine in one operation, then a THF solution of Grignard reagent was added to the above suspension to give *P*-chiral phosphinoselenoic chlorides **35–41** in good to excellent yields after work-up (Figure 5.13). The sequential addition of two different Grignard reagents to a mixture of phosphorus trichloride and elemental selenium also led to dialkyl chlorides **42** and **43** in moderate yields (Figure 5.14) [52].

The phosphinoselenoic chlorides are air- and moisture-stable and can easily be handled under air, even some chlorides can be purified by column chromatography on silica gel without any decomposition. Thus, phosphinoselenoic chlorides can be used as selenation reagents or synthetic precursors. The reactions of *P*-chiral chlorides with nitrogen nucleophiles such as metal azide [53], phosphinothioic and

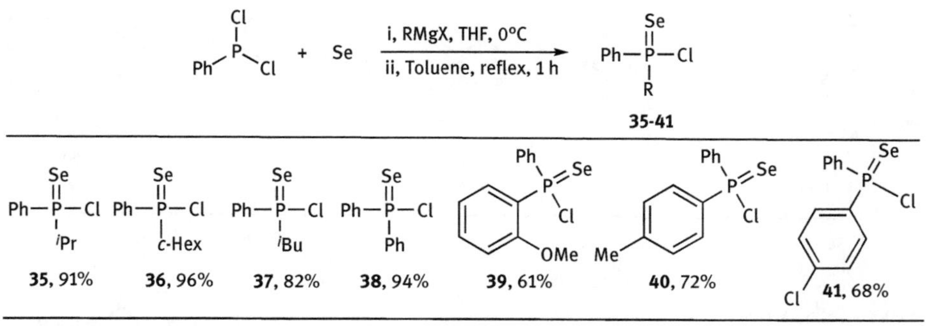

Figure 5.13: Synthesis of phosphinoselenoic chlorides **35–41**.

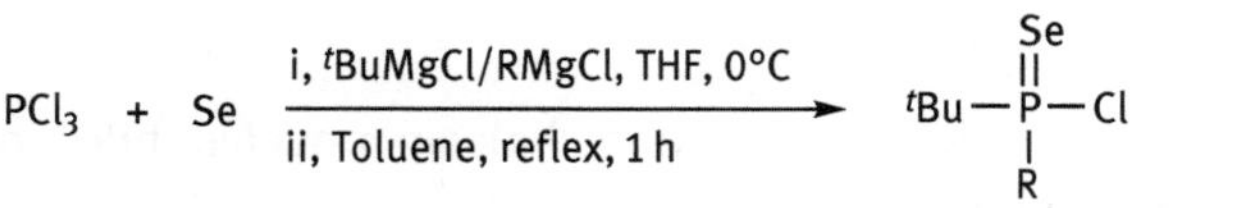

42, R = ^{i}Pr, 46%
43, R = *c*-Hex, 61%

Figure 5.14: Synthesis of phosphinoselenoic chlorides **42** and **43**.

phosphorothioic amide sodium salts [54] delivered mixed-donor unsymmetrical P-N-P ligands **44** and **45**, which are further coordinated to Co or Sn to give the corresponding metal complexes (Figure 5.15) [55].

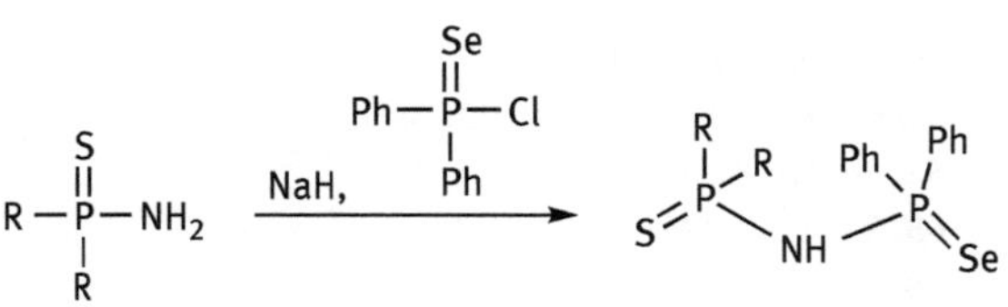

44, R = Ph; **45**, R = OPh

Figure 5.15: Synthesis of unsymmetrical P-N-P ligands **44** and **45**.

Treating diphenylphosphinoselenoic chloride **30** with optically active primary and secondary amines led to the corresponding optically active ligands **46–50** in high yields [56–58]. The ligands **46, 49,** and **50** have been used in Ti-catalyzed ethylation of aldehydes with Et_2Zn [56]. Diphenylphosphinoselenoic chloride **30** reacting with amines in the presence of KOH and 18-crown-6 under carbon dioxide atmosphere gave the products **51** and **52** with the carbon dioxide being incorporated into the bond between the phosphorus and nitrogen atoms (Figure 5.16) [59].

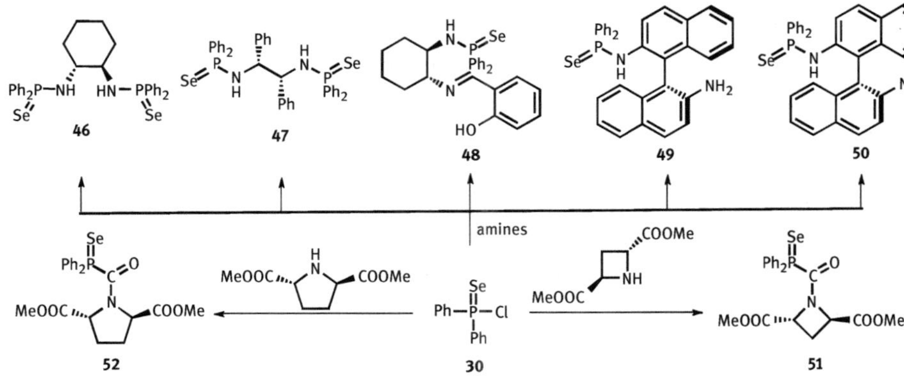

Figure 5.16: Synthesis of ligands **46–52**.

Phosphinoselenoic chlorides react with oxygen and chalcogen nucleophiles to give rise to a series of the corresponding phosphinoselenoic and phosphinoselenothioic acids E-esters (E=O, S, Se) **53–64**; alternatively, reacting with alcohols, thiols metal alkoxides and chalcogenolates led also to the above phosphinoselenoic and phosphinoselenothioic acids chalcogen-esters (Figure 5.17) [51, 60].

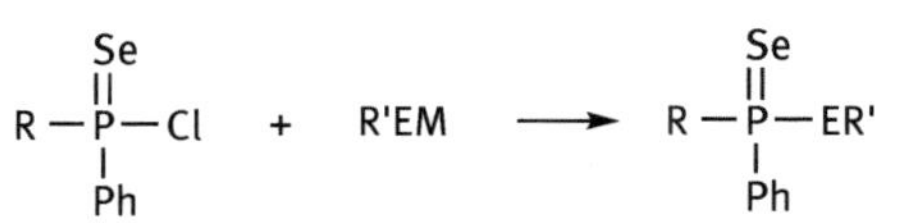

53-64

53, R = ^{t}Bu, R'EM = EtONa, 98%	**59**, R = ^{i}Pr, R'EM = ^{t}BuSLi, 99%
54, R = *c*-Hex, R'EM = EtONa, 93%	**60**, R = Ph, R'EM = ^{t}BuSLi, 81%
55, R = ^{t}Bu, R'EM = (-)-MenthylOLi, 50%	**61**, R = ^{t}Bu, R'EM = ^{t}BuSeLi, 96%
56, R = ^{t}Bu, R'EM = PhONa, 90%	**62**, R = ^{t}Bu, R'EM = ^{t}BuSeLi, 66%
57, R = ^{t}Bu, R'EM = BuSLi, 95%	**63**, R = ^{t}Bu, R'EM = PhSeLi, 99%
58, R = ^{t}Bu, R'EM = PhSLi, 79%	**64**, R = Ph, R'EM = ^{t}BuSeLi, 90%

Figure 5.17: Synthesis of phosphinochalcogenoselenoic E-esters **53–64**.

Reaction of trichlorophosphine with 1,1'-binaphthyl-2,2'-diols and elemental selenium in the presence of Et_3N gave phosphoroselenoyl chlorides **65–67** in good to high yields (Figure 5.18) [61]. All chlorides obtained are air-stable and can be purified by column chromatography on silica gel. The conversion of the chlorides into esters and amides is readily achieved by reaction with the corresponding alcohols and amines as electrophiles.

Treatment of **65** with primary and cyclic secondary amines led to the corresponding phosphoroselenoic amides **69a,b–72a,b** within 3 h. However, reaction with

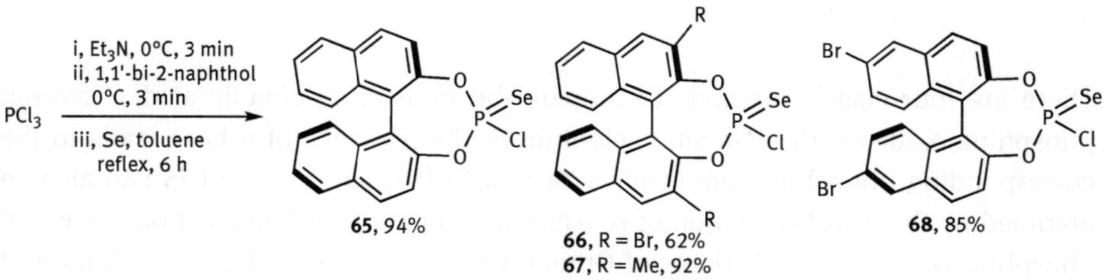

Figure 5.18: Synthesis of phosphoroselenoyl chlorides **65–68**.

acyclic secondary amines needed more than 12 h to give the phosphoroselenoic amides **73a,b** due to the steric congestion around the nucleophilic nitrogen atom (Figure 5.19). In all cases the phosphoroselenoic amides were obtained as two diastereomers in different ratios and most of them can be separated by simple recrystallization, or by column chromatography on silica gel or by HPLC [62].

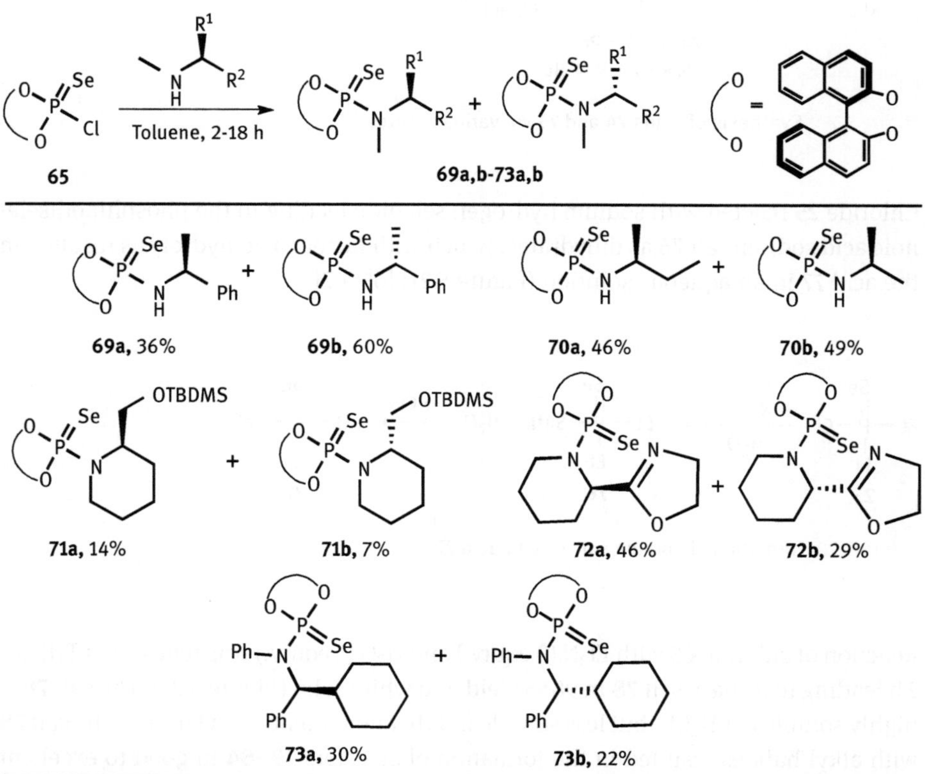

Figure 5.19: Synthesis of phosphoroselenoic amides **69a,b–73a,b**.

5.4.3 Phosphinoselenoic acids and derivatives

There are four classic methods to prepare the phosphinoselenoic acids. Treating phosphine oxides with elemental selenium in the presence of a base leads to the corresponding phosphinselenoic acids **74** and **75** [63, 64]. The acid **75** can also be obtained by the direct oxidation of phosphine selenide [63]. The reaction of diethyl phosphite with three equivalents of Grignard reagents, followed by the addition of elemental selenium gave the same acid **74**. Finally, alkaline hydrolysis of phosphinoselenoic chlorides afforded the acids **74** and **75** (Figure 5.20) [45].

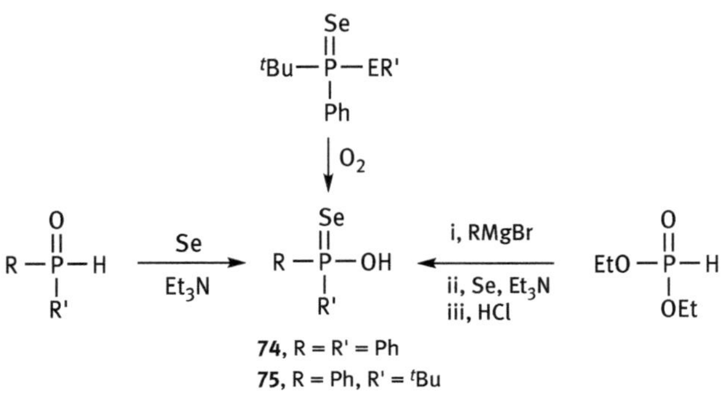

Figure 5.20: Synthesis of acids **74** and **75** via various routes.

Chloride **25** reacted with sodium hydrogen selenide leading to the phosphinodiselenoic acid sodium salt **76** as dihydrates, which with further acid hydrolysis resulted in the acid **77** in an aqueous solution (Figure 5.21) [65–67].

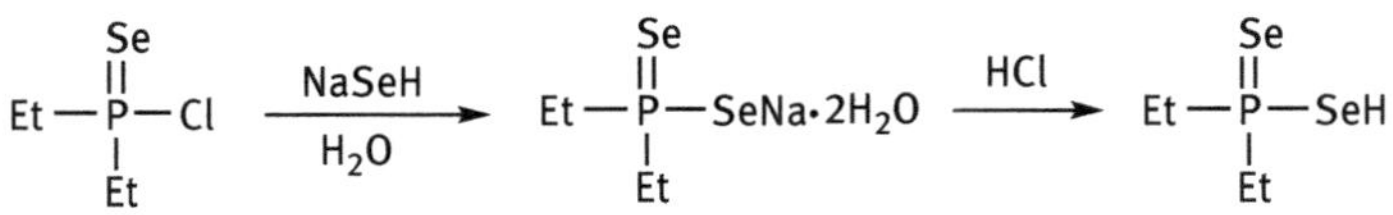

25 76 77

Figure 5.21: Formation of sodium salt **76** and acid **77**.

Reaction of chloride **65** with Et_3N (3 equiv.) and H_2O (2 equiv.) was refluxed in THF for 2 h leading to tertiary salt **78** in 98 % yield as a white solid (Figure 5.22). The salt **78** is highly soluble in CH_2Cl_2, but less soluble in THF and toluene. Alkylation of the salt **78** with alkyl halides resulted in the formation of *Se*-esters **79**–**84** in good to excellent yields. Two diastereomers were found with almost equal ratios except in the case of *Se*-ester **79** [62].

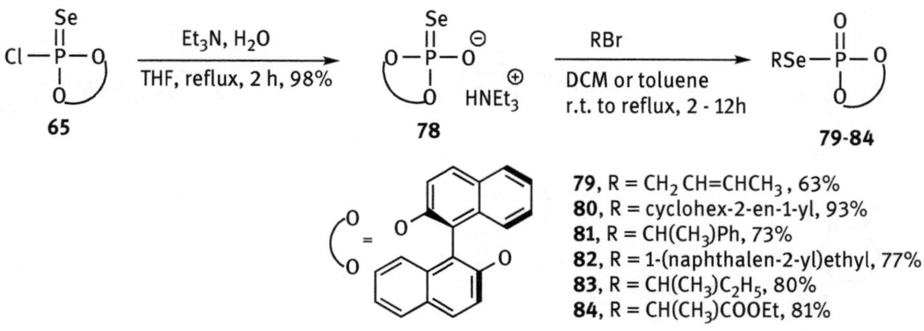

Figure 5.22: Formation of **78** and esters **79-84**.

5.4.4 *O,O,O*-Triorganorylphosphoroselenoates $(RO)_3P{=}Se$

In comparison with its sulfur analogues, *O,O,O*-triorganorylphosphoroselenoates $(RO)_3P{=}Se$ are much more difficult to obtain from the reaction of trialkyl phosphite $(RO)_3P$ with selenium. The strained phosphoroselenoates $(RO)_3P{=}Se$ **85–89** were prepared by direct treatment of the appropriate tertiary phosphine either with elemental selenium or with KSeCN (Figure 5.23) [67–70].

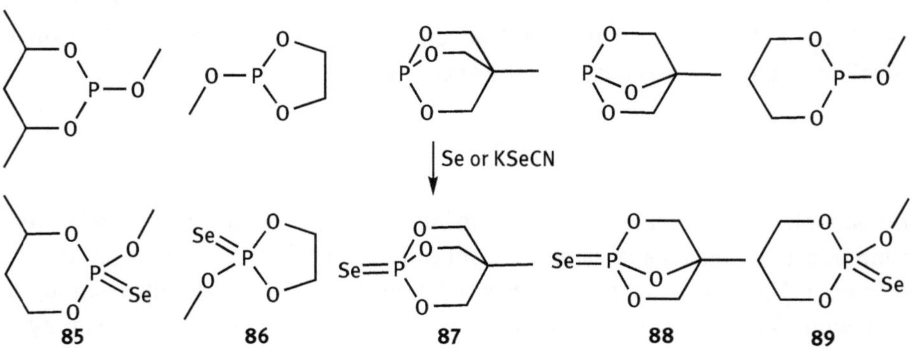

Figure 5.23: Synthesis of the strained phosphoroselenoates **85–89**.

A series of *Se*-glycosylphosphoroselenothioates **90a-93a** and *S*-glycosylphosphoroselenothioates **90b-93b** were prepared by direct glycosidation from *O,O*-dialkylphosphoroselenothioate anion and *β*-D-hexopyranoses (Figure 5.24) [71]. *β*-D-hexopyranoses: peracetylated gluco-, galacto-, 2-deoxy-arabina- and per-*O*-benzylated unprotected 1-*OH*-gluco- as glycosyl donors have different modes of activation of the anomeric position and protection of the sugar moiety for the glycosidation of the *O,O*-dialkylphosphoroselenothioate anion, promoted by $BF_3{\cdot}Et_2O$ in 1,2-dichloroethane at room temperature to give a mixture of two *Se*-

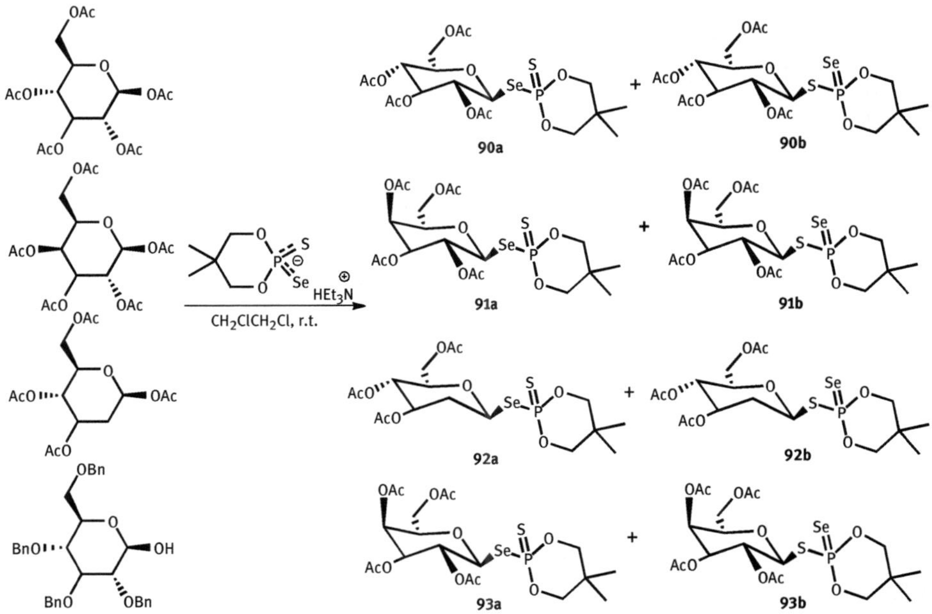

Figure 5.24: Synthesis of *Se*- and *S*- lycosylphosphoroselenothioates **90a–93a** and **90b–93b**.

and *S*-isomers. The reactants used were in equimolar amounts and the $BF_3{\cdot}Et_2O$ was three molar equivalents. Two isomers were obtained, but the *Se*-isomer prevailed in all the cases based on the spectral data.

5.4.5 Tertiary phosphine tellurides $R_3P{=}Te$

Triorganophosphine tellurides (tellurophosphoranes) R_3PTe are widely used as tellurium transfer reagents in a variety of applications. The first organophosphorus(V)-tellurium is $^nBu_3PTe$ which was synthesized more than 50 years ago by oxidation of $^nBu_3P$ with elemental tellurium in boiling toluene [72]. Since then, a wide variety of derivatives $RR^1{}_2PTe$ ($R{=}R^1{=}$Me, Et, ^nPr, Cy, ^tBu; R=Ph, $R^1{=}{}^nPr$, ^nBu) were prepared by a similar method [73, 74]. The oxidation is more difficult with the introduction of Ph substituents on the P(III) center and the yields are normally low for $PhR^1{}_2PTe$ ($R^1{=}{}^nPr$, ^nBu). For example, the bis(phosphine tellurides) $Ph_2P(Te)(CH_2)_nP(Te)Ph_2$ (n = 2, 6), and the monotellurides $Ph_2P(Te)(CH_2)_4Ph_2$ and $Ph_2MeP(Te)$ were prepared in only 6 % yield [75]. Elemental tellurium does not oxidize Ph_3P; however, to use more reactive source of Te, such as, the tellurocyanate $[Ph_4As][TeCN]$ led to a 1:1 adduct of Ph_3PTe with Ph_3P [76]. The derivatives R_3PTe (R=Cy, ^tBu) are thermally stable both in the solid and in solution; the other phosphine tellurides are isolated as air-/light-sensitive pale-yellow solids, which readily decomposed to gray tellurium in solution.

The introduction of imino substituents improve the nucleophilicity of the phosphorus(III) center. $Ph_3P{=}NP(Te)Ph_2$ [77], $(R_2N)_3PTe$ (R=Me, Et, R_2N=morpholino, piperidino, pyrrolidino [78–80], $Me(Et_2N)PTe$ and $(Et_2N)P(Te)(CH_2)_4P(Te)(NEt_2)_2$ [81] were prepared in high yields. The introduction of a saturated four-membered P_2N_2 ring changes the nucleophilicity of the phosphorus(III) center dramatically. The oxidation of $[MeP^{III}(\mu\text{-}N^tBu)]_2$ with elemental tellurium at room temperature for 48 h led to the ditelluride *cis*-$[MeP^V({=}Te)(\mu\text{-}N^tBu)]_2$ in only 8 % yield; the monotelluride $MeP^V(Te)(\mu\text{-}N^tBu)]_2P^{III}Cl$ was obtained in 17 % yield from the unsymmetrical P^{III}/P^{III} derivative $[MeP^{III}(\mu\text{-}N^tBu)]_2P^{III}Cl)$ and elemental tellurium under similar conditions [82]. The tellurides incorporating a *cyclo*-P_2N_2 framework are stable both in the solid state and in solution. Methylation of triorganophosphine tellurides R_3PTe (R=Me, ^{i}Pr, ^{n}Bu, ^{t}Bu, NMe_2) with methyl iodide in benzene led to the tellurophosphonium cations $[R_3P\text{-}TeMe]^+$ in high yields [83]. The oxidation of $^tBu_3PTe$ with a ferrocenium salt Cp_2FeX (X=BF_4, PF_6, SbF_6) gives a tritelluro dication $[^tBu_3P\text{-}Te\text{-}Te\text{-}Te\text{-}P^tBu_3]^{2+}$ [84] The dihalides Et_3PTeX_2 (X=Cl, I) were obtained from the oxidation of Et_3PTe with SO_2Cl_2 or I_2, and the corresponding dibromide (X=Br) was prepared via a halogen exchange reaction of Et_3PTeCl_2 with Me_3SiBr [85].

The coordination chemistry of the triorganophosphine tellurides has been poorly developed due to the lability of the phosphorus-tellurium bond leading to either a terminal or a bridging role in metal complexes. The first examples of R_3PTe metal complexes with terminal mode bonding were reported in 1985. The photolysis of $M(CO)_6$ in the presence of $^tBu_3PTe$ in THF generated $M(CO)_5(TeP^tBu_3)$ (M=Cr, Mo, W) [86]. The dimeric cation $[\{Hg(TeP^nBu_3)\}_3Br_5]_2^{2+}$ was prepared in good yield by reaction of elemental tellurium with $^nBu_3PTe$ and $HgBr_2$ in boiling acetone [87].

Triorganophosphine tellurides are efficient tellurium-transfer reagents that are soluble and more reactive than elemental tellurium [88, 89]. The use of $^nBu_3PTe$ for insertion of Te into the P-E (E=Si, Ge, Sn) bond of $^nBu_2P\text{-}EMe_3$ led to the formation of the acyclic $P^{III}\text{-}Te\text{-}P^{III}$ compound $^nBu_2P\text{-}Te\text{-}EMe_3$ [90]. Other examples for the phosphine tellurides as Te-transfer reagents include the synthesis of a bridging uranium telluride $[(MeC_5H_4)_3U]_2(\mu\text{-}Te)$ from $(MeC_5H_4)_3U\cdot THF$ and $^nBu_3PTe$ [91], the preparation of $R_2Ga\text{-}Te\text{-}GaR_2$ (R=$CH(SiMe_3)_2$) via insertion into metal-metal bonds from $R_2Ga\text{-}GaR_2$ and Et_3PTe [92], the formation of an $\eta^2\text{-}Te_2^{2-}$ complex $[Cp^*{}_2Ti]_2(\mu\text{-}TeTe)$ from $Cp^*{}_2Ti\text{-}Te\text{-}TiCp^*{}_2$ and $^nBu_3PTe$ [93], and the chalcogen-exchange reactions to yield the tellurometallates from selenometallates [94]. Triorganophosphine tellurides are also used as tellurium-transfer reagents for tellurium insertion into metal-carbon bonds. Elemental tellurium does not insert directly into the Mn–C bond in $RMn(CO)_5$ (R=alkyl, aryl) complexes; however, when $R^1{}_3PTe$ (R^1=Me, Et) was used, it gave the tellurolates $RTeMn(CO)_3(PR^1{}_3)_2$ via initial substitution of two CO ligands by $PR^1{}_3$, followed by Te insertion [95].

5.5 Diselenophosphinates/ditellurophosphinates

5.5.1 Diselenophosphinates

The chemistry of diselenophosphinates has been paid huge attention due to their use as efficient single-source precursors of remarkable nanomaterials, as chain transfer agents for reversible addition-fragmentation transfer polymerization, as bidentate ligands for metal complexes, as promising extractants of heavy, rare and transuranium elements, as potential drug candidates, and as building block for organic synthesis [96–104]. However, there are limited examples of well-characterized diselenophosphinates despite the regularly reported applications as single-source precursors [98, 100], synthetic building blocks [105, 106], and ligands for metal separation [107]. The diselenophosphinate complexes can be conveniently prepared from the metathesis reaction of simple metal salts with source of the free [R_2PSe_2] anions, such as alkali metal or ammonium diselenophosphinates [108].

The synthesis of the heavy metal diselenophosphinates are based on the reaction of alkylammonium diselenophosphinates with heavy metal halides. The alkylammonium diselenophosphinates are key intermediates in the synthesis of *Se*-esters of diselenophosphinoic acids [51]. The conventional synthesis of alkylammonium diselenophosphinates has multiple steps, is laborious and requires aggressive and poisonous phosphorus halides and flammable unstable organometallic reactants. For example, triethylammonium diselenophosphinates **94** were prepared by the reaction of toxic monochlorophosphines R_2PCl with triethylsilane, triethylamine, and selenium in refluxing toluene solution for *ca.* 20 h (Figure 5.25) [108, 109].

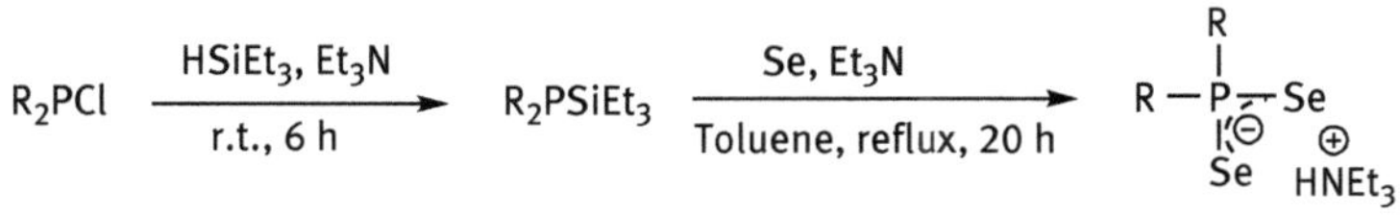

94

R = ^{i}Pr, ^{t}Bu, Ph

Figure 5.25: Formation of triethylammonium diselenophosphinates **94**.

A facile synthesis of alkylammonium diselenophosphinates from various secondary phosphines, R_2PH, elemental selenium and primary, secondary, or tertiary amine was reported [110, 111]. The three-component reaction of secondary phosphines, elemental selenium, and amines proceeds under mild conditions in ethanol leading to diselenophosphinates **95** in 85–97 % yields (Figure 5.26). The reaction was found to be of general character. For example, the reaction of secondary phosphine proceeds readily with elemental selenium and optically active (*R*)-1-aminobutan-2-ol under the similar conditions to give diselenophosphinate **96** in 86 % (Figure 5.27).

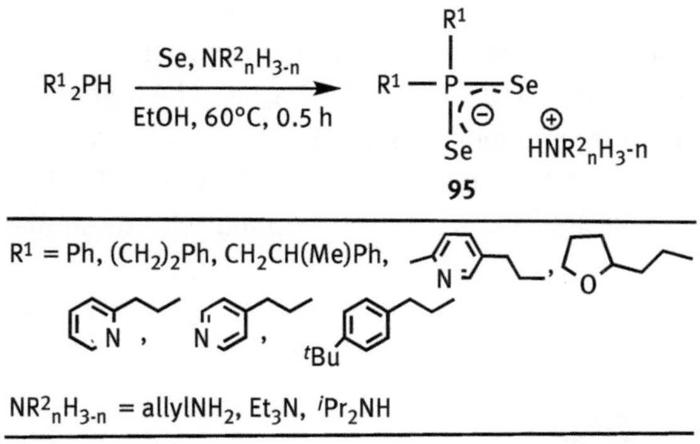

Figure 5.26: Formation of diselenophosphinates **95**.

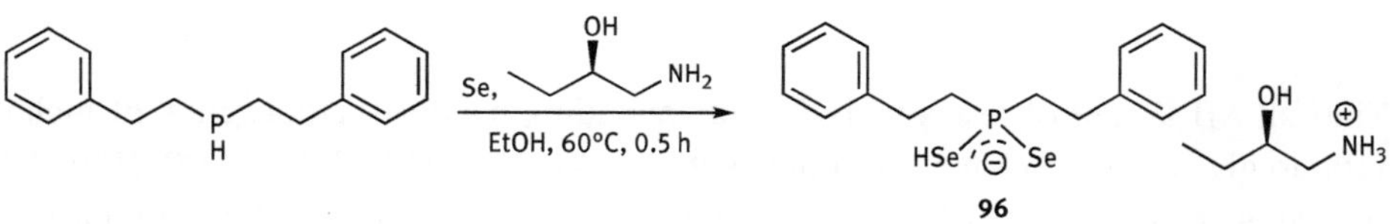

Figure 5.27: Formation of diselenophosphinates **96**.

Using secondary phosphine selenides as starting materials is another efficient method to synthesize diselenophosphinates. The reaction of the secondary phosphine selenides with the system Se/MOH (M=Li, Na, K, Rb, Cs) in the THF/EtOH media at room temperature led to almost quantitative yields (96–100 %) of alkali metal diselenophosphinates **97** (Figure 5.28) [112]. The ammonium diselenophosphinates were synthesized via the three-component reaction between secondary phosphines, elemental selenium and ammonia [112, 113]. Hydrazine can replace ammonia in the three-component reactions for the synthesis of diselenophosphinates [113, 114].

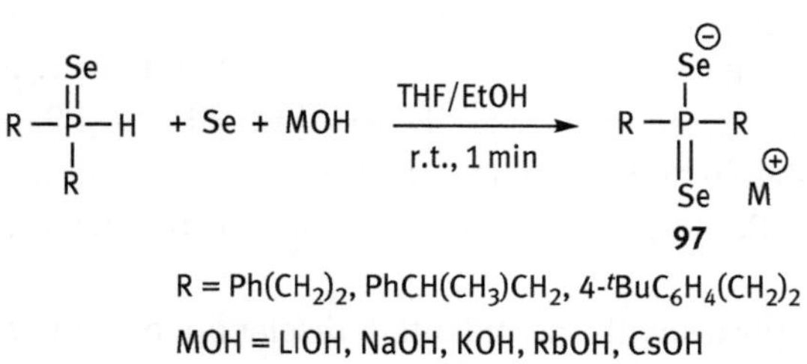

Figure 5.28: Synthesis of alkali metal diselenophosphinates **97**.

The conventional syntheses of diselenophosphinates are based on the reactions of selenophosphinic chlorides $R_2P(Se)Cl$ or monochlorophosphines R_2PCl with sodium selenide Na_2Se, sodium hydroselenide NaHSe, or polyselenides Na_xSe_y. The first example was the synthesis of sodium diethyldiselenophosphinate **98**, which was formed as the dihydrate $Et_2PSe_2Na{\cdot}2H_2O$ in 34 % yield by a multi-step route (Figure 5.29) [115]. Sodium diphenylselenophosphinate (Ph_2PSe_2Na) can be obtained via a similar approach in 65 % yield [116].

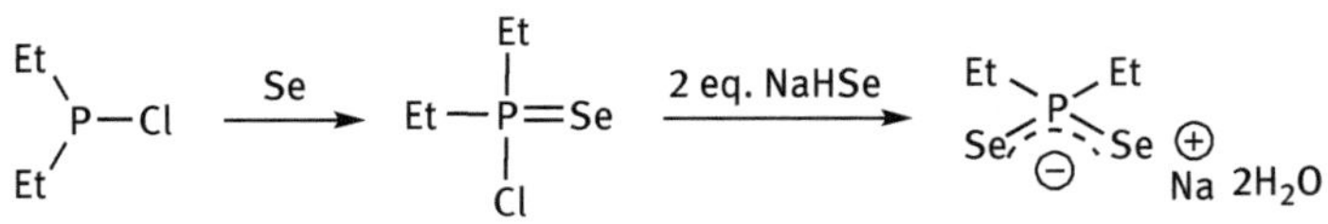

98, 34%

Figure 5.29: Synthesis of sodium diethyldiselenophosphinate **98**.

The salt $NH_4[Se_2PPh_2]$ reacts with the corresponding metal (II) oxides in a 2:1 molar ratio to give metal diselenophosphinates **99** in 64–85 % yields (Figure 5.30) [117]. All reactions proceed chemoselectively leading to the formation of only metal complexes.

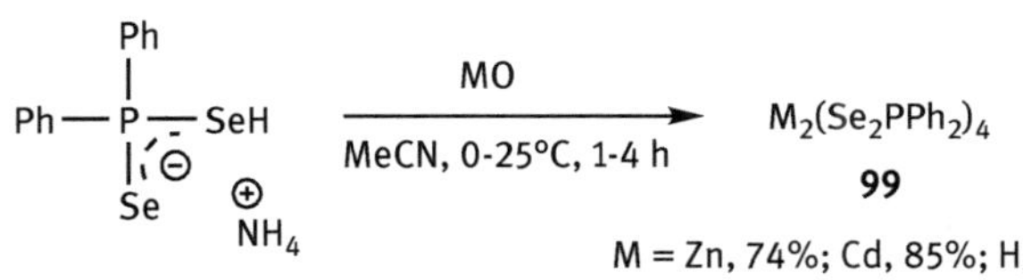

99

M = Zn, 74%; Cd, 85%; Hg, 64%

Figure 5.30: Synthesis of metal diselenophosphinates **99**.

The reaction of *N,N*-diethylammonium diselenophosphinate $[Et_2NH_2][(CH_2CH_2Ph)_2PSe_2]$ with $ZnCl_2$ or $ZnBr_2$ in a 2:1 molar ration under 25 °C in DCM for 4 h delivered a discrete *Se*-centered tetranuclear cluster **100** in 63 % isolated yield (Figure 5.31) [118]. It is apparent that the self-assembly of cluster **100** proceeds *via* initial formation of monomeric $[Zn(Se_2PR_2)_2]$ and then dimeric $[Zn_2(Se_2PR_2)_4]$ intermediates, followed attack by the free $[(CH_2CH_2Ph)_2PSe_2]^-$ resulting in cleavage of the one P-Se bond and giving a $Zn_4(Se)$ cage **100**. The reaction of diselenophosphinate $[Et_3NH_2][(CH_2CH_2Ph)_2PSe_2]$ and $ZnBr_2$ with pyridine *N*-oxide (PyNO) at 25 °C in acetone for 4 h resulted in the assembly of a helical polymer $\{ZnBr(Se_2PR_2)[PyNO]_n$ (**101**) as white powder in 84 % yield (Figure 5.32) [118].

A series of 4f and 5f metal complexes with the bidentate diselenophopshinate anion $[Se_2PPh_2]^-$ were prepared [107]. Heating the diselenophosphinate dimer $[K(Se_2PPh_2)]_2$ with 0.66 equivalents of LnX_3 (LnX_3=$La(OTf)_3$, $CeBr_3$ $NdCl_3$) in

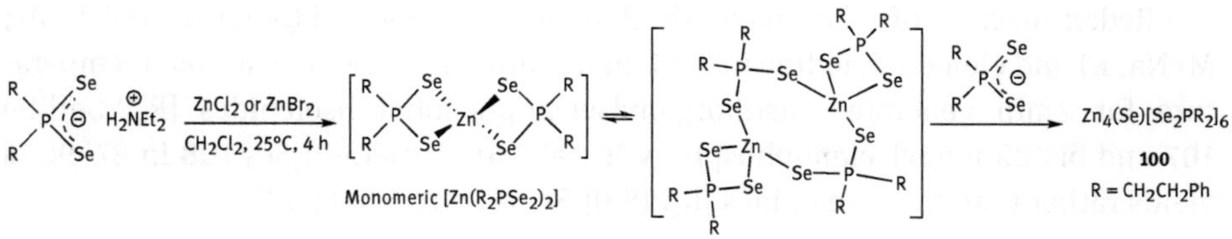

Figure 5.31: Synthesis of *Se*-centered tetranuclear cluster **100**.

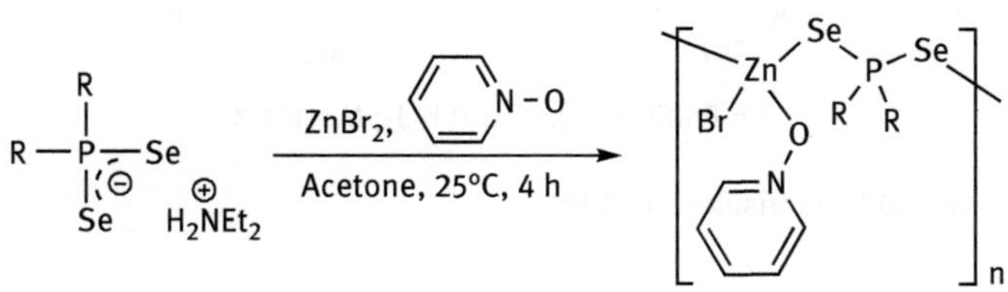

101, R = CH_2CH_2Ph

Figure 5.32: Synthesis of helical polymer **101**.

THF led to the formation of neutral complexes $[Ln(Se_2PPh_2)_3(THF)_2]$ **102–104** (Ln=La, Ce, Nd). The molecular structures of **102–104** adopt eight-coordinate metal centers with three $[Se_2PPh_2]^-$ anions bonding through both Se atoms and two coordinated THF solvent molecules. However, reacting an acetonitrile solution of either $CeBr_3$ or $[PuI_3(Py)_4]$ with two equivalents of $[K(Se_2PPh_2)]_2$ in the presence of Et_4NBr gave the homoleptic trivalent 1:4 anionic complexes $[Et_4N][Ce(Se_2PPh_2)_4]$ (**105**) and $[Et_4N][Pu(Se_2PPh_2)_4]$ (**106**) (Figure 5.33).

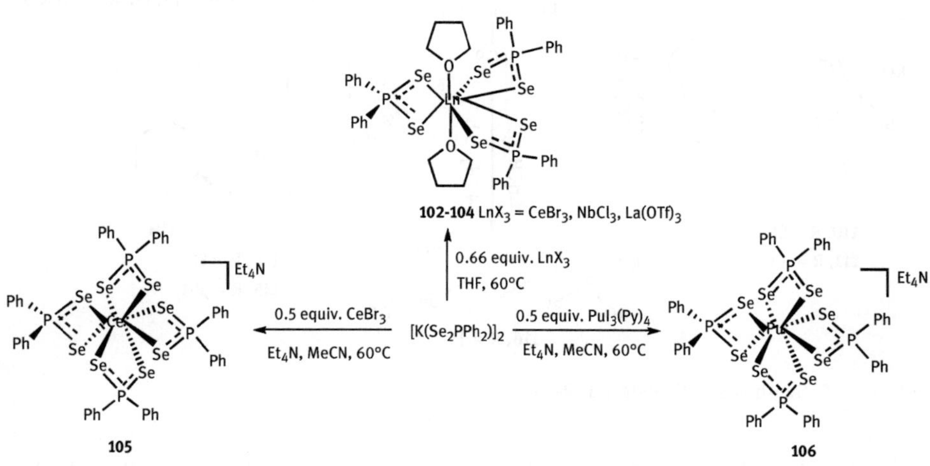

Figure 5.33: Synthesis of complexes **102–104**, and anionic complexes **105** and **106**.

Redox reaction of alkali metal diselenophosphinates $R_2P(Se)SeM$ (R=Alk, Ar; M=Na, K) and elemental iodine in a 2:1 molar ratio in 1,4-dioxane at room temperature for 2 min generated bis(diorganylselenophosphoryl)selenides $[R_2P(Se)]_2Se$ **107** and bis(diorganylselenophosphoryl)triselenides $[R_2P(Se)]_2Se_3$ **108** in 87–90 % yields rather than the diselenides $[R_2P(Se)]_2Se_2$ (Figure 5.34) [119].

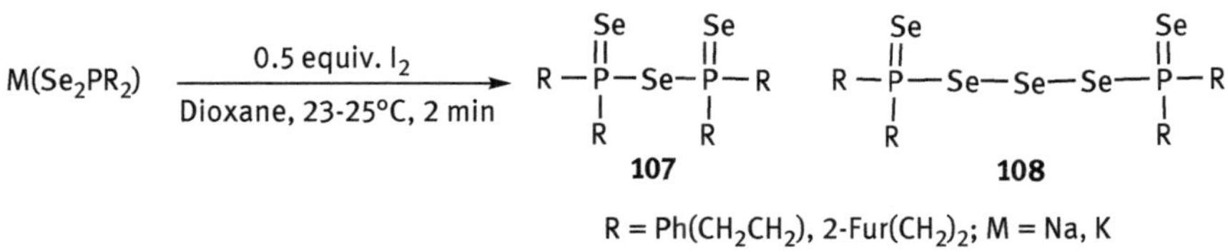

Figure 5.34: Synthesis of selenides **107** and triselenides **108**.

O,O'-Dialkyl diselenophosphates $(RO)_2PSe_2M$ can be used as the nucleophile to attack alkyl halides for the synthesis of *Se*-esters of phosphorodiselenoic acids [120, 121]. The syntheses of *Se*-esters of *O,O'*-dialkylphosphorodiselenoic acid **109–116** were achieved by the reaction of the ammonium salts. The reactions based on one of the Se atoms of the $[RO)_2PSe_2]^-$ units acting as the nucleophilic center includes Michael addition, epoxide ring opening addition, acylation, and α-alkynation (Figure 5.35) [120].

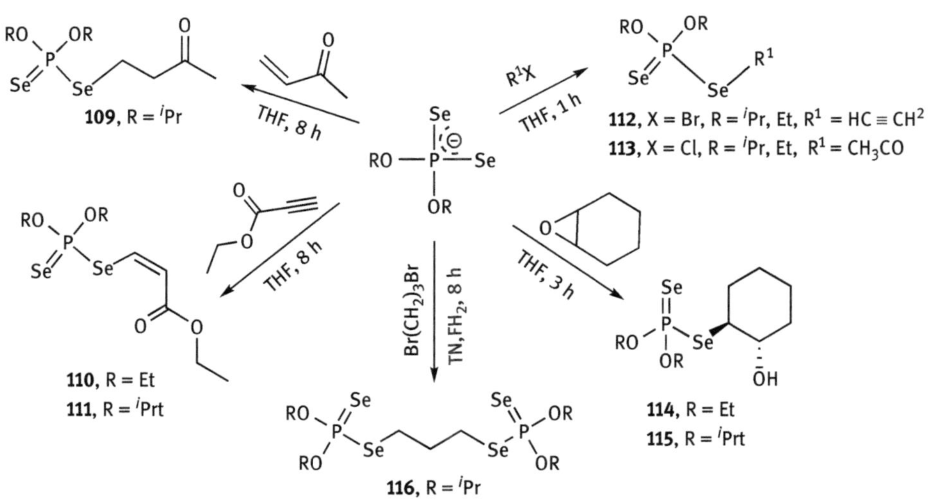

Figure 5.35: Synthesis of *Se*-esters **109–116**.

A modified method was developed by applying Na_2Se in place of sodium hydroselenide to react with (*tert*-butyl)phenylselenophosphinic chloride at 0 °C for 4 h

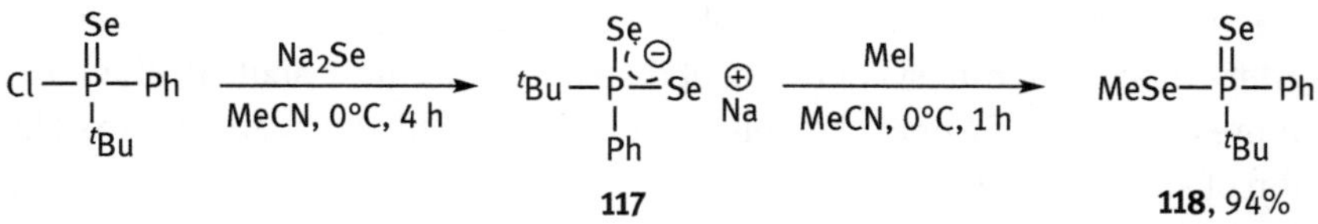

Figure 5.36: Synthesis of sodium diselenophosphinate **117** and *Se*-methyl ester **118**.

leading to sodium diselenophosphinate **117**, which reacted further *in situ* with methyl iodide to give *Se*-methyl ester **118** in 94 % yield (Figure 5.36) [51]. The method of using monochlorophosphines R_2PCl instead of selenophosphinic chlorides $R_2P(Se)Cl$ for synthesis of diselenophsophinates is preferable and has also been explored [122–124].

The reaction of alkali metals or alkylammonium diselenophosphinates with inorganic heavy metal salts led to the formation of a variety of heavy metal diselenophosphinates with some limitations. For example, the reaction of selenophosphinate anions with metal salts possessing strong oxidative properties did not or only with difficulty afforded the corresponding metal diselenophosphinates. In most cases, the major product is an oxidative dimer $(R_2PSe)_2Se_2$ of diselenophosphinate anions $[R_2PSe_2]^-$. However, the diselenide $(R_2PSe)_2Se_2$ together with the target diselenophosphinates were obtained with decreasing yield when the metal cations M^{n+} show weak oxidative properties. Thus, all synthesis of heavy metal diselenophosphinates are based on the reactions of diselenophosphinate anions $[R_2PSe_2]^-$ with metal cations M^{n+} with soft (Ag^+, Tl^+, Cd^{2+}, Hg^{2+}) or intermediate (Ni^{2+}, Co^{2+}, Zn^{2+}, Cu^{2+}, Sn^{2+}, Bi^{3+}, In^{3+}, Ga^{3+}) Lewis acids [40, 109, 125]. The cadmium salt $Cd(Se_2P^iPr_2)_2$ was applied as a single-source precursor to deposit CdSe nanoparticles, which were produced by introduction of a solution of cadmium diselenophosphinate in trioctylphosphine into a reactor composed of fused silica microcapillary at 200 °C. The resulting CdSe nanoparticles were then mixed with $Cd(S_2CNMe^nHex)_2$ and re-injected into the capillary layer to afford highly luminescent CdSe/CdS core-shell nanoparticles. The technique has opened an easy and convenient approach to high-quality semiconductor nanoparticles [98].

The preparation of nickel phosphide (NiP and $Ni_{12}P_5$) and nickel selenide (NiSe) nanoparticles by thermal decomposition of single source molecular precursor $Ni(Se_2PR_2)_2$ (R=^{i}Pr, ^{t}Bu, Ph) in solvents such as trioctylphosphine oxide (TOPO) and HAD at different temperature has been reported [125, 126]. Cobalt(II) diselenophosphinates, $Co(Se_2PR_2)_2$ (R=^{i}Pr, ^{t}Bu, Ph), can be used to grow cobalt selenide and cobalt phosphide nanoparticles by decomposition in trioctylphosphine oxide or HAD at 300 °C [96, 97]. The $Eu_{1-x}Se$ and EuSe nanoparticles were prepared by the thermal decomposition of Eu(III) diphenyldiselenophosphinate $Eu(Se_2PPh_2)_2$ in HAD at 300 °C [126, 127].

The reaction of salts of diselenophosphinic acids with organohalides provided an efficient method for synthesis of esters of diselenophosphinic acids $R_2P(Se)SeR$. The first representatives of esters of diselenophosphinic acids **119** were obtained by the reaction of sodium diethyldiselenophosphinate with ethyl or *n*-propyl bromides (Figure 5.37) [115]. Alkylation of sodium (*tert*-butyl)phenyldiselenophosphinate $^tBuPhSe_2Na$ with MeI in acetonitrile at 0 °C for 1 h gave the corresponding *Se*-methyl ester [51].

119, R = Et, ^{n}Pr

Figure 5.37: Synthesis of esters of diselenophosphinic acids **119**.

5.5.2 Tellurophosphinites and ditellurophosphinates

The chemistry of tellurophosphinites and ditellurophosphinates is less well documented despite the growing interest in using tellurolates as precursors to semiconducting metal tellurides [102, 103], and also as reagents for the synthesis of tellurium-containing organics [128]. The anionic species R_2PTe^- and $R_2PTe_2^-$ are potentially interesting ligands, which have displayed a range of coordination patterns. The anions R_2PTe^- (R=Cy, Ph) were prepared as lithium derivatives from the reaction of Li[R_2P] with elemental tellurium in THF [129]. The chalcogenotellurophosphinates $R_2P(Ch)Te^-$ (Ch=O, S, Se) can be delivered by treatment of Li[$R_2P(Ch)$] with elemental tellurium.

A tellurophosphinite [Ph_2PTe][$Li(TMEDA)_{1.33}(THF)_{1.33}$] (**121**) was prepared by the reaction of one equivalent of elemental tellurium with lithium diphenylphosphide in THF via insertion of the tellurium into the phosphorus-lithium bond. The reaction of $LiPPh_2$ with two equivalents of elemental tellurium led to ditellurophosphinate [Ph_2PTe_2][$Li(TMEDA)_{3.5}(THF)_{0.25}$] (**122**). A mixed chalcogen selenotelluro-phosphinate [$Ph_2P(Se)Te$][$Li(THF)_2(TMEDA)$] (**120**) were prepared from the reaction of lithium diphentlphopshide with elemental tellurium in a one-pot reaction; the same product **121** was able to be obtained from the reaction of **122** with one equivalent of selenium (Figure 5.38) [130]. Both compounds **121** and **122** incorporate an R_2PTe^- anion. Three tellurium containing compounds **120–122** are much more reactive than their sulfur or selenium analogues and are highly air and mositure and light sensitive, and readily decompose to deposit amorphous tellurium.

The bis(ditellurophosphinate) complex **124** was prepared in near quantitative yield by treatment of 1,2-bis(diphenylphosphanyl)ethane with elemental lithium via an intermediate 1,2-dilithio-1,2-di(phenylphosphine)ethylene (**123**), followed by

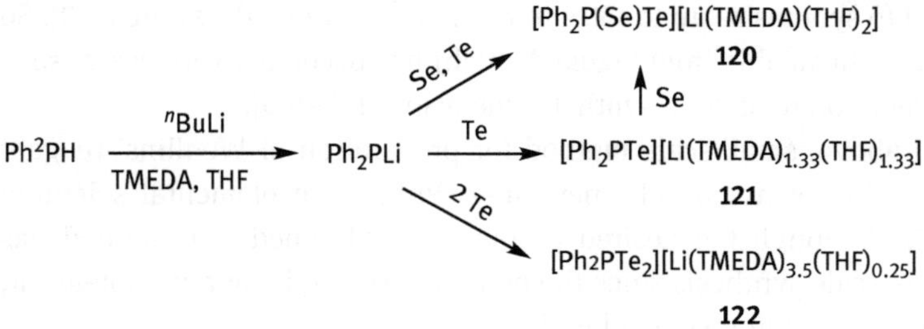

Figure 5.38: Synthesis of three tellurium containing compounds **120–122**.

subsequent reaction with powdered elemental tellurium (Figure 5.39) [131]. ^{31}P NMR studies of the reaction mixture revealed a nearly quantitative conversion of the dilithiated bisphosphine **123** into the bis(ditellurophosphinate) **124**.

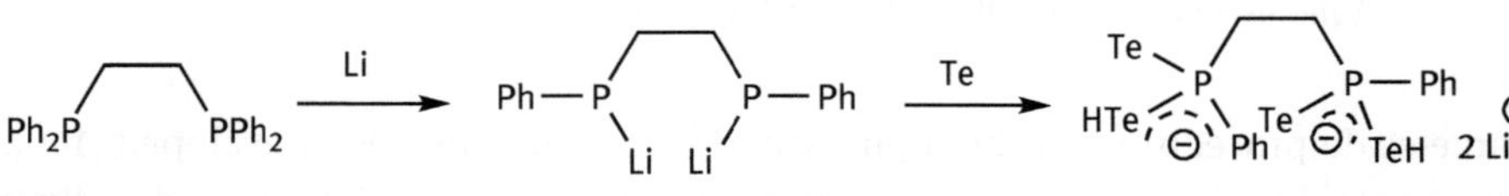

Figure 5.39: Synthesis of bis(ditellurophosphinate) **124**.

5.6 Diselenaphosphetane diselenides and Woollins' reagent

Diselenaphosphetane diselenides $(RPSe_2)_2$ have received very little attention, compared to their sulfur analogues, dithiadiphosphetane disulfides $(RPS_2)_2$. The first report was the synthesis of $(^tBuPSe_2)_2$ from the reaction of dichloro-*tert*-butylphosphine and Li_2Se_2 [132]. Since then, the same compound and its methyl and phenyl analogues have been obtained from the reaction of the silyl esters of the triselenophosphonic acids and DMSO though no detailed experimental procedure or spectroscopic data were reported [133]. A wide range of diselenaphosphetane diselenides $(RPSe_2)_2$ have been obtained from the oxidation of the corresponding homocyclic pentamers [$(RP)_5$, R=Me, Et, 4-$Me_2NC_6H_4$, An and Ph] by ten equivalents of elemental selenium [134–137]. Among them, the chemistry of 2,4-bis(phenyl)-1,3-diselena-2,4-diphosphetane-2,4-diselenide $(PhPSe_2)_2$ has received special attention since it was suggested to be named as Woollins' reagent (**WR**) [138]. It is known that Woollins' reagent is in equilibrium with a diselenaphosphorane $PhPSe_2$, which is a true reactive species in reaction [139]. However, a major obstacle to developing the chemistry of Woollins' reagent was its unwieldy synthesis from the cyclophosphine $(PhP)_5$, which could not be readily scaled up. It was not until 2005, a large-scale method for the preparation of Woollins' reagent was successfully developed, which gave a product of excellent purity and high yield and can be applied to synthesize up to 150 g in normal laboratory glassware [140]. So far, Woollins' reagent (available from Sigma-Aldrich) has become a very useful selenium source or selenation reagent in synthetic chemistry [141–149].

As mentioned above, small-scale method for preparation of Woollins' reagent was from the oxidation of homocyclic pentamer $(PhP)_5$ with elemental selenium (Figure 5.40) [135]. Although the desired product was obtained, the method has limitations for large-scale synthesis since the pentamer $(PhP)_5$ is highly unpleasant, air-sensitive precursor, and limited availability.

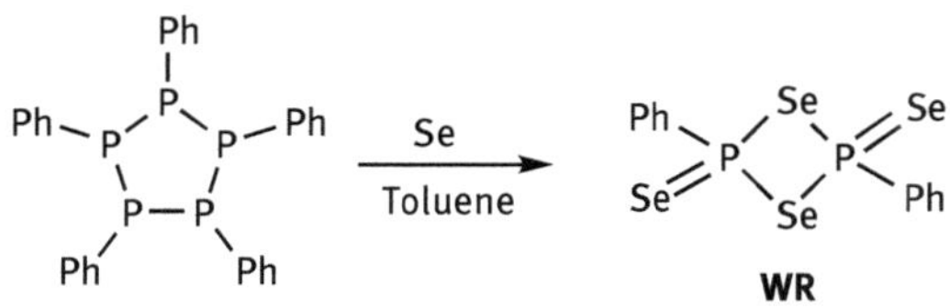

Figure 5.40: Synthesis of a small scale of Woollins' reagent.

A large-scale procedure with high purity and high yield has been developed [140]. Sodium selenide Na_2Se, which was prepared from elemental selenium and sodium metal in liquid ammonia at −78 °C, reacts with $PhPCl_2$ in refluxing toluene for 24 h. After filtering to remove NaCl, a second portion of selenium is added into the reaction mixture and the mixture is refluxed for another 24 h leading to Woollins' reagent as red crystals in ~86 % yield after cooling to room temperature (Figure 5.41). The method involves the use of the unpleasant dichlorophenylphosphine and the potentially hazardous liquid ammonia; however, it avoids the handling of the pentamer $(PhP)_5$. In addition, the method produces a product of high purity in higher yield, and the major advantage of this method is that a much larger scale up to 150 grams can be achieved by using lab glassware. Woollins' reagent is soluble in hot toluene and can be stored at room temperature for several months without apparent decomposition. Reaction of Woollins' reagent with organic substrates delivers a wide range of systems ranging from simple oxygen-selenium exchange to the formation of complex phosphorus-selenium heterocycles as well as the surprising phosphorus-selenium-free products. Nucleophilic addition of the phosphorus atom of the double P=Se bond of diselenophosphorane can readily be achieved for the synthesis of diselenophosphinates. Diselenophosphinate **125** was obtained in 27 % yield by reaction of

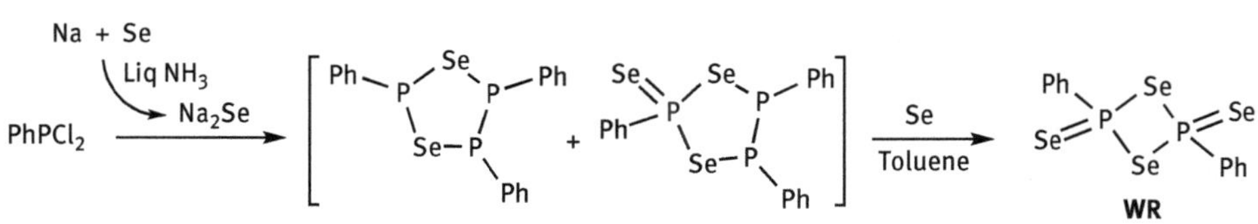

Figure 5.41: Synthesis of a large scale of Woollins' reagent.

2-phosphaindolizine with an equivalent of Woollins' reagent in benzene at ambient temperature for 2 h [150]. The reaction of 2-phosphaindolizine with Woollins' reagent in the presence of NEt_3 (in a molar ration 2:1:2) in benzene at 20 °C for 2 h gave triethylammonium diselenophosphinates **126** in 87 % yield [150]. Similarly, treating 1-aza-2-phosphaindolizine with Woollins' reagent in the presence of NEt_3 resulted in the corresponding diselenophosphinate **127** in 42 % yield (Figure 5.42).

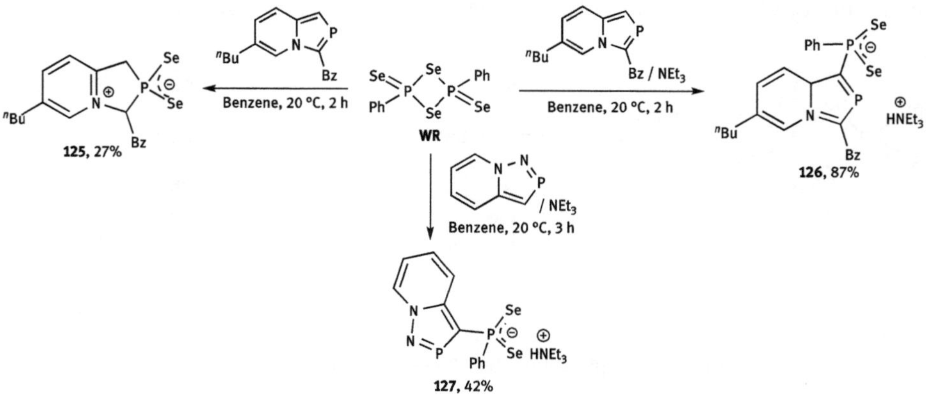

Figure 5.42: Synthesis of diselenophosphinates **125–127**.

Refluxing a mixture of equimolar amount of Woollins' reagent and cyanamides in toluene for 10 h followed treatment by water for 1 h resulted in the formation of a series of zwitterionic diselenophosphinates **128** in 29–70 % yields and cyclic selenazadiphosphoaminediselenides **129** in 11–41 % yields (Figure 5.43) [145].

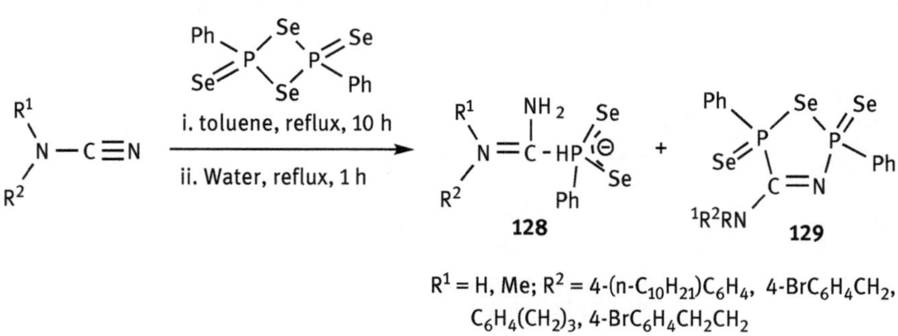

R^1 = H, Me; R^2 = 4-(n-$C_{10}H_{21}$)C_6H_4, 4-Br$C_6H_4CH_2$, $C_6H_4(CH_2)_3$, 4-Br$C_6H_4CH_2CH_2$

Figure 5.43: Synthesis of diselenophosphinates **128** and diselenides **129**.

Non-symmetric phenyldiselenophosphonates $[Ph(RO)PSe_2]^-$ **130** were synthesized *via* the cleavage of the four-membered P_2Se_2 ring in Woollins' reagent with two equivalents of sodium alkoxide [NaOR (R=Me, Et, ^{i}Pr and ^{n}Pr)] at room temperature as either white or pale-yellow solids in high yields [141, 151]. Then, sodium phenyldiselenophosphonates

130 was further treated with one equivalent of haloalkane in tetrahydrofuran at room temperature to yield the corresponding *Se*-alkyl-*O*-alkylphenylphosphonodiselenoate esters **131** in 83–99 % yields (Figure 5.44) [152]. These compounds were isolated in high yields as sticky oils or pastes soluble in both chlorinated solvents and tetrahydrofuran, and stable to air and moisture for months without obvious signals of degradation. The esters bearing bulky R^1 groups (R^1=^{i}Pr) were obtained in rather low yields, compared to the esters bearing small or medium R^1 groups (R^1=Me or Et), suggesting that steric hindrance of R^1 groups is important during the formation of the products.

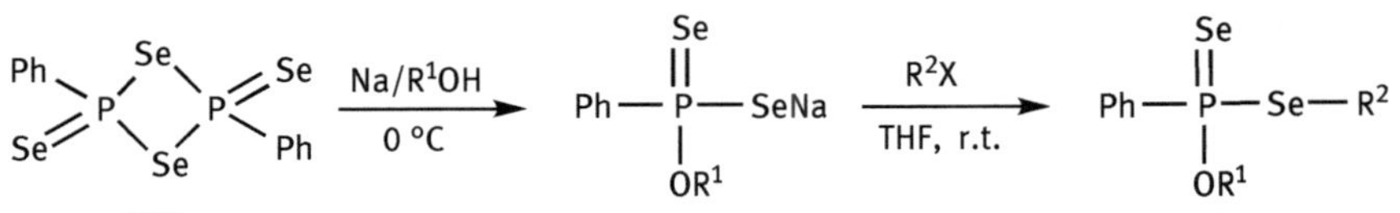

WR

130
R^1 = Me, Et, ^{i}Pr

131
R^1 = Me, Et, ^{i}Pr; X = Cl, Br, I
R^2 = 2-BrC$_6$H$_4$CH$_2$, PhCH$_2$, PhC(O)CH$_2$, CH$_3$(CH$_2$)$_6$, EtOC(O)CH$_2$C(O)CH$_2$

Figure 5.44: Synthesis of diselenophosphonates **130** and *Se*-esters **131**.

A series of alkylammonium phenylphosphonamidodiselenoates **132** were obtained in good to excellent yields by reacting two equivalents of primary or secondary amines with Woollins' reagent at room temperature (Figure 5.45) [153]. The products are soluble in polar and chlorinated solvents such as alcohols, dichloromethane, and chloroform but are insoluble in less polar solvents such as ethers and hexane, and are air stable for several weeks, after that obvious signs of degradation occur including reddening of the powders due to the expulsion of elemental selenium.

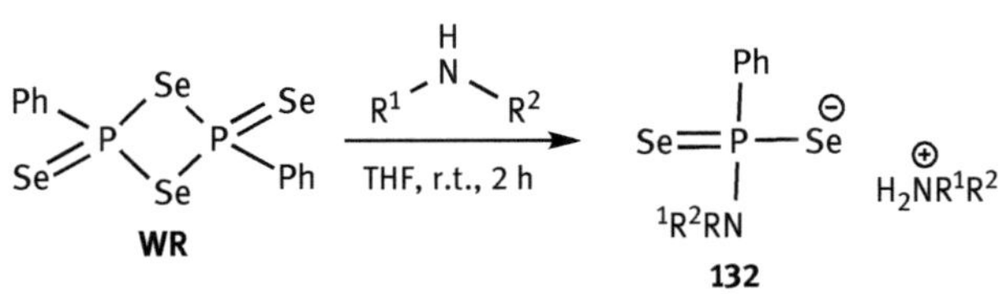

WR

132
R^1 = H, Et, ^{i}Bu, CH$_2$C$_6$H$_4$
R^2 = Et, ^{n}Bu, Ph, CH$_2$C$_6$H$_5$,^{i}Bu, H$_2$CH$_2$C$_6$H$_4$

Figure 5.45: Synthesis of alkylammonium phenylphosphonamidodiselenoates **132**.

The most attractive feature of Woollins' reagent lies in its capacity to form diverse patterns of phosphorus-selenium heterocycles. The earliest example is the reaction of Woollins' reagent with small, unsaturated molecules such as acetone or CS_2 [134, 154]. Treatment of Woollins' reagent with acetone led to colorless crystals of 4,4-

dimethyl-5-oxa-1,3-diphenyl-1,3-diseleno-2-selena-1,3-diphospholane **134**, arising from insertion of acetone in a P_2Se_2 ring. Meanwhile, Woollins' reagent reacted with CS_2 slowly resulting in the formation of a dimeric motif consisting of two five-membered CP_3Se rings being bridged *via* their trigonal carbon atoms bonding to the phosphorus atoms of a P_2Se_2 ring **133** (Figure 5.46).

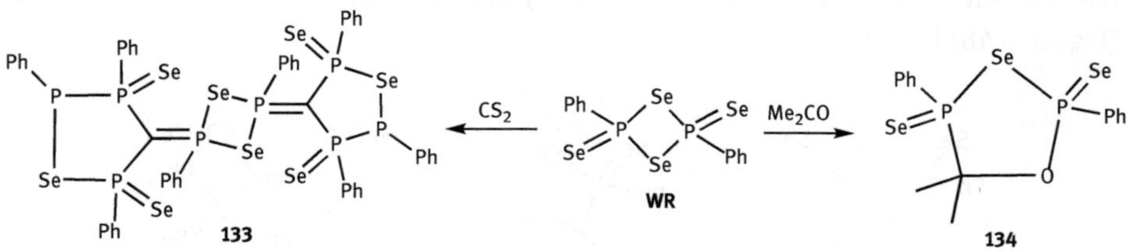

Figure 5.46: Synthesis of phosphorus-selenium heterocycles **133** and **134**.

The synthesis of a far more extensive series of P-Se heterocycles were performed based on the reaction of Woollins' reagent with a wide range of organic substrates containing reactive unsaturated C=O, C=C double and C≡C triple bonds. An unusual phosphorus-selenium spirocyclic heterocycle with a four-membered P_2SeC ring **135** (5 % yield) was achieved together with an expected selenocarbonyl compound **136** (27 % yield) from Woollins' reagent and diphenylcyclopropenone. Reaction of Woollins' reagent with methyl phenylpropiolate in refluxing toluene solution provided three five-membered phosphorus-selenium containing or selenium containing heterocyclic products **137–139**, which were found to be moderately stable in air, degrading over a period of several weeks with the obvious expulsion of red selenium (Figure 5.47) [140].

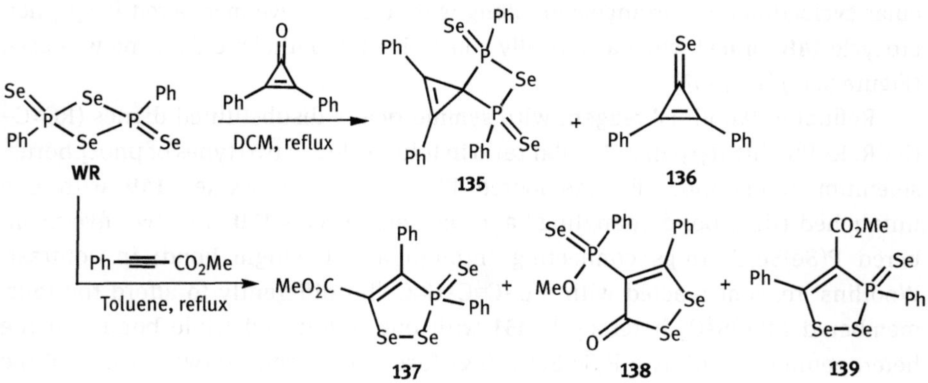

Figure 5.47: Synthesis of phosphorus-selenium heterocycles **135–139**.

A series of similarly unexpected phosphorus-selenium-containing heterocycles were also obtained from the reaction of Woollins' reagent with dimethyl but-2-ynedioate, alkynes and dialkyl cyanamides. Woollins' reagent was heated with dimethyl but-2-ynedioate in toluene leading to two five-membered P-Se heterocycles **140** and **141** and one six-membered P-Se ring **142** in 5–19 % yields [155]; however, only straightforward products **143** and **144** were harvested when Woollins' reagent reacted with ethynylbenzene or dialkyl cyanamides in the identical conditions (Figure 5.48) [155, 156].

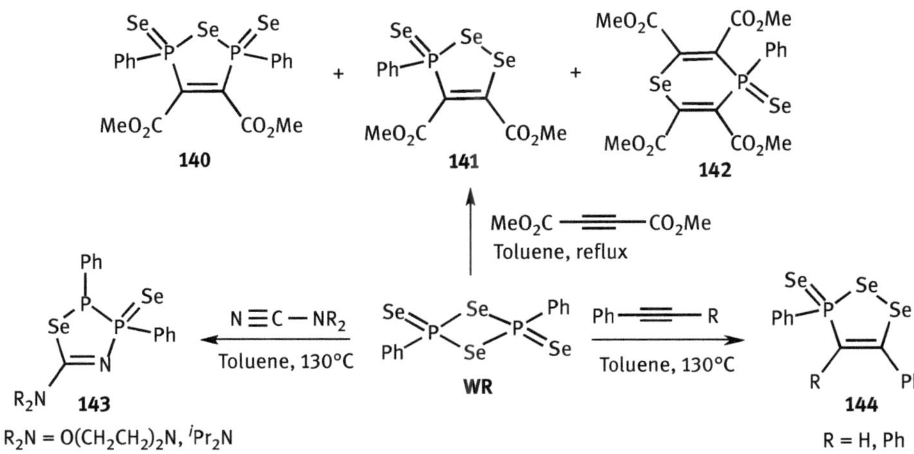

Figure 5.48: Synthesis of phosphorus-selenium heterocycles **140-144**.

Five-membered PSe_2C_2 heterocycles **145** and **146** were synthesized from the reaction of Woollins' reagent and dialkynes by formal addition of a $Ph(Se)PSe_2$ fragment to the alkyne triple bond. An unusual diselenide **147** was generated by an intramolecular cycloaddition/rearrangement along with a double five-membered PSe_2C_2 heterocycle **148** formed when a sterically constrained naphthalene dialkyne was used (Figure 5.49) [156, 157].

Refluxing Woollins' reagent with symmetrically disubstituted diynes (RC≡C–C≡CR, R=Ph, $Si(CH_3)_3$) in 2 : 1 molar ratio in toluene led to two types of phosphorus-selenium compounds, five-membered $P(Se)Se_2C_2$ heterocycles **149** with one untouched triple bond "dangling" and *bis*-heterocycles **150** with two five-membered $P(Se)Se_2C_2$ rings connecting through a C–C single bond; in contrast, Woollins' reagent reacted with ^{t}Bu–C≡C–C≡C–^{t}Bu differently to afford the four-membered $P(Se)SeC_2$ heterocycle **151** with one untouched triple bond and the heteropentalene with two $P(Se)SeC_3$ rings fused at the central two carbons of the diyne to give a heteropentalene analogue of pentalene, [3.3.0]octa-1-6-diene **152** (Figure 5.50) [157].

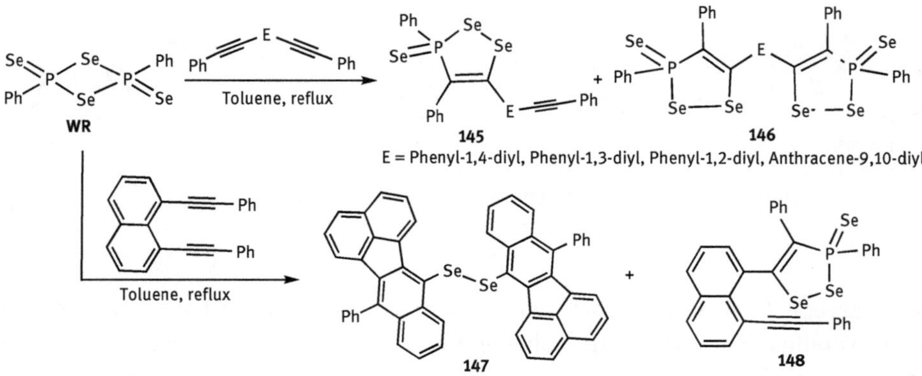

Figure 5.49: Synthesis of phosphorus-selenium heterocycles **145–148**.

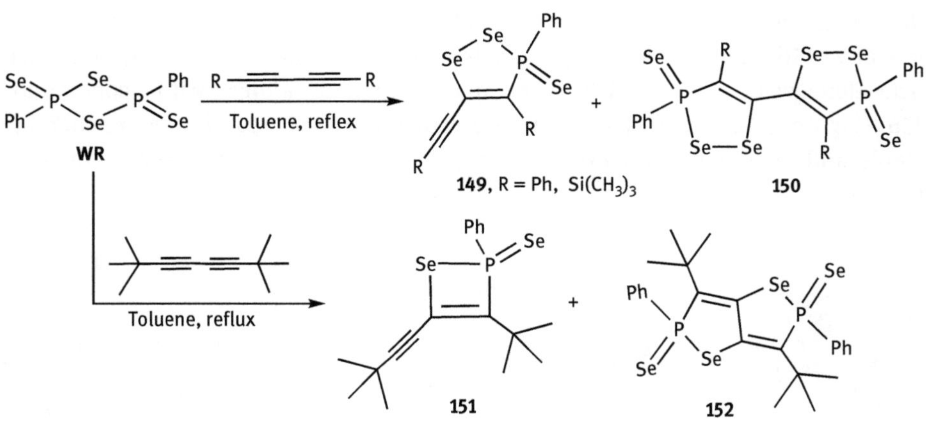

Figure 5.50: Synthesis of phosphorus-selenium heterocycles **149–152**.

Treatment of Woollins' reagent with norbornene proceeded with cleavage of the $P_2(\mu\text{-Se})_2$ ring resulting in the formation of a 1,2-diselena-3-phospholane **153** with a five-membered C_2PSe_2 ring *via* a [2 + 3] cycloaddition between the alkene and a Se–P=Se unit [157, 158], being different from the product 1,2-thiaphosphetane (C_2PS ring) of the reaction between norbornene and $[RP(S)(\mu\text{-S})]_2$ (R=ferrocenyl)] [158, 159]. Furthermore, reacting Woollins' reagent with norbornadiene did give a four-membered C_2PSe heterocycle **155** as a counterpart of 1,2-thiaphosphetane (C_2PS) together with the five-membered (C_2PSe_2) ring product **154** [158], although the reaction was carried out relatively slower than with norbornene (Figure 5.51) [160, 161].

Large ring heterocycles bearing PhP(Se)SeSeP(Se)Ph moiety could also be achieved through iodine oxidation of the ammonium salts of bisdiselenophosphonic

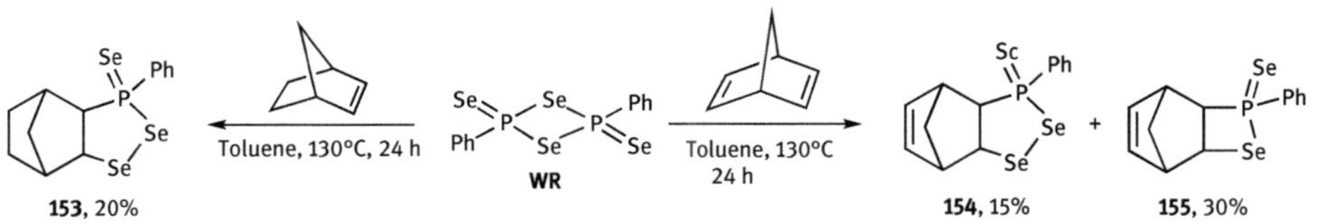

Figure 5.51: Synthesis of phosphorus-selenium heterocycles **153–155**.

acids at room temperature, which derived from diols [$HO(CH_2)_nOH$, $n = 2–6$, 8 and 10] and Woollins' reagent in dry toluene followed by treatment with butylamine in tetrahydrofuran [162]. Compound **158** is soluble in common polar organic solvents and stable to air and moisture for several months. One interesting example that should be mentioned is the remarkable reaction of Woollins' reagent with secondary amines. This reaction proceeded in the presence of dichloromethane or dibromomethane to give, predictably, bis(*N*,*N*-dialkyl-*P*-phenylphosphonamidodiselenoates) **156**. More interestingly, a twelve-membered heterocycle 1,2,4,5,7,8,10,11-octaselenacyclododecane **157** was also obtained along with compounds **156** (Figure 5.52) [163]. This new twelve-membered C_4Se_8 ring is remarkably stable and does not eliminate selenium even when refluxed in toluene for more than 24 h.

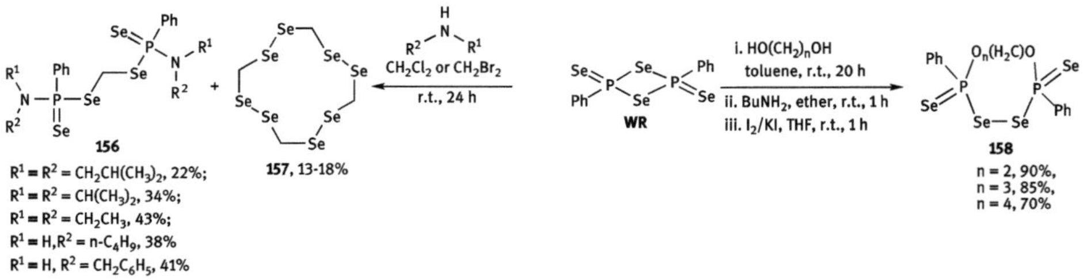

Figure 5.52: Synthesis of compounds **156–158**.

Woollins' reagent plays also an essential role for the selenation of organic compounds. For example, it reacts with secondary and tertiary amides to give selenoamides **161** by simple oxygen-selenium exchange [164]; with aryl-nitriles followed by hydrolysis with water to lead to arylselenoamides **162** [139]. It has also been used for the synthesis of selenadiazoles **163** [144, 165] and the synthesis of symmetrical and unsymmetrical (*E*)-olefins **159** and **160** via the reductive coupling reaction of aldehydes and ketones (Figure 5.53) [166].

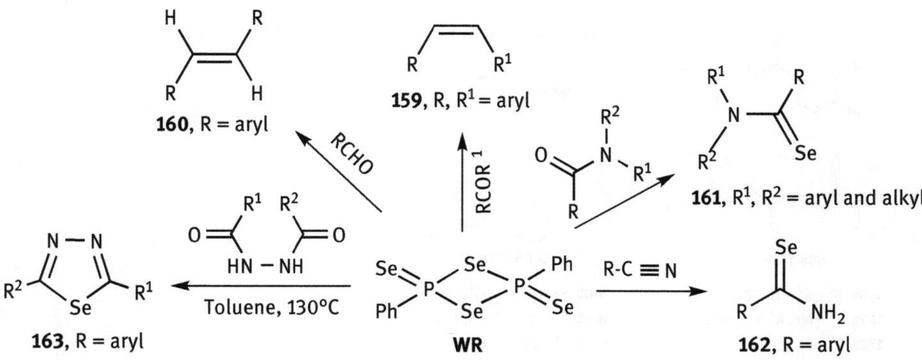

Figure 5.53: Synthesis of organic compounds **159–163**.

Furan-derived donor-acceptor-substituted cyclopropanes reacted with 0.5 equivalent of Woollins' reagent in chloromethane at room temperature to give the unprecedented selenium-containing cage-like compounds **164–167** in medium to excellent yields (Figure 5.54) [167].

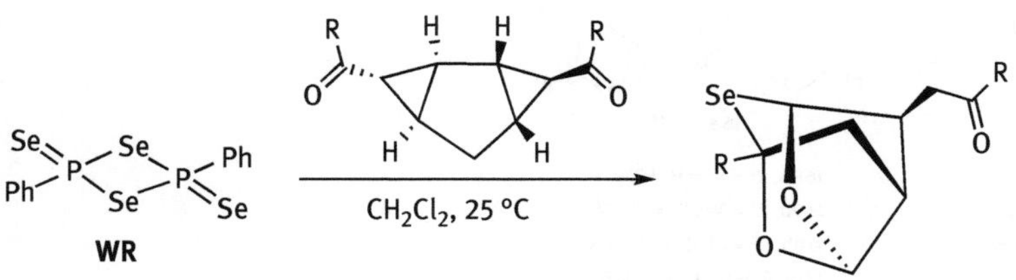

164-167

164, R = Me, 87%; **165**, R = ^{n}Pr, 70%

166, R = Ph, 84%; **167**, R = 4-$CF_3C_6H_4$, 32%

Figure 5.54: Formation of cage-like compounds **164–167**.

A selenium-assisted reduction system based on Woollins' reagent has been developed recently. Woollins' reagent was found to act as a highly chemoselective reducting reagent for the synthesis of a wide range of saturated 1,4-diketones **168a–174a** in good yields along with the corresponding cyclized products 2,5-disubstitued furans **168b–174b** in minor quantities from the corresponding 1,4-enediones and 1,4-ynediones under mild reaction conditions (Figure 5.55–Figure 5.57) [168]. The results show that the reduction of the triple bond is faster than that of the double bond to a single bond in such systems, and the reduction of arylalkyl and dialkyl enediones by Woollins' reagent is weaker than that of diaryl enediones.

The reaction of Woollins' reagent with one equivalent of porpholactone and four equivalents of $PhMe_2SiH$ in refluxing toluene afforded dihydroporpholactone **175** in 40 % isolated yield; with one equivalent of porpholactone and 100 equivalents of

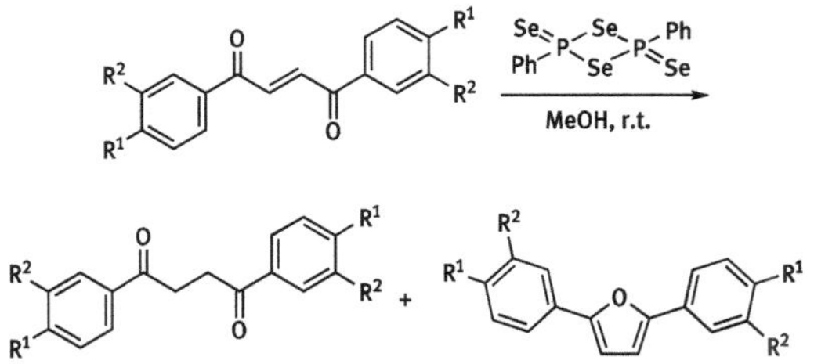

168a-172a 168b-172b

168a, $R^1 = R^2$ = H, 67%	**168b**, $R^1 = R^2$ = H, 15%
169a, R^1 = Me, R^2 = H, 64%	**169b**, R^1 = Me, R^2 = H, 16%
170a, R^1 = Cl, R^2 = H, 60%	**170b**, R^1 = Cl, R^2 = H, 15%
171a, R^1 = Br, R^2 = H, 68%	**171b**, R^1 = Br, R^2 = H, 17%
172a, R^1 = Me, R^2 = Cl, 60%	**172b**, R^1 = Me, R^2 = Cl, 16%

Figure 5.55: Reduction of 1,4-enediones.

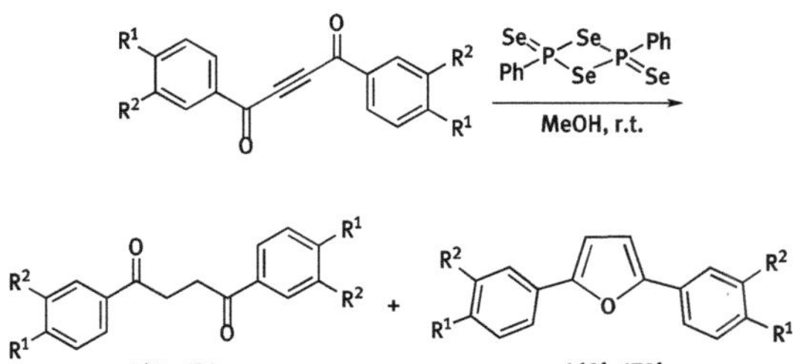

168a-172a 168b-172b

168a, $R^1 = R^2$ = H, 61%	**168b**, $R^1 = R^2$ = H, 16%
169a, R^1 = Me, R^2 = H, 65%	**169b**, R^1 = Me, R^2 = H, 15%
170a, R^1 = Cl, R^2 = H, 60%	**170b**, R^1 = Cl, R^2 = H, 18%
171a, R^1 = Br, R^2 = H, 62%	**171b**, R^1 = Br, R^2 = H, 16%
172a, R^1 = Me, R^2 = Cl, 60%	**172b**, R^1 = Me, R^2 = Cl, 16%

Figure 5.56: Reduction of 1,4-ynediones.

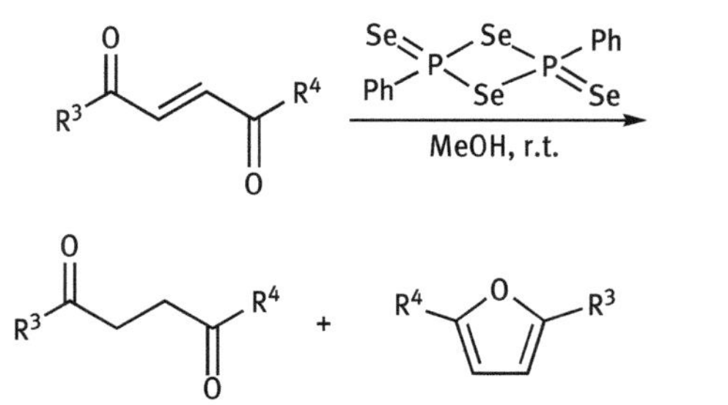

173a, R^1 = Ph, R^2 = Me, 42%	**173b**, R^1 = Ph, R^2 = Me, 18%
174a, $R^1 = R^2$ = Me, 46%	**174b**, $R^1 = R^2$ = Me, 17%

Figure 5.57: Reduction of arylalkyl and dialkyl enediones.

$PhMe_2SiH$ in refluxed toluene led to adjacent-tetrahydroporpholactone **176** in 75 % isolated yield. Similarly, tetrahydroporphyrin **177** was obtained by the reaction of Woollins' reagent with one equivalent of porphyrin and 100 equivalents of $PhMe_2SiH$ in 70 % isolated yield (Figure 5.58) [169].

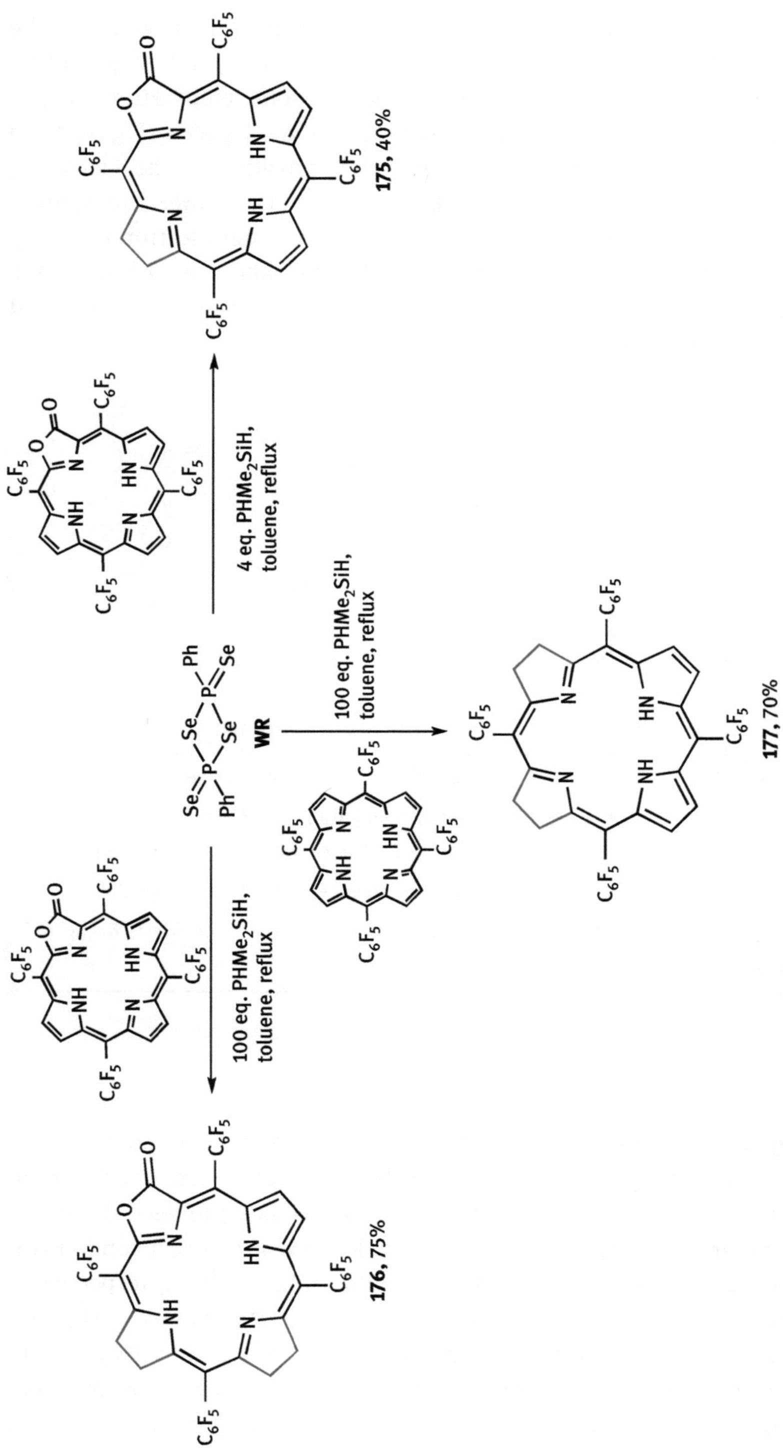

Figure 5.58: Synthesis of organic compounds **175–177**.

Woollins' reagent can act as a useful selective reducing reagent for the reduction of the double bond of 2-α,β-unsaturated thiazo and selenazolidinones. The reaction was carried out in toluene at room temperature to give the corresponding saturated heterocycles **178–188** in good yields (Figure 5.59) [170]. The results showed that the reaction is not sensitive to the electronic effects in the aromatic ring of the thiazolidinones. Therefore, the desired reduction products were prepared in good yields when either electron-donating (–OMe, –Me) or electron-withdrawing groups (–Br, –Cl), and the position of the substituent groups did not affect the reactivity of α,β-unsaturated thiazo and selenazolidinones considerably.

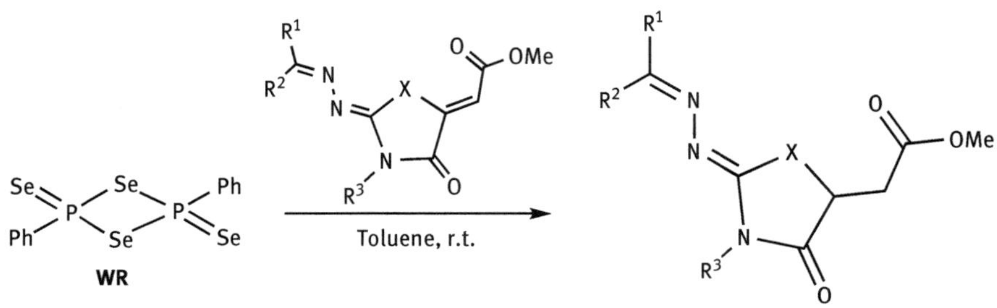

178-188

178, R^1 = Ph, R^2 = R^3 = Me, X = S, 55%
179, R^1 = 3,4-di-ClC_6H_3, R^2 = Me, R^3 = H, X = S, 72%
180, R^1 = 3-BrC_6H_4, R^2 = Me, R^3 = H, X = S, 77%
181, R^1 = 4-ClC_6H_4, R^2 = H, R^3 = Me, X = S, 68%
182, R^1 = 4-$CF_3C_6H_4$, R^2 = H, R^3 = Me, X = S, 79%
183, R^1 = 4-OMeC_6H_4, R^2 = H, R^3 = Me, X = S, 74%
184, R^1 = 4-OMeC_6H_4, R^2 = R^3 = H, X = S, 58%
185, R^1 = 4-MeC_6H_4, R^2 = R^3 = H, X = S, 25%
186, R^1 = 2,3-di-OMeC_6H_3, R^2 = R^3 = H, X = S, 89%
187, R^1 = Thiophene, R^2 = R^3 = H, X = S, 50%
188, R^1 = 3,4-di-ClC_6H_3, R^2 = Me, R^3 = H, X = Se, 67%

Figure 5.59: Selective olefin reduction of α,β-unsaturated thiazo and selenazolidinones.

Woollins' reagent can also act as both a selective reducing reagent and oxygen to selenium exchange reagent simultaneously. Treating four-component condensation adducts with two equivalents of Woollins' reagent in refluxing toluene delivered, upon workup in air, the corresponding selenoamides **189–191**, which contain two C=Se double bond groups in 68 %, 64 %, and 60 % isolated yields, respectively, rather than three C=Se double bond groups (Figure 5.60) [171]. The one C=O group conjoined phenyl ring from starting materials was reduced by Woollins' reagent into a CH_2 group, at the same time as the other two amide C=O groups were converted successfully by Woollins' reagent into the corresponding C=Se groups leading to selenoamides **189–191**.

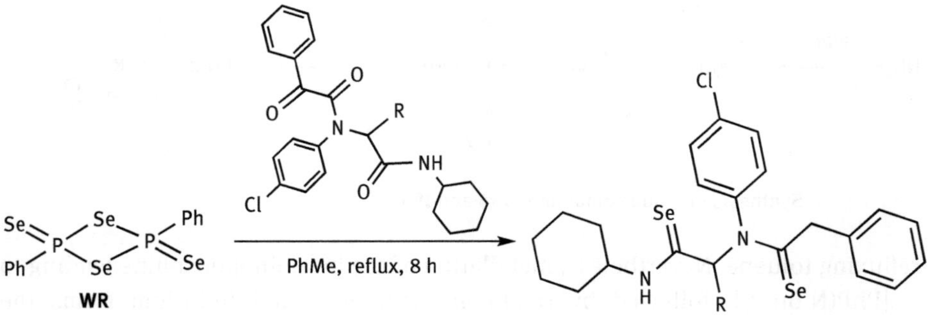

189, R = 2,3-di-OMeC_6H_3, 68%
190, R = 3-OMeC_6H_4, 64%
191, R = 4-ClC_6H_4, 60%

Figure 5.60: Selective olefin reduction of three-component condensation adducts.

5.7 Phosphorus-selenium/tellurium amides and imides

Replacement of one or more of the chalcogen atoms with an isoelectronic imido (NR) group in $PE_2(EH)$, $PE(EH)_3$, $R_2PE(EH)$, $RPE(EH)_2$ or in their anionic conjugate bases PE_3^-, PE_4^{3-}, R_2PE^{2-}, RPE_3^{2-}, (E=Se, Te) led to a new range of phosphorus-chalcogen amides and imides [172]. The chalcoamidophosphinic acids R_2PENHR' **192** (E=Se, Te), which are considered as dichalcogenophosphinic acids with one of the chalcogen atoms replaced by an NR´ group, can be prepared from the reaction of aminophosphines with chalcogen, the latter were further deprotonated with organometallic bases such as ^{n}BuLi to give the imido compounds **193** (Figure 5.61) [173].

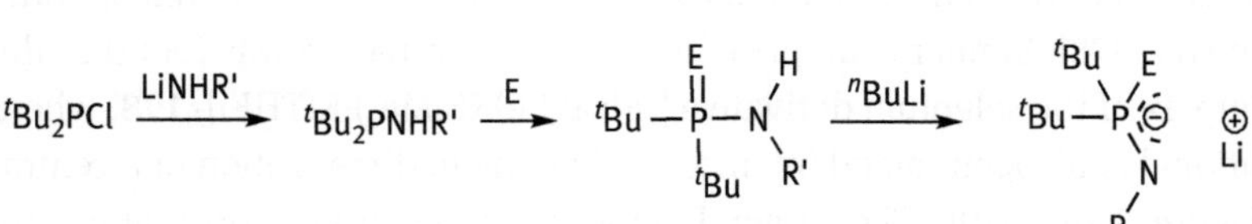

192, E = Se, Te

193, R = ^{i}Pr,*cyclo*-C_6H_{11}, E = Se, Te

Figure 5.61: Synthesis of imido compounds **192** and **193**.

Selenobisamidophosphonic acids **194** were prepared from the oxidation of the parent bisaminophosphines $RP(NHR')_2$ (which can be prepared from the aminolysis of an organodichlorophosphine) with elemental selenium [174]. Two amide protons are acidic in **194** and deprotonation by ^{n}BuLi gives the complex **195** (Figure 5.62). However, tellurobisamidophosphonic acids are not accessible from this route due to $RP(NHR')_2$ failing to react with elemental tellurium even in

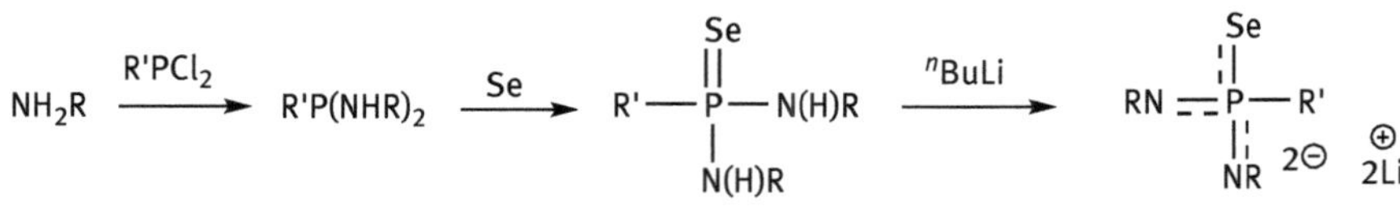

Figure 5.62: Synthesis of imido compounds **194** and **195**.

refluxing toluene. Nevertheless, metallation of the bisaminophosphine leading to $Li_2[PhP(N^tBu')_2]$, followed by treatment with elemental tellurium forms the dilithium salt **196** (Figure 5.63) [175].

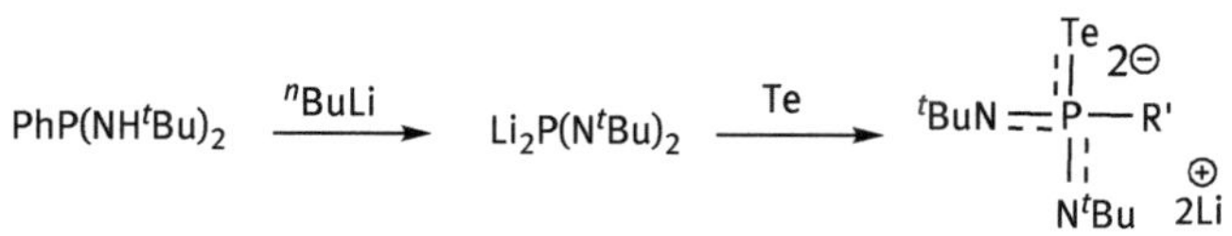

Figure 5.63: Synthesis of dilithium salt **196**.

Metal complexes of the chalcogenoimidophosphinate $[R_2P(E)(NR)]^-$ are prepared from metathesis reaction of the lithium chalocogenoimidophosphinate with metal chlorides or from protolysis of a bis(trimethylsilyl)amide metal salt $M[N(SiMe_3)]_2$ (M=Cr, Mn, Co, Zn, Cd; E=Se, Te) with chalcogenoimidophosphinic acids [172]. The chalcogenoimidophosphinate ligands act as N, E chelating ligands in metal complexes, such as $M[^tBu_2P(Te)NR]_2$ **197**, which were used as potential source of the binary metal chalcogenides. The thermolysis of transition-metal complexes produced thin films of metal ditellurides MTe_2 (M=Fe, Mn, Co) [176], and the thermally stable, sublimable group 12 complexes were discovered to produce thin films of MTe (M=Zn, Cd) free from phosphorus contamination [177, 178]. Similar example of N, E chelate formation was found in the potassium complex of the selenium derivative $[(Ph_2P(Se)NSiMe_3)\cdot K(THF)]_2$ **198**, where there are additional chalcogen–metal interactions leading to dimerization of a central K_2Se_2 ring (Figure 5.64) [179]. The larger R groups on the phosphorus atom are

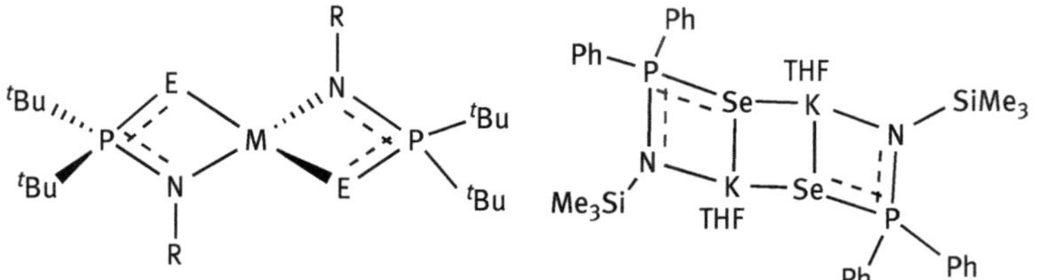

197, M = Cr, Mn, Fe, Co; E = Se, Te
R = ^{i}Pr, C_6H_{11}

198

Figure 5.64: Metal complexes **197** and **198**.

beneficial to the formation of monomeric complexes, whereas the smaller R groups prefer to the formation of dimeric complexes or larger aggregates.

Dichalcogenoimidodiphosphinates $[R_2P(E)NHP(E)R_2]$ (R=alkyl, aryl, alkoxy, aryloxy; E=Se, Te) and their deprotonated anionic compounds, dichalcogenoimindodiphosph(in)ates $[R_2P(E)NHP(E)R_2]^-$, have been widely studied due to their wide-ranging coordination chemistry which have covered almost all the metallic and semi-metallic members of the periodic table [159, 180–182]. One of the most convenient and high yield routes for the preparation of symmetric $R_2P(Se)NHP(Se)R_2$ is from the reaction of diorganochlorophosphine with hexamethyldisilazane, followed by oxidative addition of two equivalents of elemental selenium to give the diselenides **199** (Figure 5.65). Iminobisphosphine selenides **199** can theoretically exist in two different tautomeric states of **199a** and **199b**, and extensive studies have shown that in both solution and solid state the N–H tautomer predominates. Structural analysis reveals the $N(PSe)_2$ group in **199a** is planar or close to planar with P–N–P bond angles in the range of 122–133° [159].

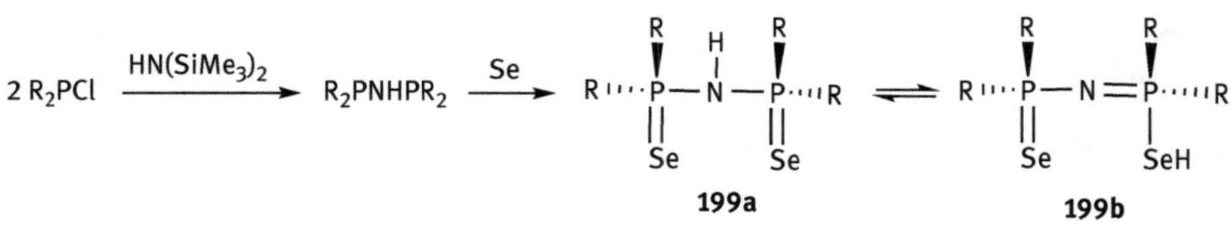

Figure 5.65: Synthesis of diselenides **199**.

Treating $Na[N(PNR_2)_2]$ with tellurium powder in hot toluene in the presence of TMEDA gave sodium salts **200** incorporating anion $[N(PR_2Te)_2]^-$ (R=Ph, ^{i}Pr, ^{t}Bu) and solvated TMEDA (Figure 5.66) [183–185]. Structural analysis shows the phenyl derivative exists as a centrosymmetric dimer $\{[Na(TMEDA)][N(PPh_2Te)_2]\}_2$ in the solid state, in which the ligand is *Te,Te'*-chelated to TMEDA-solvated Na^+ ions and the monomer units are linked by strong Na–Te interactions leading to an approximately square Na_2Te_2 ring [183].

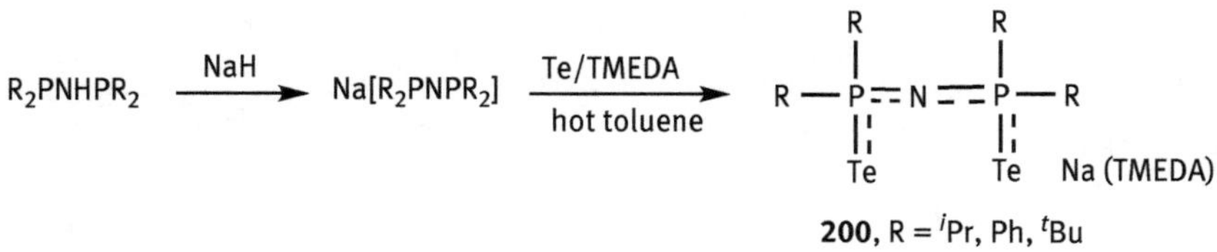

Figure 5.66: Synthesis of sodium salts **200**.

The coordination chemistry of dichalcogenoimidodiphosphinates $[R_2P(E)NHP(E)R_2]$ with almost all the metallic and semi-metallic elements has been intensively investigated [159, 180]. Five major coordination modes have been observed: the neutral ligands as the bidentate E, E' chelation (mode A) can be reacted directly with a reactive organometallic reagent; for example, the reaction of $U[N(SiMe_3)_2]_3$ with $[Ph_2P(Se)NHP(Se)Ph_2]$ gives the dichalcogenoimidodiphosph(in)ate complex $U[Ph_2P(Se)NP(Se)Ph_2]_3$ [186]; the dichalcogenoimidodiphosphinates acting as monodentate coordination patterns from one chalcogen (Mode B) are regularly observed only with the mixed chalcogen systems $[R_2P(Se)NHP(O)R_2]$; the most common coordination mode is C, in which the E, E' chelation is able to give a flexible six-membered chelate ring, which can adopt various configuration including planar, boat, and chair information; the anisobidentate chelating mode D is also known in which the negative charge is predominately localized on one of the chalcogen atoms to result in unequal pairs of P–N, P–E, and P–M bond distances in the complex; and the lanthanides and actinides have been reported to adopt tridentate coordination mode E with dichalcogenimidodiphosphinates (Figure 5.67).

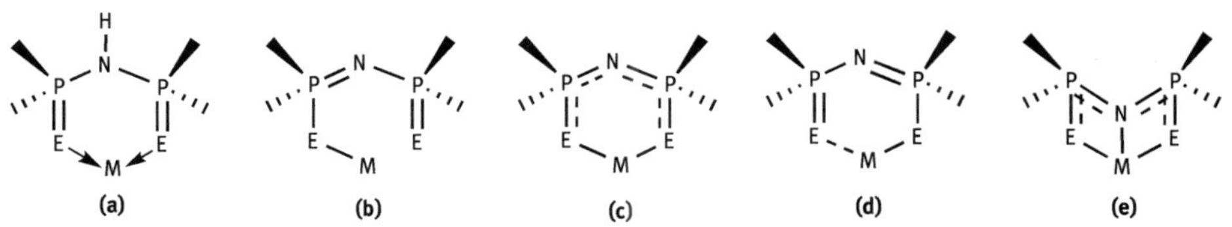

Figure 5.67: Five major coordination modes A–E.

Although these five coordination modes are the most frequently observed coordination patterns for the dichalcogenimidodiphosphinates, other coordination patterns such as multi-ligand and multi-metallic complex modes have also been reported. The metathetical reactions of alkali-metal derivatives of the ditelluroimidodiphosphinates $[R_2P(Te)NHP(Te)R_2]$ (R=^{i}Pr) with main group, transition-metal, lanthanide, and actinide halides have generated homoleptic complexes $M[N(P^iPr_2Te)_2]_n$ [187, 188] with square-planar (**201**, n = 2, M=Ni, Pd, Pt]) [189, 190], distorted tetrahedral (**202**, n = 2, M=Zn, Cd, Hg) [184], distorted octahedral (**203**, n = 3, M=Sb, Bi, La, U, Pu, Ce) [184, 191, 192], or pseudotrigonal bipyramidal (**204**, n = 2, M=Sn, Pb) geometries (Figure 5.68) [193].

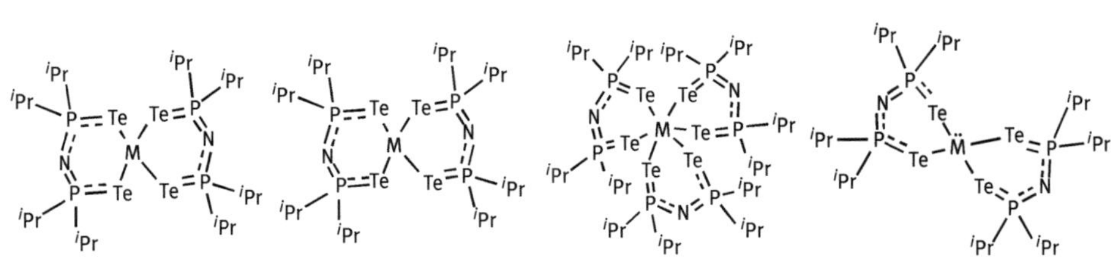

201, M = Ni, Pd, Pt **202**, M = Zn, Cd, Hg **203**, M = Sb, Bi, La, U, Pu **204**, M = Sn, Pb

Figure 5.68: Metal derivatives of the ditelluroimidodiphosphinates $[R_2P(Te)NHP(Te)R_2]$ (R=^{i}Pr).

One electron-oxidation of the anions $[N(PR_2E)_2]^-$ (R=alkyl; E=Se, Te) by iodine led to the formation of the dimeric products, which are strongly influenced by the nature of the R group and chalcogen. When R=^{i}Pr, the dichalcogenides **205** are formed [194].

However, oxidation of the *tert*-butyl derivative $[N(P^tBu_2E)_2]^-$ (E=Se, Te) by iodine afforded the diselenide **206** and the unique spirocyclic compound **207**, the latter may be viewed as a contact ion pair in which the $[N(P^tBu_2Te)_2]^-$ anion is *Te, Te'*-chelated to one tellurium atom of the cyclic $[N(P^tBu_2Te)_2]^+$ cation as a result of an internal redox process (Figure 5.69) [185].

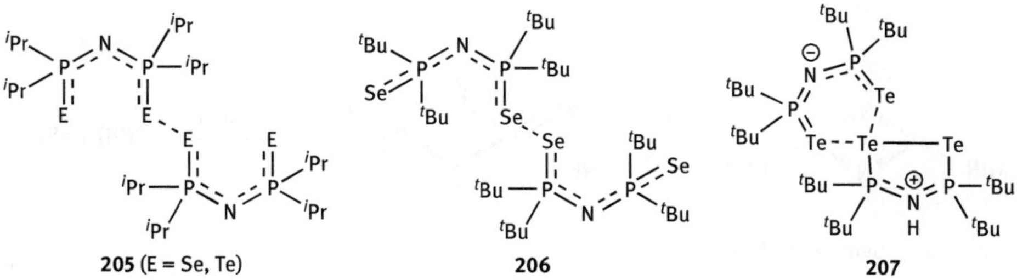

Figure 5.69: Formation of the dimeric products **205–207**.

Two electron-oxidation of the anions $[N(PR_2E)_2]^-$ (R=alkyl; E=Se, Te) by iodine generated the corresponding salts: the cyclic cations $[N(PR_2E)_2]^+$ with iodide counterions [195, 196]. The solid-state structure of these salts depends generally on the nature of the substituent on the phosphorus atoms. For instance, the isopropyl derivatives **208** are an infinite chain of nonplanar cations linked by E...I interactions [195]. Whereas, the phenyl derivative **209** forms a dimeric salt with two tellurium-containing cations [196]; by contrast, the attempted synthesis of the selenium analogue $[N(PPh_2)Se]_2]^+$ led to a 1:1 mixture of six-membered ring $[N(PPh_2Se)_2(\mu\text{-}Se)][I]$ **211** and an acyclic species $[SeP(Ph_2)N(Ph_2)PI]$ (Figure 5.70) [196].

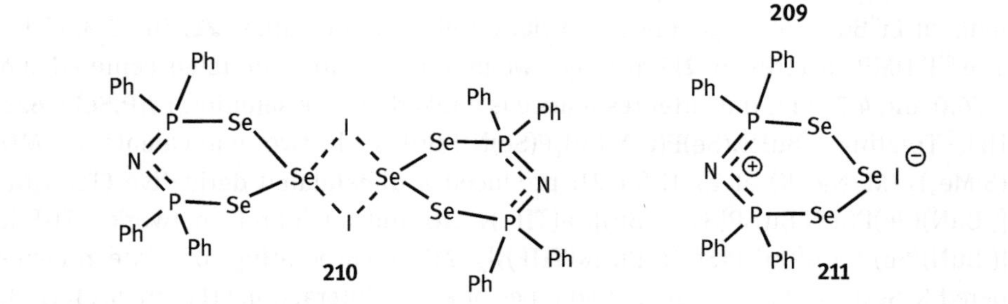

Figure 5.70: Formation of the dimeric products **208–211**.

The bisamidocyclodiphosphazane diselenides **212** are accessible from the bisamidocyclophosphazanes *via* the oxidative addition reaction with elemental selenium or alternatively from the thermolysis of triamidoselenophosphate (Figure 5.71) [197]. The tellurium analogues cannot be obtained via the same approach; however, their dianionic lithium complexes **213** can be prepared from reaction of the dilithio bis (imido)cyclodiphosphazane with elemental tellurium (Figure 5.72) [183].

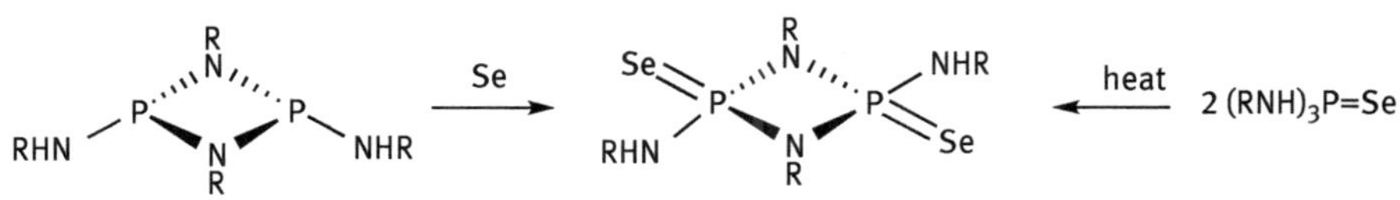

Figure 5.71: Synthesis of **212**.

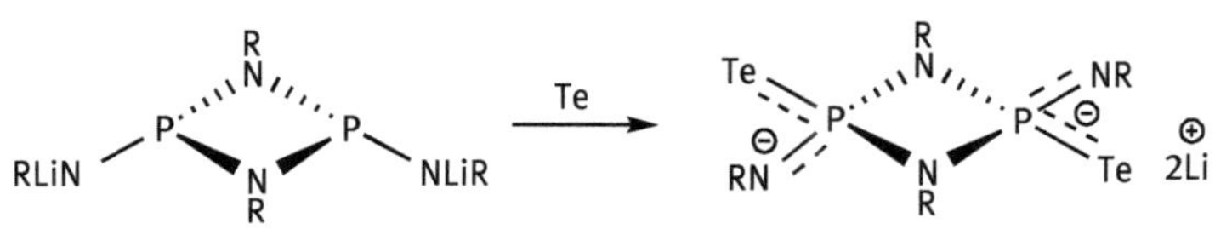

213

Figure 5.72: Synthesis of litium salts **213**.

The anions of bis(amido)cyclodiphosphazane dichalcogenides $[RN(E)P(\mu\text{-}NR)_2P(E)NR]^{2-}$ (E=Se, Te) are versatile ligands capable of a wide range of coordination modes involving two chalcogen atoms and two exocyclic and two endocyclic nitrogen atoms, which adopt a variety of bonding modes, that is *N, E; N, N' or E, E'* with metal centers [172, 197]. The reaction of $[^tBuHN(Se)P(\mu\text{-}N^tBu)_2P(Se)NH^tBu]$ with Li^nBu at 23 °C gave rise to the monolithiated complex **214** as white solid (90 % yield). The ^{31}P NMR spectrum of **214** shows two mutually coupled doublets centered at δ = 23.6 and −4.0 ppm, flanked by ^{77}Se satellites $[^1J(P,Se)$ = 860 and 695 Hz]. Interestingly, treatment of $[^tBuHN(Se)P(\mu\text{-}N^tBu)_2P(Se)NH^tBu]$ with three equivalents of Li^nBu at 65 °C generated partial deselenation complex **215** in 78 % yield. The ^{31}P NMR spectrum of **215** displays two mutually coupled doublets centered at δ = 76.0 and 4.7 ppm, the latter resonance is flanked by ^{77}Se satellites $[^1J(P,Se)$ = 622 Hz]. Treating $[^tBuHN(Se)P(\mu\text{-}N^tBu)_2P(Se)NH^tBu]$ with two equivalents of $MN(SiMe_3)_2$ (M=Na, K) at 23 °C for 2 h produced a dimetalated derivative $(THF)_2Na[(^tBuN)(Se)P(\mu\text{-}N^tBu)_2P(Se)(N^tBu)]Na(THF)_2$ **216** and an infinite network $\{(THF)K[(^tBuN)(Se)P(\mu\text{-}N^tBu)_2P(Se)(N^tBu)]K(THF)_2\}_2$ **217** incorporating of sixteen-membered $K_4Se_4P_4N_4$ ring involving two types of K… Se interactions (Figure 5.73) [198, 199]. The reaction of $RHN(Se)P(\mu\text{-}N^tBu)_2P(Se)NHR$ (R=Ph, ^tBu) with two equivalents

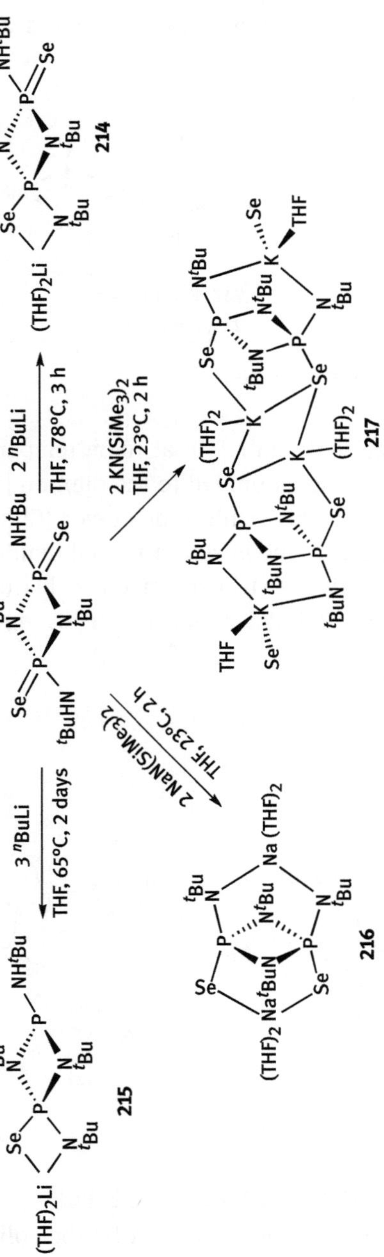

Figure 5.73: Synthesis of phosphorus-selenium complexes **214–217**.

of $AlMe_3$ at -21 °C for 24 h afforded the bis(dimethylaluminum) complexes **218** (R=^{t}Bu) and **219** (R=Ph) (Figure 5.74) [200].

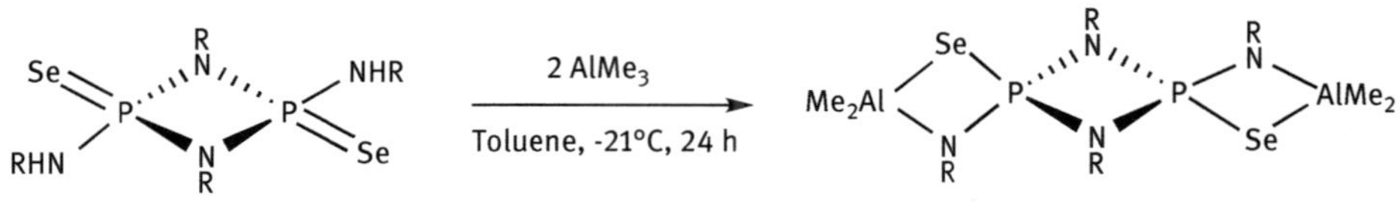

218, R = ^{t}Bu, 74%
219, R = Ph, 74%

Figure 5.74: Synthesis of phosphorus–selenium complexes **218** and **219**.

Deprotonation of [S=(H)P(μ-N^{t}Bu)]$_2$ with benzylsodium in THF at room temperature, followed by *in situ* oxidation with elemental selenium led to the dianion [S=(Se)P(μ-N^{t}Bu)]$_2^{2-}$ **220**, the latter was reacted with the dichloro-precursor [ClP(μ-N^{t}Bu)]$_2$ to give the mixed chalcogen macrocycle **221**. The *in situ* produced dianion **220** was reacted with one equivalent of iodine to give the trimeric macrocycle [{(S=(H))P(μ-N^{t}Bu)}$_2$(μ-Se-Se)]$_3$ **222** with a P–Se–Se–P linkage between P_2N_2 units (Figure 5.75) [201].

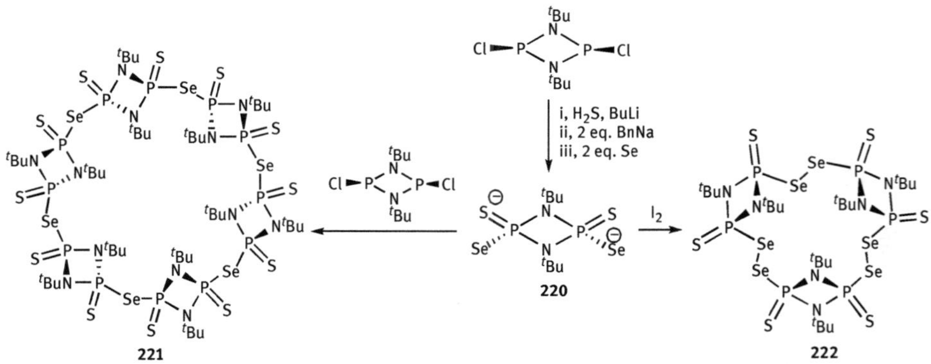

221 **220** **222**

Figure 5.75: Synthesis of phosphorus–selenium complexes **220–222**.

The tellurium-centered dianionic ligands [^{t}BuN(Te)P(μ-N^{t}Bu)$_2$P(Te)N^{t}Bu]$^{2-}$ may adopt a variety of bonding modes with metal centers as illustrated by the solid-state structure of the alkali-metal derivatives. The dilithium compound [Li(TMEDA)]$_2$[^{t}BuN(Te)P(μ-N^{t}Bu)$_2$P(Te)N^{t}Bu] **223** reveals asymmetric *Te,Te'* and *N,Te* bonding to the Li^+ ions [183]; however, the disodium analogue [Na(TMEDA)]$_2$[^{t}BuN(Te)P(μ-N^{t}Bu)$_2$P(Te)N^{t}Bu] **224** exhibits a symmetric *N,N'* and *Te,Te'* coordination arrangement [202], similar to the structures of the sulfur and selenium analogues (Figure 5.76) [198, 199].

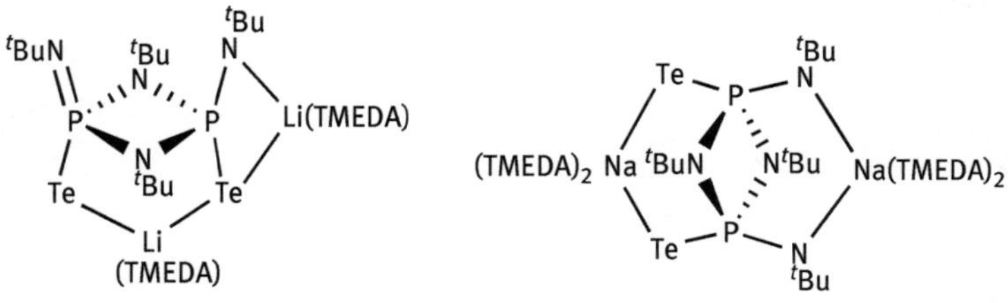

Figure 5.76: Structures of dilithium compound **223** and disodium analogue **224**.

The two-electron oxidation of **216** with I_2 gave the macrocycle **225** in which a planar P_6Se_2 framework is anchored by perpendicular $P^V{}_2N_2$ ring (Figure 5.77) [203]. By contrast, the oxidation of **224** by I_2 first led to two intermediates, the planar dianionic ditelluride **226** and the spirocyclic dianion **227** with a central Te_5 motif [204], and final product is the cyclic tritelluride **228** (Figure 5.78) [202, 205]. The preparation of the cyclic tritelluride **228** can also be obtained by the reaction of **223** with $TeCl_2$·TMTU [202].

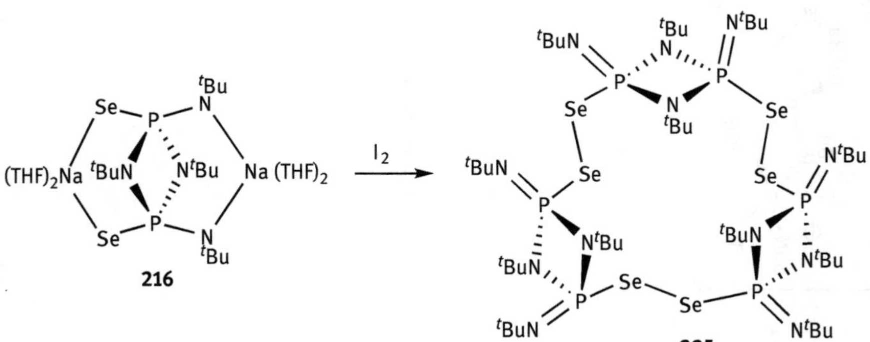

Figure 5.77: Synthesis of phosphorus–selenium macrocycle **225**.

The metathetical reaction of $[Li(TMEDA)]_2[{}^tBuN(Te)P(\mu\text{-}N^tBu)_2P(Te)N^tBu]$ **223** with *p*-block element halides provided a broader patterns of *p*-block element–phosphorus–tellurium heterocycles. In short, the reaction of **223** with Ph_2GeCl_2, $RPCl_2$ (R=Ad, ^{t}Bu, N^iPr_2), $EtAsI_2$, $PhSbCl_2$, or R_2SnCl_2 (R=^{n}Bu, Ph) leads to *Te,Te'*-chelating to the metal complexes **229–233**; with Se_2Cl_2 resulted in the products **234** and **235**, in which either one or two selenium atoms occupy the bridging position. Finally, coordinating **223** with group 13 trihalides MCl_3 (M=Ga, In) afforded the neutral spirocyclic complexes **236** in which the loss of a Te from one P atom leaves a terminal P^{III} center in one of the ligands (Figure 5.79) [205].

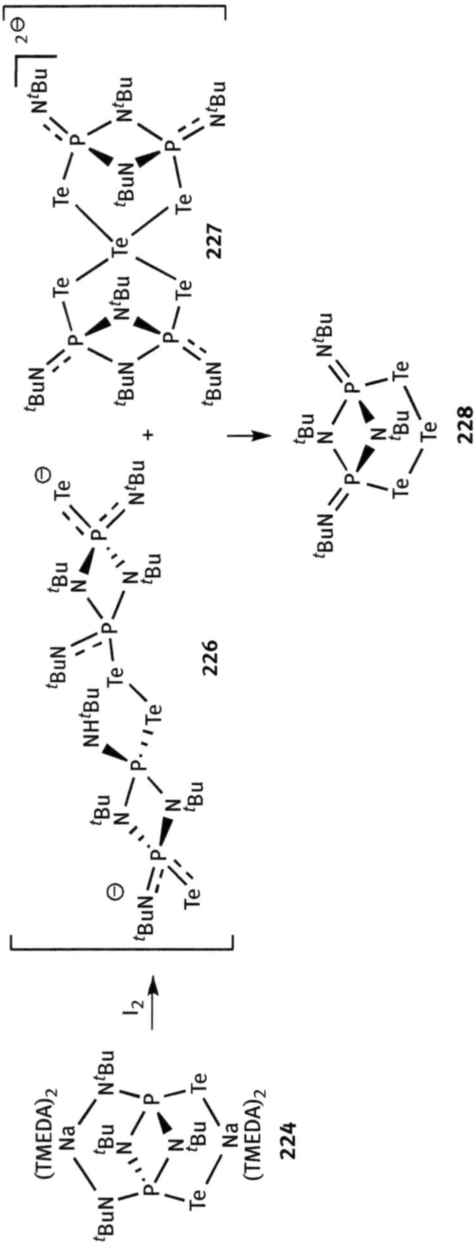

Figure 5.78: Synthesis of phosphorus–tellurium heterocycles **226–228**.

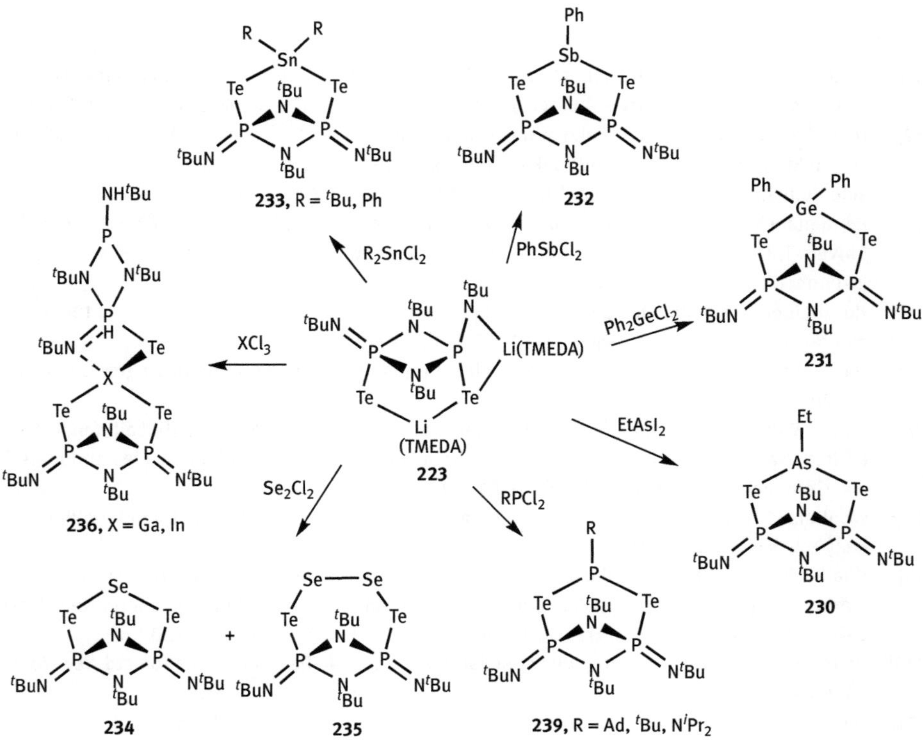

Figure 5.79: Synthesis of the phosphorus–tellurium heterocycles **229–236**.

5.8 Outlook

A variety of organic selenium- and tellurium-phosphorus compounds have found utility as reagents in synthetic inorganic and organic chemistry involving oxygen/chalcogen exchange reaction, oxidation/reduction, nucleophilic/electrophilic substitution, nucleophilic addition, free radical addition, Diel-Alder reaction, various cycloadditions, metal coordination, and so on. The products have or potentially have applications on many fields such as synthetic chemistry, biochemistry, pharmaceutical industry, agrochemical industry, and materials science. Undoubtedly, many admirable developments have been achieved through selenium- and tellurium-phosphorus reagents, such as the construction of chiral organics, the formation of hydrophobic and hydrophilic dendrimers, the synthesis of macrocycles incorporating single or multiple P–Se and P–Te scaffolds, reduction of unsaturated C≡C, C=C, and C=O bonds, the modification of nucleosides, nucleotides, DNA and RNA, application in fluorescence chemistry, etc. Therefore, we believe that the diversity of new selenium- and tellurium-phosphorus reagents will continue to expand in the future, and the new horizon of their applications will be more exciting, challenging, and fruitful.

References

[1] Liotta D. Organoselenium chemistry, New York, USA: John Wiley & Sons, 1987. (Back TG. Organoselenium chemistry – A practical approach. Oxford University Press, Oxford, UK, 1999).
[2] Davies R, Patel L. Chalcogen-phosphorus (and heavier congener) compounds. In: Devillanova FA, Du Mont WW, editors. Handbook of chalcogen chemistry: new perspectives in sulfur, selenium and tellurium. 2nd ed. UK: RSC Publishing, 2013:238–306.
[3] Petragnani N, Stefani H. Advances in organic tellurium chemistry. Tetrahedron. 2005;61:1613–79.
[4] Chivers T, Laitinen RS. Tellurium: a maverick among the chalcogens. Chem Soc Rev. 2015;44:1725–39.
[5] Nordheider N, Woollins JD, Chivers T. Organophosphorus-tellurium chemistry: from fundamentals to applications. Chem Rev. 2015;115:10378–406.
[6] Hua G, Woollins JD. Formation and reactivity of phosphorus-selenium rings. Angew Chem Int Ed. 2009;48:1368–77.
[7] Hua G, Woollins JD. Organic phosphorus-selenium chemistry. In: Woollins JD, Laitinen RS, editors, Selenium and tellurium chemistry: from small molecules to biomolecules and materials. Germany: Springer Heidelberg, 2011:1–39.
[8] Woollins JD. How not to discover a new reagent. The Evolution and Chemistry of Woollins' Reagent. Synlett. 2012;23:1154–69.
[9] Hua G, Woollins JD. Selenophosphorus compounds. In: Timperley CM, editor. Best synthetic methods: organophosphorus(V) chemistry. Amsterdam-Boston-Heidelberg-London-New York-Oxford-Paris-San Diego-San Francisco-Sydney-Tokyo: Elsevier, 2015:633–720.
[10] Housecroft C, Sharpe AG. Inorganic chemistry. 2nd ed. USA: Pearson Education Ltd New York, 2005:426–7.
[11] Leung YC, Waser J. The crystal structure of P_4S_3. Acta Cryst. 1957;10:574–82.
[12] Keulen E, Vos A. A possible explanation of the δc separations in intermediate plagioclase. Acta Cryst. 1958;12:323–4.
[13] Penny GJ, Sheldrick GM. Crystal and molecular structure of tetraphosphorus pentaselenide. J Chem Soc Inorg Phys Theor. 1971;245–8.
[14] Blachnik R, Lonnecke P, Boldt K, Engelen B. P_2S_5. Acta Cryst C Cryst Commun. 1994;50:659–61.
[15] Korak RD, Zemlyanskii NI. Intermediate esters of *O,O*-dialky(phenyl)diselenophosphoric acid. Zh Obshch Khim. 1971;41:1211–3.
[16] Zemlyanskii NI, Gorak RD. Transesterifcation and reaction of *O,O*-dialkyl diselenophosphates with hydrogen chloride. Zh Obshch Khim. 1971;41:2446–8.
[17] Baldus HP, Blachnik R. New A_4B_3 molecules: P_3SbS_3, P_4S_2Te and P_4STe_2. Z Naturforsch. 1990;45b:1605–9.
[18] Jörgens S, Johrendt D, Mewis A. BaP_4Te_2—A ternary telluride with P—Te bonds and a tructural fragment of black phosphorus. Chemistry - A European Journal. 2003;9:2405–10.
[19] Rotter C, Schuster M, Karaghiosoff K. $P_4Ch_{22}^-$ (Ch = S, Se, Te). Inorg Chem. 2009;48:7531–3.
[20] Graham CME, Valjus J, Pritchard TE, Boyle PD, Tuononen HM, Ragogna PJ. Phosphorus-chalcogen ring expansion and metal coordination. Inorg Chem. 2017;56:13500–9.
[21] Baudler M, Suchomel H, Furstenberg G, Schings U. Di- *tert*-butylthia- and -selenadiphosphirane. Angew Chem Int Ed. 1981;20:1044–5.
[22] Yoshifuji M, Shibayama K, Ando K, Inamoto N. Preparation and characterization of some unsymmetrical thiadiphosphiranes. Heterocycles. 1984;22:681–6.
[23] Yoshifuji M, Shibayama K, Inamoto N. Reaction of diphosphenes with elemental selenium. Isolation and characterization of selenadiphosphiranes and diselenoxophosphorane. Chem Lett. 1984;13:603–6.

[24] Sasamori T, Takeda N, Tokitoh N. Synthesis and reactions of new diphosphenes bearing extremely bulky substituents. J Phys Org Chem. 2003;16:450–62.

[25] Yogendra Si, Hennersdorf F, Weigand JJ. Trapping rare and elusive phosphinidene chalcogenides Nitrogen–phosphorus(III)–chalcogen macrocycles for the synthesis of polynuclear silver(I) sandwich complexes. Angew Chem Int EdInorganic Chemistry. 2017;56:62368698–40704.

[26] Lardon M. Selenium and proton nuclear magnetic resonance measurements on organic selenium compounds. J Am Chem Soc. 1970;92:5063–6.

[27] Musher JI, Corey EJ. Virtual long-range spin-spin coupling in nuclear magnetic resonance (N.M.R.). The linear 3-spin system and qualitative. Tetrahedron. 1962;18:791–809.

[28] Yogendra S, Chitnis SS, Hennersdorf F, Bodensteiner M, Fischer R, Burford N, et al. Condensation reactions of chlorophosphanes with chalcogenides. Inorg Chem. 2016;55:1854–60.

[29] Colquhoun IJ, Hc M, McFarlane W, Nash JA, Keat R, Rycroft DS, et al. Long-range nuclear spin-spin coupling between selenium-77 and phosphorus-31 in biphosphorus compounds. Org Magn Reson. 1979;12:473–5.

[30] Simirnova ES, Munoz Molina JM, Johnson A, Bandeira NA, Bo C, Echavarren AM. Polynuclear gold $[AuI]_4$, $[AuI]_8$, and bimetallic $[AuI_4AgI]$ complexes: C-H functionalization of carbonyl compounds. Angew Chem Int Ed. 2016;55:7487–91.

[31] Munoz MP. Silver and platinum-catalyzed addition of O-H and N-H bonds to allenes. Chem Soc Rev. 2014;43:3164–83.

[32] Dehnert P, Grobe J, Hildebrandt W, Le Van DZ. Perfluoromethyl elemental ligands. XXI. Preparation and characterization of organoelement-element compounds of the type R*n*E-ER*m* ($R = CH_3$, CF_3; E = P, As, S, Se, Te; n,m = 1,2). Z Naturforsch B. 1979;34:1646–52.

[33] Du Mont WW, Kubiniok S, Severengiz T. Properties of Te-Te bonds. IV. Dismutation reactions of di-*p*-tolylditelluride with *tetra-tert*-butyldiphosphane and *tetra*-isopropyldiphosphane. Z Anorg Allg Chem. 1985;531:21–5.

[34] Du Mont WW, Kroth HJ. Tellurium-125 NMR shifts and Te-P coupling constants of phosphine tellurides, tellurophosphines, and tellurophosphine complexes. Z Naturforsch B. 1981;36:332–4.

[35] Dumont WW, Severengiz T, Meyer B. Small heterocycles containing phosphorus and tellurium. Angew Chem Int Ed. 1983;22:983–4.

[36] Grimm S, Karaghiosoff K, Mayer P, Ross D. New phosphorus-tellurium heterocycles in the quasi-binary system RP/Te. Phosphorus Sulfur Silicon Relat Elem. 2001;168–169:375–8.

[37] Francis MD, Jones C, Morley CP. A novel synthetic route to chalcogen substituted diphospholes. Tetrahedron Lett. 1990;40:3815–6.

[38] Francis MD, Hibbs DE, Hitchcock PB, Hursthouse MB, Jones C, Mackewitz T, et al. First structural characterization of 1,2,4-selenadiphosphole and 1,2,4-telluradiphosphole ring systems. Crystal and molecular structures of the η^1-complexes $[M(CO)_5(P_2SeC_2{}^tBu_2)]$ (M = Cr, W) and $[W(CO)_5(P_2TeC_2{}^tBu_2)]$. J Organomet Chem. 1999;580:156–60.

[39] Mosbo JA, Verkade JG. Stereochemistry of oxidation of trivalent phosphorus and configurational assignments in 2-substituted 1,3,2-dioxaphosphorinanes. J Am Chem Soc. 1973;95:4659–65.

[40] Cross RJ, Millington D. Selenium abstraction from diethyl diselenide by tertiary phosphines. J Chem Soc Chem Commun. 1975;455–6.

[41] Brown DH, Cross RJ, Millington D. Photochemical reactions between tertiary phosphines and organic diselenides. J Chem Soc Dalton Trans. 1977;159–61.

[42] Nicpon P, Meek DW. Novel synthesis of arylphosphine selenides. Reaction of arylphosphines with potassium selenocyanate. Inorg Chem. 1966;5:1297–8.

[43] Bollmark M, Stawinski J. A new selenium-transferring reagent-triphenylphosphine selenide. Chem Commun. 2001;771–2.

[44] Stawinski J, Thelin M. Nucleoside H-phosphonates. 14. Synthesis of nucleoside phopshoroselenoates and phosphorothioselenoates via stereospecific selenization of the corresponding *H*-phosphonate and *H*-phosphonothioate diesters with the aid of new selenium-transfer reagent, 3*H*-1,2-benzothiaselenol-3-one. J Org Chem. 1994;59:130–6.
[45] Markowska A, Mickalski J. Organophosphorus compounds of sulfur and selenium. XVII. Organic seleno acids of phosphorus. Rocniki Chemii. 1960;34:1675–80.
[46] Elbel S, Tom Dieck H. Photoelectron spectra of group 5 compounds. Part V. Trifluoro- and trimethylphosphorus chalcogenide halides. J Chem Soc Dalton Trans. 1976;1762–8.
[47] Zingaro RA. Phosphine sulfides and selenides: the phosphorus-sulfur and phosphorus-selenium stretching frequencies. Inorg Chem. 1963;2:192–6.
[48] Zyablikova TA, Ilyasov AV, Ignateva SN, Malatsoin SF, Erastov OA. Selenium-77 and phosphorus-31 spin-spin interaction in cyclic organophosphorus compounds. Seriya Khim. 1990;776–80.
[49] Wieber M, Eichhorn B. Chloro(chloromethyl)methylphosphine. Chem Ber. 1973;106:2733–7.
[50] Kimura T, Murai T. P-Chiral phosphinoselenoic chlorides and optically active *P*-chiral phosphinoselenoic amides: synthesis and stereospecific interconversion with extrusion and addition reactions of the selenium atom. Chem Lett. 2004;33:878–9.
[51] Kimura T, Murai T. *P*-Chiral phosphinoselenoic chlorides and phosphinochalcogenoselenoic acid esters: synthesis, characterization, and conformational studies. J Org Chem. 2005;70:952–9.
[52] Ilyasov AV, Nuretdinov IA. NMR ^{77}Se, ^{125}Te, ^{31}P and structure of seleno- and telluro-phosphorus compounds. Phosphorus Sulfur Silicon Relat Elem. 1998;136,137&138:479–82.
[53] Schroeder HF, Mueller J. Dialkylphosphinic azides. II. Z Anorg Allg Chem. 1979;451:158–74.
[54] Necas M, Foreman MRSJ, Woollins JD, Novosad J. New mixed-donor unsymmetrical P-N-P ligands and their palladium(II) complexes. New J Chem. 2001;25:1256–63.
[55] Sekar P, Ibers JA. Imporved synthesis of $HN(SPPh_2)(SePPh_2)$ and some coordination chemistry of $[N(SPPh_2)(SePPh_2)]^-$. Inorg Chim Acta. 2001;319:117–22.
[56] Shi M, Sui WS. Chiral diphenylselenophosphoramides: a new class of chiral ligands for the titanium(IV) alkoxide-promoted addition of diethylzinc to aldehydes. Tetrahedron Asymmetry. 2000;11:835–41.
[57] Shi M, Zhang W. Asymmetric 1,4-addition of diethylzinc to α,β-unsaturated enones catalyzed by chiral imino-thiophosphoramide ligands and copper(I). Tetrahedron Asymmetry. 2004;15:167–76.
[58] Shi M, Wang CJ, Zhang W. Enantioselective conjugate addition of dialkylzinc and diphenylzinc to enones catalyzed by a chiral copper(I) binaphthylthiophosphoramide or binaphthylselenophosphoramide ligand system. Chem Eur J. 2004;10:5507–16.
[59] Shi M, Jiang JK, Shen YM, Feng YS, Lei GX. An unexpected carbon dioxide insertion in the reaction of trans-2,4-disubstitued azetidine, trans-2,5-disubstituted pyrrolidine, or trans-2,6-disubstituted piperidine with diphenylthiophosphinic chloride and diphenylselenophosphinic chlorides. J Org Chem. 2000;65:3443–8.
[60] Nuretdinov IA, Buina NA, Bayandina EV. Beckmann rearrangement of aldoximes in the presence of metallic copper and the direct synthesis of chelate complexes of group IB, IIB, and VIII metals. Zh Obshch Khim. 1978;48:1668.
[61] Murai T, Matsuoka D, Morishita K. 1,1-Binaphthyl-2,2'-diyl phosphoroselenoyl chloride as a chiral molecular tool for the preparation of enantiomerically pure alcohols and amines. J Am Chem Soc. 2006;128:4584–5.
[62] Murai T, Inaji S, Morishita K, Shibahara F, Tokunaga M, Obora Y, et al. Synthesis of 1,1'-binaphthyl-2,2'-diyl phosphoroselenoic amides and their conversion to optically pure phosphoramidites. Chem Lett. 2006;35:1424–5.
[63] Miller JA. Phosphine oxides, sulfides, and selenides. Organophosphorus Chem. 1973;5:70–82.
[64] Pilkington MJ, Slawin AM, Williams DJ, Woollins JD. The preparation and structures of $Ph_2P(E)OH$ and $Ph_2P(E)OPPh_2$ (E = S, Se). Main Group Chem. 1995;1:145–51.

[65] Kuchen W, Knop B. Derivatives of diethyl phosphinothionoselenolate. Angew Chem. 1964;76:496–7.

[66] Kuchen W, Knop B. Organophosphorus compounds. IX. Phosphinic acids $Et_2P(S)SeH$ and $Et_2P(Se)SeH$ and their derivatives. Chem Ber. 1966;99:1663–72.

[67] Kuchen W, Knop B. Oxidative phosphorylation, by phosphate adducts of compounds with a $C{=}N^+$ double bond. Angew Chem. 1965;77:259–60.

[68] Kroshefsky RD, Weiss R, Verkade JD. Constraint and electronegativity effects on selenium-77-phosphorus-31 spin-spin couplings in selenophosphoryl compounds. Inorg Chem. 1979;18:469–72.

[69] Kudchadker MV, Zingaro RA, Irgolic KJ. Chemistry of phosphorus pentaselenide. I. Its reaction with alcohols. Can J Chem. 1968;46:1415–24.

[70] Liu CW, Shang IJ, Hung CM, Wang JC, Keng TC. Novel silver diselenophosphate clusters: structures of $Ag_{10}(\mu_{10}\text{-}Se)[Se_2P(OEt)_2]_8$ and $(Ag[Se_2P(O^iPr)_2])_6$. Dalton Trans. 2002;1974–9.

[71] Kudelska W. Glycosidation of an ambident organic phosphoroselenothioate in the presence of a Lewis acid. Heteroatom Chem. 1999;10:259–62.

[72] Zingaro RA. Tributylphosphine telluride. J Organomet Chem. 1963;1:200.

[73] Zingaro RA, Steeves BH, Irgolic K. Phosphine telluride. J Organomet Chem. 1965;4:320–3.

[74] Brodie AM, Rodley GA, Wilkins CJ. Cobalt(II) complexes with trimethylphosphine selenide. J Chem Soc A. 1969;2927–9.

[75] Jeremias L, Babiak M, Kubat V, Calhodra MJ, Travnicek Z, Novosad J. Successful oxidation of $Ph_2P(CH_2)nPPh_2$ (n = 2,4,6) by tellurium leading to $Ph_2P(Te)(CH_2)nP(Te)Ph_2$. RSC Adv. 2014;4:15428–30.

[76] Austad T, Rod T, Ase K, Songstad J, Norbury AH. Adduct between triphenyl phosphine and triphenyl phosphine telluride. Tellurium(0) compound. Acta Chem Scand. 1973;27:1939–49.

[77] Mardersteig HG, Noth H. Aminophosphanes. XI. Reaction of *N*-(diphenylphosphino)triphenyl-phosphazene. Z Anorg Allg Chem. 1970;375:272–80.

[78] Dean PA. Nuclear magnetic resonance studies of the solvation of phosphorus(V) selenides, 1,2-bis(diphenylphosphino)ethane, and tris(dimethylamino)phosphine telluride by sulfur dioxide. Can J Chem. 1979;57:754–61.

[79] Rømming C, Iversen AJ, Songstad J. Structural studies on the phosphorus-nitrogen bond. III. The crystal structure of tri(morpholino)phosphine telluride. The tellurium basicity of tervalent phosphorus species. Acta Chem Scand. 1980;34A:333–42.

[80] Rømming C, Maartmann-Moe K, Songstad J. Structural studies on the phosphorus-nitrogen bond. VIII. The crystal structure of tris(pyrrolidino)phosphine selenide and tris(pyrrolidino) phosphine telluride. Acta Chem Scand. 1984;38A:349–57.

[81] Diemert K, Kuchen W, Kutter J. Organophosphorus compounds. XX. Bifunctional aminophosphines polymethylenebis[bis(diethylamino)phosphines] $(Et_2N)_2P[CH_2]_nP(NEt_2)_2$ and $Et_2N(R)P[CH_2]_nP(R)NEt_2$. Phosphorus Sulfur Relat Elem. 1983;15:155–64.

[82] Scherer OJ, Schnabl G. Element-organic amine/imine compounds, XVI. New diazadiphosphetidines. Chem Ber. 1976;109:2996–3004.

[83] Kuhn N, Schumann H. Phosphorus tellurium compounds. II. Tellurophosphonium cations. Phosphorus Sulfur Relat Elem. 1986;26:199–201.

[84] Kuhn N, Schumann H, Boese R. Oxidative coupling of tellurophosphoranes, a route to phosphine stabilized tritellurium dications. J Chem Soc Chem Commun. 1987;1257–8.

[85] Konu J, Chivers T. Synthesis, spectroscopic and structural characterization of tertiary phosphine tellurium dihalides Et_3PTeX_2 (X = Cl, Be, I). Dalton Trans. 2006;3941–6.

[86] Kuhn N, Schumann H, Wolmershauser GM. $M(CO)_5(R_3PTe)$ (M = Cr, Mo, W; R = CMe_3): the first stable tellurophosphorane complexes. J Chem Soc Chem Commun. 1985;1595–7.

[87] Huang L, Zingaron RA, Meyers EA, Reibenspies JH. Reaction of mercury(II) dibromide with tris(*n*-butyl)phosphine telluride: formation of an unusual $(HgTe)_3$ ring system. Heteroat Chem. 1996;7:57–65.
[88] Sadekov ID, Maksimenko AA, Nivorozhkin VL. Organic derivatives of monocoordinated tellurium. Russ Chem Rev. 1998;67:193–208.
[89] Chivers T. Tellurium compounds of the main-group elements: progress and prospects. J Chem Soc Dalton Trans. 1996;1185–94.
[90] Ww DM. Bis(di-butylphosphino)tellurium(II): a derivative of phosphinous anhydride exhibiting umpolung. Angew Chem Int Ed. 1980;19:554–5.
[91] Brennan JG, Andersen RA, Zalkin A. Chemistry of trivalent uranium metallocenes: electron-transfer reactions. Synthesis and characterization of $[(MeC_5H_4)_3U]_2E$ (E = S, Se, Te) and the crystal structures of hexakis(methylcyclopntadienyl)sulfidiuranium and tris(methylcyclopentadienyl)(triphenylphosphine oxide)uranium. Inorg Chem. 1986;25:1761–5.
[92] Uhl W, Schutz U, Hiller W, Heckel M. Synthesis and crystal structure of a monomeric tetraalkyldigallium telluride with a bent Ga-Te-Ga group. Organometallics. 1995;14:1073–5.
[93] Fischer JM, Piers W, MacGillivray LR, Zawarotko MJ. Penmethyltitanocene tellurides and ditellurides. X-ray structures of $[(C_5Me_5)_2Ti]_2(\mu\text{-Te})$ and $(C_5Me_5)_2Ti(\eta_2\text{-Te}_2)$. Inorg Chem. 1995;34:2499–500.
[94] McCoie JM, Bollinger JC, Ibers JA. Chalcogenide substitution reactions. Inorg Chem. 1993;32:3923–7.
[95] Coleman AP, Dickson RS, Deacon GB, Fallon GD, Ke M, McGregor K, et al. The synthesis of some alkyltelluride-manganese(I) complexes, and an assessment of their suitability for MOCVD applications. Polyhedron. 1994;13:1277–90.
[96] Moon J, Nam H, Kim S, Ryu J, Han C, Lee C, et al. Synthesis of phosphinodiselenoic acid esters and their application as RAFT agents in styrene polymerization. Tetrahedron Lett. 2008;49:5137–40.
[97] Maneeprakorn W, Malik MA, O'Brien P. The preparation of cobalt phosphide and cobalt chalcogenide (CoX, X = S, Se) nanoparticles from single source precursors. Mater Chem. 2010;20:2329–35.
[98] Malik MA, Afzaal M, O'Brien P. Precursor chemistry for main group elements in semiconducting materials. Chem Rev. 2010;110:4417–46.
[99] Hasegawa Y, Adachi T, Tanaka A, Afzaal M, O'Brien P, Doi T, et al. Remarkable magneto-optical properties of europium selenide nanoparticles with wide energy gaps. J Am Chem Soc. 2008;130:5710–5.
[100] Fan D, Afzaal M, Malik MA, C Q N, O'Brien P, Thomas PJ. Using coordination chemistry to develop new routes to semiconductor and other materials. Coord Chem Rev. 2007;251:1878–88.
[101] Harris LM, Tam EC, Cummins SJ, Coles MP, Fulton JR. The reactivity of germanium phosphinides with chalcogens. Inorg Chem. 2017;56:3087–94.
[102] Behrke AC, Kerridge A, Walensky JR. Dithio- and diselenophosphinate thorium(IV) and uranium (IV) complexes: molecular and electronic structures, spectroscopy, and transmetalation reactivity. Inorg Chem. 2015;54:11625–36.
[103] Chauhan HPS. Chemistry of diorganodithiophosphate (and phosphinate) derivatives with arsenic, antimony and bismuth. Coord Chem Rev. 1998;173:1–30.
[104] Devillanova FA. Handbook of chalcogen chemistry: new perspectives in sulfur, selenium and tellurium. Cambridge UK: RSC Publishing, 2007:286–343.
[105] Artem'ev AV, Chemysheva NA, Gusarova NK, S F M, Yasrko SV, Albanov AL, et al. An efficient and general synthesis of *Se*-esters of diselenophosphinic acids via reaction of alkali metal diselenophosphinates with organic halides. Synthesis. 2011;1309–13.

[106] Latouche C, Lee YC, Liao JH, Furer E, Saillard JY, Liu CW, et al. Structure and spectroscopic preperties of gold (I) diselenophosph(in)ate complexes: a joint experimental and theoretical study. Inorg Chem. 2012;51:11851–9.

[107] Jones MB, Gaunt AJ, Cordon JC, Kaltsoyannis N, Neu MP, Scott BL. Uncovering *f*-element bonding differences and electronic structure in a series of 1:3 and 1:4 complexes with a diselenophosphinate ligand. Chem Sci. 2013;4:1189–203.

[108] Artem'ev AV, Gusarova NK, Malysheva SF, Trofimov BA. Diselenophosphinates. Synthesis and Applications. Org Prep Proc Int. 2011;43:381–449.

[109] Nguyen CQ, Afzaal M, Malik MA, Helliwell M, Raftery J, O'Brien P. Novel inorganic rings and materials deposition. J Organomet Chem. 2007;692:2669–77.

[110] Trofimov BA, Artem'ev AV, Gusarova NK, Malysheva SF, Fedorov SV, Kazheva ON, et al. One-pot reaction of secondary phosphine selenides with selenium and nitrogen bases: a novel synthesis of diorganodiselenophosphinates. Synthesis. 2009;3332–8.

[111] Artem'ev AV, Gusarova NK, Malysheva SF, Kraikivskii PB, Belogorlova NA, Trofimov BA. Efficient general synthesis of alkylammonium diselenophosphinates via multicomponent one-pot reaction of secondary phosphines with elemental selenium and amines. Synthesis. 2010;3724–30.

[112] Trofimov BA, Artem'ev AV, Malysheva SF, Gusarova NK. Reaction of secondary phosphine selenides with the system Se/MOH (M = Li, Na, K, Rb, Cs): a novel three-component synthesis of diorganodiselenophosphinates. J Organomet Chem. 2009;694:4116–20.

[113] Artem'ev AV, Malysheva SF, Gusarova NK, Belogorlova NA, Trofimov BA. Facile atom-economic synthesis of ammonium diselenophosphinates via three-component reaction of secondary phosphines, elemental selenium, and ammonia. Synthesis. 2010;1777–80.

[114] Artem'ev AV, Gusarova NK, Malysheva SF, Trofimov BA. Selective synthesis of hydrazinium diselenophosphinates from secondary phosphines, elementary selenium, and hydrazine. Russ Chem Bull. 2010;59:1671–3.

[115] KuChen W, Knop B. Synthesis of halogenoalkyl hydroperoxides by "bromine-induced autoxidation" of allyl halides. Angew Chem Int Ed. 1965;4:244.

[116] Muller A, Christophliemk P, Rao VV. Transition metal chalcogen compounds. Preparation, electronic, and vibrational spectra of diphenyldiselenophosphinato complexes. Chem Ber. 1971;104:1905–14.

[117] Jhang RY, Liao JH, Liu CW, Kuimov VA, Gusarova NK, Artem'ev AV. A new convenient synthetic route to metal diselenophosphinates: synthesis and characterization of $[M_2(Se_2PPh_2)_4]$ (M = Zn, Cd and Hg) complexes. J Organmet Chem. 2014;758:60–4.

[118] Jhang RY, Liao JH, Liu CW, Artem'ev AV. Synthesis of the first chalcogen-centered diselenophosphinato Zn(II) clusters, $[Zn_4(\mu_4\text{-}X)\{Se_2PR2\}_6]$ (X = S or Se), and a zigzag polymer $\{ZnBr(\mu\text{-}Se_2PR_2)[PyNO]\}_n$. J Organomet Chem. 2015;781:72–6.

[119] Artem'ev AV, Malysheva SF, Sukhov BG, Belogorlova NA, Gatilov YV, Mamatyuk VI, et al. Unexpected redox reaction of alkali metal diselenophosphinates with elemental iodine. Mendeleev Commun. 2012;22:18–20.

[120] Sarkar B, Fang CS, You LY, Wang JC, Liu CW. The chemistry of phosphorodiselenoates: structure, catalysis and formation of *Se*-esters. New J Chem. 2009;33:626–33.

[121] Krawczyk E, Skowronska A, Michalski J. Studies on the interaction of phosphine selenides and their structural analogues with dihalogens and sulfuryl chloride. J Chem Soc Dalton Trans. 2002;4471–8.

[122] Williams DJ, Wynne KJ. Chalcogen chemistry. 10. Synthesis and characterization of triarylphosphine adducts of selenium(II) bromide. Inorg Chem. 1976;15:1449–51.

[123] Pilkington MJ, Slawin AMZ, Williams DJ, Woollins JD. The X-ray crystal structure of Na_2 $[PhPSe_2]_2$.5H2O – a polymer containing six-membered sodium-water rings and sodium coordinated by selenium. Polyhedron. 1991;10:2641–5.
[124] Du Mont WW, Hensel R, McFarlane W, Ij C, Ml Z, Serhadli O. Telluration and selenation of *tetra-tert*-butyldiphosphine. Chem Ber. 1989;122:37–41.
[125] Gelmini L, Stephan DW. Synthesis, characterization, and chemistry of titanium(IV), zirconium(IV), and hafnium(IV) complexes of phosphine sulfides and selenides. The crystal and molecular structures of $Cp_2Ti(SPCy_2)_2$, $Cp_2Ti(S_2PCy_2)$, and $Cp_2Ti(Se_2PPh_2)$. Organometallics. 1987;6:1515–22.
[126] Maneeprakorn W, Nguyen CQ, Malik MA, O'Brien P, Raftery J. Synthesis of nickel selenophosphinates $[Ni(Se_2PR_2)_2]$ (R = iPr, tBu and Ph) and their use as single source precursors for the deposition of nickel phosphide or nickel selenide nanoparticles. Dalton Trans. 2009;2103–8.
[127] Tanaka A, Adachi T, Hasegawa Y, Kawai T. Crystal growth of nanoscales europium selenide having characteristic crystal shapes. J Alloy Compd. 2009;488:538–40.
[128] Silvestru C, Haiduc I. Structural patterns in inorganic and organoantimony derivatives of oxo- and thiodiorganophosphorus ligands. Coord Chem Rev. 1996;147:117–46.
[129] Bildstein B, Sladky F. Novel anionic chalcogen ligands. Tellurophosphinites R_2PTe^- and chalcogenotellurophopshinates $R_2P(X)Te^-$ [X = oxygen, sulfur, selenium, tellurium]. Phosphorus Silicon Relat Elem. 1990;47:341–7.
[130] Davies RP, Matinelli MG, Patel L, Wheatley AEH, White AJP, Williams DJ. Structural studies of lithium telluro- and seleno-phosphorus compounds. Eur J Inorg Chem. 2003;3409–16.
[131] Davies RP, Matinelli MG, Patel L, White AJ. Facile synthesis of bis(dichalcogenophosphinate)s and a remarkable $[Li_8(OH)_6]^{2+}$ polyhedron. Inorg Chem. 2010;49:4626–31.
[132] Shore JT, Pennington WT, Noble MC, Cordes AW. Crystal structure of *tert*-butylperthiophosphonic anhydride, (*tert*-$C_4H_9PS_2)_2$, and its selenium analog (*tert*-$C_4H_9PSe_2)_2$. Phosphorus Sulfur Relat Elem. 1988;39:153–7.
[133] Hahn J, Hopp A, Borkowsky A. Sulfur-phosphorus and selenium-phosphorus heterocycles synthesis and structure elucidation by modern NMR techniques. Phosphorus Sulfur Relat Elem. 1992;64:129–36.
[134] Wood PT, Woollins JD. Phosphorus-selenium heterocycles. J Chem Soc Chem Commun. 1988;1190–1.
[135] Pilkington MJ, Slawin AM, Williams DJ, Woollins JD. The preparation and characterization of binary phosphorus-selenium rings. Heteroatom Chem. 1990;1:351–5.
[136] Karaghiosoff K, Eckstein K. Phosphorus-selenium heterocycles $(RP)_nSe_m$: syntheses and structures. Phosphorus Sulfur Relat Elem. 1993;75:257–60.
[137] Karaghiosoff K, Eckstein K, Motzer R. Phosphorus-selenium heterocycles in the quasi-binary system RP/Se. Phosphorus Sulfur Relat Elem. 1994;93/94:185–8.
[138] Baxter I, Hill AF, Malget JM, White AJP, Williams DJ. Selenoketenyl and selenoalkyne complexes via the reactions of ketenyl complexes with Woollins' reagent. Chem Commun. 1997;2049–50.
[139] Hua G, Li Y, Slawin AM, Woollins JD. Synthesis of primary arylselenoamides by reaction of aryl nitriles with Woollins' reagent. Org Lett. 2006;8:5251–4.
[140] Gray IP, Bhattacharyya P, Slawin AM, Woollins JD. A new synthesis of $(PhPSe_2)_2$ (Woollins' reagent) and its use in the synthesis of novel P-Se heterocycles. Chem Eur J. 2005;11:6221–7.
[141] Rothenberger A, Shi W, Shafaei-Fallah M. The anhydride route to group 15/16 ligands in oligomeric and polymeric environments: from metal complexes to supraionic chemistry. Chem Eur J. 2007;13:5974–81.
[142] Shi W, Zhang L, Shafaei-Fallah M, Rothenberrger A. Synthesis and crystal structures of octahedral metal complexes containing the new dianion $[PhP(Se,O)SeSe(O,Se)PPh]^{2-}$. Z Anorg Allg Chem. 2007;633:440–2.

[143] Shi W, Shafaei-Fallah M, Zhang L, Matern E, Rothenberrger A. Polymeric organometallic architectures of novel P-Se anions. Chem Eur J. 2007;13:598–603.
[144] Hua G, Li Y, Fuller AL, Slawin AM, Woollins JD. Facile synthesis and structure of novel 2,5-disubstituted 1,3,4-selenadiazoles. Eur J Org Chem. 2009;1612–8.
[145] Hua G, Zhang Q, Li Y, Slawin AM, Woollins JD. Novel heterocyclic selenazadiphospholoamine-diselenides, zwitterionic carbamidoyl(phenyl)phosphinodiselenoic acids and selenoureas derived from cyanamides. Tetrahedron. 2009;65:6074–2.
[146] Hua G, Fuller AL, Li Y, Slawin AM, Woollins JD. Synthesis and X-ray structures of new phosphorus-selenium heterocycles with an E-P(Se)-E' (E, E' = N, S, Se) linkage. New J Chem. 2010;34:1665–71.
[147] Hua G, Henry JB, Li Y, Mount AR, Slawin AM, Woollins JD. Synthesis of novel 2,5-diarylselenophenes from selenation of 1,4-diarylbutane-1,4-diones or methanol/'arylacetylenes. Org Biomol Chem. 2010;8:1655–60.
[148] Hua G, Fuller AL, Slawin AM, Woollins JD. Novel five- to ten-membered organoselenium heterocycles from the selenation of aromatic diols. Eur J Org Chem. 2010;2607–15.
[149] Hua G, Fuller AL, Slawin AM, Woollins JD. Formation of new organoselenium heterocycles and ring reduction of 10-membered heterocycles into seven-membered heterocycles. Polyhedron. 2011;30:805–8.
[150] Bansal RK, Gupta N, Bharatiya N, Gupta G, Surana A, Hackenbracht G, et al. Sulfur and selenium derivatives of 2-phosphaindolizines. Heteroat Chem. 1998;9:445–52.
[151] Gay IP, Slawin AM, Woollins JD. Synthesis and structure of novel $[Ph(RO)PSe_2]^-$ complexes. Dalton Trans. 2005;2188–94.
[152] Hua G, Woollins JD. Sodium phenyldiselenophosphonate salts and *Se*-alkyl-*O*-alkylphenylphosphonodiselenoates and *Se,Se'*-dialkyl-*O,O'*-dialkyl-bis(phenylphosphonodiselenoate)s. Polyhedron. 2012;23:1170–5.
[153] Hua G, Randall RA, Slawin AM, Woollins JD. Ammonium phenylphosphonamidodiselenoates and phenylphosphonamidodiselenoic diamides from the selenation of primary and secondary amines. Z Anorg Allg Chem. 2011;637:1800–6.
[154] Fitzmaurice JD, Williams DJ, Wood PT, Woollis JD. Organophosphorus-selenium heterocycles. J Chem Soc Chem Commun. 1988;741–3.
[155] Bhattacharyya P, Slawin AM, Woollins JD. The reactivity of $[PhP(Se)(\mu\text{-}Se)]_2$ and $(PhP)_3Se_2$ towards acetylenes and cyanamides: X-ray crystal structures of some P-Se-C and P-Se-C-N heterocycles. Chem Eur J. 2002;8:2705–11.
[156] Hua G, Li Y, Slawin AM, Woollins JD. Synthesis of novel vinylic P–Se heterocycles from selenation of alkynes by $[PhP(Se)(\mu\text{-}Se)]_2$. Eur J Inorg Chem. 2007;891–7.
[157] Hua G, Li Y, Slawin AM, Woollins JD. Unexpected four- and eight-membered organo P-Se heterocycles. Chem Commun. 2007;1465–7.
[158] Bhattacharyya P, Slawin AM, Woollins JD. Heterocycle formation using $[PhP(Se)(\mu\text{-}Se)]_2$. The crystal structures of $[PhP(Se)(\mu\text{-}Se)]_2$, $PhP(Se)Se_2(C_7H_{10})$ and $PhP(Se)(\mu\text{-}Se)(\mu\text{-}NPh)P(Se)Ph$. J Chem Soc Dalton Trans. 2001;300–3.
[159] Silvestru C, Drake JE. Tetraorganodichalcogenoimidodiphosphorus acids and their main group metal derivatives. Coord Chem Rev. 2001;223:117–46.
[160] Foreman MR, Slawin AM, Woollins JD. The reaction of dithiadiphosphetane disulfide with dienes, alkenes and thioaldehydes. J Chem Soc Dalton Trans. 1999;1175–84.
[161] Foreman MR, Slawin AM, Woollins JD. Novel 1,2-thiaphosphetanes from diferrocenyldithiadiphosphetane disulfide. J Chem Soc Chem Commun. 1997;855–6.
[162] Hua G, Li Y, Slawin AM, Woollins JD. Synthesis and structure of eight-, nine-, and ten-membered rings with P-Se-Se-P linkages. Angew Chem Int Ed. 2008;47:2857–9.

[163] Hua G, Griffin JM, Ashbrook SE, Slawin AM, Woollins JD. Octaselenocyclodecane. Angew Chem Int Ed. 2011;50:4123–6.
[164] Bhattacharyya P, Woollins JD. One-pot deracemization of an enol acetate derived from a prochiral cyclohexanone. Tetrahedron Lett. 2001;42:5949–51.
[165] Ostrowski W, Gierczyk B, Franski RJ. Synthesis of monosubstituted 1,3,4-selenadiazoles using Woollins' reagent. Heterocycl Chem. 2012;49:1266–8.
[166] Hua G, Slawin AM, Woollins JD. Stereoselective synthesis of olefins by a reductive coupling reaction. Dalton Trans. 2007;1477–80.
[167] Kaschel J, Schmidt CD, Mumby M, Kratzert D, Stalke D, Werz DB. Donor-acceptor cyclopropanes with Lawesson' and Woollins' reagents: formation of bisthiophenes and unprecedental cage-like molecules. Chem Commun. 2013;49:4403–5.
[168] Mandal M, Chatterjee S, Jaisankar O. Woollins' reagent: a chemoselective reducing agent for 1,4-enediones and 1,4-ynediones to saturated 1,4-diones. Synlett. 2012;2615–8.
[169] Yu Y, Furuyama T, Tang J, Wu ZY, Chen JZ, Kobayashi N, et al. Stable iso-bacteriochlorin mimics from porpholactone: effect of *a,β*-oxazolone moiety on the frontier *n*-molecular orbitals. Inorg Chem Front. 2015;2:671–7.
[170] Pizzo C, Mahler G. Woollins' reagent promotes selective reduction of *α,β*-unsaturated thiazo and selenazolidinones. Tetrahedron Lett. 2017;58:1445–7.
[171] Hua G, Du J, Fuller AL, Arachchige KS, Cordes DB, Slawin AM, et al. Synthesis and selenation of tandem multicomponent condensation adducts. Synlett. 2015;26:839–45.
[172] Chivers T. Imido analogues of phosphorus oxo and chalcogenido anions. Top Curr Chem. 2003;229:143–59.
[173] Bochmann M, Bwembya GC, Whilton N, Song XJ, Hursthouse MB, Coles SJ, et al. Synthesis of selenophosphinic and tellurophosphinic amides and amidato complexes. Crystal structures of $^{t}Bu_2P(Te)NH(C_6H_{11})$, $[Ti(\eta\text{-}C_5H_5)Cl_2\{^{t}Bu_2P(Se)NCHMe_2\}]$ and $[TiCl_2\{^{t}Bu_2P(Se)N(C_6H_{11})\}_2]\cdot C7H_8$. J Chem Soc Dalton Trans. 1995;1887–92.
[174] Eichhorn B, Noth H, Seifert T. (*N*-Lithioamino)diorganophosphines and bis(*N*-lithioamino) organophosphines. Synthesis and structures. Eur J Inorg Chem. 1999;2355–68.
[175] Briand GG, Chivers T, Krahn M, Parez M. Synthesis and X-ray structures of dilithium complexes of the phosphonate anions $[PhP(E)(N^{t}Bu)_2]^{2-}$ (E = O, S, Se, Te) and dimethylaluminum derivatives of $[PhP(E)(N^{t}Bu)(NH^{t}Bu)]^{-}$ (E = S, Se). Inorg Chem. 2002;41:6808–15.
[176] Song X, Bochmann M. Synthesis of phosphinochalcogenoic amidato complexes of divalent transition metals and their thermolysis to metal selenide and telluride phases. J Chem Soc Dalton Trans. 1997;2688–92.
[177] Bochmann M. Metal chalcogenide materials. Chalcogenolato complexes as "single-source" precursors. Chem Vap Deposition. 1996;2:85–96.
[178] Bwembya GC, Song X, Bochmann M. Phosphinochalcogenoic amidato complexes of zinc and cadmium as novel single-source precursors for the deposition of metal selenide and telluride films. Chem Vap Deposition. 1995;1:78–80.
[179] Chivers T, Parvez M, Seay MA. An acyclic nitrogen-phosphorus-selenium anion: preparation, structures, and reactions of $(K[Ph_2P(Se)NSiMe_3].THF)_2$ with iodine and chlorodiphosphine sulfide. Inorg Chem. 1994;33:2147–50.
[180] Haiduc I. Thiophopshorus and related ligands in coordination, organometallic and supramolecular chemistry. A personal account. J Organomet Chem. 2001;623:29–42.
[181] Bhattacharyya P, Woollins JD. Bis(diphenylphosphino)amine and related chemistry. Polyhedron. 1995;14:3367–73.
[182] Woollins JD. P-N-S/Se-contained metallacycles. J Chem Soc Dalton Trans. 1996;2893–901.
[183] Briand GG, Chivers T, Parvez M. A new approach to metalated imido and amido tellurophosphoranes. Angew Chem Int Ed. 2002;41:3468–70.

[184] Chivers T, Eiler DJ, Ritch JS. Synthesis and structures of $M[N(TePPr^i_2)_2–Te,Te']_n$ (n = 2, M = Zn, Cd, Hg; n = 3, M = Sb, Bi): the first ditelluroimidodiphosphinato *p*- and *d*-block metal complexes. Dalton Trans. 2005;2675–7.
[185] Ritch JS, Chivers T, Eiler DJ, Tuononen HM. Experimental and theoretical investigations of structural isomers of dichalcogenoimidodiphosphinate dimers: dichalcogenides or spirocyclic contact ion pairs. Chem Eur J. 2007;13:4643–53.
[186] Gaunt AJ, Scott BL, Neu MP. Homoleptic uranium(III) imidodiphosphinochalcogenides including the first structurally characterized molecular trivalent actinide-Se bond. Chem Commun. 2005;3215–7.
[187] Chivers T, Ritch JS, Robertson SD, Konu J, Tuononen HM. New insight into the chemistry of imidodiphosphinates from investigations of tellurium-centered systems. Acc Chem Res. 2010;43:1053–62.
[188] Chivers T, Konu J, Ritch JS, Copsey MC, Eisler DJ, Tuononen HM. New tellurium-containing ring systems. J Organomet Chem. 2007;692:2658–68.
[189] Levasanos N, Robertson SD, Maganas D, Raptopoulou CP, Terzis A, Kyritsis P, et al. Ni $[(EPiPr_2)_2N]_2$ complexes: stereoisomers (E = Se) and square-planar coordination (E = Te). Inorg Chem. 2008;47:2949–51.
[190] Robertson SD, Ritch JS, Chivers T. Palladium and platinum complexes of tellurium-containing imidodiphosphinate ligands: nucleophilic attack of $Li[(P^iPr_2)(TeP^iPr_2)N]$ on coordinated 1,5-cyclooctadiene. Dalton Trans. 2009;8582–92.
[191] Gaunt AJ, Scott BL, Neu MP. A molecular actinide-tellurium bond and comparison of bonding in $[M\{N(TeP^iPr_2)_3\}$ (M = U, La). Angew Chem Int Ed. 2006;45:1638–41.
[192] Gaunt AJ, Reilly SD, Enriquez AE, Scott BL, Ibers JA, Sekar P, et al. Experimental and theoretical comparison of actinide and lanthanide bonding in $M[N(EPR_2)_2]_3$ complexes (M = U, Pu, La, Ce; E = S, Se, Te; R = Ph, iPr, H). Inorg Chem. 2008;47:29–41.
[193] Ritch JS, Chivers T, Ahmad K, Afzaal M, O'Brien P. Synthesis, structures, and multinuclear NMR spectra of Tin(II) and Lead(II) complexes of tellurium-containing imidodiphoshinate ligands: preparation of two morphologies of phase-pure PbTe from a single-source precursor. Inorg Chem. 2010;49:1198–205.
[194] Chivers T, Eisler DJ, Ritch JS, Tuononen HM. An unusual ditelluride: synthesis and molecular and electronic structures of the dimer of the tellurium-centered radical $[TeP^iPr_2NiPr_2PTe]$. Angew Chem Int Ed. 2005;44:4953–6.
[195] Konu J, Chivers T, Tuononen HM. The cyclic $[N(P^iPr_2E)_2]^+$ (E = Se, Te) cations: a new class of inorganic ring system. Chem Commun. 2006;1634–6.
[196] Konu J, Chivers T, Tuononen HM. Synthesis, spectroscopic, and structural investigation of the cyclic $[N(PR_2E)_2]^+$ cations (E = Se, Te; R = ^{i}Pr, Ph): the effect of anion and R-group exchange. Inorg Chem. 2006;45:10678–87.
[197] Briand GG, Chivers T, Krahn M. Coordination complexes of bis(amido)cyclodiphosph(III/V and V/V)azane imides and chalcogenides. Coord Chem Rev. 2002;233:237–54.
[198] Chivers T, Krahn M, Parvez M. An extended network of twenty-membered $K_6Se_6P_4N_4$ rings: X-ray structure of $\{[(THF)K[^tBuN(Se)P(\mu\text{-}N^tBu)_2P(Se)N^tBu]K(THF)_2]_2\}$. Chem Commun. 2000;463–4.
[199] Chivers T, Krahn M, Parvez M, Schatte G. Preparation and X-ray structures of alkali-metal derivatives of the ambidentate anions $[^tBuN(E)P(\mu\text{-}N^tBu)_2P(E)N^tBu]^{2-}$ (E = S, Se) and $[^tBuN(Se)P(\mu\text{-}N^tBu)_2PN(H)^tBu]^-$. Inorg Chem. 2001;40:2547–53.
[200] Lief GR, Carrow CJ, Stahl L. Trispirocyclic bis(dimethylaluminum)bis(amido)cyclodiphosph(V) azanes. Organometallics. 2001;20:1629–35.
[201] Plajer AJ, Garcia-Rodriguez R, Benson CG, Matthews PD, Bond AD, Singh S, et al. A modular approach to inorganic phosphazane macrocycles. Angew Chem Int Ed. 2017;56. DOI: 10.1002/anie.201702558.

[202] Nordheider A, Chivers T, Thirumoorthi R, Athukorala Arachige KS, Slawin AM, Woollins JD, et al. A planar dianionic ditelluride and a cyclic tritelluride supported by P_2N_2 rings. Dalton Trans. 2013;42:3291–4.
[203] Nordheider A, Chivers T, Thirumoorthi R, Vargas-Baca I, Woollins JD, Vargas-Baca IA. Planar P_6E_6 (E = Se, S) macrocycles incorporating P_2N_2 scaffolds. Chem Commun. 2012;48:6346–8.
[204] Nordheider A, Athukorala Arachige KS, Slawin AM, Woollins JD, Chivers T. Sodium and rhodium complexes of a spirocyclic Te_5 dianion supported by P_2N_2 rings. Dalton Trans. 2015;8781–83.
[205] Nordheider A, Hull K, Prentis JK, Athukorala Arachige KS, Slawin AM, Woollins JD, et al. Main group tellurium heterocycles anchored by a $P^V{}_2N_2$ scaffold and their sulfur/selenium analogues. Inorg Chem. 2015;54:3043–54.

Jamie S. Ritch

6 Synthesis and coordination chemistry of cyclic seleno- and telluroureas

Abstract: Chalcogenated derivatives of N-heterocyclic carbene ligands have received increasing attention due to their diverse chemical reactivity and potential applications in fields such as medicine and materials chemistry. This chapter summarizes the synthetic methods for the preparation of cyclic heavy chalcogenoureas featuring heterocyclic cores and explores their diverse coordination chemistry with p- and d-block metals.

Keywords: chalcogenoureas, nitrogen heterocycles, coordination chemistry

6.1 Introduction

Selenium- and tellurium-containing analogues of urea and its *N*-substituted derivatives have received increasing attention in the last few decades. These heavy chalcogenoureas (HCUs) have been investigated in many areas from fundamental structural and coordination chemistry to applications in material science [1, 2], medicine [3], and biochemistry [4–6]. A large number of compounds fall into this distinction, and this chapter will focus in particular on *cyclic* seleno- and telluroureas, which are formally chalcogen derivatives of the popular *N*-heterocyclic carbene (NHC) ligands and which feature flanking tertiary nitrogen centres. In the following sections, the preparation and physical properties of cyclic HCUs will be described, followed by a review of their coordination chemistry. Literature up to late 2017 is covered.

The chemistry of ureas and thioureas is of course very well developed in its own right, as are HCUs featuring less organo-substitution. There are many studies featuring acyclic HCUs such as *N,N*-dimethylselenourea, or those which are cyclic but featuring NH substituents (i. e. imidazolidine-2-selone). To provide a focused review, this contribution will only feature selenium and tellurium analogues featuring rings, and with full substitution at nitrogen. Many comprehensive reviews on related areas of chemistry are available, including main group carbene complexes (including non-chalcogens) [7, 8] and the redox chemistry of HCUs [3].

It should be noted that there are several names for this class of compound to be found in the literature, with no particular term being universally adopted. Popular examples include chalcone (selone or selenone, tellone or tellurone) and chalcogenourea (selenourea, tellurourea).

This article has previously been published in the journal *Physical Sciences Reviews*. Please cite as: Ritch, J. Synthesis and Coordination Chemistry of Cyclic Seleno- and Telluroureas. *Physical Sciences Reviews* [Online] **2018** DOI: 10.1515/psr-2017-0128

https://doi.org/10.1515/9783110529340-006

6.2 Structural variety of cyclic HCUs

Within this class of compounds, wherein the C = E (E = Se, Te) carbon atom is within a ring, and each nitrogen atom contains one other organic substituent, there is considerable structural variation possible. Figure 6.1 illustrates the most common molecular architectures featuring the HCU functional group. The ring is usually five membered, though six- and seven-membered rings have been achieved as well. Flanking this ring are the two *N*-substituents which are usually the same group, though asymmetric variants are known. In particular, HCUs with more than one chalcogenourea group (Figure 6.2) often feature an alkyl or aryl spacer. The most commonly encountered HCU is 1,3-dimethylimidazole-2-selone (dmise), perhaps due to the ease of its preparation (see Section 6.3.3).

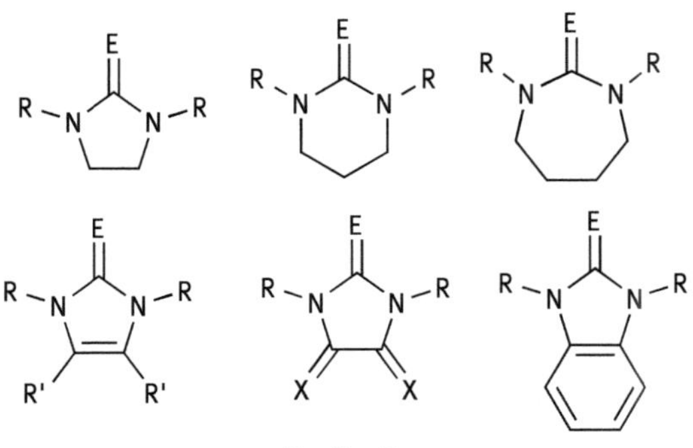

E = Se, Te

R = alkyl, aryl; R' = H, Me, Cl; X = O, NR, CR_2

Figure 6.1: Common HCU compounds with one C = E functional group.

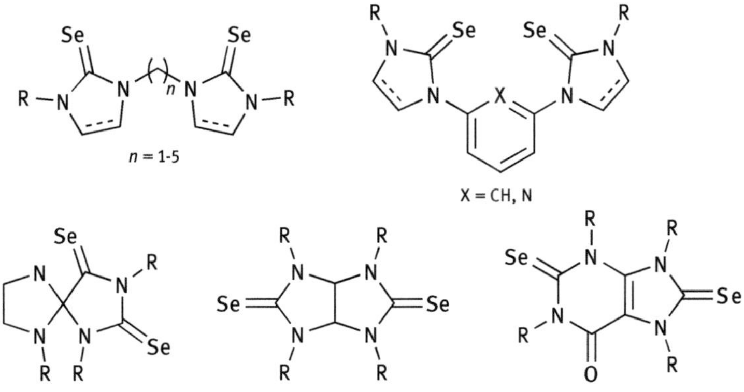

Figure 6.2: HCU compounds with multiple C = E functional groups.

6.3 Synthesis of heavy chalcogenoureas

Cyclic HCUs can be considered derivatives of NHC ligands, and most reported synthetic routes to these compounds begin by assembling the *N*-heterocyclic ring, i. e. an imidazolium or imidazolinium salt, prior to introduction of the chalcogen centre. The numerous routes to prepare carbene precursors have been reviewed thoroughly [9], and this section will focus on several synthetic routes to cyclic HCUs not including these prior steps.

6.3.1 Direct chalcogenation

If one has already prepared or purchased an NHC, then a direct reaction of the carbene with elemental chalcogen is quite facile, generally proceeding cleanly in any solvent compatible with the NHC, and without the need to elevate temperature (eq. (6.1)). Soon after reports of stable NHC ligands in the early 1990s [10, 11], telluroureas [12, 13] and then subsequently selenoureas [14] were prepared in this manner.

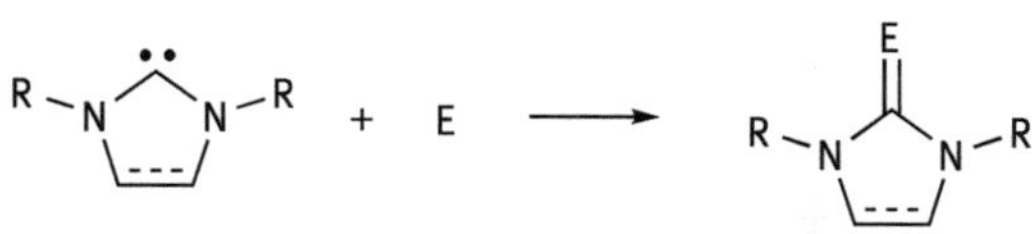

(6.1)

Biimidazolidines (carbene dimers) have been shown to react with elemental chalcogen (eq. (6.2)) to form a selenourea (E = Se, R = Ph) [15], and the first well-characterized example of a tellurourea (E = Te, R = Et) [16]. The selenium derivative formed in high yield in tetrahydrofuran (THF) solution at 0 °C, while tellurourea formation required refluxing toluene. Interestingly, the latter method predates the advent of stable, "bottleable" carbenes [10] by nearly a decade. The tellurium derivative was reported to extrude tellurium and revert back to the biimidazolidine upon photolysis or thermolysis [16].

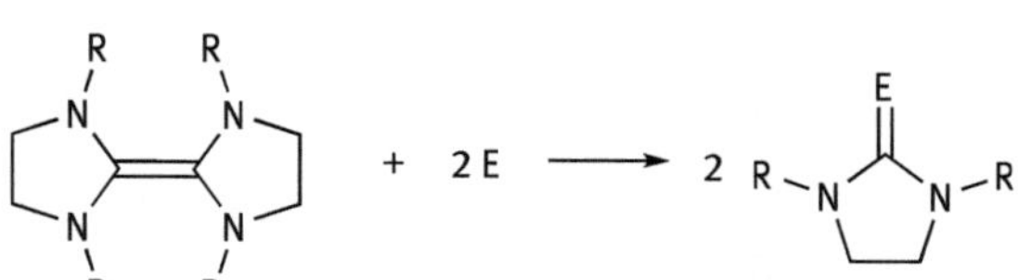

(6.2)

6.3.2 Deprotonation-chalcogenation

Most preparations of cyclic HCUs feature deprotonation of the imidazolium or imidazolinium salt with a base, and concomitant reaction with a chalcogen source (eq. (6.3). and Table 6.1). The base deprotonates the carbene precursor to yield an

Table 6.1: Selected deprotonation–chalcogenation protocols for cyclic seleno- and telluroureas.

Entry	Heterocycle	Chalcogen source	Base	Solvent	Temperature, time	Product (% yield)	Reference
1	 E = Se: X = Cl E = Te, X = PF_6	Se or Te	K_2CO_3	MeOH	24 h (Se), 96 h (Te)	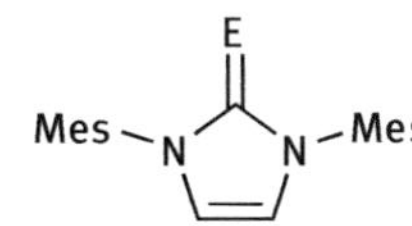 (68–70 %)	[42]
2	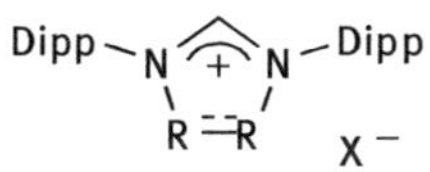 R = CH, X = Cl R = CH_2, X = BF_4	Se	KO^tBu	THF	RT, overnight	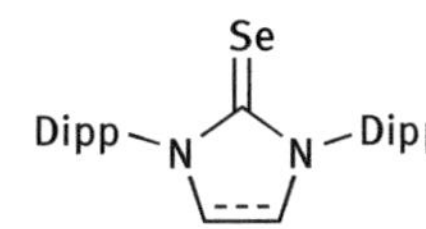 (81–86 %)	[56]
3	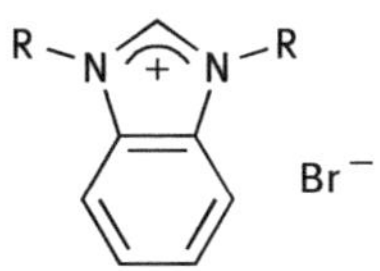 R = ^nBu, ^iPr	Na_2Se_2 or Na_2Te_2	KO^tBu	THF	1) Na_2E_2, 8 h; 2) KO^tBu, 10 h	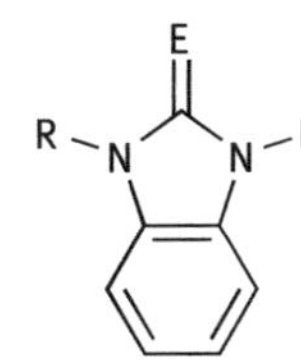 (41–48 %)	[41]

(continued)

Table 6.1 (continued)

Entry	Heterocycle	Chalcogen source	Base	Solvent	Temperature, time	Product (% yield)	Reference
4	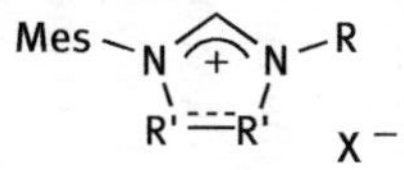 $R = C_6F_5$, $R' = CH_2$, X = Br R = Cy, $R' = CH_2$, X = Cl $R = {}^nBu$, R' = CH, X = Br	Se	$NaN(SiMe_3)_2$	Et_2O	1) −78 °C, 30 min 2) RT, overnight	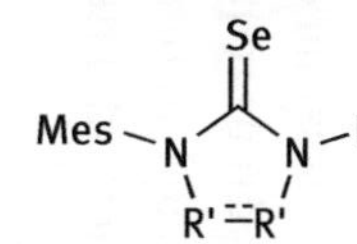 (61–77 %)	[70]
5	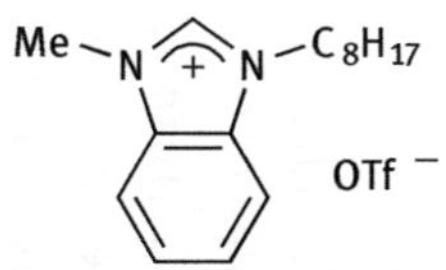	Se	DBU	MeCN	150 °C (MW), 1 h	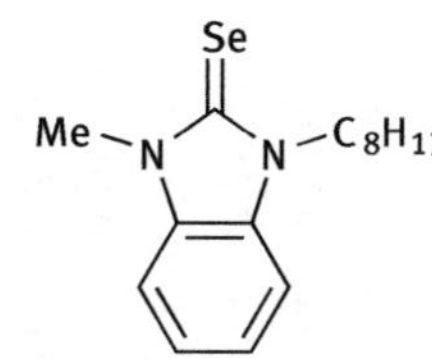 (95 %)	[71]
6	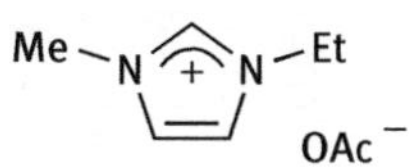	Se	none	neat	75 °C, several days	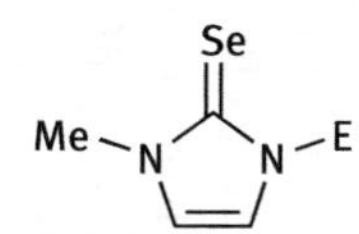 (91 %)	[72]

NHC *in situ*, which then reacts with the chalcogen source to form the desired HCU. These methods vary in the base and chalcogen source used, as well as other reaction conditions such as solvent and temperature. The two most commonly encountered protocols make use of powdered elemental chalcogen and either methanolic K_2CO_3 (entry 1) or KO^tBu in THF (entry 2) as the base/solvent system. The use of hindered alkoxide has the advantage of proceeding cleanly at room temperature, but requires anhydrous conditions. Methanolic potassium carbonate generally requires reflux for the reaction to progress, but anhydrous requirements are less strict and the reactions can sometimes be conducted under ambient conditions without a significant loss in product yield. This method dates back to 1970, when it was first used to generate a cyclic thiourea [17].

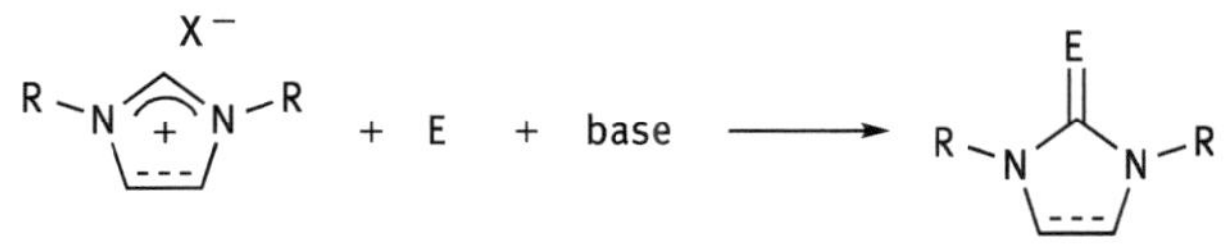

(6.3)

The use of Na_2E_2 salts has been reported as giving higher yields for benzimidazole derivatives (entry 3), versus the elemental chalcogens as powders. This requires the additional *in situ* preparation of the sodium dichalcogenide using a procedure such as sodium reduction of the chalcogen in THF in the presence of naphthalene as an electron donor catalyst [18]. Additionally, when used in an equimolar ratio with the carbenic species, the chalcogen is present in twofold excess. Nonetheless, this strategy has been reported to improve yields for some systems by up to 50 % compared to using elemental chalcogen for some difficult transformations. Na_2Se has been used as a stand-in for Se, eliminating the excess chalcogen, but this method has not been explored thoroughly and its scope is unknown.

Strong amide bases such as $NaN(SiMe_3)_2$ have been utilized (entry 4), though inert atmosphere and low temperature (−78 °C) conditions are required. The organic base DBU has been reported to effect high-yielding selenylation of a benzimidazolium triflate salt in acetonitrile under microwave irradiation (entry 5). Another interesting variation involves the anion of an imidazolium salt acting as the proton acceptor, without the need for an external base. The ionic liquid 1-ethyl-3-methylimidazolium acetate (mp 30 °C) reacts with selenium in the absence of solvent to form the imidazole selone product (entry 6).

Most bidentate chalcogenourea ligands are prepared using the method from entry 1, using bis(imidazolium) bromide or iodide salts. Alkyl bridged bis(selone) ligands (n = 1 and 2 [19], 3 and 4 [20], 5 [21]) can be synthesized in this way (eq. (6.4)). In a study featuring the n = 2 to 5 series, it was noted that the yield of the ethylene bridged (n = 2) bis(selone) ligand was noticeably lower (34 % versus 53–57 %). A vinyl-substituted monoselone side product was also obtained from this reaction in 16 % yield [6]. This suggests that a competing elimination pathway, such as that in eq. (6.5), is operating in this instance. Bis(benzimidazole chalcones) follow the general conditions of entry 3 (eq. (6.6)) [22].

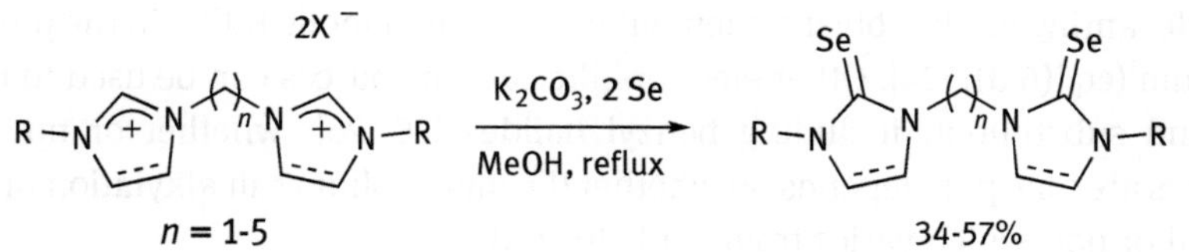

(6.4)

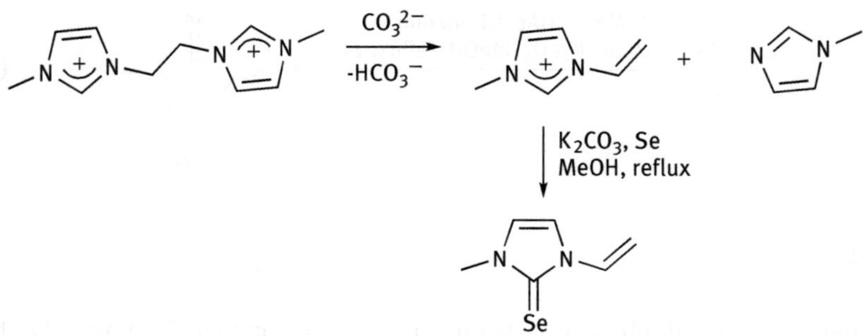

(6.5)

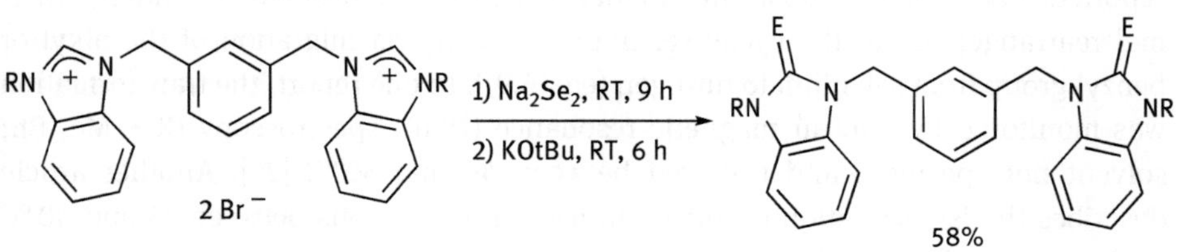

(6.6)

6.3.3 One-pot reactions

So-called one-pot syntheses combine the chalcogen addition step with one or more steps in the synthesis of imidazolium or imidazolinium precursors, *in situ*. Examples of this strategy include the use of *N,N′*-disubstituted diaminoethanes and triethylorthoformate in a high-temperature sealed bomb reaction with selenium, to form selenoureas with an imidazoline backbone (eq. (6.7)). The *tert*-butyl substituted product was only obtained in 39 %, but the other derivatives afforded yields of 85–95 % [6, 23].

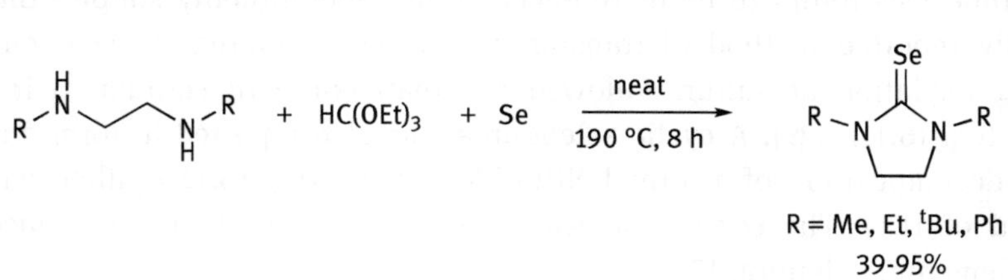

(6.7)

Other reports have combined the chalcogenation with alkylation of an imidazole or benzimidazole molecule to yield HCUs. For example, the popular HCU dmise was first reported in 1993, made by alkylating 1-methylimidazole with methyl iodide in ethyl acetate, decanting off the solvent, then refluxing in methanolic K_2CO_3 in the presence of selenium (eq. (6.8)) [24]. Other electrophilic carbon sources can be used to install the second substituent, including benzyl halides [25, 26]; whether or not these procedures are 'one pot' depends on whether the imidazolium salt alkylation product is isolated or not, which varies from study to study.

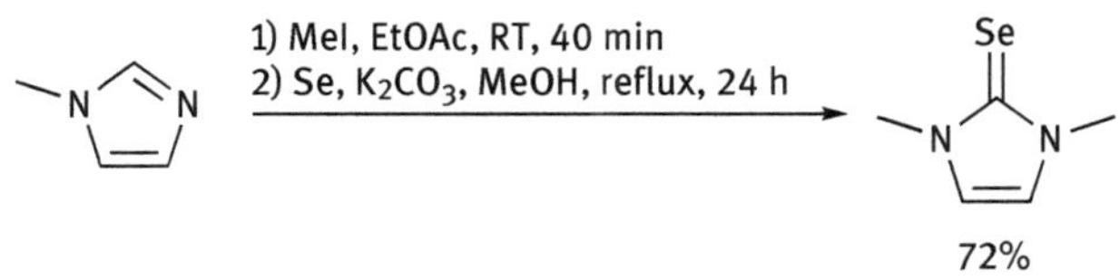

(6.8)

6.3.4 Other methods

A few other methods which do not fit into the previous categories have also been reported. 2-Selenoether-substituted imidazoles have been observed to undergo thermal rearrangement to the cyclic selenourea isomer, via migration of the alkyl or benzyl group from selenium to nitrogen (eq. (6.9)). In one report, the transformation was monitored by nuclear magnetic resonance (NMR) spectroscopy (R = Me, Bn; solvent not specified) and occurred between 40 and 50 °C [26]. Another article describes the isomerization occurring in methanol solutions between 35 and 40 °C for R = Me, *m*-oMeBn [27].

(6.9)

Diorganosubstituted acyclic selenoureas have been cyclized to yield oxamide-type heterocycles via reaction with oxalyl chloride and base (eq. (6.10)) [28]. This method was found to be more effective and operationally simpler than a previously reported method of transforming acyclic thioureas to selenoureas involving akylation at sulfur, followed by treatment with sodium hydrogen selenide (eq. (6.11)) [29]. A cyclic selenourea was also reported to form during thermal decomposition of a formal NHC-CSe_2 adduct (eq. (6.12)), albeit with a low yield of 11 %. The conversion was increased to 21 % in the presence of excess elemental selenium [30].

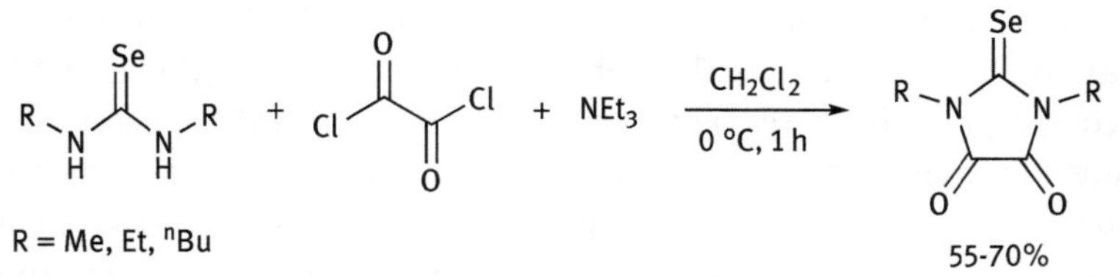

(6.10)

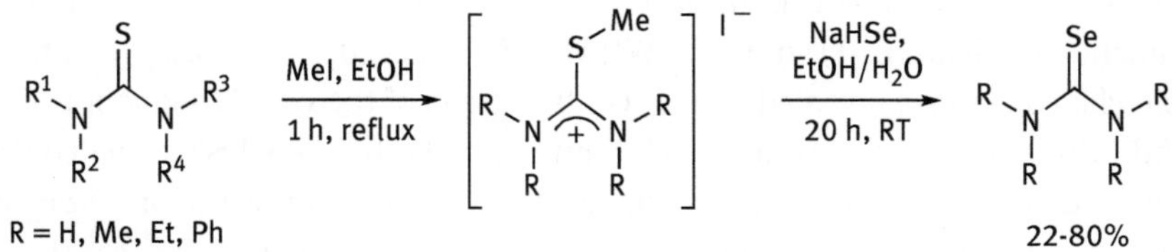

(6.11)

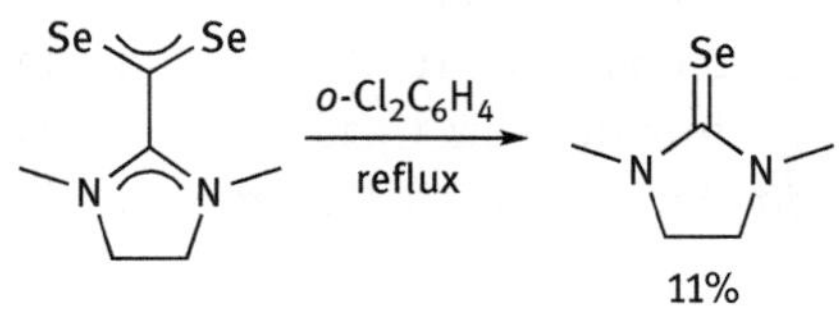

(6.12)

Five-membered selenohydantoin derivatives are accessible via the reaction of alkyl- or aryl-substituted isoselenocyanates with α-amino acids (eq. (6.13)) [31]. This method does not require an added base, but conversion depends strongly on the substitution of the amino acid substrate. β-Amino acids did not give high conversion to six-membered heterocycles. For instance, *N*-methylanthranilic acid yielded only 25 % of the desired product, while β-alanine yielded an acyclic disubstituted selenourea with no ring-closure product observed.

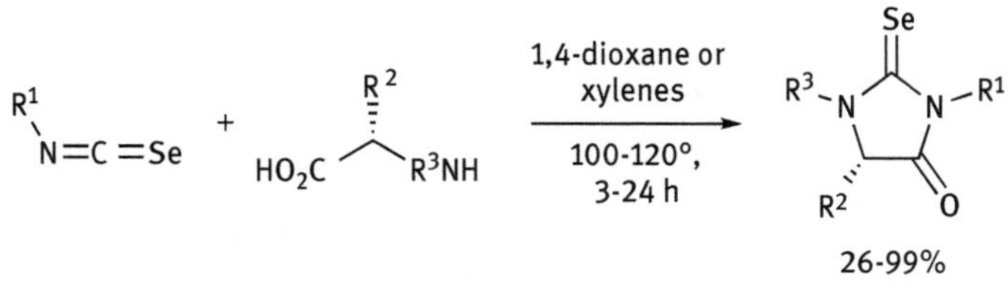

(6.13)

6.4 General properties of HCUs

Cyclic selenoureas, unlike their NHC counterparts, are generally air- and moisture-stable crystalline solids which are stable to column chromatography and can be stored under ambient conditions. However, telluroureas are prone to detelluration and have been reported as being light and moisture sensitive, and slowly depositing tellurium mirrors when in solution [32]. Solubility in organic solvents is a function of substitution pattern; typically, selenoureas are more soluble than telluroureas, and those with more aliphatic groups are more soluble than those with more aromatic substituents.

The frontier molecular orbitals of dmise (M06-2X/6-31G(d,p)), a typical HCU compound, are illustrated in Figure 6.3. The highest occupied molecular orbital (HOMO) and HOMO−1 consist primarily of selenium-centred p orbitals, and these, along with the aforementioned partial negative charge, highlight the ability of the chalcogen centre of HCUs to coordinate to metal centres (see Section 6.6). The HOMO−2 shows π-bonding character for the C = Se fragment and the C = C unit of the imidazole ring, while the LUMO orbital is π-antibonding with respect to C = Se and the imidazole N–C–N unit. A density functional theory (DFT) study (MPW1PW91/LANL2DZ(dp)/6-311G(d,p)) on benzimidazole chalcogenones revealed similar contributions of the C = E fragment to the HOMO and LUMO of these compounds [32]. Natural bond orbital analysis revealed the chalcogen centres bear a partial negative charge (approximate averages of S: −0.25; Se: −0.20; Te: −0.15), while the directly bound carbon atom is positive (ca. +0.0.13 to +0.30). Wiberg bond indices for the chalcogenone functionality also revealed a decrease in double-bond character going from S to Te. A DFT study on related 2-imidazolones (B3LYP/LAV3P/6-31G(d,p)) attributes this trend to an increased contribution to the resonance hybrid by a zwitterionic form [33]. (Figure 6.4).

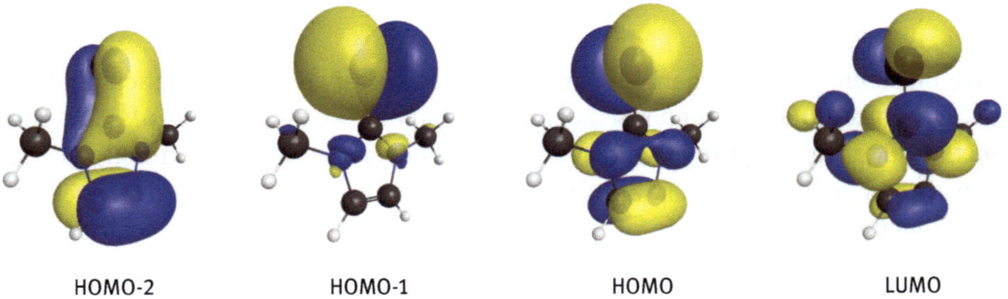

Figure 6.3: Frontier orbitals of dmise.

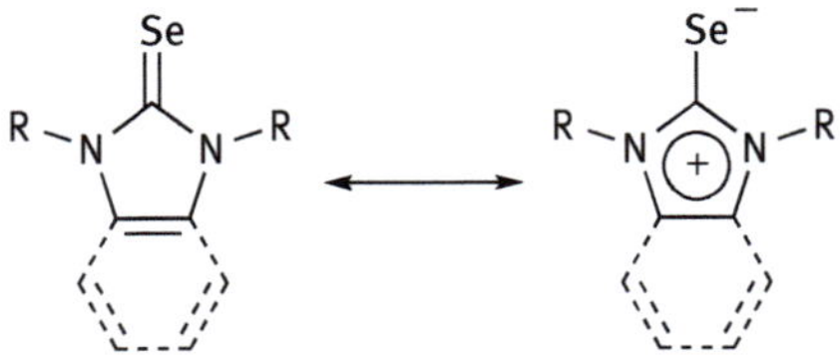

Figure 6.4: Resonance contributors to the electronic structure of HCUs.

Many cyclic HCUs have been characterized by single-crystal X-ray diffraction. These studies reveal several characteristic structural features. The N–C(=E)–N framework is planar, as expected for a functional group featuring lone pair delocalization. The average C = Se and C = Te bond distances found in the Cambridge Structural Database represent contractions of ca. 5 % and 2 %, respectively, compared to the sum of the covalent radius values (1.833(15) Å from 88 hits, 2.068(12) Å from 12 hits, respectively) [34]. This supports the presence of some multiple bond character between the carbon and chalcogen atoms. The average NCN bond angles of HCUs are found to be 106(2)°; large deviations from this average can be found with atypical ring sizes e. g. a seven-membered ring (118°, Figure 6.5) [35].

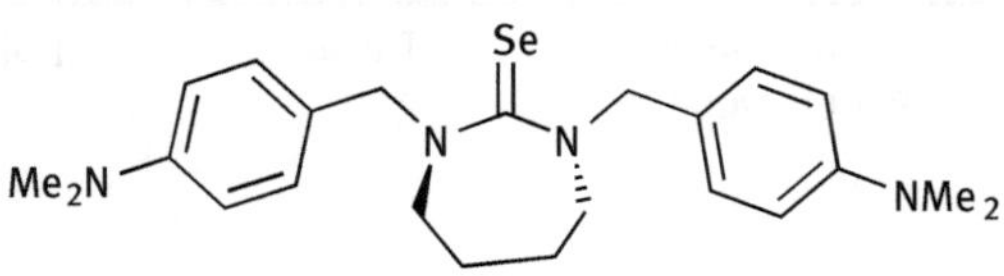

Figure 6.5: Structure of an HCU with a large NCN bond angle.

6.5 ^{77}Se and ^{125}Te NMR spectroscopic properties

These compounds have been used as probes of the electronic properties of parent NHC ligands [36–38], as has been done for phosphinidenes [39]. In particular, ^{77}Se NMR (I = ½, 7.6 %) resonances for various cyclic selenoureas have been associated with the π-accepting ability of the parent NHC ligand. The chemical shift values were found to be correlated to the energy gap of an Se(n)→(C = Se)π* transition, while no correlation between the σ-bonding energy (as determined by energy decomposition analysis) and chemical shift was found.

Though fewer in number, several cyclic telluroureas have been characterized by ^{125}Te NMR (I = ½, 7.0 %) spectroscopy (see Table 6.2 for a survey of relevant compounds). The range of chemical shifts for the species listed spans from δ −168 to −4 and seems to reflect similar trends as has been observed for the Se-containing analogues. Those telluroureas expected to be more π-accepting, e. g. a chlorosubstituted imidazole-based HCU (δ −4, entry 1) and the imidazoline (δ −55.5, entry 3) are found at the higher frequency end of the range, while less π-accepting HCUs including imidazole-based derivatives are found at the lower frequency end (δ −168, entry 2; δ −150, entry 1). However, the delineations are not all clear, as the similarly substituted TeIMes and TeIPr exhibit very disparate chemical shifts of δ −150 (entry 1) and −52 (entry 3), respectively. The benzimidazole selenoureas in entries 4–6, which feature alkyl or benzyl substituents, resonate in the fairly narrow range δ −143 to 126. Further study in this area would be warranted to examine if there is any correlation between bonding tendencies and the ^{125}Te chemical shift, as has been done for the selenium derivatives [36–38].

6.6 Coordination chemistry

A plethora of metal complexes have been prepared using cyclic HCUs. This section is categorized by block and group of the periodic table, and will focus on salient structural features of those complexes whose crystal structures have been determined. Exhaustive structural parameters and features such as co-crystallized solvent are omitted in favour of a general discussion of trends in aggregation type and coordination number. For multistep syntheses, yields (when available) are given are for the last step in the preparation.

Table 6.2: Selected ^{125}Te NMR data for some cyclic telluroureas.

Entry	Compound	$\delta(^{125}Te)$ (ppm)	Solvent	Reference
1	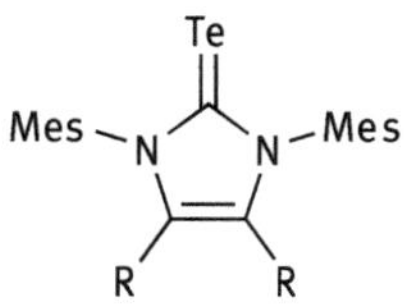 R = H, Cl	R = H: −149.8; R = Cl: −4.08	THF-d_8	[13]
2	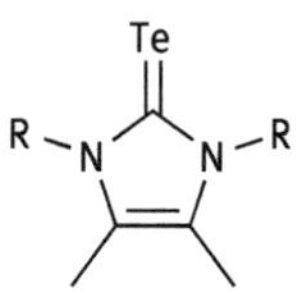 R = Me, ^{i}Pr	R = Me: −168.13 R = ^{i}Pr: −167.82	C_6D_6	[73]
3	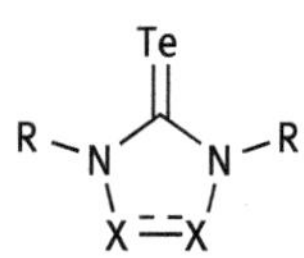 R = Dipp, X = CH; R = Mes, X = CH_2	R = Dipp, X = CH: −52.4; R = Mes, X = CH_2: −55.5	$CDCl_3$	[74]
4	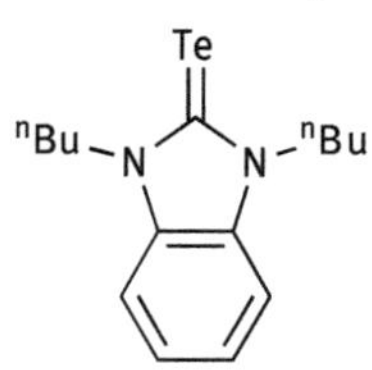	−143	$CDCl_3$	[41]
5	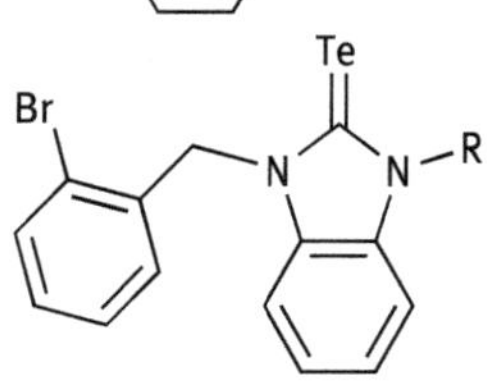 R = ^{n}Bu, ^{i}Pr	R = ^{n}Bu: −130; R = ^{i}Pr: −126	$CDCl_3$	[32]
6	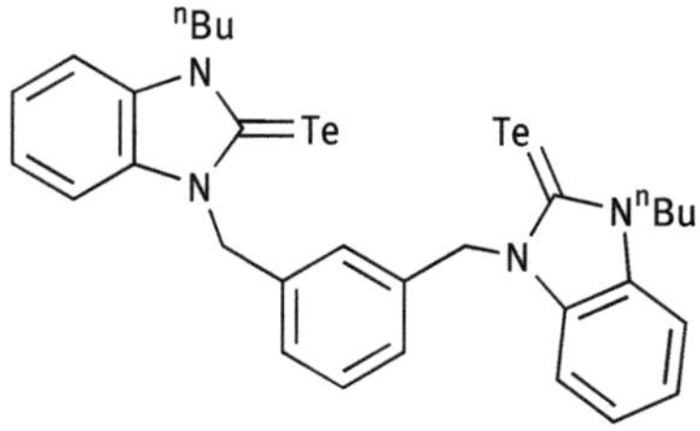	−129.1	$CDCl_3$	[22]

6.6.1 p-block (groups 13–16)

Among the group 13 elements, thallium(I) complexes with two bis(selenoureas) exhibit two motifs (eq. (6.14)) [40]. Upon reaction with thallium(I)hexafluorophosphate in acetonitrile, a bridging coordination mode is adopted within the resulting coordination polymer. Treating the ligand with a thallium(I) diarylaurate salt in THF yields a chelated thallium(I) centre engaged in metal–metal bonding with the gold(I) centre. The bimetallic units are linked by Tl ··· Se and CH ··· Se interactions to afford coordination polymers.

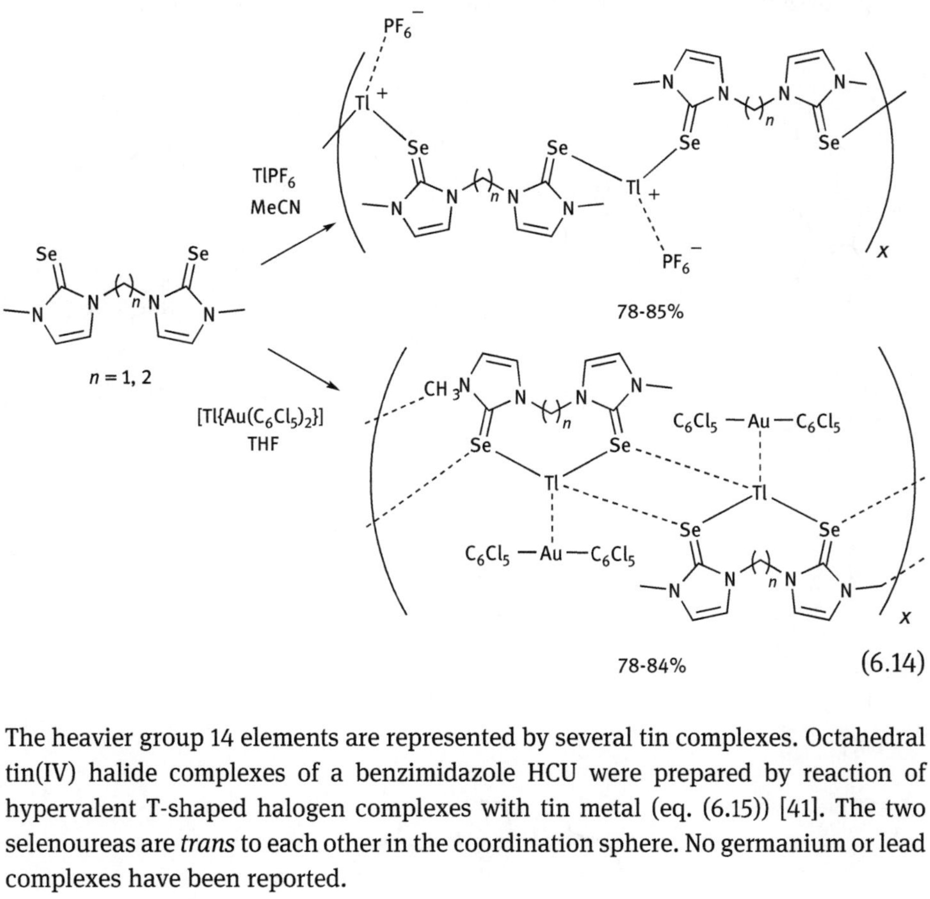

(6.14)

The heavier group 14 elements are represented by several tin complexes. Octahedral tin(IV) halide complexes of a benzimidazole HCU were prepared by reaction of hypervalent T-shaped halogen complexes with tin metal (eq. (6.15)) [41]. The two selenoureas are *trans* to each other in the coordination sphere. No germanium or lead complexes have been reported.

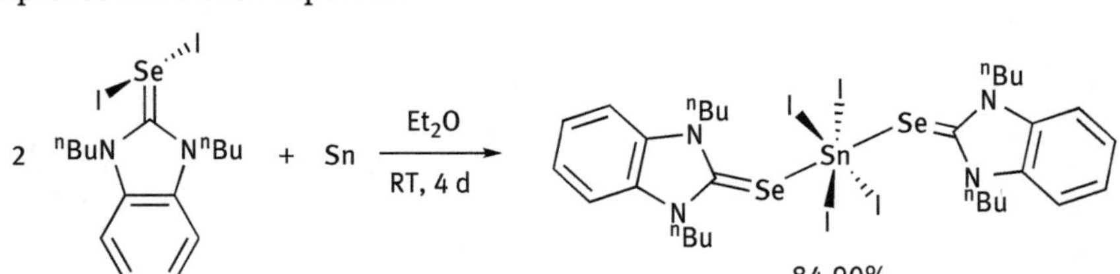

(6.15)

The reaction of this hypervalent selenourea derivative with bismuth(III) iodide yielded a tetrameric cluster featuring a double *seco*-cubane motif (eq. (6.16)). In another study, the use of bulkier aryl-substituted selenoureas afforded two different types of complexes, depending on the size of the pendent aromatic groups. For diisopropylphenyl substitution, monometallic [$LBiX_3$] complexes were obtained, while in the case of smaller mesityl groups, bimetallic complexes of type [{$LBiX_2(\mu_2$-X)}$_2$] formed (eq. (6.17)) [42]. In both cases, bismuth ⋯ aryl interactions are observed. Analogous tellurourea complexes were synthesized but were not able to be studied by X-ray diffraction or NMR spectroscopy, due to very low solubility. A phosphorus adduct of a selenourea, via reaction with the phosphadiazonium salt [Mes*N ≡ P]OTf, has been reported (eq. (6.18)) [43].

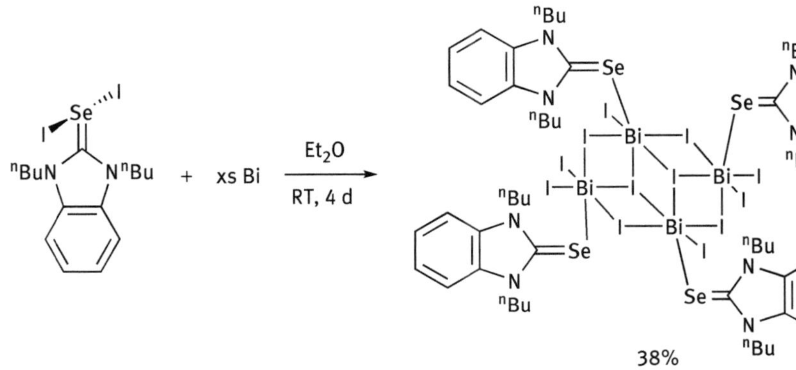

(6.16)

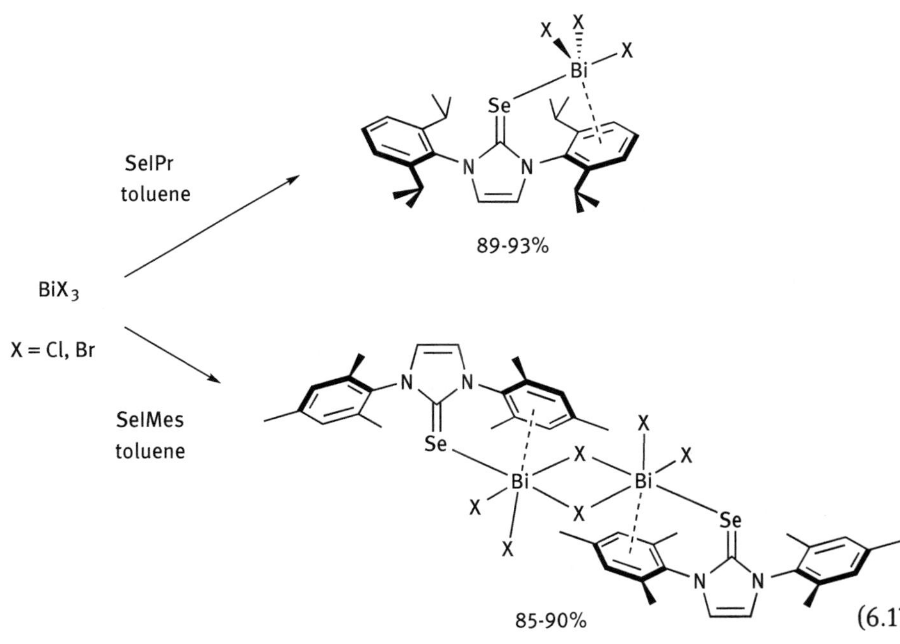

(6.17)

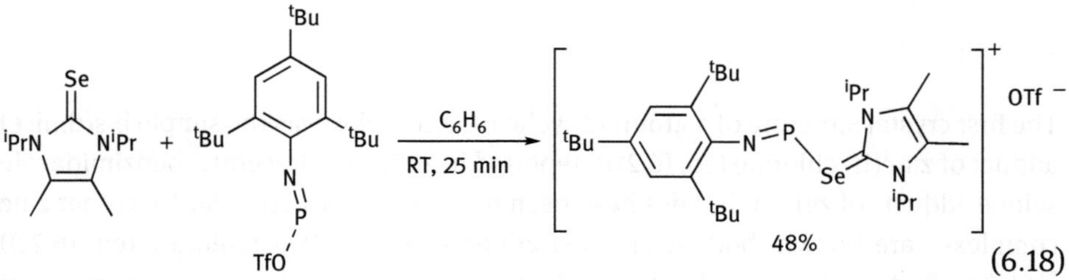

(6.18)

Complexes of the heavier group 16 elements with cyclic HCUs have been reported. The reaction of benzimidazole selenourea dihalides with a diaryl ditelluride resulted in the formation of HCU-stabilized T-shaped aryltellurenyl halides (eq. (6.19)). Halide abstraction with silver tetrafluoroborate in acetonitrile afforded the bent HCU-stabilized aryltellurenyl cation. When reacting various benzimidazole selone dihalides with elemental tellurium, a slow conversion to bis-adducts of Te^{IV} tetrahalides was observed (eq. (6.20)) [41]. The two selone ligands are *trans* to each other in the coordination environment of tellurium. The reactions give moderate (50–60 %) conversion to product when conducted with excess powdered tellurium, and the yields do not improve when a 2:1 ratio of ligand:tellurium is utilized. In one instance (R = ⁿBu, X = Br), a square-planar Te^{II} complex was obtained as a minor by-product in 10 % yield.

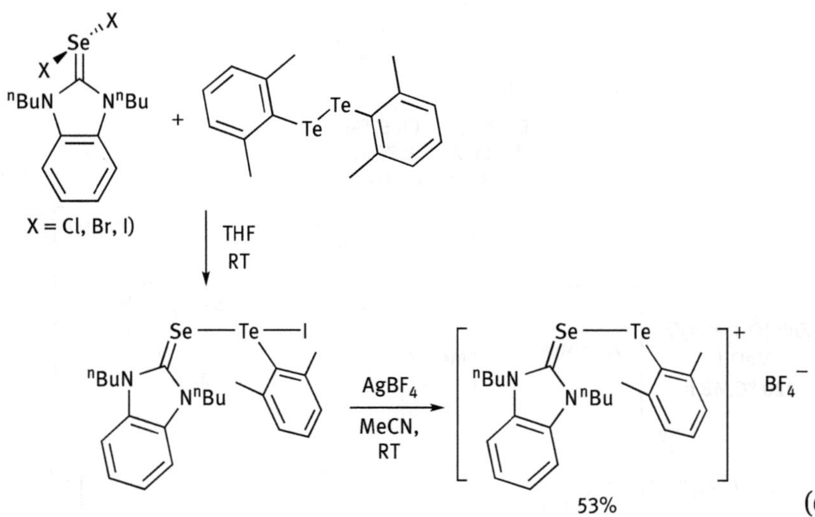

(6.19)

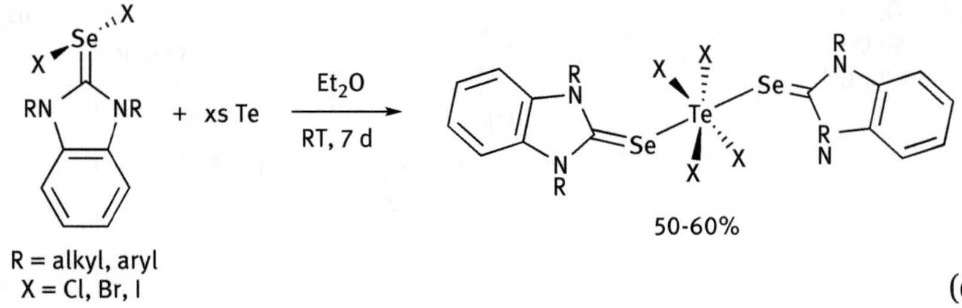

(6.20)

6.6.2 Group 12

The first crystal structure of a group 12 cyclic HCU complex was the simple bis(dmise) adduct of zinc(II) chloride (eq. (6.21)), reported in 2002 [44]. Recently, benzimidazole selone adducts of zinc(II) halides have been prepared (eq. (6.21)) [45]. Two other zinc complexes are known, both selenourea adducts of zinc(II) perchlorate (eq. (6.22)) [46]. The first is a tetracoordinate structure with a distorted tetrahedral geometry at zinc, while the second is an unusual bent two-coordinate motif (Se–Zn–Se bond angle of ca. 103°). Interestingly, the original ligands feature a free carboxylic acid, but under the reaction conditions these functional groups esterify.

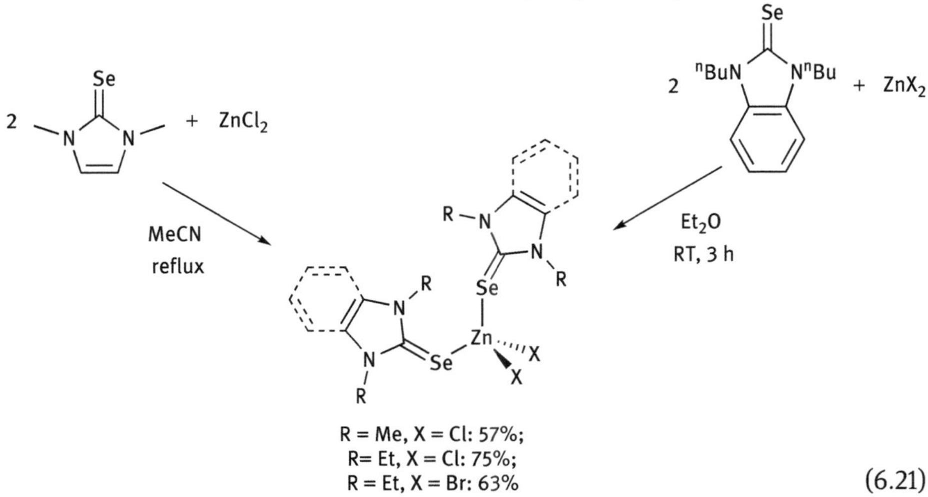

(6.21)

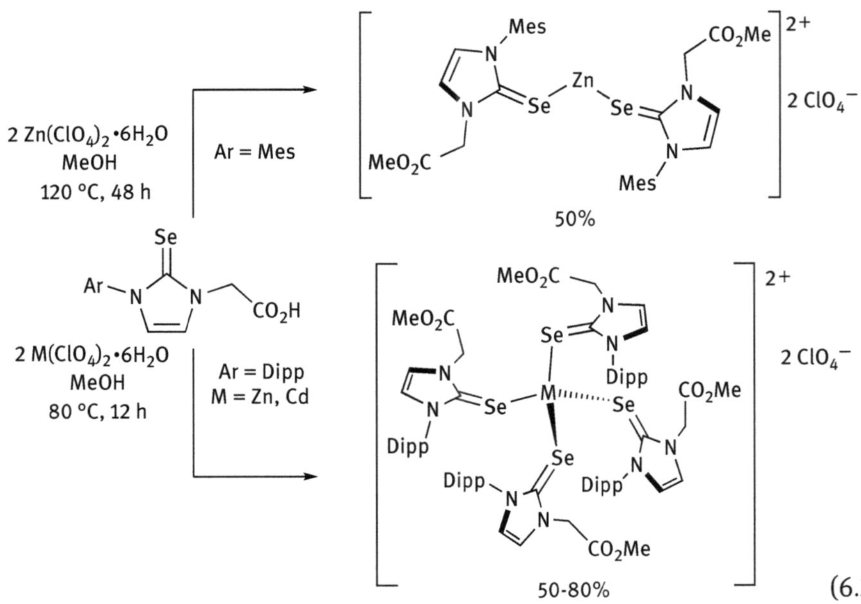

(6.22)

A dmise adduct of cadmium(II) hexafluorophosphate – the first homoleptic selenourea–cadmium complex, was reported in 2007 (eq. (6.23)) [47]. Using mixed aryl/alkyl ester ligands (as in eq. (6.22)), three unique structural motifs are obtainable when reacted with cadmium salts in methanolic HCl (eq. (6.24)) [46]. When the aryl substituents are Mes, a cationic complex with one chloride and three selenourea ligands is obtained in the reaction with cadmium perchlorate and when cadmium nitrate is used, a coordination polymer of the form $(LCdCl_2)_x$ is obtained, with distorted trigonal bipyramidal geometry at cadmium. In the case of 2,6-diisopropylphenyl (Dipp) substituents on the nitrogen centres, a neutral complex featuring two selenourea ligands, one nitrate and one chloride, is observed to form.

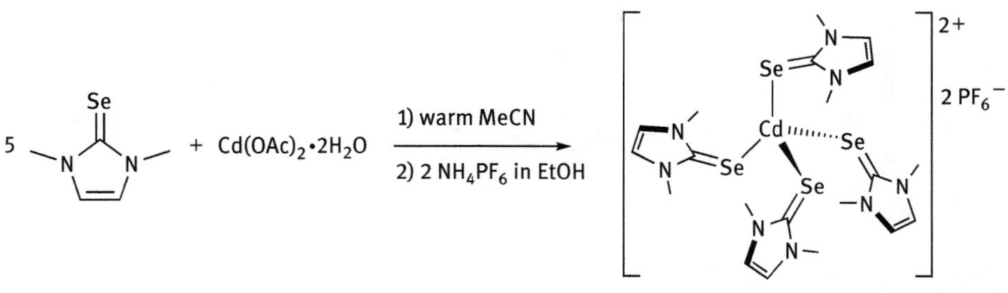

(6.23)

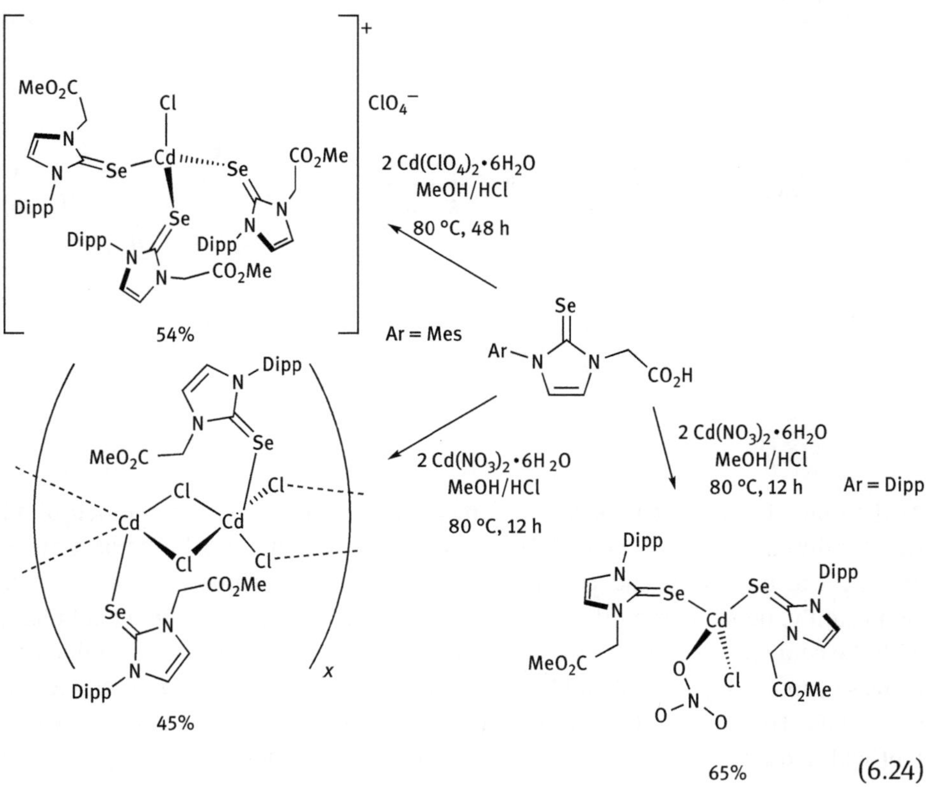

(6.24)

By reacting a benzimidazole selonourea ligand with ether-containing alkyl substituents on nitrogen with cadmium iodide, a bimetallic complex was afforded (eq. (6.25)) [45]. The metal centres are bridged by two iodide ligands. The analogous reaction with mercury(II) halides affords two bimetallic mercury complexes. The dimeric motif has also arisen from the reaction of $HgCl_2$ with a dicationic diselenide, via cleavage of the Se–Se bond (eq. (6.25)) [48]. One monomeric mercury complex with distorted tetrahedral geometry has been reported, via a 1:1 reaction of a bidentate selenourea/pyridine ligand with HgI_2 (eq. (6.26)).

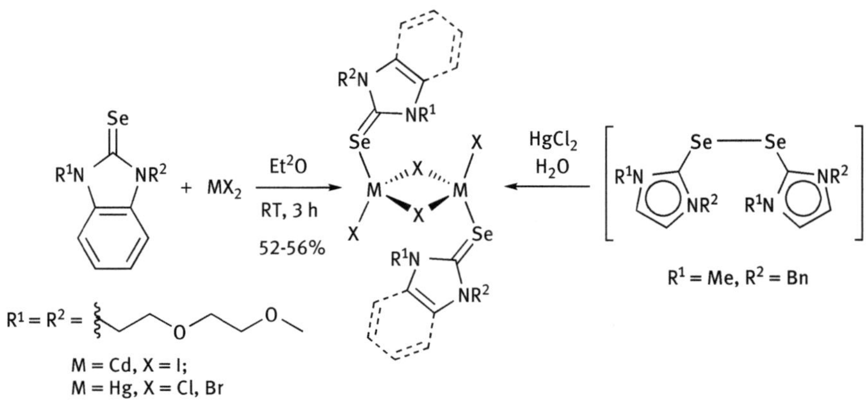

(6.25)

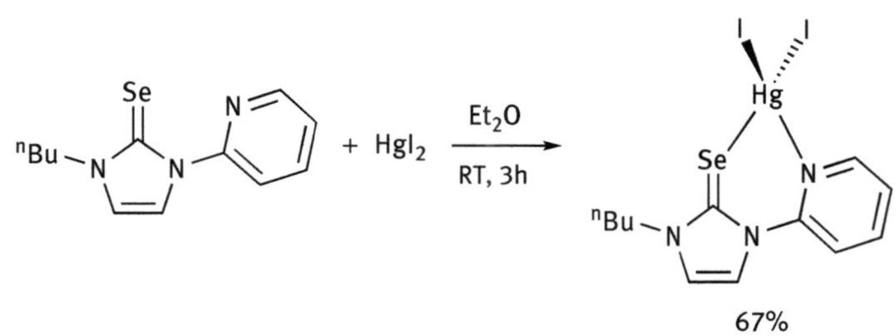

(6.26)

6.6.3 Group 11

At the time of writing, complexes of coinage metals make up over one half of all structurally characterized cyclic HCU–metal adducts, and of these the majority are of copper. The nuclearity and coordination number of the copper complexes vary significantly, depending on the steric and electronic properties of the ligand along with the counterion. Cationic two-coordinate Cu^{I} complexes have been obtained by using selenourea ligands with bulky aryl groups installed on the nitrogen centres (eq. (6.27)) [49]. The metal salts used contain Cu^{II}, and so the complexation is concomitant with reduction of the metal by some of the chalcogenone. The complexes are centrosymmetric in the solid state, featuring 180° Se–Cu–Se bond angles, except for one derivative, which crystallized in another space group and is pseudo-centrosymmetric (Ar = Mes, X = ClO_4; 176° bond angle). In another report, a neutral two-coordination copper(I) complex was obtained by the reaction of a very bulky selenourea with CuBr in toluene (eq. (6.28)) [50].

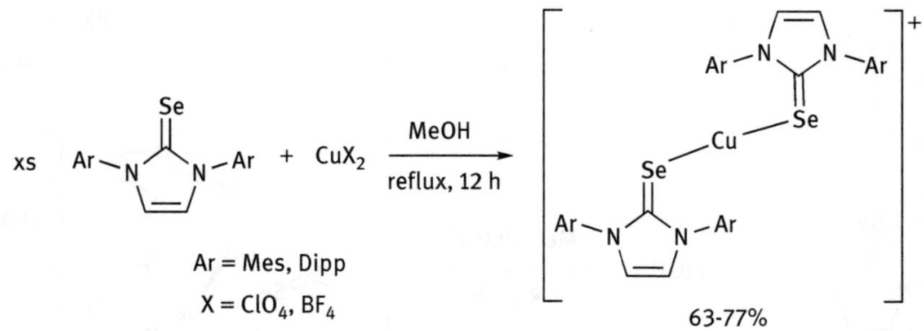

(6.27)

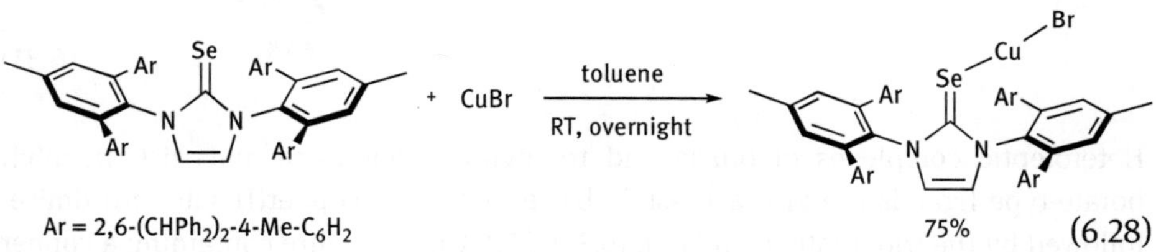

(6.28)

Nearly all of the higher nuclearity/coordination number complexes of Cu feature the small *N,N′*-dimethylimidazole selone (dmise) ligand or the methylene bridged bis (selone) analogue. An exception is a dimeric copper(I) iodide complex of a benzimidazole selone with *n*-butyl groups on the nitrogen centres (eq. (6.29)) [45]. The 2:1 reaction of dmise with copper(I) halides (eq. (6.30)) affords trigonal planar adducts where the two ligands are either *syn* (X = Br, I) or *anti* (X = Cl) [51]. The *syn* conformer exhibits intramolecular π–π interactions between the imidazole rings. A homoleptic cationic complex is afforded by the reductive coordination of dmise with copper(II) triflate (eq. (6.31)) [52]. In this reaction, an oxidation product of the HCU, [(dmise)$_3$] [OTf]$_2$, was also isolated in 15 % yield.

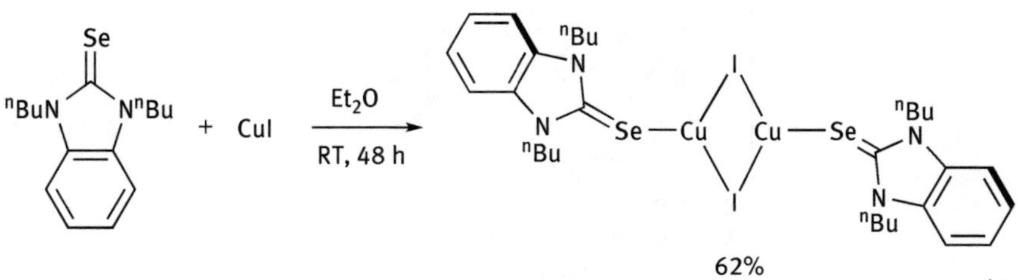

(6.29)

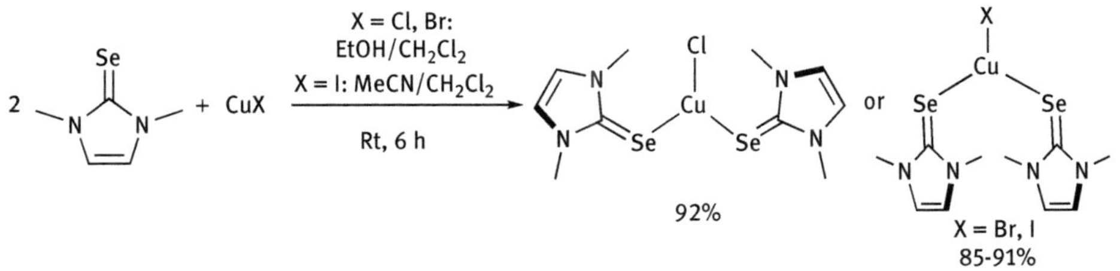

(6.30)

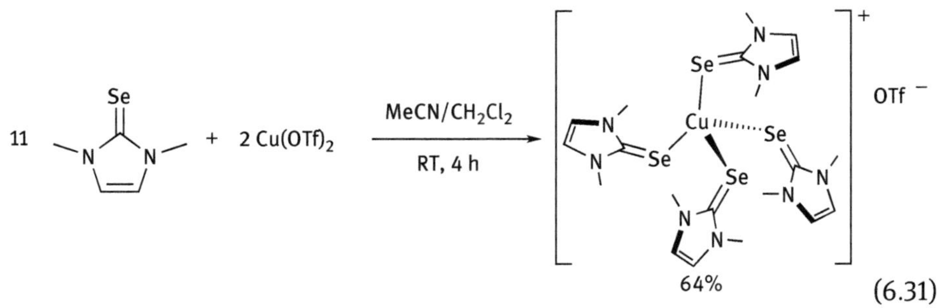

(6.31)

Heteroleptic complexes of dmise and tris(pyrazolyl)methane and tris(pyrazolyl) borate-type ligands are also accessible by reaction of a copper(I) salt with dmise, followed by the tridentate ligand (eq. (6.32)) [53]. One structure containing a copper (II) centre has been reported. The strategy used to stabilize the +2 oxidation state in the presence of the reducing selenourea ligand was to bind Cu^{II} with π-acidic phenanthroline ligands prior to coordination with dmise (eq. (6.33)) [54]. The copper(II) centre exhibits a distorted trigonal bipyramidal geometry, with the selenium donor atom in one of the equatorial sites.

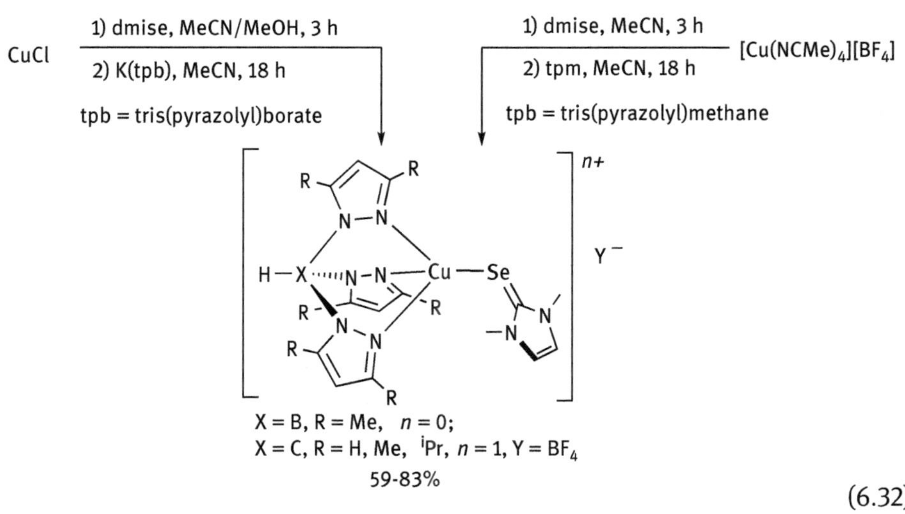

(6.32)

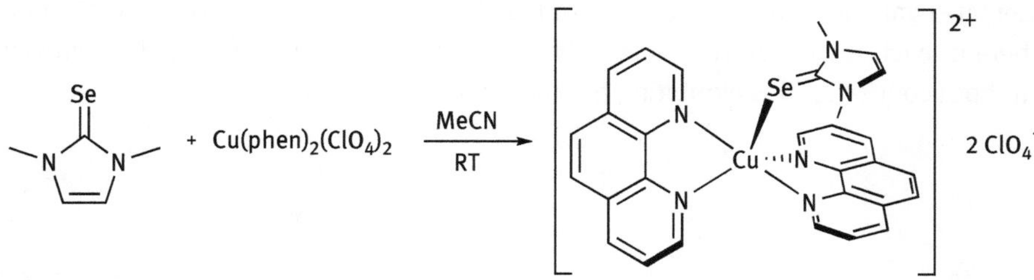

77% (6.33)

Four dinuclear complexes featuring copper(I) have been synthesized and structurally characterized [55]. The 1:1 reaction of bmise with tetrakis(acetonitrile)copper(I) tetrafluoroborate afforded a 3:2 ligand:metal complex featuring two copper centres, each chelated by one bmise and bridged together by a third ligand (eq. (6.34)). The central Cu_2Se_2 ring is butterfly-shaped.

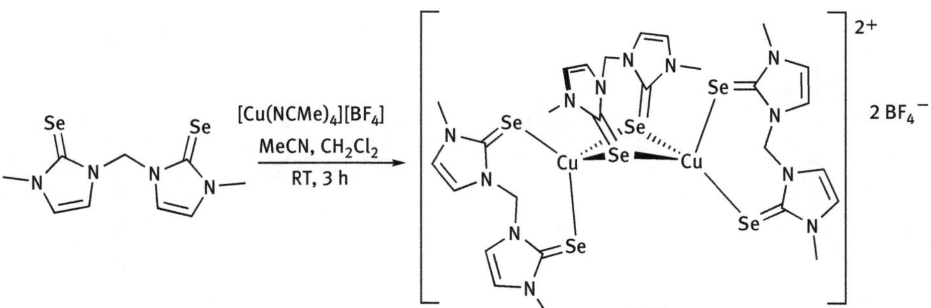

45% (6.34)

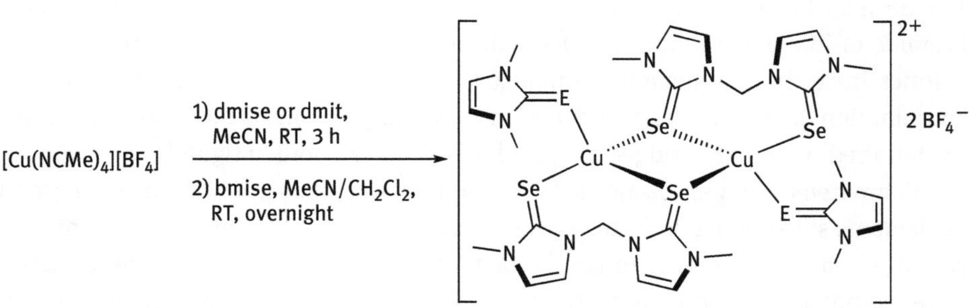

E = S, Se
38-46% (6.35)

When using a mixture of monodentate and bidentate chalcogenourea ligands in a stepwise fashion (eq. (6.35)), another bimetallic complex with tetrahedral coordination is formed, but in this case the central Cu_2Se_2 ring is planar, with each bridging selenium donor belonging to a different bmise ligand. The other selenium centre on each bidentate ligand is terminal, and one terminal dmise ligand completes the coordination sphere of each metal centre. Using only dmise afforded another bimetallic complex, with both copper centres exhibiting trigonal planar geometry (eq. (6.36)).

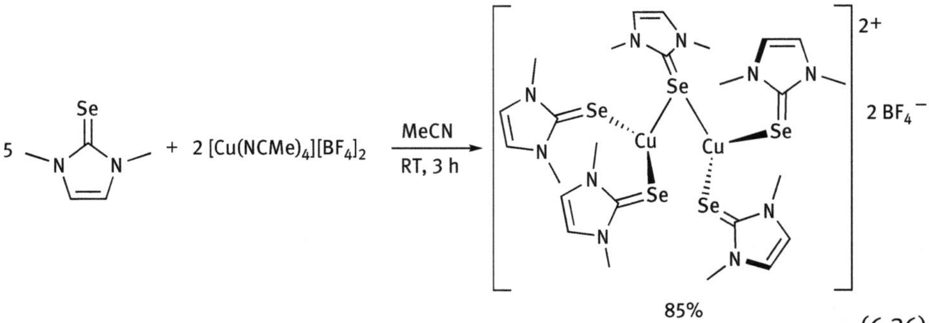

(6.36)

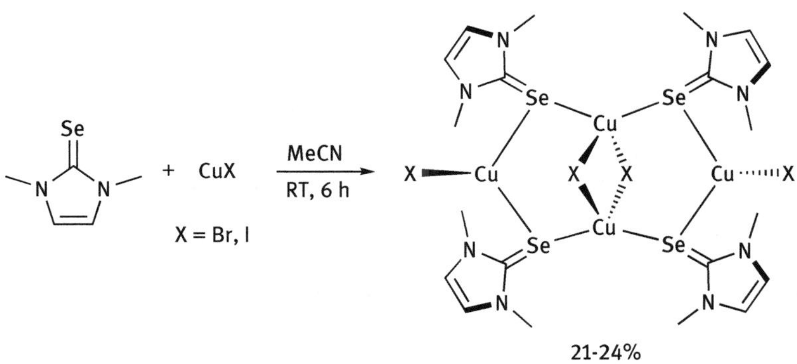

(6.37)

A neutral tetrametallic complex was prepared by the simple 1:1 reaction of copper(I) bromide or iodide with dmise, in acetonitrile (eq. (6.37)) [51]. The structure is quite distinct from the tetrabismuth complex in eq. (6.16), which featured octahedral coordination with only halide ligands in bridging positions. The copper(I) centres are tetrahedral and are bridged by both halide and selenourea ligands.

Compared to copper, the cyclic HCU coordination chemistry of the heavier group 11 elements silver and gold is much less thoroughly explored. However, a number of complexes have been structurally characterized in recent years, and these metals have the potential to equal the structural diversity of copper. Heavy group 11 analogues of the linear copper complexes in eq. (6.27) can be prepared in an equivalent fashion, using aryl-substituted cyclic selenoureas (eq. (6.38)). There are, however, noteworthy differences between all the elements. In the case of gold, two structure types were seen: neutral selenourea-gold(I) chloride adducts and cationic bis(selenourea) complexes with a dichloroaurate counterion [56]. For silver, ionic complexes were observed (featuring triflate or dinitritoargentate anions) when a Dipp-substituted ligand was used, but neutral bimetallic complexes with three-coordinate metal centres were seen for the smaller mesityl-substituted analogues [57]. The aggregation mode of gold was surmised to arise from the π-accepting nature of the selenourea ligands: ionic complexes were observed for more π-accepting selenoureas (as judged by the ^{77}Se chemical shift value) [36–38]. In the case of silver, the bulkier Dipp groups seemed to promote the monometallic ionic form, while the less sterically encumbered Mes group yielded bimetallic complexes.

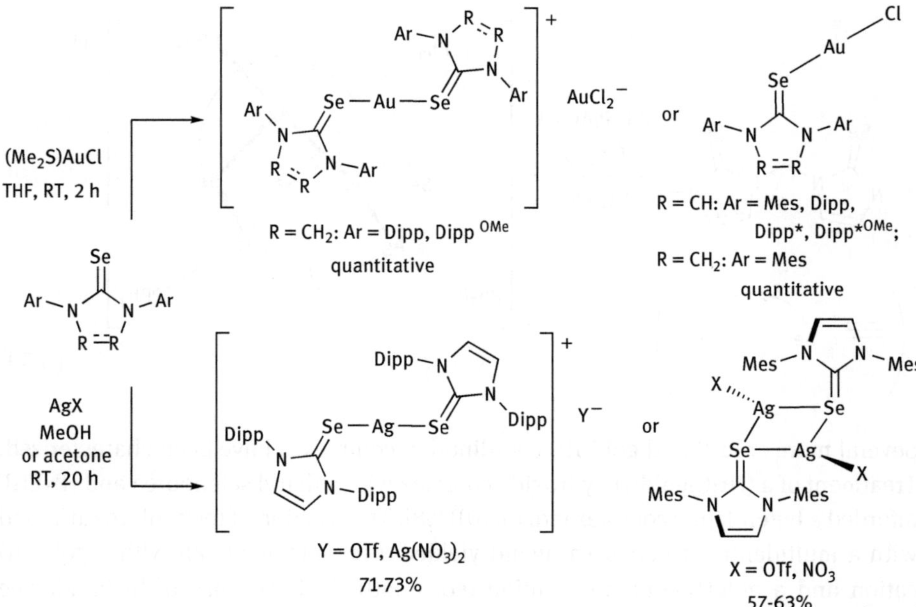

(6.38)

Recrystallization of one bimetallic silver complex (Ar = Mes, X = NO_3) from hot methanol resulted in the formation of a coordination polymer with three crystallographically distinct silver(I) environments (Figure 6.6). In this product, the ligand:

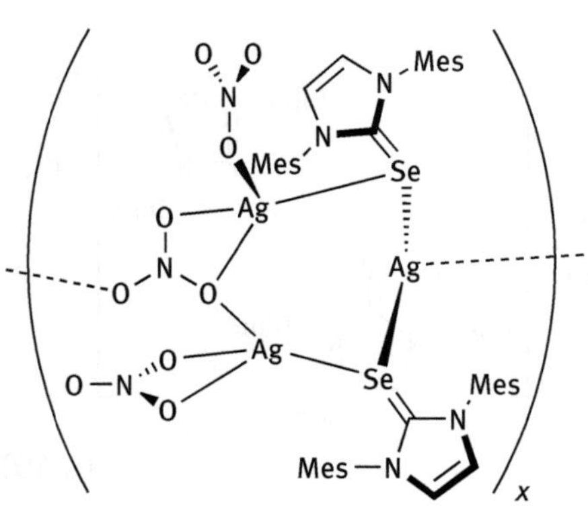

Figure 6.6: A silver(I) coordination polymer featuring HCU ligands.

metal ratio has been altered from 1:1 to 2:3. Two of the silver sites are four coordinate and one is three coordinate, including the weak Ag ··· O contact which links the monomers together.

The reaction of triphenylphosphine silver(I) triflate and bmise afforded a tetrametallic cluster featuring an Ag_4Se_6 diamondoid core, with the four silver centres capped by terminal PPh_3 ligands and doubly bridging bmise ligands (eq. (6.39)) [21].

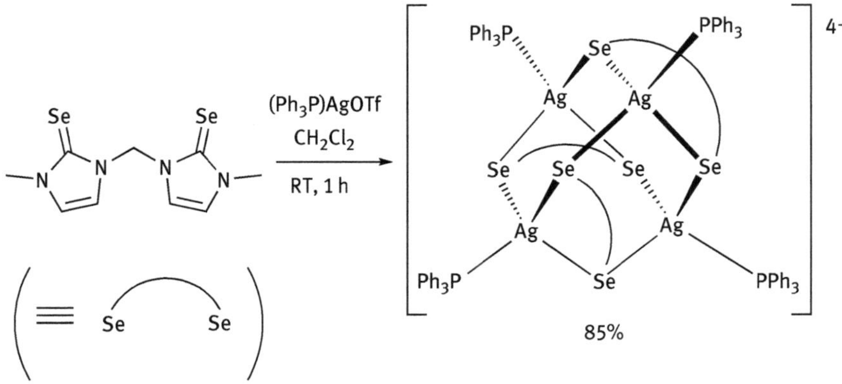

(6.39)

Several more gold(I) and gold(III) coordination complexes have been characterized. Treatment of a (NHC)gold(I) hydroxide complex with a diaryl selenourea and $HNEt_3F$ afforded a heteroleptic complex (eq. (6.40)) [58]. The reaction of tetrachloroauric acid with a multidentate selenourea ligand yielded a mixed-valent salt with a gold(III) cation and a gold(I)-containing anion (eq. (6.40)) [59]. The use of bis(imidazole selone) ligands in combination with pentafluorophenyl-substituted gold(I) and gold(III) salts resulted in two bimetallic salts in which the metal centres are bridged by the whole ligand (eq. (6.42)), rather than brought into close proximity, c.f. the copper(I) complexes in Equations 34–35 [21]. As expected, the gold(I) complex features linear coordination, while the gold(III) complex exhibits square-planar metal centres.

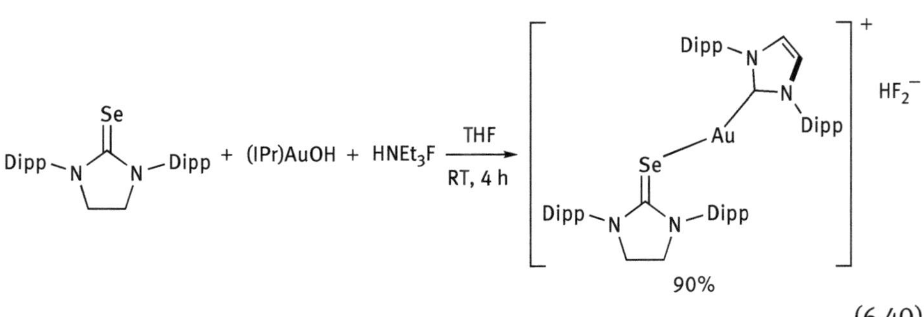

(6.40)

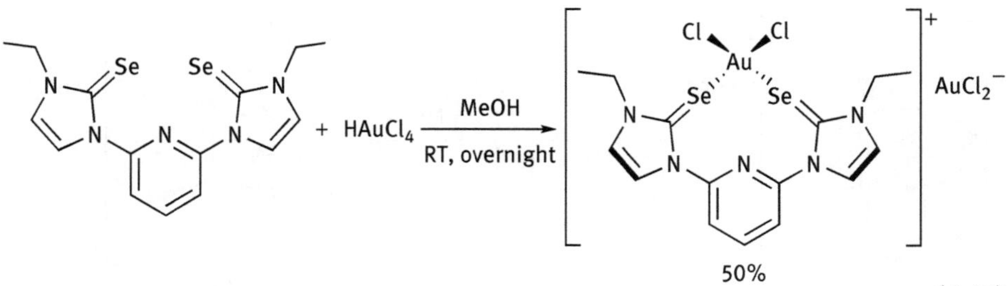

(6.41)

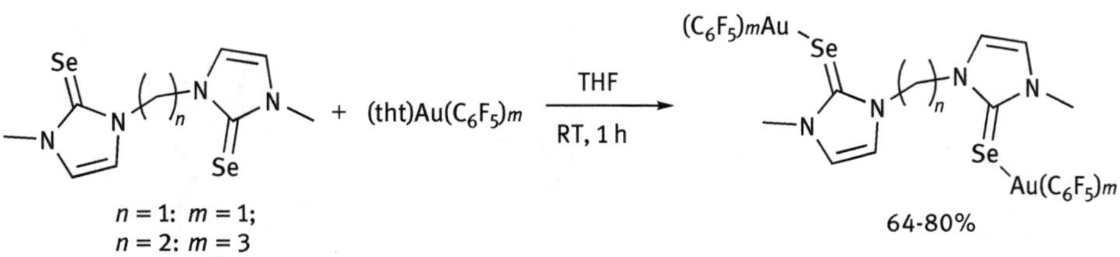

(6.42)

6.6.4 Group 10

Several M^{II} (d^8) complexes of cyclic HCUs have been synthesized and characterized by X-ray diffraction. The lack of M^{IV} complexes is to be expected, given the reducing nature of the chalcogenoureas. Two nickel(II) complexes have been reported: one neutral $NiBr_2$ adduct with tetrahedral geometry (eq. (6.43)) [60] and one cationic, distorted octahedral selenourea adduct featuring a macrocyclic ligand (eq. (6.44)) [61]. Palladium(II) is also represented by one cationic and one neutral complex. Starting from a potential SeCSe pincer ligand precursor, the reaction with bis(benzonitrile)dichloropalladium(II) under anion exchange conditions afforded a dicationic complex with a square-planar, $PdSe_4$ core (eq. (6.45)) [22]. The same bis(selenourea) in reaction with a dipalladium complex afforded a neutral, monodechalcogenated product. The low yield and missing selenium atom in the product complex indicate formation of palladium selenide phases as a likely side reaction.

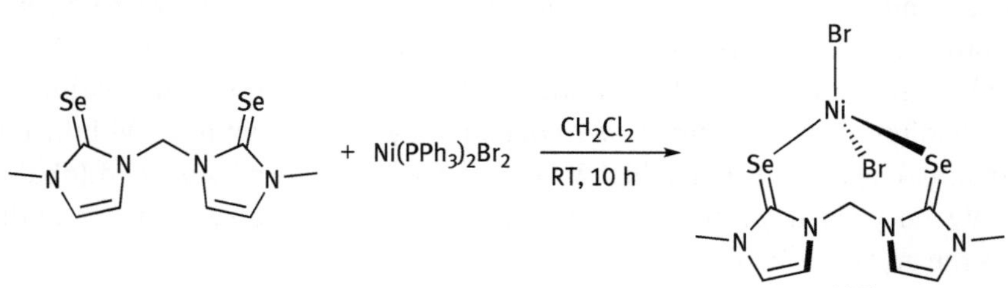

(6.43)

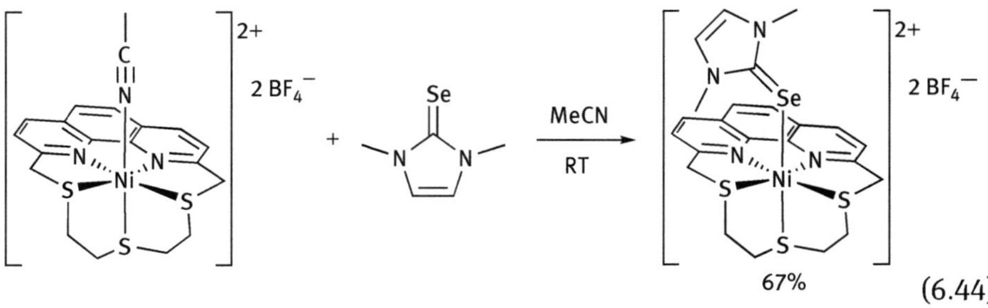

(6.44)

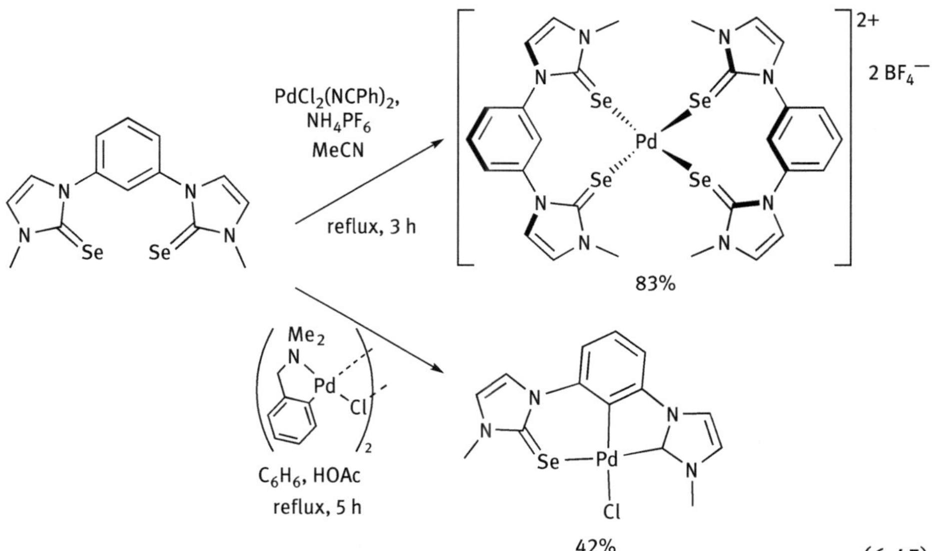

(6.45)

Several dinuclear complexes of palladium, and one of platinum, are also known. A crystal growth experiment featuring an imidazole selenourea and palladium(II) chloride yielded a bimetallic complex featuring bridging selenium atoms in *cis* orientations (eq. (6.46)) [62]. Unlike the silver complexes in eq. (6.38), the central ring is not square but rather a butterfly-shaped motif, with the two *cis* chloride ligands completing each square-planar palladium(II) metal centre being *syn* to each other. The imidazole rings are also nearly in the same plane, with the smaller methyl substituents of each ligand oriented towards each other. One platinum and one palladium dimeric complex were obtained by reaction with a aryl-bridged bis (selenourea) ligand (eq. (6.47)) [63]. When reacted with the Pd^0 source $Pd_2(dba)_3$, the same ligand yielded an intractable mixture with no detectable oxidative addition across the aryl C–Br bond.

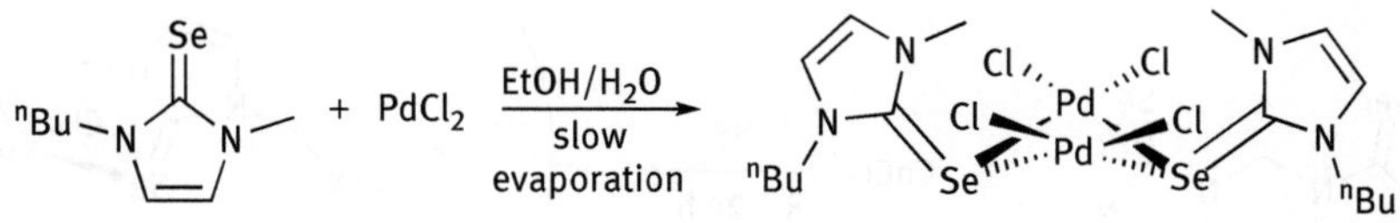

(6.46)

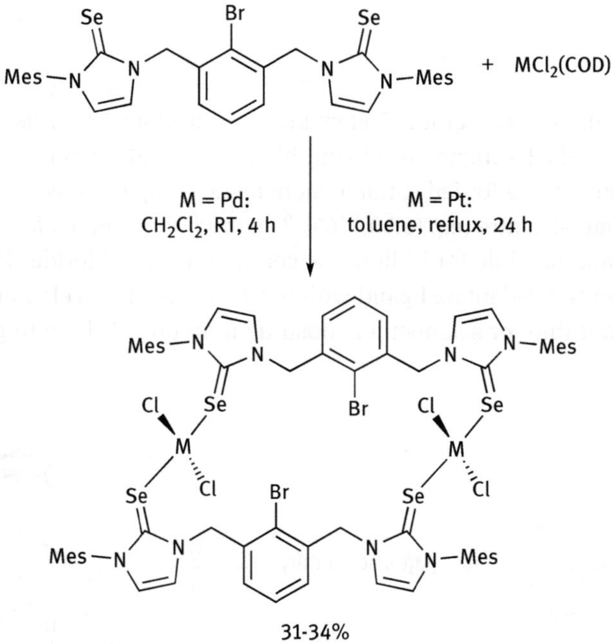

(6.47)

6.6.5 Group 9

Two cobalt complexes have been structurally characterized. A monomeric bis(selenourea) adduct similar to $ZnCl_2(dmise)_2$ (eq. (6.21)) was prepared by boiling down a methanolic of the ligand and metal salt (eq. (6.48)) [64]. A coordination polymer of a bis(selenourea) was obtained upon reaction of the ligand with $CoCl_2$ in THF (eq. (6.49)) [60]. The structure features tetrahedral coordination at cobalt, and the packing reveals a racemic mixture of left- and right-handed helical chains.

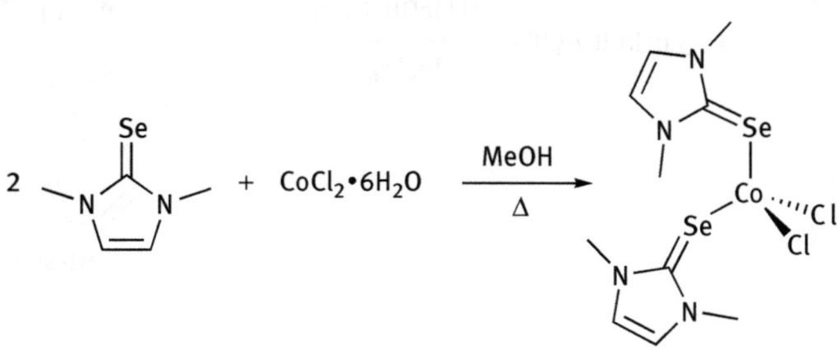

(6.48)

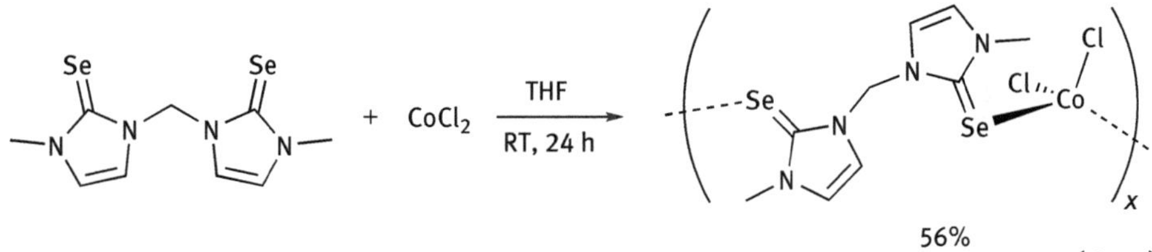

(6.49)

Complexes of the heavier group 9 elements are all represented as adducts to the $[Cp^{*}MCl]^{+}$ (M = Rh, Ir) fragment. Mixing bis(selenourea) ligands with the dimeric $[Cp^{*}MCl_2]_2$ complexes afforded cationic monomeric complexes featuring the ligands chelating the metal centre (eq. (6.50)) [65]. The rhodium complex featured a rhodate (III) complex anion, while for iridium the counterion was chloride. Two related Ir^{III} complexes feature a bidentate ligand with one benzimidazole selenourea donor and either a pendant thio- or selenoether substituent on one of the nitrogen atoms (eq. (6.51)) [66].

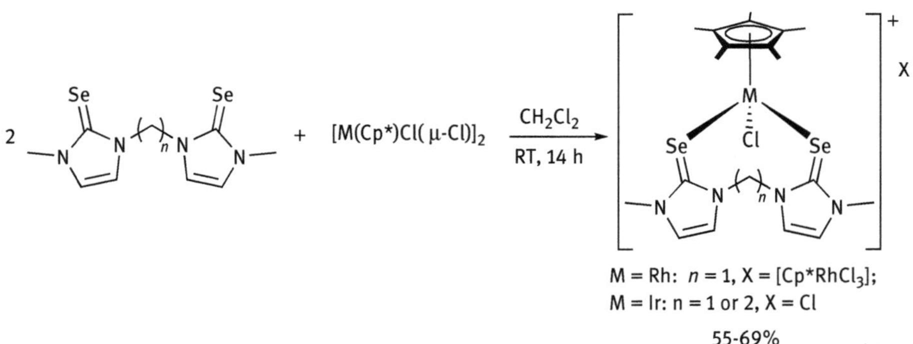

(6.50)

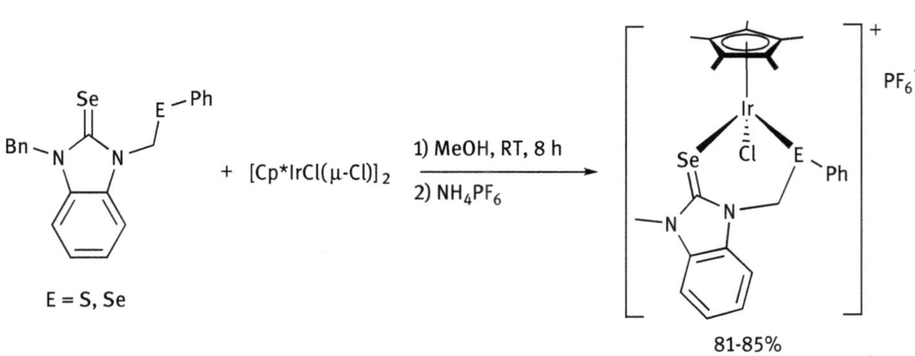

(6.51)

6.6.6 Group 8

Two dmise adducts of iron(II) salts are reported. The simple adduct $(dmise)_2FeCl_2$ was prepared by mixing the ligand and iron(II) chloride tetrahydrate in $CH_2Cl_2/HC(OEt)_3$ solvent system (eq. (6.52)) [67]. Under these conditions, the triethyl orthoformate acts as both a solvent and a water scavenger via its hydrolysis reaction. The same report discloses that the 4:1 reaction of dmise with iron(II) tetrafluoroborate yields a cationic complex similar to the copper(I) complex in eq. (6.31). Two cationic ruthenium(II) complexes are generated using the same procedure shown in eq. (6.53) [68].

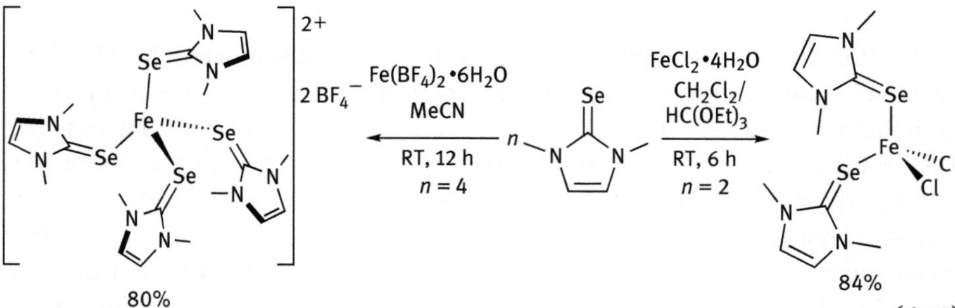

(6.52)

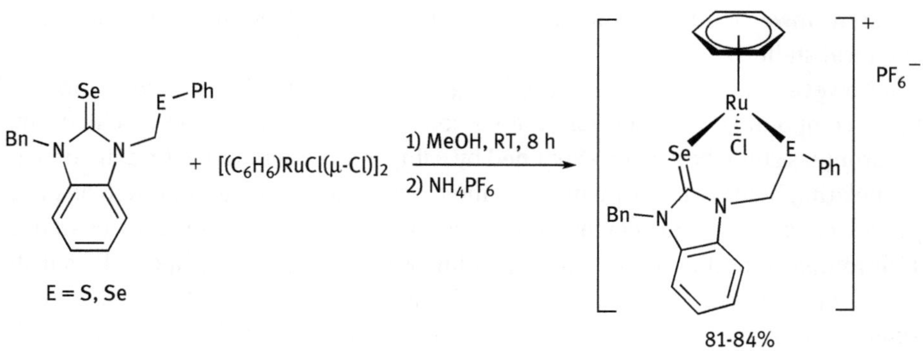

(6.53)

6.6.7 Group 6

The one and only crystal structure of a group 6 complex of a cyclic HCU dates back to the 1980 report of the first well-characterized tellurourea [16]. Treatment of the imidazoline tellurourea with a Cr^0 source afforded the 1:1 adduct of ligand to the $Cr(CO)_5$ fragment (eq. (6.54)). This is also the only structurally characterized cyclic tellurourea complex. It slowly decomposes at 20 °C in toluene solution via detelluration.

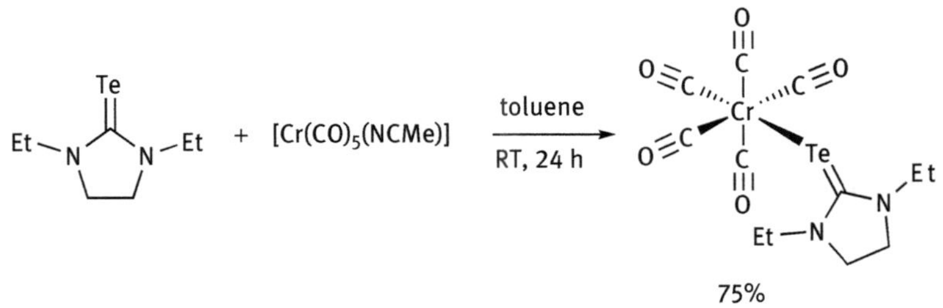

(6.54)

6.7 Conclusions and outlook

Cyclic HCUs featuring selenium and tellurium are accessible by a number of reliable synthetic routes, building upon the well-explored chemistry of NHCs. They can be designed with different heterocyclic cores and organic substituents on nitrogen. The coordination chemistry of HCUs clearly shows a preference for softer metals, particularly late d-block metals. There remain many metals for which HCU complexes are not known. In particular, crystallographic studies of s-block and f-block metal complexes are completely absent at the time of writing. The tellurium derivatives are barely represented in crystallographic studies of coordination chemistry, with only single reported example. This paucity of data is understandable, given the lower solubility of telluroureas compared to selenoureas, and their lower chemical stability.

However, this represents an exciting opportunity for future explorations of these compounds. It is possible the coordination chemistry of both seleno- and telluroureas could be greatly expanded by using multifunctional hybrid ligands (e. g. chelating bidentate or tridentate ligands) featuring strong donors such as phosphines or carbenes, to increase the range of stable metal complexes available. Tellurourea solubility issues could be addressable by installing lipophilic substituents on nitrogen. The lower chemical stability of the C = Te bond may be exploitable in the design of new non-innocent ligands, which may transfer tellurium to chalcophilic metals. For instance, the reaction of 1,3-diisopropylimidazole tellone with a Mn^0 precursor was recently shown to yield insertion products in which $(MnTe)_n$ clusters formed cubane and dicubane motifs, ligated by the remaining NHC fragment [69]. This type of chemistry has the potential to generate other new metal chalcogenide clusters.

Overall, the chemistry of cyclic HCUs remains very active, with many recent studies elaborating new aspects of the synthetic and structural chemistry of these ligands and their resultant metal complexes. Given this pace, the future looks very promising for the discovery of even more new structure types and modes of reactivity.

Funding: Financial support from the Natural Sciences and Engineering Research Council (Canada) and the University of Winnipeg is gratefully acknowledged. Prof. Neil Burford is thanked for hosting JSR's sabbatical visit to the University of Victoria in summer 2017, during which a portion of this chapter was written.

Abbreviations

DBU	1,8-diazabicyclo[5.4.0]undec-7-ene
Mes	2,4,6-trimethylphenyl (mesityl)
Dipp*^{OMe}	2,6-bis(diphenylmethyl)-4-methoxyphenyl
Dipp*	2,6-bis(diphenylmethyl)-4-methylphenyl
Dipp	2,6-diisopropylphenyl
Dipp^{OMe}	2,6-diisopropyl-4-methoxyphenyl
Bn	benzyl
Bmise	bis(*N*-methylimidazole selone)methane
dba	dibenzylideneacetone
HCU	heavy chalcogenourea
MW	microwave
NHC	*N*-heterocyclic carbene
dmise	*N,N′*-dimethylimidazole selone
IPr	*N,N′*-bis(2,6-diisopropylphenyl)imidazol-2-ylidene
IMes	*N,N′*-bis(2,4,6-trimethylphenyl)imidazol-2-ylidene
THF	tetrahydrofuran
OTf	trifluoromethanesulfonate (triflate)

References

[1] García-Rodríguez R, Hendricks MP, Cossairt BM, Liu H, Owen JS. Conversion reactions of cadmium chalcogenide nanocrystal precursors. Chem Mater. 2013;25:1233–49.

[2] Campos MP, Hendricks MP, Beecher AN, Walravens W, Swain RA, Cleveland GT, et al. A library of selenourea precursors to PbSe nanocrystals with size distributions near the homogeneous limit. J Am Chem Soc. 2017;139:2296–305.

[3] Lippolis V. The relevant chemistry of imidazoline-2-selone donors with potential biological applications. In: Jain V.K., Priyadarsini K.I., editor(s). Organoselenium compounds in biology and medicine. London, UK: Royal Society of Chemistry, 2017:122–49.

[4] Stadelman BS, Brumaghim JL. Thione- and selone-containing compounds, their late first row transition metal coordination chemistry, and their biological potential. In: Bayse C.A., Brumaghim J.L., editor(s). Biochalcogen chemistry: The biological chemistry of sulfur, selenium, and tellurium. Washington, DC, USA: American Chemical Society, 2013:33–70.

[5] Hussain RA, Badshah A, Shah A. Synthesis and biological applications of selenoureas. Appl Organomet Chem. 2014;28:61–73.

[6] Roy G, Jayaram PN, Mugesh G. Inhibition of lactoperoxidase-catalyzed oxidation by imidazole-based thiones and selones: A mechanistic study. Chem – Asian J. 2013;8:1910–21.

[7] Kuhn N, Al-Sheikh A. 2, 3-dihydroimidazol-2-ylidenes and their main group element chemistry. Coord Chem Rev. 2005;249:829–57.

[8] Kirmse W. Carbene complexes of nonmetals. Eur J Org Chem. 2005;237–60.
[9] Benhamou L, Chardon E, Lavigne G, Bellemin-Laponnaz S, César V. Synthetic routes to N-heterocyclic carbene precursors. Chem Rev. 2011;111:2705–33.
[10] Arduengo AJ, Harlow RL, Kline M. A stable crystalline carbene. J Am Chem Soc. 1991;113: 361–3.
[11] Kuhn N, Kratz T. Synthesis of imidazol-2-ylidenes by reduction of imidazole-2(3H)-thiones. Synthesis. 1993;561–2.
[12] Kuhn N, Henkel G, Kratz T. Beiträge zur chemie des Imidazols, III. 2-telluroimidazoline – Stabile tellurocarbonyl-verbindungen. Chem Ber. 1993;126:2047–9.
[13] Arduengo AJ, Davidson F, Dias HV, Goerlich JR, Khasnis D, Marshall WJ, et al. An air stable carbene and mixed carbene "dimers". J Am Chem Soc. 1997;119:12742–9.
[14] Kuhn N, Henkel G, Kratz T. 2-selenoimidazoline/2-selenoimidazolines. Z Für Naturforschung B. 1993;48:973–7.
[15] Schönberg A, Singer E, Stephan W. C=C-doppelbindungen mit extremer reaktivität, II. 1,3-diphenyl-2-imidazolidinselenon aus 1,1′,3,3′-tetraphenyl-Δ2,2′-biimidazolidin und selen. Chem Ber. 1983;116:2068–73.
[16] Lappert MF, Martin TR, McLaughlin GM. Telluroureas and derived transition metal complexes: The crystal and molecular structure of $[Cr(CO)_5\{TeCN(Et)CH_2CH_2NEt\}]$. J Chem Soc Chem Commun. 1980;635–7.
[17] Ansell GB, Forkey DM, Moore DW. The molecular structure of 1,3-dimethyl-2(3H)-imidazolethione ($C_5H_8N_2S$). J Chem Soc Chem Commun 1970 56b–7.
[18] Thompson DP, Boudjouk P. A convenient synthesis of alkali metal selenides and diselenides in tetrahydrofuran and the reactivity differences exhibited by these salts toward organic bromides. Effect of ultrasound. J Org Chem. 1988;53:2109–12.
[19] Bigoli F, Pellinghelli MA, Deplano P, Devillanova FA, Lippolis V, Mercuri ML, et al. Reaction of imidazole-2-selone derivatives with diiodine: Synthesis, structural and spectroscopic characterization of the adduct 1,1′-bis(3-methyl-4-imidazoline-2-selone)methane bis(diiodine) and of the first examples of I-Se-I hypervalent selenium compounds: 1,3-dimethyl-4-imidazolin-2-ylium diiodo selenanide and 1,2-bis(3-methyl-4-imidazolin-2-ylium diiodo selenanide)ethane bis(dichloromethane). Gazzetta Chim Ital. 1993;124:445–54.
[20] Bhabak KP, Satheeshkumar K, Jayavelu S, Mugesh G. Inhibition of peroxynitrite- and peroxidase-mediated proteintyrosine nitration by imidazole-based thiourea and selenourea derivatives. Org Biomol Chem. 2011;9:7343–50.
[21] Aroz MT, Gimeno MC, Kulcsar M, Laguna A, Lippolis V. Group 11 complexes with imidazoline-2-thione or selone derivatives. Eur J Inorg Chem. 2011;2884–94.
[22] Ghavale N, Manjare ST, Singh HB, Butcher RJ. Bis(chalcogenones) as pincer ligands: isolation and heck activity of the selone-ligated unsymmetrical C,C,Se–Pd pincer complex. Dalton Trans. 2015;44:11893–900.
[23] Zhou Y, Denk MK. Synthesis and reactivity of subvalent compounds. Part 13: Reaction of triethyl orthoformate with amines and selenium—A convenient one-step three-component synthesis for selenoureas. Tetrahedron Lett. 2003;44:1295–9.
[24] Williams DJ, Fawcett-Brown MR, Raye RR, VanDerveer D, Pang YT, Jones RL, et al. Synthesis, characterization, and X-ray crystallographic structure of 1,3-dimethyl-2(3H)-imidazoleselone. Heteroat Chem. 1993;4:409–14.
[25] Banerjee M, Karri R, Rawat KS, Muthuvel K, Pathak B, Roy G. Chemical detoxification of organomercurials. Angew Chem Int Ed. 2015;54:9323–7.
[26] Roy G, Das D, Mugesh G. Bioinorganic chemistry aspects of the inhibition of thyroid hormone biosynthesis by anti-hyperthyroid drugs. Inorganica Chim Acta. 2007;360:303–16.

[27] Bhabak KP, Mugesh G. Antithyroid drugs and their analogues protect against peroxynitrite-mediated protein tyrosine nitration—A mechanistic study. Chem – Eur J. 2010;16:1175–85.

[28] Arca M, Demartin F, Devillanova FA, Isaia F, Lelj F, Lippolis V, et al. An experimental and theoretical approach to the study of the properties of parabanic acid and related compounds: Synthesis and crystal structure of diethylimidazolidine-2-selone-4,5-dione. Can J Chem. 2000;78:1147–57.

[29] Klayman DL, Shine RJ. Synthesis of selenoureas and selenothiocarbamic esters from thioureas. J Org Chem. 1969;34:3549–51.

[30] Nakayama J, Kitahara T, Sugihara Y, Sakamoto A, Ishii A. Isolable, stable diselenocarboxylate and selenothiocarboxylate salts: Syntheses, structures, and reactivities of 2-(1,3-dimethylimidazolidinio)diselenocarboxylate and 2-(1,3-dimethylimidazolidinio) selenothiocarboxylate. J Am Chem Soc. 2000;122:9120–6.

[31] Segi M, Maeda H, Sakata K, Takashima M, Honda M, Watanabe T. Synthesis of selenohydantoins from isoselenocyanates and α-amino acids. Heterocycles. 2010;82:1709–17.

[32] Manjare ST, Sharma S, Singh HB, Butcher RJ. Facile synthesis of benzimidazolin-2-chalcogenones: Nature of the carbon–chalcogen bond. J Organomet Chem. 2012;717:61–74.

[33] Rong Y, Al-Harbi A, Kriegel B, Parkin G. Structural characterization of 2-imidazolones: Comparison with their heavier chalcogen counterparts. Inorg Chem. 2013;52:7172–82.

[34] Cordero B, Gómez V, Platero-Prats AE, Réves M, Echeverría J, Cremades E, et al. Covalent radii revisited. Dalton Trans. 2008;2832–8.

[35] Aygun M, Çetinkaya E, Gök Y, Kendi E, Çetinkaya B. Synthesis and crystal structure of hexahydrobis[(1,3-p-dimethylaminobenzyl)-1,3-diazepine]-2-selenone, $C_{23}H_{32}N_4Se$. Anal Sci. 2003;19:1093–4.

[36] Vummaleti SV, Nelson DJ, Poater A, Gómez-Suárez A, Cordes DB, Slawin AMZ, et al. What can NMR spectroscopy of selenoureas and phosphinidenes teach us about the π-accepting abilities of N-heterocyclic carbenes?. Chem Sci. 2015;6:1895–904.

[37] Liske A, Verlinden K, Buhl H, Schaper K, Ganter C. Determining the π-acceptor properties of N-heterocyclic carbenes by measuring the ^{77}Se NMR chemical shifts of their selenium adducts. Organometallics. 2013;32:5269–72.

[38] Verlinden K, Buhl H, Frank W, Ganter C. Determining the ligand properties of N-heterocyclic carbenes from ^{77}Se NMR parameters. Eur J Inorg Chem. 2015;2416–25.

[39] Back O, Henry-Ellinger M, Martin CD, Martin D, Bertrand G. ^{31}P NMR chemical shifts of carbene–phosphinidene adducts as an indicator of the π-accepting properties of carbenes. Angew Chem Int Ed. 2016;52:2939–43.

[40] Arca M, Aroz T, Concepción Gimeno M, Kulcsar M, Laguna A, Lasanta T, et al. Homopolynuclear TlI and heteropolynuclear AuI–TlI complexes with organodiselone ligands: Activation of luminescence by intermetallic interactions. Eur J Inorg Chem. 2011;2288–97.

[41] Manjare ST, Yadav S, Singh HB, Butcher RJ. Redox reaction between main-group elements (Te, Sn, Bi) and N-heterocyclic-carbene-derived selenium halides: A facile method for the preparation of monomeric halides. Eur J Inorg Chem. 2013;5344–57.

[42] Srinivas K, Suresh P, Babu CN, Sathyanarayana A, Prabusankar G. Heavier chalcogenone complexes of bismuth(III)trihalides: Potential catalysts for acylative cleavage of cyclic ethers. RSC Adv. 2015;5:15579–90.

[43] Burford N, Phillips AD, Spinney HA, Robertson KN, Cameron TS, McDonald R. Chalcogeno-urea ligands on a phosphadiazonium Lewis acceptor: A new synthetic approach to Ch–P bonds (Ch = O, S, Se). Inorg Chem. 2003;42:4949–54.

[44] Williams DJ, White KM, VanDerveer D, Wilkinson AP. Dichlorobis[1,3-dimethyl-2(3H)-imidazoleselone]zinc(II): A potential zinc selenide synthon. Inorg Chem Commun. 2002; 5:124–6.

[45] Yadav S, Singh HB, Butcher RJ. Synthesis and reactivity of selones and dihaloselones: Complexation of selones with d^8- and d^{10}-metal ions. Eur J Inorg Chem. 2017;2968–79.
[46] Babu CN, Srinivas K, Prabusankar G. Facile access to zinc and cadmium selones: Highly active catalysts for barbier reactions in aqueous media. Dalton Trans. 2016;45:6456–65.
[47] Williams DJ, McKinney BJ, Baker B, Gwaltney KP, VanDerveer D. The preparation, characterization and X-ray structural analysis of tetrakis[1,3-dimethyl-2(3H)-imidazoleselone]cadmium (II) hexafluorophosphate. J Chem Crystallogr. 2007;37:691–4.
[48] Choi J, Park SY, Yang HY, Kim HJ, Ihm K, Nam JH, et al. Submicro-polymer particles bearing imidazoline-2-selenone: Dual mode adsorbents with color-sensing for halogens and mercury ions. Polym Chem. 2011;2:2512–7.
[49] Srinivas K, Babu CN, Prabusankar G. Linear Cu(I) chalcogenones: synthesis and application in borylation of unsymmetrical alkynes. Dalton Trans. 2015;44:15636–44.
[50] Parvin N, Pal S, Khan S, Das S, Pati SK, Roesky HW. Unique approach to copper(I) silylene chalcogenone complexes. Inorg Chem. 2017;56:1706–12.
[51] Kimani MM, Bayse CA, Brumaghim JL. Synthesis, characterization, and DFT studies of thione and selone Cu(I) complexes with variable coordination geometries. Dalton Trans. 2011;40:3711–23.
[52] Kimani MM, Wang HC, Brumaghim JL. Investigating the copper coordination, electrochemistry, and Cu(II) reduction kinetics of biologically relevant selone and thione compounds. Dalton Trans. 2012;41:5248–59.
[53] Kimani MM, Brumaghim JL, VanDerveer D. Probing the antioxidant action of selenium and sulfur using Cu(I)-chalcogenone tris(pyrazolyl)methane and -borate complexes. Inorg Chem. 2010;49:9200–11.
[54] Blake AJ, Lippolis V, Pivetta T, Verani G. (1,3-Dimethylimidazolidine-2-selone-κSe)bis(1,10-phenanthroline-κ^2N,N′)copper(II) bis(perchlorate) and bis(2,2′-bipyridyl-κ^2N,N′)(imidazolidine-^{2}-thione-κS)copper(II) bis(perchlorate). Acta Crystallogr C. 2007;63:m364–7.
[55] Kimani MM, Watts D, Graham LA, Rabinovich D, Yap GP, Brumaghim JL. Dinuclear copper(I) complexes with N-heterocyclic thione and selone ligands: synthesis, characterization, and electrochemical studies. Dalton Trans. 2015;44:16313–24.
[56] Nelson DJ, Nahra F, Patrick SR, Cordes DB, Slawin AM, Nolan SP. Exploring the coordination of cyclic selenoureas to gold(I). Organometallics. 2014;33:3640–5.
[57] Perras JH, Mezibroski SM, Wiebe MA, Ritch JS. Diverse silver(I) coordination chemistry with cyclic selenourea ligands. Dalton Trans. 2018;47:1471–8.
[58] Nahra F, Patrick SR, Bello D, Brill M, Obled A, Cordes DB, et al. Hydrofluorination of alkynes catalysed by gold bifluorides. ChemCatChem. 2015;7:240–4.
[59] Zhang H-N, Jia W-G, Xu Q-T, Ji C-C. Synthesis, characterization and catalytic activity of gold complexes with pyridine-based selone ligands. Inorg Chim Acta. 2016;450: 315–20.
[60] Jia W-G, Huang Y-B, Lin Y-J, Wang G-L, Jin G-X. Nickel complexes and cobalt coordination polymers with organochalcogen (S, Se) ligands bearing an N-methylimidazole moiety: Syntheses, structures, and properties. Eur J Inorg Chem. 2008;4063–73.
[61] Blake AJ, Casabò J, Devillanova FA, Esriche L, Garau A, Isaia F, et al. Mixed aza–thia crowns containing the 1,10-phenanthroline sub-unit. Substitution reactions in $[NiL(MeCN)][BF_4]_2$ {L = 2,5,8-trithia[9](2,9)-1,10-phenanthrolinophane}. J Chem Soc Dalton Trans. 1999; 1085–92.
[62] Wang C, Tong Y, Huang Y, Zhang H, Yang Y. Selone behavior towards palladium(II) extraction with hydrophobic ionic liquids and mechanism studies. RSC Adv. 2015;5:63087–94.
[63] Rani V, Singh HB, Butcher RJ. Bis(selone) complexes of palladium(II), platinum(II), and gold(III): Synthesis and structural studies. Eur J Inorg Chem. 2017;3720–8.

[64] Williams DJ, Jones TA, Rice ED, Davis KJ, Ritchie JA, Pennington WT, et al. Dichlorobis[1,3-dimethylimidazole-2(3H)selone-Se]cobalt(II). Acta Crystallogr C. 1997;53:837–8.
[65] Jia W-G, Huang Y-B, Lin Y-J, Jin G-X. Syntheses and structures of half-sandwich iridium(III) and rhodium(III) complexes with organochalcogen (S, Se) ligands bearing N-methylimidazole and their use as catalysts for norbornene polymerization. Dalton Trans. 2008;5612–20.
[66] Sharma AK, Joshi H, Bhaskar R, Singh AK. Complexes of (η^5-Cp*)Ir(III) with 1-benzyl-3-phenylthio/selenomethyl-1,3-dihydrobenzoimidazole-2-thione/selenone: Catalyst for oxidation and 1,2-substituted benzimidazole synthesis. Dalton Trans. 2017;46:2228–37.
[67] Stadelman BS, Kimani MM, Bayse CA, McMillen CD, Brumaghim JL. Synthesis, characterization, DFT calculations, and electrochemical comparison of novel iron(II) complexes with thione and selone ligands. Dalton Trans. 2016;45:4697–711.
[68] Sharma AK, Joshi H, Sharma KN, Gupta PL, Singh AK. 2-Propanol vs glycerol as hydrogen source in catalytic activation of transfer hydrogenation with (η^6-benzene)ruthenium(II) complexes of unsymmetrical bidentate chalcogen ligands. Organometallics. 2014;33:3629–39.
[69] Choi B, Paley DW, Siegrist T, Steigerwald ML, Roy X. Ligand control of manganese telluride molecular cluster core nuclearity. Inorg Chem. 2015;54:8348–55.
[70] Engl PS, Santiago CB, Gordon CP, Liao W-C, Fedorov A, Copéret C, et al. Exploiting and understanding the selectivity of Ru-N-heterocyclic carbene metathesis catalysts for the ethenolysis of cyclic olefins to α,ω-dienes. J Am Chem Soc. 2017;139:13117–25.
[71] Wonner P, Vogel L, Düser M, Gomes L, Kniep F, Mallick B, et al. Carbon–halogen bond activation by selenium-based chalcogen bonding. Angew Chem Int Ed. 2017;56:12009–12.
[72] Rodríguez H, Gurau G, Holbrey JD, Rogers RD. Reaction of elemental chalcogens with imidazolium acetates to yield imidazole-2-chalcogenones: direct evidence for ionic liquids as protocarbenes. Chem Commun. 2011;47:3222–4.
[73] Kuhn N, Fawzi R, Kratz T, Steimann M, Henkel G. Heterocyclen Als liganden Xxii. 2-chalkogenoimidazoline Als liganden in pentacarbonylmetall-komplexen. Phosphorus Sulfur Silicon Relat Elem. 1996;108:107–19.
[74] Liu L, Zhu D, Cao LL, Stephan DW. N-heterocyclic carbene stabilized parent sulfenyl, selenenyl, and tellurenyl cations (XH^+, X = S, Se, Te). Dalton Trans. 2017;46:3095–9.

Fateh V. Singh and Thomas Wirth

7 Selenium and Tellurium Electrophiles in Organic Synthesis

Abstract: This chapter highlights the utility of electrophilic achiral and chiral organoselenium reagents in organic synthesis. A range of reactions from alkene functionalizations, the functionalization of aliphatic and aromatic C–H bonds using stoichiometric and catalytic approaches as well as rearrangement reactions are described. In addition, the utility of organotellurium reagents in organic synthesis is covered in this chapter.

Keywords: selenium electrophiles, selenenylation, selenium-catalyzed reactions, chiral selenium electrophiles, tellurium electrophiles

7.1 Introduction

The chemistry of organoselenium compounds has been identified as an important portfolio in synthetic and medicinal chemistry [1]. In the beginning, the development of organoselenium chemistry was quite limited but several synthetically important organoselenium reagents have been discovered after the selenoxide elimination in the early 1970s [2]. In past few decades, various synthetic transformations including selenenylations, selenocyclizations, selenoxide eliminations and 2,3-sigmatropic rearrangements have been successfully achieved using organoselenium reagents [2, 3]. Furthermore, the application of these reagents as catalysts and ligands provides an additional asset to the organic chemists [4]. The chemistry of organoselenium reagents has now become a well-established research area and books [5], book chapters [6] and review articles [7] have appeared to describe their chemical potential. This chapter highlights the utility of electrophilic achiral and chiral organoselenium reagents in organic synthesis. In addition, the utility of organotellurium reagents in organic synthesis is also covered in this chapter.

7.2 Selenium electrophiles

Organoselenium reagents are mainly known for their electrophilic nature but there are several reports in which these reagents have been used as nucleophile. Variety of selenium electrophiles can be generated by the cleavage of Se–Se bond of diselenides.

This article has previously been published in the journal *Physical Sciences Reviews*. Please cite as: Singh, F.V., Wirth, T. Selenium and Tellurium Electrophiles in Organic Synthesis. *Physical Sciences Reviews* [Online] **2018** DOI: 10.1515/psr-2017-0131

https://doi.org/10.1515/9783110529340-007

Selenium electrophiles are quite powerful reagents and react with olefins to form three-membered seleniranium ion intermediate. Furthermore, seleniranium ion intermediate reacts with different nucleophiles to undergo various selenenylation reactions. The scope of these selenium electrophiles is not limited to selenylation of alkenes but selenation of other aliphatic and aromatic species has been successfully achieved using these electrophiles.

7.2.1 Selenenylation reactions

7.2.1.1 Selenenylation of Alkenes

The addition of selenium electrophiles to alkenes is one of the oldest reactions in the area of organoselenium chemistry. Initially, commercially available PhSeCl was successfully used to achieve the selenenylation of alkenes but this reaction was suffered with some side reactions due to nucleophilic chlorine anion [8]. In 1989, Tiecco and coworkers developed methoxyselenenylation of various acyclic and cyclic alkenes in good yields [9]. In this report, active selenium electrophilic species was generated by the oxidation of diphenyl diselenide using ammonium peroxydisulfate as an oxidant. Furthermore, *N*-phenyl selenophthalimide (*N*-PSP) and *N*-phenyl selenosuccinimide were successfully used as selenenylating agents in oxyselenation of alkenes [10, 11]. In 1991, Yoshshida and coworkers developed a new selenenylating agent benzeneselenenyl *m*-nitrobenzenesulfonate **3** by oxidation of diphenyl diselenide **2** using benzeneselenenyl *m*-nitrobenzenesulfonyl peroxide **1** as oxidant (Figure 7.1) [12].

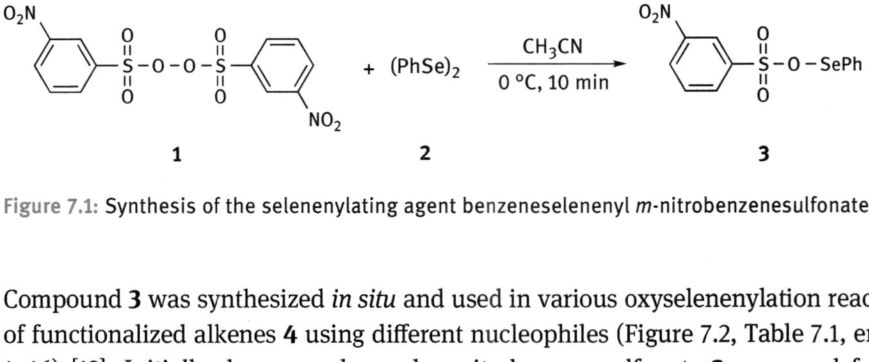

Figure 7.1: Synthesis of the selenenylating agent benzeneselenenyl *m*-nitrobenzenesulfonate **3**.

Compound **3** was synthesized *in situ* and used in various oxyselenenylation reactions of functionalized alkenes **4** using different nucleophiles (Figure 7.2, Table 7.1, entries 1–16) [12]. Initially, benzeneselenenyl *m*-nitrobenzenesulfonate **3** was used for the methoxyselenenylation of alkenes **4** using methanol as nucleophile (Figure 7.2, Table 7.1, entries 1–3). All reactions were performed in acetonitrile and selenenylated products **6** were isolated in excellent yields.

After achieving the methoxyselenenylations, acetoxyselenenylation of similar alkenes was achieved using organoselenium reagent **3** and AcOH as a sources of acetoxy ions (Figure 7.2, Table 7.1, entries 4–7) [12]. Initially, acetoxyselenylation was performed in acetonitrile but the reaction did not proceed well (Figure 7.2, Table 7.1,

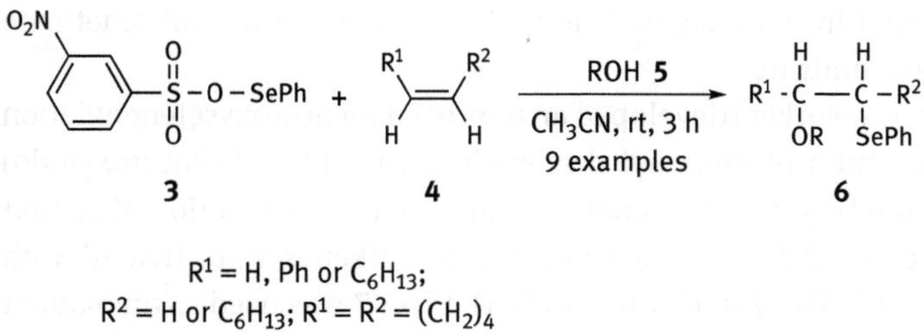

Figure 7.2: Oxyselenenylation of alkenes **4** using benzeneselenenyl *m*-nitrobenzenesulfonate **3**.

Table 7.1: Oxyselenenylation of alkenes **4** using benzeneselenenyl *m*-nitrobenzenesulfonate **3**.

Entry	4		5	Solvent	6
	R^1	R^2	R		Yied (%)
1	Ph	H	Me	CH_3CN	98
2	Hex	H	Me	CH_3CN	83
3	R^1=R^2=$(CH_2)_4$		Me	CH_3CN	92
4	R^1=R^2=$(CH_2)_4$		Ac	CH_3CN	25
5	R^1=R^2=$(CH_2)_4$		Ac	CH_3NO_2	70
6	Ph	H	Ac	CH_3NO_2	11
7	Hex	H	Ac	CH_3NO_2	73
8	Ph	H	H	CH_3NO_2	61
9	Ph	H	H	CH_3NO_2	77
10	R^1=R^2=$(CH_2)_4$		H	CH_3CN	33
11	R^1=R^2=$(CH_2)_4$		H	CH_3NO_2	72
12	Hex	H	H	CH_3NO_2	–
13	Hex	H	Ph	CH_3CN	25
14	Hex	H	Ph	CH_3NO_2	62
15	R^1=R^2=$(CH_2)_4$		Ph	CH_3CN	38
16	R^1=R^2=$(CH_2)_4$		Ph	CH_3NO_2	64

entries 4). The reaction products **6** were obtained in moderate yields when reaction was performed in nitromethane (Figure 7.2, Table 7.1, entries 5 and 7). Poor yields were observed when styrene was treated with electrophile **3** under similar reaction conditions (Figure 7.2, Table 7.1, entry 6).

The same reagent **3** was successfully used for the hydroxyselenenylation of similar alkenes (Figure 7.2, Table 7.1, entries 8–11) [12]. Hydroxyselenenylations were working in both acetonitrile and nitromethane but nitromethane was found to be a more effective solvent than acetonitrile. Interestingly, hydroxyselenenylation did not proceed when 1-octene was used as substrate (Figure 7.2, Table 7.1, entry 12). Finally, phenoxyselenenylation of 1-octene and cyclohexene was also achieved successfully using selenium electrophile **3** (Figure 7.2, Table 7.1, entries 13–16). The reactions were performed in both acetonitrile and nitromethane but nitromethane was well suited over acetonitrile.

In 1998, Tingoli and coworkers developed an approach for acetoxyselenenylation of alkenes **8** by the reaction of diphenyl diselenide **2** with PIDA [(diacetoxyiodo) benzene] **7** in acetonitrile (Figure 7.3). In addition, hydroxyselenenylation of similar alkenes **8** was also achieved in moderate yields when alkenes were treated with diphenyl diselenide **2** and PIDA **7** in MeCN/H_2O (5:1). PIDA **7** was used as an oxidant to oxidize diphenyl diselenide **2** to more electrophilic selenium species [13].

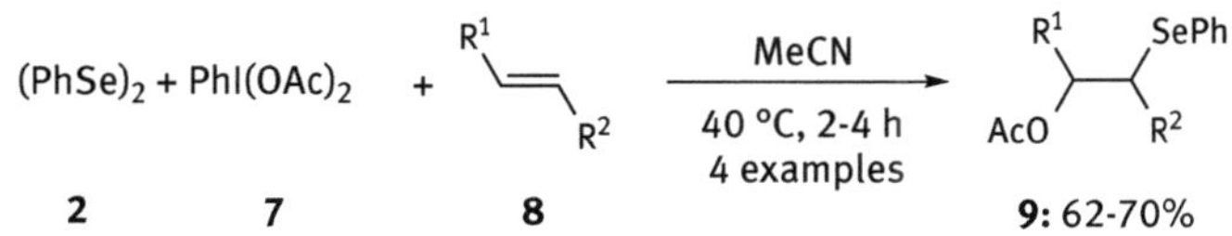

R^1 = Ph, C_6H_{13}, C_3H_7;
R^2 = H, C_3H_7; R^1 = R^2 = cyclohexene

Figure 7.3: Acetoxyselenenylation of alkenes **8** by the reaction with diphenyl diselenide **2** and PIDA **7**.

In 2006, Tingoli and coworkers synthesized a new selenium electrophile *N*-phenyl selenosaccharin (NPSSac) **12** by the reaction of commercially available phenyl selenium halide **11** and silver saccharin (AgSac) **10** in dichloromethane at room temperature (Figure 7.4). The synthesized selenium reagent **12** was further used for methoxyselenenylation of olefins **13**. All the reactions were performed in dichloromethane at room temperature and reaction products **14** were isolated in moderate-to-high yields (Figure 7.4) [14].

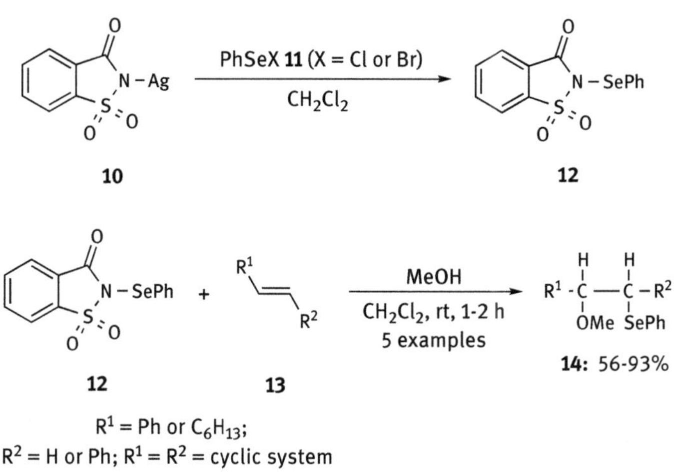

R^1 = Ph or C_6H_{13};
R^2 = H or Ph; R^1 = R^2 = cyclic system

Figure 7.4: Methoxyselenenylation of alkenes **13** using selenium electrophile *N*-phenyl selenosaccharin (NPSSac) **12**.

In 2012, Thomas and coworkers reported an iodine catalyzed addition of functionalized styrenes **15** using *in situ*-generated electrophilic species PhSeI. In this reaction, 20 mol% of iodine was used as catalyst for the addition of styrenes **15** with diphenyl diselenide **2** in dichloromethane at 70°C and excess of styrene was used as nucleophile (Figure 7.5) [15].

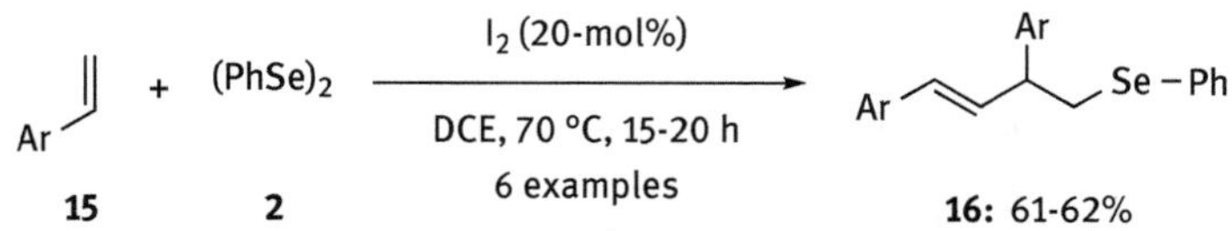

Ar = Ph, 2-MeC_6H_4, 3-MeC_6H_4,
4-MeC_6H_4, 4-PhC_6H_4, 2-Nap

Figure 7.5: Iodine-catalyzed addition of styrenes **15** using *in situ*-generated electrophilic species PhSeI.

The catalytic cycle for iodine-mediated addition of styrenes is described in Figure 7.6. According to that, the catalytic cycle was initiated by the reaction of diphenyl diselenide **2** with iodine to forms active catalytic species phenyl selenium iodide **17** which further reacts with styrene **15** and form three-membered seleniranium ion intermediate **18**. Furthermore, intermediate **18** could be attacked by the styrene nucleophile to form addition product **19**. After that, the reaction intermediate **19** undergoes elimination reaction and HI with final product **16**. Finally, HI could further react with diphenyl diselenide to phenyl selenyl iodide **17** to continue the catalytic cycle.

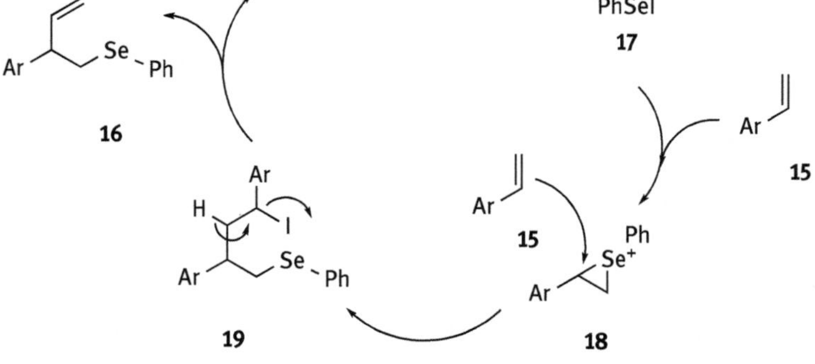

Figure 7.6: Catalytic cycle for iodine-mediated addition of styrenes **15** using *in situ*-generated electrophilic species PhSeI **17**.

In 2015, Yan and coworkers developed an approach for the acetoxyselenenylation of alkenes **15** by the reaction with diorganyl diselenide **20** and catalytic amount of KBr using *m*-CPBA as an oxidant. All the reactions were performed in AcOH and reaction products **21** were isolated in good-to-high yields (Figure 7.7) [16]. Various aliphatic and aromatic olefins were successfully tolerated under mild reaction conditions. In addition, NaCl was also used as catalyst under similar reaction conditions.

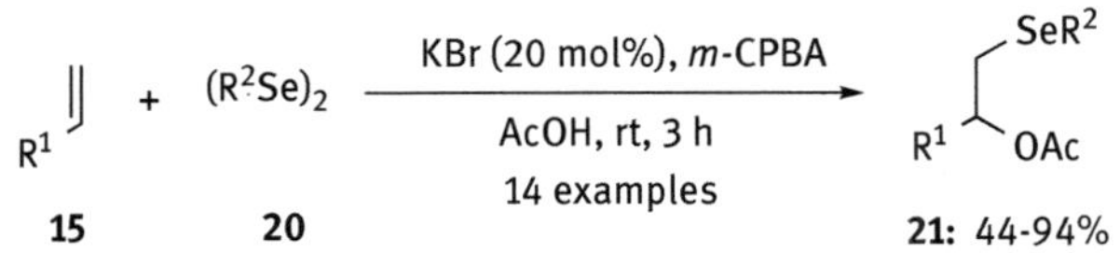

R^1 = *n*-Bu, $Me_2(COH)$, Ph, 4-MeC_6H_4, 4-BrC_6H_4, 4-ClC_6H_4, 4-FC_6H_4, 4-$AcOC_6H_4$, 4-*t*-BuC_6H_4, Py; R^2 = Ph or Bn

Figure 7.7: KBr-catalyzed acetoxyselenylation of alkenes **15** by the reaction of diorganyl diselenides **20** with *m*-CPBA as catalyst.

The catalytic cycle for KBr-catalyzed acetoxyselenylation of alkenes **15** is shown in Figure 7.8. The catalytic cycle initiates with oxidation of bromide ion to bromine using *m*-CPBA as an oxidant. Bromine was further reacted with diselenide **20** to form more electrophilic species phenyl selenium bromide **22.** Electrophilic selenium species **22** activates the double bond of styrene **15** to the three-membered seleniranium ion intermediate **23** along with the formation of bromide ion. Finally, intermediate **23** reacts with acetic acid to yield the final product **21** while the bromide ion was further oxidized to continue the catalytic cycle.

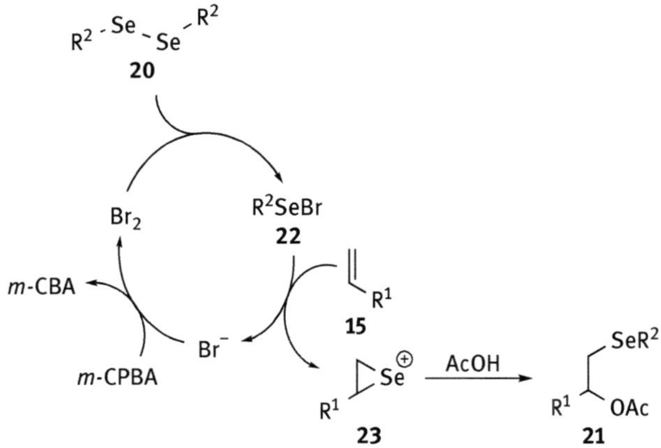

Figure 7.8: The catalytic cycle for KBr-catalyzed acetoxyselenylation of alkenes **15**.

In 2015, Braga and coworkers developed an iodine-catalyzed microwave-assisted methoxyselenenylation of alkenes **15** by the reaction of diorganyl diselenide **20**, catalytic amount of iodine in DMSO using methanol as source of nucleophile. All the reactions were completed in very short reaction time while reaction products **24** were isolated in poor-to-excellent yields (Figure 7.9) [17]. It was observed that methoxyselenenylated products **24** were obtained in high yields when diorganyl diselenides having electron-donating substituents were used. Various styrenes were successfully employed as substrates but the methoxyselenenylation of aliphatic alkenes was not successful. Molecular iodine reacts with diphenyl diselenide to form more electrophilic species PhSeI **17** which was used to activate the double bond of styrenes **15**. In addition, various other oxyselenenylations of alkenes were successfully achieved under similar reaction and conditions.

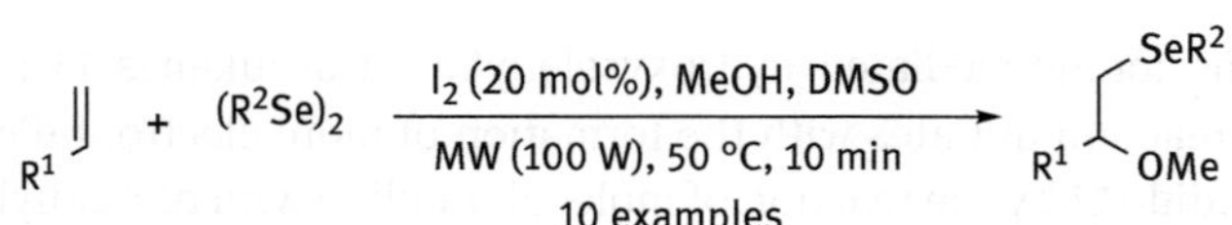

15 **20** **24:** 21-96%

R^1 = Ph, 4-MeC_6H_4, 4-ClC_6H_4, 4-(Me)$_3$COC_6H_4, R^2 = *n*-Bu, Ph, 4-MeC_6H_4, 4-OMeC_6H_4, 4-ClC_6H_4, 3-$CF_3C_6H_4$

Figure 7.9: Iodine-catalyzed microwave-assisted methoxyselenenylation of alkenes **15**.

Recently, Yan and coworkers developed an iodine-mediated hydroxyselenenylation of alkenes **15** with the reaction of diorganyl diselenide **20** in the presence of molecular iodine in solvent mixture of EtOH and H_2O (1:1) under oxygen atmosphere. β-Hydroxyselenides **25** were isolated in good-to-excellent yields (Figure 7.10) [18]. Various electron withdrawing and donating functionalities at aromatic ring in styrenes were successfully tolerated under mild reaction conditions. The same approach was successfully applied for the hydroxyselenenylation of aliphatic and cyclic alkenes.

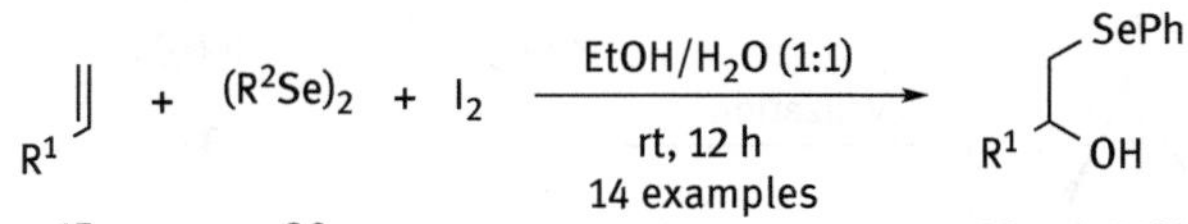

15 **20** **25:** 70-92%

R^1 = *n*-Bu, Ph, 4-MeC_6H_4, 4-BrC_6H_4, 4-ClC_6H_4, 4-$NO_2C_6H_4$, 4-AcOC_6H_4, 4-*n*-BuC_6H_4, 4-*t*-BuC_6H_4; R^2 = Ph or Bn

Figure 7.10: Iodine-mediated hydroxyselenenylation of styrenes **15** with diorganyl diselenide **20** in EtOH/H_2O (1:1).

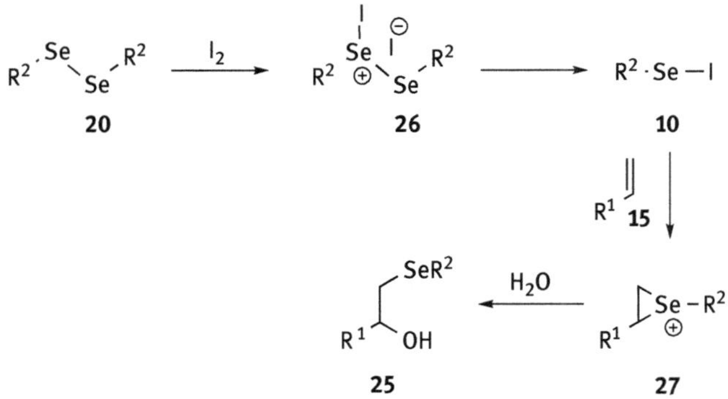

Figure 7.11: Mechanism for the iodine-mediated hydroxyselenenylation of styrenes **15** with diorganyl diselenide **20** in EtOH/H_2O (1:1).

The mechanism for the iodine-mediated hydroxyselenylation of alkenes **15** is shown in Figure 7.11. The reaction initiates with the formation of more electrophilic species phenyl selenium iodide **17** by the reaction of molecular iodine with diorganyl diselenide **20**. Electrophilic selenium species **22** activates the double bond of styrene **15** to form three-membered selenarium ion intermediate **27** along with the formation of iodide. Finally, intermediate **27** reacts with water to yield the product **25**.

7.2.1.2 Selenocyclizations

Selenocyclization is an approach for the synthesis of various biologically active heterocyclic compounds (3f, 3g) [19]. The selenocyclization process is a quite similar process to oxyselenenylations of alkenes as shown in Figure 7.12. Initially, the selenium electrophile **29** activates the double bond of alkene **28** to form seleniranium intermediate **30**. Furthermore, seleniranium intermediate is then opened by an intramolecular nucleophilic attack of the internal nucleophile resulting in an *anti*-addition of the selenium moiety to yield cyclization product. Depending on ring size

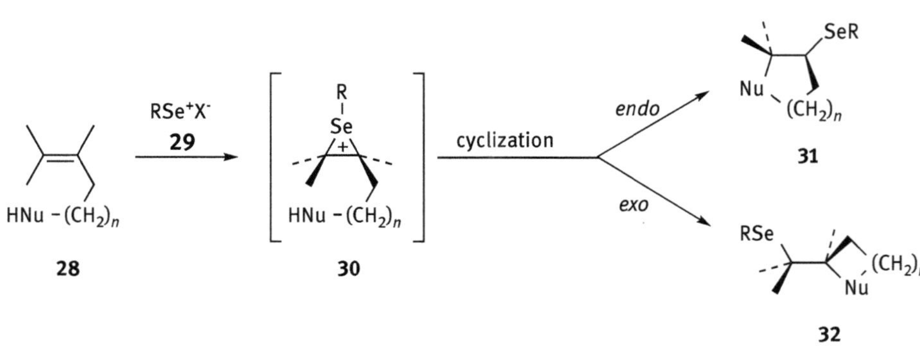

Figure 7.12: Selenocyclization of alkenes **28** to *endo*- and *exo*-cyclization products **31** and **32** using selenium electrophiles **29**.

and reaction conditions, the seleniranium intermediate **30** can undergo either *endo*-cyclization to product **31** or *exo*-cyclizations yielding other heterocyclic derivatives such as compounds of type **32** (Figure 7.12). The selenium moiety in the products **31** and **32** can be used to develop selenium-catalyzed cyclization reactions.

Various selenium electrophiles have been successfully used to achieve the selenocyclization of alkenes having internal nucleophile. Initially, different benzene selenenyl sulfates were used as electrophiles to achieve the selenocyclization of unsaturated alcohols and unsaturated carboxylic acids under mild reaction conditions [20–22]. In 1991, benzeneselenenyl *m*-nitrobenzenesulfonate **3** was synthesized *in situ* and used in the selenocyclization of unsaturated alcohols **33** and **34** in acetonitrile at 0°C (Figure 7.13) [12]. Both reactions were completed in 1 h and *exo*-cyclic products **35** and **36** were isolated in excellent yields (Figure 7.13). Interestingly, *endo*-cyclic products were also observed during the selenocyclization of unsaturated alcohol **33** (n = 1) while *exo*-cyclic product **35** formed at –40°C.

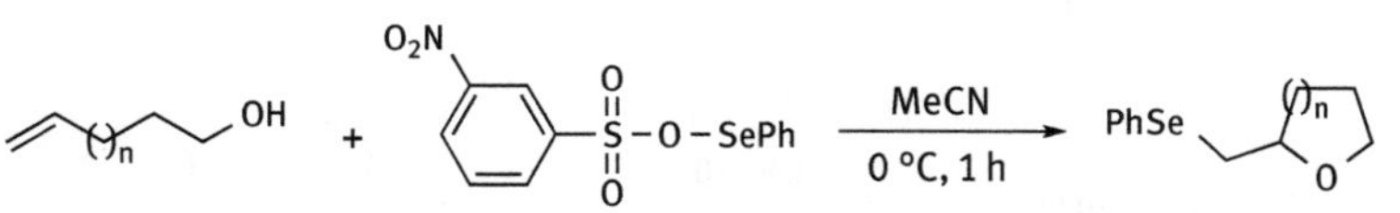

33: n = 1
34: n = 2

3

35: n = 1; 85%
36: n = 2; 100%

Figure 7.13: Selenocyclization of unsaturated alcohols **33** and **34** to *O*-heterocyclic compounds **35** and **36** using selenium electrophiles **3**.

Selenium electrophile **3** was also used for the selenocyclization of unsaturated carboxylic acids **37** and **38** in acetonitrile at 0°C (Figure 7.14) [12]. *exo*-Cyclizations were observed and selenolactones **39** and **40** were isolated in excellent yields (Figure 7.14). The *endo*-cyclic product was not observed during these selenocyclizations.

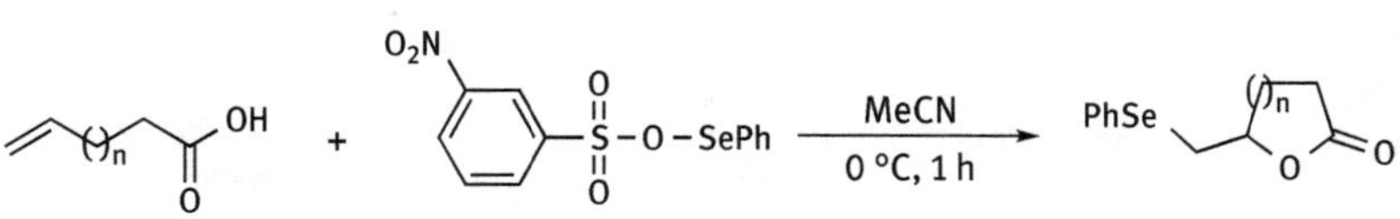

37: n = 1
38: n = 2

3

39: n = 1; 93%
40: n = 2; 96%

Figure 7.14: Selenocyclization of unsaturated carboxylic acids **37** and **38** to selenolactones **39** and **40** using selenium electrophile **3**.

In 1998, Tingoli and coworkers developed an approach for the selenocyclization of unsaturated alkenes **33** and **34** by the reaction of diphenyl diselenide **2** with PIDA **7** in acetonitrile (Figure 7.15). The cyclic products **35** and **36** were isolated in good yields.

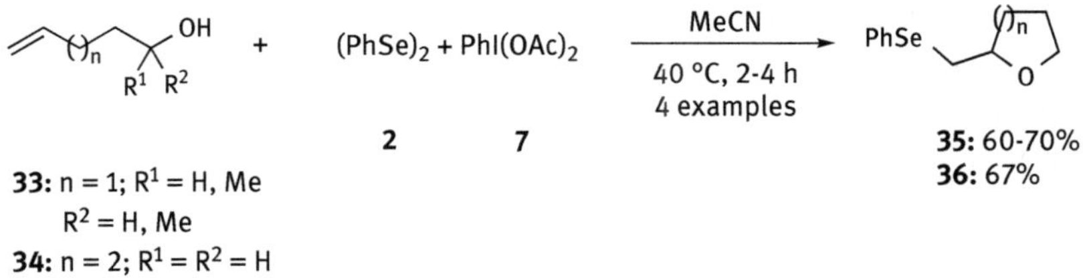

Figure 7.15: Iodine(III)-mediated selenocyclization of unsaturated alcohols **33** and **34** to selenoethers **35** and **36** by the reaction of diphenyl diselenide **2** with PIDA **7** as an oxidant.

The electrophilic selenium species was generated *in situ* by the oxidation of diphenyl diselenide **2** using PIDA **7** as an oxidant [13].

In addition, similar reaction conditions were used for selenocyclization of unsaturated carboxylic acids **37** and **41** in acetonitrile at 40°C (Figure 7.16) [13]. Both cyclic and acyclic unsaturated carboxylic acids were successfully used as substrates but better yields were observed with acyclic substrates **37**.

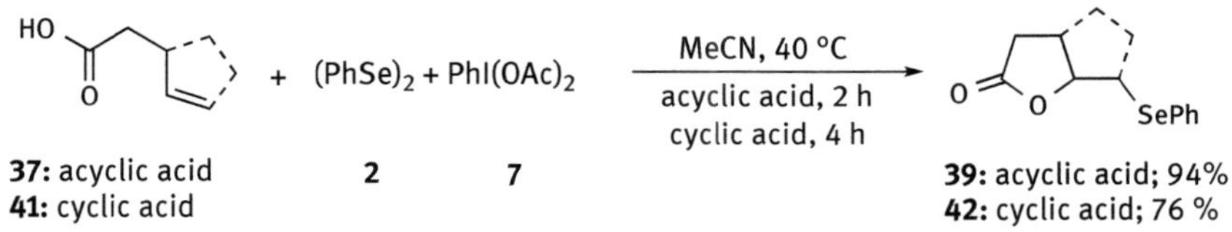

Figure 7.16: Iodine(III)-mediated selenocyclization of unsaturated carboxylic acids **37** and **41** to selenolactones **39** and **42** by the reaction of diphenyl diselenide **2** with PIDA **7** as an oxidant.

Functionalized ketones **43** and **44** were cyclized to corresponding 2,3-dihydrofurans **45** and **46** under similar reaction conditions. Both reactions proceeded well and 2,3-dihydrofurans **45** and **46** were isolated in moderate yields (Figure 7.17) [13].

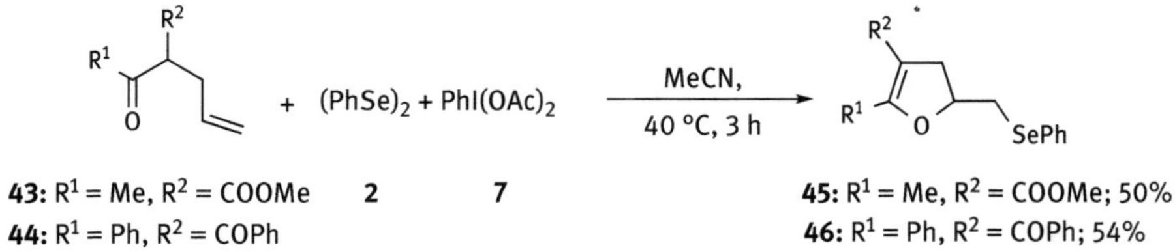

Figure 7.17: Iodine(III)-mediated selenocyclization of ketones **43** and **44** to 2,3-dihydrofurans **45** and **46**, respectively, by the reaction of diphenyl diselenide **2** with PIDA **7** as an oxidant.

Finally, substituted benzamide **47** was cyclized to the corresponding 4,5-dihydrooxazole **48** under similar reaction conditions. The product was isolated in 65 % yield (Figure 7.18) [13].

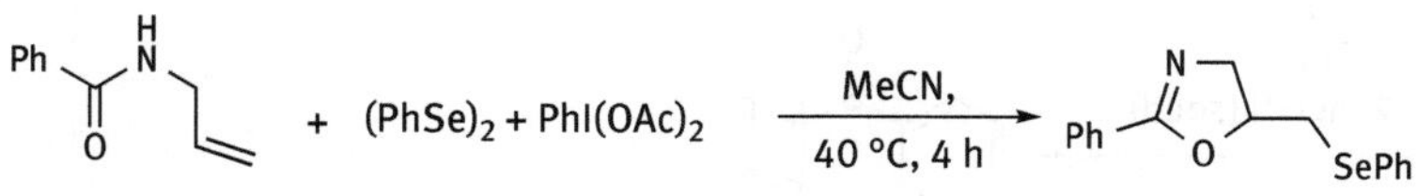

47 **2** **7** **48:** 65%

Figure 7.18: Iodine(III)-mediated selenocyclization of amide **47** to 4,5-dihydrooxazole derivative **48** by the reaction of diphenyl diselenide **2** with PIDA **7** as an oxidant.

In 2006, Tingoli and coworkers used another selenium electrophile NPSSac **12** for selenocyclization of unsaturated alcohol **33.** The cyclization reaction was performed in dichloromethane at room temperature for 1 h and cyclic ether **35** was isolated in 92 % yield (Figure 7.19) [14].

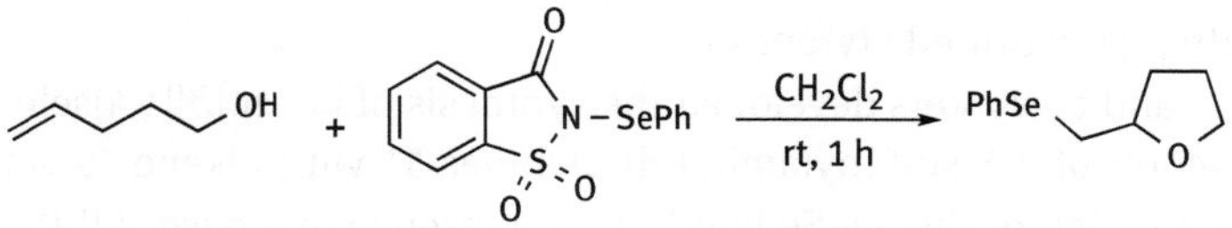

33 **12** **35:** 92%

Figure 7.19: Selenocyclization of unsaturated alcohol **33** to cyclic ether **35** using *N*-phenylseleno-saccharin (NPSSac) **12** as an electrophile.

In addition, a similar approach was used to cyclize the unsaturated carboxylic acid **37** to selenolactone **39** under similar reaction conditions. The selenolactone **39** was isolated in 89 % yield (Figure 7.20) [14].

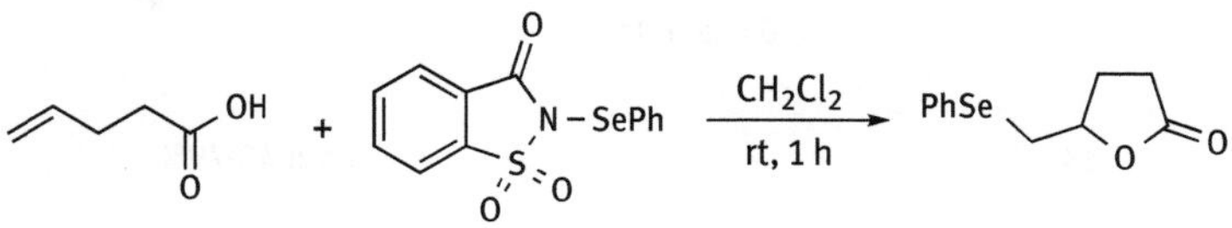

37 **12** **39:** 89%

Figure 7.20: Selenocyclization of unsaturated carboxylic acid **37** to corresponding selenolactone **39** using *N*-phenyl selenosaccharin (NPSSac) **12** as an electrophile.

In 2010, Wirth and coworkers developed the intermolecular selenocyclization of alkene **15** and *in situ* aryl selenium triflate by the reaction of diselenide **49** and bromine followed by the addition of silver triflate in THF at 0°C (Figure 7.21) [23].

This reaction was performed without using any external nucleophile and the formation of cyclic product occurs by the activation of double bond to form selenarium ion followed by the cyclization involving sulfoxide moiety of the *in situ*-generated electrophile. The cyclic products **50** were only formed in poor yields

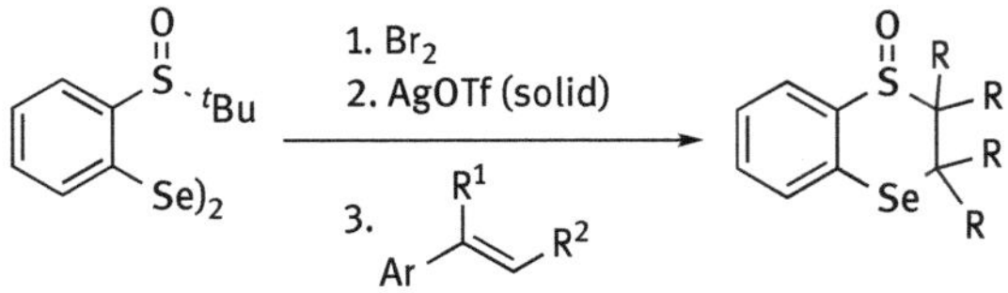

49 15 50: 9-16%

THF, °C, 2h
4 examples

R^1 = Ph, 2-ClC_6H_4, 2-Napthyl;
R^2 = H, Me, COOMe; R^2 = H, Me

Figure 7.21: Intermolecular selenocyclization of alkenes **15** and electrophilic species aryl selenium triflate.

(Figure 7.21). Notably, the cyclization reaction was proceeding only with α-substituted styrenes not with β-substituted styrenes [23].

In 2012, Menichetti and coworkers developed the synthesis of benzo[*b*][1,4]selenazines **52** by the reaction of 2-*N*-sulfonylamino diselenides **51** with alkene **15** *via* copper(II)-catalyzed activation of the Se–Se bond in the presence of a base. All the reactions were performed in chloroform and benzo[*b*][1,4]selenazines **52** were isolated moderate-to-good yields (Figure 7.22) [24]. The course of reaction was quite slow and most of reactions were completed in 2–3 days.

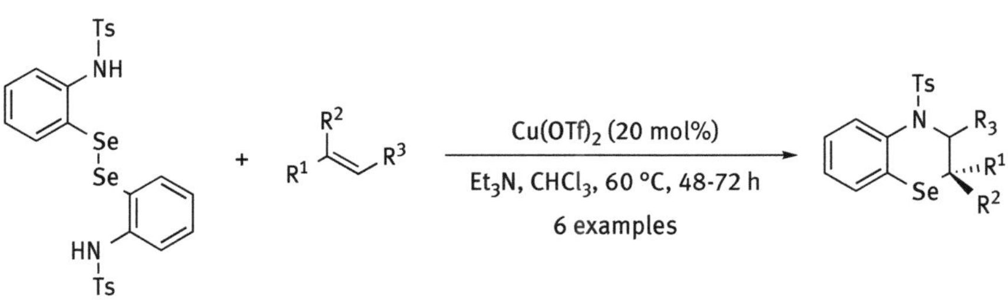

51 15 52: 43-79%

R^1 = H, Me; R^2 = H, Me; R^3 = 4-OMeC_6H_4,
4-$NH_2C_6H_4$, OPh, *N*-2-pyrrolidone etc.

Figure 7.22: Cu(II)-catalyzed synthesis of benzo[*b*][1,4]selenazines **52** by the reaction of 2-*N*-sulfonylamino diselenides **51** with alkene **15**.

Additionally, similar cyclization reactions were performed with cyclic alkenes **53** and **54** under similar reaction conditions. The cyclic reaction was working well and Se-containing tricyclic compound **55** and **56** were isolated in 60 % and 59 % yield, respectively (Figure 7.23) [24].

Furthermore, Viglianisi and coworkers synthesized various types of 2-*N*-sulfonylamino diselenides **57** having different electron-withdrawing and donating

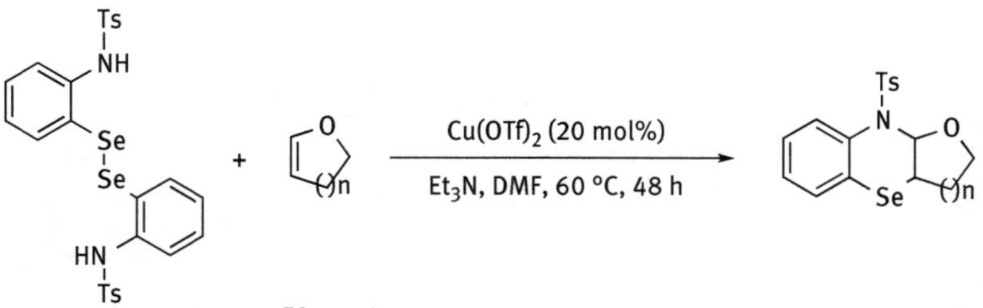

51 **53:** n = 1; **54:** n = 2 → **55:** n = 1; 60%; **56:** n = 2; 59%

Figure 7.23: Cu(II)-catalyzed synthesis of Se-containing tricyclic compounds **55** and **56** by the reaction of 2-*N*-sulfonylamino diselenides **51** with cyclic alkenes **53** and **54**.

functionalities. The synthesized diselenides **57** were treated with 4-methoxystyrene using $Cu(OTf)_2$ as catalyst and triethyl amine as base in DMF. The reactions were completed in 1–10 days and benzo[*b*][1,4]selenazines **58** were isolated good-to-high yields (Figure 7.24) [25]. It was noted that the course of reaction was slow when diselenides with electron-withdrawing group was used in the aromatic ring in substrates. Recently, Supuran and coworkers reported the synthesis of benzo[*b*][1,4]selenazines of type **58** and identified as carbonic anhydrase inhibitors [26].

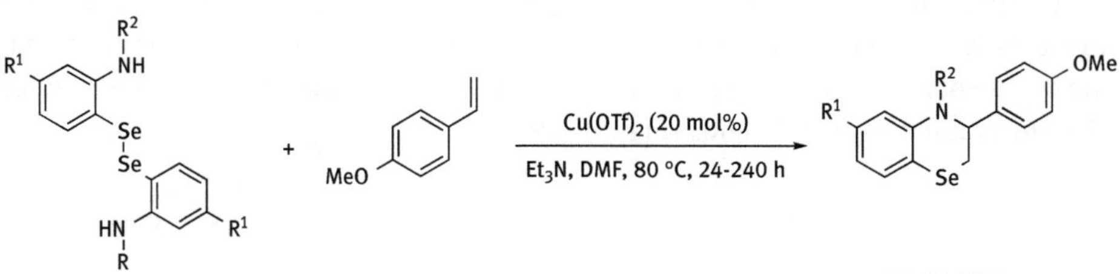

57 **15** **58:** 58-83%

R^1 = H, Me, Cl, Br, CF_3;
R^2 = Ts, Ns

Figure 7.24: Cu(II)-catalyzed synthesis of benzo[*b*][1,4]selenazines **58** by the reaction of 2-*N*-sulfonylamino diselenides **57** with 4-methoxystyrene **15**.

The catalytic cycle for the synthesis of Se-containing heterocyclic compounds **52** by the reaction of diaryl diselenide **51** and alkenes **15** is shown in Figure 7.25 [25]. The catalytic cycle was initiated with the formation of intermediate **59** by reaction of Cu $(OTf)_2$ in the presence of a base. Intermediate **59** reacts with alkene **15** and forms selenarium ion intermediate **61** along with the formation of selenolate ion intermediate **60.** The seleniranium ion intermediate **61** converts into another intermediate **62,** which undergoes intramolecular cyclization to form the final product **52.** Finally, selenolate ion intermediate **60** was oxidized to diaryl diselenide **51** to restart the catalytic cycle.

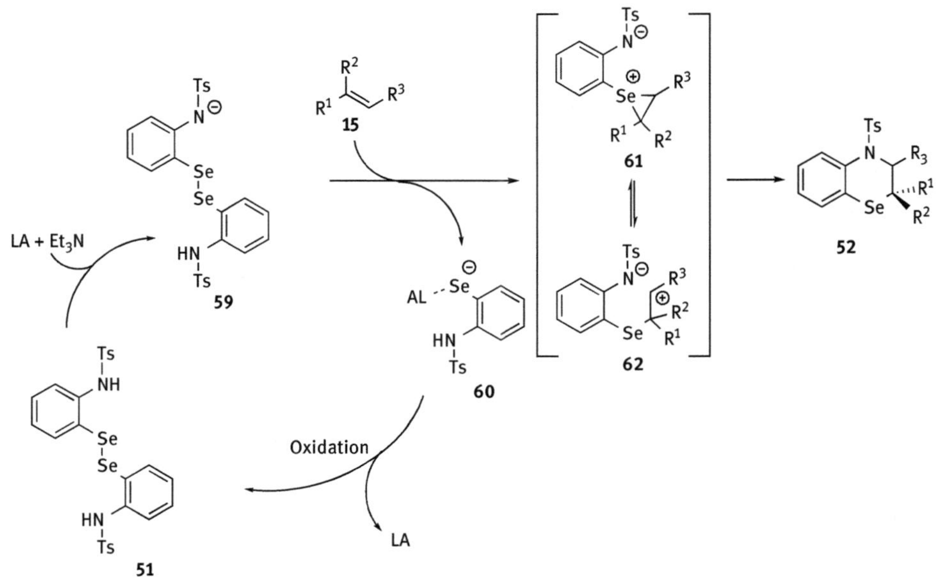

Figure 7.25: Cu(II)-catalyzed synthesis of benzo[*b*][1,4]selenazines **52** by the reaction of 2-*N*-sulfonylamino diselenides **51** with alkene **15**.

Furthermore, the synthesized benzo[*b*][1,4]selenazines **58** were treated with thioacetic acid and LiOH in dry DMF and a clean denosylation reaction was observed. The course of reaction was slightly slow once again but *N*-unsubstituted selenazines **63** were isolated in useful yields (Figure 7.26) [25].

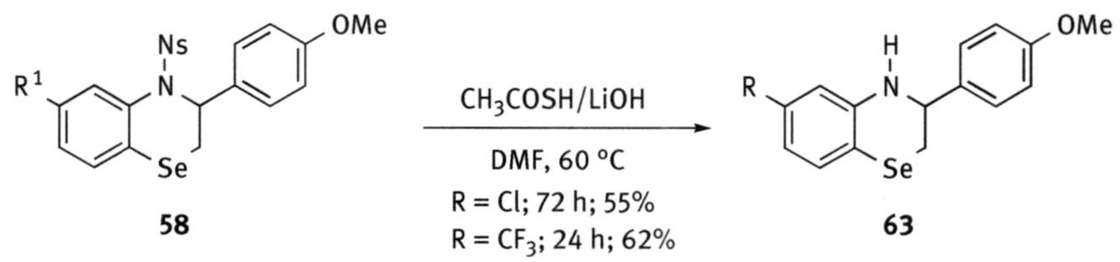

Figure 7.26: Denosylation of compound **58** with thioacetic acid and LiOH in DMF.

Recently, Wang and Bates used the selenocyclization approach for the synthesis of one intermediate during the total synthesis of naturally occurring compound allahabadolactone **A** [27].

7.2.2 Selenenylation of aliphatic C–H bonds

There are only few reports in which selenium electrophiles have been used for the selenenylation of aliphatic C–H bonds. In 2006, Tingoli and coworkers used selenium

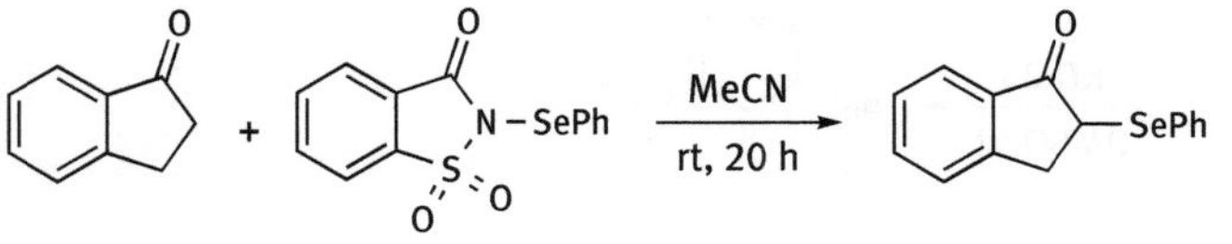

64 12 65: 72%

Figure 7.27: Functionalization of 1-indanone **64** at α-position using *N*-phenyl selenosaccharin (NPSSac) **12** as an electrophile.

electrophile NPSSac **12** for the functionalization of 1-indanone **64** in the α-position. The reaction was performed in acetonitrile at room temperature for 20 h and reaction product **65** was isolated in 72 % yield (Figure 7.27) [14].

A similar reaction was used for the functionalization of heptanal **66** in the α-position with selenium electrophile **12** and α-selenenylated heptanal **67** was isolated in 91 % yield (Figure 7.28) [14].

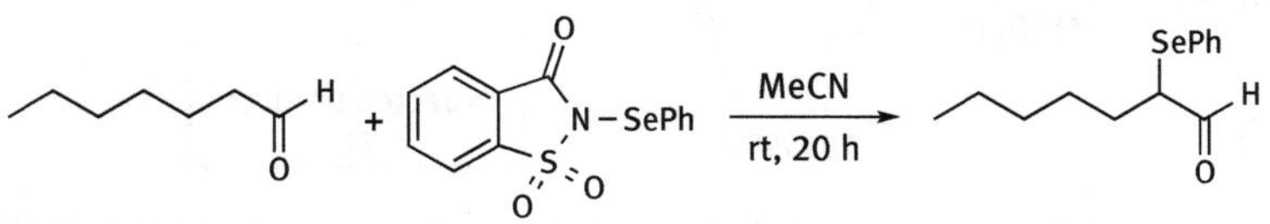

66 12 67: 91%

Figure 7.28: Functionalization of heptanal **66** at α-position using *N*-phenyl selenosaccharin (NPSSac) **12** as an electrophile.

Recently, Kumar and coworkers developed another approach for the α-functionalization of cyclic amides using diorganyl diselenide **20** and potassium *tert*-butoxide in DMSO at room temperature. The reaction products **69** were isolated in moderate to good yields (Figure 7.29) [28].

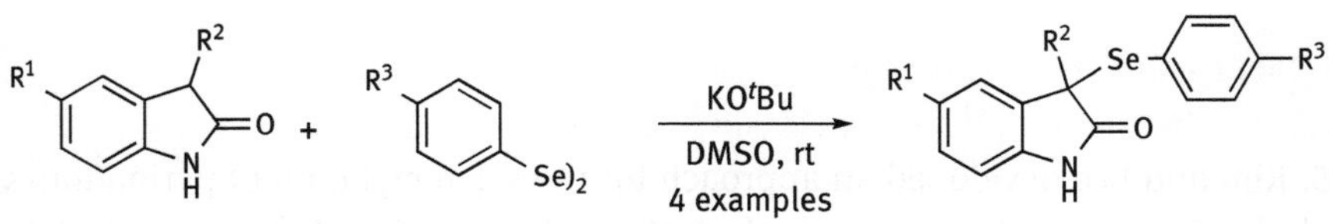

68 20 69: 46-79%

R^1 = H, Cl; R^2 = H, Bn; R^3 = H, OMe

Figure 7.29: Functionalization of cyclic amides **68** in the α-position using diorganyl diselenide **20** and potassium *tert*-butoxide in DMSO.

In addition, acyclic amide **70** was used as a precursor under similar reaction conditions and reaction product **71** was isolated in 75 % yield (Figure 7.30). It was observed that the course of reaction was quite similar to cyclic amides [28].

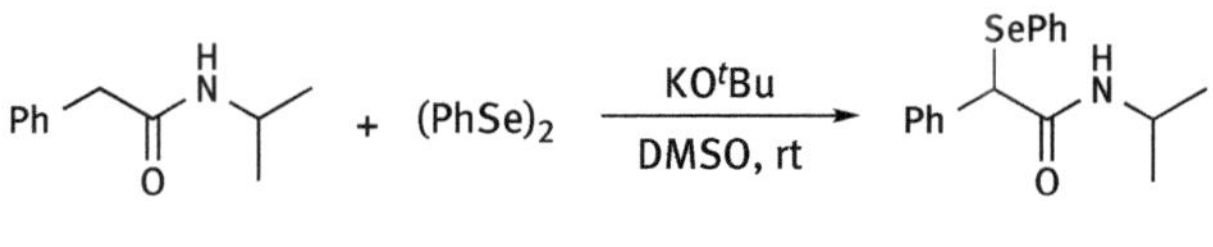

70 **2** **71:** 75%

Figure 7.30: Functionalization of acyclic amides **70** at α-position using diphenyl diselenide **2** and potassium *tert*-butoxide in DMSO.

Furthermore, the reaction was performed for the functionalization of α-tetralone **72** at α-position but the expected reaction product **74** was not observed. The isolated product was characterized as 2-phenylselanyl-1-naphthol **73** by its spectroscopic analysis and obtained in 73 % yield (Figure 7.31) [28].

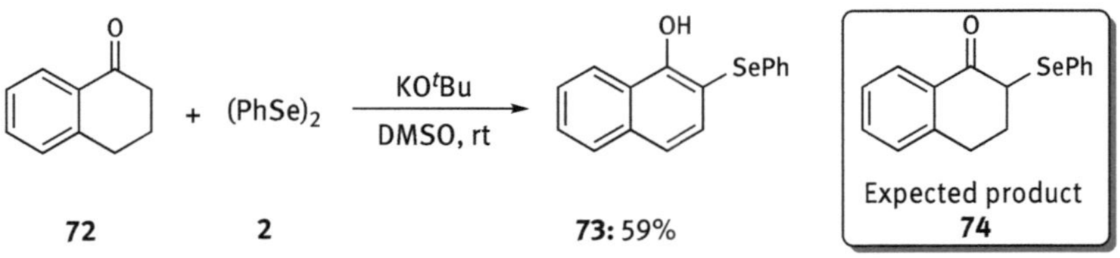

72 **2** **73:** 59%

Figure 7.31: Conversion of α-tetralone **72** to 2-phenylselanyl-1-naphthol **73** using diphenyl diselenide **2** and potassium *tert*-butoxide in DMSO.

7.2.3 Selenenylation of aromatic C–H bonds

There are some reports in which selenium electrophiles have been used for the selenenylation of aromatic C–H bonds. In few approaches, transition metals have been used to initiate the reaction while few reactions proceeded without using any transition metal.

7.2.3.1 Metal-mediated selenenylation of aromatic C–H bonds

7.2.3.1.1 Copper-mediated selenenylation of aromatic C–H bonds

In 1995, Kim and Lee developed an approach for the selenenylation of pyrimidones using diphenyl diselenide as source of electrophile and $Mn(OAc)_3$ as catalyst in DMSO but this approach suffered from poor reaction yields [29]. In 2014, Shibahara and coworkers reported Cu(I)-catalyzed selenenylation of 3-(4-methoxyphenyl)imidazo[1,5-*a*]pyridine **75** using diphenyl diselenide **2** as source for the electrophile and CuBr as catalyst in DMSO at 30°C. The selenenylated product phenylimidazopyridyl selenide **76** was obtained in quantitative yield (Figure 7.32) [30].

In addition, a Cu(I)-catalyzed approach was used for the selenenylation of *N*-methylindole **77** under similar reaction conditions. The selenenylated product phenyl indolyl selenide **78** was isolated in 38 % yield (Figure 7.33) [30].

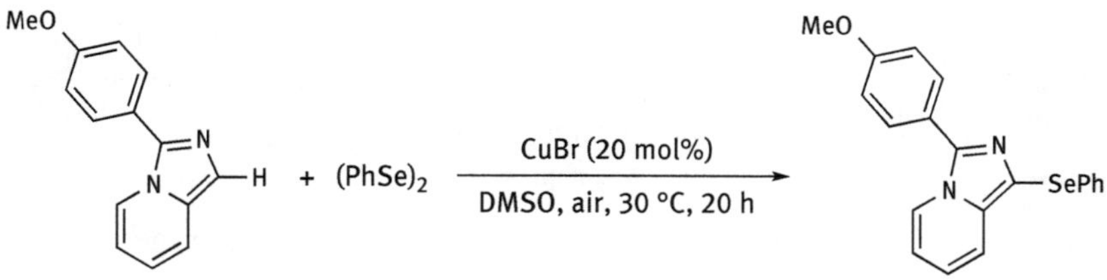

75 **2** **76:** 100%

Figure 7.32: Cu(I)-catalyzed selenenylation of 3-(4-methoxyphenyl)imidazo[1,5-*a*]pyridine **75** using diphenyl diselenide **2** as an electrophile in DMSO.

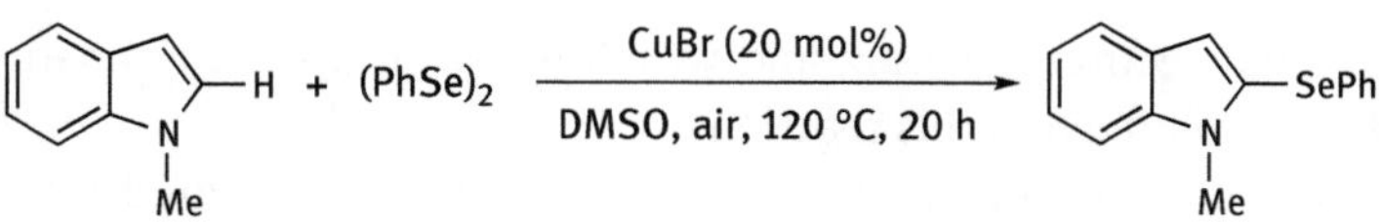

77 **2** **78:** 38%

Figure 7.33: Cu(I)-catalyzed selenenylation of *N*-methylindole **77** using diphenyl diselenide **2** as an electrophile in DMSO.

Furthermore, a Cu(II)-mediated approach was developed for the selenenylation of quinolones [31]. In this approach, 8-aminoquinolones **79** were reacted with various diorganyl diselenides **20** using 1.5 equivalents of $CuBr_2$ in DMSO under oxygen atmosphere at 160°C leading to the corresponding diaryl selenides **80** in good yields (Figure 7.34) [31]. Various electron-donating and withdrawing groups at the aromatic ring in diselenide substrates were successfully tolerated in this reaction.

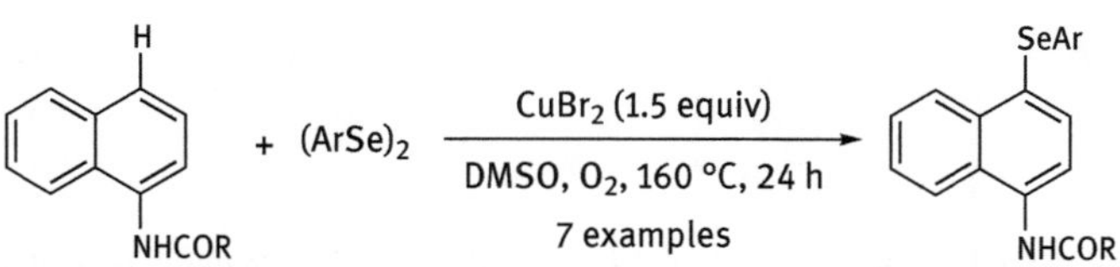

79 **20** **80:** 68-81%

R = Me, ^{t}Bu, cyclopentyl; Ar = Ph, 4-ClC_6H_4, 2,6-$Me_2C_6H_3$, 2,4,6-$Me_3C_6H_2$

Figure 7.34: Cu(II)-mediated selenenylation of 8-aminoquinolones **79** to the corresponding diaryl selenides **80** using diorganyl diselenide **20** in DMSO.

In 2016, Baidya and coworkers developed another Cu(II)-mediated approach for the diselenenylation of benzamides **81** using diorganyl diselenide **20** and $Cu(OAc)_2$ in DMSO at 80°C. The reaction products **82** were isolated in moderate-to-good yields

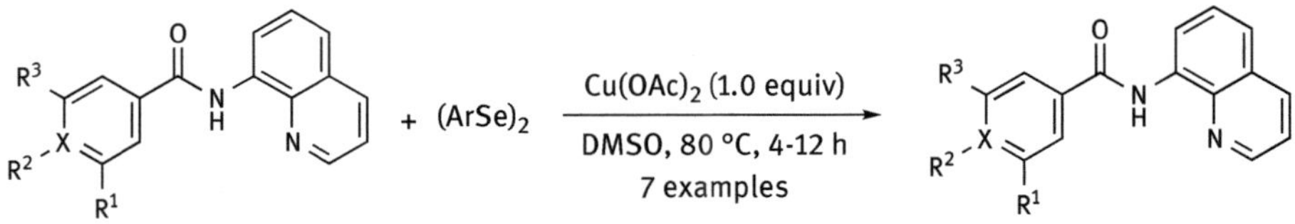

81 **20** **82:** 52-70%

R^1 = H, Me; R^2 = H, Me, OMe, Cl; X = CH, N;
R^3 = H, Me, OMe; Ar = Ph, 4-$(Me)_3CC_6H_4$

Figure 7.35: Cu(II)-mediated diselenenylation of benzamides **81** using diorganyl diselenides **20** and $Cu(OAc)_2$ in DMSO.

(Figure 7.35) [32]. Various electron-donating and withdrawing groups on aromatic ring in substrates **81** were tolerated in this reaction.

In addition, *N*-(quinolin-8-yl)thiophene-2-carboxamide **83** was treated with diorganyl diselenides **20** under similar reaction conditions. The selenenylation of the thiophene functionality was observed at C-3 position and reaction products **84** were isolated in moderate-to-excellent yields (Figure 7.36) [32]. The selenenylation reaction was working well with diaryl diselenides having both electron-donating and withdrawing groups at the aromatic ring but better yields were obtained with diorganyl diselenides having electron-donating functionalities.

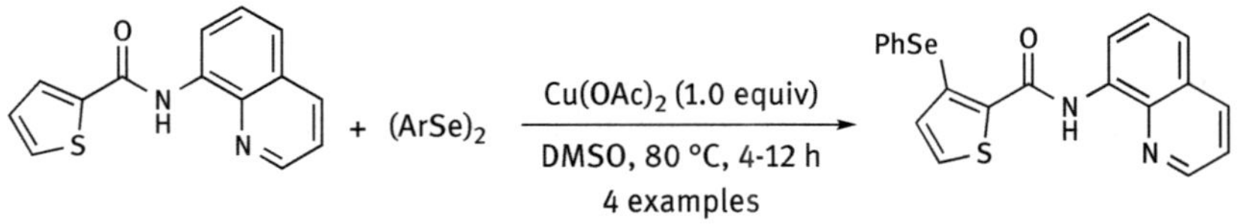

83 **20** **84:** 55-92%

Ar = Ph, 4-$OMeC_6H_4$,
4-ClC_6H_4, 4-BrC_6H_4

Figure 7.36: Cu(II)-mediated diselenenylation of *N*-(quinolin-8-yl)thiophene-2-carboxamide **83** using diorganyl diselenides **20** and $Cu(OAc)_2$ in DMSO.

The catalytic version of the same selenenylation reaction was developed by reacting benzamides **81** with diselenide **20** using $Cu(OAc)_2$ as catalyst and KF as an additive in the presence of silver carbonate in DMSO. The reaction products **82** were isolated in moderate-to-good yields (Figure 7.37) [32]. Silver carbonate served as an oxidant to reoxidize the Cu(I) to the Cu(II) species. Substrates **81** bearing various electron-donating and withdrawing groups on phenyl, pyridyl and quinoline rings were successfully used in this catalytic reaction.

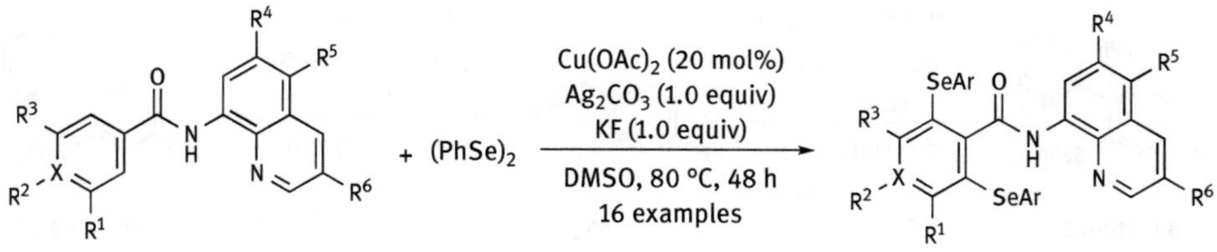

81 **2** **82:** 45-80%

R^1 = H, Me, OMe; R^2 = H, Me, OMe, Cl, F, Br; X = CH, N; R^3 = H, Me, OMe; R^4 = H, OMe; R^5 = H, I; R^6 = H, Br

Figure 7.37: Cu(II)-catalyzed mono-/diselenenylation of benzamides **81** using diphenyl diselenides **2** and catalytic amounts of $Cu(OAc)_2$ in DMSO.

7.2.3.1.2 Palladium-mediated selenenylation of aromatic C–H bonds

Recently, some selenium electrophiles have been successfully used to achieve selenenylation of aromatic species in the presence of palladium catalysts. In 2015, Law and others developed a Pd-catalyzed approach for the selenenylation of arenes **85** using *N*-PSP **86** in the presence of 10 mol% of $[PdCl_2(MeCN)_2]$ **87** in water at 110°C. Unsymmetrical diaryl selenides **88** were isolated in poor-to-excellent yields (Figure 7.38) [33]. In addition, diselenenylated products **89** were isolated in reactions with up to 29 % yield (Figure 7.38) [33].

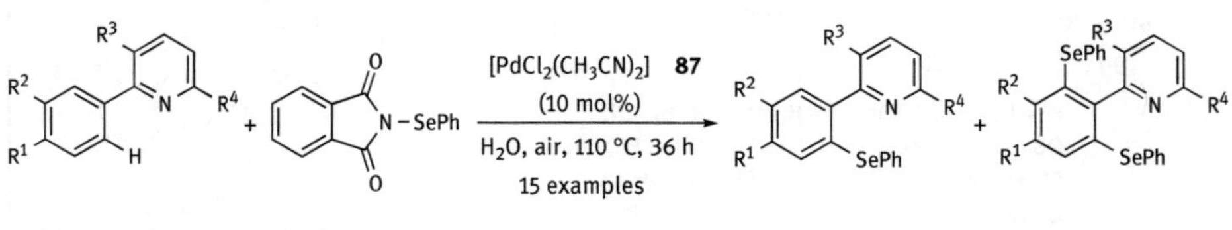

85 **86** **88:** 19-90% **89:** 12-29%

R^1 = H, Me, ^{t}Bu, F, Br, COMe; R^2 = H, F, Cl, NO_2; R^3 = H, Me; R^4 = H, Me

Figure 7.38: Pd-catalyzed mono-/diselenenylation of arenes **85** using *N*-(phenylseleno)phthalimide (*N*-PSP) **86** and 10 mol% of $[PdCl_2(MeCN)_2]$ **87**.

The same research group developed a Pd-catalyzed approach for the mono- or diselenenylation of arenes selectively using *N*-PSP **86**. Monoselenenylation of arenes **85** was achieved by the reaction with selenium electrophile *N*-PSP **86** in the presence of 10 mol% $[PdCl_2(CH_3CN)_2]$ **87** in DMSO:H_2O (1:1) at 100 C. The reaction products **88** were isolated in good-to-excellent yields (Figure 7.39). Additionally, diselenenylation of similar arenes **85** was achieved in excellent yields by using electrophilic species **86** and catalyst **87** in a similar solvent combination DMSO:H_2O in a different ratio (4:1) (Figure 7.39) [34].

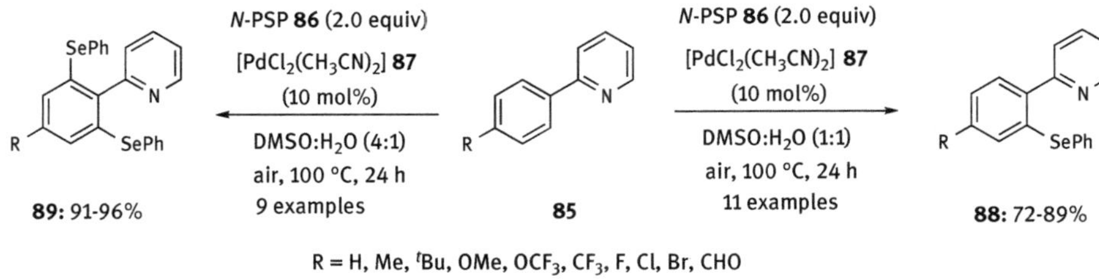

Figure 7.39: Pd-catalyzed mono-/diselenenylation of arenes **85** using *N*-(phenylseleno)phthalimide (*N*-PSP) **86** and 10 mol% of $[PdCl_2(MeCN)_2]$ **87**.

7.2.3.1.3 Iridium-mediated selenenylation of aromatic C–H bonds

Recently, Liu and coworkers reported a selenenylation of (hetero)arenes **90** by the reaction with diorganyl diselenide **20** using 2.0 mol% of FlrPic **91** as catalyst in the presence of visible light in acetonitrile under open atmosphere. Unsymmetrical diaryl selenides **92** were isolated (Figure 7.40) [35]. Various electron-donating and withdrawing substituents are tolerated in both, (hetero)arene and diorganyl diselenide substrates.

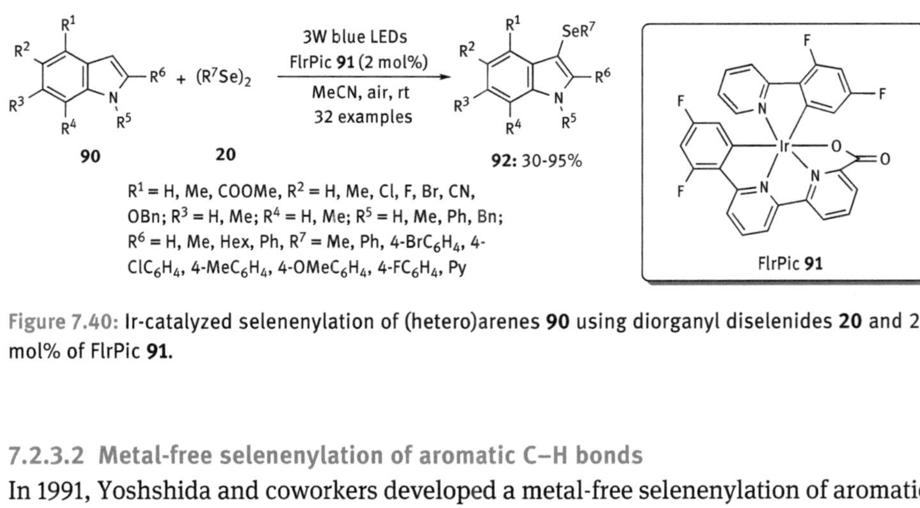

Figure 7.40: Ir-catalyzed selenenylation of (hetero)arenes **90** using diorganyl diselenides **20** and 2 mol% of FlrPic **91**.

7.2.3.2 Metal-free selenenylation of aromatic C–H bonds

In 1991, Yoshshida and coworkers developed a metal-free selenenylation of aromatic species using benzeneselenenyl *m*-nitrobenzenesulfonate **3** as selenium electrophile [12]. In 2006, Tingoli and coworkers developed another metal-free reaction for the selenenylation of electron-rich arenes **93a** and **93b** using selenium electrophile NPSSac **12** in acetonitrile at room temperature. The reaction products **94a** and **94b** were isolated in 80 % and 87 % yield, respectively (Figure 7.41) [14].

In 2016, Braga and coworkers reported a molecular iodine-catalyzed selenenylation of imidazo[1,2-*a*]pyridines **95** using diorganyl diselenides **20** as source of

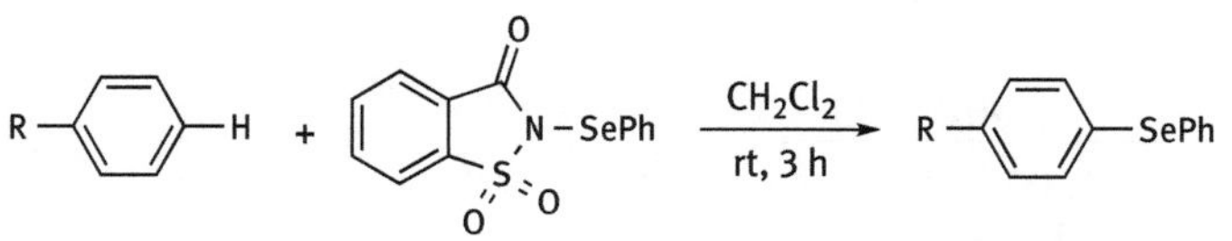

93a: R = OMe
93b: R = NMe_2

12

94a: R = OMe; 80%
94b: R = NMe_2; 87%

Figure 7.41: Metal-free selenenylation of electron-rich arenes **93a** and **93b** using *N*-phenylselenosaccharin (NPSSac) **12** as an electrophile in acetonitrile.

electrophile and DMSO as an oxidant under solvent-free conditions. The selenenylation of substrates **95** occurred at the C-3 position selectively and unsymmetrical diaryl selenides **96** were obtained in high yields (Figure 7.42) [36]. Various electron-donating and withdrawing substituents are tolerated in arene and diorganyl diselenide substrates. Additionally, dibutyl diselenide was also used successfully as source of electrophile in this reaction.

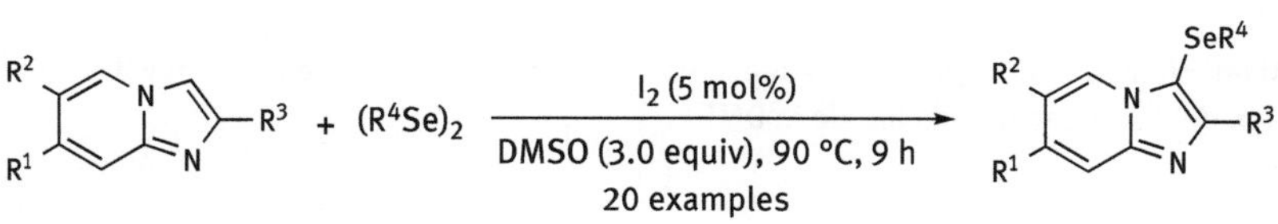

95 **20** **96:** 69-91%

R^1 = H, Me; R^2 = H, Me; R^3 = H, Ph, 3-BrC_6H_4, 4-FC_6H_4, 4-ClC_6H_4, 4-CNC_6H_4, 4-$OMeC_6H_4$, R^4 = ^{n}Bu, Ph, 2-MeC_6H_4, 4-MeC_6H_4, 2-$OMeC_6H_4$, 4-$OMeC_6H_4$, 4-FC_6H_4, 4-ClC_6H_4, 4-$CF_3C_6H_4$, Np, Thio

Figure 7.42: Iodine-catalyzed selenenylation of imidazo[1,2-*a*]pyridines **95** using diorganyl diselenides **20** as source of electrophile and DMSO as an oxidant.

Recently, another metal-free approach for the selenenylation of chromones **97** was developed by the reaction with diorganyl diselenide **20** and ammonium iodide in DMF at 135°C under open atmosphere [37].

The selenenylation of chromones **97** takes place at the C-3 position selectively and unsymmetrical diaryl selenides **98** are obtained in moderate-to-good yields (Figure 7.43) [37]. The selenenylation reaction was proceeding well when electron-donating and withdrawing substituents were used in both arene and diaryl diselenide substrates.

Additionally, the same approach was used for the selenenylation of quinolone **99** under similar reaction and conditions. Similar to chromones, the selenenylation of quinolone **99** occurred selectively at the C-3 position and 3-(phenylselanyl)quinolin-4(1*H*)-one **100** was isolated in 53 % yield (Figure 7.44) [37].

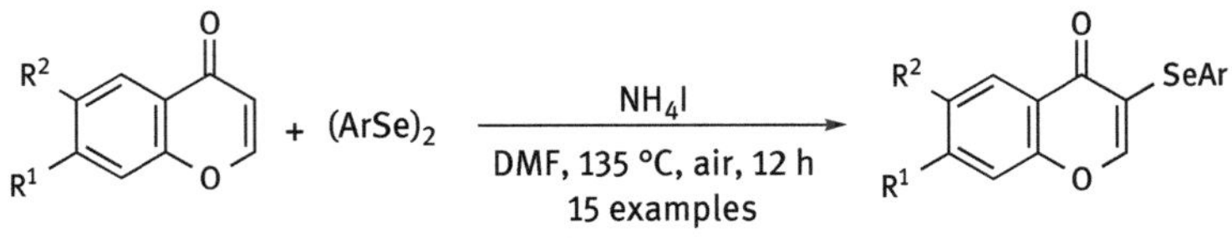

97 20 98: 45-73%

R^1 = H, Me, OMe, Br; R^2 = H, Me, Cl, F, Br, NO_2; Ar = Ph, 4-BrC_6H_4, 4-ClC_6H_4, 4-MeC_6H_4, 4-OMeC_6H_4, 4-FC_6H_4, 1-Nap

Figure 7.43: Ammonium iodide-mediated selenenylation of chromones **97** by the reaction with diorganyl diselenide **20** and ammonium iodide in DMF.

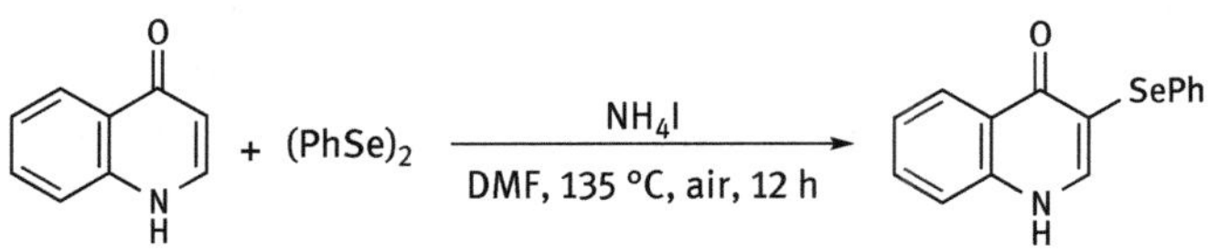

99 2 100: 53%

Figure 7.44: Ammonium iodide-mediated selenenylation of quinolone **99** by the reaction with diphenyl diselenide **2** and ammonium iodide in DMF.

7.2.4 Cyclizations

7.2.4.1 Metal-free cyclizations

Selenium electrophiles have been used to achieve various cyclization reactions with or without transition metals. In 2009, Amosova and coworkers reported the synthesis of the cyclic compound 2,6-dichloro-1,4-thiaselenane **103** in quantitative yield by the reaction of divinyl sulfide **101** with selenium dichloride **102** in chloroform at −50°C without using any metal salt (Figure 7.45) [38]. In addition, the synthesized compound **103** was further used as substrate for the synthesis of other selenium-containing heterocycles.

7.2.4.2 Metal-mediated cyclizations

7.2.4.2.1 Iron-mediated cyclizations

Other cyclization reactions have been achieved using selenium electrophiles in the presence of transition metals in either stoichiometric or catalytic amounts. In 2015, Zeni and coworkers developed an Fe(III)-mediated approach for the cyclization of functionalized 1,3-diynes **104** by the reaction of two equivalents of dibutyl diselenide **20** and 1.5 equivalents of $FeCl_3$ in dichloromethane at 40°C. The isolated products were characterized as selenophenes **105** and obtained in moderate-to-good yields

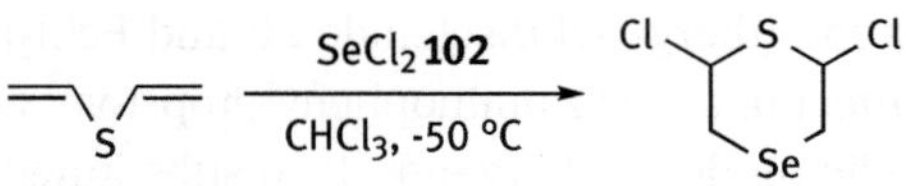

101 **103:** 100%

Figure 7.45: Synthesis of 2,6-dichloro-1,4-thiaselenane **103** by the reaction of divinyl sulfide **101** with selenium dichloride **102**.

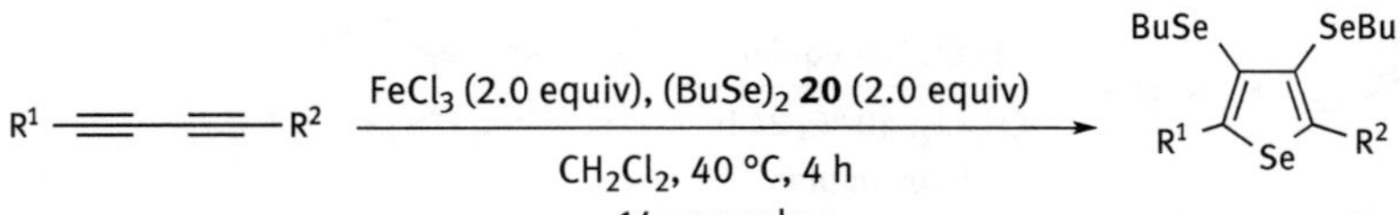

104 **105:** 39-74%

R^1 = Bu, Ph, 4-FC_6H_4, 4-ClC_6H_4, 4-MeC_6H_4, 3-MeC_6H_4, 2-MeC_6H_4, 4-$OMeC_6H_4$, 1-Nap, 2-Nap; R^2 = Bu, Ph, 4-FC_6H_4, 4-ClC_6H_4, 4-MeC_6H_4, 3-MeC_6H_4, 2-MeC_6H_4, 4-$OMeC_6H_4$, 1-Nap, 2-Nap

Figure 7.46: The cyclization of functionalized 1,3-diynes **104** to selenophenes **105** by the reaction of dibutyl diselenide **20** and $FeCl_3$ in dichloromethane.

(Figure 7.46) [39]. This approach was used to prepare of various symmetrical and unsymmetrical selenophenes using easily accessible precursors.

The same research group developed another Fe(III)-mediated approach for the cyclization of functionalized *o*-alkynylbenzamides **106** by the reaction of 0.5 equivalents of diorganyl diselenide **20** and 2.0 equivalents of $FeCl_3$ in dichloromethane at room temperature under oxygen atmosphere. The cyclic products 4-(phenylselanyl)-1*H*-isochromen-1-imines **107** were isolated in moderate-to-good yields as major reaction products (Figure 7.47) [40]. In some reactions, functionalized isobenzofuran-1(3*H*)-imines **108** were observed as minor reaction products.

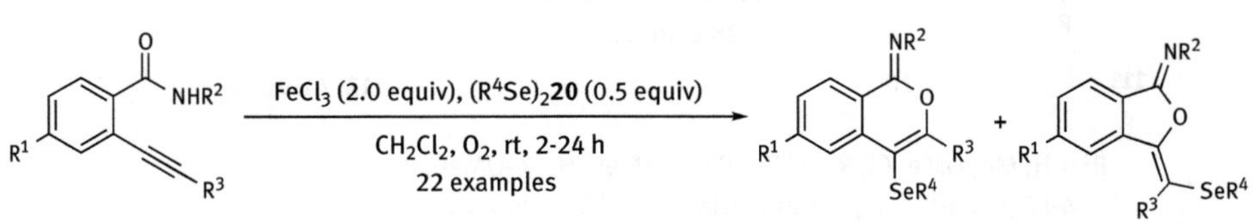

106 **107** major **108** minor

R^1 = H, 4-FC_6H_4, 4-$OMeC_6H_4$; R^2 = ^{t}Bu, Ph, 4-$OMeC_6H_4$, 4-ClC6H4; R^3 = ^{n}Bu, Ph, 4-FC_6H_4, 4-$OMeC_6H_4$, 4-ClC_6H_4; R^4 = ^{n}Bu, Ph, 4-FC_6H_4, 4-$OMeC_6H_4$, 3-CF3C_6H_4, 4-ClC_6H_4

Figure 7.47: The cyclization of functionalized *o*-alkynylbenzamides **106** by the reaction of diorganyl diselenide **20** and $FeCl_3$ in dichloromethane.

Furthermore, a similar reagent combination (diorganyl diselenide **20** and $FeCl_3$) was used to achieve the aminocyclization of functionalized 2-aminophenylprop-1-yn-3-ols **109** in dichloromethane at 40°C. The cyclic products, 3-organoseleno-substituted quinolines **110**, were isolated in poor-to-good yields (Figure 7.48) [41]. The cyclization is proceeding well when electron-donating and withdrawing substituents were used in both arene and diorganyl diselenide substrates.

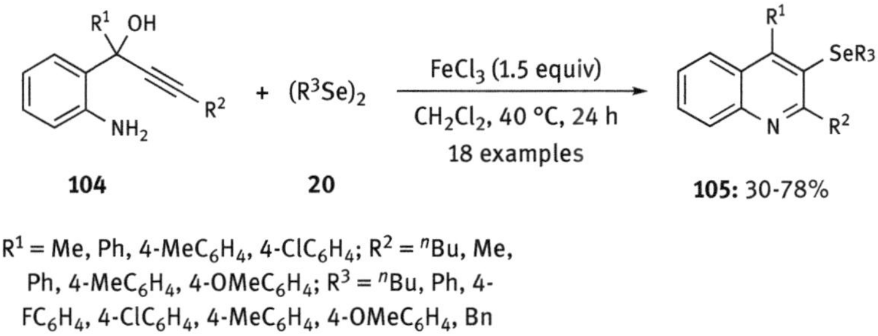

Figure 7.48: The cyclization of functionalized 2-aminophenylprop-1-yn-3-ols **109** to 3-organoseleno-substituted quinolines **110** by the reaction of diorganyl diselenides **20** and $FeCl_3$ in dichloromethane.

Recently, Zeni and coworkers used a similar reagent combination (diorganyl diselenide **20** and $FeCl_3$) to achieve the carbocyclization of benzylic-substituted propargyl alcohols **111** in dichloroethane (DCE) at 70°C under oxygen atmosphere. The cyclic products 2-organoselenyl-naphthalenes **112** were obtained in moderate-to-excellent yields (Figure 7.49) [42]. Various electron-donating and withdrawing substituents were successfully tolerated in both alcohol and diorganyl diselenide substrates.

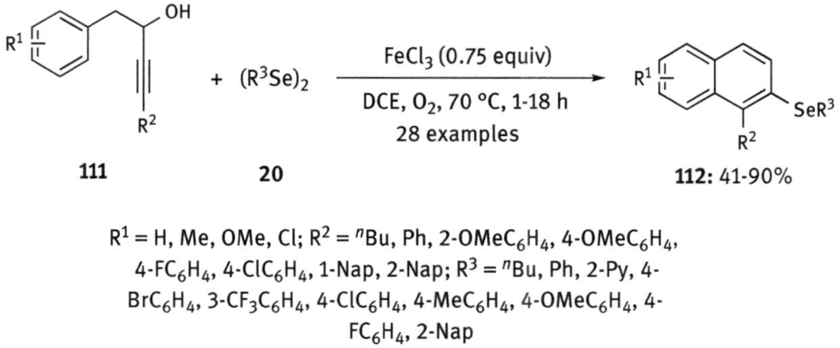

Figure 7.49: The cyclization of benzylic-substituted propargyl alcohols **111** to 2-organoselenyl-naphthalenes **112** by the reaction of diorganyl diselenides **20** and $FeCl_3$ in DCE.

In 2016, Zeni and coworkers developed an Fe(III)-mediated cyclization of 1,3-diynyl chalcogen derivatives **113** with diorganyl diselenides **20** and $FeCl_3{\cdot}6H_2O$ in dichloromethane at reflux temperature. The cyclic products benzo[*b*]furan-fused selenophenes (**114**: X=O) were isolated in moderate-to-excellent yields (Figure 7.50) [43]. In addition, the synthesis of benzo[*b*]thiophene-fused selenophenes (**114**: X=S) or benzo[*b*]seleno-fused selenophenes (**114**: X=Se) was also achieved using a similar approach.

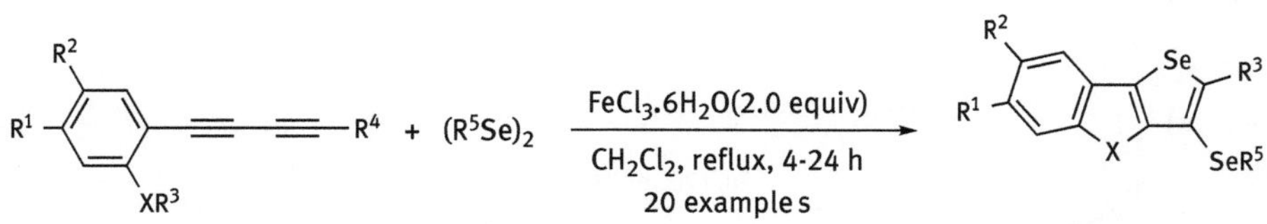

113 **20** **114:** 35-89%

R^1 = H, Me; R^2 = H, F, Me; R^3 = Me, Et, ^{n}Bu; R^4 = ^{n}Bu, Ph, 4-MeC_6H_4, 4-ClC_6H_4; 2-ClC_6H_4; 4-BrC_6H_4; 2-Nap; R^5 = Et, ^{n}Pr, ^{n}Bu, ^{n}Oct; X = O, S, Se

Figure 7.50: The cyclization of 1,3-diynyl chalcogen derivatives **113** to benzo[*b*]chalcogen-fused selenophenes **114** by the reaction of diorganyl diselenides **20** and $FeCl_3$ in dichloromethane.

Another Fe(III)-catalyzed approach was developed by Zeni and coworkers for the cyclization of 1,4-butyne-diols **115** with diorganyl diselenides **20** using 20 mol% of $FeCl_3{\cdot}6H_2O$ in DCE at room temperature under oxygen atmosphere. The 3,4-bis (organoselanyl)-2,5-dihydrofuran products **116** were isolated in poor-to-excellent yield (Figure 7.51) [44]. In addition, the same catalytic approach was also applied to the synthesis of 3,6-dihydro-2*H*-pyrans and 2,5-dihydro-1*H*-pyrroles.

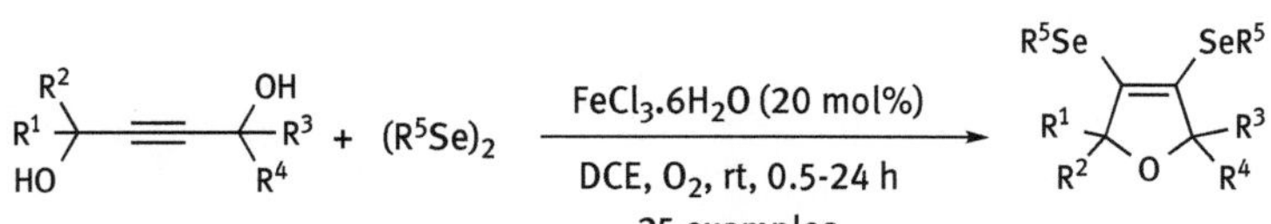

115 **20** **116:** 15-97%

R^1 = H, Me; R^2 = H, Me; R^3 = H, Me; R^4 = Me, Ph, 4-$OMeC_6H_4$, 2-ClC_6H_4, 2-Nap; R^5 = Ph, 4-FC_6H_4, 4-MeC_6H_4

Figure 7.51: Fe(III)-catalyzed cyclization of 1,4-butyne-diols **115** to 3,4-bis(organoselanyl)-2,5-dihydrofurans **116**.

7.2.4.2.2 Copper-mediated cyclizations

Recently, Zeni and coworkers reported a Cu(I)-catalyzed approach for the cyclization of functionalized propargylpyridines **117** with diorganyl diselenides **20** using 20 mol% of KI in the presence of Na_2CO_3 base in DMF at 60 C. The functionalized 2-(phenylselanyl) indolizines **118** were isolated in poor-to-excellent yields (Figure 7.52) [45]. Various electron-donating and withdrawing substituents are tolerated in both alcohol and diorganyl diselenide substrates. In addition, a similar cyclization reaction was investigated with dibutyl diselenide **20** (R=^{n}Bu) as source of electrophile but no cyclization could be observed.

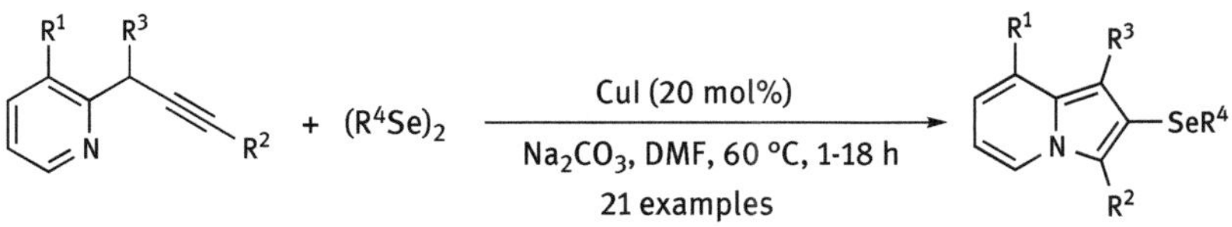

117 **20** **118:** 32-84%

R^1 = H, Me; R^2 = H, Ph, 4-MeC_6H_4, 4-$OMeC_6H_4$; R^3 = OAc, OCOOEt; R^4 = Ph, 2-MeC_6H_4, 4-MeC_6H_4, 4-ClC_6H_4, 2-$OMeC_6H_4$, 4-$OMeC_6H_4$, 3-$CF_3C_6H_4$

Figure 7.52: Cu(I)-catalyzed cyclization of functionalized propargylpyridines **117** to 2-(phenylselanyl) indolizines **118**.

7.2.4.2.3 Lewis acid-mediated cyclizations

In 2009, Shahzad and Wirth developed an approach for the carbocyclization of stilbenes **119** to dihydronaphthalenes **120** with the combination of a Lewis acid and phenylselenenyl chloride. The cyclic products **120** were isolated in good-to-excellent yields (Figure 7.53) [46]. Both $SnCl_4$ and $BF_3{\cdot}OMe_2$ can be used as Lewis acid in this reaction but cyclic product **120** was obtained in higher yields with $BF_3{\cdot}OMe_2$.

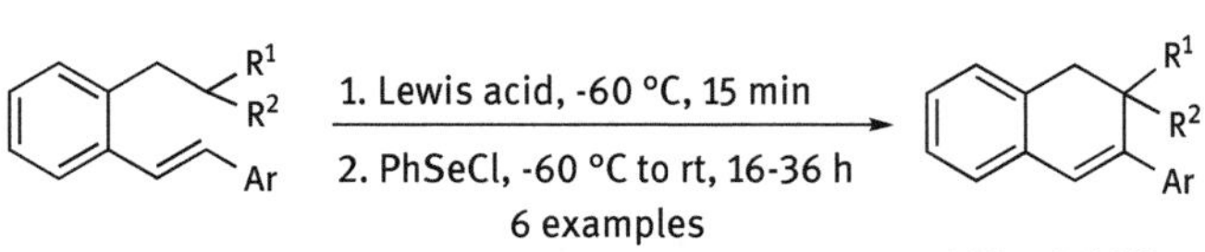

119 **120:** 50-90%

R^1 = COOMe, COOEt; R^2 = COMe, COOMe; Ar = Ph, 2-ClC_6H_4, 3-ClC_6H_4, 4-ClC_6H_4, 2,4-$Cl_2C_6H_4$

Figure 7.53: Selenium-mediated approach for the carbocyclization of stilbenes **119** to dihydronaphthalenes **120**.

In addition, the double carbocyclization of stilbenes **121** was achieved by using the combination of the Lewis acid $BF_3{\cdot}OMe_2$ and phenylselenenyl chloride under similar reaction conditions. The cyclic products benzo[*b*]fluorenes **122** were isolated in reasonable yields (Figure 7.54) [46].

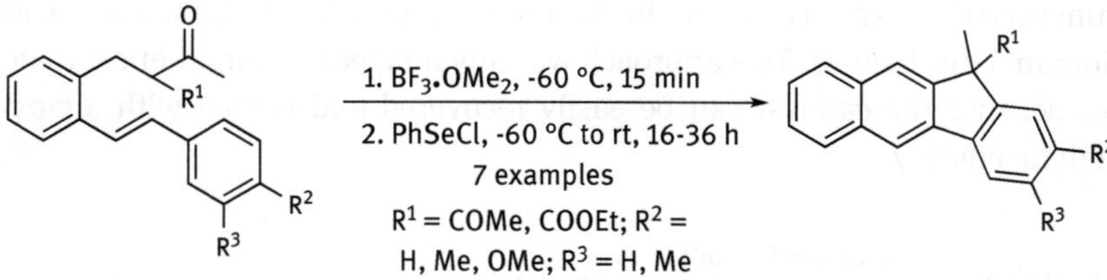

121 **122:** 30-90%

Figure 7.54: Selenium-mediated approach for the double carbocyclization of stilbenes **121** to benzo [*b*]fluorenes **122**.

7.2.5 Metal-catalyzed coupling reactions

Metal-catalyzed coupling reactions of organonucleophilic species with selenium electrophiles have been used to construct new C–Se bonds. The synthesis of various symmetrical and unsymmetrical diorganoselenides has been achieved using these coupling reactions.

7.2.5.1 Copper-catalyzed coupling reactions

In 2007, Taniguchi reported a Cu(I)-catalyzed coupling reaction of diorganyl diselenides **20** with alkyl- or arylboronic acids **123** using 5.0 mol% of CuI–bpy (1:1) in DMSO:H_2O (2:1) at 100°C under oxygen atmosphere. Symmetrical and unsymmetrical selenides **124** were isolated in good-to-excellent yields (Figure 7.55) [47]. In this reaction, organoboronic acid and diorganyl diselenide species were used as nucleophilic and electrophilic partners, respectively. Various electron-donating and withdrawing functionalities are tolerated at the aromatic ring of the organoboronic acid substrates.

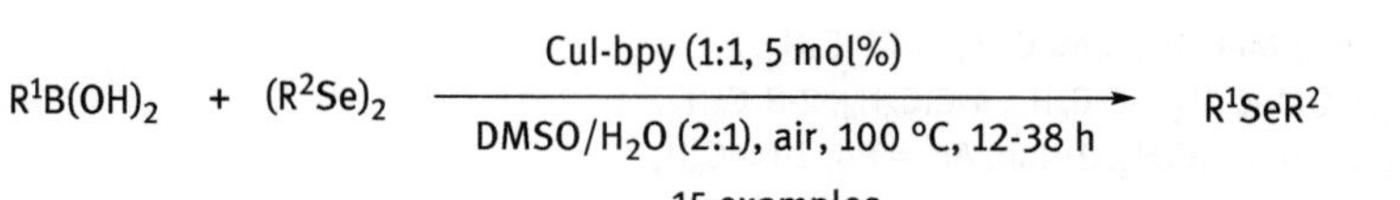

123 **20** **124:** 63-99%

R^1 = Me, ^{n}Bu, Ph, 2-MeC_6H_4, 4-MeC_6H_4, 4-FC_6H_4, 2-$OMeC_6H_4$, 4-$OMeC_6H_4$, 2-ClC_6H_4, 4-ClC_6H_4, 4-OHC_6H_4, 4-$CHOC_6H_4$, 4-OHC_6H_4, 4-$MeO_2CC_6H_4$; R^2 = Ph, Bn

Figure 7.55: Cu(I)-catalyzed coupling reaction of diorganyl diselenides **20** with alkyl- or arylboronic acids **123** using 5 mol% of CuI–bpy (1:1) in DMSO:H_2O (2:1).

CuO nanoparticles (CuO Nps) have been used to catalyze similar coupling reactions of diorganyl diselenides **20** and arylboronic acids **123**. The coupling reaction of substrates **20** and **123** was performed in DMSO using 3 mol% CuO Nps at 100°C under open atmosphere. Symmetrical and unsymmetrical diaryl selenides **124** were isolated in excellent yields (Figure 7.56) [48]. Various electron-donating and withdrawing functionalities are acceptable in both substrates, the organoboronic acid and the diorganyl diselenides. This approach was advantageous over other catalytic approaches because the catalyst can be easily recovered and reused with almost similar catalytic efficacy.

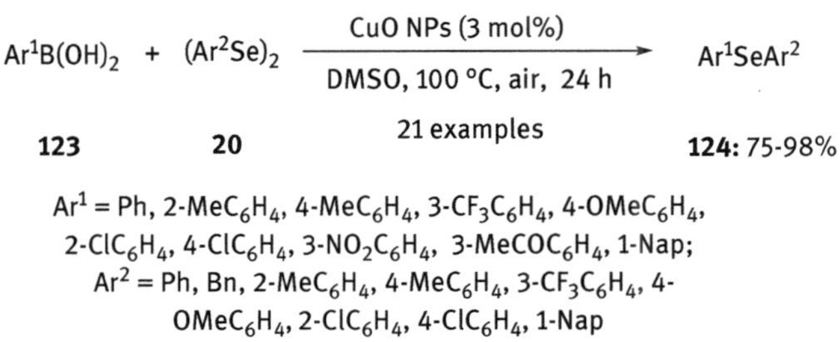

Figure 7.56: CuO nanoparticle-catalyzed coupling reaction of diorganyl diselenides **20** with arylboronic acids **123** using 3 mol% of CuO Nps in DMSO.

In 2012, Alves and others reported a CuI-catalyzed coupling reaction of diaryl diselenides **20** with arylboronic acids **123** using CuI as catalyst and DMSO as an additive in glycerol at 100°C under oxygen atmosphere. Diaryl selenides **124** bearing electron-donating and withdrawing group were obtained in good-to-excellent yields (Figure 7.57) [49]. Various substituents are tolerated on the aromatic ring in organoboronic acid substrates. Additionally, the glycerol–CuI mixture was reused in the cross-coupling reactions.

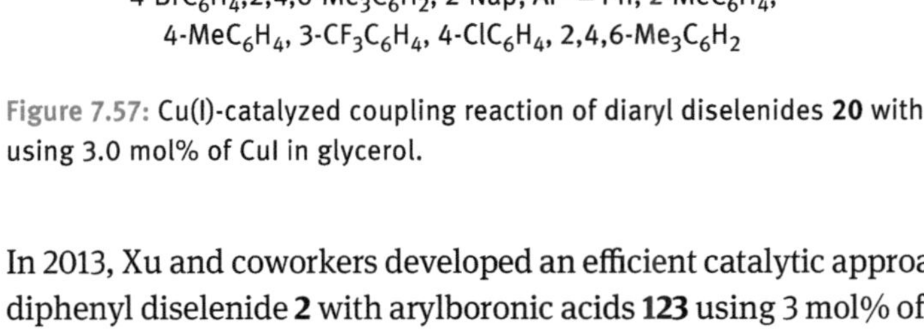

Figure 7.57: Cu(I)-catalyzed coupling reaction of diaryl diselenides **20** with arylboronic acids **123** using 3.0 mol% of CuI in glycerol.

In 2013, Xu and coworkers developed an efficient catalytic approach for the coupling of diphenyl diselenide **2** with arylboronic acids **123** using 3 mol% of $CuSO_4$ and 3 mol% of

1,10-Phen·H_2O in the presence of sodium carbonate in ethanol at 100°C under open atmosphere. Diaryl selenides **124** bearing electron-donating or -withdrawing groups were obtained in good-to-excellent yields (Figure 7.58) [50].

$$\underset{\mathbf{123}}{ArB(OH)_2} + \underset{\mathbf{2}}{(PhSe)_2} \xrightarrow[\substack{Na_2CO_3\ (5\%\ aq.,\ 0.1\ mL) \\ \text{Ethanol, rt, air, 5 h} \\ \text{26 examples}}]{\substack{CuSO_4\ (3\ mol\%) \\ \text{1,10-Phen.}H_2O\ (3\ mol\%)}} \underset{\mathbf{124}:\ 50\text{-}98\%}{ArSePh}$$

Ar = Ph, 2-MeC_6H_4, 4-MeC_6H_4, 3-$OCF_3C_6H_4$, 3-CNC_6H_4, 4-$OMeC_6H_4$, 2-ClC_6H_4, 4-ClC_6H_4, 2-BrC_6H_4, 4-BrC_6H_4, 4-$EtO_2CC_6H_4$, 3-$CH_2OHC_6H_4$, 3-$CH_3COC_6H_4$, 1-Nap, 2-Nap

Figure 7.58: Cu(II)-catalyzed coupling reaction of diphenyl diselenide **2** with arylboronic acids **123** using 3 mol% of each $CuSO_4$ and 1,10-Phen·H_2O in ethanol.

7.2.5.2 Iron-catalyzed coupling reactions

In 2009, Wang and coworkers developed coupling reactions of diorganyl diselenides **20** and arylboronic acids **123** using 10 mol% of Fe powder in DMSO at 130°C. Symmetrical and unsymmetrical aryl selenides **124** were obtained in high yields (Figure 7.59) [51]. In addition, dibutyl diselenide **20** (R=Bu) was also used successfully in this reaction but coupling products were isolated in lower yields in the comparison with diaryl diselenides. Various electron-donating and withdrawing functionalities are tolerated in both substrates organoboronic acid and diaryl diselenide.

$$\underset{\mathbf{123}}{ArB(OH)_2} + \underset{\mathbf{20}}{(RSe)_2} \xrightarrow[\substack{\text{DMSO, 130 °C, 20 h} \\ \text{19 examples}}]{\text{Fe powder (10 mol\%)}} \underset{\mathbf{124}:\ 62\text{-}97\%}{ArSeR}$$

Ar = Ph, 2-MeC_6H_4, 3-MeC_6H_4, 4-MeC_6H_4, 2-$OMeC_6H_4$, 3-$OMeC_6H_4$, 4-$OMeC_6H_4$, 2,4-$(OMe)_2C_6H_3$, 2,6-$(OMe)_2C_6H_3$, 4-$MeSC_6H_4$, 4-$MeO_2CC_6H_4$, 4-FC_6H_4, 4-ClC_6H_4, 4-BrC_6H_4; R = ^{n}Bu, Ph, Bn, 4-$OMeC_6H_4$

Figure 7.59: Fe(0)-catalyzed coupling reaction of diorganyl diselenides **20** with arylboronic acids **123** using 10 mol% of Fe powder in DMSO at 130°C.

7.2.5.3 Indium-catalyzed coupling reactions

An $InBr_3$-catalyzed approach for the coupling of diorganyl diselenides **20** and arylboronic acids **123** has been reported using 10 mol% of $InBr_3$ in DMSO at 130°C. Symmetrical and unsymmetrical diaryl selenides **124** were obtained in high yields

$$\underset{\mathbf{123}}{Ar^1B(OH)_2} + \underset{\mathbf{20}}{(Ar^2Se)_2} \xrightarrow[\text{DMSO, 130 °C, 20 h}]{InBr_3\text{ (10 mol\%)}} \underset{\mathbf{124}\text{: 71-98\%}}{Ar^1SeAr^2}$$

19 examples

Ar^1 = Ph, 2-MeC_6H_4, 3-MeC_6H_4, 4-MeC_6H_4, 2-$OMeC_6H_4$, 3-$OMeC_6H_4$, 4-$OMeC_6H_4$, 2,4-$(OMe)_2C_6H_3$, 2,6-$(OMe)_2C_6H_3$, 4-$MeSC_6H_4$, 4-$MeO_2CC_6H_4$, 4-FC_6H_4, 4-ClC_6H_4, 4-BrC_6H_4; Ar^2 = Ph, Bn, 4-$OMeC_6H_4$

Figure 7.60: $InBr_3$-catalyzed coupling reaction of diorganyl diselenides **20** with arylboronic acids **123** using 10 mol% of $InBr_3$ in DMSO at 130°C.

(Figure 7.60) [52]. Additionally, dibutyl diselenide **20** (R=Bu) was also used successfully in this reaction but coupling product was isolated comparatively in low yield.

7.2.5.4 Silver-catalyzed coupling reactions

In 2016, Alves and coworkers achieved the same coupling reaction of diorganyl diselenides **20** and arylboronic acids **123** using 10 mol% of $AgNO_3$ in DMSO at 130 °C. The coupling products **124** were isolated in moderate-to-excellent yields (Figure 7.61) [53]. This catalytic approach was equally working with both electron-withdrawing and donating substituents at the aromatic ring in both substrates.

$$\underset{\mathbf{123}}{Ar^1B(OH)_2} + \underset{\mathbf{20}}{(Ar^2Se)_2} \xrightarrow[\text{DMSO, 130 °C, 20 h}]{AgNO_3\text{ (10 mol\%)}} \underset{\mathbf{124}\text{: 55-96\%}}{Ar^1SeAr^2}$$

21 examples

Ar^1 = Ph, 2-MeC_6H_4, 3-MeC_6H_4, 4-MeC_6H_4, 2-$OMeC_6H_4$, 3-$OMeC_6H_4$, 4-$OMeC_6H_4$, 3-$CF_3C_6H_4$, 4-FC_6H_4, 2-ClC_6H_4, 4-ClC_6H_4, 2-BrC_6H_4, 4-BrC_6H_4, 2-Nap; Ar^2 = Ph, 2-Py, 2-MeC_6H_4, 4-MeC_6H_4, 4-FC_6H_4, 4-ClC_6H_4, 3-$CF_3C_6H_4$, 2,4,6-$Me_3C_6H_2$, 4-$OMeC_6H_4$

Figure 7.61: Ag(I)-catalyzed coupling reaction of diorganyl diselenides **20** with arylboronic acids **123** using 10 mol% of $AgNO_3$ in DMSO at 130°C.

7.2.5.5 Coupling reactions using ionic liquids

In 2011, the same research group achieved a coupling reaction of aryl selenenyl chlorides **125** and arylboronic acids **123** in ionic liquids [54]. The coupling of aryl selenenyl chlorides **125** and arylboronic acids **123** was performed in imidazolium ionic liquid [bmim][PF_6] without using any metal catalyst. Unsymmetrical and symmetrical diorganyl selenides **124** were isolated in moderate-to-excellent yields (Figure 7.62) [54]. Other imidazolium ionic liquids such as [bmim][BF_4] and [bmim][NTf_2] were used but lower yields of coupling products **124** were obtained in

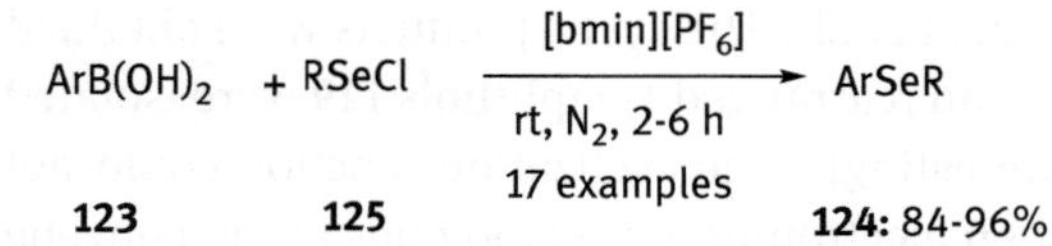

Ar = Ph, 2-MeC_6H_4, 3-MeC_6H_4, 4-MeC_6H_4, 2-OMeC_6H_4, 3-OMeC_6H_4, 4-OMeC_6H_4, 3-$CF_3C_6H_4$, 4-FC_6H_4, 2-ClC_6H_4, 4-ClC_6H_4, 2-BrC_6H_4, 4-BrC_6H_4, 2-Nap; R = ^{n}Bu, Ph, 2-Py, 2-MeC_6H_4, 4-MeC_6H_4, 4-FC_6H_4, 4-ClC_6H_4, 3-$CF_3C_6H_4$, 2,4,6-$Me_3C_6H_2$, 4-OMeC_6H_4

Figure 7.62: Synthesis of diorganyl selenides **124** by the coupling of aryl selenenyl chlorides **125** with arylboronic acids **123** in ionic liquid.

comparison with [bmim][PF_6]. The coupling reaction was working well with both alkyl- and aryl selenyl chlorides. In addition, aryl selenyl bromides can be used as electrophilic substrates under similar reaction conditions.

7.2.6 Carbene insertion reaction

In 2016, Arunprasatha and Sekar reported a metal-free approach for carbene insertion in Se–Se bonds [55]. In this report, *N*-tosylbenzylidenehydrazines **126** were treated with diphenyl diselenide **2** in the presence of potassium *tert*-butoxide in DMSO at 100°C. The carbene insertion reaction proceeded well and bis(phenylseleno)acetals **127** were isolated in good yields (Figure 7.63) [55]. In addition, a similar approach was applied for the carbene insertion in S–S bonds where thioacetals were obtained as reaction products. It was noted that the course reaction was quite similar when both electron-withdrawing and donating groups were used at the aromatic ring in substrates **126.**

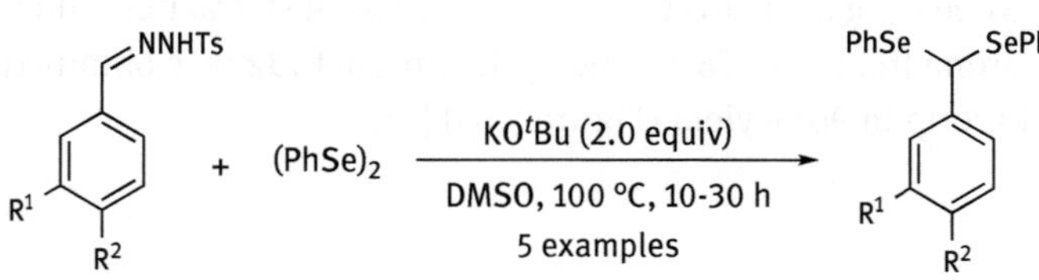

126 **2** **127:** 67-79%

R^1 = H, OMe, Br; R^2 = H, ^{i}Pr, CF_3

Figure 7.63: Metal-free and base-mediated approach for carbene insertion in Se–Se bonds.

7.2.7 Rearrangements

In 2010, Wirth and coworkers developed a new approach for the cyclization of stilbenes **128** (R=Ar) having β-keto ester functionality using phenyl selenenyl chloride in the presence of the Lewis acid $FeCl_3$. The cyclic products were obtained with a 1,2-migration of the aryl group and rearranged 1-naphthols **129** were isolated in good yields (Figure 7.64) [56]. Interestingly, the cyclization reaction could not proceed under similar conditions when electron-donating moieties at the benzene ring in stilbene **128** (R=4-MeO-C_6H_4) were used. The cyclization of stilbene **128** (R=4-MeO-C_6H_4) was achieved by the reaction with *in situ* generated more reactive selenium electrophilic species phenyl selenenyl trifluoracetate by the reaction of bis(trifluoroacetoxy)iodobenzene with stoichiometric amounts of diphenyl diselenide. In addition, styrene **128** (R=Me) was also used as substrate and the mixture of cyclic product with 1,2-methyl migration **130** (R=Me) and without migration **129** (R=Me) was obtained in overall 50 % yield with 2:1 ratio (Figure 7.64) [56]. During the cyclization of **128** (R=Ar), a cyclic product without 1,2-aryl migration **130** (R=Ar) was not observed.

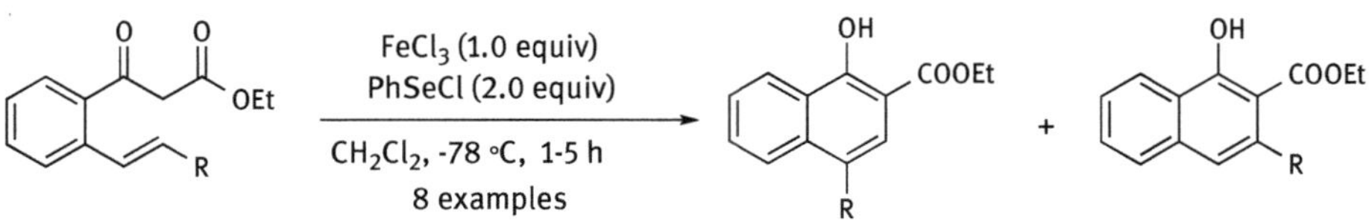

128 **129:** 68-96% **130:** R = Me

R = Me, 2-ClC$_6$H$_4$, 3-ClC$_6$H$_4$, 4-ClC$_6$H$_4$, 4-MeC$_6$H$_4$, 4-OMeC$_6$H$_4$, 1-Nap, 2-Nap

Figure 7.64: The cyclization of stilbenes **128** to cyclic products **129** with 1,2-alkyl- or aryl migration in dichloromethane.

The identical reagent combination ($FeCl_3$ and PhSeCl) was used by Tancock and Wirth to cyclize stilbene **131** bearing a β-keto ester moiety to yield the tetrasubstituted naphthalene **132** in dichloromethane at −78°C. The cyclic product **132** was obtained with 1,2-migration of aryl group in 96 % yield (Figure 7.65) [57].

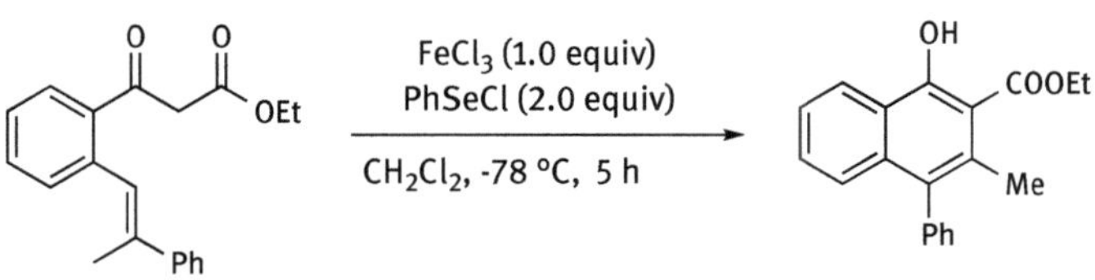

131 **132:** 96%

Figure 7.65: The cyclization of stilbenes **131** to cyclic products **132** with 1,2-aryl migration in dichloromethane.

7.2.8 Catalytic reactions

The scope of organoselenium electrophiles is not limited to their stoichiometric reactions as various catalytic protocols have been developed using organoselenium catalysts [4, 58–63]. In this section, catalytic transformations using organoselenium electrophiles developed in last decade are covered.

7.2.8.1 Selenium-catalyzed lactonization reactions

In 2007, Wirth and coworkers developed the catalytic lactonization of β,γ-unsaturated carboxylic acids **133** to butenolides **135** using 5 mol% of diphenyl diselenide **2** and [bis(trifluoroacetoxy)iodo]benzene (PIFA) **134** as stoichiometric oxidant. The reaction products **135** were isolated in moderate-to-excellent yields (Figure 7.66) [64]. Both aliphatic and aromatic substituents on alkene functionality in substrates **133** are tolerated in this catalytic reaction.

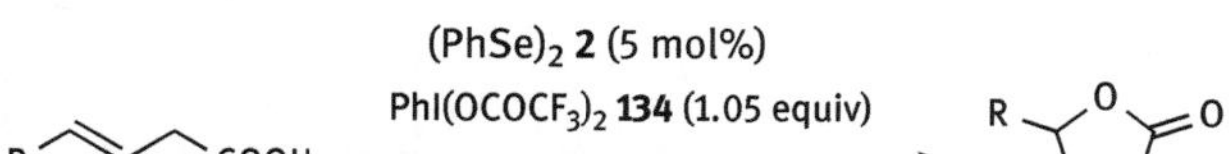

133 **135**: 49-96%

R = ^{n}Pr, ^{n}Bu, ^{n}Pen, ^{n}Dec, Ph, Bn, 4-MeC$_6$H$_4$, 4-BrC$_6$H$_4$, 2-Nap

Figure 7.66: Selenium-catalyzed lactonization of β,γ-unsaturated carboxylic acids **133** to butenolides **135** using diphenyl diselenide **2** as catalyst and [bis(trifluoroacetoxy)iodo]benzene (PIFA) **134** as oxidant.

In 2011, Singh and Wirth extended the same selenium-catalyzed approach for the cyclization γ,δ-unsaturated carboxylic acids **136** to 3,6-dihydro-2*H*-pyran-2-ones **137** using 10 mol% of diphenyl diselenide **2** and [bis(trifluoroacetoxy)iodo]benzene (PIFA) **134** as stoichiometric oxidant in acetonitrile under argon atmosphere. All the lactonization reactions were proceeding well and 3,6-dihydro-2*H*-pyran-2-ones **137** were isolated in good yields (Figure 7.67) [65].

In addition, the lactonization of carboxylic acids **138** was performed under similar reaction conditions. Interestingly, the expected seven-membered lactone was not observed but six-membered lactones **139** were obtained in good yields via an *exo*-cyclization process (Figure 7.68) [65].

The possible reaction mechanism for the selenium-catalyzed lactonization of γ,δ-unsaturated carboxylic acids **136** to 3,6-dihydro-2*H*-pyran-2-ones **137** is shown in Figure 7.69 [65]. According to the catalytic cycle, the reaction is initiated by the formation of the more electrophilic species phenyl selenenyl trifluoroacetate **141** by reaction of diphenyl diselenide **2** with PIFA **134**. Phenyl selenenyl trifluoroacetate **141**

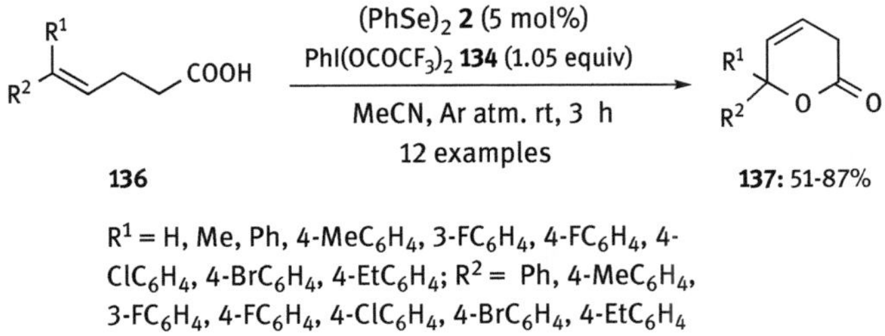

Figure 7.67: Selenium-catalyzed lactonization of γ,δ-unsaturated carboxylic acids **136** to 3,6-dihydro-2*H*-pyran-2-ones **137** using diphenyl diselenide **2** as catalyst and [bis(trifluoroacetoxy)iodo]benzene (PIFA) **134** as oxidant.

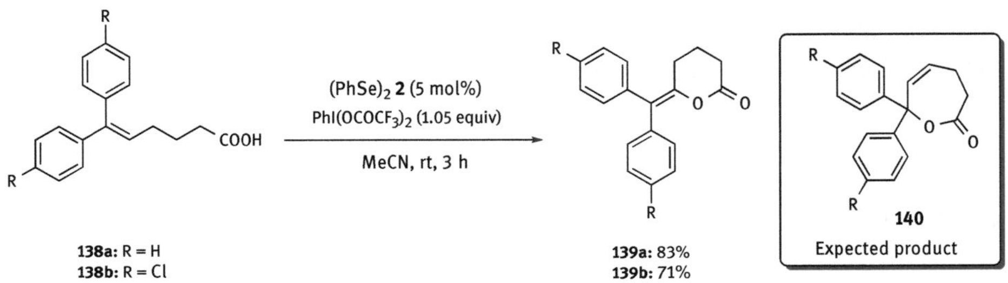

Figure 7.68: Selenium-catalyzed lactonization of carboxylic acids **138** to the corresponding 6-(diarylmethylene)tetrahydro-2*H*-pyran-2-ones **139** using selenium electrophile **2** as catalyst and PIFA **134** as oxidant.

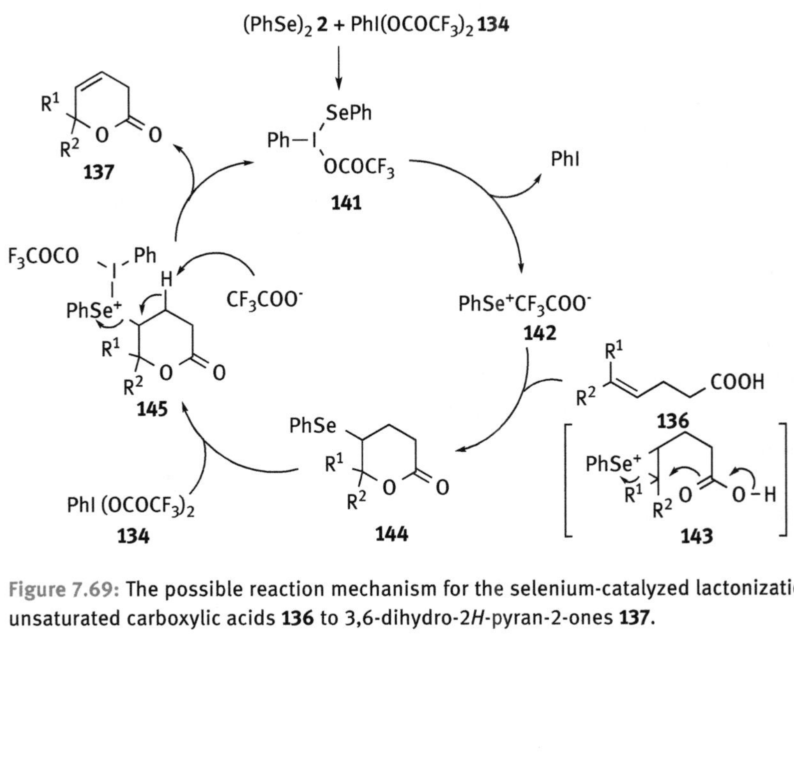

Figure 7.69: The possible reaction mechanism for the selenium-catalyzed lactonization of γ,δ-unsaturated carboxylic acids **136** to 3,6-dihydro-2*H*-pyran-2-ones **137**.

then reacts with the γ,δ-unsaturated carboxylic acid **136** to form selenolactone **144**. The selenide functionality of lactone **144** is then activated for elimination by [bis(trifluoroacetoxy)iodo]benzene *via* intermediate **145** and forms 3,6-dihydro-2*H*-pyran-2-ones **137**. The selenium electrophile **141** is regenerated by this process to continue the catalytic cycle.

In 2010, Wirth and coworkers used a similar selenium-catalyzed approach for the lactonization of (*E*)-2-styrylbenzoic acids **146** to 3-aryl-1*H*-isochromen-1-ones **147** with 5 mol% diphenyl diselenide **2** and [bis(trifluoroacetoxy)iodo]benzene (PIFA) **134** as stoichiometric oxidant. These lactonizations were proceeding well and 3-aryl-1*H*-isochromen-1-ones **147** could be isolated in good yields (Figure 7.70) [66].

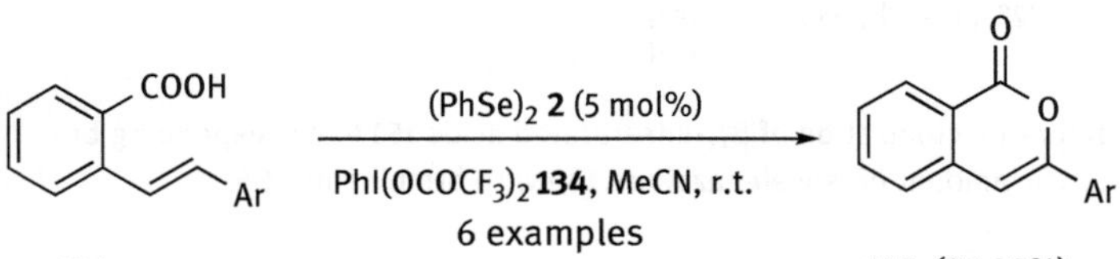

146 **147:** (81-99%)

Ar = Ph, 2-MeC_6H_4, 3-MeC_6H_4, 4-MeC_6H_4, 4-$OMeC_6H_4$, 4-PhC_6H_4, 1-Nap, 2-Nap

Figure 7.70: Selenium-catalyzed lactonization of (*E*)-2-styrylbenzoic acids **146** to corresponding 3-aryl-1*H*-isochromen-1-ones **147** using selenium electrophile **2** as catalyst and PIFA **134** as oxidant.

In 2012, Kumar and coworkers reported another selenium-catalyzed approach for the bromolactonization of functionalized alkenoic acids **148** using catalytic amounts of isoselenazolone **149** with bromine or *N*-bromosuccinimide (NBS) in the presence of potassium carbonate in dichloromethane at room temperature. The lactonization proceeded via an *exo*-cyclization and lactones **150** with different ring sizes were isolated in excellent yields (Figure 7.71) [67].

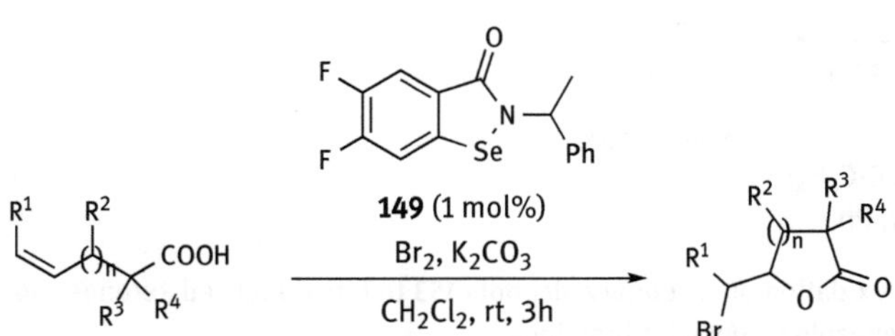

148 **150:** 88-95%

R^1 = H, ^{n}Hex; R^2 = H, Me; R^3 = H, Me; R^4 = H, Me; n = 1,2

Figure 7.71: Selenium-catalyzed lactonization of alkenoic acids **148** to corresponding 3-lactones **150** using catalytic amount of isoselenazolone **149** with bromine or NBS.

The same approach was used in the bromolactonization of β,γ-unsaturated acids **151** under similar reaction conditions. The lactonization proceeded via an *endo*-cyclization process and lactones **152a** and **152b** were isolated in 72% and 95% yield, respectively (Figure 7.72) [67].

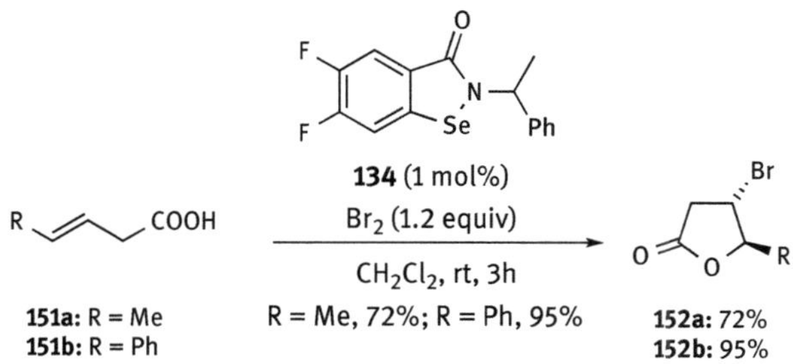

Figure 7.72: Selenium-catalyzed lactonization of β,γ-unsaturated acids **151** to corresponding bromolactones **152** using catalytic amount of isoselenazolone **149** with bromine or NBS.

7.2.8.2 Selenium-catalyzed oxidations of alcohols

In addition, isoselenazolone **149** was used to catalyze the oxidation of secondary alcohols **153**. In this approach, secondary alcohols **153** were reacted with bromine in the presence of catalytic amount of isoselenazolone **149** in dichloromethane. The oxidation reactions were proceeding well and functionalized ketones **154** are obtained as oxidation products in moderate-to-excellent yields (Figure 7.73) [67]. Both cyclic and acylic alcohols were successfully used as substrates in this isoselenazolone-catalyzed oxidation reaction.

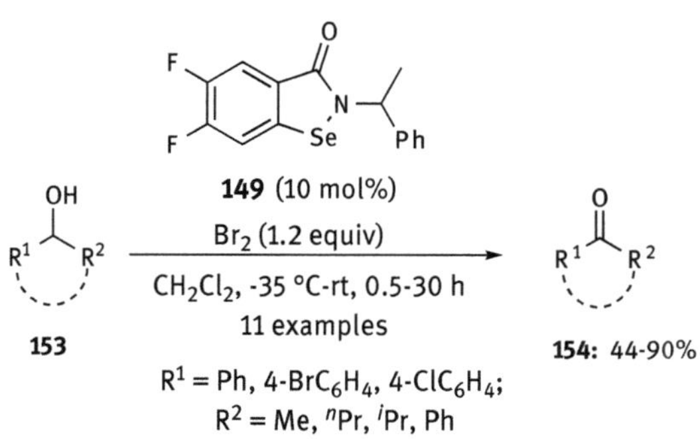

Figure 7.73: Selenium-catalyzed oxidation of secondary alcohols **153** to functionalized ketones **154** using catalytic amount of isoselenazolone **149** with bromine.

7.2.8.3 Selenium-catalyzed aminocyclizations

In 2015, Ortgies and Breder developed a new selenium-catalytic approach for the aminocyclization of styrenes or stilbenes **155** having an amino functionality in the

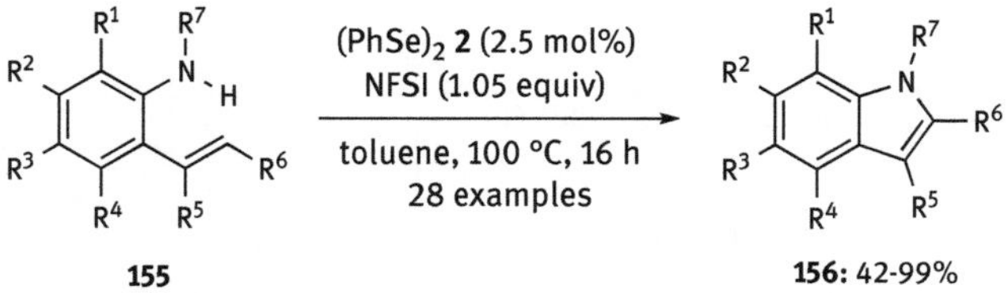

R^1 = H, Br; R^2 = H, CF_3; R^3 = H, Me, Cl, ^{n}Pr, CN, COOMe, R^4 = H, F; R^5 = H, 4-OMeC_6H_4; R^6 = Me, ^{n}Pr, Cy, CH_2OMe, CH_2^{c}Pent, ^{c}Pr, Ph, Bn; R^7 = Ts, Ns, Ms, Ac, $COCCl_3$

Figure 7.74: Selenium-catalyzed aminocyclization of styrenes or stilbenes **155** to functionalized indoles **156** using diphenyl diselenide **2** as catalyst and NFSI as oxidant.

ortho-position using 2.5 mol% of diphenyl diselenide **2** and *N*-fluorobenzenesulfonimide as an oxidant in toluene at 100°C. Various functionalized indoles **156** were obtained as reaction products in high yields (Figure 7.74) [68]. The aminocyclization reaction was tolerant to various electron-donating and withdrawing groups at different positions in the substrates **155**.

In 2015, Zhao and coworkers extended this approach to the aminocyclization of styrenes or stilbenes **155** using 10 mol% of diphenyl diselenide **2** and *N*-fluorobenzenesulfonimide as an oxidant in dioxane at 30°C. Similar functionalized indoles **156** were isolated in poor-to-excellent yields (Figure 7.75) [69]. The aminocyclization reaction was tolerant to various electron-donating and withdrawing groups R at the double bond in the substrates **155**. Additionally, this approach was also applicable for the synthesis of *N*-tosyl-2-substituted indoles under similar reaction conditions.

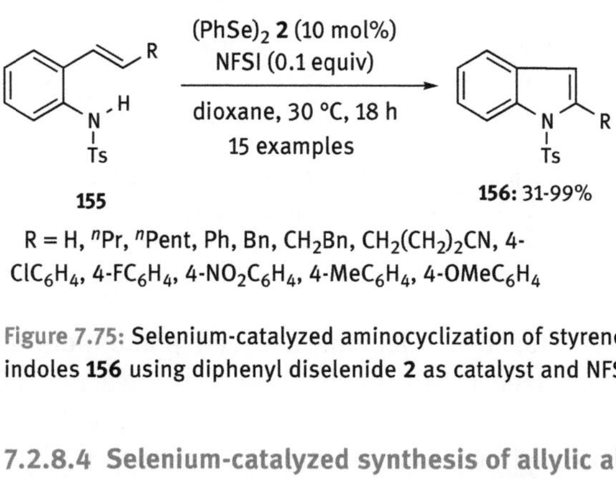

R = H, ^{n}Pr, ^{n}Pent, Ph, Bn, CH_2Bn, $CH_2(CH_2)_2$CN, 4-ClC_6H_4, 4-FC_6H_4, 4-$NO_2C_6H_4$, 4-MeC_6H_4, 4-OMeC_6H_4

Figure 7.75: Selenium-catalyzed aminocyclization of styrenes or stilbenes **155** to functionalized indoles **156** using diphenyl diselenide **2** as catalyst and NFSI as oxidant.

7.2.8.4 Selenium-catalyzed synthesis of allylic alcohols

In 2015, Zhao and coworkers developed a selenium-catalyzed approach for the synthesis of 3-amino allylic alcohols **158** by reaction of terminal alkenes **157** using catalytic amounts of diphenyl diselenide **2** and NFSI as oxidant in the presence of base in THF at room temperature. The reaction products **158** were isolated in good-to-excellent yields

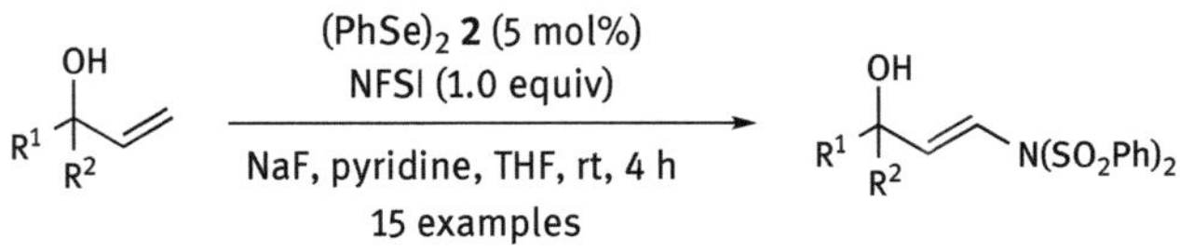

157 158: 63-95%

R^1 = ^{n}Pent, Ph, Bn, CH_2Bn, 4-ClC_6H_4, 4-BrC_6H_4, 4-$CF_3C_6H_4$, 2-$OMeC_6H_4$, 3-$OMeC_6H_4$, 4-MeC_6H_4, 4-$OMeC_6H_4$, 2-Nap; R^2 = H, Me, Et

Figure 7.76: Selenium-catalyzed synthesis of 3-amino allylic alcohols **158** by the reaction of terminal alkenes **157** using diphenyl diselenide **2** as catalyst and NFSI as oxidant.

(Figure 7.76) [70]. Notably, the presence of hydroxyl group for coordinating the selenium reagent seems to be crucial in this reaction.

Additionally, terminal alkenes **157** were reacted with the catalyst diphenyl diselenide **2** and oxidant NFSI in EtOAc without using any base and additive. Interestingly, 3-amino allylic alcohols **158** were not obtained and the isolated compounds were characterized as α,β-unsaturated aldehydes **159** by spectroscopic analysis. The reaction products **159** were obtained in moderate-to-high yields (Figure 7.77) [70].

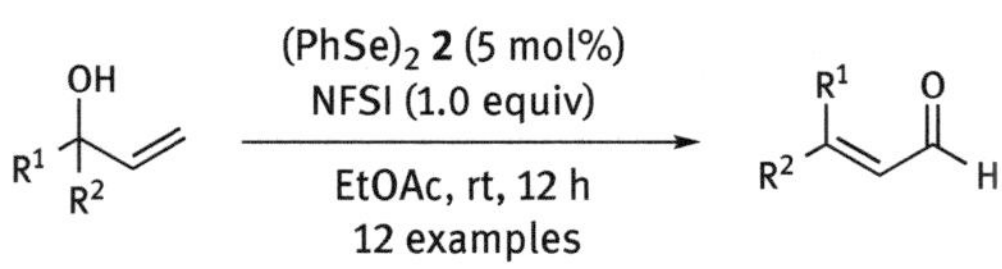

157 159: 43-97%

R^1 = H, Me; R^2 = Ph, Bn, CH_2Bn, 4-ClC_6H_4, 4-BrC_6H_4, 4-$CF_3C_6H_4$, 2-$OMeC_6H_4$, 3-$OMeC_6H_4$ 4-MeC_6H_4, 4-$OMeC_6H_4$, 2-Nap

Figure 7.77: Selenium-catalyzed synthesis of α,β-unsaturated aldehydes **158** by the reaction of terminal alkenes **157** using diphenyl diselenide **2** as catalyst and NFSI as oxidant.

7.2.8.5 Selenium-catalyzed allylic esterification

Recently, Breder and coworkers developed a selenium-catalyzed photo-induced allylic esterification of disubstituted alkenes **160.** In this report, the reaction of alkenes **160**, carboxylic acid **161,** 10 mol% diphenyl diselenide **2** and 5 mol% of photo-catalyst 2,4,6-tris(4-methoxyphenyl)pyrylium tetrafluoroborate **162** in acetonitrile was irradiated at 465 nm for 16–24 h. Allylic esters **163** were obtained as reaction products in reasonable yields (Figure 7.78) [71]. Actually, the photo-catalyst **162** was activated after irradiation at 465 nm and initiated the process of forming the electrophilic selenium species.

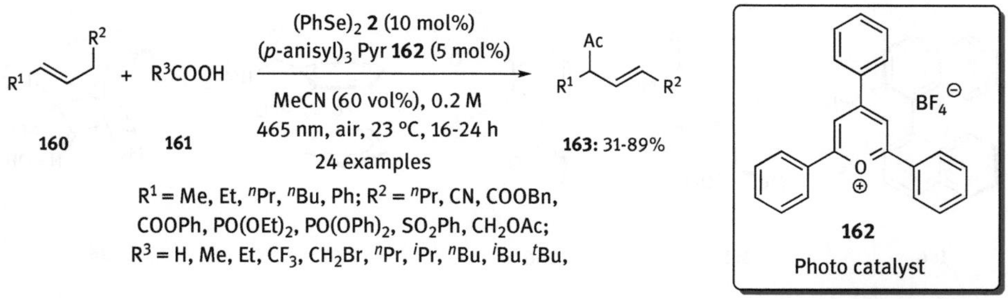

Figure 7.78: Selenium-catalyzed photo-induced allylic esterification of disubstituted alkenes **160** to allylic esters **163**.

7.3 Chiral selenium electrophiles in stereoselective reactions

In last decades, different chiral organoselenium electrophiles have been developed and used to achieve stereoselective selenenylations of alkenes including selenocyclizations [3f, 3g, 3j, 3m, 6e]. Additionally, several chiral selenium electrophiles have been successfully used as catalysts to develop catalytic stereoselective transformations [4a, 4d]. Various chiral organoselenium electrophiles **164–181** have been synthesized and used in asymmetric selenenylation reactions (Figure 7.79). Some new chiral binaphthyl diselenides **164** were synthesized by Fujita and his research group and used in stereoselective methoxyselenenylation of styrene **15** (R=H) (Figure 7.80) [72–77]. The methoxyselenenylated product **184** was obtained with up to 49 % diastereomeric excess (Table 7.2, entry 1) [74]. Déziel and coworkers reported the synthesis of C_2 symmetric chiral diselenides **165** and **166** and used in methoxyselenenylation of styrene **15** (R=H) in diethyl ether at −78°C using methanol as source of nucleophile [78, 80, 81]. The methoxyselenenylated product **184** was obtained in 77 % and 73 % diastereomeric ratio with chiral diselenides **165** and **166**, respectively (Table 7.2, entries 2 and 3) [79, 81].

Furthermore, Uemura and coworkers developed the synthesis of ferrocenyl-cored chiral diselenide of type **167** and used successfully in stereoselective reactions [82–88]. The methoxyselenenylation of styrene **15** (R=H) was performed in dichloromethane at room temperature and selenenylated product **184** was obtained with up to 97 % de (Table 7.2, entry 4) [85].

The synthesis of camphor-based chiral diselenide **168** and its derivatives was developed by Back and others and successfully used in asymmetric selenenylation reactions [89–97]. The methoxyselenenylated product **184** was obtained in 92 % diastereomeric excess using chiral diselenide **168** (Table 7.2, entry 5) [98]. Tomoda and his research group developed the synthesis of chiral diselenides of type **169** containing cyclic amines as chiral moieties and used in different stereoselective reactions [98–101]. These diselenides have a nitrogen atom at the position

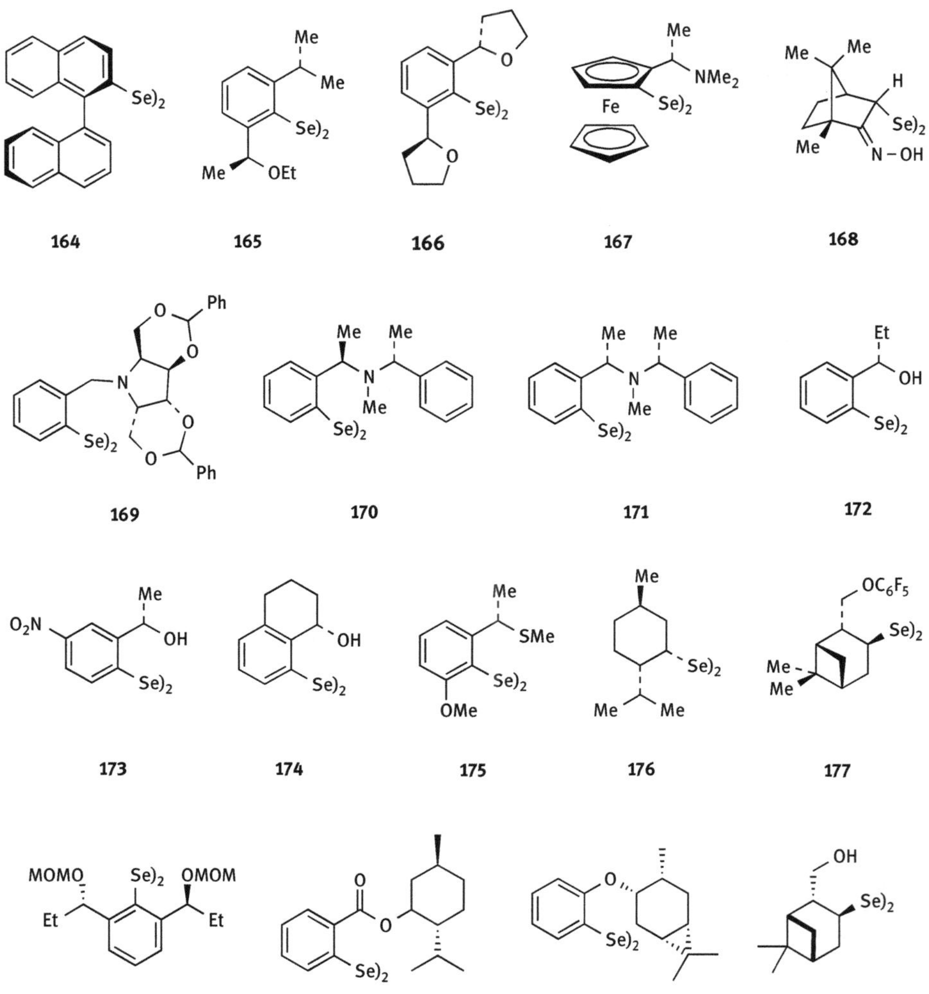

Figure 7.79: The structures of selective chiral selenium electrophiles **164–181**.

segregated by four bonds from the selenium atom. Chiral diselenide **169** showed the best selectivity in asymmetric methoxyselenenylation of styrene **15** (R=H) (Table 7.2, entry 6) [101].

In addition, Wirth and others have developed the synthesis of nitrogen and oxygen-containing chiral diselenides **170, 171** and **172–174**, respectively [102–108]. Asymmetric methoxyselenylation of styrene **15** (R=H) was obtained with high selectivities using nitrogen and oxygen containing chiral diselenides **170, 171** and **172–174** (Table 7.2, entries 7–11). The selenenylation reactions were performed in different solvent systems at different temperatures and best result was observed at –114°C [107]. Furthermore,

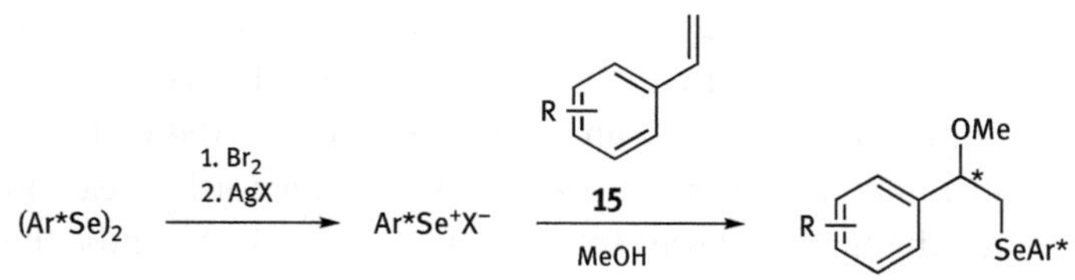

164-181 182 183

Figure 7.80: Asymmetric methoxyselenenylation of styrene using chiral electrophiles **164–181**.

Table 7.2: Stereoselective methoxyselenenylation of styrenes **15**.

Entry	Chiral diselenide	Counterion X	Reaction conditions	184 de (%)	Reference
1	**164**	Br	MeOH, 25°C	49	[74]
2	**165**	OTf	Ether, −78°C	77	[81]
3	**166**	OTf	Ether, −78°C	73	[79]
4	**167**	Br	CH_2Cl_2, 25°C	97	[82]
5	**168**	OTf	CH_2Cl_2/MeOH, −78°C	92	[96]
6	**169**	PF_6	CH_2Cl_2/MeOH, −78°C	42	[101]
7	**170**	OTf	MeOH, −78°C	94	[107]
8	**171**	OSO_3H	MeOH, 25°C	62	[103]
9	**172**	OTf	MeOH, −114°C	92	[103]
10	**173**	OTf	MeOH, −114°C	95	[103]
11	**174**	OTf	MeOH, −100°C	93	[103]
12	**175**	OTf	CH_2Cl_2/MeOH, −78°C	92	[109]
13	**176**	OTf	MeOH, −78°C	40	[111]
14	**177**	OTf	MeOH, −78°C	72	[112]c
15	**178**	OTf	MeOH, −100°C	72	[113]
16	**179**	Br	MeOH, rt	40	[114]
17	**180**	OTf	MeOH, CH_2Cl_2, −78°C	36	[115]
18	**181**	OTf	MeOH, −78°C	44	[116]

Tiecco and coworkers synthesized sulfur-containing chiral diselenide of type **175** and asymmetric methoxyselenenylation reaction styrene was obtained with up to 96% diastereomeric excess [109]. In 2010, Wirth and coworkers synthesized new sulfoxide-containing diselenides and used in asymmetric methoxyselenenylation reactions of alkenes [23]. In addition, Cox and Wirth have reported the synthesis of selenium-stabilized diselenides and used them in stereoselective selenenylation reactions [110].

Furthermore, Scianowski and his research group reported the synthesis of menthol- and terpene-cored chiral diselenides **176** and **177** and used them in asymmetric selenenylation reactions (Table 7.2, entries 13 and 14) [111, 112]. In 2009, Cox and Wirth developed the synthesis of C-2 symmetric chiral diselenide **178** and asymmetric selenenenylation of styrenes **15** was achieved with up to 95% *de* by employing this diselenide **178** (Table 7.2, entry 15) [113]. In 2012, Santi and coworkers achieved asymmetric methoxyselenenylation of styrene with up to 40 % *de* by using chiral diselenide **179** (Table 7.2, entry 16) [114]. In 2016, Scianowski and coworkers reported the synthesis of optically active *ortho*-substituted diaryl diselenide **180** which was further used in similar asymmetric reactions with up to 36 % *de* (Table 7.2, entry 17) [115]. Recently, the same research group developed the synthesis of various dipinanyl diselenides of type **181** which were used in asymmetric methoxyselenenylations with moderate diastereoselectivities (Table 7.2, entry 18) [116].

The scope of chiral organoselenium reagents is not only limited to asymmetric selenenylation reactions but also several other synthetically important asymmetric transformations such as selenocyclizations [117], aminoselenocyclizations [118], carboselenocyclizations [119] and α-functionalizations of carbonyl compounds [120] have been successfully achieved using these electrophiles. Additionally, stereoselective catalytic reactions have also been developed using selenium electrophiles as catalysts [4, 67, 121].

7.4 Tellurium electrophiles in organic synthesis

The chemistry of tellurium electrophiles is by far not as well explored as other chalcogen electrophiles and the applications of these electrophiles in organic synthesis are quite limited [122]. In 1996, Uemura and coworkers reported the application of tellurium electrophiles in coupling reactions [123]. Furthermore, Zeni and Comasseto synthesized few *Z*-vinylic tellurides which were used as electrophiles in coupling reactions with alkynes [124]. In 2000, some unsaturated organotellurium electrophiles were applied in the palladium-catalyzed C–C bond formation with nucleophilic species such as diethylzinc [125]. Furthermore, Zeni and others reported the synthesis of geminal enediynes by palladium-catalyzed coupling reaction of alkynes with electrophilic ketene butyl telluroacetals [126].

In 2003, Braga and others developed Sonogashira cross-coupling reactions of terminal alkynes **184** using vinylic tellurodichlorides **185** in the presence of a catalytic combination of $PdCl_2/CuI$ (1:1) in methanol. Enynes **186** were isolated as cross-coupled products and obtained in good yields (Figure 7.81) [127].

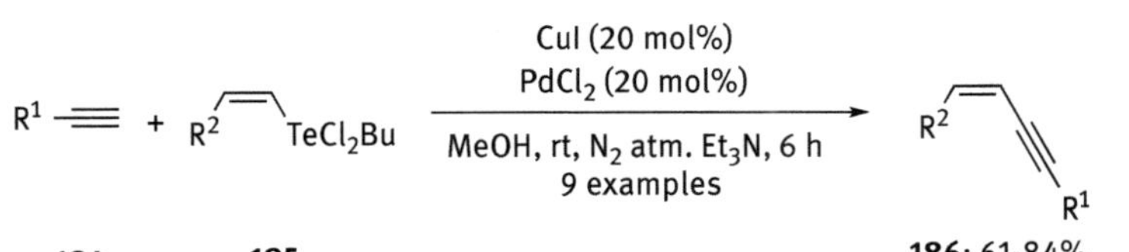

184 **185** **186**: 61-84%

R^1 = Ph, C_8H_{17}, $P(O)(OEt)_2$; R^2 = CH_2OH, CH_2NH_2, $(CH_2)_2CH_2OH$, $(Me)_2CHOH$

Figure 7.81: Pd-catalyzed coupling of alkynes **184** with vinylic tellurodichlorides **185** in methanol.

Furthermore, Stefani and coworkers developed a Suzuki-type coupling reaction of (*Z*)-vinylic tellurides **187** with potassium alkynyltrifluoroborate salts **188** using catalytic combination Pd(acac)$_2$/CuI (1:2) in methanol. Enynes **186** were isolated as cross-coupled products in good yields (Figure 7.82) [128].

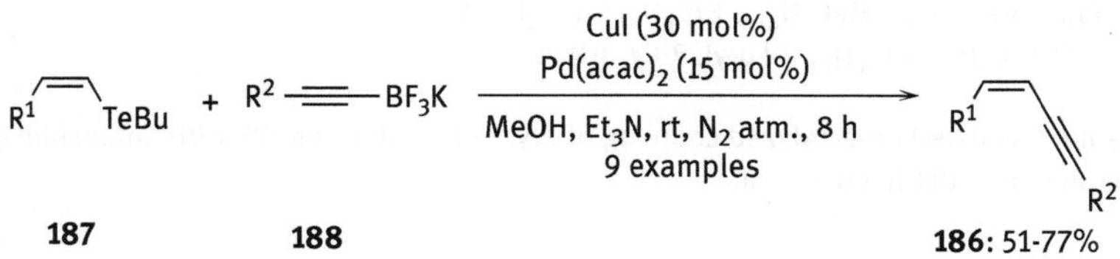

R^1 = Ph, CH$_2$OH, COOEt,CH$_2$NEt$_2$;
R^2 = C$_8$H$_{17}$, CH$_2$OMe, (Me)$_3$C

Figure 7.82: Pd-catalyzed coupling of potassium alkynyltrifluoroborate salts **188** with vinylic tellurides **187** in methanol.

Also, vinylic tellurides **187** were used for the synthesis of 1,3-dienes by Pd-catalyzed coupling with potassium vinylic trifluoroborates [129]. In 2006, Cella and Stefani used similar (*Z*)-vinylic tellurides **189** for the synthesis of (*Z*)-stilbenes **191** by Pd-catalyzed Suzuki coupling with potassium organotrifluoroborate salts **190.** All the coupling reactions were performed in ultrasonic bath at room temperature and (*Z*)-stilbenes **191** were obtained in good-to-excellent yields (Figure 7.83) [130]. Both electron-donating and withdrawing groups at the aromatic ring in substrates **189** and **190** were tolerated in the reaction conditions shown in Figure 7.83.

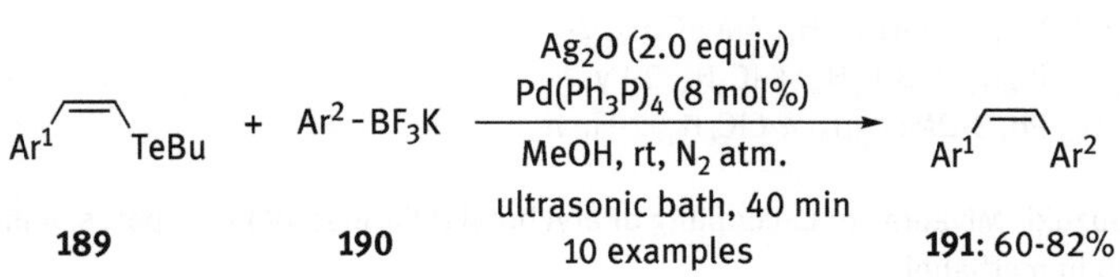

Ar^1 = Ph, 4-MeC$_6$H$_4$, 4-BrC$_6$H$_4$; Ar^2 = Ph, 4-ClC$_6$H$_4$, 4-OMeC$_6$H$_4$, 4-MeC$_6$H$_4$, 2-MeC$_6$H$_4$, 2-Furyl, 3-Py, 1-Nap

Figure 7.83: Ultrasound-assisted Pd-catalyzed coupling of vinylic tellurides **189** with potassium aryltrifluoroborate salts **190** in methanol.

In addition, aryl butyl tellurides **193** were synthesized and used as an electrophilic species in Pd-catalyzed Suzuki coupling with potassium (*E*)-vinyl trifluoroborate salts **192** under the reaction conditions mentioned in Figure 7.83. The coupling products (*E*)-stilbenes **194** were isolated in good-to-excellent yields (Figure 7.84) [130]. Both electron-donating and withdrawing groups at the aromatic ring in

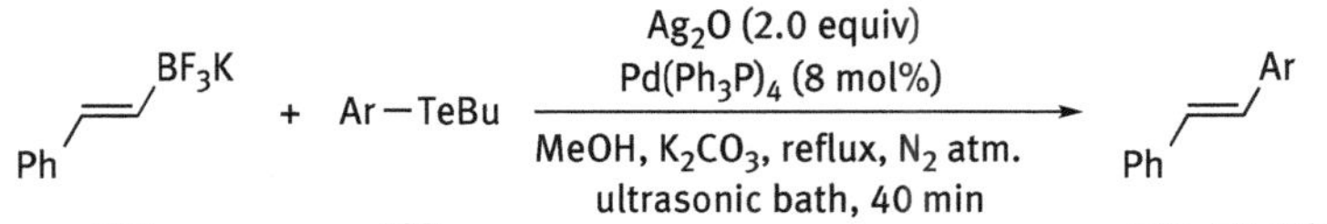

192 193 194: 59-91%

Ar = Ph, 4-MeC$_6$H$_4$, 2-MeC$_6$H$_4$, 4-BrC$_6$H$_4$, 4-ClC$_6$H$_4$, 4-OMeC$_6$H$_4$, 4-IC$_6$H$_4$, 2-Furyl, 3-Py, 1-Nap

Figure 7.84: Ultrasound-assisted Pd-catalyzed coupling of aryl butyl tellurides **193** with potassium (*E*)-vinyl trifluoroborate salts **192** in methanol.

substrates **192** and **193** are tolerated in this coupling reaction. The role of Ag$_2$O was to reoxidize the Pd(0) to Pd(II) species.

In 2006, Stefani and others developed the palladium-catalyzed Suzuki–Miyaura cross-coupling of aryl butyl tellurides **193** with potassium aryltrifluoroborate salts **190** using 10 mol% of Pd(Ph$_3$P)$_4$ in methanol at reflux temperature. Symmetrical and unsymmetrical biaryls **195** were isolated in moderate-to-excellent yields (Figure 7.85) [131]. The reaction was working efficiently when coupling was performed with the substrates having both electron-donating and withdrawing substituents on the aromatic ring.

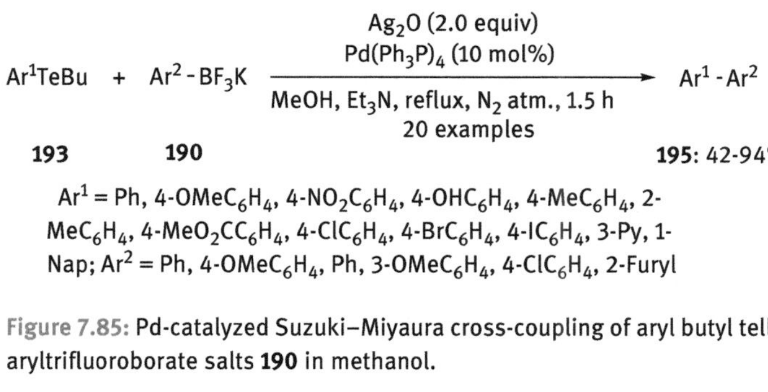

193 190 195: 42-94%

Ar^1 = Ph, 4-OMeC$_6$H$_4$, 4-NO$_2$C$_6$H$_4$, 4-OHC$_6$H$_4$, 4-MeC$_6$H$_4$, 2-MeC$_6$H$_4$, 4-MeO$_2$CC$_6$H$_4$, 4-ClC$_6$H$_4$, 4-BrC$_6$H$_4$, 4-IC$_6$H$_4$, 3-Py, 1-Nap; Ar^2 = Ph, 4-OMeC$_6$H$_4$, Ph, 3-OMeC$_6$H$_4$, 4-ClC$_6$H$_4$, 2-Furyl

Figure 7.85: Pd-catalyzed Suzuki–Miyaura cross-coupling of aryl butyl tellurides **193** with potassium aryltrifluoroborate salts **190** in methanol.

In 2008, a different organotellurium electrophile, α-*n*-butyl tellurostyrene **196**, was synthesized and used in palladium-catalyzed coupling reaction with potassium alkynyltrifluoroborate salts **188** using 10 mol% of Pd(Ph$_3$P)$_4$ in methanol at room temperature in an ultrasonic bath. The coupling products 1,3-enynes **197** were isolated in high yields (Figure 7.86) [132].

Alkynyl butyl tellurides **199** were used in similar coupling reactions with potassium α-styryltrifluoroborate salt **198**. Functionalized 1,3-enynes **197** were obtained in good yields (Figure 7.87) [132].

The electrophilic species **193** was also employed in palladium-catalyzed coupling reactions with potassium α-styryltrifluoroborate salt **198** with 10 mol% of Pd(Ph$_3$P)$_4$

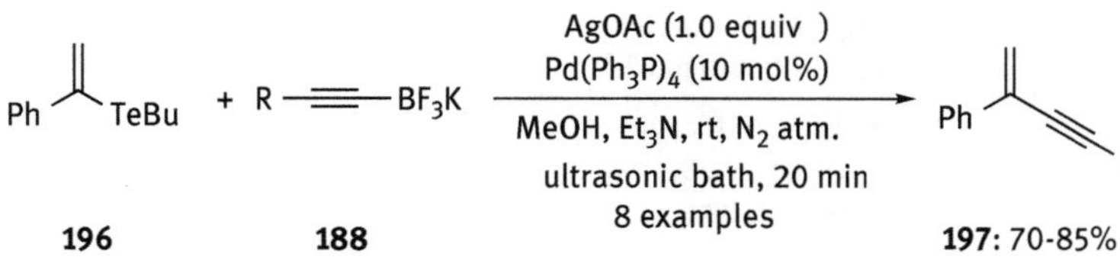

R = Ph, $(Me)_3Si$, C_6H_{13}, (Me)(Et)(OMe)C, cyclohexene and 2-Nap etc

Figure 7.86: Pd-catalyzed Suzuki–Miyaura cross-coupling of α-*n*-butyltellurostyrene **196** with potassium alkynyltrifluoroborate salts **188** in methanol.

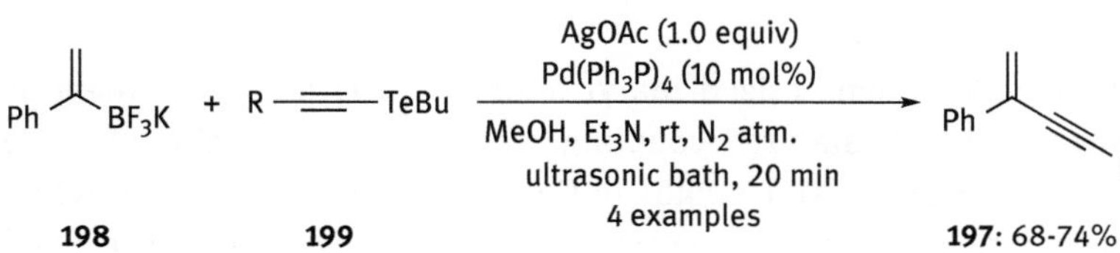

R = Ph, C_5H_{11}, C_6H_{13}, cyclohexene

Figure 7.87: Pd-catalyzed Suzuki–Miyaura cross-coupling of alkynyl butyl tellurides **199** with potassium α-styryltrifluoroborate salt **198** in methanol.

and 1 equiv of AgOAc as an additive in methanol at room temperature. The reaction mixture was irradiated in an ultrasonic bath and functionalized 1,1-diarylethenes **200** were isolated in high yields (Figure 7.88) [133]. Both electron-withdrawing and donating groups at the aromatic ring in tellurides **193** were tolerated.

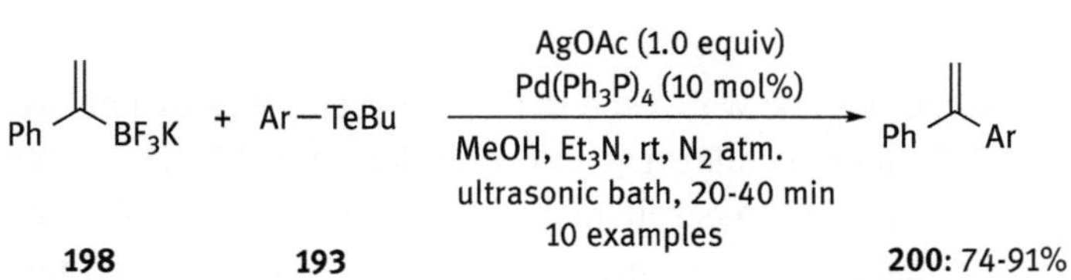

Ar = Ph, 3-$CF_3C_6H_4$, 2-MeC_6H_4, 4-$OMeC_6H_4$, 4-MeC_6H_4, 4-OHC_6H_4, 4-ClC_6H_4, 4-BrC_6H_4, 4-FC_6H_4

Figure 7.88: Pd-catalyzed Suzuki–Miyaura cross-coupling of aryl butyl tellurides **193** with potassium α-styryltrifluoroborate salt **198** in methanol.

Furthermore, Stefani and coworkers synthesized a new tellurium electrophile **201** having tellurium and chlorine moiety and used in Pd-catalyzed coupling reaction with potassium aryltrifluoroborate salts **190** using 10 mol% of $Pd(Ph_3P)_4$ and 1 equiv of AgOAc as an additive in methanol at room temperature. The coupling products **202** were obtained in good-to-excellent yields (Figure 7.89) [134]. Interestingly, the

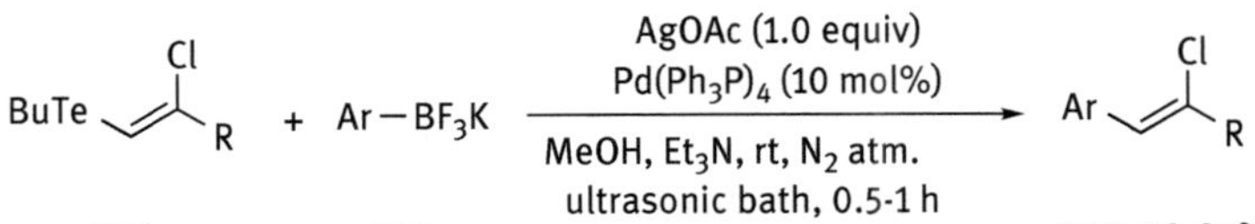

201 190 202: 58-85%

R = Ph, C_3H_7, C_5H_{11}, C_6H_{13}, CH_2OMe, $(Me)_3C$, 1-cyclohexene, Ar = 4-OMeC_6H_4, 4-ClC_6H_4

Figure 7.89: Pd-catalyzed Suzuki–Miyaura cross-coupling of alkenyl butyl tellurides **201** with potassium aryltrifluoroborate salts **190** in methanol.

chlorine substituent remained in the coupling products while the tellurium moiety was participating in the reaction.

Stefani and coworkers synthesized diaryl tellurides **203** and used them as electrophilic partners in Pd-catalyzed coupling reactions with arylzinc chloride **204** with 10 mol% of $Pd(dppf)_4$ and 2 equiv of CuI as an additive in THF at reflux temperature. Symmetrical and unsymmetrical biaryls **197** were isolated in good yields (Figure 7.90) [135].

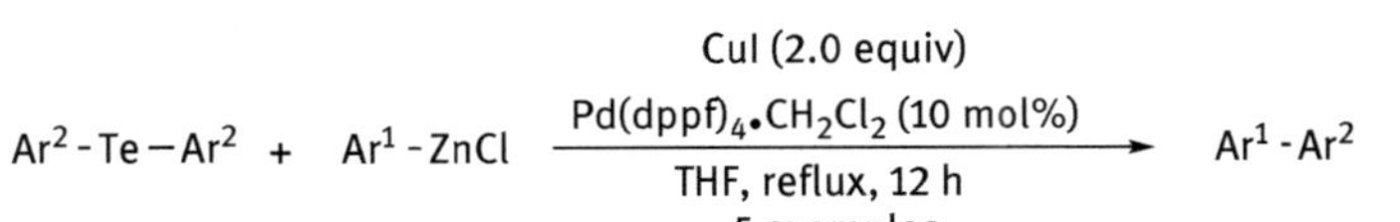

203 204 195: 32-88%

Ar^1 = Ph, 4-OMeC_6H_4; Ar^2 = Ph, 4-FC_6H_4, 4-MeC_6H_4, 4-OMeC_6H_4, 4-ClC_6H_4, thionyl

Figure 7.90: Pd-catalyzed Suzuki–Miyaura cross-coupling of diaryl tellurides **203** with arylzinc chloride **204** in THF.

In the same report, diaryl tellurides **203** were used in Pd-catalyzed coupling reaction with alkynylzinc chloride **205** under similar reaction conditions and 1,3-substituted alkynes **206** were obtained in good-to-excellent yields (Figure 7.91) [135]. Both electron-withdrawing and donating substituents at the aromatic ring in diaryl tellurides **206** are tolerated in the coupling with organozinc reagents **204** and **205.**

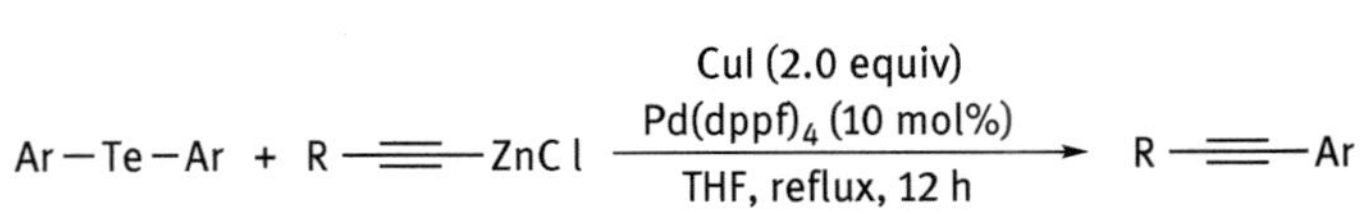

203 205 206: 59-90%

R = Ph, ^{n}Bu; Ar = Ph, 4-FC_6H_4, 4-MeC_6H_4, 4-OMeC_6H_4, 4-ClC_6H_4

Figure 7.91: Pd-catalyzed Suzuki–Miyaura cross-coupling of diaryl tellurides **203** with alkynylzinc chloride **205** in THF.

Stefani and coworkers also developed palladium-catalyzed homo-coupling reactions successfully using aryl butyl tellurides **193** and alkynyl butyl tellurides **199** under mild reaction conditions [136]. Additionally, few organotellurium electrophiles have been successfully used in cycloaddition reactions with both cyclic and acyclic olefins [137, 138].

References

[1] (a) Flohe L, Gunzler EA, Hh S. Glutathione peroxidase: a selenoenzyme. FEBS Lett. 1973;32:132–4; (b) Rotruck JT, Pope AL, Ganther HE, Swanson AB, Hafeman DG, Hoekstra WG. Selenium: biochemical role as a component of glutathione peroxidase. Science. 1973;179:588–90; (c) Behne D, Kyriakopoulos A, Meinhold H, Kohrle J. Identification of type I iodothyronine 5′-deiodinase as a selenoenzyme. Biochem Biophys Res Commun. 1990;173:1143–9; (d) Tamura T, Stadtman TC. A new selenoprotein from human lung adenocarcinoma cells: purification, properties, and thioredoxin reductase activity. Proc Natl Acad Sci USA. 1996;93:1006–10; (e) Thomas SP, Satheeshkumar K, Mugesh G, Row TNG. Unusually short chalcogen bonds involving organoselenium: insights into the Se–N bond cleavage mechanism of the antioxidant ebselen and analogues. Chem Eur J. 2015;21:6793–800; (f) Wirth T. Small organoselenium compounds: more than just glutathione peroxidase mimics. Angew Chem. 2015;127:10212–4; Angew Chem Int Ed. 2015;54:10074–6; (g) Akhoon SA, Naqvi, T Nisar S, Rizvi MA. Synthetic organo-selenium compounds in medicinal domain: a Review. Asian J Chem. 2015;27:2745–52; (h) Bhowmick D, Srivastava S, D'Silva P, Mugesh G. Highly efficient glutathione peroxidase and peroxiredoxin mimetics protect mammalian cells against oxidative damage. Angew Chem Int Ed. 2015;54:8569–73.

[2] (a) Sharpless KB, Lauer RF. Mild procedure for the conversion of epoxides to allylic alcohols. First organoselenium reagent. J Am Chem Soc. 1973;95:2697–9; (b) Sonnet PE. Olefin inversion. Tetrahedron 1980;36:557–604; (c) Clive DLJ, Kale VN. Generation of carbon-carbon double bonds from β-oxygenated phenylseleno, phenylthio, and iodo species. A new use for the chlorotrimethylsilane-sodium iodide reagent. J Org Chem. 1981;46:231–4; (d) Schmid H. A standard definition of the term "thiranium ion" phosphorus sulfur silicon relat elem. 1988;36:197–200; (e) Mihelich ED. Stereoselective synthesis of highly substituted tetrahydrofurans through acid-catalyzed ring closure of selenyl diols. J Am Chem Soc. 1990;112:8995–7. (f) Wirth T. Chiral selenium compounds in organic synthesis. Tetrahedron. 1999;55:1–28.

[3] (a) Song CE, Oh CR, Roh EJ, Choo DJ. Cr(salen) catalysed asymmetric ring opening reactions of epoxides in room temperature ionic liquids. Chem Commun. 2000;1743–4; (b) Wirth T. Organoselenium chemistry in stereoselective reactions. Angew Chem Int Ed. 2000;39:3740–9; (c) Yang M, Zhu C, Yuan F, Huang Y, Pan Y. Enantioselective ring-opening reaction of meso-epoxides with ArSeH catalyzed by heterometallic Ti–Ga–Salen system. Org Lett. 2005;7:1927–30; (c) Browne DM, Wirth T. New developments with chiral electrophilic selenium reagents. Curr Org Chem. 2006;10:1893–903. (d) Santi C, Santoro S, Testaferri L, Tiecco M. A simple zinc-mediated preparation of selenols. Synlett. 2008:1471–4; (e) Freudendahl DM, Shahzad SA, Wirth T. Recent advances in organoselenium chemistry. Eur J Org Chem. 2009;1649–64; (f) Shahzad SA, Wirth T. Fast synthesis of benzofluorenes by selenium-mediated carbocyclizations. Angew Chem Int Ed. 2009;48:2588–91; (g) Rodrigues OED, Souza D, Soares LC, Dornelles L, Burrow RA, Appelt HR, et al. Stereoselective synthesis of selenosteroids. Tetrahedron Lett. 2010;51:2237–40; (h) Waskow B, Mano RA, Giacomini RX, Oliveira DH, Schumacher RF, Wilhelm EA, et al. Synthesis and beckmann rearrangement of novel (*Z*)-2-organylselanyl ketoximes:

promising agents against grapevine anthracnose infection. Tetrahedron Lett. 2016;57:5575–80; (i) Wilkins LC, Günther BAR, Walther M, Lawson JR, Wirth T, Melen RL, et al. Contrasting frustrated Lewis pair reactivity with selenium- and boron-based Lewis acids. Angew Chem Int Ed. 2016;55:11292–5.

[4] (a) Braga AL, Ludtke DS, Vargas F, Braga RC. Catalytic applications of organoselenium compounds in asymmetric synthesis. Synlett. 2006;1453–66. (b) Santi C, Santaro S, Battistelli B. Organoselenium compounds as catalysts in nature and laboratory. Curr Org Chem. 2010;14:2442–62; (c) Freudendahl DM, Santoro S, Shahzad SA, Santi C, Wirth T. Green chemistry with selenium reagents: development of efficient catalytic reactions. Angew Chem Int Ed. 2009;48:8409–11; (d) Singh FV, Wirth T. Selenium compounds as ligands and catalysts. In: Wirth T, editor. Organoselenium chemistry: synthesis and reactions. Ist ed. Weinheim, Germany: Wiley-VCH, 2011:321–56; (e) Santi C, Lorenzo RD, Tidei C, Bagnoli L, Wirth T. Stereoselective selenium catalysed dihydroxylation and hydroxymethoxylation of alkenes. Tetrahedron 2012;68:10530–5; (f) Curran SP, Connon SJ. Selenide ions as catalysts for homo- and crossed tishchenko reactions of expanded scope. Org Lett. 2012;14:1074–7.

[5] (a) Nicolaou KC, Petasis NA. Selenium in natural products synthesis. Philadelphia: CIS, 1984; (b) Paulmier C. Selenium reagents and intermediates in organic synthesis. Pergamon Press: Oxford, 1986; (c) Patai S, Rappoport Z. The chemistry of organic selenium and tellurium compounds, vol. 1. 1986; (d) Patai S, Rappoport Z. The chemistry of organic selenium and tellurium compounds, vol. 2. Wiley, 1987; (e) Liotta D, Organoselenium chemistry. New York: John Wiley & Sons, 1987; (f) Krief A, Hevesi L, Organoselenium chemistry I. Berlin: Springer, 1988; (g) Back TG. Organoselenium chemistry. Oxford: Oxford University Press, 1999; (h) Wirth T. Organoselenium chemistry: modern developments in organic synthesis. Top Curr Chem. Springer: Berlin, 2000: 208; (j) Wirth T. Organoselenium chemistry: synthesis and reactions. Germany: Wiley-VCH, 2011; (k) Patai S, Rappoport Z. The chemistry of organic selenium and tellurium compounds, vol. 3, Wiley, 2012.

[6] (a) Nishibayashi Y, Uemura S. Selenoxide elimination and [2,3]-sigmatropic rearrangement. In: Wirth T editor. Topics in current chemistry, vol. 208, 1st ed. Berlin, Germany: Springer, 2000:201–34; (b) Beaulieu PL, Déziel R. In: Back TG, editor. Organoselenium chemistry: a practical approach, Ist ed. Oxford: Oxford University Press, 1999:35–66; (c) Freudendahl DM, Wirth T. New selenium electrophiles and their reactivity. In: Woolins JD, Laitinen RS, editor. Selenium and Tellurium Chemistry, Ist ed. Berlin, Germany: Springer, 2011:41–56; (d) Bhujan BJ, Mugesh G. Biological and biochemical aspects of selenium compounds. In Wirth T, Organoselenium chemistry. Weinheim, Germany: Wiley-VCH, 2011:361–92; (e) Singh FV, Wirth T. Stereoselective reactions of organoselenium reagents including catalysis. In: Rappoport Z, editor. Patai Series: Organic Selenium and Tellurium Compounds, Vol. 3, 2nd ed. West Sussex, England: John Wiley & Sons, 2012:303–56; (f) Nogueira CW, Rocha JBT. Organoselenium and organotellurium compounds: toxicology and pharmacology. In: Rappoport Z, editor, Patai Series: Organic Selenium and Tellurium Compounds, Vol. 3, West Sussex, England: John Wiley & Sons, 2012:1277–358; (g) Bhuyan BJ, Lamani DS, Mugesh G, Wirth T. Current research on mimics and models of selenium-containing antioxidants. In: Devillanova FA, du Mont W-W, Handbook of chalcogen chemistry: new perspectives in sulfur, selenium and tellurium, Vol 2, 2nd ed. Cambridge, UK: RSC, 2013:25–46; (h) Bhowmick D, Mugesh G. Enzyme mimetic chemistry of organoselenium compounds. In: Rappoport Z, editor. Patai Series: Organic Selenium and Tellurium Compounds, vol. 4, 3rd ed. West Sussex, England: John Wiley & Sons, 2013:1175–236; (i) Braga AL, Rafique J. Synthesis of biologically relevant small molecules containing selenium. Part C. Miscellaneous biological activities. In: Rappoport Z, editor. Patai series: organic selenium and tellurium compounds. vol. 4, 3rd ed. West Sussex,

England: John Wiley & Sons, 2014:1175-236; (j) Singh FV, Wirth T. Synthesis of organoselenium compounds with potential biological activities. In: Jain VK, editor, Priyadarsini KI, editors. Organoselenium compounds in biology and medicine: synthesis, biological and therapeutic treatments. 1st ed. Cambridge, UK: RSC, 2018:77–121.

[7] (a) Reich HJ. Functional group manipulation using organoselenium reagents. Acc Chem Res. 1979;12:22–30; (b) Tiecco M, Testaferri L, Tingoli M, Bagnoli L, Marini F, Santi C, Temperini A. Production of reactivity of new organoselenium intermediates. Formation of carbon-oxygen and carbon-nitrogen bonds. Gazz Chimi Ital. 1996;126:635-43; (c) Nishibayashi Y, Uemura S. Novel synthesis of chiral differrocenyl dichalcogenides and their application to asymmetric reactions. Heteroatom Chem. 1996;14:83–118; (d) Wessjohann LA. Benzeneselenenyl reagents in organic synthesis. J Prakt Chem. 1998;340:189–203; (e) Uemura S. Chiral diferrocenyl dichalcogenides in asymmetric synthesis. Phosphorus Sulfur. 1998;136–138:219–34; (f) Mugesh G, Singh HB. Synthetic organoselenium compounds as antioxidants: glutathione peroxidase activity. Chem Soc Rev. 2000;29:347–57; (g) Tiecco M. Electrophilic selenium, selenocyclizations. Top Curr Chem. 2000;208:7–54; (h) Ren X, Yang L, Liu J, Su D, You D, Liu C, et al. A novel glutathione peroxidase mimic with antioxidant activity. Arch Biochem Biophys. 2001;387:250–6; (i) Petragnani N, Stefani HA, Valduga CJ. Recent advances in selenocyclofunctionalization reactions. Tetrahedron. 2001;57:1411–48; (j) Mugesh G, du Mont WW, Sies H. Chemistry of biologically important synthetic organoselenium compounds. Chem Rev. 2001;101:2125–80; (k) Tiecco M, Testaferri L, Marini F, Bagnoli L, Santi C, Temperini A, et al. Asymmetric syntheses promoted by organoselenium reagent. Phosphorus Sulfur. 2005;180:729–40; (l) Guillena G, Ramon DJ. Enantioselective α-heterofunctionalization of carbonyl compounds: organocatalysis is the simplest approach. Tetrahedron: Asymmetry. 2006;17:1465–92; (m) Narajji C, Karvekar MD, Das AK. Biological importance of organoselenium compounds. Indian J Pharm Sci. 2007;69:344–51; (n) Sharma BK, Mugesh G. Thiol cofactors for selenoenzymes and their synthetic mimics. Org Biomol Chem. 2008;6:965–74; (o) Bhabak KP, Mugesh G. Functional mimics of glutathione peroxidase: Bioinspired synthetic antioxidant. Acc Chem Res. 2010;43:1408–19; (p) Alberto EE, do Nascimento V, Braga AL. Catalytic application of selenium and tellurium compounds as glutathione peroxidase enzyme mimetics. J Braz Chem Soc. 2010;21:2032–41; (q) Mukherjee AJ, Zade SS, Singh HB, Sunoj RB. Organoselenium chemistry: role of intramolecular interactions. Chem Rev. 2010;110:4357–414; (r) Rafique J, Canto RFS, Saba S, Barbosa FAR, Braga AL. Recent advances in the synthesis of biologically relevant selenium-containing 5-membered heterocycles. Curr Org Chem. 2016;20:166–88; (s) Reich HJ, Hondal RJ. Why nature chose selenium. ACS Chem Biol. 2016;11:821–41.

[8] (a) Toshimitsu A, Aoai T, Owada H, Uemura S, Okano M. Phenylselenenyl chloride in acetonitrile-water: a highly convenient reagent for hydroxyselenation of olefins and preparation of cyclic ethers from dienes. Tetrahedron. 1985;41:5301–6; (b) Toshimitsu A, Aoai T, Owada H, Uemura S, Okano M. Highly convenient method for hydroxyselenation of olefins. J Chem Soc Chem Commun. 1980:412–3.

[9] Tiecco M, Testaferri L, Tingoli M, Chianelli D, Bartoli D. The reaction of diphenyl diselenide with peroxydisulphate ions in methanol a convenient procedure to effect the methoxyselenenylation of alkenes. Tetrahedron Lett. 1989;30:1417–20.

[10] Nicolaou KC, Claremon DA, Barnette WE, Seitz SP. *N*-Phenylselenophthalimide (*N*-PSP) and *N*-phenylselenosuccinimide (*N*-PSS). Two versatile carriers of the phenylseleno group. Oxyselenation of olefins and a selenium-based macrolide synthesis. J Am Chem Soc. 1979;101:3704–6.

[11] Nicolaou KC, Petasis NA, Claremon DA. *N*-phenylselenophthalimide (NPSP): A valuable selenenylating agent. Tetrahedron. 1985;41:4835–41.

[12] Yoshshida M, Sasage S, Kawamura K, Suzuki T, Kamigata N. Oxidative cleavage of diselenide by m-nitrobenzenesulfonyl peroxide: novel method for the electrophilic benzeneselenenylations of olefins and aromatic rings. Bull Chem Soc Jpn. 1991;64:416–22.

[13] Tingoli M, Tiecco M, Testafem L, Temperini A. Iodosobenzene diacetate and diphenyl diselenide: an electrophilic selenenylating agent of double bonds. Synth Commun. 1998;28:1769–78.

[14] Tingoli M, Diana R, Panunz B. N-Phenylselenosaccharin (NPSSac): a new electrophilic selenium-containing reagent. Tetrahedron Lett. 2006;47:7529–31.

[15] Gabriele E, Singh FV, Freudendahl DM, Wirth T. Selenenylations of alkenes with styrene nucleophiles. Tetrahedron. 2012;68:10573–6.

[16] Shi H-W, Yu C, Yan J. Potassium bromide or sodium chloride catalyzed acetoxyselenenylation of alkenes with diselenides and *m*-CPBA. Chin Chem Lett. 2015;26:1117–20.

[17] Vieira AA, Azeredo JB, Godoi M, Santi C, Da Silva EN, Braga AL. Catalytic chalcogenylation under greener conditions: A solvent-free sulfur- and seleno-functionalization of olefins via I_2/DMSO oxidant system. J Org Chem. 2015;80:2120–7.

[18] Wang X-L, Li H-J, Yan J. Iodine-mediated regioselective hydroxyselenenylation of alkenes: facile access to β-hydroxy selenides. Chin Chem Lett. 2018;29:479–81 .

[19] Tiecco M, Testaferri L, Marini F, Sternativo F, Santi C, Bagnoli L, et al. Conjugated additions of selenium containing enolates to enones–enantioselective synthesis of δ-oxo-α-seleno esters and their facile transformations. Eur J Org Chem. 2005;543–51.

[20] Murata S, Suzuki T. Stereoselective selenolactonization by superelectrophilic benzeneselenenyl triflate. Chem Lett. 1987;849–52.

[21] Murata S, Suzuki T. Cyclization of olefinic alcohol by benzeneselenenyl triflate. Tetrahedron Lett. 1987;28:4297–8.

[22] Tiecco M, Testaferri L, Tingoli M, Bartoli D, Balducci R. Ring-closure reactions initiated by the peroxydisulfate ion oxidation of diphenyl diselenide. J Org Chem. 1990;55:429–34.

[23] Freudendahl DM, Iwaoka M, Wirth T. Synthesis of new sulfoxide-containing diselenides and unexpected cyclization reactions to 2,3-dihydro-1,4-benzoselenothiine 1-oxides. Eur J Org Chem. 2010;3934–44.

[24] Viglianisi C, Simone L, Menichetti S. Copper-mediated one-pot transformation of 2-n-sulfonylaminoaryl diselenides into benzo[*b*][1,4]selenazines. Adv Synth Catal. 2012;354:77–82.

[25] Menichetti S, Capperucci A, Tanini D, Braga AL, Botteselle GV, Viglianisi C. A one-pot access to benzo[*b*][1,4]selenazines from 2-aminoaryl diselenides. Eur J Org Chem. 2016;3097–102.

[26] Angeli A, Tanini D, Viglianisi C, et al. Evaluation of selenide, diselenide and selenoheterocycle derivatives as carbonic anhydrase I, II, IV, VII and IX inhibitors. Bioorg Med Chem. 2017;25:2518–23.

[27] Wang K, Bates RW. Synthesis of Allahabadolactone A. J Org Chem. 2017;82:12624–30.

[28] Prasad CD, Sattar M, Kumar S. Transition-metal-free selective oxidative $C(sp^3)$–S/Se coupling of oxindoles, tetralone, and arylacetamides: synthesis of unsymmetrical organochalcogenides. Org Lett. 2017;19:774–7.

[29] Lee DH, Kim YH. Regioselective phenylselenenylation at the 5-position of pyrimidine nucleosides mediated by manganese(III) acetate. Synlett. 1995;349–50.

[30] Shibahara F, Kanai T, Yamaguchi E, Kamei A, Yamauchi T, Murai T. Copper-catalyzed C–H bond direct chalcogenation of aromatic compounds leading to diaryl sulfides, selenides, and diselenides by using elemental sulfur and selenium as chalcogen sources under oxidative conditions. Chem Asian J. 2014;9:237–44.

[31] Zhu L, Qiu R, Cao X, Copper-mediated remote C–H bond chalcogenation of quinolines on the C5 position. Org Lett. 2015;17:5528–31.

[32] Mandal A, Sahoo H, Baidya M. Copper-catalyzed 8-aminoquinoline-directed selenylation of arene and heteroarene C–H bonds. Org Lett. 2016;18:3202–5.

[33] Jin W, Zheng P, Wong W-T, Law W-L. Efficient palladium-catalyzed direct C–H phenylselenylation of (hetero)arenes in water. Asian J Org Chem. 2015;4:875–8.
[34] Jin W, Zheng P, Law G-L, Wong W-T. Palladium(II)-catalyzed switchable mono-/diselenylation of arenes controlled by solvent effects. J Organomet Chem. 2016;812:66–73.
[35] Zhang Q-B, Ban Y-L, Yuan P-F, et al. Visible-light-mediated aerobic selenation of (hetero)arenes with diselenides. Green Chem. 2017;19:5557–63.
[36] Rafique J, Saba S, Rosrio AR, Braga AL. Regioselective, solvent- and metal-free chalcogenation of imidazo[*1,2-a*]pyridines by employing I_2/DMSO as the catalytic oxidation system. Chem Eur J. 2016;22:11854–62.
[37] Guo T. Ammonium iodide-mediated regioselective chalcogenation of chromones with diaryl disulfides and diselenides. Synth Commun. 2017;47:2053–61.
[38] Amosova SV, Penzik MV, Albanov AI, Potapov VA. The reaction of selenium dichloride with divinyl sulfide. J Organomet Chem. 2009;694:3369–72.
[39] Bilheri FN, Stein AL, Zeni G. Synthesis of chalcogenophenes via cyclization of 1,3-diynes promoted by iron(III) chloride and dialkyl dichalcogenides. Adv Synth Catal. 2015;357:1221–8.
[40] Neto JSS, Back DF, Zeni G. Nucleophilic cyclization of *o*-alkynylbenzamides promoted by iron (III) chloride and diorganyl dichalcogenides: synthesis of 4-organochalcogenyl-1*H*-isochromen-1-imines. Eur J Org Chem. 2015;1583–90.
[41] Stein AL, Rosário AR, Zeni G. Synthesis of 3-organoseleno-substituted quinolines through cyclization of 2-aminophenylprop-1-yn-3-ols promoted by iron(III) chloride with diorganyl diselenides. Eur J Org Chem. 2015;5640–8.
[42] Recchi AMS, Back DF, Zeni G. Sequential carbon–carbon/carbon–selenium bond formation mediated by iron(III) chloride and diorganyl diselenides: synthesis and reactivity of 2-organoselenyl-naphthalenes. J Org Chem. 2017;82:2713–23.
[43] Neto JSS, Iglesias BA, Back DF, Zeni G. Iron-promoted tandem cyclization of 1,3-diynyl chalcogen derivatives with diorganyl dichalcogenides for the synthesis of benzo[*b*]furan-fused selenophenes. Adv Synth Catal. 2016;358:3572–85.
[44] Casola KK, Back DF, Zeni G. Iron-Catalyzed cyclization of alkynols with diorganyl diselenides: synthesis of 2,5-dihydrofuran, 3,6-dihydro-2*H*-pyran, and 2,5-dihydro-1*H*-pyrrole organoselanyl derivatives. J Org Chem. 2015;80:7702–12.
[45] Goulart TAC, Back DF, Zeni G. Copper-catalyzed carbon-nitrogen/carbon-selenium bonds formation: synthesis of 2-(organochalcogenyl)-indolizines. Adv Synth Catal. 2017;359:1901–11.
[46] Shahzad SA, Wirth T. Fast synthesis of benzofluorenes by selenium-mediated carbocyclizations. Angew Chem Int Ed. 2009;48:2588–91.
[47] Taniguchi N. Convenient synthesis of unsymmetrical organochalcogenides using organoboronic acids with dichalcogenides via cleavage of the S-S, Se-Se, or Te-Te bond by a copper catalyst. J Org Chem. 2007;72:1241–5.
[48] Alves D, Santos CG, Paixãod MW, et al. CuO nanoparticles: an efficient and recyclable catalyst for cross-coupling reactions of organic diselenides with aryl boronic acids. Tetrahedron Lett. 2009;50:6635–8.
[49] Ricordi VG, Freitas CS, Perin G, et al. Glycerol as a recyclable solvent for copper-catalyzed cross-coupling reactions of diaryl diselenides with aryl boronic acids. Green Chem. 2012;14:1030–4.
[50] Zheng B, Gong Y, Xu H-J. Copper-catalyzed Ce–se coupling of diphenyl diselenide with arylboronic acids at room temperature. Tetrahedron. 2013;69:5342–7.
[51] Wang M, Ren K, Wang L. Iron-catalyzed ligand-free carbon-selenium (or tellurium) coupling of arylboronic acids with diselenides and ditellurides. Adv Synth Catal. 2009;351:1586–94.
[52] Ren K, Wang M, Wang L. Lewis acid $InBr_3$-catalyzed arylation of diorgano diselenides and ditellurides with arylboronic acids. Org Biomol Chem. 2009;7:4858–61.

[53] Goldani B, Ricordi VG, Seus N, Lenardao EJ, Schumacher RF, Alves D. Silver-catalyzed synthesis of diaryl selenides by reaction of diaryl diselenides with aryl boronic acids. J Org Chem. 2016;81:11472–6.
[54] Freitas CS, Barcellos AM, Ricordi VG, et al. Synthesis of diaryl selenides using electrophilic selenium species and nucleophilic boron reagents in ionic liquids. Green Chem. 2011;13:2931–8.
[55] Arunprasatha D, Sekar G. A transition-metal-free and base-mediated carbene insertion into sulfur-sulfur and selenium-selenium bonds: an easy access to thio- and selenoacetals. Adv Synth Catal. 2016;358:1–12.
[56] Shahzad SA, Vivant C, Wirth T. Selenium-mediated synthesis of biaryls through rearrangement. Org Lett. 2010;12:1364–7.
[57] Tancock J, Wirth T. Selenium-mediated synthesis of tetrasubstituted naphthalenes through rearrangement. Molecules. 2015;20:10866–72.
[58] Hori T, Sharpless KB. Selenium-catalyzed nonradical chlorination of olefins with *N*-chlorosuccinimide. J Org Chem. 1979;44:4204–8.
[59] Tunge JA, Mellegaard SR Selective selenocatalytic allylic chlorination. Org Lett. 2004;6:1205–7.
[60] Barrero AF, del Moral Jfq, Herrador MM, et al. Solid-phase selenium-catalyzed selective allylic chlorination of polyprenoids: facile syntheses of biologically active terpenoids. J Org Chem. 2006;71:5811–4.
[61] Mellegaard SR, Tunge JA. Selenium-catalyzed halolactonization: nucleophilic activation of electrophilic halogenating reagents. J Org Chem. 2004;69:8979–81.
[62] Carrera I, Brovetto MC, Seoane GA. Selenium-catalyzed iodohydrin formation from alkenes. Tetrahedron Lett. 2006;47:7849–52.
[63] Ehara H, Noguchi M, Sayama S, Onami T. Bis[2-(2-pyridyl)phenyl] diselenide, a more effective catalyst for oxidation of alcohols to carbonyl compounds. J Chem Soc Perkin Trans. 2000;1:1429–31.
[64] Browne DM, Niyomura O, Wirth T. Catalytic use of selenium electrophiles in cyclizations. Org Lett. 2007;9:3169–71.
[65] Singh FV, Wirth T. Selenium-catalyzed regioselective cyclization of unsaturated carboxylic acids using hypervalent iodine oxidants. Org Lett. 2011;13:6504–7.
[66] Shahzad SA, Venin C, Wirth T. Diselenide- and disulfide-mediated synthesis of isocoumarins. Eur J Org Chem. 2010;465–72.
[67] Balkrishna SJ, Prasad CD, Panini P, Detty MR, Chopra D, Kumar S, et al. Isoselenazolones as catalysts for the activation of bromine: bromolactonizationof alkenoic acids and oxidation of alcohols. J Org Chem. 2012;77:9541–52.
[68] Ortgies S, Breder A. Selenium-catalyzed oxidative C(sp^2)–H amination of alkenes exemplified in the expedient synthesis of (aza-)indoles. Org Lett. 2015;17:2748–51.
[69] Zhang X, Guo R, Zhao X. Organoselenium-catalyzed synthesis of indoles through intramolecular C–H amination. Org Chem Front. 2015;2:1334–7.
[70] Deng Z, Wei J, Liao L, Huang H, Zhao X. Organoselenium-catalyzed, hydroxy-controlled regio- and stereoselective amination of terminal alkenes: efficient synthesis of 3-amino allylic alcohols. Org Lett. 2015;17:1834–7.
[71] Ortgies S, Depken C, Breder A. Oxidative allylic esterification of alkenes by cooperative selenium-catalysis using air as the sole oxidant. Org Lett. 2016;18:2856–9.
[72] Tomoda S, Iwaoka M, Yakushi K, Kawamoto A, Tanaka J. Synthesis of optically active 2,2'-diselenocyanato-1,1'-binaphthyl and its molecular structure. Hypervalent nature of divalent selenium atom in crystal state. J Phys Org Chem. 1988;1:179–84.
[73] Tomoda S, Iwaoka M. Synthesis of selenium-containing binaphthyls and their application to the asymmetric ring-opening of cyclohexene oxide. J Chem Soc Chem Commun. 1988;1283–4.

[74] Tomoda S, Iwaoka M. Asymmetric trans-addition reactions of a chiral selenobinaphthyl with prochiral olefins. The case of methoxyselenenylation. Chem Lett. 1988;17:1895–8.
[75] Tomoda S, Fujita K, Iwaoka M. Asymmetric oxyselenenylation of olefins using optically active selenobinaphthyls and *d*-menthol as a nucleophile. Chem Lett. 1990;19:1123–4.
[76] Tomoda S, Fujita K, Iwaoka M. Double differentiation in asymmetric methoxyselenenylation of *trans*-β-methylstyrene. J Chem Soc Chem Commun. 1990;129–31.
[77] Tomoda S, Fujita K, Iwaoka M. Asymmetric *trans*-addition reactions using chiral selenobinaphthyls. Phosphorus, Sulfur Silicon Relat Elem. 1992;67:247–52.
[78] Déziel R, Goulet S, Grenier L, Bordeleau J, Bernier J. Asymmetric selenomethoxylation of olefins involving a chiral C2 symmetrical electrophilic organoselenium reagent. J Org Chem. 1993;58:3619–21.
[79] Déziel R, Malenfant E. Asymmetric Ring Closure Reactions Mediated by a Chiral C2 Symmetrical Organoselenium Reagent. J Org Chem. 1995;60:4660–2.
[80] Déziel R, Malenfant E, Bélanger G. Practical synthesis of (*R,R*)- and (*S,S*)-bis[2,6-bis(1-ethoxyethyl)phenyl] diselenide. J Org Chem. 1996;61:1875–6.
[81] Déziel R, Malenfant E, Thibault C, Frechette S, Gravel M. 2,6-Bis[(2S)-tetrahydrofuran-2-yl] phenyl diselenide: an effective reagent for asymmetric electrophilic addition reactions to olefins. Tetrahedron Lett. 1997;38:4753–6.
[82] Nishibayashi Y, Jd S, Uemura S, Fukuzawa S. Synthesis of chiral diferrocenyl diselenides and their application to asymmetric reactions. Tetrahedron Lett. 1994;35:3115–8.
[83] Fukuzawa S, Kasugahara Y, Uemura S. A highly selective asymmetric methoxyselenylation of alkenes with a chiral ferrocenylselenium reagent. Tetrahedron Lett. 1994;35:9403–6.
[84] Nishibayashi Y, Singh JD, Fukuzawa S, Uemura S. Synthesis of [R,S;R,S]- and [S,R;S,R]-bis[2-[1-(dimethylamino)ethyl]ferrocenyl] diselenides and their application to asymmetric selenoxide elimination and [2,3]-sigmatropic rearrangement. J Org Chem. 1995;60:4114–20.
[85] Nishibayashi Y, Srivastava SK, Takada H, Fukuzawa S, Uemura S. Highly selective asymmetric intramolecular selenocyclisation. J Chem Soc Chem Commun. 1995;2321–2.
[86] Uemura S. Chiral diferrocenyl dichalcogenides in asymmetric synthesis. Phosphorus Sulfur Silicon Relat Elem. 1998;136-138:219–34.
[87] Fukuzawa S, Takahashi K, Kato H, Yamazaki H. Asymmetric methoxyselenenylation of alkenes with chiral ferrocenylselenium reagents. J Org Chem. 1997;62:7711–6.
[88] Bolm C, Kesselgruber M, Grenz A, Hermanns N, Hildebrand JP. A novel ferrocenyl diselenide for the catalytic asymmetric aryl transfer to aldehydes. New J Chem. 2001;25:13–5.
[89] Back TG, Dyck BP, Parvez M. Unexpected formation of 1,3-diselenetanes from the reaction of camphor enolate with selenium. J Chem Soc Chem Commun. 1994;515–6.
[90] Back TG, Dyck BP, Parvez M. 1,3-Diselenetanes and 1,3-dithietanes derived from camphor. Formation, structure, stereochemistry, and oxidation to selenoxide and sulfoxide products. J Org Chem. 1995;60:703–10.
[91] Back TG, Dyck BP. Asymmetric cyclization of unsaturated alcohols and carboxylic acids with camphor-based selenium electrophiles. Chem Commun. 1996;2567–8.
[92] Back TG, Nan S. Asymmetric methoxyselenenylations with camphor-based selenium electrophiles. J Chem Soc Perkin Trans. 1998;1:3123–4.
[93] Tiecco M, Testaferri L, Santi C, Marini F, Bagnoli L, Temperini A, et al. Asymmetric selenohydroxylation of alkenes with camphorselenenyl sulfate. Eur J Org Chem. 1998;2275–7.
[94] Tiecco M, Testaferri L, Santi C, Marini F, Bagnoli L, Temperini A. Asymmetric selenomethoxylation of alkenes with camphorselenenyl sulfate. Tetrahedron Lett. 1998;39:2809–12.
[95] Back TG, Dyck BP, Nan S. Asymmetric electrophilic methoxyselenenylations and cyclizations with 3-camphorseleno derivatives. Tetrahedron. 1999;55:3191–208.

[96] Back TG, Moussa Z. New chiral auxiliaries for highly stereoselective asymmetric methoxyselenenylations. Org Lett. 2000;2:3007–9.
[97] Back TG, Moussa Z, Parvez M. Asymmetric methoxyselenenylations and cyclizations with 3-camphorseleno electrophiles containing oxime substituents at C-2: formation of an unusual oxaselenazole from an oxime-substituted selenenyl bromide. J Org Chem. 2002;67:499–509.
[98] Fujita K, Murata K, Iwaoka M, Tomoda S. Asymmetric intramolecular selenoetherification and selenolactonization using an optically active diaryl diselenide derived from D-mannitol. J Chem Soc Chem Commun. 1995;1641–2.
[99] Fujita K, Iwaoka M, Tomoda S. Synthesis of diaryl diselenides having chiral pyrrolidine rings with c_2 symmetry. Their application to the asymmetric methoxyselenenylation of *trans*-β-methylstyrenes. Chem Lett. 1994;923–6.
[100] Fujita K, Murata K, Iwaoka M, Tomoda S. Asymmetric methoxyselenenylation of olefins using an optically active diaryl diselenide derived from d-mannitol. Tetrahedron Lett. 1995;36:5219–22.
[101] Fujita K, Murata K, Iwaoka M, Tomoda S. Design of optically active selenium reagents having a chiral tertiary amino group and their application to asymmetric inter- and intramolecular oxyselenenylations. Tetrahedron. 1997;53:2029–48.
[102] Wirth T. Asymmetric reaction of arylalkenes with diselenides. Angew Chem Int Ed Engl. 1995;34:1726–8.
[103] Wirth T. A new stereoselective synthesis of chiral tetrahydrothiopyrans and their corresponding sulfoxides. Tetrahedron Lett. 1995;36:1849–52.
[104] Wirth T, Kulicke KJ, Fragale G. Chiral diselenides from benzylamines: catalysts in the diethylzinc addition to aldehydes. Helv Chim Acta. 1996;79:1957–66.
[105] Wirth T, Fragale G. Asymmetric addition reactions with optimized selenium electrophiles. Chem Eur J. 1997;3:1894–902.
[106] Santi C, Fragale G, Wirth T. Synthesis of a new chiral nitrogen containing diselenide as a precursor for selenium electrophiles. Tetrahedron: Asymmetry. 1998;9:3625–8.
[107] Tiecco M, Testaferri L, Santi C, Tomassini C, Marini F, Bagnoli L, et al. New nitrogen containing chiral diselenides: synthesis and asymmetric addition reactions to olefins. Tetrahedron: Asymmetry. 2000;11:4645–50.
[108] (a) Uehlin L, Fragale G, Wirth T. New and efficient chiral selenium electrophiles. Chem Eur J. 2002;8:1125–33; (b) Santi C, Lorenzo RD, Tidei C, Bagnoli L, Wirth T. Stereoselective selenium catalyzed dihydroxylation and hydroxymethoxylation of alkenes. Tetrahedron. 2012;68:10530–5.
[109] (a) Tiecco M, Testaferri L, Bagnoli L, Marini F, Temperini A, Tomassini C, et al. Efficient asymmetric selenomethoxylation and selenohydroxylation of alkenes with a new sulfur containing chiral diselenide. Tetrahedron Lett. 2000;41:3241–5; (b) Tiecco M, Testaferri L, Santi C, Tomassini C, Marini F, Bagnoli L, et al. Preparation of a new chiral non-racemic sulfur-containing diselenide and applications in asymmetric synthesis. Chem Eur J. 2002;8:1118–24.
[110] Cox M, Wirth T. Synthesis of a selenium-substituted diselenide. Phosphorus Sulfur Silicon Relat Elem. 2005;180:659–66.
[111] (a) Scianowski J, Rafinski Z, Wojtczak A. Syntheses and reactions of new optically active terpene dialkyl diselenides. Eur J Org Chem. 2006;3216–25. (b) Rafinski Z, Scianowski J. Synthesis and reactions of enantiomerically pure dialkyl diselenides from the *p*-menthane group. Tetrahedron: Asymmetry. 2008;19:1237–44.
[112] (a) Scianowski J. Convenient route to dialkyl diselenides from alkyl tosylates. Synthesis of di (cis-myrtanyl) diselenide. Tetrahedron Lett. 2005;46:3331–4; (b) Rafinski Z, Scianowski J, Wojtczak A. Asymmetric selenocyclization with the use of dialkyl monoterpene diselenides. Tetrahedron: Asymmetry. 2008;19:223–30; (c) Scianowski J, Rafinski Z, Szuniewicz A, Wojtczak A, New chiral selenium electrophiles derived from functionalized terpenes. Tetrahedron. 2009;65:10162–74; (d) Scianowski J, Rafinski Z, Szuniewicz A, Burczynski K, Syntheses and

reactions of terpene β-hydroxyselenides and β-hydroxydiselenides. Tetrahedron: Asymmetry. 2009;20:2871–9.
[113] Uehlin L, Wirth T. Synthesis of chiral acetals by asymmetric selenenylations. Phosphorus Sulfur Silicon Relat Elem. 2009;184:1374–85.
[114] Incipini L, Rongoni E, Bagnoli L, Marini F, Santi C. New chiral electrophilic selenium reagents: synthesis and structural investigation. 16th International Electronic Conference on Synthetic Organic Chemistry 2012, 1–6.
[115] Scianowski J, Pacuła AJ, Zielinska-Błajetb M, Wojtczakc A. New diphenyl diselenides *O*-substituted by an O(S,Se)-caranyl skeleton-synthesis and application in asymmetric reactions. New J Chem. 2016;40:6697–705.
[116] Ścianowski J, Szumera J, Pacuła AJ, Rafiński Z. Reactivity of dipinanyl diselenides functionalized at the C-10-position with -CH_2O(Se)Ph, -OH and -$OCPh_3$ substituents. Arkivoc. 2017;ii:272–84.
[117] (a) Deziel R, Malenfant E. Asymmetric ring closure reactions mediated by a chiral c2 symmetrical organoselenium reagent. J Org Chem. 1995;60:4660–2; (b) Wirth, T, Kulicke KJ, Fragale G. Chiral diselenides in the total synthesis of (+)-Samin. J Org Chem. 1996;61:2686–9; (c) Fragale G, Wirth T, Fragale G, Neuburger M. New and efficient selenium reagents for stereoselective selenenylation reactions. Chem Commun. 1998:1867–8; (d) Tiecco M, Testaferri L, Bagnoli L, Purgatorio V, Temperini A, Marini F, et al. Efficient asymmetric selenocyclizations of alkenyl oximes into cyclic nitrones and 1,2-oxazines promoted by sulfur containing diselenides. Tetrahedron: Asymmetry. 2001;12:3297–304; (e) Tiecco M, Testaferri L, Santi C, Tomassini C, Marini F, Bagnoli L, et al. Asymmetric synthesis of thioamido selenides. A simple synthetic route to enantiopure thiazolines. Tetrahedron: Asymmetry. 2002;13:429–35; (f) Tiecco M, Testaferri L, Temperini A, Bagnoli L, Marini F, Santi C, et al. A New synthesis of α-phenylseleno γ- and δ-lactones from terminal alkynes. Synlett. 2003;655–8; (g) Aprile C, Gruttadauria M, Amato ME, Anna FD, Meo PL, Riela S, et al. Studies on the stereoselective selenolactonization, hydroxy and methoxy selenenylation of α- and β-hydroxy acids and esters. Synthesis of δ- and γ-lactones. Tetrahedron. 2003;59:2241–51; (h) Khokhar SS, Wirth T. Selenocyclizations: control by coordination and by the counterion. Angew Chem Int Ed. 2004;43:631–3 (i) Khokhar SS, Wirth T. Nucleophile-selective selenocyclizations. Eur J Org Chem. 2004:4567–81; (j) Tiecco M, Testaferri L, Santi C, Tomassini C, Marini F, Bagnoli L, et al. Synthesis of enantiomerically pure perhydrofuro[2,3-b]furans. Tetrahedron: Asymmetry. 2005;16:2429–35; (j) Tiecco M, Testaferri L, Bagnoli L, Scarponi C, Temperini A, Marini F, et al. Organoselenium mediated asymmetric cyclizations. Synthesis of enantiomerically pure 1,6-dioxaspiro[4.4]nonanes. Tetrahedron: Asymmetry. 2006;17:2768–74.
[118] (a) Wirth T, Fragale G. Stereoselective isoquinoline alkaloid synthesis with new diselenides. Synthesis. 1998;162–6; (b) Tiecco M, Testaferri L, Santi C, Tomassini C, Marini F, Bagnoli L, et al. Asymmetric amidoselenenylation of alkenes promoted by camphorselenenyl sulfate: a useful synthetic route to enantiopure oxazolines. Eur J Org Chem. 2000:3451–7; (c) Tiecco M, Testaferri L, Marini F, Sternativo S, Santi C, Bagnoli L, et al. Optically active isoxazolidines and 1,3-amino alcohols by asymmetric selenocyclization reactions of O-allyl oximes. Tetrahedron: Asymmetry. 2001;12:3053–9; (d) Tiecco M, Testaferri L, Santi C, Tomassini C, Marini F, Bagnoli L, et al. Asymmetric azidoselenenylation of alkenes: a key step for the synthesis of enantiomerically enriched nitrogen-containing compounds. Angew Chem Int Ed. 2003;42:3131–3; (e) Tiecco M, Testaferri L, Santi C, Tomassini C, Santoro S, Marini F, et al. Synthesis of enantiomerically pure β-azidoselenides starting from natural terpenes. Tetrahedron. 2007;63:12373–8.
[119] (a) Déziel R, Malenfant E, Thibault C. Asymmetric arene-alkene cyclizations mediated by a chiral organoselenium reagent. Tetrahedron Lett. 1998;39:5493–6; (b) Okamoto K, Nishibayashi Y, Uemura S, Toshimitsu A. Stereospecific carbon–carbon bond formation by the reaction of a chiral episelenonium ion with aromatic compounds. Tetrahedron Lett. 2004;45:6137–9;

(c) Toshimitsu A. stereo- and site-selectivity in the reaction of chiral episelenonium ion with carbon nucleophile. Phosphorus Sulfur Silicon Relat Elem. 2005;180:935–7; (d) Okamoto K, Nishibayashi Y, Uemura S, Toshimitsu A. Asymmetric carboselenenylation reaction of alkenes with aromatic compounds. Angew Chem Int Ed. 2005;44:3588-91.

[120] (a) Shirahata M, Yamazaki H, Fukuzawa S. Highly diastereoselective reduction of chiral (ferrocenylseleno)methyl aryl and alkyl ketones. Chem Lett. 1999;245–6; (b) Wang W, Wang, J, Li H. A simple and efficient l-prolinamide-catalyzed α-selenenylation reaction of aldehydes. Org Lett. 2004;6:2817–20; (c) Wang J, Li H, Mei Y, Lou B, Xu D, Xie D, et al. J Org Chem. 2005;70:5678–87; (d) Giacalone F, Gruttadauria M, Marculescu AM, Noto R, Polystyrene-supported proline and prolinamide. Versatile heterogeneous organocatalysts both for asymmetric aldol reaction in water and α-selenenylation of aldehydes. Tetrahedron Lett. 2007;48:255–9; (e) Tiecco M, Carlone A, Sternativo S, Marini F, Bartoli G, Melchiorre P, et al. Organocatalytic asymmetric α-selenenylation of aldehydes. Angew Chem Int Ed. 2007;46:6882–5; (f) Sunden H, Rios R, Cordova A. Organocatalytic highly enantioselective α-selenenylation of aldehydes. Tetrahedron Lett. 2007;48:7865–9; (g) Marini F, Sternativo S, Del Verme F, Testaferri L, Tiecco, M. Enantioselective organocatalytic Michael addition of α-substituted cyanoacetates to α,β-unsaturated selenones. Adv Synth Catal. 2009;351:103–6.

[121] (a) Wirth T, Häuptli S, Leuenberger M. Catalytic asymmetric oxyselenenylation–elimination reactions using chiral selenium compounds. Tetrahedron: Asymmetry. 1998;9:547–50; (b) Browne DM, Niyomura O, Wirth T. Catalytic addition-elimination reactions towards butenolides. Phosphorus Sulfur Silicon Relat Elem. 2008;183:1026–35; (c) Tang E, Zhao Y, Li W, Wang W, Zhang M, Dai X, et al. Catalytic selenium-promoted intermolecular Friedel–Crafts alkylation with simple alkenes. Org Lett. 2016;18:912–5.

[122] (a) Comasseto JV, Ling LW, Petragnani N, Stefani HA. Vinylic selenides and tellurides - preparation, reactivity and synthetic applications. Synthesis. 1997;373–403; (b) Zeni G, Braga, AL, Stefani, HA. Palladium-catalyzed coupling of sp^2-hybridized tellurides. Acc Chem Res. 2003;36:731–8; (c) Petragnani N, Stefani HA. Advances in organic tellurium chemistry. Tetrahedron. 2005;61:1613–79; (d) Potapov VA, Musalov MV, Musalova MV, Amosova SV. Recent Advances in organochalcogen synthesis based on reactions of chalcogen halides with alkynes and alkenes. Curr Org Chem. 2016;20:136–45; (e) Singh FV, Stefani HA. Ultrasound-assisted synthesis of symmetrical biaryls by palladium-catalyzed detelluration of 1,2-diarylditellanes. Tetrahedron Lett. 2010;51:863–7.

[123] (a) Nishibayashi Y, Cho CS, Uemura S. Palladium-catalyzed cross-coupling reactions between organic tellurides and alkenes. J Organomet Chem. 1996;507:197–200; (b) Nishibayashi Y, Cho CS, Ohe K, Uemura S. Palladium-catalyzed homocoupling reactions of organic tellurides. J Organomet Chem. 1996;526:335–9.

[124] Zeni G, Comasseto JV. Coupling of *Z*-vinylic tellurides with alkynes catalysed by $PdCl_2CuI$: synthesis of *Z*-enynes and *Z*-enediynes. Tetrahedron Lett. 1999;40:4619–22.

[125] Dabdoub MJ, Dabdoub VB, Marino JP. Palladium-catalyzed cross-coupling reactions of dimethyl- and diethylzinc with unsaturated organotellurium compounds. Tetrahedron Lett. 2000;41:433–6.

[126] Zeni G, Perin G, Cella R, Jacob RG, Braga AL, Silveira CC, et al. Synthesis of cross-conjugated geminal enediynes via palladium catalyzed cross-coupling reaction of ketene butyltelluroacetals. Synlett. 2002;975–7.

[127] Braga AL, Ludtke DS, Vargas F, Donato RK, Silveira CC, Stefani HA, et al. Sonogashira cross-coupling reaction of organotellurium dichlorides with terminal alkynes. Tetrahedron Lett. 2003;44:1779–81.

[128] Stefani HA, Cella R, Dorr FA, Pereira CMP, Zeni G, Gomes M. Synthesis of 1,3-enynes via Suzuki-type reaction of vinylic tellurides and potassium alkynyltrifluoroborate salt. Tetrahedron Lett. 2005;46:563–7.

[129] Cella R, Orfao ATG, Stefani HA. Palladium-catalyzed cross-coupling of vinylic tellurides and potassium vinyltrifluoroborate salt: synthesis of 1,3-dienes. Tetrahedron Lett. 2006;47:5075–8.
[130] Cella R, Stefani HA. Ultrasound-assisted synthesis of *Z* and *E* stilbenes by Suzuki cross-coupling reactions of organotellurides with potassium organotrifluoroborate salts. Tetrahedron. 2006;62:5656–62.
[131] Cella R, Cunha RLOR, Reis AES, Pimenta DC, Klitzke CF, Stefani HA. Suzuki-Miyaura cross-coupling reactions of aryl tellurides with potassium aryltrifluoroborate salts. J Org Chem. 2006;71:244–50.
[132] Singh FV, Weber M, Guadagnin RC, Stefani HA. Ultrasound-assisted synthesis of functionalized 1,3-enynes by palladium-catalyzed cross-coupling reaction of α-styrylbutyltelluride with alkynyltrifluoroborate salts. Synlett. 2008;1889–93.
[133] Singh FV, Milagre HMS, Eberlin MN, Stefani HA. Synthesis of benzophenones from geminal biaryl ethenes using *m*-chloroperbenzoic acid. Tetrahedron Lett. 2009;50:2312–6.
[134] Guadagnin RC, Suganuma CA, Singh FV, Vieira AS, Cella R, Stefani HA, et al. Chemoselective cross-coupling Suzuki–miyaura reaction of (*Z*)-(2-chlorovinyl)tellurides and potassium aryltrifluoroborate salts. Tetrahedron Lett. 2008;49:4713–6.
[135] Stefani HA, Pena JM, Manarin F, Ando RA, Petragnani N. Negishi Cross-coupling of organotellurium compounds: synthesis of biaryls, aryl-, and diaryl acetylenes. Tetrahedron Lett. 2011;52:4398–401.
[136] Singh FV, Amaral MFZJ, Stefani HA. Synthesis of symmetrical 1,3-diynes via homocoupling reaction of *n*-butyl alkynyltellurides. Tetrahedron Lett. 2009;50:2636–9.
[137] Borisov AV, Matsulevich ZV, Osmanov VK, Borisova FN, Naumov VI, Mammadova GZ, et al. Cycloaddition of 2-pyridinetellurenyl chloride to alkenes. Russ Chem Bull Int Ed. 2012;61:91–4.
[138] Borisov AV, Matsulevich ZV, Osmanov VK, Borisova GN. Polar cycloaddition of 2-pyridinetellurium trichloride to multiple bonds. Russ Chem Bull Int Ed. 2013;62:1042–3.

Peter C. Ho, Jin Wang and Ignacio Vargas-Baca

8 Reagents that Contain Se-H or Te-H Bonds

Abstract: Species that contain bonds between hydrogen and selenium or tellurium have a characteristic high reactivity, which can be harnessed in the synthesis of valuable organic compounds. This overview includes the synthesis of dihydrides, alkali metal hydrochalcogenides, chalcogenols, chalcogenocarboxylic and chalcogenocarbamic acids, and their application in reactions of reduction, addition to unsaturated compounds, and nucleophilic substitution.

Keywords: Organic synthesis, Organochalcogen compounds, Selenium, Tellurium

8.1 Introduction

Organotellurium and organoselenium compounds are increasingly important in fundamental (e. g. supramolecular [1] and biomimetic [2]) chemistry and applied (e. g. soft materials [3], anticancer agents [4] and biological probes [5]) research. Atoms of selenium and tellurium (here denoted as Ch) are usually introduced into an organic structure by reactions of addition or substitution. Other than the elemental forms of these chalcogens, their anions and their halides, the most important precursors contain reactive Ch-H bonds. Here we provide an overview of the properties and applications of such reagents. In many instances, we note that the use of a Ch-H reagent is followed by one or more steps that accomplish a complex transformation, e. g. cyclization, further substitution, or dissociation. There are also examples of reactions take advantage of the reducing ability of these compounds.

8.2 Chalcogen dihydrides (H_2Ch)

Both H_2Se and H_2Te are fairly reactive species due to a combination of factors, the most important of which are compiled in Table 8.1. Given the elemental electronegativities (Cf. 2.2 for H), Ch-H bonds have a greater covalent character than the O-H bonds of water. The H_2Ch molecules are weakly polar, yet both dihydrides are stronger Brønsted acids than H_2S. The HCh^- anions that result from their first deprotonation are nucleophilic enough to achieve the displacement of good leaving groups. Ch-E bond energies are small, consequently, H_2Se has a modest heat of formation and H_2Te actually is an endothermic compound. While H_2Se can be obtained by a combination of the elements, H_2Te is prepared either by electrolysis or hydrolysis of Al_2Te_3.

This article has previously been published in the journal *Physical Sciences Reviews*. Please cite as: Ho, P.C., Wang, J., Vargas-Baca, I. Reagents that Contain Se-H or Te-H Bonds. *Physical Sciences Reviews* [Online] **2018** DOI: 10.1515/psr-2017-0130

https://doi.org/10.1515/9783110529340-008

Table 8.1: Selected properties relevant to the reactivity of the dihydrides of the heavy chalcogens.

	H_2Se	H_2Te
x^{Ch} (Pauling)	2.6	2.1
b.p. (°C)	−41.25	−2.2
μ (D)	0.24 [6]	
BE HCh-H, Ch-H (kcal/mol)	78.99±0.18 74.27±0.23[7]	65.0±0.1 63.8±0.4[8]
BE Ch-Ch[[9]]	320	258.3
ΔH_f^0(kcal/mol)	-6.3 ± 0.1[10]	23.83 ± 0.20[11, 12]
pKa_1, pKa_2[13]	3.81, 15.00	2.64, 11.80
E^0 Ch,$2H^+/H_2Ch$ (V)	-0.399	0.793

In general, these reagents are handled in anhydrous media, under inert atmospheres and occasionally at low temperatures and in the dark. The dihydrides themselves are used either as reducing agents or in reactions of addition to unsaturated molecules.

H_2Se inserts into nitriles forming selenoamides, which can be used in the synthesis of 1,3 selenazoles and selenoxo-isoquinolines [14]. Selenopyrans can be obtained by the reaction of diketones or pyrans with H_2Se generated *in situ* from Al_2Se_3 to HCl (Figure 8.1(a)) [15]. The reaction with dicyclohexyl carbodiimide yields a selenourea that is a valuable selenium transfer agent, suitable for ^{75}Se radiolabeling (Figure 8.1(b)) [16].

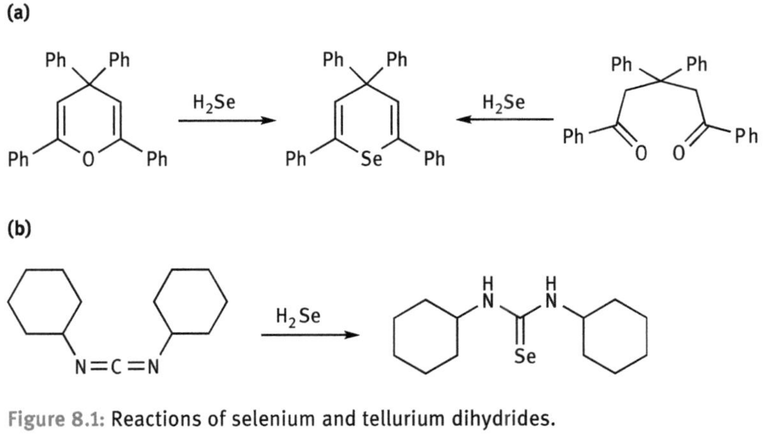

Figure 8.1: Reactions of selenium and tellurium dihydrides.

H_2Te reduces *α,β*-unsaturated carbonyl compounds to the corresponding ketones at mild temperatures, higher temperatures, and longer times or an excess of H_2Te result in alcohols. Aromatic nitro compounds are reduced to the corresponding anilines, the reaction is more efficient in THF and hampered by electron withdrawing substituents. Similarly, imines, hydroxylamines, nitroso, azo, and azo *N*-oxide groups are reduced with H_2Te. In the case of nitrosobenzene and azo compounds, the predominant product is the hydrazine. Epoxyethyl benzene is reduced to styrene, which is further reduced to ethyl benzene, albeit inefficiently [17].

There are, however, important constraints to working with the dihydrides, both are gases in ambient conditions and are notorious because of their toxicities and stench. The reagents decribed in the following sections have been developed as more convenient options.

8.3 Sodium Hydrochalcogenides (NaHCh)

The authentic salts are prepared by introducing the gaseous dihydride into the aqueous base. While NaHSe has been crystallographically characterized [18], the structure of the Te analogue is not known but the structure of $[PPh_4]^+[HTe]^-$ [19] has been determined. Because these reagents are readily oxidized, most frequently they are prepared *in-situ* either by bubbling H_2Ch into a base solution or by reduction of the elemental chalcogen with $NaBH_4$. These procedures accommodate a variety of solvents, ranging from water to DMF.

In general, reduction and addition reactions that work with H_2Se are also feasible with NaHSe [20]. The inherent reactivity of the Se-H bond leads to the formation of dialkyl selenides from the nucleophilic substitution of alkyl halides with NaHSe [21]. Iminium salts, obtained from phenols and Vilsmeier reagent, react with NaHSe forming selenocarbamates, these undergo a Newman-Kwart rearrangement that is useful to form transient phenylselenols and stable diselenides upon hydrolysis (Figure 8.2(a)) [22].

A 2-selenouridine is obtained when NaHSe is used to displace a geranyl sulfide group by nucleophilic substitution and proton transfer [23]. Sulfur can be displaced from thioureas to form selenourea, with the assistance of a methylating agent [24]. Vilsmeier reagent forms diselenocarbamates that later react with 3-bromobutan-2-one forming selenoketones; further reaction with H_2Se gives a selone and coupling assisted by $P(OEt)_3$ gives a tetraselenafulvalene [25]. NaHSe is also used in the synthesis of selenolo[2,3-*d*]-1,3-dithiole-2-thione, which facilitates C-C coupling to form the corresponding diselenolotetrathiafulvalene (Figure 8.2(b)) [26].

NaHTe hydrogenates olefins. Depending on the substrate and conditions, the products conform to either hydride-ion, hydrogen-atom or electron transfer mechanisms [27]. Reduction of terminal alkenes is preferred [28]. Electron withdrawing groups usually are necessary to accomplish alkyne reduction. The reduction of α,β-unsaturated carbonyls often spares the C=O bond but occasionally leads to C-C coupling [29]. Similarly, it is possible to selectively reduce the C=C bond of 1-cyanoethenyl acetamides [30]. Electron withdrawing groups promote the reduction of the carbonyl of substituted benzaldehydes to the benzyl alcohols [31]. Similar to carbonyl compounds, reduction or addition are possible with Schiff bases [32]. Less-common selective reductions effected with NaHTe include the *N*-oxide to the

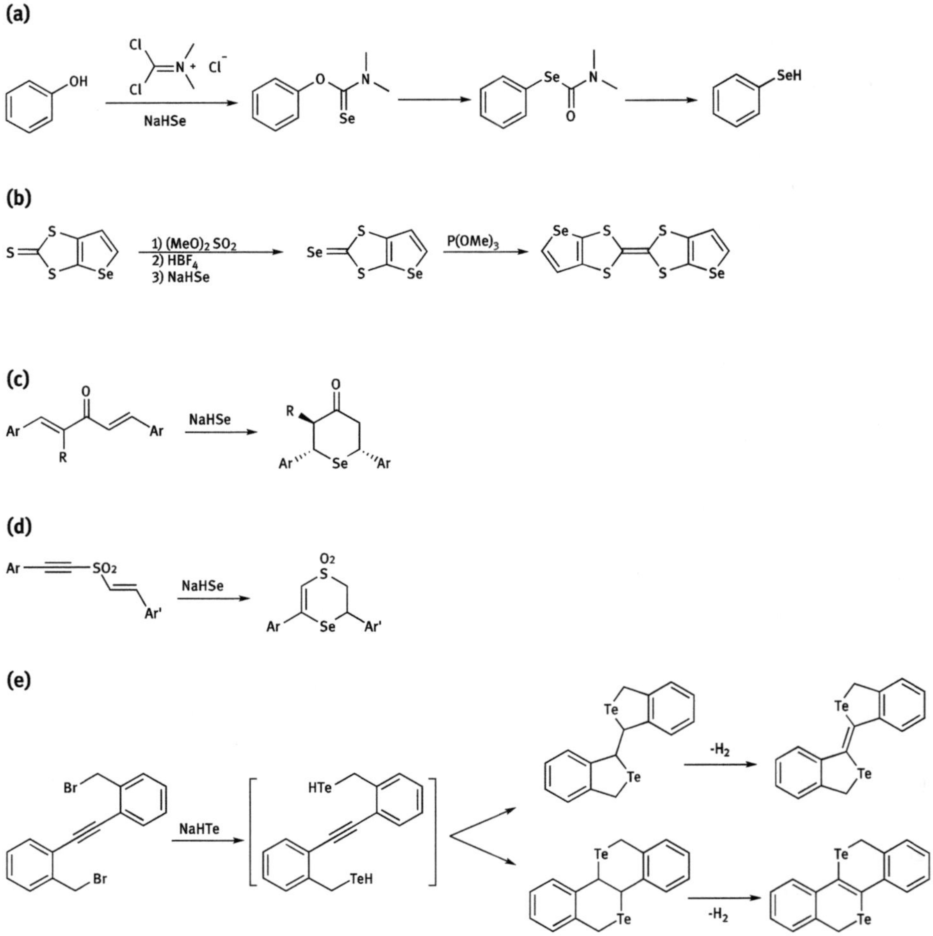

Figure 8.2: Reactions of sodium hydrochalcogenides.

corresponding amine of a stemofoline [33], desulfonation [29, 34], epoxide reduction [35] and the removal of the NO_2 group from a tertiary allylic carbon [36].

Markovnikov NaHTe addition to terminal alkenes generates dialkyltellurides, the reaction is facilitated by electron withdrawing groups and small amount of reduction products is observed [37]. Acetal deprotection of carbonyls with NaHTe produces transient telluroketones [38].

Nucleophilic aromatic substitution to form tellurophenols and ditellurides is possible with good leaving halides (I^-, Br^-) and ancillary electron withdrawing substituents such as Cl, Br, I, and carbonyl functional groups (C(=O)R, C(=O)H, and C(=O)OR) [39].

Tellurophenes can be conveniently prepared adding butadiynes to a mixture of elemental Te and $NaBH_4$ [40], although the reactant is assumed to be Na_2Te, it likely is NaHTe given the reaction conditions. Similar double additions produce tetrahydroselenopyran-4-ones from 1,4-pentadien-3-ones [41] (Figure 8.2(c)) and 1-thia-4-selenin-1,1-dioxides from 2-arylethenyl-2-arylethynyl sulfones [42] (Figure 8.2 (d)). From bioxirane, double addition produces tetrahydroselenophene-3,4-diol [43]. Double substitution of bromide followed by addition to olefin yields bicyclic compounds. For example, bi-isobenzotellurophene and ditellurachrysene, which are simultaneously formed (Figure 8.2(e)) [44].

8.4 Lithium dihydroselenoxo aluminate Li(AlH$_2$Se)

This is the assumed composition of the product of the reaction between Se and LiAlH$_4$. It has been used to produce cyclic and acyclic acyl selenides and diselenides, selenolactones (Figure 8.3(a)), selenoureas (Figure 8.3(b)) and selenoamides. Reaction with phenyl isocyanate yields a phenylcarbamate that is isolated after methylation (Figure 8.3(c)). *N,N*-dimethyl selenocarbamoyl chloride is formed by reaction with Vilsmeier reagent; in turn, this grants access to *S*-alkyl selenothiocarbamates and *Se*-alkyl diselenocarbamates (Figure 8.3(d)) [45].

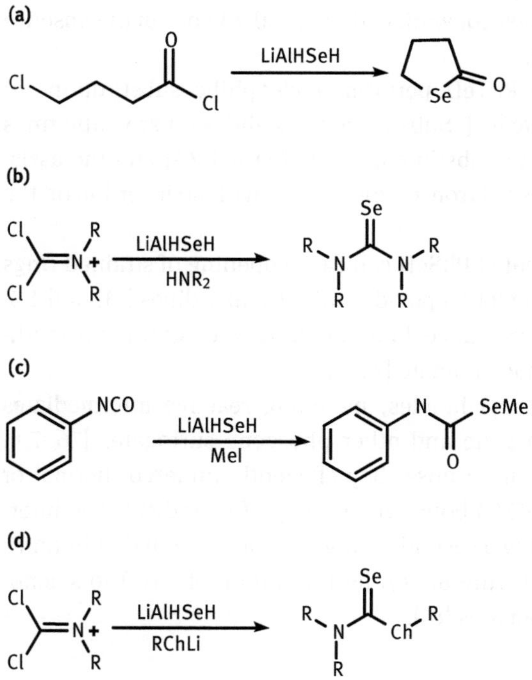

Figure 8.3: Reactions of sodium Li(AlH$_2$Se).

8.5 Chalcogenols RChH

Although PhSeH is commercially available, this type of compounds is more easily oxidized than their sulfur analogues. Thus, it is more convenient to prepare chalcogenols *in situ* by reduction of the corresponding dichalcogenide with reagents such as $NaBH_4$, $LiAlH_4$ or phosphinic acid in the case of tellurols. Alternatively, tellurophenol is generated by the reaction of PhTe-$SiMe_3$ with alcohol [46].

Reductions carried out with chalcogenols include the reduction of nitroaromatics to anilines [46] and imines to amines; this method can be extended to perform the reductive amination of carbonyl compounds to primary and secondary amines [47].

Hydrochalcogenation of alkynes preferentially gives anti-adducts and is favored by electron withdrawing groups (-C(=O)H, -CHR(OH), -SO_2-R, -P(=O)$(OEt)_2$) moderate heating and catalysts such as $Pd(OAc)_2$, $Ni(acac)_2$ [48, 49, 51], however, with RhCl $(PPh_3)_3$-Ph_2P(O)H, the syn-adducts were observed [49]. Hydroselenation of 1,3-butadiynes proceeds with highly regio-, and stereo-selectivity, on the terminal triple bond (Figure 8.4(a)) [52]. Hydroselenation of the terminal alkene of α-, β-unsaturated carbonyls is applicable to produce amides and ketones [53, 54].

Isotopic labeling experiments suggest that hydrotelluration reactions involve protons from protic solvents [55]. However, RTeNa and RTeH have different reactivities towards olefins in nonprotic solvent, addition proceeds with yields over 80 % at room temperature with RTeH but requires heating and gives lower yields (<; 40 %) with RTeNa [56]. Hydrotelluration does not work with internal alkynes in the absence of electron withdrawing groups [57].

Assisted by base, PhSeH and PhTeH perform nucleophilic substitutions of halides [58–60] and methyl sulfonate [61]. Substitution of halides on aromatic rings is possible with electron withdrawing substituents (NO_2, COOH) [62] and the assistance of ^{t}BuOK [63]. PhSeH substitutes Cl from 6-chlorouracil with no insertion on the double bond (Figure 8.4(b)) [64].

Nucleophilic attack by deprotonated PhSeH can cause opening of strained rings such as oxetane [65], γ-butyrolactone [66], epoxides [67–70], aziridines [71] and 1,3-propanesultone [72]. With $BF_3.OEt_2$ assistance, PhSeH performs substitution of methoxide [73], acetate [74] and trichloroacetimidate [75] groups.

PhSeH and PhTeH react with acyl halides, including reactive intermediates generated by reactions with triphosgene and other phosgene surrogates [76–79]. Acyl tellurides are particularly useful because they efficiently undergo thermal or photochemical homolysis of the Te-C(O) bond, generating a free radical. The intermediate readily eliminates CO, leaving a secondary organic radical [80] that initiates a useful radical reaction pathway (Figure 8.4(c)) [60]. Addition of RTeH to selenocyanates forms bis-chalcogenocarbamates [81].

8.6 Monochalcogenocarboxylic acids RC(O)ChH

The heavy congeners of organic acids are prepared by protonation of the product of reaction of diacyl chalcogenides with bases such as NaOEt (Figure 8.5(a)) [82]. Selenocarboxylic acids can also be obtained from the reaction of carboxylic acids

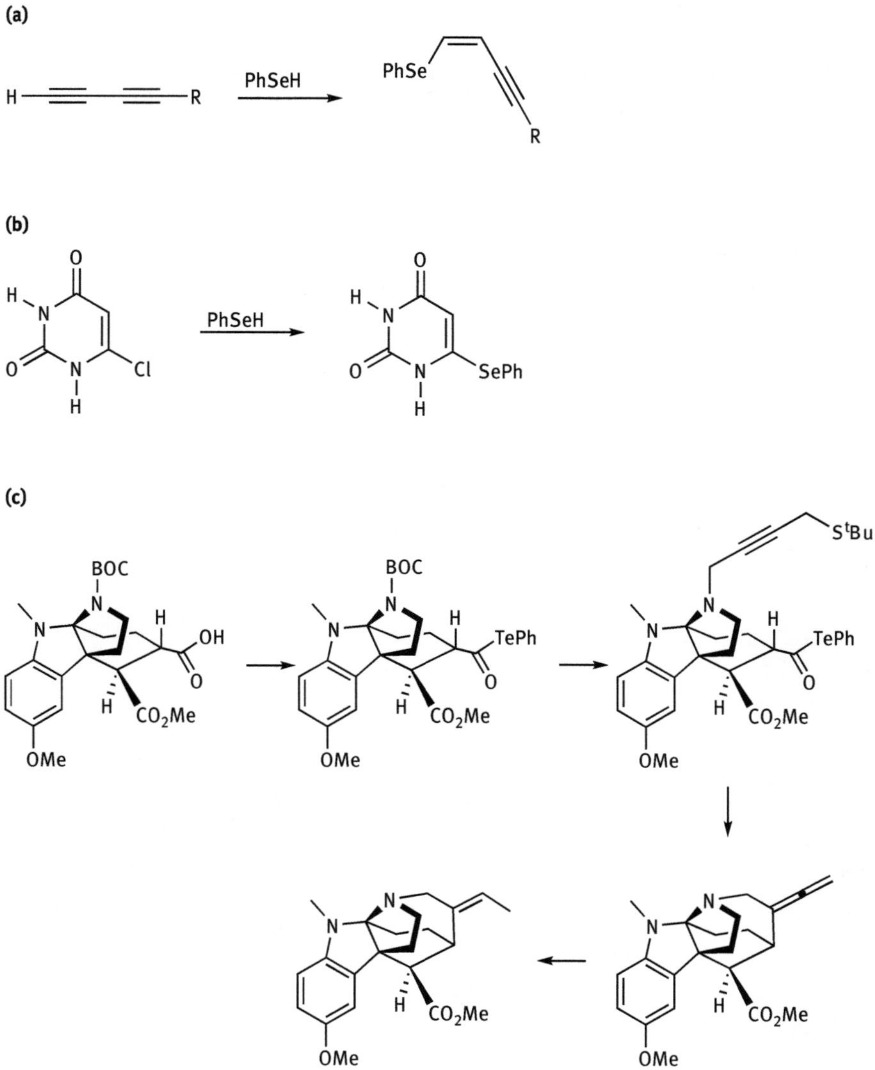

Figure 8.4: Reactions of chalcogenols.

with Woollins' reagent ($(PhP)_2Se_4$) (Figure 8.5(a)) [83]. Chalcogenocarboxylic acids exist as the O-H and Ch-H tautomers in equilibrium. The former is favored by polar solvents at low temperature while the latter is predominant in nonpolar media and the solid state [84]. Demonstrated reactions of these species include addition to alkenes (Figure 8.5(b)), acid-promoted ring-opening by addition to carbon bonded to oxygen (Figure 8.5(c)), alkylation by condensation with alcohols or substitution of good leaving groups such as halides and triflate, amidation by reaction with azides, and addition to aryl isocyanates forming the corresponding acyl carbamyl chalcogenides (Figure 8.5(d)) [83–85].

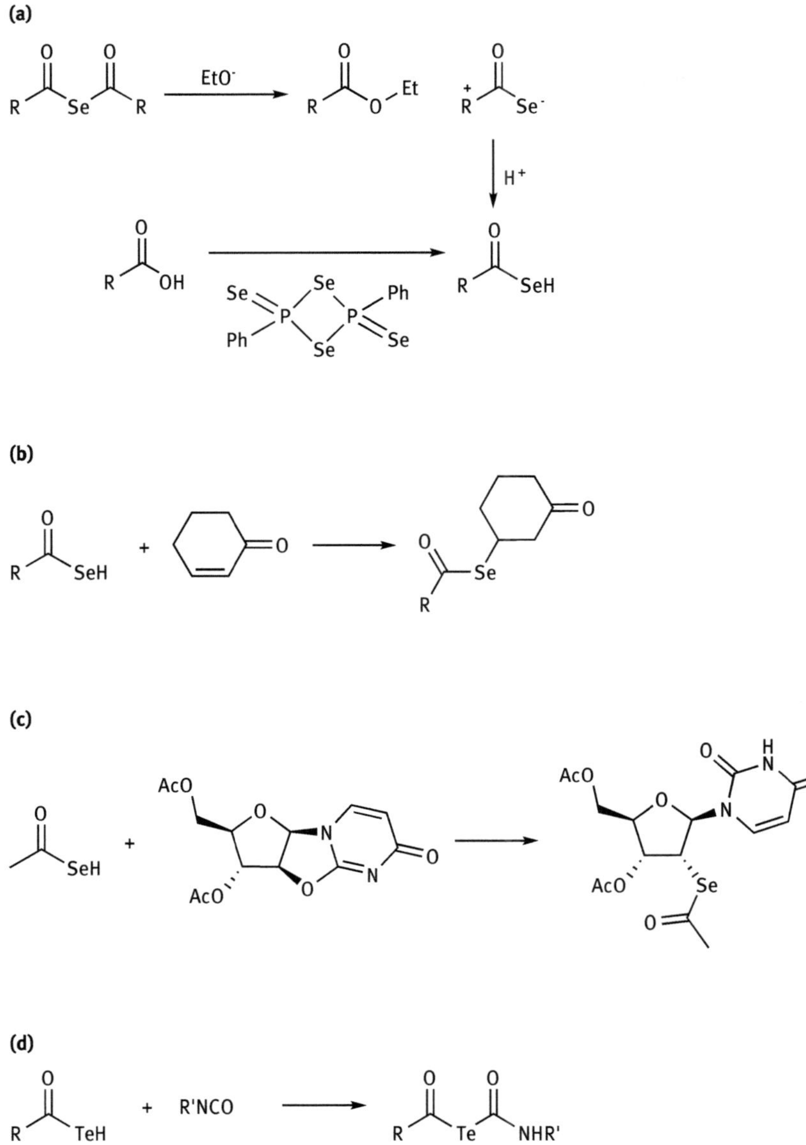

Figure 8.5: Reactions of monochalcogenocarboxylic acids.

8.7 N,N-dimethyl Chalcogenocarbamic acids $Me_2NC(O)ChH$

As they are not isolable in pure form [86], these reagents are prepared in-situ by reduction of the carbamyl dichalcogenides with $NaBH_4$, [87] or reaction of NaHCh with a carbamyl chloride (Figure 8.6(a)) [1]. The latter method, followed by oxidation, is suitable to prepare multigram stocks of the carbamyl dichalcogenides. An earlier preparation of dichalcogenides relied on heating a mixture of DMF, sodium metal and the chalcogen in elemental form, followed by aerobic oxidation [88].

The N,N-dimethyl chalcogenocarbamic acids have been used in the preparation of Ch-N heterocycles. Their addition to ynones predominantly yields the Z isomer. Condensation with hydroxylamine-*O*-sulfonic acid followed by hydrolysis yield iso-chalcogenazole N-oxides and the corresponding iso-chalcogenazoles upon reduction (Figure 8.6(b)) [1, 87].

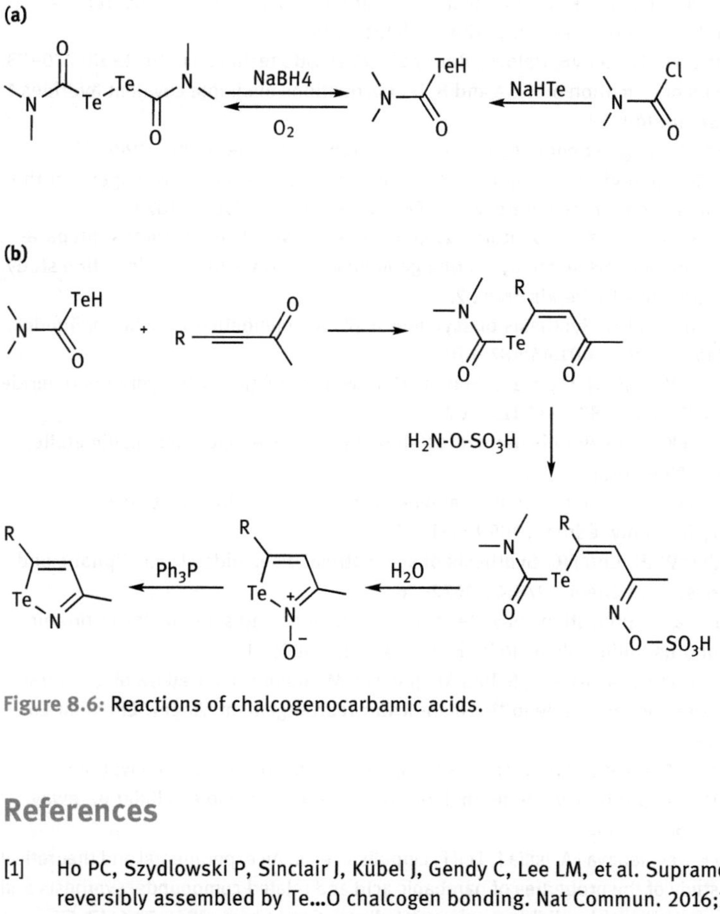

Figure 8.6: Reactions of chalcogenocarbamic acids.

References

[1] Ho PC, Szydlowski P, Sinclair J, Kübel J, Gendy C, Lee LM, et al. Supramolecular macrocycles reversibly assembled by Te...O chalcogen bonding. Nat Commun. 2016;7:11299

[2] Manna D, Mugesh G. Regioselective deiodination of thyroxine by iodothyronine deiodinase mimics: an unusual mechanistic pathway involving cooperative chalcogen and halogen bonding. J Am Chem Soc. 2012;134:4269–79.

[3] Carrera EI, Seferos DS. Semiconducting polymers containing tellurium: perspectives toward obtaining high-performance materials. Macromolecules. 2015;48:297–308.

[4] Plano D, Karelia DN, Pandey MK, Spallholz JE, Amin S, Sharma AK, et al. Biological evaluation of novel selenium (Se-NSAID) molecules as anticancer agents. J Med Chem. 2016;59:1946–59.

[5] Edgar LJ, Vellanki RN, Halupa A, Hedley D, Wouters BG, Nitz M. Identification of hypoxic cells using an organotellurium tag compatible with mass cytometry. Angew Chem Int Ed. 2014;53:11473–77.

[6] Jache AW, Moser PW, Gordy W. Millimeter wave spectrum, molecular structure, and dipole moment of hydrogen selenide. J Chem Phys. 1956;25:209–10.

[7] Gibson ST, Greene JP, Berkowitz J. A photoionization study of SeH and H_2Se. J Chem Phys. 1986;85:4815–24.

[8] Underwood J, Chastaing D, Lee S, Wittig C. Heavy hydrides: H_2Te ultraviolet photochemistry. J Chem Phys. 2005;123:084312.

[9] Douglas B, McDaniel DH, Alexander JJ. Concepts and models of inorganic chemistry. 3rd ed. New York: Wiley; 1994.

[10] Barin I, Knacke O, Kubaschewski O. Thermochemical properties of inorganic substances [Internet]. Berlin, Heidelberg: Springer Berlin Heidelberg; 1977.

[11] Wagman DD, Evans WH, Parker VB, Halow I, Bailey SM, et al. NBS technical note . 1968: 270–73.

[12] Gunn SR. The heats of formation of H_2Se and H_2Te. Correlations of simple covalent hydrides 1. J Phys Chem. 1964;68:949–52.

[13] Schröke H. Die Entstehung der endogenen Erzlagerstätten. Berlin: deGruyter, 1986:597.

[14] Dyachenko IV, Dyachenko VD. Ethyl 3-amino-3-selenoxopropanoate as a new reagent for the synthesis of selenium-containing heterocycles. Russ J Gen Chem. 2015;85:1673–76.

[15] Kroulík J, Čejka J, Böhm S, et al. Substituted 2,4,4,6-tetraphenyl-4H-selenopyrans: preparation, photocolouration and 4H-selenopyran ring geometry; an X-ray and DFT calculation study. J Chem Soc. 2002;0:1909–16. Perkin Trans 2.

[16] Blum T, Ermert J, Coenen HH. Synthesis of asymmetric [75Se]selenoethers via carbodiimides. J Label Compd Radiopharm. 2001;44:587–601.

[17] Kambe N, Inagaki T, Miyoshi N, Ogawa A, Sonoda N, New A. Reduction with hydrogen telluride TI. Nippon Kagaku Kaishi. 1987;1987:1152–62.

[18] Teichert VW, Klemm W. Zur kenntnis der hydrosulfide und hydroselenide der alkalimetalle. Z Anorg Allg Chem. 1939;243:86–98.

[19] Huffman JC, Haushalter RC. The tellurium analogue of the hydroxide ion: syntheses and structure of $(PPh_4)TeH$. Polyhedron. 1989;8:531–32.

[20] Lewicki JW, Gunther WHH, Chu JYC. Synthesis of symmetrical diselenides from aliphatic and aromatic aldehydes. J Org Chem. 1978;43:2672–76.

[21] Iwaoka M, Kumakura F. Applications of water-soluble selenides and selenoxides to protein chemistry. Phosphorus Sulfur Silicon Relat Elem. 2008;183:1009–17.

[22] Sørensen A, Rasmussen B, Agarwal S, Schau-Magnussen M, Sølling TI, Pittelkow M. Conversion of phenols into selenophenols: seleno Newman–Kwart rearrangement. Angew Chem Int Ed. 2013;52:12346–49.

[23] Bartos P, Maciaszek A, Rosinska A, Sochacka E, Nawrot B. Transformation of a wobble 2-thiouridine to 2-selenouridine via S-geranyl-2-thiouridine as a possible cellular pathway. Bioorg Chem. 2014;56:49–53.

[24] Arca M, Demartin F, Devillanova FA, Isaia F, Lelj F, Lippolis V, et al. An experimental and theoretical approach to the study of the properties of parabanic acid and related compounds: synthesis and crystal structure of diethylimidazolidine-2-selone-4,5-dione. Can J Chem. 2000;78:1147–57.

[25] Chiang L-Y, Poehler TO, Bloch AN, Cowan DO. A modified synthesis of tetraselenafulvalenes. J Chem Soc, Chem Commun. 1980;0:866b–867.

[26] Jigami T, Takimiya K, Otsubo T, Novel Selenocycle-Fused AY. TTF-Type of Electron Donors Forming Conducting Molecular Complexes: bis(ethyleneseleno)tetrathiafulvalene (BES-TTF), Diselenolotetrathiafulvalene (DS-TTF), and Bis(ethyleneseleno)tetraselenafulvalene (BES-TSF). J Org Chem. 1998;63:8865–72.

[27] Barton DHR, Bohé L, Lusinchi X. Sodium hydrogen telluride: a mechanistic chameleon. Tetrahedron Lett. 1987;28:6609–12.

[28] Barton DHR, Bohé L, Lusinchi X. The action of sodium hydrogen telluride on olefins. Tetrahedron. 1990;46:5273–84.

[29] Yamashita M, Tanaka Y, Arita A, Nishida M. The reaction of sodium hydrogentelluride with α, β-unsaturated carbonyl compounds. J Org Chem. 1994;59:3500–02.

[30] Zeng P, Hu Y, Hu H, Convenient Reduction A. Of N-(2-Substituted-1-cyanoethenyl) acetamides with NaTeH. Synth Commun. 1997;27:939–44.

[31] Yamashita M, Tanaka Y. Organic reactions using sodium hydrogentelluride: part 4. The selective reduction of aromatic aldehydes or ketones with electron-withdrawing groups on the benzene ring by sodium hydrogentelluride. Appl Organomet Chem. 1996;10:791–94.

[32] Barton DHR, Bohé L, Lusinchi X. A mechanistic approach to the reaction between imines and sodium hydrogen telluride. Tetrahedron Lett. 1988;29:2571–74.

[33] Ye Y, Velten RF. Semi-syntheses of new stemofoline derivatives. Tetrahedron Lett. 2003;44:7171–73.

[34] Huang X, Pi J-H, Huang -Z-Z. Stereoselective synthesis of (2Z)-αβ-unsaturated nitriles via tandem condensation-desulfonylation of α-cyanosulfones with aldehydes. Phosphorus Sulfur Silicon Relat Elem. 2006;67:177–82.

[35] Osuka A, Taka-Oka K, Suzuki H. Chemo- and regioselective reduction of α,β-epoxy ketones to β-hydroxy ketones by sodium hydrogentelluride. Chem Lett. 1984;13:271–72.

[36] Anderson DA, Hwu JR Synthesis and application of tertiary allylic nitro compounds. J Org Chem. 1990;55:511–16.

[37] Barton DHR, Bohé L, Lusinchi X. The addition of sodium hydrogen telluride to unactivated carbon-carbon double bonds. Tetrahedron Lett. 1990;31:93–96.

[38] Lue P, Fan W-Q, Zhou X-J, Novel A. Cleavage of five-membered cyclic acetals using sodium hydrogen telluride. Synthesis (Mass). 1989;1989:692–93.

[39] Liu J, Qiu M, Zhou X. Direct synthesis of some diaryl ditellurides from aryl halides. Synth Commun. 1990;20:2759–67.

[40] Sweat DP, Stephens CE. A modified synthesis of tellurophene using $NaBH_4$ to generate sodium telluride. J Organomet Chem. 2008;693:2463–64.

[41] Reddy DB, Reddy AS, Padmavathi V. 1,4-pentadien-3-ones. A source for selenano 1,2,3- selena and thiadiazoles. Synth Commun. 2006;31:3429–37.

[42] Reddy DB, Babu NC, Padmavathi V, Padmaja A. 2-arylethenyl-2″-arylethynyl sulfones: a potential source for new heterocycles. Phosphorus Sulfur Silicon Relat Elem. 2000;165:237–42.

[43] Iwaoka M, Tomoda S. trans-3,4-dihydroxy-1-selenolane oxide: a new reagent for rapid and quantitative formation of disulfide bonds in polypeptides. Chem Lett. 2000;29:1400–01.

[44] Sashida H, Kaname M, Nakayama A, Suzuki H, Minoura M. A straightforward double intramolecular cyclization of dibenzyl dichalcogenols into a triple bond. Tetrahedron. 2010;66:5149–57.

[45] Ishihara H, Koketsu M, Fukuta Y, Nada F. Reaction of lithium aluminum hydride with elemental selenium: its application as a selenating reagent into organic molecules. J Am Chem Soc. 2001;123:8408–09.

[46] Ohira N, Aso Y, Otsubo T, Ogura F. Reduction of aromatic nitro compounds to amines by benzenetellurol. Chem Lett. 2006;13:853–54.

[47] Fujimori K, Yoshimoto H, Oae S. Reduction with organic selenium compounds II. reduction of schiff bases with selenophenol reductive alkylation of amines with carbonyl compounds. Tetrahedron Lett. 1980;21:3385–88.

[48] Dos Santos AA, Castelani P, Bassora BK, Fogo Junior JC, Costa CE, Comasseto JV. Chemoselective reduction of β-butyltellanyl α,β-unsaturated carbonyl compounds to allylic alcohols. Tetrahedron. 2005;61:9173–79.

[49] Kawaguchi S-I, Kotani M, Atobe S, Nomoto A, Sonoda M, Ogawa A. Rhodium-catalyzed highly stereoselective hydroselenation of internal alkynes bearing an electron-withdrawing group. Organometallics. 2011;30:6766–69.
[50] Comasetto JV, Brandt CA. ChemInform abstract: selenium in organic synthesis: a novel route to 1-phenylselenobutadienes and 1,4-dicarbonyl compounds. Synythesis. 1987;1987:146–49.
[51] Huang X, Liang C-G, Xu Q, He Q-W. Alkyne-based, highly stereo- and regioselective synthesis of stereodefined functionalized vinyl tellurides. J Org Chem. 2001;66:74–80.
[52] Dabdoub MJ, Baroni ACM, Lenardão EJ, Gianeti TR, Hurtado GR. Synthesis of (Z)-1-phenylseleno-1,4-diorganyl-1-buten-3-ynes: hydroselenation of symmetrical and unsymmetrical 1,4-diorganyl-1,3-butadiynes. Tetrahedron. 2001;57:4271–76.
[53] Ds M, Simpkins NS, J Begley M, K Terrett N. Synthesis of spiroethers using radical cyclisations. Tetrahedron. 1990;46:545–64.
[54] Middleton DS, Simpkins NS, Terrett NK. Synthesis of N-protected spiroamines related to natural products using radical cyclisations. Tetrahedron Lett. 1989;30:3865–68.
[55] Marino JP, Nguyen HN. Electrotelluration: a new approach to Tri- and Tetrasubstituted alkenes. J Org Chem. 2002;67:6291–96.
[56] Barrientos-Astigarraga RE, Castelani P, Sumida CY, Zukerman-Schpector J, Comasseto JV. A general method of synthesis of functionalized Z-vinylic tellurides starting from β-dicarbonyl compounds. Tetrahedron. 2002;58:1051–59.
[57] Petragnani N, Stefani HA. Advances in organic tellurium chemistry. Tetrahedron. 2005;61:1613–79.
[58] Cox DJ, Singh GP, Watson AJA, Fairbanks AJ. Neighbouring group participation during glycosylation: do 2-substituted ethyl ethers participate?. Eur J Org Chem. 2014;2014:4624–42.
[59] Kobayashi S, Kinoshita T, Kawamoto T, Wada M, Kuroda H, Masuyama A, et al. Stereocontrolled synthesis of substituted bicyclic ethers through Oxy-Favorskii rearrangement: total synthesis of (±)-communiol E. J Org Chem. 2011;76:7096–103.
[60] Horning BD, MacMillan DWC. Nine-step enantioselective total synthesis of (–)-vincorine. J Am Chem Soc. 2013;135:6442–45.
[61] Fukui H, Shiina I. Asymmetric total synthesis of botcinins C, D, and F. Org Lett. 2008;10: 3153–56.
[62] Lambert C, Christiaens L. Stabilisation of condensed selenium heterocycles by a nitro group in the ortho-position to the chalcogen. Tetrahedron. 1991;47:9053–60.
[63] Chan S-C, Jang J-P, Cherng Y-J. Microwave-assisted synthesis of substituted phenanthrenes, anthracenes, acenaphthenes, and fluorenes. Tetrahedron. 2009;65:1977–81.
[64] Fang W-P, Cheng Y-T, Cheng Y-R, Cherng Y-J. Synthesis of substituted uracils by the reactions of halouracils with selenium, sulfur, oxygen and nitrogen nucleophiles under focused microwave irradiation. Tetrahedron. 2005;61:3107–13.
[65] Wang Y, Bekolo H, Howell AR. Ring opening reactions of 2-methyleneoxetanes. Tetrahedron. 2002;58:7101–07.
[66] Mori T, Tohmiya H, Satouchi Y, Higashibayashi S, Hashimoto K, Nakata M. Synthetic studies on thiostrepton family of peptide antibiotics: synthesis of the cyclic core portion containing the dehydropiperidine, dihydroquinoline, l-valine, and masked dehydroalanine segments. Tetrahedron Lett. 2005;46:6423–27.
[67] Günther WHH, Acid H. A novel reagent for the reduction of diselenides and the selenol-catalyzed reduction of disulfides. J Org Chem. 1966;31:1202–05.
[68] Ericsson C, Microwave-Assisted Group-Transfer EL. Cyclization of organotellurium compounds. J Org Chem. 2004;69:5143–46.
[69] Jung ME, Sun DL. Trapping of Payne rearrangement intermediates with arylselenide anions. Tetrahedron Lett. 2015;56:3082–85.

[70] Tschöp A, Nandakumar MV, Pavlyuk O, Schneider C. Scandium–bipyridine-catalyzed, enantioselective selenol addition to aromatic meso-epoxides. Tetrahedron Lett. 2008;49:1030–33.
[71] Besev M, Engman L. Diastereocontrol by a hydroxyl auxiliary in the synthesis of pyrrolidines via radical cyclization. Org Lett. 2002;4:3023–25.
[72] Kanda T, Engman L, Cotgreave IA, Novel Water-Soluble PG. Diorganyl tellurides with thiol peroxidase and antioxidant activity. J Org Chem. 1999;64:8161–69.
[73] Martins MAP, Bastos GP, Sinhorin AP, Zimmermann NEK, Bonacorso HG, Zanatta N, et al. Preparation of 4-methyl- and 4-phenylseleno-1,1,1-trihalo-3-alken-2-ones and their usefulness in the synthesis of 3-trihalomethylisoselenazoles. Synthesis (Mass). 2002;2002:2220–24.
[74] Viso A, Poopeiko N, Castillón S. Stereoselective synthesis of nucleosides from 1-thio and 1-seleno glycosides through consecutive 1,2-migration and glycosylation under Mitsunobu conditions. Tetrahedron Lett. 2000;41:407–11.
[75] Nicolaou KC, Mitchell HJ, Fylaktakidou KC, Suzuki H, Rodríguez RM. 1,2-Seleno migrations in carbohydrate chemistry: solution and solid-phase synthesis of 2-deoxy glycosides, orthoesters, and allyl orthoesters. Angew Chem Int Ed. 2000;39:1089–93.
[76] Rigby JH, Danca DM, Horner JH. Carbamoyl radicals from Se-phenylselenocarbamates: intramolecular additions to alkenes. Tetrahedron Lett. 1998;39:8413–16.
[77] Evans PA, Murthy VS. Enantioselective synthesis of the 4-hydroxy buteneolide terminus of mucocin and related annonaceous acetogenins. Tetrahedron Lett. 1998;39:9627–28.
[78] Álvarez C, Peláez R, Medarde M. New dicyclopentadiene-based scaffolds. Tetrahedron. 2007;63:2132–41.
[79] Singh R, Singh GC, Ghosh SK, Novel Approach A. Towards dibenzylbutyrolactone lignans involving heck and radical reactions: application to (±)-matairesinol synthesis. Eur J Org Chem. 2007;2007:5376–85.
[80] Chen C, Crich D, Papadatos A. The chemistry of acyl tellurides: generation and trapping of acyl radicals, including aryltellurium group transfer. J Am Chem Soc. 1992;114:8313–14.
[81] Kanda T, Aoki H, Mizoguchi K, Shiraishi S, Murai T, Kato S. Selenotellurocarbamic acid Te-Alkyl Esters: first isolation and characterization. Organometallics. 1996;15:5753–55.
[82] Kageyama H, Murai T, Kanda T, Kato S. Tautomeric equilibrium between selenol and selenoxo forms of selenocarboxylic acids. J Am Chem Soc. 1994;116:2195–96.
[83] Knapp S, Darout E. New reactions of selenocarboxylates. Org Lett. 2005;7:203–06.
[84] Kato S, Kawahara Y, Kageyama H, Yamada R, Niyomura O, Murai T, et al. Thion (RCSOH), selenon (RCSeOH), and Telluron (RCTeOH) acids as predominant species. J Am Chem Soc. 1996;118:1262–67.
[85] Kageyama H, Tani K, Kato S, Kanda T. Acyl carbamoyl selenides and related sulfur isologues: synthesis and X-Ray structural analyses. Heteroatom Chem. 2001;12:250–58.
[86] Ibi K, Kawai K, Kato S, Ando F, Koketsu J. O-Triorganosilyl carbamoselenoates – synthesis and Reactions. Z Anorg Allg Chem. 2007;633:621–24.
[87] Shimada K, Moro-Oka A, Maruyama A, Fujisawa H, Saito T, Kawamura R, et al. Synthesis of isochalcogenazole rings by treating β-(N,N-Dimethylcarbamoylchalcogenenyl)alkenyl ketones with hydroxylamine-O-sulfonic Acid. Bull Chem Soc Jpn. 2007;80:567–77.
[88] Zingaro RA, Herrera C, Meyers EA. Isolation and crystal structure of an unusual ditelluride: bis(N,N-dimethylaminoformyl) ditelluride. J Organomet Chem. 1986;306:C36–C40.

Claudio Santi, Lucia Capoccia and Bonifacio Monti

9 Zinc-Selenium reagents in organic synthesis

Abstract: Organoselenolates, due to the high polarizability of the chalcogen atoms, are generally weak bases and soft nucleophiles used to introduce in stereoselective and mild way a selenium functionality through substitution or addition reactions. Among several methods reported for their preparation, recently the reduction of Se-Se or Se-Halogen bond mediated by elemental zinc becomes particularly attractive for the simplicity and the efficiency of the protocols. An overview on the most recent developments in the field is here reported.

Keywords: selenium, zinc, nucleophile

9.1 Introduction

Organoselenium compounds are playing an important role in organic synthesis as well as in biological and medicinal chemistry. The introduction of a selenium moiety in an organic substrate confers it peculiar characteristics from a biological and synthetic point of view, considering its redox properties and enabling a series of further transformations that can be ended with the easy removal of the selenium either via selenoxide syn-elimination or [2, 3]-sigmatropic rearrangement. Furthermore, a series of examples were reported in which, carbon-hydrogen, carbon-halogen, carbon-lithium, or carbon-carbon bonds replace the carbon-selenium bond, representing an extraordinary versatile tool for synthetic organic chemists.

Interestingly the introduction, manipulation and removal of selenium moieties can be realized under mild conditions, are generally characterized by high chemo regio and stereoselectivities and recently several aspects of their involvement in new green protocols have been discussed and highlighted [1–6].

Introduction of organoselenium groups can be easily achieved through radical, electrophilic and nucleophilic selenium centered reagents that can be easily prepared starting from the corresponding diselenides via homolytic, oxidative and reductive selenium-selenium bond cleavage, respectively. Among these, the use of selenolates (selenide anions) is particularly convenient and common having some valuable characteristics respect to the similar sulfur analogues. The high polarizability of selenium atom is responsible of the higher nucleophilicity of selenols and selenolates respect to the corresponding thiol/thiolates as well as of the weakness of the carbon-selenium bond according to the experimental evidences that selenoxide

This article has previously been published in the journal *Physical Sciences Reviews*. Please cite as: Santi, C., Capoccia, L., Monti, B. Zinc-Selenium Reagents in Organic Synthesis. *Physical Sciences Reviews* [Online] **2018** DOI: 10.1515/psr-2017-0129

https://doi.org/10.1515/9783110529340-009

eliminations typically occur easily and faster than those of sulfoxides [7–12]. Furthermore based on the hard-soft theory selenolates are considered good soft nucleophiles (low ionization potential and high polarizability) able to bring easily the formation of the carbon-selenium bond and at the same time having a lower basicity respect to the thiolates they are also better leaving groups than sulfur analogues [13]. Selenols and selenolates can be prepared starting from stable and easily accessible diselenides by reduction or oxidative metal insertion as well as by the reductive treatment of other intermediates like selenocyanates or seleninic acids [14–39] (Figure 9.1).

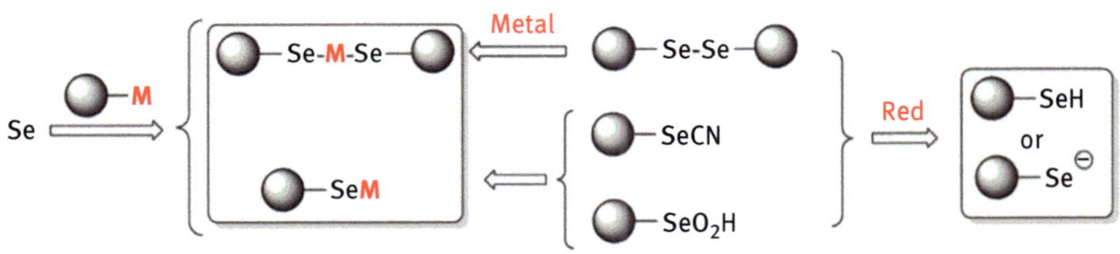

Figure 9.1: General Scheme for selenol/selenolates preparation.

Some of these methods were recently collected in reviews and book chapters [13, 40, 41]. In this chapter, we will focus our discussion on the use of zinc for the preparation of selenols and zinc selenolates due a growing recent interest on these classes of reagents from a synthetical point of view but also as mimetic of the glutathione peroxidase endowed with some unexpected biological activities. In order to address a broad panel of readers the chapter will be constructed on selected examples in literature that better explain the state of art and it will be not a fully comprehensive review in the topic.

9.2 Oxidative zinc insertion in Se-Se and Se-X bond

In a general overview the zinc-selenium bond can be formed through the oxidative insertion of metal into selenium-selenium bond **A** or into selenium-halogen bond **B** affording zinc selenates that are generally prepared and used in situ and only in few cases were isolated as simple Se-Zn complexes. Better stability can be obtained by additional coordination of zinc with nitrogen containing pincers **C**. Some examples are described below (Figure 9.2).

9.2.1 Synthesis and reactivity of RSeZnSeR complexes

Due to their stability, diselenides are very common reagents in organic synthesis also because some of them are commercially available and/or can be prepared according to

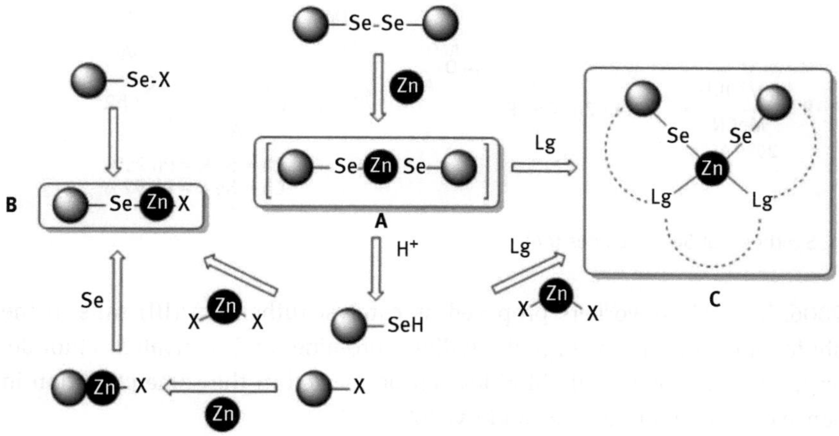

Figure 9.2: General scheme for the preparation of zinc selenates.

robust and efficient procedures [41]. The oxidative insertion of elemental zinc into Se-Se bond to afford the *in-situ* formation of a [RSeZnSeR] reactive intermediate was performed for the first time by Movassagh *et al.* using $AlCl_3$ as a Lewis Base. The reduction required the use of a large excess of zinc and aluminium trichloride (4.6 and 2.0 equivalents, respectively) in DMF at 65 °C affording, in the presence of acyl chlorides, the corresponding selenol esters in good yield [42]. More recently, the same author described that diselenide in the presence of Zn/$AlCl_3$ can be used for the synthesis of selenocarbamates from the corresponding isocyanates. In this case, the role of the Lewis acid was demonstrated both in the reduction of the Se-Se bond and in the activation of the isocyanates by the coordination with the oxygen atom [43] (Figure 9.3).

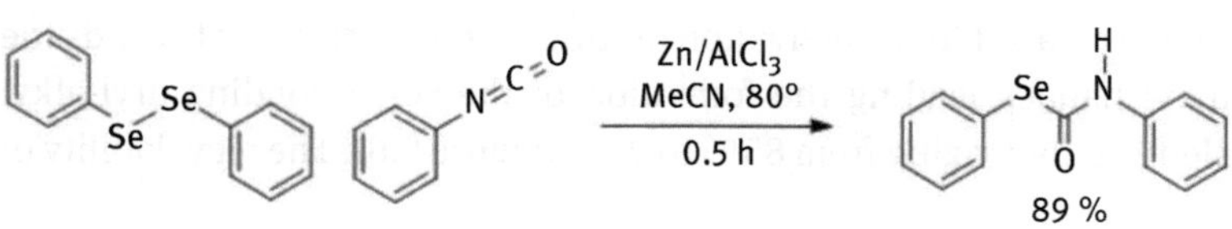

Figure 9.3: Synthesis of selenol esters catalysed by Lewis acids [43].

A similar insertion was obtained using catalytic amount of $ZnCl_2$ as Lewis acid in presence of zinc powder in refluxing acetonitrile. This procedure was demonstrated to be efficient both for the reduction of Se-Se and S-S bonds and it has been successfully used for the synthesis of thio- and selenoglycosides, starting from the corresponding bromo or chloro derivatives. The *in situ* formation of the corresponding zinc chalcogenide as a reactive intermediate was proposed also in this case [44] (Figure 9.4).

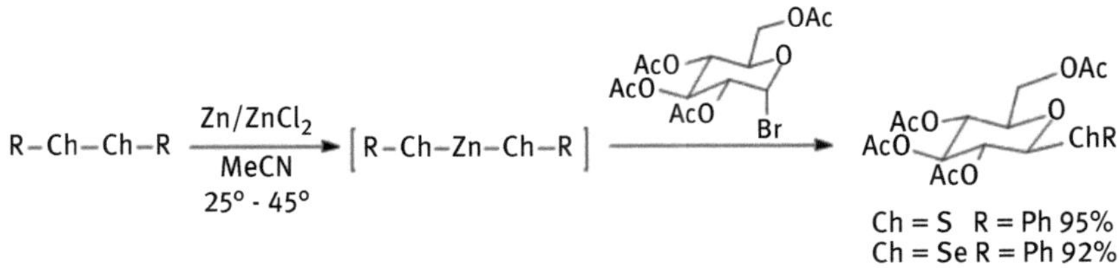

Figure 9.4: Synthesis of Se-glycosides [44].

In 2006, Yu and co-workers proposed as catalyst ruthenium (III) salts in the nucleophilic substitution or organic iodine, bromide and activated chloride. Interestingly, the non-activated chlorides can be reacted in the same condition in the presence of sodium bromide as additive [45].

A similar procedure was used later by other authors to promote a Michael-type addition of three different zinc selenates [RSeZnSeR] (R = Ph, 4-ClC_6H_5, Bn) in aqueous media ($MeCN/H_2O$ 4:1) (Figure 9.5). Esters, ketones, aldehydes and nitriles were reported as Michael acceptors [46].

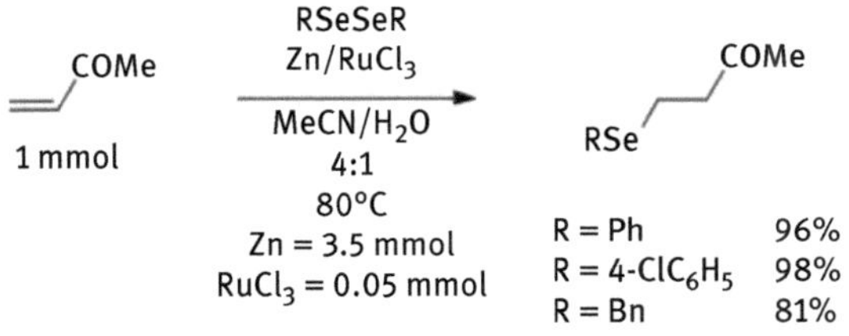

Figure 9.5: Michael type addition under $RuCl_3$ catalysis [46].

MCM-412-immobilized bidentate nitrogen Ru(III) was demonstrated to be an efficient and easily recyclable catalyst in the activation of ArSeZnSeAr complexes towards the reaction with alkyl halides leading the formation of the corresponding aryl-alkyl halides in excellent yields ranging from 85 % to 95 % (Figure 9.6). The recyclability of

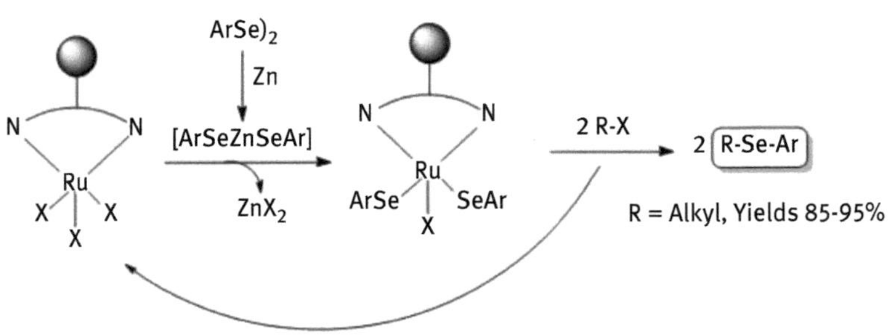

Figure 9.6: A recyclable ruthenium catalyst [47].

the Ruthenium catalyst has been demonstrated for eight consecutive cycles, without any appreciable loss of reactivity [47].

Se-Se reductive cleavage mediated by zinc was catalysed also by In (III) for the formation of unsymmetrical diorganyl selenide in DMF at 100 °C, starting from alkyl and benzyl halides. As catalyst, Indium bromide ($InBr_3$) resulted to be superior compared with the corresponding chloride or triflate and the [RSeZnSeR] was supposed to react with InX_3 to afford the reactive specie $(RSe)_2InX$ as intermediate [48].

Krief and coworkers demonstrated that when a diselenide, a tertiary alkyl halides and zinc powder were reacted at room temperature for 5 h, the corresponding tertiary selenide can be formed only in non-polar solvents (like dichloromethane) suggesting the involvement of a radical mechanism [49]. Radical intermediates are not involved using polar medium (e. g. water and MeCN) as indicated by the higher reactivity of primary halides.

It was demonstrated that the oxidative insertion to Se-Se bond can be achieved also in basic conditions in the presence of NaOH. Comparison of these results with those obtained using the above mentioned $Zn/AlCl_3$ in the ring opening of epoxides, basic conditions afforded higher yields and higher level of regioselectivity (Figure 9.7) [50]. Similarly, in basic conditions, Barros and coworkers reported a regio and stereoselective procedure to perform Michael type addition of several propiolic acids and esters. Authors demonstrated that the reaction mechanism involves a competition of ionic and a radical pathway and that this latter is responsible of a lost in stereoselectivity [51].

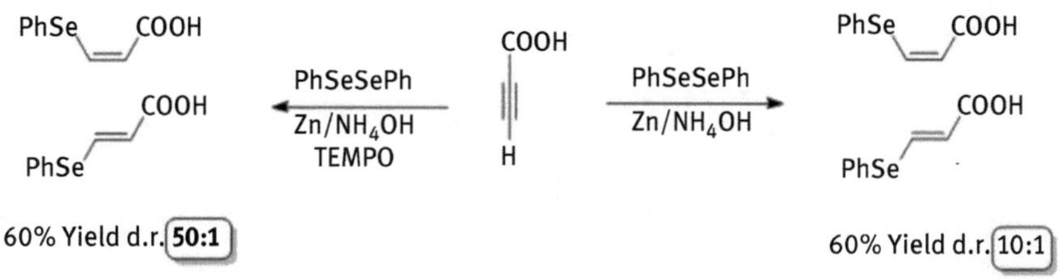

Figure 9.7: Stereoselective conjugated addition [50].

The oxidative insertion of zinc into Se-Se bond of a diselenide was reported by Braga and coworkers using ionic liquids as a recyclable medium. The reduction of diaryl diselenides occurred at room temperature in BMIM-BF_4 affording at the presence of alkyl, allyl and benzyl halides the corresponding selenides in good yields (ranging from 52 % to 98 %) in very short reaction time (10–45 minutes) [52]. The same authors reported also that the versatility and efficiency of the method can be improved by introducing a catalytic amount of ZnO nanopowder (3 %) in the synthesis of β-amino selenides starting from the corresponding mesylates and tosylates. The exploration of the scope demonstrated that this protocol results efficient also in variously functionalized systems allowing the synthesis of the biologically active selenocysteine derivatives reported in Figure 9.8 in 78 % yield [53].

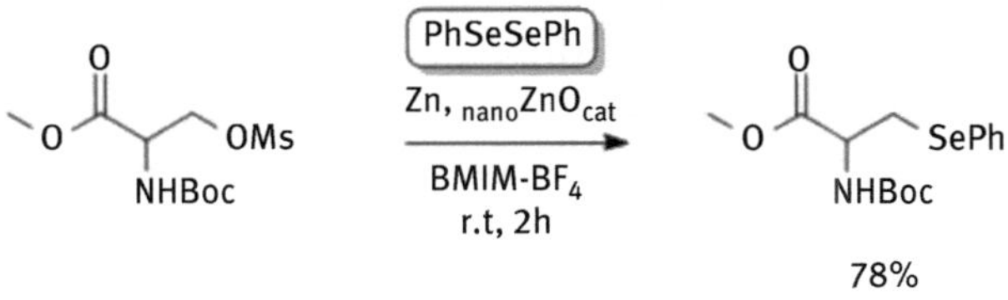

Figure 9.8: Synthesis of selenocysteine derivatives [53].

Later, Ranu and co-workers optimized the conditions for the synthesis of unsymmetrically substituted diaryl chalcogenides (sulfides, selenides and tellurides) by the reaction of aryl diazonium fluoroborates dichalcogenides reduced *in situ* by zinc dust in the presence of dimethyl carbonate under heating (80 °C) or microwave activations. The mechanism proposed by the authors involve the intermediate formation of the zinc dichalcogenate [ArChZnChAr] that reacts with the diazonium salts releasing a molecule of nitrogen affording the desired products [54] (Figure 9.9).

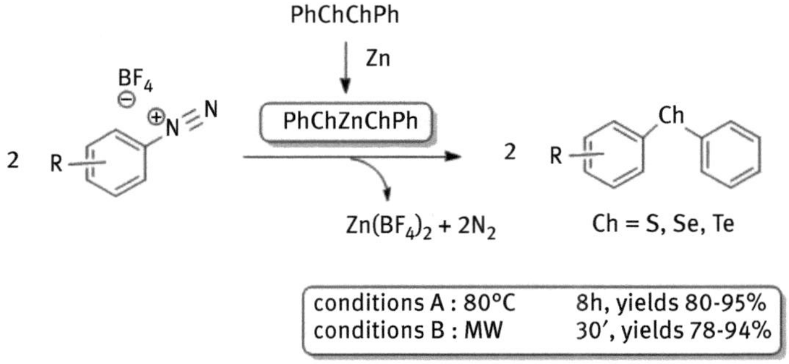

Figure 9.9: Synthesis of non-symmetric diaryl chalcogenides [54].

9.2.2 A versatile and recyclable biphasic system

We recently reported an efficient recyclable and simple method for the zinc mediated reduction of dichalcogenides and the generation of chalcogenols, which can be isolated or reacted *in situ* with the desired electrophiles [55]. The reaction is performed using zinc in a biphasic acidic medium composed by diethyl ether and aqueous HCl (10 %). As reported in Figure 9.10, the zinc activated by the acidic conditions reduces the Ch-Ch bond and we supposed the intermediate formation of a zinc dichalcogenate that, due to the acidic conditions can decompose leading the

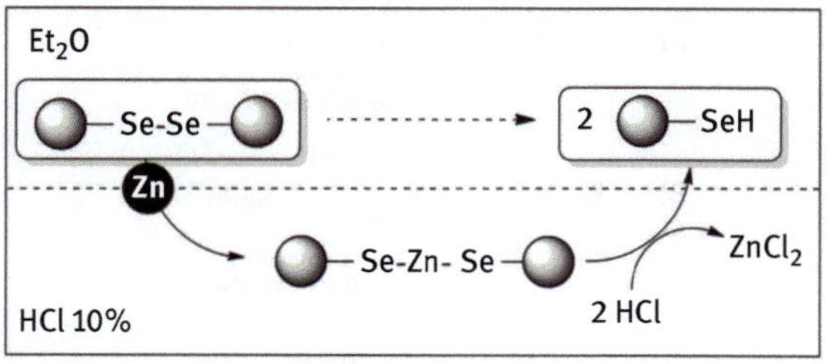

Figure 9.10: General scheme for the Se-Se reduction in a biphasic system.

formation of the chalcogenols. When the dichalcogenide is a diselenide (or a ditelluride) the reaction can be easily followed by the discoloration of the organic phase. The procedure is very simple allowing to produce and use selenols and thiols in a closed vial, overcoming the problems connected with the characteristic smell of these derivatives.

The broad applicability of this protocol was demonstrated by us and other authors. Using as electrophiles alkyl halides or epoxides the corresponding alkyl aryl selenides and β-hydroxy selenides were obtained in good to excellent yields (Figure 9.11a and Figure 9.11b, respectively). Braga demonstrated that this method can be efficiently used also for the ring opening of unprotected aziridines, a reaction that cannot be obtained using the other known protocols for nucleophilic selenenylation (Figure 9.11c) [56].

The biphasic system was also used to perform the hydrochalcogenation of alkynes leading to the stereoselective of the corresponding *Z* vinyl chalcogenides (Figure 9.11d). Both diselenides and disulfides were demonstrated to be efficient as chalcogenol precursors but the reaction with selenium-containing reagents resulted to be more stereoselective in comparison with those with sulfur analogues. Interestingly in this case it was demonstrated that the possibility to reuse the aqueous part of the biphasic system together with the excess of unreacted zinc for 10 consecutive cycles allows the easy gram scale scale-up in the synthesis of (*Z*)-phenyl(styryl)selane [57]. Moderate yields were obtained when a series of acyl chloride was converted into the corresponding selenol esters using the same protocol (Figure 9.11e). The prevalent formation of the carboxylic acid, evidenced a limit in the use of strong hydrolytic conditions. Attempting to avoid this undesirable side reaction, THF and triflic acid were used in the place of the biphasic system and an unprecedented reaction of ring opening of tetrahydrofuran was reported [58] (Figure 9.12).

The simplicity and versatility of the biphasic system was claimed by Flamer, who affirmed that it was the most effective method (among a large number of other reported in literature) for the synthesis of cysteine and selenocysteine derivatives to be used as peptide building blocks [59] and in peptide chemistry [60] (Figure 9.11f).

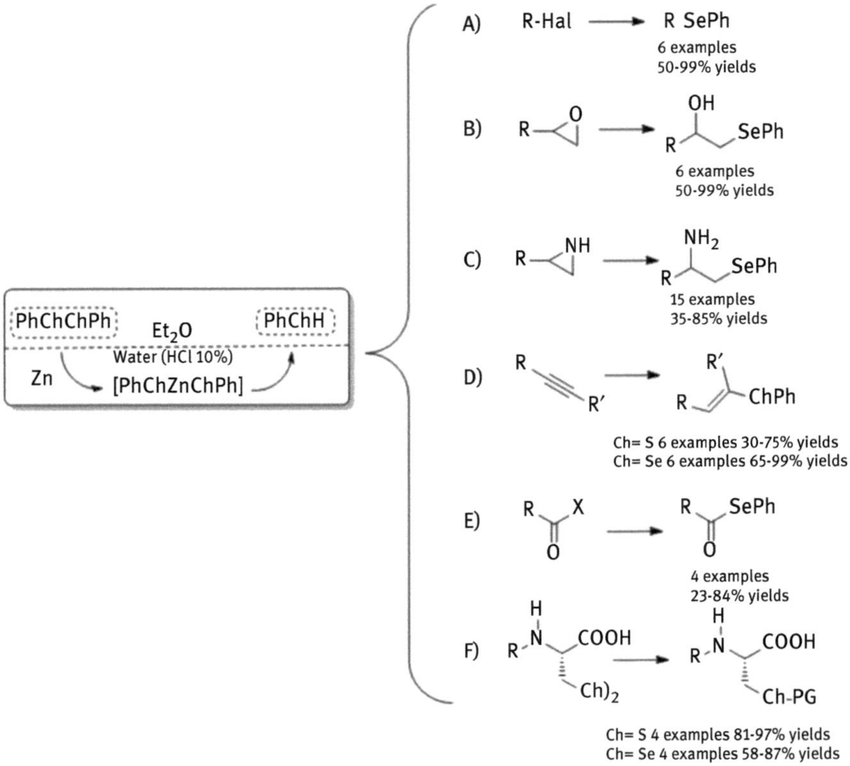

Figure 9.11: Synthetic application of the biphasic system.

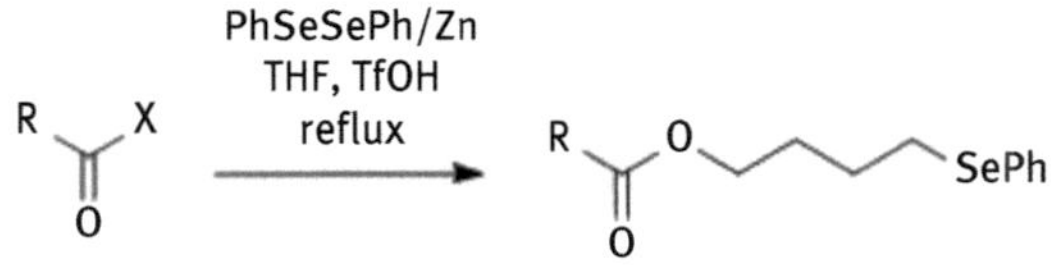

Figure 9.12: Ring opening of THF [58].

9.2.3 Synthetic applications of TMEDA stabilized PhSeZnSePh

The bis(phenylselenolato)zinc, $Zn(SePh)_2$, described above is normally polymeric and is characterized by a poor solubility in organic solvents. For these reasons, it is normally generated *in situ* and the reaction is activated by mild acidic or basic conditions. It was demonstrated that this polymer can be also generated by the reaction of phenylselenol (PhSeH) with $Zn[N(SiMe_3)_2]_2$ in toluene and, after precipitation from heptane, the resulting white solid reacts with *N,N,N',N'*-tetramethylethylenediamine (TMEDA) affording a soluble, air-stable, monomeric compound that was characterized as $Zn(SePh)_2$-(TMEDA) [61].

This compound has been used for the preparation of ZnSe quantum dots by thermal decomposition and having a size tuneable by the temperature. Due to the particular stability and the reduced nucleophilicity of selenium $Zn(SePh)_2$-(TMDEA) was never used in organic synthesis until 2017 when we compared the reactivity of Zn $(SePh)_2$ and $Zn(SePh)_2$-(TMEDA) in the conversion of acyl chlorides into the corresponding selenol esters using water as medium [62]. We demonstrated that the reactivity of TMEDA derivative is only slightly reduced and interestingly its stability enabled to setup a unique new protocol for the one pot reaction-purification process that allows to avoid the usual workup of the reaction (Figure 9.13).

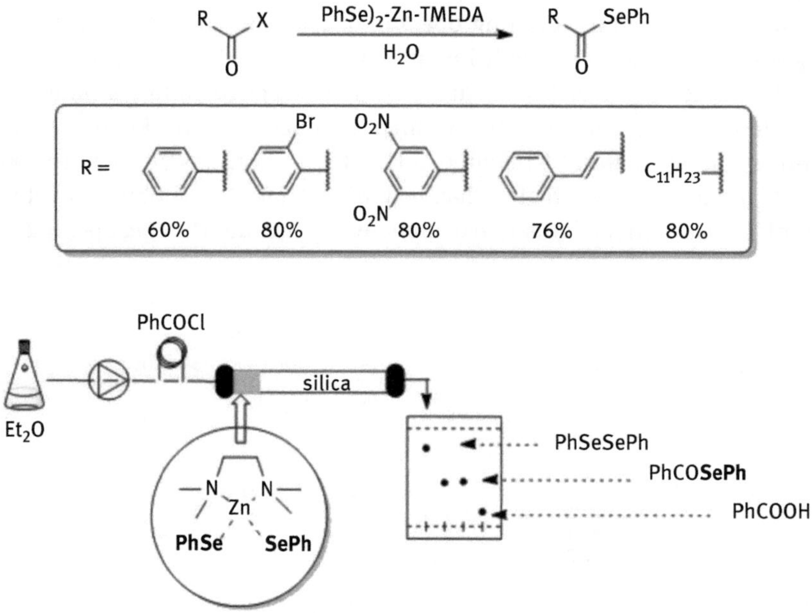

Figure 9.13: Acyl substitution using TMEDA stabilized $(PhSe)_2Zn$ complex.

9.2.4 PhSeZn-halides and the unexpected on-water reactivity

The formation of RSeZnCl was for the first time proposed by Huang and co-workers in 1998 as an intermediate formed *in situ* during the reaction of alkyl or aryl-zinc halides with elemental selenium for the formation of a nucleophilic selenium reagents in the synthesis of allyl halides and α-selenocarbonyl derivatives.

In the first case allylzinc halides reacted with elemental selenium and subsequently with diaryliodonium salts affording the corresponding aryl selenides in good yields [63]; in the latter, alkyl and aryl zin halides, generated by the reaction of elemental zinc and the corresponding Grignard reagents, were subjects to the selenium insertion into the zinc-carbon bond to generate the nucleophilic selenium cantered reagent for the nucleophilic substitution of α-halocarbonyl substrates [64].

Ten years later, for the first time we reported a novel protocol for the synthesis of PhSeZn-halides starting from the commercially available PhSeCl and PhSeBr by oxidative insertion of elemental zinc [65]. The main novelty of this synthesis is that, in refluxing THF, after the consumption of the zinc and the complete umpolung of the selenium atoms (clearly evidenced by the discoloration of the solution) a white amorphous and bench stable solid with stoichiometric composition of PhSeZnCl or PhSeZnBr can be precipitated by diethyl ether or heptane, characterized by ^{77}SeNMR, stored for long periods under ambient conditions and used as nucleophilic selenenylating reagents in several reactions.

Among all these reactions, the ring opening of epoxides for the synthesis of β-hydroxyselenides was the most deeply investigated by us and by other groups. The reactivity of PhSeZnX (X = Cl or Br) is usually strongly accelerated under on water conditions affording quantitatively the desired product in 2 h (Figure 9.14a, Figure 9.14b) as opposed to the 24 h required in THF. In addition, in water suspension the stability of the complex was higher than in THF solution were a photoactivated oxidation rapidly results in the formation of diphenyl diselenide. The ring opening of

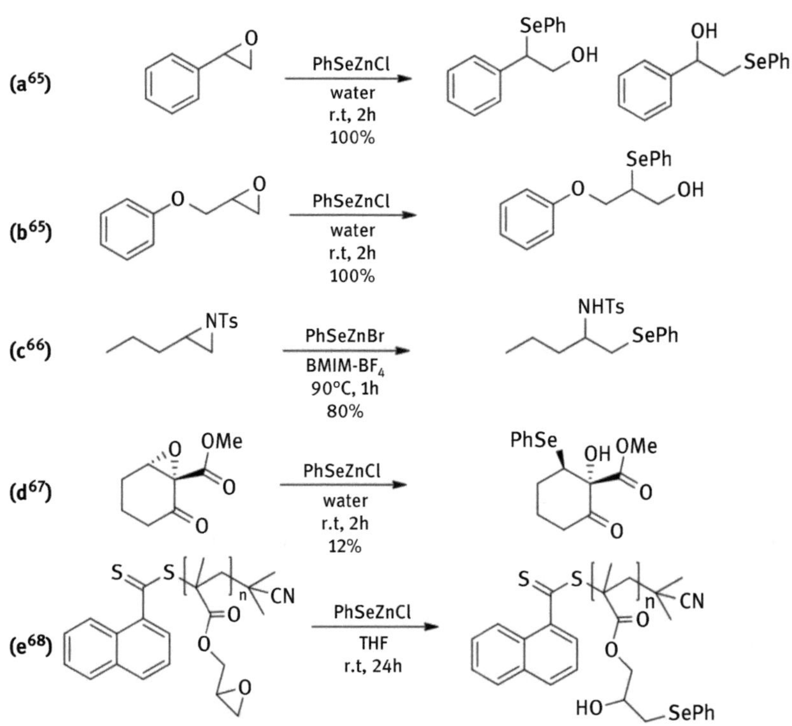

Figure 9.14: Ring opening of epoxides and aziridines.

styrene oxide afforded the prevalent formation of the electronically controlled isomer over the sterically controlled one indicating that the reagent combines the nucleophilic properties of selenium with the Lewis acid character of the zinc. The zinc (II) atom, coordinating to the oxygen of the heterocycle, activates the ring opening driving the regiochemistry of the process when a strong stabilization of an incipient positive charge is possible (Figure 9.14a).

A regioselective process has been reported also by Braga and co-workers for the ring opening of protected and unprotected aziridines under neutral conditions. The reaction affords β-seleno amines using PhSeZnBr in ionic liquids at 90 °C leading better yields in the case of the N-Tosyl derivatives [66] (Figure 9.14c). In a recent study on the total synthesis of idesolide, it was reported that the functionalized epoxide depicted in Figure 9.14d can be converted in the corresponding β-seleno alcohol using PhSeZnCl in water. Even if the yield is particularly poor, the classical protocol, involving $NaBH_4$/EtOH and diselenide failed affording to a mixture of unidentified compounds probably due to the presence of functional groups that are sensitive to reductive conditions [67].

A further interesting application was reported in 2017 by Zhu and co-workers that used the ring opening of epoxide promoted by PhSeZnCl in THF as a highly efficient post-synthetic modification of poly(glycidyl methacrylate) preparing a series of polymers at different concentration of selenium having an high refractive index [68] (Figure 9.14e).

Several examples of nucleophilic substitutions using PhSeZnCl or PhSeZnBr have been reported and are summarized in Figure 9.15, evidencing the broad applicability of these reagents under on water conditions. As electrophile alkyl and vinyl halides resulted to be particularly reactive affording in 2 h from good to excellent yields. In the first case it was demonstrated that the yield is negatively affected by the grade of substitution on the carbon bearing the halogen atom, indicating a mechanism that proceed through a S_N2 mechanism and not through a radical pathway (Figure 9.15a) [65].

When the substrate is a vinyl halide, the substitution normally proceeds with retention of configuration (Figure 9.15b) with the exception of the olefins bearing a substituent able to coordinate to the zinc atom. In this case, it was demonstrated by DFT calculations that the coordination drives the substitution toward the Z-isomer when starting from both the E- and Z- halides [69] (Figure 9.15c).

Churchill et al. applied this reaction to the synthesis of a selenium functionalized coumarin as a fluorescent probe of glutathione. In this case the nucleophilic reagent was PhSeZnBr used at room temperature in aqueous condition with a reaction time of 2 h [70] (Figure 9.15d). Very interestingly, the reaction can be performed in water suspension also when the substrate is an acyl halide or a chloroformiate (Figure 9.15e) and, in the presence of PhSeZnBr, the corresponding selenoesters or selenoformiate can be obtained in good yield indicating a favourable rate of the selenenylation over the hydrolysis [71]. Also in this case the presence of the zinc was

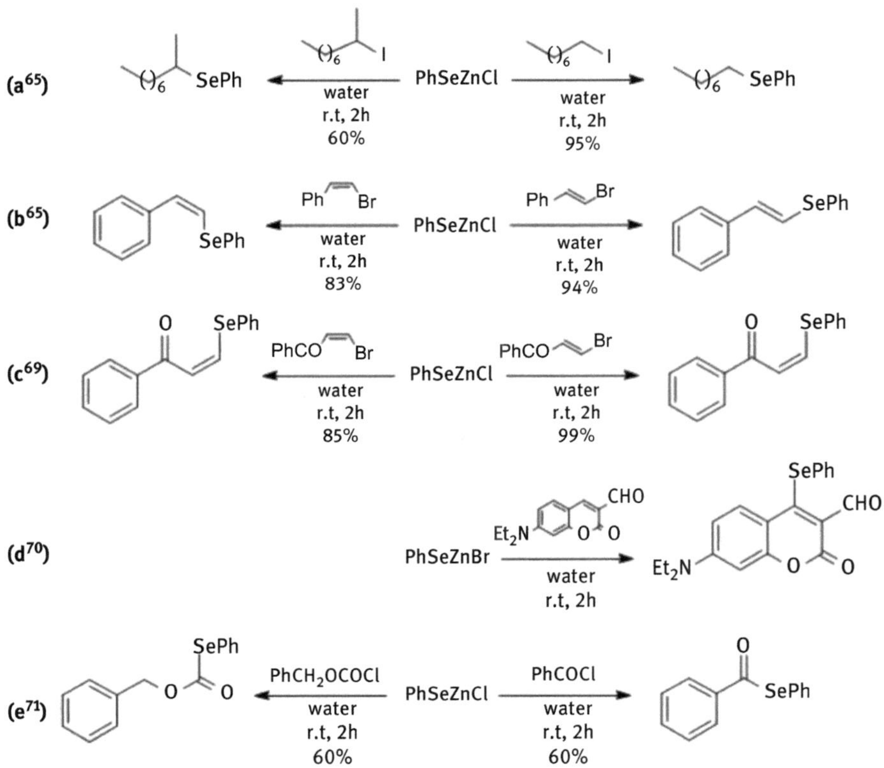

Figure 9.15: Nucleophilic substitutions.

supposed to be responsible of the chemo selectivity and it was suggested that it coordinates to the oxygen atom of the carbonyl moiety bringing the nucleophilic selenium atom near to the reactive centre and facilitating the formation of the selenoesters. In this case in order to claim the eco-friendly aspect of the use of PhSeZn-halides in on water condition, the recyclability of the aqueous medium was studied and demonstrated for four subsequent cycles.

Finally, it was also demonstrated that PhSeZnCl can be used in seleno-Michael addition reaction starting from conjugated alkenes or alkynes [72]. In the case of alkenes, it was observed a severe limitation in the choice of the electron withdrawing group (only ketones afforded desired products) and the reaction was not accelerated by the on-water conditions. On the contrary, when the addition involves carbon-carbon triple bonds the reaction resulted accelerated on water affording the stereoselective formation of the vinyl selenide Z either starting from ketone, aldehydes and esters.

In 2016 PhSeZnCl was for the first-time named as *Santi's reagent* [73] and nowadays it is also commercially available.

References

[1] Wirth T, editor. Organoselenium chemistry. Modern developments in organic synthesis. Berlin, Heidelberg: Springer; 2000.
[2] Browne DM, Wirth T. New developments with chiral electrophilic selenium reagents. Curr Org Chem. 2006;10:1893–903.
[3] Freudendahl DM, Santoro S, Shahzad SA, Santi C, Wirth T. Green chemistry with selenium reagents: development of efficient catalytic reactions. Angew Chemie - Int Ed. 2009;48:8409–11.
[4] Singh FV, Wirth T. Stereoselective reactions of organoselenium reagents including catalysis. In: PATAI'S chemistry of functional groups. Chichester, UK: John Wiley & Sons, Ltd; 2011.
[5] Santoro S, Azeredo JB, Nascimento V, Sancineto L, Braga AL, Santi C. The green side of the moon: ecofriendly aspects of organoselenium chemistry. RSC Adv. 2014;4:31521–35.
[6] Jain VK, Priyadarsini KI, editors. Organoselenium compounds in biology and medicine: Synthesis, biological and therapeutic treatment. Cambridge: Royal Society of Chemistry; 2018.
[7] Emerson DW, Korniski TJ. Substituent effects in the pyrolysis of aryl n-propyl sulfoxides. J Org Chem. 1969;34:4115–18.
[8] Sharpless KB, Young MW, Lauer RF. Reactions of selenoxides: thermal-elimination and $H_2{}^{18}O$ exchange. Tetrahedron Lett. 1973;14:1979–82.
[9] Sharpless KB, Young MW. Olefin synthesis. Rate enhancement of the elimination of alkyl aryl selenoxides by electron-withdrawing substituents. J Org Chem. 1975;40:947–49.
[10] Grieco PA, Gilman S, Nishizawa M. Organoselenium chemistry. A facile one-step synthesis of alkyl aryl selenides from alcohols. J Org Chem. 1976;41:1485–86.
[11] Reich HJ, Wollowitz S, Trend JE, Chow F, Wendelborn DF. Syn elimination of alkyl selenoxides. Side reactions involving selenenic acids. Structural and solvent effects on rates. J Org Chem. 1978;43:1697–705.
[12] Nishibayashi Y, Uemura S. Selenoxide elimination and [2,3]sigmatropic rearrangement. In: Wirth T, editor. Organoselenium chemistry. Springer-V; 2000. 201–33.
[13] Movassagh B, Navidi M. Nucleophilic selenium: nucleophilic substitution. In: Santi C, editor. Organoselenium chemistry between synthesis and biochemistry. Bentham Science Publishers; 2014. p. 61–93.
[14] Bird ML, Challenger F. Potassium alkaneselenonates and other alkyl derivatives of selenium. J Chem Soc. 1942;570–72.
[15] Klayman DL. The synthesis of aminoethyl-substituted selenium compounds. J Org Chem. 1965;30:2454–56.
[16] Ley SV, O'Neil IA, Low CMR. Ultrasonic formation and reactions of sodium phenylselenide. Tetrahedron. 1986;42:5363–68.
[17] Fukuzawa S, Fujinami T, Sakai S. Samarium(II) di-iodide induced synthesis of allylic phenyl selenides from allylic acetates and diphenyl diselenide in the presence of palladium catalyst. Chem Lett. 1990;19:927–30.
[18] Sakakibara M, Katsumata K, Watanabe Y, Toru T, Ueno Y, Convenient A. Procedure for the preparation of organic selenides. Synthesis (Stuttg). 1992;377–9.
[19] Krief A, Trabelsi M, Dumont W. Syntheses of alkali selenolates from diorganic diselenides and alkali metal hydrides: scope and limitations. Synthesis (Stuttg). 1992;933–35.
[20] Pedersen ML, Berkowitz DB. Formal alpha-vinylation of amino acids. Use of a new benzene-selenolate equivalent. J Org Chem. 1993;58:6966–75.
[21] Zhang Y, Bao W, Zheng Y, Zhou J, Novel A. Synthesis of allyl sulfides in aqueous media promoted by indium. Tetrahedron Lett. 1996;37:9333–34.
[22] Krief A, Delmotte C, Dumont W. Chemoselective reduction of organoselenocyanates to diselenides and selenolates. Tetrahedron Lett. 1997;38:3079–80.

[23] Liao P, Bao W, Zhang Y, Novel A. Synthesis of allyl and prop-2-ynyl selenides promoted by tin in the presence of water. J Chem Res. 1998;3:150–51.
[24] Zheng Y, Bao W, Zhang Y. Synthesis of allyl-type selenides and α-selenoesters in Aqueous Media Promoted by Cadmium. Synth Commun. 2000;30:1731–36.
[25] Kundu A, Roy S. Copper(II)/Tin(II) reagent for allylation, propargylation, alkynylation, and benzylation of diselenides: a novel bimetallic reactivity. Organometallics. 2000;19:105–07.
[26] Günther WHH. Hypophosphorous acid, a novel reagent for the reduction of diselenides and the selenol-catalyzed reduction of disulfides. J Org Chem. 1966;31:1202–05.
[27] Guo H, Zheng Y, Zhang Y. Convenient one-pot synthesis of α-selenonitriles, α-selenoesters and asymmetrical selenides by a $Sm/ZnCl_2$ system in DMF-H_2O. J Chem Res. 2001;160–61.
[28] Zheng Y, Guo H-Y, Zhang Y. Synthesis of allyl selenides promoted by an $Sm/ZnCl_2$ bimetal system in the presence of water. Chinese J Chem. 2001;19:2000–02.
[29] Nishino T, Okada M, Kuroki T, Watanabe T, Nishiyama Y, Sonoda N. One-pot synthetic method of unsymmetrical diorganyl selenides: reaction of diphenyl diselenide with alkyl halides in the presence of lanthanum metal. J Org Chem. 2002;67:8696–98.
[30] Ranu BC, Mandal T, Samanta S. Indium(I) iodide-mediated cleavage of diphenyl diselenide. An efficient one-pot procedure for the synthesis of unsymmetrical diorganyl selenides. Org Lett. 2003;5:1439–41.
[31] Dan W, Deng H, Chen J, Liu M, Ding J, Wu H. A new odorless one-pot synthesis of thioesters and selenoesters promoted by Rongalite®. Tetrahedron. 2010;66:7384–88.
[32] Su W-K, Li Y-S, Zhang Y-M, Zheng Y-F. Facile formation of ytterbium diiodide and its use in the synthesis of Allyl selenides. Chinese J Chem. 2010;20:174–77.
[33] Schmidt M, Block HD. Notiz zur vereinfachten Darstellung von Alkanselenolen. Chem Ber. 1970;103:3348–49.
[34] Klayman DL, Griffin TS. Reaction of selenium with sodium borohydride in protic solvents. A facile method for the introduction of selenium into organic molecules. J Am Chem Soc. 1973;95:197–99.
[35] Sharpless KB, Lauer RF, Mild A. Procedure for the conversion of epoxides to Allylic Alcohols. First organoselenium reagent. J Am Chem Soc. 1973;95:2697–99.
[36] Liotta D, Markiewicz W, Santiesteban H. The generation of uncomplexed phenyl selenide anion and its applicability to S_N2-type ester cleavages. Tetrahedron Lett. 1977;18: 4365–67.
[37] Dowd P, Kennedy P. Convenient preparation of sodium and potassium phenyl selenides. Reduction of the Selenium-Selenium Bond by Sodium Hydride and Potassium Hydride. Synth Commun. 1981;11:935–41.
[38] Maruoka K, Miyazaki T, Ando M, et al. Organoaluminum-promoted Beckmann rearrangement of oxime sulfonates. J Am Chem Soc. 1983;105:2831–43.
[39] Syper L, Młochowski J. The convenient syntheses of organoselenium reagents. Synthesis (Stuttg). 1984;439–42.
[40] Iwaoka M, Tomoda S. Nucleophilic Selenium. In: Wirth T, editor. Topics in current chemistry. Weinheim, Germany: Springer-Verlag; 2000. 53–109.
[41] Potapov VA. Organic diselenides, ditellurides, polyselenides and polytellurides. synthesis and reactions. In: Zvi Rappoport, editor. PATAI'S chemistry of functional groups. Chichester, UK: John Wiley & Sons, Ltd; 2013. p. 765–844.
[42] Movassagh B, Mirshojaei F. Synthesis of selenol esters from acid chlorides and organic diselenides in the presence of the $Zn/AlCl_3$ system. Monatshefte Fur Chemie. 2003;134: 831–35.
[43] Movassagh B, Moradi M. One-pot synthesis of selenocarbamates from isocyanates and diselenides using the $Zn/AlCl_3$ system. Chinese Chem Lett. 2013;24:192–94.

[44] Mukherjee C, Tiwari P, Misra AK. Synthesis of thio- and selenoglycosides by cleavage of dichalconides in the presence of zinc/zinc chloride and reaction with glycosyl bromides. Tetrahedron Lett. 2006;47:441–45.

[45] Zhao X, Yu Z, Yan S, et al. Ruthenium(III) chloride catalyzed efficient synthesis of unsymmetrical diorganyl selenides via cleavage of dibenzyl and diphenyl diselenides in the presence of zinc. J Org Chem. 2005;70:7338–41.

[46] Movassagh B, Tatar A. Zn/$RuCl_3$ -promoted cleavage of diselenides: an efficient Michael addition of zinc selenolates to conjugated alkenes in aqueous media. Synlett. 2007;2007:1954–56.

[47] Chen Q, Wang P, Yan T, Cai M. A highly efficient heterogeneous ruthenium(III)-catalyzed reaction of diaryl diselenides with alkyl halides leading to unsymmetrical diorganyl selenides. J Organomet Chem. 2017;840:38–46.

[48] Braga AL, Schneider PH, Paixão MW, Deobald AM. Efficient synthesis of diorganyl selenides via cleavage of Se-Se bond of diselenides by indium(III) catalyst and zinc. Tetrahedron Lett. 2006;47:7195–98.

[49] Krief A, Derock M, Lacroix D. Zinc-mediated synthesis of tertiary alkyl selenides from tertiary alkyl halides. Synlett. 2005;2005:2832–34.

[50] Movassagh B, Shamsipoor M. Stereo- and regioselective zinc-mediated ring-opening of epoxides with diselenides. Synlett. 2005;1316–18.

[51] Lóren Nunes V, De Oliveira IC, Soares Do Rêgo Barros O. Organylzinc chalcogenolate promoted Michael-type addition of α,β-unsaturated carbonyl compounds. European J Org Chem. 2014;2014:1525–30.

[52] Narayanaperumal S, Alberto EE, Gul K, Rodrigues OED, Braga AL. Synthesis of diorganyl selenides mediated by zinc in ionic liquid. J Org Chem. 2010;75:3886–89.

[53] Narayanaperumal S, Gul K, Kawasoko CY, et al. Transition metal oxide nanopowder and ionic liquid: an efficient system for the synthesis of diorganyl selenides, selenocysteine and derivatives. J Braz Chem Soc. 2010;21:2079–87.

[54] Kundu D, Ahammed S, Ranu BC. Microwave-assisted reaction of aryl diazonium fluoroborate and diaryl dichalcogenides in dimethyl carbonate: a general procedure for the synthesis of unsymmetrical diaryl chalcogenides. Green Chem. 2012;14:2024–30.

[55] Santi C, Santoro S, Testaferri L, Tiecco M. A simple zinc-mediated preparation of selenols. Synlett. 2008;2008:1471–74.

[56] Braga AL, Schwab RS, Alberto EE, Salman SM, Vargas J, Azeredo JB. Ring opening of unprotected aziridines by zinc selenolates in a biphasic system. Tetrahedron Lett. 2009;50:2309–11.

[57] Tidei C, Sancineto L, Bagnoli L, Battistelli B, Marini F, Santi C. A recyclable biphasic system for stereoselective and easily handled hydrochalcogenations. European J Org Chem. 2014;2014:5968–75.

[58] Bellino G, Scisciani M, Vargas JP, et al. Reaction of Acyl chlorides with in situ formed zinc selenolates: synthesis of selenoesters versus ring-opening reaction of tetrahydrofuran. J Chem. 2016;1–8.

[59] Flemer S. A comprehensive one-pot synthesis of protected cysteine and selenocysteine SPPS derivatives. Protein Pept Lett. 2014;21:1257–64.

[60] Flemer S. Fmoc-Sec(Xan)-OH: synthesis and utility of Fmoc selenocysteine SPPS derivatives with acid-labile sidechain protection. J Pept Sci. 2015;21:53–59.

[61] Jun Y, Koo J-E, Cheon J. One-step synthesis of size tuned zinc selenide quantum dots via a temperature controlled molecular precursor approach. Chem Commun. 2000;2000:1243–44.

[62] Sancineto L, Pinto Vargas J, Monti B, et al. Atom efficient preparation of zinc selenates for the synthesis of selenol esters under "On Water" Conditions. Molecules. 2017;22:953–62.

[63] Huang X, Xu X-H, New A. Method for the synthesis of Allyl arylselenides via the reaction of the Zinc Allyl selenoates with diaryliodonium salts. Synth Commun. 1998;28:801–05.
[64] Huang X, Xu X-H. Insertion of elemental selenium into zinc carbon bond and application in synthesis of α-selenocarbonyl compound. Synth Commun. 1998;28:807–11.
[65] Santi C, Santoro S, Battistelli B, Testaferri L, Tiecco M. Preparation of the first bench-stable phenyl selenolate: an interesting "On Water" nucleophilic reagent. European J Org Chem. 2008;32:5387–90.
[66] Salman S, Schwab R, Alberto E, et al. Efficient ring opening of protected and unprotected aziridines promoted by stable zinc selenolate in ionic liquid. Synlett. 2011;1:69–72.
[67] Nagasawa T, Shimada N, Torihata M, Kuwahara S. Enantioselective total synthesis of idesolide via $NaHCO_3$-promoted dimerization. Tetrahedron. 2010;66:4965–69.
[68] Jiang H, Pan X, Li N, Zhang Z, Zhu J, Zhu X. Selenide-containing high refractive index polymer material with adjustable refractive index and Abbe's number. React Funct Polym. 2017;111:1–6.
[69] Santoro S, Battistelli B, Testaferri L, Tiecco M, Santi C. Vinylic substitutions promoted by PhSeZnCl: synthetic and theoretical investigations. European J Org Chem. 2009;74:4921–25.
[70] Kim Y, Mulay SV, Choi M, Yu SB, Jon S, Churchill DG. Exceptional time response, stability and selectivity in doubly-activated phenyl selenium-based glutathione-selective platform. Chem Sci. 2015;6:5435–39.
[71] Santi C, Battistelli B, Testaferri L, Tiecco M. On water preparation of phenylselenoesters. Green Chem. 2012;14:1277–80.
[72] Battistelli B, Lorenzo T, Tiecco M, Santi C. "On-Water" Michael-type addition reactions promoted by PhSeZnCl. European J Org Chem. 2011;2011:1848–51.
[73] Perin G, Alves D, Jacob RG, Barcellos AM, Soares LK, Lenardão EJ. Synthesis of organochalcogen compounds using non-conventional reaction media. ChemistrySelect. 2016;1:205–58.

Alexander M. Polgar and John F. Corrigan

10 Recent advances in the self-assembly of polynuclear metal–selenium and –tellurium compounds from 14–16 reagents

Abstract: The use of reagents containing bonds between group 14 elements and Se or Te for the self-assembly of polynuclear metal–chalcogen compounds is covered. Background material is briefly reviewed and examples from the literature are highlighted from the period 2007–2017. Emphasis is placed on the different classes of 14–16 precursors and their application in the targeted synthesis of metal–chalcogen compounds. The unique properties arising from the combination of specific 14–16 precursors, metal atoms, and ancillary ligands are also described. Selected examples are chosen to underline the progress in (i) controlled synthesis of heterometallic (ternary) chalcogen clusters, (ii) chalcogen clusters with organic functionalized surfaces, and (iii) crystalline open-framework metal chalcogenides.

Keywords: cluster, synthesis, heterometallic, complexes, surface derivatization, extended, networks

10.1 Introduction

Polynuclear metal compounds of the heavier chalcogen elements selenium and tellurium are among the most structurally diverse assemblies in modern synthetic chemistry. They range in size from small clusters of metal atoms up to nanoscale "cut-outs" of bulk metal chalcogenides [1–6] and infinite three-dimensional network solids [7–9]. An impressive set of physical properties may be accessed in these systems as a function of their composition and structure. Accordingly, metal–chalcogen compounds occupy a special position in materials science. Early interest in metal chalcogenide clusters was generated because of their relevance as model compounds for the cofactors of chalcogen-containing metalloproteins [10–13] or the active sites of heterogeneous catalysts used in the petroleum industry (e. g. for hydrodesulfurization) [14, 15]. Far from being solely models, polynuclear metal chalcogen compounds are themselves technologically useful. They exhibit unique optical and electronic properties and are commonly used as superior photoluminescent [16, 17], semiconducting [18], non-linear optical [19–21], and thermoelectric materials [22–25] and may possess more advanced hybrid functional properties

This article has previously been published in the journal *Physical Sciences Reviews*. Please cite as: Polgar, A.M., Corrigan, J.F., Recent Advances in The Self-Assembly of Polynuclear Metal-Selenium and -Tellurium Compounds from 14-16 Reagents. *Physical Sciences Reviews* [Online] **2018** DOI: 10.1515/psr-2017-0126

https://doi.org/10.1515/9783110529340-010

[26, 27]. Porous metal chalcogenides based on the self-assembly of polynuclear chalcogenide clusters combine the adsorbent and ion-exchanging properties of zeolites with cutting-edge chemical and electronic properties [28–31].

Numerous approaches exist for the controlled synthesis of polynuclear metal chalcogen compounds [32]. The use of reagents containing bonds between group 14 and group 16 elements for their self-assembly has become one of the most popular and variegated approaches over the past few decades. Modern synthetic methods incorporating 14–16 precursors can be broadly classified into two major categories. The first employs precursors with labile bonds between a chalcogen (E = Se, Te) and a silyl group (SiR_2R') [33, 34]. Reaction of silylchalcogen precursors ($-ESiR_2R'$) with metal (pseudo)halide (MX_n) complexes results in the delivery of anionic chalcogen ligands that initiate cluster self-assembly. Formation of a strong Si–X bond via the elimination of a silane drives the generation of new bonds between the metal and bridging chalcogen ligands. The second major method involves the coordination chemistry of neutral or anionic 14–16 ligands [35–37]. These binary 14–16 ligands may be classified as either chalcogenidotetrelates $[T_xE_y]^{n-}$ (T = Ge, Sn, Pb; E = Se, Te) or organotetrel chalcogenides $(R_xT)_yE_z$ (R = organic group). Anionic chalcogenidotetrelates combine with metal cations in solution or react under solvothermal conditions to form molecular or extended M/14/16 (M = d-block or main-group metal) structures [38]. The reactions of neutral organotetrel chalcogenides with metal complexes have given rise to discrete binary and ternary chalcogenide clusters of significant research interest.

This chapter will provide selected highlights of the rational design of polynuclear metal–chalcogen compounds from 14–16 precursors during the period from 2007 to 2017. Special emphasis will be placed on recent developments in the field: (i) the controlled synthesis of heterometallic chalcogen clusters, (ii) the "functionalization" of cluster surfaces with useful organic moieties, and (iii) the preparation of crystalline open-framework (porous) metal chalcogenides from solution.

10.2 Silane elimination methods

Modern silane elimination methods for metal chalcogen cluster synthesis originate from the systematic study of organotin–chalcogen compounds and their reactivity with Lewis acids [39–41]. In 1965, Abel extended these studies to metal complexes. This resulted in the synthesis of a discrete nickel(II) thiolate hexamer $[Ni(SEt)_2]_6$ from $NiCl_2$ and $Me_2Sn(SEt)_2$ under elimination of Me_2SnCl_2 [42]. Here, a metal chalcogen cluster was synthesized via fission of the bond between heavier elements of group 14 and 16. Subsequent work would show the analogous silylated precursors to be more convenient for metal cluster synthesis [43–45]. Both the higher volatility and stability of halosilane elimination products relative to halostannanes drive the reaction and facilitate ease of reaction work-up. Silylated chalcogen reagents are, however, less stable with respect to hydrolysis than their stannylated counterparts [39].

The steric properties of the silyl group (SiR_2R') may be tuned to impart greater stability on these reagents, albeit with a corresponding decrease in reactivity [33]. Certain bulky silanes favour terminal coordination of the chalcogen, rather than metal bridging modes needed for cluster formation [46, 47]. Small trimethylsilylchalcogen ($-ESiMe_3$) reagents appear to give the best overall reactivity towards cluster formation [48–50]. Silyl groups of intermediate sizes such as $SiPh_3$ are desirable for the preparation of low-nuclearity metal complexes with terminal or μ-bridging silylchalcogenolate ligands, although cleavage of the E–Si bond with F^- prior to metalation affords larger clusters [51, 52]. The remainder of this section will deal exclusively with trimethylsilylchalcogen reagents and their reactivity patterns.

10.2.1 Bis(trimethylsilyl)chalcogenide reagents

10.2.1.1 Binary metal chalcogenide clusters

Reagents of general formula $E(SiMe_3)_2$ (E = S, Se, Te) react in organic solvents with metal (pseudo)halide complexes (MX_n) in the presence of a stabilizing ancillary ligand (L; usually in excess). This results in the self-assembly of ligand-protected binary metal chalcogenide clusters (e. g. Figure 10.1). $E(SiMe_3)_2$ thus acts as a soluble source of chalcogenide ions (E^{2-}) in organic solvents. The ancillary ligands prevent uncontrolled condensation to bulk or nanocrystalline phases and solubilize the metal chalcogenides in organic solvents [53, 54]. Early uses for silylated chalcogenide reagents by Holm and Fenske were concerned with synthesis of binary transition metal chalcogenide clusters, for example $[Fe_2S_2Cl_4]$ and $[Ni_{34}Se_{22}(PPh_3)_{10}]$ [43–45].

$$[NiCl_2(PPh_3)_2] + 2\ Se(SiMe_3)_2 \xrightarrow[\text{PhMe, 4d, 298 K}]{} [Ni_{34}Se_{22}(PPh_3)_{10}]$$

Figure 10.1: Fenske's synthesis of a nickel(II) selenide cluster from a phosphine-ligated metal complex and bis(trimethylsilyl)selenide [45].

This approach is general for early and late 3d and 4d metals. The most success, in terms of the quantity and diversity of cluster frameworks, has come from the combination of $E(SiMe_3)_2$ with d^{10} group 11 metal salts, particularly Ag(I) and Cu(I) carboxylates [3–5]. Group 11 cations are ideal choices for cluster growth, due to their flexible coordination requirements and attractive dispersive forces between the metals which aid in aggregation. Copper selenide (Cu_2Se) clusters are perhaps the most thoroughly studied of these systems. The transformation from small, spherical polynuclear assemblies up to layered triangular nanoclusters such as $[Cu_{146}Se_{73}(PPh_3)_{30}]$ and $[Cu_{140}Se_{70}(PEt_3)_x]$ (x = 34, 36) is extensively documented [2]. Recently, investigations of the luminescence from copper(I)-chalcogenide systems have demonstrated their promise as low-cost triplet emitting elements. For example, the dodecanuclear copper selenide cluster [$Cu_{12}Se_6$(dppo)$_4$] (dppo = 1,8-bis(diphenylphosphino)octane; Figure 10.2) emits red light (λ_{max} = 638 nm) with an impressive quantum yield of 0.53 in the solid state [55]. The favourable photoluminescence properties of metal chalcogen clusters are a consequence of the structural rigidity of the frameworks. This simultaneously reduces emission quenching by vibrational deactivation of the excited state and increases photostability.

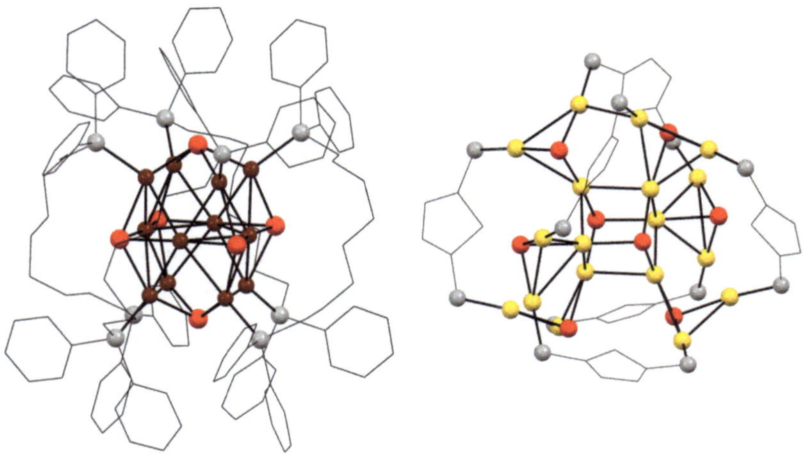

Figure 10.2: Molecular structures of the binary copper(I)-selenide [$Cu_{12}Se_6$(dppo)$_4$] (*left*) [55] and the dicationic gold(I)–selenide [$Au_{18}Se_8$(dppthph)$_6$]$^{2+}$ (*right*) [56]. Atoms are represented by spheres, bonds by solid lines, organic ligands as wireframe drawings, and all hydrogen atoms are omitted throughout. Phenyl groups are omitted in [$Au_{18}Se_8$(dppthph)$_6$]$^{2+}$. Colour scheme: Cu, *brown*; Au, *yellow*; Se, *red*; P, *grey*.

The analogous chemistry of the 5d metals has developed more slowly. For example, the heaviest coinage metal, gold, is usually resistant to cluster formation with chalcogenide ligands compared to Ag(I) and Cu(I), owing to the strong preference of Au(I) centres for terminal, linear coordination. However, reports of luminescent, high nuclearity gold(I) selenide clusters, such as [$Au_{18}Se_8$(dppthph)$_6$](Cl)$_2$ [dppthph = 2,5-bis(diphenylphosphanyl)thiophene; Figure 10.2] prepared from Se($SiMe_3$)$_2$ and phosphinogold(I) chlorides, have emerged over the past decade [56–58]. The preparative chemistry of the f-block element chalcogenides is less amenable to 14–16 precursors, instead being dominated by oxidative addition of diaryl dichalcogenide reagents followed by ligand-based redox reactions [59].

10.2.1.2 Ternary metal chalcogenide clusters

Heterometallic chalcogenide clusters may be obtained from reactions of the corresponding metal salts with E($SiMe_3$)$_2$ in the presence of stabilizing phosphine ligands. The high reactivity of silylated chalcogen reagents enables simultaneous activation of both metal salts, rather than binary metal chalcogenide formation. This method is effective for preparing discrete chalcogenide clusters containing both early and late metals. For example, Fenske has shown that the presence of middle and late 3d metals Fe(II), Ni(II), Co(II), and Cu(I) can result in the formation of discrete ternary chalcogenide clusters with Nb(V) and Ta(V) (Figure 10.3) [60, 61].

$$4\,[CoCl_2(PPh_3)_2] + NbCl_5 \xrightarrow[\text{MeCN, 273 K, 2d}]{4\,Se(SiMe_3)_2} [NbCo_3Se_4(PPh_3)(CH_3CN)_3][CoCl_3(PPh_3)]$$

Figure 10.3: Synthesis of a mixed Co(II)–Nb(V) selenide heterocubane cluster from the respective metal chlorides and $Se(SiMe_3)_2$ [61].

This provides an alternative to more traditional approaches to early/late heterometallic chalcogenides, based on purely inorganic chalcogenometalate anions such as $[ME_4]^{2/3-}$ (M = V, Mo, W) or $[Nb_6S_{17}]^{6-}$ as a chalcogenide source [19, 20, 62, 63]. Consequently, this approach enables the synthesis of new ternary chalcogenides. $[NbCo_3Se_4(PPh_3)(CH_3CN)_3]^+$, for example, is the first structurally characterized Nb(V)/Co(II)/Se cluster [61].

Heterometallic chalcogenide clusters incorporating main group metals are of interest in part because of their relation to ternary metal chalcogenide semiconductors. An important theme of research in this area is the design of molecular systems that mimic the composition and structure of the bulk solids. For example, the semiconducting Cu_2SnS_3 system ($E_g > 0.9$ eV) is a promising absorber material for thin film photovoltaic applications [18, 64]. Molecular Cu(I)/Sn(IV)/S clusters have been prepared from silylated chalcogenides and exhibit complementary properties such as a narrow optical bandgap resulting in near-infrared (NIR) photoluminescence [65, 66]. This has been extended to general Cu(I)/Sn(IV)/E (E = S, Se, Te) clusters $[Cu_6E_6(SnPh)_2(PPh_2Et)_6]$. The isostructural clusters were prepared from the reaction of CuOAc with $PhSnCl_3$ and $E(SiMe_3)_2$ in the presence of PPh_2Et [67]. The core structure contains a nonbonded $[E_6]$ octahedron capped by μ_3-SnPh along two opposing faces and μ_3-$CuPPh_2Et$ on the remaining six faces [68]. Heavier chalcogen atoms reduce the absorption energy onset from 2.83 eV (S) to 2.32 eV (Se) and 1.71 eV (Te). The clusters are photoluminescent in the NIR region with bathochromically shifted luminescence as a function of the chalcogenide (λ_{em} = 800, 920, and 1280 nm for E = S, Se, and Te, respectively). This is attributed to the destabilization of the highest occupied molecular orbital as more electron-rich chalcogen atoms are introduced from silylated sources.

The “inverse” situation of a heterometallic chalcogen cluster containing a single metal and two distinct chalcogenide ligands has been reported. This uncommon situation is observed in the ternary Cu(I)/Se/S cluster $[Cu_{84}Se_{42-x}S_x(PEt_2Ph)_{24}]$ ($x \approx 15$) [69]. The cluster is prepared from the bis-silylated selenyl sulfide Me_3Si–SeS–$SiMe_3$ with PEt_2Ph solubilized CuOAc. Me_3Si–SeS–$SiMe_3$ is synthesized by insertion of S_8 into the Li–Se bond of $Li[SeSiMe_3]$, followed by metathesis of $Li[SSeSiMe_3]$ with

$ClSiMe_3$. The core of the cluster consists of three cubic layers of chalcogenide ligands (Figure 10.4), typical of larger copper selenide clusters, with the holes filled by Cu(I) centres. Mixed Se/S site occupation and Cu atom disorder make precise quantification of chalcogen atom distribution difficult, although the proposed formula is supported by energy dispersive X-ray spectroscopy (EDX) analysis. The overall structure is intermediate between smaller, spherical copper sulfide clusters and large trigonal copper selenide arrangements.

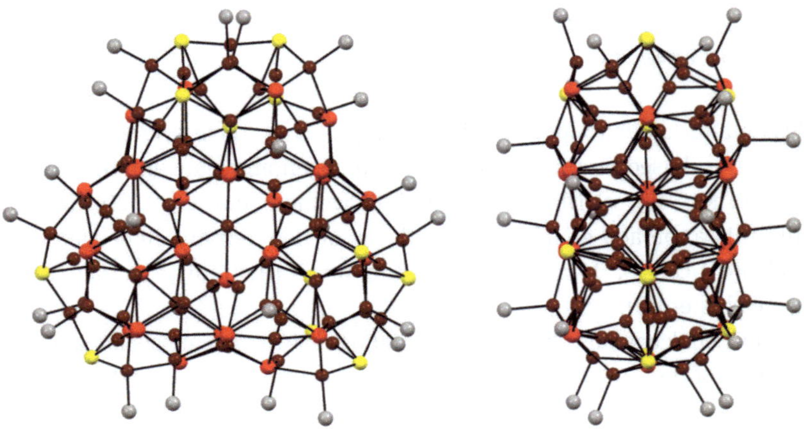

Figure 10.4: Molecular structure of the ternary copper(I) sulfide/selenide nanocluster $[Cu_{84}Se_{42-x}S_x(PEt_2Ph)_{24}]$ ($x \approx 15$) viewed along *c* (*left*). An alternate view approximately parallel to [1 1 0] is shown (*right*) [69]. Organic groups are omitted. Colour scheme: Cu, *brown*; Se, *red*; disordered S/Se, *yellow*; P, *grey*.

10.2.1.3 Metal chalcogenide clusters with functionalized surfaces

Variation of the ligands used to stabilize metal chalcogenide clusters is mostly restricted to changes in the electron-donating ability and steric properties (i.e. Tolman cone angle and electronic parameter [70]). These parameters are crucial in guiding the results of the self-assembly process. Ligands with advanced reactivity and/or functionality have not been studied to the same degree. Recent interest has been stimulated by the use of electronically conjugated ancillary phosphine ligands for metal chalcogenide clusters, which are components in single-molecule junctions [71, 72]. Corrigan and co-workers prepared copper(I) selenide and telluride clusters stabilized by the redox-active ligand 1,1′-bis(diphenylphosphino)ferrocene (dppf) [73]. The combination of [Cu(OAc)(dppf)] and half an equivalent of $E(SiMe_3)_2$ gave rise to $[Cu_8(\mu_4\text{-Se})_4(\mu\text{-dppf})_3]$ and a mixture of $[Cu_4(\mu_4\text{-Te})(\mu_4\text{-}\eta^2\text{-Te}_2)(\mu\text{-dppf})_2]$ and $[Cu_{12}(\mu_5\text{-Te})_4(\mu_8\text{-}\eta^2\text{-Te}_2)(\mu\text{-dppf})_4]$ (Figure 10.5). Importantly, it was possible to oxidize the ferrocene moiety and phosphorous atoms of dppf ligands on the copper selenide and telluride clusters without decomposition of the metal chalcogenide core.

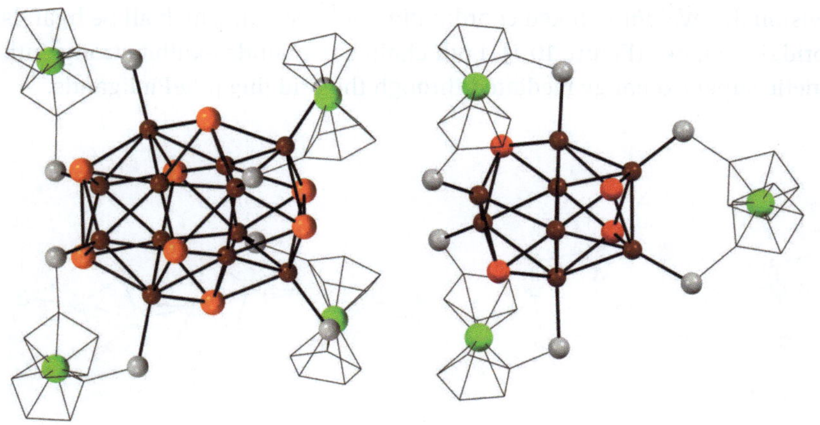

Figure 10.5: Ferrocenyl-functionalized copper telluride cluster $[Cu_{12}(\mu_5\text{-}Te)_4(\mu_8\text{-}\eta^2\text{-}Te_2)(\mu\text{-}dppf)_4]$ (*left*) and copper selenide $[Cu_8(\mu_4\text{-}Se)_4(\mu\text{-}dppf)_3]$ (*right*) [73]. The phenyl groups of dppf have been omitted for clarity. Colour scheme: Cu, *brown*; Fe, *green*; Se, *red*; Te, *orange*; P, *grey*.

10.2.2 Organic trimethylsilylchalcogenides

10.2.2.1 Metal chalcogenolate coordination polymers

Replacing one silyl group of $E(SiMe_3)_2$ with an organic moiety (R) yields organo trimethylsilylchalcogenide reagents $RESiMe_3$. These serve as organic-soluble sources of RE^- for the self-assembly of metal chalcogenolate clusters. For silylthiolate ligands, a wide scope of potential R groups may be implemented [74, 75]. Selenolate and tellurolate ligands, on the other hand, mainly employ either simple aromatics or bulky silylated groups, due to their lower overall stability [76, 77]. Reaction of $RESiMe_3$ with a metal halide (MX_n) produces oligomeric species of general formula $[M(ER)_n]_m$ or metal chalcogenolate coordination polymers [78]. Eichhöfer and co-workers prepared one-dimensional tin(II) phenylchalcogenolate complexes $[Sn(EPh)_2]_n$ (E = Se, Te) by this route (Figure 10.6) [79]. The polymer consists of planar $Sn(EPh)_2$ units assembled into a one-dimensional chain by weak Sn(II)–E interactions between adjacent units.

$$SnCl_2 + 2\ PhESiMe_3 \xrightarrow[THF]{} [Sn(EPh)_2]_n$$

Figure 10.6: Synthesis of a one-dimensional homoleptic Sn(II) phenylchalcogenolate coordination polymer from $PhESiMe_3$ (E = Se, Te) [79].

Related Fe(II) and Mn(II) diselenolates and $[Bi(SePh)_3]_n$ have been isolated from the combination of the corresponding metal salts and $PhSeSiMe_3$ [80–82]. In contrast to the Sn(II)/E system, the solid-state structures of $[M(SePh)_2]_n$ (M = Fe, Mn) show one-dimensional, covalently linked coordination polymers in which all Se ligands adopt μ-bridging modes (Figure 10.7). Both chain compounds exhibit strong antiferromagnetic super exchange mediated through the bridging μ-SePh ligands.

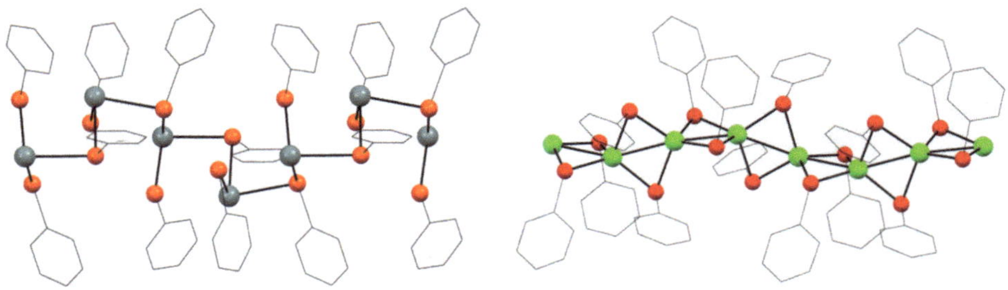

Figure 10.7: Segments of one-dimensional metal chalcogenolate coordination polymers [Sn(TePh)$_2$]$_n$ viewed parallel to *b* (*left*) [79] and [Fe(SePh)$_2$]$_n$ parallel to *c* (*right*) [81]. Colour scheme: Sn, *dark grey*; Fe, *green*; Se, *red*; Te, *orange*.

10.2.2.2 Molecular metal chalcogenolate aggregates

When reactions between metal salts and silylated chalcogenolates are carried out in the presence of ancillary ligands, aggregation into polymers is suppressed and ligand-stabilized metal chalcogenolate clusters result. The size and structure of the resultant cluster is dependent on the reaction stoichiometry, the silylchalcogenide ligand, the ancillary ligand, and the metal centre. An equivalent method uses ancillary ligands to coordinate to preformed metal chalcogenolate polymer, in which case the product spectrum is mostly sensitive to the nature and amount of the solubilizing ligand. Krautscheid provided a systematic demonstration of these effects for the copper(I) chalcogenolate system in the presence of trialkyl phosphines of various sizes (Figure 10.8) [83].

$$m\ R_3P + n\ CuOAc + n\ Me_3SiEPh \longrightarrow [(R_3P)_m(CuEPh)_n]$$

Figure 10.8: Synthesis of phosphine-stabilized copper(I) chalcogenolate clusters with varying ratios of phosphine and trimethylsilylchalcogenide reagents [83].

In the presence of excess phosphine ($PR_3/RE^- = m/n = 3/1$), terminal copper(I) chalcogenolates like [$(Et_3P)_3$CuSePh] are preferred. These may also be prepared from the reaction of larger clusters such as [$(Et_3P)_5(CuEPh)_6$] with the addition of excess phosphine. When the m/n ratio is reduced to 2/1 for the small alkyl phosphine Me_3P, a dimeric structure [$(Me_3P)_2(CuEPh)]_2$ (E = Se, Te) is obtained in which all copper(I) atoms are tetrahedrally coordinated. If the sterically demanding phosphine, ^{t}Bu$_3$P is used with $m/n = 1/1$, the dimer [(^{t}Bu$_3$P)(CuSePh)]$_2$ is obtained with trigonal planar Cu(I) centres.

Substoichiometric amounts of phosphine lead to the formation of polynuclear clusters with more complex geometries. With a 3/4 ratio of m/n with bulky phosphines $^{i}Pr_3P$ and $^{t}Bu_3P$, the clusters $[(R_3P)_3(CuTePh)_4]$ are obtained. The structure is that of a Cu (I) centred $[\{(R_3P)CuTePh\}_3]$ ring completed by an additional μ-TePh ligand. A 5/6 m/n ratio yields hexanuclear copper(I) selenolate and tellurolate clusters $[(Et_3P)_5(CuEPh)_6]$. The structures consist of an octahedral arrangement of EPh ligands with three trigonal faces capped by $CuPEt_3$, one edge-bridging $Cu(PEt_3)_2$ and two trigonal Cu(I) centring opposite faces. Another hexanuclear cluster $[(^{i}Pr_3P)_4(CuSePh)_6]$ is obtained using a bulkier phosphine in a 4/6 m/n ratio. The cluster contains only trigonal planar copper centres and consists of two stacked $[(CuSePh)_3]$ rings. With a 3/7 m/n ratio, the related copper selenolate $[(^{i}Pr_3P)_3(CuSePh)_7]$ is obtained. This can be constructed by formally replacing a terminal phosphine in $[(^{i}Pr_3P)_4(CuSePh)_6]$ with a μ_3-bridging CuSePh moiety. Structures of the clusters are given in Figure 10.9.

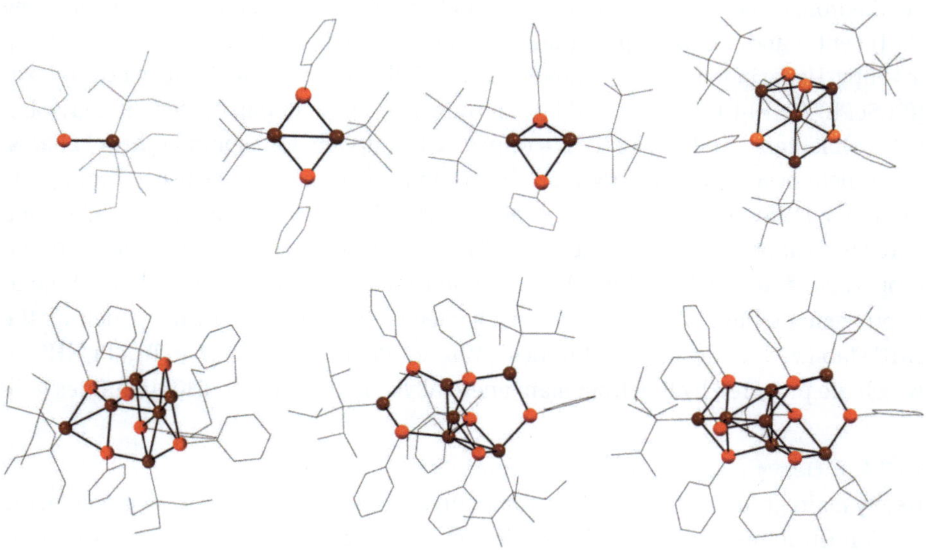

Figure 10.9: Core structures of phosphino-copper(I) phenylchalcogenolates prepared from $RESiMe_3$ with varying ratios of PR_3/RE^- (m/n) [83]. *Top row*: $[(Et_3P)_3CuSePh]$ (3/1), $[(Me_3P)_2(CuSePh)]_2$ (2/1), $[(^{t}Bu_3P)(CuSePh)]_2$ (1/1), $[(^{i}Pr_3P)_3(CuTePh)_4]$ (3/4); *bottom row*: $[(Et_3P)_5(CuSePh)_6]$ (5/6), $[(^{i}Pr_3P)_4(CuSePh)_6]$ (4/5), $[(^{i}Pr_3P)_3(CuSePh)_7]$ (3/7). Colour scheme: Cu, *brown*; Se, *red*; Te, *orange*.

Different stabilizing ligands beyond phosphines are useful for varying the structural properties of metal chalcogenolates. The use of 4,4′-bipyridine (bpy) in conjunction with CuOAc, PPh_3, and $PhSeSiMe_3$ has given rise to discrete copper(I) selenolate clusters bridged into a one-dimensional coordination polymer of $[(CuSePh)_6(PPh_3)_3]$ units linked through 4,4′-bpy (Figure 10.10) [84]. The cluster units are based on a copper-bridged nonbonded $[Se_6]$ octahedron. Two Cu(I) centre opposing faces of the

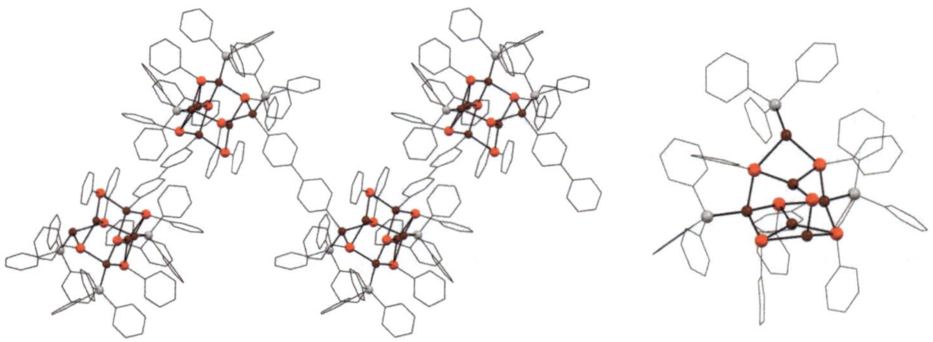

Figure 10.10: Segment of the one-dimensional zigzag chain of $[(CuSePh)_6(PPh_3)_3(4,4'\text{-bpy})]_n$ viewed along *a* (*left*). Expansion of the monomeric copper(I)–selenolate cluster unit (*right*) [84]. Colour scheme: Cu, *brown*; Se, *red*; P, *grey*.

octahedron, three cap the trigonal faces, and one is edge-bridging. The edge-bridging Cu(I) centre and a Cu(I) capping an adjacent face are ligated by the cluster-bridging 4,4′-bpy. Using bidentate phosphines instead of PPh_3 gives rise to dinuclear complexes $[(CuSePh)_2(P^\wedge P)_2]$ linked by 4,4′-bpy into linear (P^P = dppm; 1,1-bis(diphenylphosphino)methane) and zigzag (P^P = dppe; 1,2-bis(diphenylphosphino)ethane) chains.

N-heterocyclic carbenes (NHCs) have recently been shown to be useful for improving the stability and solubility of copper(I) and silver(I) selenolate aggregates compared to their phosphine analogues [85]. This is a consequence of the excellent σ donor properties of NHCs which limit cluster condensation promoted by ligand loss. Similar improvements have also been shown for metal–chalcogenide clusters, such as the NHC-stabilized manganese tellurides $[Mn_4Te_4(^iPr_2NHC)_4]$ and $[Mn_8Te_8(^iPr_2NHC)_6]$, which are prepared by Roy from manganese allyl complexes and NHC–tellurides [86].

10.2.2.3 Heavy main group metal tellurolates

Heavy main group metal tellurium compounds are suitable single-source precursors for nanostructured and thin-film semiconducting materials [25, 87]. Nanostructures are typically prepared by chemical processes (e. g. solvothermal synthesis, atomic layer deposition, thermolysis) from single- or multi-source precursors, whereas thin films involve deposition of the precursors by physical means (chemical vapour deposition, solution-based deposition) followed by an annealing step [88, 89]. In either case, using a single precursor containing a preformed M–Te bond can improve the quality of the resulting materials in terms of its compositional and morphological uniformity [90–92]. Metal tellurolate precursors offer one route into metal telluride phases by thermal elimination of ER_2 (Figure 10.11).

Well-defined bismuth(III), antimony (III), tin(II), and lead(II) tellurolate precursors are of special interest because of the importance of their respective tellurides as thermoelectric materials [22, 25]. Molecular compounds of these systems are rare due

$$(L)_nM(TeR)_m \xrightarrow{\Delta} n\,L + MTe_{m/2} + m/2\,R_2Te$$

Figure 10.11: Idealized thermal decomposition of a mononuclear metal tellurolate in the deposition of a bulk metal telluride phase.

to the instability of bonds between heavy main group elements. For the most part, only mononuclear compounds have been prepared [93–96]. von Hänisch and co-workers have recently reported Sn(II) and Pb(II) tellurolate clusters prepared from sterically encumbered silyltelluride reagents $^tBu_2PhSiTeSiMe_3$, $(Me_3Si)_3CTeSiMe_3$, and $(Me_3Si)_3SiTeSiMe_3$ and tin(II) or lead(II) chlorides (Figure 10.12) [97]. This chemistry is facilitated by the reactive nature of the silyltellurides even at the low temperatures required to kinetically stabilize Sn(II)–Te and Pb(II)–Te bonds in solution.

$$MCl_2 + 2\,{}^tBu_2PhSiTeSiMe_3 \xrightarrow[Et_2O,\ 199\ K]{} [({}^tBu_2PhSiTe)_4M_2]\ (M = Sn, Pb)$$

$$(Me_3Si)_3CPbCl + 2\,{}^tBu_2PhSiTeSiMe_3 \xrightarrow[Et_2O,\ 199\ K]{} [{}^tBu_2PhSiTePbC(SiMe_3)_3]_2$$

$$SnCl_2 + 2\,(Me_3Si)_3SiTeSiMe_3 \xrightarrow[Et_2O,\ 199\ K]{} [\{(Me_3Si)_3SiTe\}_4Te_2Sn_4]$$

Figure 10.12: Synthesis of a tin(II) and lead(II) tellurolate clusters by silane elimination chemistry [97].

Dimeric species $[({}^tBu_2PhSiTe)_2M]_2$ (M = Sn, Pb) and $[{}^tBu_2PhSiTePbC(SiMe_3)_3]_2$ were synthesized and structurally characterized [96]. They adopt a folded $[M_2Te_2]$ ring with *trans*-disposed silyl ligands. Because of the weak nature of the bonds present, these complexes were dynamic in solution. $[({}^tBu_2PhSiTe)_2Sn]_2$ underwent Te–Si bond fission and subsequent scrambling of the silyl ligands above −3 °C. $[({}^tBu_2PhSiTe)_2Pb]_2$ is thought to engage in a monomer–dimer equilibrium initiated by cleavage of the weak Pb–Te bond. All compounds decomposed between 100 and 300 °C to give the corresponding MTe phase. Unexpectedly, reaction of $SnCl_2$ with $(Me_3Si)_3SiTeSiMe_3$ gave rise to a mixed tin(II) telluride/tellurolate cluster $[\{(Me_3Si)_3SiTe\}_4Te_2Sn_4]$. The cluster consists of a planar $[Sn_2Te_2]$ ring with two *trans*-disposed $\{(Me_3Si)_3SiTe\}_2Sn$ fragments bound to Te. The system has structural similarities to cubic SnTe and may represent an intermediate molecular species in the decomposition of $[Sn\{TeSi(SiMe_3)_3\}_2]_2$ to SnTe as demonstrated by Arnold [98].

10.2.2.4 Ternary metal chalcogenolate clusters

Heterometallic chalcogenolate clusters can be prepared from mixtures of the respective metal complexes in the presence of $RESiMe_3$ and ancillary ligands. As was the case with mixed-metal chalcogenides (Section 10.2.1.2), the reactive nature of silylchalcogen reagents is exploited for the simultaneous activation of two different metal complexes under mild conditions. Krautscheid and co-workers used this method for the first synthesis of molecular copper(I)–zinc(II) selenolate and tellurolate complexes (Figure 10.13) [99].

$$[(^{i}Pr_3P)_4(CuSePh)_6] + {}^{i}Pr_3P + ZnEt_2 \xrightarrow[\text{EtOH, 298 K}]{PhSeSiMe_3} [(^{i}Pr_3PCu)_3(ZnEt)(SePh)_4]$$

$$[(^{i}Pr_3P)_3(CuTePh)_4] + {}^{i}Pr_3P + ZnEt_2 \xrightarrow[\text{EtOH, 273 K}]{PhTeSiMe_3} [(^{i}Pr_3PCu)_3Cu(^{i}Pr_3PZn)(TePh)_6]$$

Figure 10.13: Synthesis of ternary Cu(I)/Zn(II)/Se and Cu(I)/Zn(II)/Te chalcogenolate clusters from the mixture of the respective metal complexes and $PhESiMe_3$ [99].

These reactions are conducted in EtOH and the reactivity is a combination of silane elimination chemistry and chalcogenolysis (see also Section 10.2.3.2). $[(^{i}Pr_3PCu)_3(ZnEt)(SePh)_4]$ consists of a pyramidal $[(CuP^{i}Pr_3)_3(\mu_3\text{-}SePh)]$ unit capped by $[(ZnEt)(\mu\text{-}SePh)_3]$. Each selenolate bridges the Zn(II) and one Cu(I). $[(^{i}Pr_3PCu)_3Cu(^{i}Pr_3PZn)(TePh)_6]$ has an adamantane-like structure: a six-membered ring with alternating Cu(I) and Te^{2-} centred by Cu(I) is capped by a tetrahedral $[(^{i}Pr_3PZn)(\mu\text{-}TePh)_3]$ unit. $[Cu_2Sn(SePh)_6(PPh_3)_2]$ is another representative of Cu/M selenolate clusters derived from a ternary semiconductor phase [66]. The cluster was prepared from $PhSeSiMe_3$ and a mixture of CuCl and $[SnCl_4(THF)_2]$ and its structure is based on an octahedral $[SnSe_6]$ arrangement capped on two opposite faces by $CuPPh_3$. At temperatures below 100 K, it emits brightly in the NIR region by phosphorescence (λ_{em} = 1100 nm) in the solid state. Increasing the temperature decreases the emission intensity and induces hypsochromic shifting.

Silylated organoselenolate precursors have enabled access into some less common ternary metal chalcogen systems. For example, Fenske prepared rare Ag(I)/Bi/Se compounds $[Bi_4Ag_3(SePh)_6X_3(PR_3)_3]_2$ (X = Cl, Br, PR_3 = PPh_3, $PPh^{i}Pr_2$) using $PhSeSiMe_3$ as a source of $PhSe^-$ [100]. The structure contains a trigonal pyramidal $Bi_4{}^{6+}$ unit linked through the basal Bi centres to three $AgPR_3$ through μ-SePh bridges. The μ_3-halide ligands bridge two Bi and one Ag at distances indicative of non-covalent interactions (Figure 10.14). This unit dimerizes in the solid state through weak Se···Bi interactions, as also observed in $Bi(SePh)_3$ which forms a linear polymer through these bonding modes [81]. Electronic structure calculations show the $Bi_4{}^{6+}$ unit contains a neutral Bi^0 at the apex of the trigonal pyramid, bound to three Bi^{2+} which engage in 3 centre–4 electron bonding to form Se–Bi–Se moieties.

10.2.2.5 Functionalized chalcogenolate surface ligands

The tunability of the organic group in trimethylsilylchalcogenide reagents has provided access to metal chalcogenolate clusters with functionalized ligand shells [34]. Fuhr showed that the electronic properties of dinuclear copper(I) complexes with

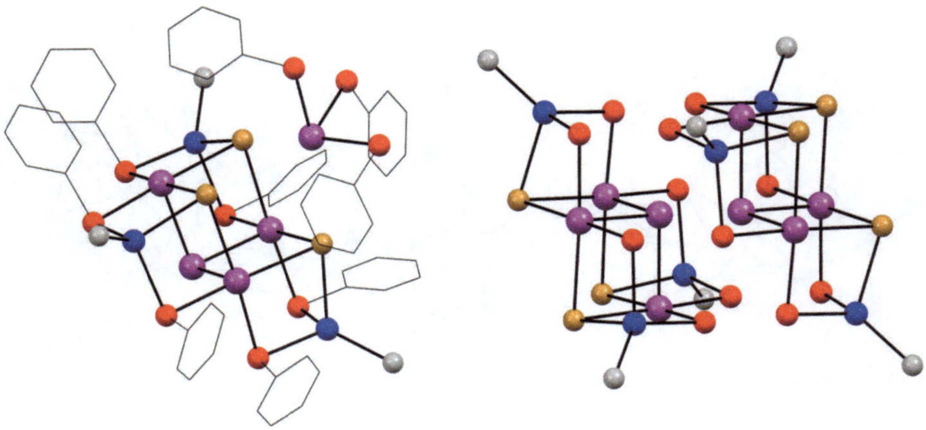

Figure 10.14: Core structure of ternary chalcogenolate $[Bi_4Ag_3(SePh)_6Br_3(PPh_3)_3]$ with one co-crystallizing molecule of $Bi(SePh)_3$ and the phenyl ligands of the phosphine omitted (*left*). Its aggregation into an intermolecular dimer in the crystal is shown (*right*) with all organic substituents omitted [100]. Colour scheme: Ag, *blue*; Bi, *purple*; Se, *red*; Br, *beige*; P, *grey*.

phenylselenolate ligands may be tuned by installing electron donating groups about the aromatic ring, without changing the core cluster geometry [101]. On the other hand, the self-assembly properties of selenolates with distinctive steric and geometric properties are underdeveloped. Polydentate aryl trimethylsilylselenide reagents with several positions of the aromatic ring substituted are promising candidates in this respect. Preliminary work includes the development of 1,4-, 1,3,5-, and 1,2,4,5-phenylchalcogenosilanes $Ar(CH_2ESiMe_3)_n$ (n = 2, 3, 4; E = S, Se) for polyferrocenyl assemblies [102]. The use of these reagents as precursors to polynuclear metal chalcogen compounds was demonstrated by the synthesis of a dinuclear palladium complex $[(dppp)_2Pd_2\text{-}\mu\text{-}\kappa^2Se\text{-}\{1,2\text{-}(SeCH_2)_2C_6H_4\}](Cl)_2$ (dppp = 1,3-bis(diphenylphosphino)propane) [103].

Polyferrocenyl arrangements on inorganic supports are attractive targets for multifunctional nanomaterials with potential applications as electrochemical and/or luminescent sensors [27, 104], electrode materials [105], and multielectron catalysts [106]. Group 14–16 reagents have been used as precursors to both mono- and polydentate ferrocenylchalcogen ligands for metal chalcogen clusters with redox-active ligand shells. Simple ferrocenylchalcogenolates $FcESiMe_3$ (Fc = $[Fe(\eta^5\text{-}C_5H_5)(\eta^5\text{-}C_5H_4)]$) and $fc(SeSiMe_3)_2$ (fc = $[Fe(\eta^5\text{-}C_5H_4)_2]$) are prepared by insertion of elemental Se or Te into the C–Li bond of [FcLi] or $[fcLi_2]$, followed by metathesis with $ClSiMe_3$. They react with ligand solubilized metal acetate salts to yield molecular metal chalcogenolate clusters, for example $[Ag_8(Se_2fc)_4(P^nPr_3)_4]$ and $[Ag_{16}(Se_2fc)_8(PPh_2Et)_6]$ (Figure 10.15) [107]. Cyclic voltammetry of metal chalcogenolate clusters prepared in this manner usually show one irreversible oxidation followed by two quasi-reversible oxidations [108]. In these cases, oxidation of Fe^{2+} results in oxidation of the ferrocenylselenolate ligand to the corresponding diselenide which is

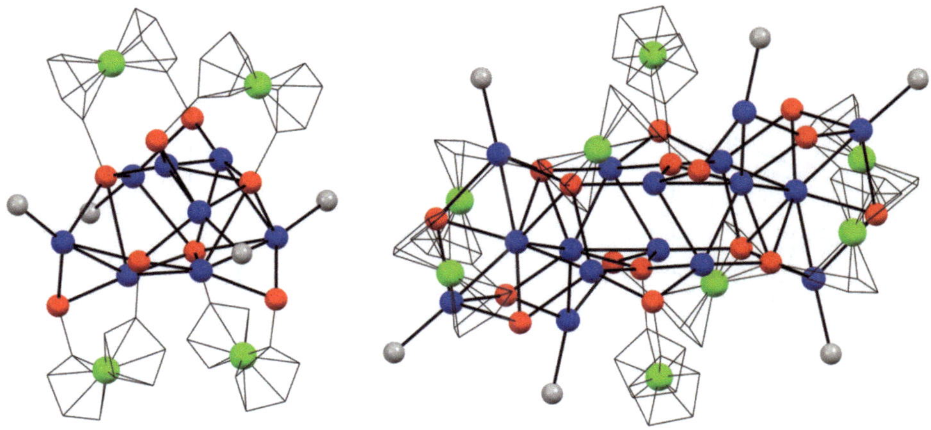

Figure 10.15: Molecular structures of ferrocenyl-functionalized silver–selenolate clusters $[Ag_8(Se_2fc)_4(P^nPr_3)_4]$ (*left*) and $[Ag_{16}(Se_2fc)_8(PPh_2Et)_6]$ (*right*) [107]. All organic substituents on the phosphine ligands are omitted. Colour scheme: Ag, *blue*; Fe, *green*; Se, *red*; P, *grey*.

responsible for the ensuing reversible redox chemistry. This causes the decomposition of the metal chalcogen framework. Oxidative degradation may be avoided by the incorporation of a methylene spacer between ferrocene and the chalcogen centre. Silylated ferrocenylmethylchalcogenolates are thus used as effective precursors to metal chalcogenolate clusters such as $[Ag_8(SeCH_2Fc)_8(PPh_3)_4]$ [109]. The redox stability of the resulting frameworks has been demonstrated for the case of a silver sulfide/thiolate nanocluster [110].

Nanomaterials functionalized with polyferrocenyl ligands may act as sensitive electrochemical ion probes. Having many redox probes per metal particle and the ability to trap polar analytes with non-covalent interactions (e. g. hydrogen bonding) make these ideal sensing materials. A first attempt involved the preparation of silylated ferrocenoylselenolate reagents $FcC(O)SeSiMe_3$ from FcC(O)Cl and $Li[SeSiMe_3]$ [111]. Reaction with $(Ph_3P)_2M(OAc)$ (M = Cu, Ag) gave rise to tetramers $[M\{Se(O)CFc\}(Ph_3P)]_4$. Electrochemical oxidation of the ferrocene units resulted in decomposition of the cluster framework and generation of $\{FcC(O)\}_2Se$, with a mechanism probably like that of $FcSe^-/fcSe_2^{2-}$ stabilized clusters. Introducing an ester linkage with a flexible ethyl spacer in $FcC(O)OCH_2CH_2SeSiMe_3$ has enabled the synthesis of metal chalcogen nanoclusters with both redox stability and superior ion-sensing capabilities (see Section 10.2.3.4). A smaller cluster $[Ag_7Br(dppe)_3\{SeCH_2CH_2O(O)CFc\}_6]$ was also prepared from the corresponding silylated selenolate ligand and [Ag(dppe)OAc] in the presence of NPr_4Br [112]. The cluster retains its structural integrity upon repeated electrochemical cycling. A novel silver ferrocenylselenolate coordination polymer $[\{AgSeCH_2CH_2O(O)CFc\}_8]_n$ was prepared from AgOAc and $FcC(O)OCH_2CH_2SeSiMe_3$ in the absence of stabilizing ligands and adopts a one-dimensional helical chain structure in the solid state (Figure 10.16) [112].

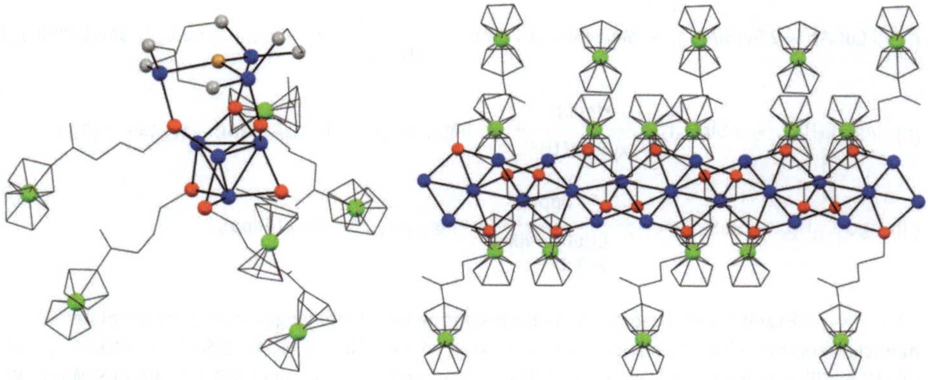

Figure 10.16: Molecular structure of the ferrocenylester-functionalized silver selenolate cluster $[Ag_7Br(dppe)_3\{SeCH_2CH_2O(O)CFc\}_6]$ (*left*) and a portion of the coordination polymer $[\{AgSeCH_2CH_2O(O)CFc\}_8]_n$ (*right*) viewed parallel to *c* [112]. The organic groups of the phosphine ligands are omitted. Colour scheme: Ag, *blue*; Fe, *green*; Se, *red*; Br, *beige*; P, *grey*.

10.2.3 Metal chalcogenide/chalcogenolate clusters

10.2.3.1 Nanoscopic silver and copper chalcogenide/chalcogenolate clusters

The largest crystallographically characterized discrete metal chalcogenide (non-oxide) clusters were reported by Fenske in 2008, measuring ca. 3 nm in core diameter [113]. They consist of Ag_2S cores enveloped in AgSR shells, further stabilized against condensation with phosphine ancillary ligands. In other words, they belong to the class of coinage metal chalcogenide/chalcogenolate nanoclusters. Silylated chalcogen reagents can be used to prepare these clusters according to three approaches: (i) reaction of MOAc (M = Cu, Ag) with $E(SiMe_3)_2$, $RESiMe_3$, and ancillary phosphines, (ii) treating metal chalcogenolate coordination polymers $[M(ER)]_n$ with $E(SiMe_3)_2$ and phosphines, and (iii) reaction of MX (X = Cl, OAc) with $RESiMe_3$ under E–C bond cleavage (Figure 10.17), whereas methods (i) and (ii) are rather general for M = Cu, Ag and E = S, Se, Te, method (iii) only becomes important as the strength of the E–C bond weakens, i.e. for silylated selenolate and tellurolate precursors.

The size and structure of the products that crystallize are highly dependent on the specific reaction conditions; the steric properties of ancillary phosphine ligands are crucial; however, variation of the chalcogenolate ligand, reaction solvent, and temperature influence cluster formation and crystallization. When silylselenolate or silyltellurolate ligands are employed, redox chemistry is commonly observed either in the formation of polychalcogenide ligands or in the preparation of mixed valence clusters, such as $[Cu^{I}_{36}(Cu^{II}Cl)_2Se_{13}(SePh)_{12}(dppe)_6]$ reported by Duan and co-workers [114]. Other considerations include the presence of additives in the reaction mixture to induce C–E bond cleavage in method (iii). For example, Dehnen prepared $[(Ph_3PAg)_8Ag_6(\mu_6\text{-}Se)_{1-x/2}(SePh)_{12}]^{x+}$ (x = 0, 1) from $[(Ph_3P)Ag(NO_3)]$ using $PhSeSiMe_3$

$$\text{(i)}\quad 5\,\mathrm{CuOAc} + 2\,\mathrm{Se(SiMe_3)_2} + \mathrm{Me_3SiSe\text{-}C_6H_4\text{-}SMe} \xrightarrow[\text{THF, 243 K}]{\text{10 PPh}_3} [\mathrm{Cu_{93}Se_{42}(Se\text{-}C_6H_4\text{-}SMe)_9(PPh_3)_{18}}]$$

$$\text{(ii)}\quad [\mathrm{AgS}^t\mathrm{C_5H_{11}}]_n + \mathrm{S(SiMe_3)_2} \xrightarrow[\text{xylene/THF}]{\text{dppbp}} [\mathrm{Ag_{490}S_{188}(S}^t\mathrm{C_5H_{11})_{114}}] + [\mathrm{Ag_{352}S_{128}(S}^t\mathrm{C_5H_{11})_{96}}]$$

$$\text{(iii)}\quad \mathrm{AgO_2C_{11}H_{23}} + {}^n\mathrm{BuSeSiMe_3} \xrightarrow[\text{CHCl}_3\text{ / PhMe, 253 K}]{\text{dppp}} [\mathrm{Ag_{172}Se_{40}(Se}^n\mathrm{Bu)_{92}(dppp)_4}]$$

Figure 10.17: Examples of three methods for preparing metal–chalcogenide/chalcogenolate nanoclusters from silylated chalcogen reagents: (i) combination of MOAc, $E(SiMe_3)_2$, $RESiMe_3$, and PR_3 [118], (ii) reaction of $[M(ER)]_n$ with $E(SiMe_3)_2$ [113], and (iii) reaction of MO_2CR with $RESiMe_3$ [119]. dppbp = 4,4′-bis(diphenylphosphino)biphenyl, dppp = 1,3-bis(diphenylphosphino)propane.

as the only selenium source [115]. It was necessary to include organotin chloride reagents in the reaction mixture for crystalline products to be obtained. Another common approach involves photolysis of C–E bonds, particularly in the conversion of metal tellurolates to metal tellurolate/telluride clusters via photoelimination of R_2Te [116]. Thermal elimination of R_2Se from $(Me_4N)_2[Cd(SePh)_4]$ via C–Se bond cleavage yields highly monodisperse cadmium selenide/selenolate nanoparticle systems, including a superlattice based on the T5 cluster $[Cd_{54}Se_{32}(SePh)_{48}(dmf)_4]^{4-}$ (*dmf* = *N*, *N*′-dimethylformamide) [117].

Coinage metal–chalcogenide/chalcogenolate clusters prepared from silylated sources display structural relationships to bulk copper or silver chalcogenide phases. For example, $[Cu_{93}Se_{42}(Se–C_6H_4–SMe)_9(PPh_3)_{18}]$ belongs to a class of triangular copper selenide nanoclusters containing 70–146 Cu atoms (Figure 10.18) [118]. In common to all clusters in this series is a layered A–B–A packing of selenium ligands, corresponding to a hexagonal modification of cubic $Cu_{2-x}Se$. The largest crystallographically characterized silver selenide/selenolate clusters $[Ag_{112}Se_{32}(Se^nBu)_{48}(P^tBu_3)_{12}]$, $[Ag_{114}Se_{34}(Se^nBu)_{46}(P^tBu_3)_{14}]$, and $[Ag_{172}Se_{40}(Se^nBu)_{92}(dppp)_4]$ (all prepared as of method (iii)) mimic a body-centred cubic-type packing of Se atoms, with Ag occupying the voids within and between the layers [119].

10.2.3.2 Metal chalcogenide/chalcogenolate clusters of intermediate sizes

Metal chalcogen cluster compounds of intermediate sizes (<50 metal atoms) can be prepared from silylchalcogen reagents by the methods outlined in Section 10.2.3.1. The flexibility of this reaction chemistry, in terms of the number of variable reaction conditions, facilitates access into metal chalcogenide systems with a diverse product spectrum. For example, two tetrahedral manganese(II) selenide clusters were prepared from the one-dimensional coordination polymer $[Mn(SePh)_2]_n$ and $Se(SiMe_3)_2$ in the presence of either P^nPr_3 or [Na(benzo-15-crown-5)][SePh] (Figure 10.19) [120]. The clusters $[Mn_{32}Se_{14}(SePh)_{36}(P^nPr_3)_4]$ and $[Na(benzo\text{-}15\text{-}crown\text{-}5)(C_4H_8O)]_2[Mn_8Se$

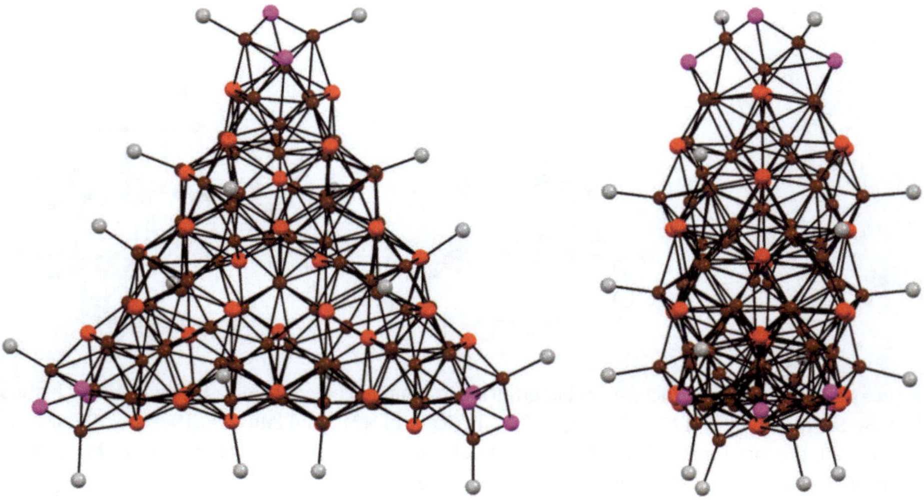

Figure 10.18: Core structure of the triangular copper(I) selenide/selenolate nanocluster $[Cu_{93}Se_{42}(Se{-}C_6H_4{-}SMe)_9(PPh_3)_{18}]$ viewed along *a* (*left*) and viewed parallel to [0 1 1] (*right*) [118]. All organic substituents in the phosphines and selenolates are omitted for clarity. Colour scheme: Cu, *brown*; Se^{2-}, *red*; RSe^{-}, *magenta*; P, *grey*.

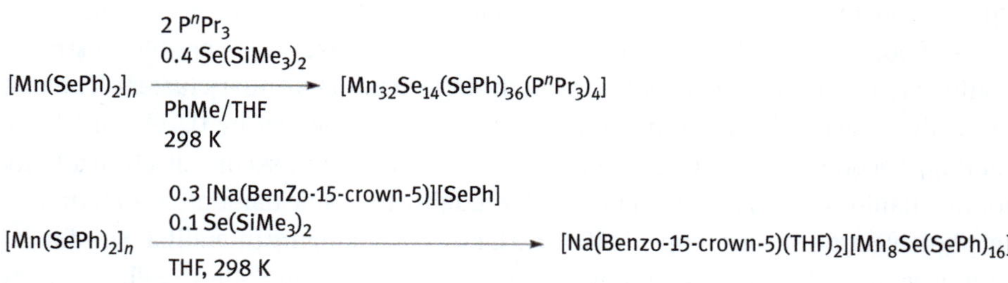

Figure 10.19: Synthesis of tetrahedral manganese(II) selenide/selenolate clusters from silylated chalcogen precursors [120].

$(SePh)_{16}]$ are structurally related to the class of supertetrahedral 12–16 clusters [29]. $[Mn_{32}Se_{14}(SePh)_{36}(P^nPr_3)_4]$ is composed of 13 fused adamantoid MnSe cages and 4 barrelanoid MnSe cages at the corners capped by one phosphine ligand. It is the largest discrete manganese selenide cluster structurally characterized to date. $[Mn_8Se(SePh)_{16}]^{2-}$ is a P1-type cluster, consisting of a $[MnSe_4]$ tetrahedron capped by four $[Mn(SePh)_4]$ tetrahedra (Figure 10.20). Despite the large number of Mn(II) atoms in these clusters, they have large, negative Weiss constants of –339(2)K and –162(2)K, respectively, indicating strong antiferromagnetic coupling.

Bis(trimethylsilylamide) metal complexes react with chalcogenols (REH) and $E(SiMe_3)_2$ to yield metal chalcogenide/chalcogenolate clusters by combining silane elimination chemistry with chalcogenolysis. This method has proven beneficial for

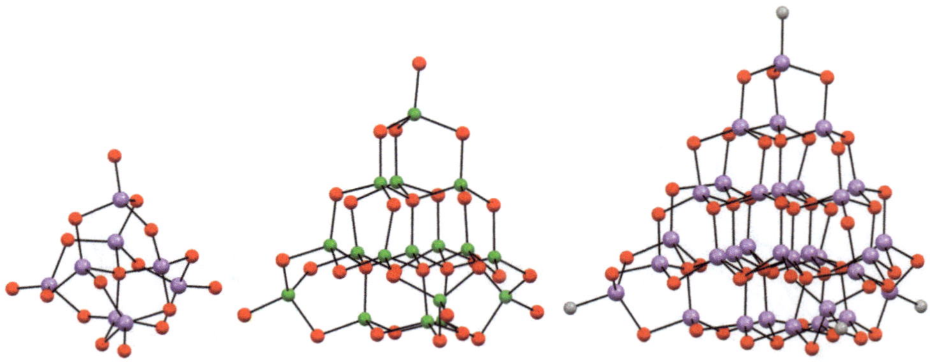

Figure 10.20: Molecular structures of supertetrahedral metal chalcogenide/chalcogenolate clusters $[Mn_8Se(SePh)_{16}]^{2-}$ (*left*) [120], $[Cd_{17}Se_4(SePh)_{28}]^{2-}$ (*centre*) [121], and $[Mn_{32}Se_{14}(SePh)_{36}(P^nPr_3)_4]$ (*right*) [120] prepared from silylated chalcogenolate sources. All surface organic groups have been omitted. Colour scheme: Mn, *violet*; Cd, *green*; Se, *red*; P, *grey*.

the generation of halide-free tetrahedral 12–16 clusters, such as the dianions $[Cd_8Se(SePh)_{16}]^{2-}$ and $[Cd_{17}E_4(E'Ph)_{28}]^{2-}$ (E, E′ = S or Se; Figure 10.20) [121]. Furthermore, the ability to generate homoleptic clusters by this method prevents surface isomerization driven by scrambling of ligands and counterions. With the molecular structure "pinned-down" in this way, a detailed comparison of the valence electronic structures of homologous chalcogenide clusters of different sizes was possible. Quantum confinement effects are noticeable, even in these small molecular clusters. The first optically allowed electronic transition shifts to higher energy by 0.5 eV for $[Cd_8Se(SePh)_{16}]^{2-}$ compared to $[Cd_{17}Se_4(SePh)_{28}]^{2-}$ and the adiabatic second electron affinity for the dianions shifts to lower energies by roughly the same amount, as determined by photoelectron spectroscopy. In $[Cd_{17}E_4(E'Ph)_{28}]^{2-}$, changing interstitial Se^{2-} for S^{2-} produces little change, but varying the surface ligands from SePh to SPh produces shifts of ca. 0.1 eV to higher energy absorption onsets and lower electron affinity.

10.2.3.3 Ternary chalcogenide/chalcogenolate clusters

Cluster complexes containing two distinct metals, chalcogenolate, and chalcogenide ligands are highly tunable systems with advanced functional properties. The use of 14–16 reagents in their synthesis has broadened the range of tenable ternary systems. In the simplest case, such compounds are prepared from the direct mixture of two metal salts, $RESiMe_3$, and $E(SiMe_3)_2$. This was the method used to prepare a copper(I)/iron(III) selenide/selenolate cluster $[Cu_{30}Fe_2Se_6(SePh)_{24}(dppm)_4]$ by inclusion of small amounts of $FeCl_3$ in the reaction mixture [122]. This elemental combination is related to that of the important semiconducting materials $CuFeS_2$ and $CuFeSe_2$ [123]. The core structure consists of a central $[Fe_2Se_6]$ unit encapsulated within a three-layer distorted cubic network of 24 selenolate ligands (Figure 10.21). The 30 Cu(I) centres occupy the tetrahedral holes inside this network and trigonal sites on the surface. The

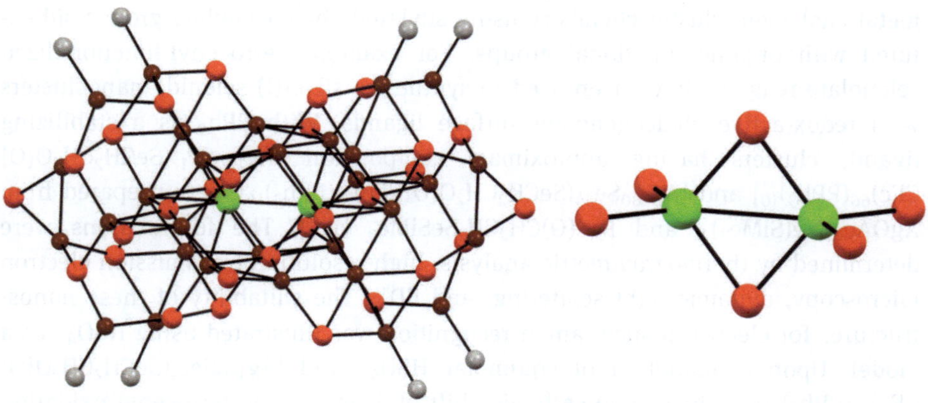

Figure 10.21: Core structure of the heterometallic Cu(I)/Fe(III) selenide/selenolate cluster $[Cu_{30}Fe_2Se_6(SePh)_{24}(dppm)_4]$ (*left*) and the central $[Fe_2Se_6]$ units (*right*) [122]. All organic groups are omitted. Colour scheme: Cu, *brown*; Fe, *green*; Se, *red*; P, *grey*.

presence of Fe^{3+} was confirmed by X-ray absorption near-edge spectroscopy, Mössbauer spectroscopy, and synchrotron X-ray crystallography. The dopant Fe^{3+} ions result in broad absorption bands at lower energies compared to other copper selenide clusters of comparable size [124].

Organic-soluble metal bis(trimethylsilyl)amide complexes are attractive reagents for the synthesis of heterometallic chalcogen clusters in combination with 14–16 precursors. The reaction of bis(trimethylsilyl)amide complexes of Cd(II) and Mn(II) with $PhSeSiMe_3$ and $E(SiMe_3)_2$ in the presence of P^nPr_3 gives rise to ternary clusters $[Cd_4Mn_6Se_4(SePh)_{12}(P^nPr_3)_4]$ and $[Cd_4Mn_4S(SePh)_{14}(P^nPr_3)_2]$ [125]. These adopt known tetrahedral $[M_{10}E_4(EPh)_{14}]^{2-}$ and $[M_8E(EPh)_{16}]^{2-}$ motifs from 12/16 cluster chemistry [30]. Tetrahedral Cd(II)–Mn(II) systems like these are of special interest as they are complementary to dilute magnetic semiconductor nanomaterials based on CdE and Mn(II) [26]. Crystallographic and mass spectral data indicate that facile Mn(II)–Cd(II) exchange occurs, so the formulae may be better described as $[Cd_{4+x}Mn_{6-x}Se_4(SePh)_{12}(P^nPr_3)_4]$ and $[Cd_{4+x}Mn_{4-x}S(SePh)_{14}(P^nPr_3)_2]$. This impacts the observed temperature-dependent magnetic susceptibility measurements. While both clusters have large, negative Weiss constants consistent with antiferromagnetic coupling between Mn(II) centres, paramagnetic impurities need to be introduced to fully model the magnetic properties. These could reasonably arise from structurally similar clusters bearing an odd number of Mn(II) ions. Moreover, the clusters are luminescent in solution with emission maxima red-shifted by comparison to Mn(II)-doped CdSe nanoparticles [26].

10.2.3.4 Metal chalcogenide clusters passivated with functional chalcogenolates

A popular approach for synthesizing nanomaterials with advanced functionality is to devise methods of surface derivatization. This has been demonstrated in metal chalcogen cluster chemistry using silylated chalcogenolate groups substituted with organic functional groups. For example, ferrocenyl-functionalized selenolate reagents have been used to synthesize silver(I) selenide nanoclusters with redox-active chalcogenolate surface ligands. With PPh_3 as a stabilizing ligand, clusters having approximate composition $[Ag_{100}Se_{17}(SeCH_2CH_2O\{O\}CFc)_{66}(PPh_3)_{10}]$ and $[Ag_{180}Se_{54}(SeCH_2CH_2O\{O\}CFc)_{72}(PPh_3)_{14}]$ were prepared from AgOAc, $Se(SiMe_3)_2$, and $FcC(O)CH_2CH_2SeSiMe_3$ [126]. The formulations were determined by thermogravimetric analysis, high-resolution transmission electron microscopy, dynamic light scattering, and EDX. The suitability of these nanostructures for electrochemical anion recognition was illustrated using HSO_4^- as a model. Upon combination of equimolar HSO_4^- and $[Ag_{180}Se_{54}(SeCH_2CH_2O\{O\}CFc)_{72}(PPh_3)_{14}]$, a distinctive cathodic shift of 50 mV in the ferrocenyl oxidation wave is observed by cyclic voltammetry, indicating the formation of an ion pair between ferrocenium and HSO_4^-.

The applicability of metal chalcogenide/chalcogenolate clusters as advanced functional materials can be enhanced through the design of "core–shell" clusters. These commonly contain a metal selenide core stabilized by a metal thiolate shell. These systems benefit from the greater stability of the thiolate precursors and their complexes compared to selenolates and tellurolates. A high degree of tunability is inherent to these systems and can be exploited in synthesis for targeted applications. A total of 14–16 reagents enable access into core–shell clusters by using silylated reagents of distinct chalcogens [typically $Se(SiMe_3)_2$ and $RSSiMe_3$].

The group of Fenske has produced copper(I) and silver(I) selenide clusters with surface functionalities including trimethylsiloxy, nitro, thiazole, cyano, and thioether – introduced from silylated arylthiolate reagents [127–131]. Importantly, the reactive functional groups of the surface ligands do not ligate the metal centres of the cluster. Their capacity for subsequent coordination could be exploited for directing reactions between functionalized cluster and metal atoms or surfaces. A preliminary result in this direction is the preparation of $[Ag_{14}S(S{-}C_6H_4{-}CN)_{12}(PPh_3)_8][FeCl_2(THF)_4]_3$ from $[Ag_{14}S(S{-}C_6H_4{-}CN)_{12}(PPh_3)_8]$ and $FeCl_2$ in THF. The two neutral molecules co-crystallize in the same unit cell; however, the cyano moieties remain unbound and the coordination sphere of iron is occupied by THF [129]. Trimethylsiloxyphenyl-thiolate and -selenolate ($Me_3SiO{-}C_6H_4{-}E{-}$) functionalized copper(I) and silver(I) selenide clusters, such as, $[Cu_{22}Se_6(Se{-}C_6H_4{-}OSiMe_3)_{10}(PPh_3)_8]$,[130] $[Cu_{52}Se_{20}(S{-}C_6H_4{-}OSiMe_3)_{12}(PPh_3)_{11}]$ [130] and $[Ag_{40}Se_4(S{-}C_6H_4{-}OSiMe_3)_{32}(PPh_3)_8]$ [129], are of special interest (Figure 10.22). The Me_3SiOR moiety can be deprotected by F^- to produce surface phenolate groups, which could form covalent links between semiconducting metal chalcogenide clusters and positively charged surfaces.

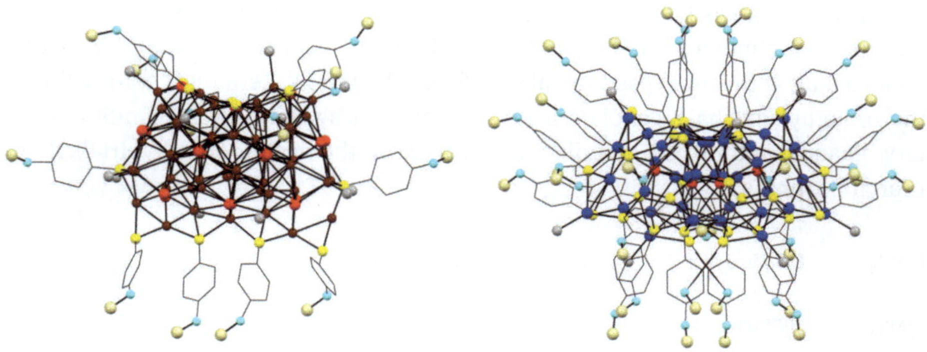

Figure 10.22: Molecular structure of core–shell metal selenide/thiolate nanoclusters functionalized with trimethylsiloxy moieties: $[Cu_{52}Se_{20}(S{-}C_6H_4{-}OSiMe_3)_{12}(PPh_3)_{11}]$ (*left*) [130] and $[Ag_{40}Se_4(S{-}C_6H_4{-}OSiMe_3)_{32}(PPh_3)_8]$ (*right*) [129]. All organic groups in the phosphines are omitted. Colour scheme: Cu, *brown*; Ag, *blue*; S, *yellow*; Se, *red*; P, *grey*; Si, *cream*; O, *cyan*.

10.2.4 Metal trimethylsilylchalcogenolates

10.2.4.1 Synthesis of metal trimethylsilylchalcogenolates

The chemistry of metal trimethylsilylchalcogenolate precursors for metal chalcogenide cluster synthesis was preceded some 25 years by the analogous trimethylstannyl derivatives. Vahrenkamp showed in 1970 that stoichiometric reaction of $[Re(CO)_5Cl]$ with $S(SnMe_3)_2$ gave a rhenium trimethylstannylthiolate dimer $[Re(CO)_4SSnMe_3]_2$, containing a reactive $Re{-}S{-}SnMe_3$ moiety [132]. Combining this with $[Mo(\eta^5{-}C_5H_5)(CO)_3Cl]$ resulted in the elimination of $ClSnMe_3$ and formation of a mixed transition metal sulfide cluster $[Re_2Mo(\eta^5\text{-}C_5H_5)(CO)_8](S)[SMo(\eta^5\text{-}C_5H_5)(CO)_3]$ [133].

The first example of a metal trimethylsilylchalcogenolate complex (containing an $Mo{-}S{-}SiMe_3$ motif) was reported by Yu in a serendipitous discovery in 1992 [134]. In 1996, Dehnen prepared two related clusters $[Cu_{31}Se_{15}(SeSiMe_3)(P^tBu_2Me)_{12}]$ and $[Cu_{70}Se_{35}(P^tBu_2Me)_{21}]$, which co-crystallize from the same reaction mixture of $Se(SiMe_3)_2$, CuOAc, and P^tBu_2Me in Et_2O [135]. Due to structural similarities between these two clusters, their crystallization from the same reaction mixture, and the reactive nature of the terminal $SeSiMe_3$ ligand in $[Cu_{31}Se_{15}(SeSiMe_3)(P^tBu_2Me)_{12}]$, it was postulated that this was an intermediate stage in the formation of the larger cluster. From these, insights evolved the systematic study of reactive mononuclear complexes $(L)_nM(ESiMe_3)_m$ (L = phosphine, amine; E = S, Se, Te) as reactive precursors *en route* to binary and ternary cluster compounds. The methodology exploits the mild and selective nature of silane elimination reactions from these complexes for the delivery of "metallachalcogenolate" ME^- for (ternary) cluster formation.

Metal trimethylsilylchalcogenolate reagents are generally prepared either by the partial silane elimination of $E(SiMe_3)_2$ reagents or by salt metathesis of $A[ESiMe_3]$ (A = alkali metal) with metal (pseudo)halides (Figure 10.23) [33]. Terminal coordination of the trimethylsilylchalcogenolate moiety is enforced by sterically demanding ancillary ligands within the coordination sphere of the metal, usually trialkyl- or triarylphosphines.

$$L_nMX_m + m\,E(SiMe_3)_2 \longrightarrow L_nM(ESiMe_3)_m + m\,XSiMe_3$$

$$L_nMX_m + m\,Li[ESiMe_3] \longrightarrow L_nM(ESiMe_3)_m + m\,LiX$$

Figure 10.23: Synthesis of metal trimethylsilylchalcogenolate reagents from silylated chalcogenides and lithium chalcogenides.

10.2.4.2 Controlled routes to molecular ternary chalcogenides

Reaction of metal trimethylsilylchalcogenolate complexes ($L_nM(ESiMe_3)_m$) with a (pseudo)halide complex of a secondary metal ($M'X_m$) gives rise to ternary M/M′/E systems. Zinc(II) and copper(II) trimethylsilylchalcogenolates were the first of these reagents for which systematic studies of this reactivity were undertaken [49, 50, 136, 137]. In these cases, the focus was on the preparation of molecular analogues of ternary semiconducting chalcogenides CuHgE, $CuInE_2$, and $Cd_xZn_{1-x}E$, from the metal chlorides or acetates. When more electron-rich metal complexes Me_3M (M = In, Ga) are combined with the copper(I) trimethylsilylchalcogenolate reagent $[(^iPr_3P)_2CuESiMe_3]$, silane elimination is suppressed and the adduct $[(^iPr_3P)_2CuESiMe_3(MMe_3)]$ (E = S, Se) forms instead [138].

Krautscheid has prepared main group metal trimethylsilylchalcogenolates $[(Me_2M)_6Se(SeSiMe_3)_4]$ (M = Ga, In) which serve as alternative entry points into ternary semiconductor systems [139]. Main group metal alkyl precursors $Me_3Ga{\cdot}2Et_2O$ and $Me_3In{\cdot}2Et_2O$ react with silylated chalcogen reagents with elimination of tetramethylsilane (Figure 10.24). MeOH is also present in the reaction mixture to generate H_2Se *in situ* via proteolysis of $Se(SiMe_3)_2$. $[(Me_2M)_6Se(SeSiMe_3)_4]$ has a spirocyclic structure in which all M atoms are tetrahedrally coordinated and are incorporated into a contiguous cluster through bridging μ-$SeSiMe_3^-$ ligands and

$$2\,Me_3M{\cdot}2Et_2O + 3\,Se(SiMe_3)_2 \xrightarrow[MeOH]{} [(Me_2M)_6Se(SeSiMe_3)_4]$$

$$4\,CuOAc + 2/3\,[(Me_2M)_6Se(SeSiMe_3)_4] \xrightarrow[MeOH/THF]{8/3\,Se(SiMe_3)_2,\ 4\,R_3P} [(R_3PCu)_4(MeM)_4Se_6]$$

Figure 10.24: Synthesis of indium(III) and gallium(III) trimethylsilylselenolates $[(Me_2M)_6Se(SeSiMe_3)_4]$ [139] and their reaction with CuOAc to form ternary Cu/M(III)/E clusters $[(R_3PCu)_4(MeM)_4Se_6]$ [140] (M = Ga, In).

one μ_4-Se^{2-}. Reaction of these precursors with CuOAc and $Se(SiMe_3)_2$ in the presence of tertiary phosphines in methanol/tetrahydrofuran results in Cu(I)/M(III)/Se clusters $[(R_3PCu)_4(MeM)_4Se_6]$ (M = Ga, In; Figure 10.24) [140].

The structure of the ternary clusters $[(R_3PCu)_4(MeM)_4Se_6]$ can be described as a cube of eight metal atoms (Cu and either Ga or In) face-capped by an octahedral array of Se^{2-} ligands (Figure 10.25). Constitutional isomerism occurs whereby the positions of the Cu(I) and M(III) centres in the cube can take two distinct arrangements with either near T_d or D_{2h} symmetry. These clusters were examined as precursors to semiconducting solids by thermolysis. In contrast to most ternary chalcogen systems, these do not contain chalcogenolate ligands which require high temperature to form chalcogenide ligands and chalcogenoethers (R_2E; Section 10.2.2.3). Thus, full thermal decomposition of the clusters in thermogravimetric analysis experiments occurs at temperatures as low as 225°C. The solids remaining after pyrolysis are composed principally of the ternary semiconductors $CuMSe_2$ in yields exceeding 85 % by mass. In contrast, thermolysis of a mixed sulfide/thiolate cluster $[(^iPr_3PCuSPh)_3(MeGaS)_4]$ required temperatures of 342°C to complete thermal decomposition by loss of methane, $^iPr_3P$ and Ph_2S.

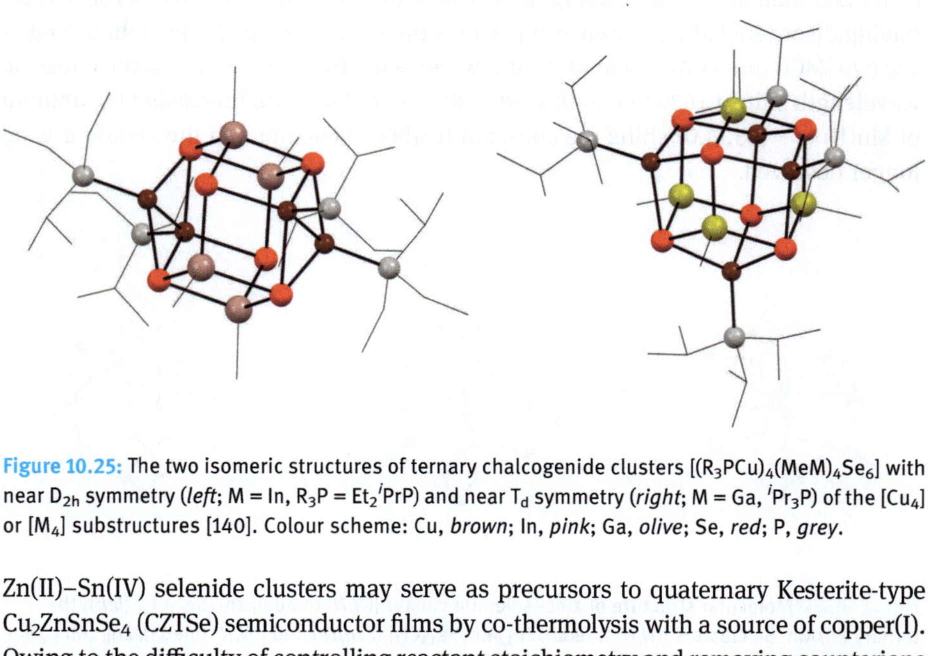

Figure 10.25: The two isomeric structures of ternary chalcogenide clusters $[(R_3PCu)_4(MeM)_4Se_6]$ with near D_{2h} symmetry (*left*; M = In, $R_3P = Et_2{}^iPrP$) and near T_d symmetry (*right*; M = Ga, $^iPr_3P$) of the $[Cu_4]$ or $[M_4]$ substructures [140]. Colour scheme: Cu, *brown*; In, *pink*; Ga, *olive*; Se, *red*; P, *grey*.

Zn(II)–Sn(IV) selenide clusters may serve as precursors to quaternary Kesterite-type $Cu_2ZnSnSe_4$ (CZTSe) semiconductor films by co-thermolysis with a source of copper(I). Owing to the difficulty of controlling reactant stoichiometry and removing counterions in dual-precursor approaches, suitable molecular precursors should be neutral and have the corresponding elements present at levels near the stoichiometric ratio. The clusters $[(N,N'\text{-tmeda})Zn(SnR_2)_2Se_3]$ (R = Ph, ^{t}Bu; tmeda = tetramethylethylenediamine) are candidate precursors, which are prepared in a single pot starting from *N,N'*-tmeda, $Zn(OAc)_2$ and $Se(SiMe_3)_2$ [141]. This combination gives rise to the previously reported zinc(II) trimethylsilylchalcogenolate $[(N,N'\text{-tmeda})Zn(SeSiMe_3)_2]$ [136]. $R_2Sn(OAc)_2$ is then added to give the trinuclear cyclic complexes. Co-thermolysis of [(*N,N′*-tmeda)Zn($Sn^tBu_2)_2Se_3$] with [($^iPr_3PCu)_2SeC_2H_4Se)]_2$ yields CZTSe in a yield of ca. 43 wt%, albeit contaminated with other phases including SnSe and Cu_2SnSe_4.

Metal trimethylsilylchalcogenolate compounds additionally enable the synthesis of binary chalcogenide systems doped with certain metal impurities. Manganese (II) and cobalt(II) trimethylsilylselenolates have been synthesized which could serve as entry points into dilute magnetic semiconductor clusters or complementary systems via combination with Zn(II) or Cd(II) salts [142]. Proof of this principle has already been shown for the ZnSe system substituted with Mn(II), using zinc(II) trimethylsilylchalcogenolate precursors [143]. Combination of [(*N,N′*-tmeda)Zn($SeSiMe_3)_2$] with varying ratios of $MnCl_2$ and $Zn(OAc)_2$ yields clusters of general formula [(*N,N′*-tmeda)$_6Zn_{14-x}Mn_xSe_{13}Cl_2$] ($x = 0$, 2.0, 3.2, 3.3, or 5.6). For $x = 0$, the unique ZnSe cluster [(*N,N′*-tmeda)$_6Zn_{14}Se_{13}Cl_2$] was obtained. The cluster has a central [$Zn_6(\mu_6$-Se)] unit of distorted octahedral geometry with four face-capping and two edge-bridging [(*N,N′*-tmeda)$ZnSe_2$], and two face-capping [ZnCl] (Figure 10.26). This binary cluster is luminescent at 298 K in solution with an emission maximum at 450 nm. Doping the cluster with Mn(II) gives analogous structures having disordered Mn/Zn occupancy within the central [$M_6(\mu_6$-Se)] octahedron and the two [MCl] units (M = Zn or Mn). At low ratios of Mn(II) ($x = 2.0$, 3.2), the emission wavelength shifts to 620 nm, with a weak shoulder at 560 nm. Increasing the amount of Mn(II) ($x = 3.3$, 4.0) shifts the emission further to 640 nm and the shoulder is no longer observed.

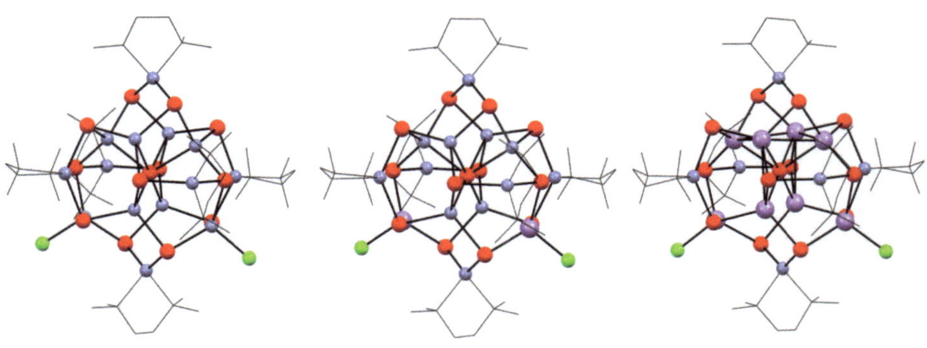

Figure 10.26: Molecular structure of zinc–selenide cluster [(*N,N′*-tmeda)$_6Zn_{14}Se_{13}Cl_2$] (*left*), the ternary Zn/Mn/Se clusters [(*N,N′*-tmeda)$_6Zn_{12}Mn_2Se_{13}Cl_2$] (*centre*) and [(*N,N′*-tmeda)$_6Zn_{8.4}Mn_{5.6}Se_{13}Cl_2$] (*right*) [143]. Colour scheme: Zn, *blue*; Mn, *violet*; Se, *red*; Cl, *green*.

10.2.4.3 Coinage metal trimethylsilylchalcogenolates

Suitable Cu(I) and Ag(I) reagents for the preparation of ternary metal chalcogen (nano) cluster assemblies are of substantial interest. A total of 14–16 reagents with preformed

M(I)–E bonds have special application in this area as they offer controlled routes into (ternary) metal chalcogenide systems of defined composition and size. However, difficulties arise for phosphine-ligated precursors ([$(R_3P)_n MESiMe_3$]) because of their thermal instability. In addition to their high moisture and oxygen sensitivity, selenolate and tellurolate reagents decomposed at temperatures below 0 °C [144]. Replacement of phosphine ancillary ligands with NHCs as stronger σ donor ligands provides a significant improvement to the thermal stability of copper(I) and silver(I) trimethylsilylchalcogenolates [145, 146]. For example, the NHC-ligated silver(I) trimethylsilylselenolate [(IPr)$AgSeSiMe_3$] (IPr = 1,3-bis(2,6-diisopropylphenyl)imidazole-2-ylidene; Figure 10.27) is thermally stable up to the melting point 170 °C.

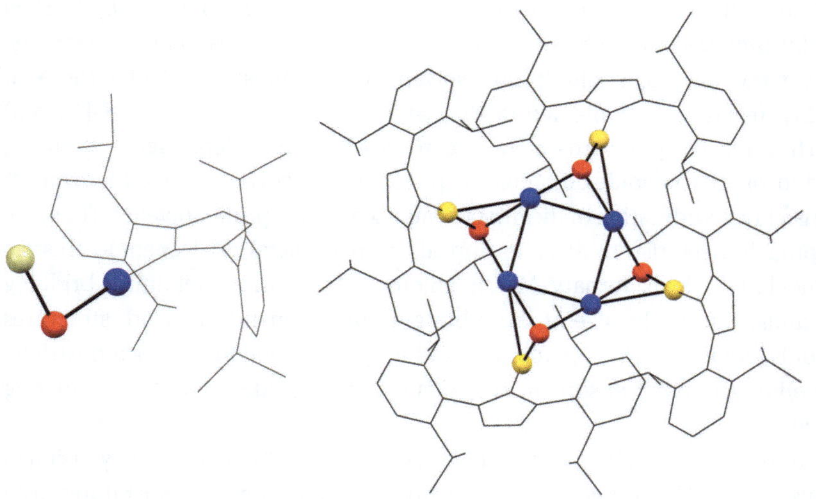

Figure 10.27: Molecular structures of carbene-ligated silver(I) trimethylsilylselenolate [(IPr) $AgSeSiMe_3$] *(left)* [146] and ternary Au/Ag/Se cluster [$Au_4Ag_4Se_4(IPr)_4$] *(right)* [147]. Colour scheme: Ag, *blue*; Au, *yellow*; Se, *red*; Si, *cream*.

Analogous NHC-ligated gold(I) trimethylsilylchalcogenolates ([(IPr)$AuESiMe_3$]) were stable in chlorinated solvents for days (Te) to weeks (Se) in room-temperature solution [147]. These reacted with phosphine or phosphite solubilized AgOAc or AuOAc, respectively, to yield a series of isostructural octanuclear clusters [$Au_4M_4(\mu_3\text{-}E)_4(IPr)_4$] (M = Ag, E = S, Se; M = Au; E = S, Se, Te). Four [(IPr)Au] units shield the [M_4E_4] core via μ_3-E ligands (Figure 10.27). Independently varying the identity of the E and M atoms changes the photoluminescence emission energies in a frozen glass matrix at 77 K. More electron-rich metal and chalcogen centres cause a bathochromic shift of the emission maximum from 562 nm (Au/Ag/S) to 657 nm (Au/Se). [$Au_8Te_4(IPr)_4$] emits at 694 nm in the solid state at 77 K.

10.3 Coordination of 14–16 ligands

Up to this point, self-assembly reactions that operate via cleavage of a single bond between silicon and a group 16 element have been considered. Reagents containing bonds between the heavier tetrel atoms (Ge, Sn, and Pb) and the chalcogens are more likely to direct self-assembly by coordination and/or fission, rather than elimination. A review published in 2012 gives an account of this chemistry [36]. The focus in the subsequent sections is on recent findings involving the heavier chalcogen atoms.

10.3.1 Chalcogenidotetrelates

Chalcogenidotetrelate ligands are complex anions of the general formula $[T_xE_y]^{n-}$. They are generally prepared by high-temperature fusion of the corresponding elements, followed by extraction into a polar (protic) solvent [35]. The most prevalent of these for self-assembly reactions are the tetrahedral anions $[TE_4]^{4-}$, $[T_2E_6]^{4-}$, $[T_2E_7]^{6-}$, and $[T_4E_{10}]^{4-}$. Their reactivity towards metal complexes is solvent dependent, favouring oxidation and/or chalcogenide donation in aprotic organic solvents and either chalcogenide transfer or coordination of the intact complex anion in protic solvents. Thus, the former approach gives rise to discrete (mixed valence) metal chalcogenide clusters whereas the latter yields ternary M/T/E anionic frameworks containing bridging $[T_xE_y]^{n-}$ ligands. Due to the size of these bridging units, open-framework structures are commonly obtained. Their porosity draws a natural comparison to industrially relevant zeolites and other adsorbent siliceous materials made from semiconducting scaffolds [29].

Early synthesis of crystalline open-framework chalcogenides proceeded by fusion of the elements E, T, and M or their alkali(ne earth) metal salts at high temperatures or in polychalcogenide fluxes [7, 8, 148]. More recently, solvothermal techniques and their variants have been popularized because they operate at lower temperature and allow for highly tunable synthetic protocols [5]. In these cases, chalcogenidotetrelate-bridging ligands arise in the metal chalcogenide phases serendipitously and are in no way guaranteed by the synthetic protocol. A few examples exist in which binary 14–16 compounds such as GeSe and SnSe are employed as precursors for solvothermal treatment [149, 150]. This takes advantage of preformed T–E bonds for the generation of bridging $[T_xE_y]^{n-}$ ligands.

Yaghi showed that open-framework chalcogenides could be prepared under ambient conditions and in aqueous solution if chalcogenidotetrelate anions were deliberately introduced as precursors [151]. Starting from the thiogermanate precursor $(Me_4N)_4[Ge_4Se_{10}]$ and $Mn(OAc)_2$ in water, a crystalline microporous lattice of composition $\{(Me_4N)_2[MnGe_4S_{10}]\}_n$ was formed. Tetrahedrally coordinated Mn(II) atoms at the corners of adamantane $[Ge_4Se_{10}]^{4-}$ units serve to bridge these anions into a Sphalerite-type network solid penetrated by channels containing the alkylammonium counterions (see Figure 10.30). A general and versatile coordination chemistry approach to the synthesis of crystalline M/14/16 can be extended from this basic protocol [36]. Aggregation of metal chalcogenidotetrelates into porous, non-crystalline networks in solution is an important feature of emerging materials like chalcogels [152–154] but will not be covered herein.

10.3.1.1 Self-assembly of ternary metal chalcogenidotetrelates in protic solvents

When high-temperature fusion methods are used to prepare metal chalcogenidotetrelates, extended network solids are most commonly observed. In contrast, discrete anionic chalcogenide clusters arise frequently from the coordination chemistry of chalcogenidotetrelate ligands in protic solvents. To date, three different types of ligand-free ternary anions containing Se and Te have been prepared: $[M_4(\mu_4\text{-}E)(SnE_4)_4]^{10-}$, $[Zn_5Sn(\mu_3\text{-}Se)_4(SnSe)_4]^{10-}$ [155], and $[Mn_6Ge_4Se_{17}]^{6-}$ [156]. P1-type $[M_4(\mu_4\text{-}E)(SnE_4)_4]^{10-}$ clusters (E = Se, M = Mn, Co, Cd, Zn, Hg; E = Te, M = Mn, Zn, Cd, Hg) are formed by the combination of the corresponding metal chloride salts and $A_4[SnE_4]$ (A = Na, K, Rb) in H_2O or H_2O/MeOH mixture [155, 157–162]. $[Zn_5Sn(\mu_3\text{-}Se)_4(SnSe)_4]^{10-}$ is a T3-type cluster: a regular tetrahedral cut-out of the Sphalerite-type lattice with three metal layers. This remains the only T3 cluster derived from the M/14/16 system. The Mn/Se core of $[Mn_6Ge_4Se_{17}]^{6-}$ is a cut-out of the α-MnSe (halite-type) lattice. A central $[Mn_6(\mu_6\text{-}Se)]$ octahedron is capped by four $[GeSe_4]^{4-}$ units to yield an overall tetrahedral ternary cluster ion. The structures of these ions are given in Figure 10.28.

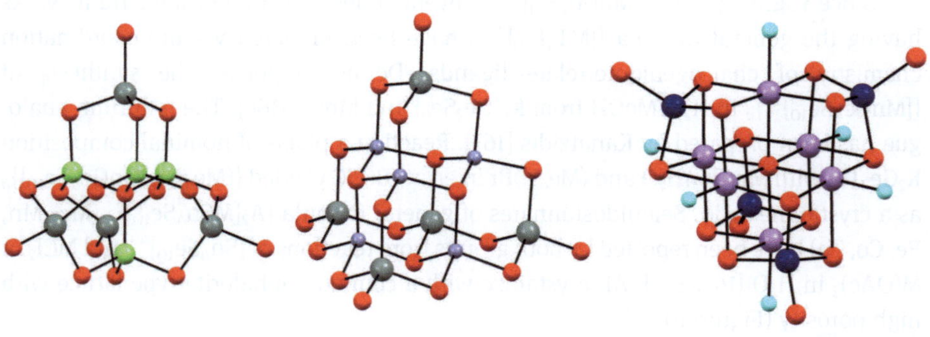

Figure 10.28: Structure of discrete ternary tetrahedral clusters prepared from chalcogenidotetrelate precursors in protic solvents: $[Co_4(\mu_4\text{-}Se)(SnSe_4)_4]^{10-}$ (*left*) [157], $[Zn_5Sn(\mu_3\text{-}Se)_4(SnSe)_4]^{10-}$ (*centre*) [155], and $[Mn_6Ge_4Se_{17}]^{6-}$ (*right*) [156]. Cations and solvent molecules are omitted. Colour scheme: Co, *green*; Mn, *violet*; Zn, *blue*; Sn, *grey*; Ge, *dark blue*; Se, *red*; O, *cyan*.

One-dimensional metal chalcogenide chains or "strands" $\{[MSnE_4]^{2-}\}$ are produced from the reaction between $[SnE_4]^{4-}$ salts and MCl_2 in H_2O or in ethylenediamine (*en*) at 100 °C [159, 163–165]. The related compound $\{[InGeTe_4]^-\}_n$ is prepared from $InCl_3$ and $[Ge_4Te_{10}]^{4-}$, with PPh_4Br in *en* at 100 °C [163]. The chains are formed by edge-sharing of nonbonded $[Te_4]$ tetrahedra centred by M and T atoms that may be

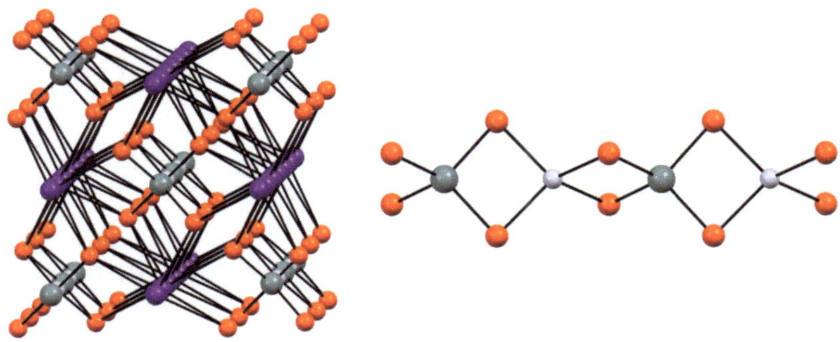

Figure 10.29: Extended three-dimensional structure of the ternary chalcogenidometallate $\{Rb_2[HgSnTe_4]\}_n$ viewed approximately along the *c* axis (*left*). View of the polymeric anionic unit along *a*, showing an alternating pattern of Sn(II) and Hg(II) centres (*right*) [165]. Colour scheme: Hg, *silver*; Sn, *grey*; Te, *orange*; Rb, *purple*.

alternating or statistically distributed along the polymer (Figure 10.29). Voids between individual strands are occupied by the counterions which align in a linear fashion. This unique dual alignment of cations and metal chalcogenide coordination polymers, as well as the lack of co-crystallizing solvent molecules, has generated interest in the charge- and ion-conducting capabilities of the solids. For instance, $\{Cs_2[MnSnTe_4]\}_n$ exhibits fast charge transport via a polaronic mechanism – indicating that charge is mobilized along the metal chalcogenide chain [165].

Since Yaghi reported $\{[MnGe_4S_{10}]^{2-}\}_n$, several other three-dimensional frameworks having the general formula $\{[MT_4E_{10}]^{2-}\}_n$ have been prepared via the coordination chemistry of chalcogenidotetrelate ligands. Dehnen reported the synthesis of $\{[MnGe_4Se_{10}]^{2-}\}_n$ in H_2O/MeOH from $K_4[Ge_2Se_6]$ and $MnCl_2$ [166]. The tellurium analogue has been prepared by Kanatzidis [167]. Reacting a phase of nominal composition $K_2Ge_2Te_5$ with $MnCl_2{\cdot}4H_2O$ and $(Me_4N)Br$ in *en* at 100 °C yielded $\{(Me_4N)_2[MnGe_4Te_{10}]\}_n$ as a crystalline solid. Selenidostannates of general formula $\{A_2[MSn_4Se_{10}]\}_n$ (M = Mn, Fe, Co, Zn) have been reported by both groups from reactions of $[Sn_4Se_{10}]^{4-}$ and MCl_2 or $M(OAc)_2$ in H_2O [167, 168]. All crystallize with a common Sphalerite-type lattice with high porosity (Figure 10.30).

Combining K or Cs salts of $[SnE_4]^{4-}$ and HgX_2 (X = Cl, OAc) gives rise to three-dimensional lattices composed of linked $[Hg_4(\mu_4\text{-}E)(SnE_4)_3]^{6-}$ (E = Se, Te) substructures [159, 160]. Such arrangements were first described by Kanatzidis in the hydrothermal preparation of $\{K_6[Cd_4Sn_3Se_{13}]\}$ [169]. The building blocks of this network consist of a P1 cluster fragment in which one $[SnSe_4]$ capping unit has been formally removed. The coordination sphere about Hg is completed by three μ-bridging Se^{2-} ligands that interconnect the cluster units into a network solid (Figure 10.31). In this way, all individual cluster units are octahedrally surrounded and organized into a porous cubic framework. The diameter of the pores is slightly sensitive to the nature of the ternary metal chalcogenide system, typically 5–7 Å [159, 169].

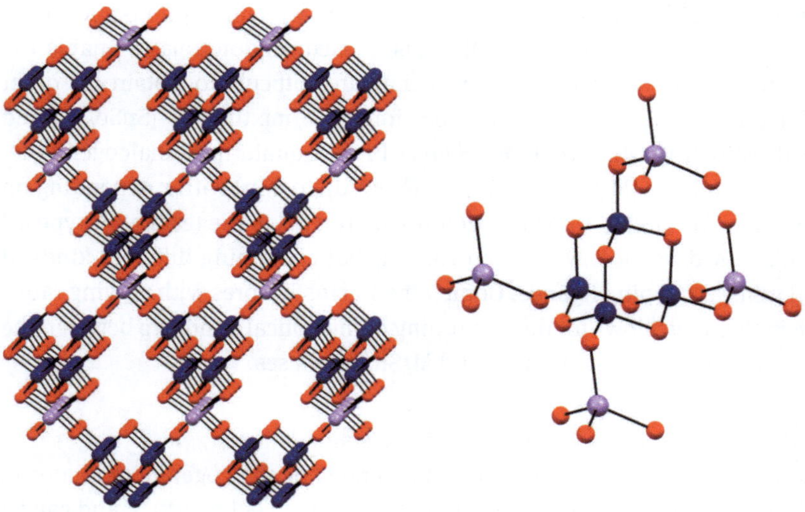

Figure 10.30: Extended three-dimensional structure of the ternary selenidogermanate $\{(Me_4N)_2[MnGe_4Se_{10}]\}_n$ viewed down *a* (*left*). The fundamental repeating unit of an adamantane $[Ge_4Se_{10}]$ cluster which is corner linked to four $[MnSe_4]$ tetrahedra is shown (*right*) [166]. All counterions and co-crystallizing solvent molecules are omitted for clarity. Colour scheme: Mn, *violet*; Ge, *dark blue*; Se, *red*.

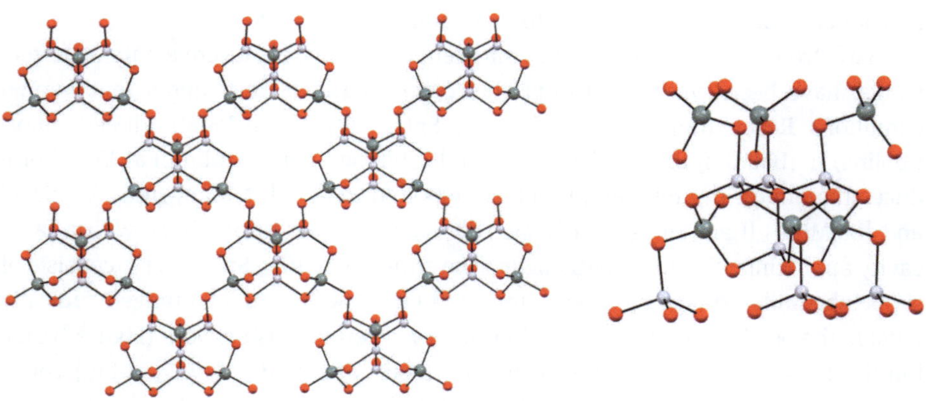

Figure 10.31: Extended three-dimensional structure of the ternary chalcogenidometallate $\{[Hg_4(\mu_4\text{-}Se)(SnSe_4)_3]^{6-}\}_n$ viewed approximately parallel to $[\bar{1}11]$ (*left*). Also shown is the fundamental repeating unit of a partial P1 cluster $[Hg_4(\mu_4\text{-}Se)(SnSe_4)_3]$ corner linked through three tetrahedral $[HgSe_4]$ units and three $[SnSe_4]$ (*right*) [159]. Colour scheme: Hg, *silver*; Sn, *grey*; Se, *red*.

Quaternary M/T/E/E′ anionic frameworks are obtained by the simple mixture of two distinct chalcogenidotetrelates with a metal complex. The dynamic nature of these mixtures in aqueous solutions is supported by multinuclear nuclear magnetic resonance spectroscopy and the characterization of ternary T/E/E′ anions such as $K_4[SnSe_2Te_2]$ and $K_2(PPh_4)_2[Sn_2Se_2Te_4]$ [170]. Thus, chalcogenidotetrelates enable the synthesis of molecular quaternary systems that are difficult to obtain by other methods, including silane elimination. Reaction of a mixture of $[SnSe_4]^{4-}$ and $[SnTe_4]^{4-}$ with MCl_2 in MeOH or H_2O/MeOH gives P1-type quaternary chalcogenidotetrelate anions formulated $[M_4Sn_4Se_{17-x}Te_x]^{10-}$ (M = Cd, Zn, Mn) with x depending on reaction stoichiometry [168, 170]. The structures are related to the ternary P1-type M/T/E anions discussed previously (Section 10.3.1.1) but containing different kinds of $[SnSe_{4-x}Te_x]$ units as capping ligands. Doping the P1 cluster cores with varying ratios of Se and Te in this way allows for the fine tuning of the optical band gap between the two extremes of the respective M/Sn/Se and M/Sn/Te phases.

10.3.1.2 Self-assembly of metal chalcogenidotetrelates in ionic liquids

Dehnen and co-workers are pursuing the self-assembly of chalcogenidotetrelates in ionic liquids. Ionic liquids are salt-like compounds that melt below 100 °C and can be used as solvents in molecular transformations at ambient conditions or at elevated temperatures (ionothermal synthesis). Ionic liquids with (silylated) chalcogen anions were recently reported by Sundermeyer and their application in the preparation of new chalcogenide phases is actively being explored [171–173]. A review of the progress in self-assembly of metal chalcogenide phases in ionic liquids has recently been published [38]. In this section, the preparation of anionic frameworks involving the heavier chalcogen atoms from 14–16 precursors is summarized.

Two "zero-dimensional" chalcogenidotetrelate solids, i.e. discrete anionic frameworks, have been synthesized from chalcogenidotetrelate salts under ionothermal conditions. Reaction of $[K_4(H_2O)_3][Ge_4Se_{10}]$, $SnCl_4 \cdot 5H_2O$, and DMMP (dimethylmorpholine) in $[BMMIm][BF_4]$ (BMMIm = 1-butyl-2,3-dimethylimidazolium) at 150 °C produces the largest discrete ternary anions reported to date $[BMMIm]_{24}[Ge_{24}Sn_{36}Se_{132}]$ and $[BMMIm]_{24}[Ge_{32.5}Sn_{27.5}Se_{132}]$ [174]. The anion is spherical in shape with a large cavity approximately 11.6 Å in diameter. The structure of $[Ge_{24}Sn_{36}Se_{132}]^{24-}$ consists of six octahedrally arranged $[Sn_6Se_{18}]$ units and eight $[Ge_3Se_6]$ units arranged cubically outside the octahedron, linked together to form a contiguous network (Figure 10.32). $[Sn_6Se_{10}]$ is a "double-defect heterocubane" motif: it consists of two $[Sn_3Se_4]$ defect heterocubane cores bridged through two μ-Se^{2-} ligands and bridged to adjacent $[Ge_3Se_9]$ units through eight further μ-Se^{2-} ligands [38]. $[Ge_3Se_9]$ is based on a hexagonal $[Ge_3Se_3]$ cycle with all Ge(II) tetrahedrally coordinated to Se^{2-} ligands. Due to its open-framework, the anion was studied for its molecular trapping potential using I_2 as a model compound. In addition to reversibly taking up 11 equivalents of I_2, the anion heterolytically activates the I–I bond, forming charge-transfer complexes $I^+\cdots I^-$ and $I^+\cdots I_3^-$, as determined by electronic absorption spectroscopy.

One-dimensional frameworks commonly result from the ionothermal conversion of chalcogenidotetrelate precursors. The simplest example of this is $\{[BMIm]_2$

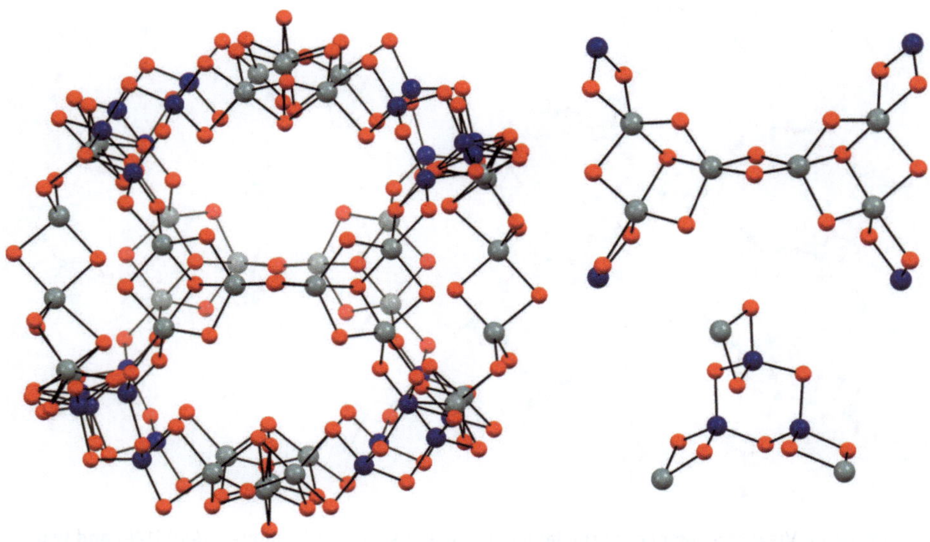

Figure 10.32: Ternary T/T′/E anion $[Ge_{24}Sn_{36}Se_{132}]^{24-}$ viewed approximately parallel to [0 1 $\bar{1}$] (*left*). The two fundamental repeating units are shown (*left*): double-defect heterocubane $[Sn_6Se_{10}]$ linked through eight μ-Se^{2-} to four Ge atoms (*upper*) and hexagonal $[Ge_3Se_6]$ units linked through six μ-Se^{2-} to three Sn atoms (*lower*) [174]. Colour scheme: Ge, *dark blue*; Sn, *grey*; Se, *red*.

$[Ge_4Se_9]\}_n$ (BMIm = 1-butyl-3-methylimidazolium) which is prepared from $K_4[Ge_4Se_{10}]$ and $SnCl_4{\cdot}5H_2O$ in the presence of DMMP in [BMIm]$[BF_4]$ [175]. The solid-state structure of the anion consists of an infinite one-dimensional zigzag chain of corner sharing $[Ge_4Se_{10}]$ adamantane units with two terminal Se^{2-} ligands. A similar reaction carried out with $SnCl_2$ in [BMIm]$[BF_4]$ yields a related one-dimensional chain $\{[BMMIm]_2[SnGe_4Se_{10}]\}_n$ [175]. In contrast to $\{[MGe_4Se_{10}]^{2-}\}_n$ based on a tetrahedral M(II) cations which give the Sphalerite-type structure (Figure 10.30), the stereochemically active lone pair on Sn(II) results instead in a double chain structure of $[Ge_4Se_{10}]$ units linked through trigonal pyramidal centres, each having one terminal Se^{2-} ligand.

Top-down strategies for the ionothermal preparation of chalcogenidotetrelate phases have been demonstrated as effective for preparing low-dimensional solids. For example, treatment of the three-dimensional network $\{K_2[Sn_2Se_5]\}_n$ in [BMMIm] $[BF_4]$ for two days yields at 150 °C the one-dimensional solid $\{[BMMIm]_4[Sn_6Se_{14}]\}_n$ [176]. This can formally be considered a one-dimensional cut-out of the lamellar $\{[Sn_3Se_7]^{2-}\}_n$ phase comprising hexagonal arrays of double-defect heterocubane units [177] (Figure 10.33).

The two-dimensional material $\{[BMMIm]_{16}[Sn_{24}Se_{56}]\}_n$ was prepared by the same top-down approach as used for $\{[BMMIm]_4[Sn_6Se_{14}]\}_n$, using lower temperature and longer reaction time and is structurally similar to $\{[Sn_3Se_7]^{2-}\}_n$ phases [176]. Other two-dimensional anions have been prepared by conventional "bottom-up" ionothermal

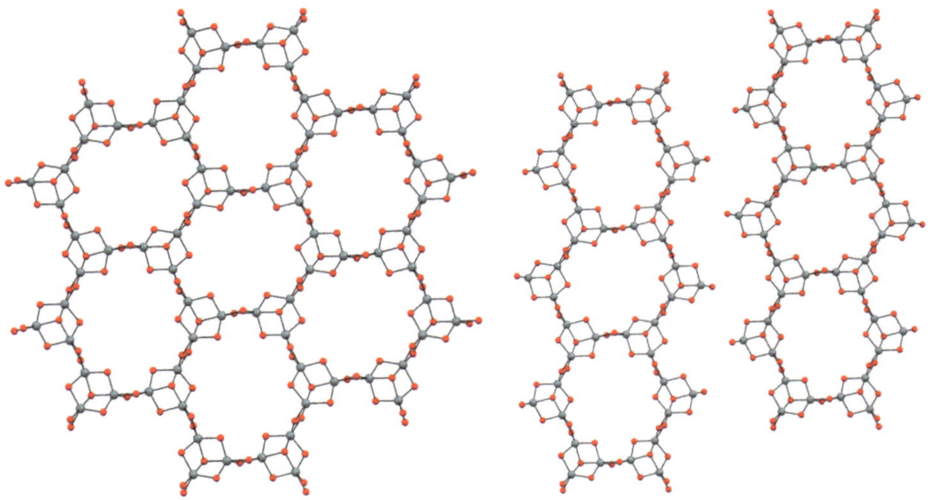

Figure 10.33: View of a segment of the lamellar phase $\{Cs_2[Sn_3Se_7]\}_n$ along *c* (*left*) [178] and two strands of the structurally derived one-dimensional selenidostannate(IV) $\{[BMMIm]_4[Sn_6Se_{14}]\}_n$ viewed along *c* (*right*) [176]. All counterions and co-crystallization solvent molecules have been omitted. Colour scheme: Sn, *grey*; Se, *red*.

syntheses. Reaction of $K_4[Ge_4Se_{10}]$ and $SnCl_4{\cdot}5H_2O$ in the presence of DMMP or *en* gives a mesoporous ternary chalcogenide formulated as $\{[BMMIm]_2[Ge_{0.83}Sn_{3.17}Se_{9.06}]\}_n$ [175]. The anionic substructure has $\{[GeSn_3Se_9]^{2-}\}_n$ layers constructed from $[Sn_6Se_{10}]$ double-defect heterocubane units bridged through two μ-Se to two $[MSe_4]$ tetrahedra. The structural refinement indicates a mixed Ge/Sn occupancy of this metal site of about 17 % Sn. The overall arrangement is of distorted 8-unit rings which stack as layers of highly porous corrugated sheets (Figure 10.34). The voids within the sheets have dimensions of approximately 19.9 × 9.6 Å^2.

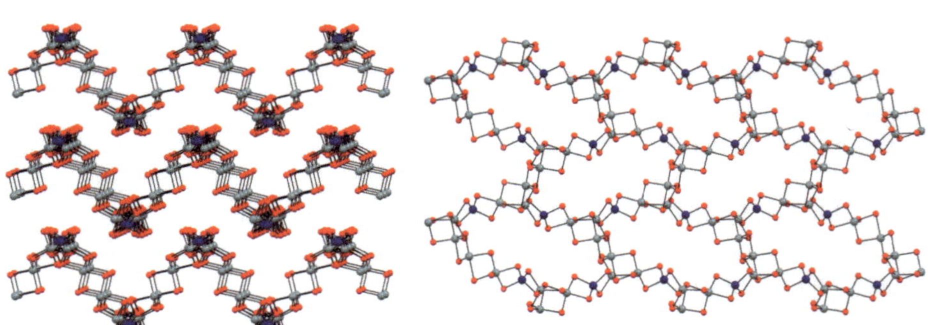

Figure 10.34: Extended structure of the anionic framework of $\{[BMMIm]_2[Ge_{0.83}Sn_{3.17}Se_{9.06}]\}_n$ viewed parallel to the [1 0 1] plane (*left*) and along *a* (*right*) [175]. Colour scheme: Sn, *grey*; disordered Sn/Ge, *dark blue*; Se, *red*.

The three-dimensional anionic framework $\{[BMIm]_4[Sn_9Se_{20}]\}_n$ has been obtained by ionothermal conversion of chalcogenidotetrelate precursors by two different routes. Treatment of $K_4[SnSe_4]$ in [BMIm][BF_4] at 130 °C for 7 days yields $\{[BMIm]_4[Sn_9Se_{20}]\}_n$ [179]. The combination of $[Ge_4Se_{10}]^{4-}$, $SnCl_4 \cdot 5H_2O$ and *en* under ionothermal conditions also produces this same binary framework and the related phase $\{[BMMIm]_8[Sn_{18}Se_{40}]\}_n$ [175]. The basic building block of the two networks is the double-defect heterocubane $[Sn_6Se_{10}]$ with all Sn atoms surrounded by Se^{2-} in a trigonal bipyramidal coordination geometry. Bridging $[SnSe_4]$ and $[Sn_2Se_6]$ motifs connect these units into a contiguous three-dimensional network (Figure 10.35). $\{[BMIm]_4[Sn_9Se_{20}]\}_n$ is highly porous, with about 56.8 % of the crystal lattice being comprised of solvent or counterion accessible voids. $\{[BMMIm]_8[Sn_{18}Se_{40}]\}_n$ has a different global connectivity which allows even larger pores (ca. 16.2 × 7.7 Å^2 at most) compared to $\{[BMIm]_4[Sn_9Se_{20}]\}_n$.

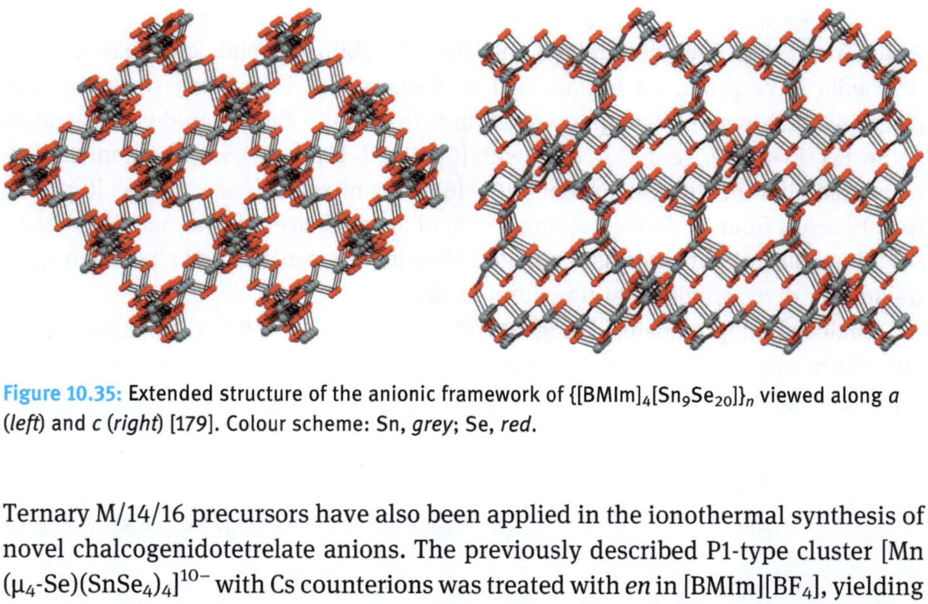

Figure 10.35: Extended structure of the anionic framework of $\{[BMIm]_4[Sn_9Se_{20}]\}_n$ viewed along *a* (*left*) and *c* (*right*) [179]. Colour scheme: Sn, *grey*; Se, *red*.

Ternary M/14/16 precursors have also been applied in the ionothermal synthesis of novel chalcogenidotetrelate anions. The previously described P1-type cluster $[Mn(\mu_4\text{-}Se)(SnSe_4)_4]^{10-}$ with Cs counterions was treated with *en* in [BMIm][BF_4], yielding a two-dimensional network solid formulated as $\{[Mn(en)_{2.5}(en\text{-}Me)_{0.5}][Sn_3Se_7]\}_n$ [162]. The anionic substructure is a distorted honeycomb network of defect heterocubanes $[Sn_3Se_4]$ units linked through two μ-Se bridges per tin atom to three adjacent units. This composition and distorted hexagonal topology was previously reported to result from the solvothermal synthesis of $\{[Mn(peha)_2][Sn_3Se_7]\}_n$ (*peha* = pentaethylenehexamine) from Sn, Se, and $MnCl_2 \cdot 4H_2O$ under surfactant-thermal conditions in triethylenetetramine/glycol [180]. The cavities of the hexagonally layered sheets are filled by two octahedral Mn^{2+} amine complexes, where one ligand is generated through methylation of *en* by the ionic liquid cation (Figure 10.36).

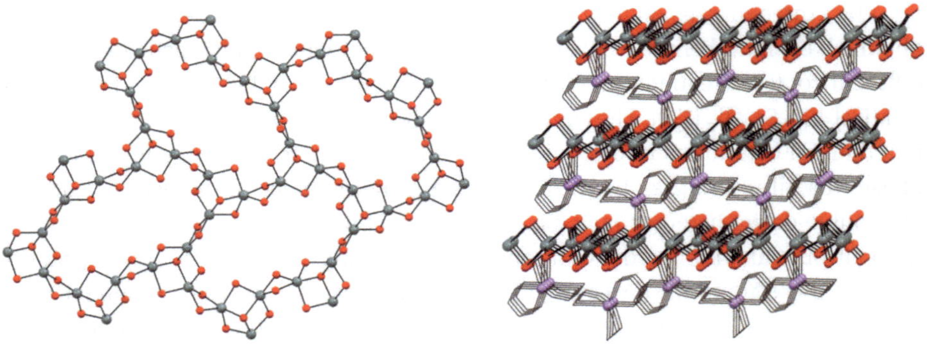

Figure 10.36: Extended structure of the anionic framework of $\{[Mn(en)_{2.5}(en\text{-}Me)_{0.5}][Sn_3Se_7]\}_n$ viewed along *a* with the cationic complexes removed from the channels (*left*) and along *b* with the cationic complexes present (*right*) [162]. Colour scheme: Sn, *grey*; Mn, *violet*; Se, *red*.

10.3.1.3 Reactions of chalcogenidotetrelates in polar aprotic solvents

Reactions of chalcogenidotetrelates with ligand-stabilized metal complexes in aprotic media have produced ternary M/T/E clusters with interesting structural and electronic features. An exemplary reaction is that of the chalcogenidostannates(III) $A_6[Sn_2E_6]$ (E = S, Se, Te; A = K, *en*H) with $[Cp^{\star}CoCl]_2$ ($Cp^{\star}$ = 1,2,3,4,5-pentamethylcyclopentadienyl) in ethereal solvents [181, 182]. The mixed-valence clusters $[(Cp^{\star}Co)_3(\mu_3\text{-}E)_2]$ result from oxidative decomposition of the chalcogenidostannate(III) precursor with concomitant chalcogen transfer, showing the versatility of chalcogenidotetrelates for diverse molecular transformations.

Chalcogenidoplumbate(II) reagents $[Pb_2E_3]^{2-}$ (E = Se, Te) have been shown to be effective mildly oxidizing and chalcogenide-donating reagents for cluster self-assembly (Figure 10.37).

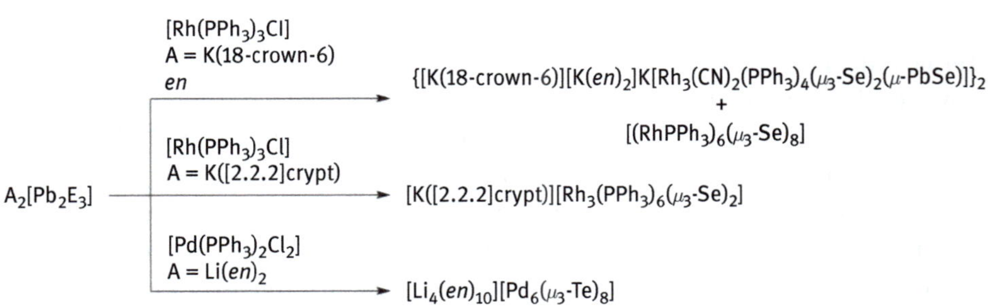

Figure 10.37: Reactions of chalcogenidoplumbates(II) with phosphine-stabilized Rh(I) and Pd(II) chloride complexes [183, 184].

Reaction of $[K(18\text{-crown-6})]_2[Pb_2Se_3]$ with $[Rh(PPh_3)_3Cl]$ in THF results in Se^{2-} transfer to and oxidation of the rhodium centres, yielding the mixed-valence $Rh^{II}{}_2Rh^{III}{}_4$

selenide cluster $[(RhPPh_3)_6(\mu_3\text{-}Se)_8]$ [183]. The molecular structure is reminiscent of condensed Chevrel phases, with a Rh_6 octahedron face-capped by six $\mu_3\text{-}Se^{2-}$ in a cubic arrangement (Figure 10.38). Co-crystallizing with this was a ternary Rh(I)/Pb (II)/Se dimer with unprecedented structural features $\{[K(18\text{-crown-6})][K(en)_2]K[Rh_3(CN)_2(PPh_3)_4(\mu_3\text{-}Se)_2(\mu\text{-}PbSe)]\}_2$ (Figure 10.38) [184]. The monomeric cluster unit is based on a Rh_3Se_2 trigonal bipyramid. One equatorial Rh atom is square planar, ligated by two phosphines and two μ_3-bridging Se^{2-} ligands. The other two Rh atoms have a distorted square pyramidal geometry. The four basal ligands are two $\mu_3\text{-}Se^{2-}$, one CN^- (generated from *en*), and one PPh_3. The axial positions are occupied by a μ-bridging PbSe ligand. The local geometry about lead is trigonal pyramidal, consistent with Pb(II). This was the first instance in which PbSe has been observed as a ligand and the authors characterized it as a heavy-atom homologue of CO with distinct electronic and coordination characteristics.

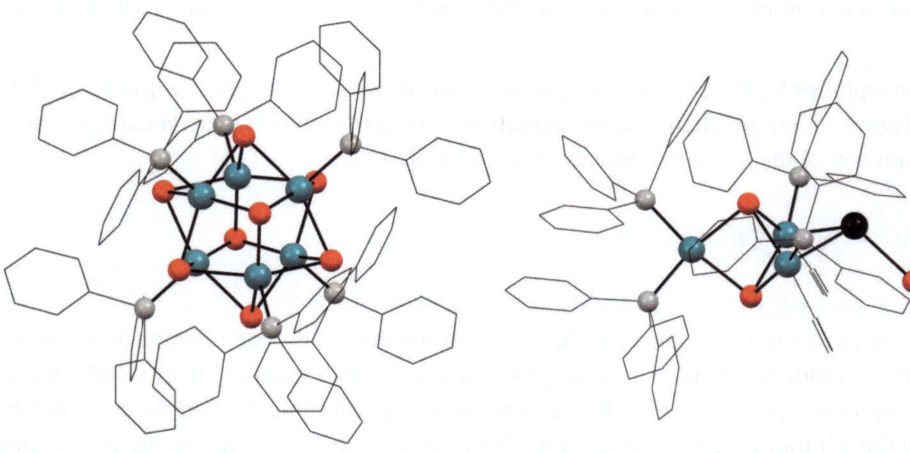

Figure 10.38: Molecular structures of $[(RhPPh_3)_6(\mu_3\text{-}Se)_8]$ (*left*) [183] and the monomeric anion $[Rh_3(CN)_2(PPh_3)_4(\mu_3\text{-}Se)_2(\mu\text{-}PbSe)]^{3-}$ (*right*) [184]. Colour scheme: Rh, *turquoise*; Pb, *black*; Se, *red*; P, *grey*.

Replacing the cation of $[Pb_2Se_3]^{2-}$ with [K([2.2.2]crypt)] in the reaction with $[Rh(PPh_3)_3Cl]$ results in the formation of a binary rhodium(I) selenide cluster $[K([2.2.2]crypt)][Rh_3(PPh_3)_6(\mu_3\text{-}Se)_2]$ [184]. All Rh centres are coordinated in a square planar geometry with two PPh_3 and two $\mu_3\text{-}Se^{2-}$. The difference in the formation of this cluster and $\{[K(18\text{-crown-6})][K(en)_2]K[Rh_3(CN)_2(PPh_3)_4(\mu_3\text{-}Se)_2(\mu\text{-}PbSe)]\}_2$ is clearly related to both the availability of CN^- and is influenced by the macrocyclic ethers used to extract Pb_2Se_3 from the solid phase. In this case, the selenidoplumbate(II) has only acted as a Se^{2-} source. The same can be said of the reaction of $[Li(en)]_2[Pb_2Te_3]$ with $[Pd(PPh_3)_2Cl_2]$. This yields the Chevrel-type cluster $[Li_4(en)_{10}][Pd_6(\mu_3\text{-}Te)_8]$, having similar geometry to $[(RhPPh_3)_6(\mu_3\text{-}Se)_8]$ without any terminally bound

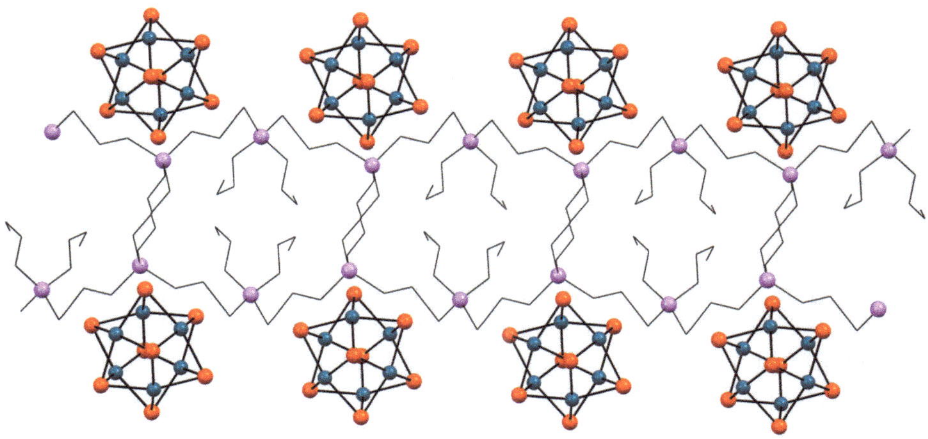

Figure 10.39: Extended structure of $[Li_4(en)_{10}][Pd_6(\mu_3\text{-}Te)_8]$ running parallel to [1 0 1] showing one double layer of anionic and cationic units [183]. Colour scheme: Pd, *turquoise*; Li, *violet*; Te, *orange*.

phosphines [183]. The anion is instead stabilized within a network of $\{[Li_4(en)_{10}]^{4+}\}_n$ (Figure 10.39). No change in the oxidation state at the metal is observed and the sole function of the 14–16 reagents is as a source of Te^{2-}.

10.3.2 Organotetrel chalcogenides

10.3.2.1 Synthesis of organotetrel chalcogenide clusters

Binary organotin chalcogenide clusters are accessible via the combination of organotin (IV) chloride reagents and chalcogenide sources. Traditionally, alkali metal chalcogenides or H_2E were used as E^{2-} sources and this significantly limited the accessibility of the tellurides [148]. Recently, it has been shown that $Te(SiMe_3)_2$ enables access into the poorly studied Sn/Te system (Figure 10.40) [185]. Reaction of organotin chlorides (R^1SnCl_3; R^1 = $Me_2CH_2C(Me)O$) with $Te(SiMe_3)_2$ in a 4:6 ratio yields the binary Sn/Te cluster $[(R^1Sn)_4Te_6]$. The cluster adopts a "double-decker" structure of two $[(R^1Sn)_2Te_2]$ units bridged at the Sn atoms by two μ-Te^{2-} ligands. Similar structures are well known for the Sn/S and Sn/Se systems [186]. Varying the ratio of R^1SnCl_3 to $Te(SiMe_3)_2$ gave

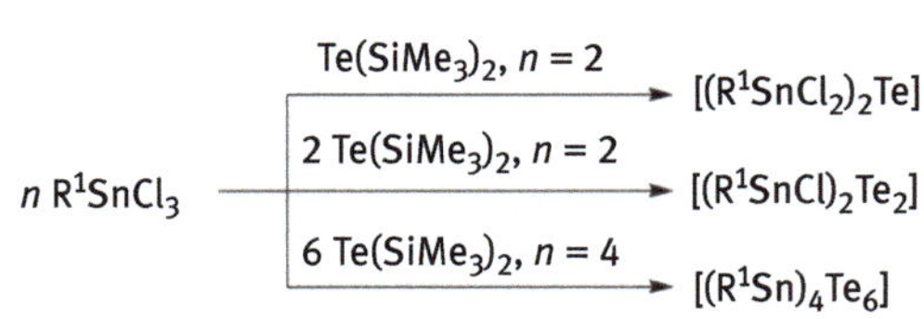

Figure 10.40: Synthesis of organotin(IV) tellurides from organotin chlorides and bis(trimethylsilyl) telluride [185].

mixed organotin chloro-tellurides $[(R^1SnCl_2)_2Te]$ and $[(R^1SnCl_2)_2Te_2]$ which displayed further reactivity towards Sn/Te cluster formation (*vide infra*).

Organotetrel chalcogenide compounds react under a variety of conditions to yield new binary 14/16 frameworks. The two common approaches used are (i) direct manipulation of the inorganic core, or (ii) reaction at the organic shell with concomitant reorganization of the cluster core. A typical example of (i) is the rearrangement of T/E cores by photolysis. Jurkschat reported an example of an unusually stable organotin telluride $[\{R^2Sn(Te)\}_2Te]$ (R^2 = 2,6-$(Me_2NCH_2)_2C_6H_3$), containing two terminally bound tellurium ligands [187]. Upon photolysis, this binary 14/16 reagent rearranges to the cluster $[(R^2Sn)_2SnTe_4]$. Dehnen and co-workers prepared the double-decker-type organogermanium(IV) selenide $[(R^1Ge)_4Se_6]$, having a keto-functionalized organic shell, from R^1GeCl_3 and Na_2Se [188]. Crystals of this highly reactive framework were found to undergo photolysis in the mother liquor to afford the mixed valence anion $[(R^1Ge^{IV})_2(R^1Ge^{III})(Ge^{III}Se)Se_5]^-$. The anion has a noradamantane-type structure containing a Ge^{III}–Ge^{III} single bond generated upon photoreduction of the Ge^{IV} precursor.

Organotetrel chalcogenides with reactive surface moieties can undergo post-synthetic modification at the cluster surface, frequently resulting in changes to the core structure. This chemistry has been extensively developed for the Sn/S system [189–191]. Only recently the parallel developments have been made for the heavier chalcogen-containing systems [185]. One productive methodology is the decoration of organotin(IV) chalcogenides with keto functionalities which, upon exposure to hydrazines, form C,N-chelating hydrazone ligands. For example, $[(R^1SnCl_2)_2Te_2]$ reacts with phenylhydrazine to yield the defect heterocubane salt $[(R^3Sn)_3Te_4][SnCl_3]$ (R^3 = $CMe_2CH_2C(Me)NNHPh$). This same precursor reacts with a multidentate carbohydrazide $H_2N_2C(O)N_2H_3$ to generate a complicated reaction mixture containing three Sn/Te clusters simply through variation of the surface moieties. The major product is $[(HR^4)Sn_2TeCl_3]$ (R^4 = $\{CMe_2CH_2C(Me)NN\}_2C(O)$) which first crystallizes from solution. Smaller amounts of $[(HR^4)(HR^5)Sn_3TeCl_5]$ (R^5 = $CMe_2CH_2CH_2C(Me)NNC(O)NNH_2$) and $[(R^1)(HR^5)(H_2R^5)_2Sn_7Te_8Cl_4]$ crystallize 3 days later. Structurally, the most interesting of these is $[(R^1)(HR^5)(H_2R^5)_2Sn_7Te_8Cl_4]$. Ignoring the organic scaffolding, the cluster core can be considered a contiguous chain of two μ-Te bridged $[Sn_2Te_2]$ cycles which are capped at either end by organotin chloro-tellurides (Figure 10.41). Intramolecular self-complementary hydrogen-bonding interactions force the Sn/Te chain to fold inwards and form a "molecular capsule" which dimerizes through intermolecular N–H···O hydrogen bonds.

10.3.2.2 Reactions of organotetrel chalcogenides with metal complexes

As was the case with chalcogenidotetrelates, organotetrel chalcogenide reagents react with ligand-supported metal complexes in organic solvent to yield ternary M/14/16 systems with unusual structures. For example, rare examples of ternary Sn/Pd/Se clusters adopting a trigonal bipyramidal structure were recently reported to arise

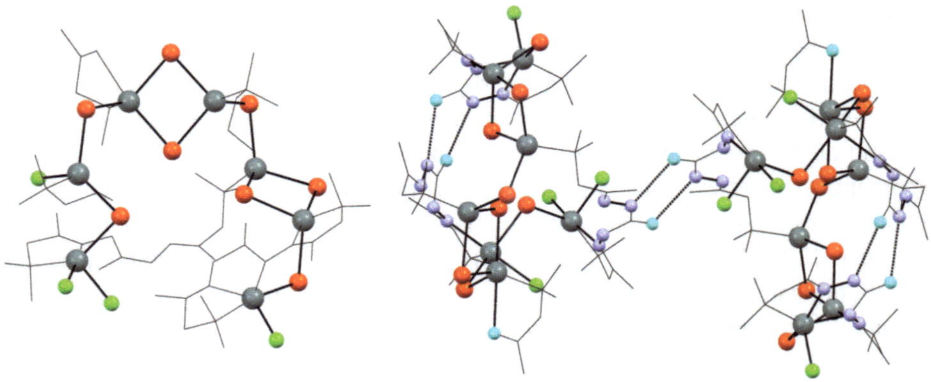

Figure 10.41: Core metal–chalcogen structure of $[(R^1)(HR^5)(H_2R^5)_2Sn_7Te_8Cl_4]$ (*left*) and the extended structure formed via dimerization through hydrogen bonding interactions (*right*) [185]. Colour scheme: Sn, *grey*; Te, *orange*; Cl, *green*; O, *cyan*; N, *blue*.

from 14/16 precursors [192]. Direct combination of the "double-decker" organotin selenide cluster $[(R^1Sn)_4Se_6]$ with five equivalents of $[Pd(PPh_3)_2Cl_2]$ resulted in fission and partial hydrolysis of the Sn/Se framework, yielding $[Pd_3(PPh_3)_5(SeH)(\mu_3\text{-}Se)_2][SnCl_3]$. The addition of hydrazine hydrate to this reaction mixture gave rise to two additional clusters $[Pd_3(PPh_3)_5(SnCl_3)(\mu_3\text{-}Se)_2][SbCl_3]$ and $[Pd_2(PPh_3)_3Cl\{SnR^6Cl_2\}\{\mu_3\text{-}Se)_2]$ ($R^6 = CMe_2CH_2C(Me)NNH_2$).

Ternary Ag/Sn/Se clusters have been isolated from the reaction of $[Ag(PPh_3)_3Cl]$ with $Se(SiMe_3)_2$ and either $[(SnR^1)_3Se_4Cl]$ or $[(SnR^1)_4Se_6]$ (Figure 10.42) [193]. When this reaction is conducted in the presence of light, the Ag/Sn/Se cluster $[Ag_6(\mu_6\text{-}Se)(Ag_8Se_{12})\{(R^1Sn)_2Se_2\}_6]$ is isolated. In the absence of light, this reactivity is different, instead yielding the mixed-valence cluster $[(AgPPh_3)_2(Sn^{II}Cl)_2\{(R^1Sn^{IV})_2Se_2\}_2]$. In the presence of hydrazine hydrate, the keto functionalities in $CMe_2CH_2C(Me)O$ react to form the corresponding hydrazones. This change in the coordination environment about the cluster causes reorganization of the cluster core and the formation of a new cluster $[Ag_7(\mu_7\text{-}Se)(Ag_7Se_{12})\{(R^6Sn)_2Se_2\}_6]$ with the C,N-coordinating hydrazone R^6 =

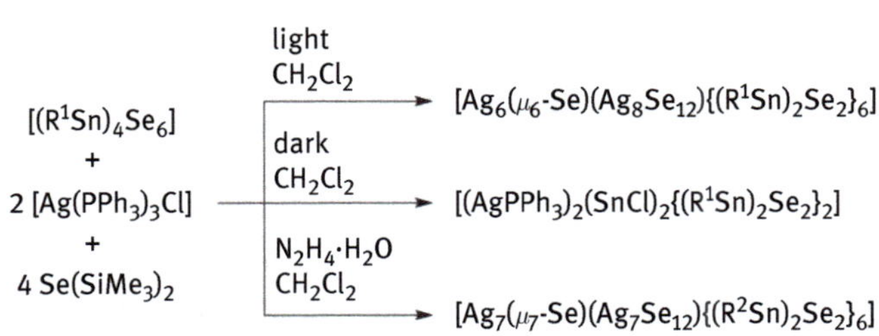

Figure 10.42: Synthesis of silver(I) selenide clusters stabilized with organotin(IV) selenide moieties from organotin(IV) selenides under different experimental conditions [193].

$CMe_2CH_2C(Me)NNH_2$. The preparation of such clusters is proof of principle for the use of (RSn)/Se stabilizing units in metal selenide clusters instead of the currently ubiquitous tertiary phosphines.

$[Ag_6(\mu_6\text{-Se})(Ag_8Se_{12})\{(R^1Sn)_2Se_2\}_6]$ is composed of a binary $[Ag_{14}Se_{13}]$ core stabilized by six $[(SnR^1)_2Se_2]$ surface moieties. $[Ag_{14}Se_{13}]$ cores are known from the chemistry of silver selenide/selenolate clusters such as $[(Ph_3PAg)_8Ag_6(\mu_6\text{-Se})(SePh)_{12}]$ (see Section 10.2.3.1). A central μ_6-Se ligand is centred in a nonbonded Ag_6 octahedron, surrounded by an Ag_8 cube and finally an Se_{12} icosahedron (Figure 10.43). The μ_3-Se atoms bridging the edges of the cube are bound to one tin atom from $[(R^1Sn)_2Se_2]$. Each Ag atom of the octahedron is additionally bound by one selenium atom within $[Sn_2Se_2]$. The structure of $[(AgPPh_3)_2(Sn^{II}Cl)_2\{(R^1Sn^{IV}Se)_2Se_2\}_2]$ has been previously observed in the sulfur homologues $[(MPPh_3)_2(Sn^{II}Cl)S_4\{(R^1Sn^{IV})_2S_2\}_2]$ (M = Cu, Ag) and consists of two $[R^1Sn(\mu\text{-E})(\mu_3\text{-E})]_2$ units linked through two $[(MPPh_3)(SnCl)]$ [194]. The core of $[Ag_7(\mu_7\text{-Se})(Ag_7Se_{12})\{(R^6Sn)_2Se_2\}_6]$ is related to that of $[Ag_6(\mu_6\text{-Se})(Ag_8Se_{12})\{(R^1Sn)_2Se_2\}_6]$ by structural isomerism induced from changes in the ligand sphere. A central μ_7-bridging selenide is surrounded by a pentagonal star-shaped unit of ten silver atoms capped by $[\{(R^6Sn)Se\}_2Se_2]$ units.

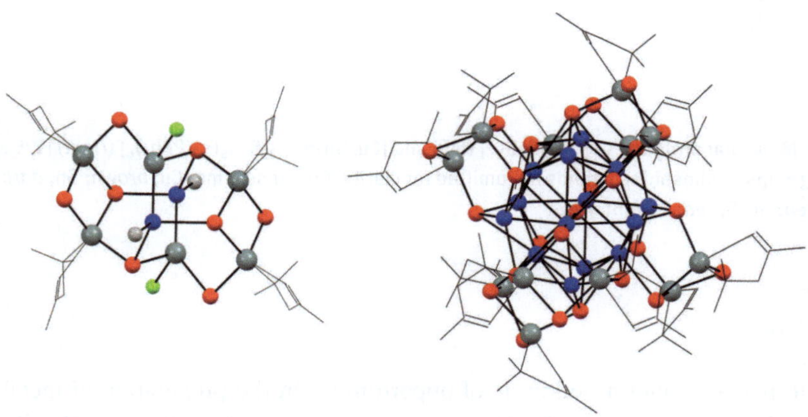

Figure 10.43: Molecular structure of $[(AgPPh_3)_2(Sn^{II}Cl)_2\{(R^1Sn^{IV}Se)_2Se_2\}_2]$ (*left*) and $[Ag_6(\mu_6\text{-Se})(Ag_8Se_{12})\{(R^1Sn)_2Se_2\}_6]$ (*right*) [193]. The organic groups of phosphine ligands are omitted for clarity. Colour scheme: Ag, *blue*; Sn, *dark grey*; Se, *red*; P, *light grey*; Cl, *green*.

Ferrocenyl functionalized Sn/Se and Sn/Te clusters have been prepared from $FcSnCl_3$ and K_2E [195]. When E = Se, an adamantane-type $[Sn_4Se_6]$ unit is stabilized by four Fc moieties at Sn. Conversely, for E = Te, a five-membered $[Sn_3Te_2]$ ring with a Sn–Sn single bond is produced and stabilized by two Fc units per tin atom. The reactivity of the clusters differs from that of the related ferrocenylchalcogenolate precursors (Section 10.2.2.5) because they act as sources of "SnFc" and E^{2-} rather than FcE^-. This was illustrated in the synthesis of a ferrocenyl-passivated quaternary chalcogenide $[Cu_6Sn_2Se_xS_{6-x}Fc_2(PPh_3)_6]$ (x = 3.25) from $[(FcSn)_4Se_6]$, $Na_2S{\cdot}9H_2O$, and $[CuCl(PPh_3)_3]$. The core structure consists of the well-known $[(R_3PM)_6E_6]$ subunit [68, 196, 197], capped by two [FcSn] units (Figure 10.44). An analogy between this chemistry may be made to the use of functionalized silylchalcogenolate reagents for cluster formation (Sections 10.2.2.5 and 10.2.3.4).

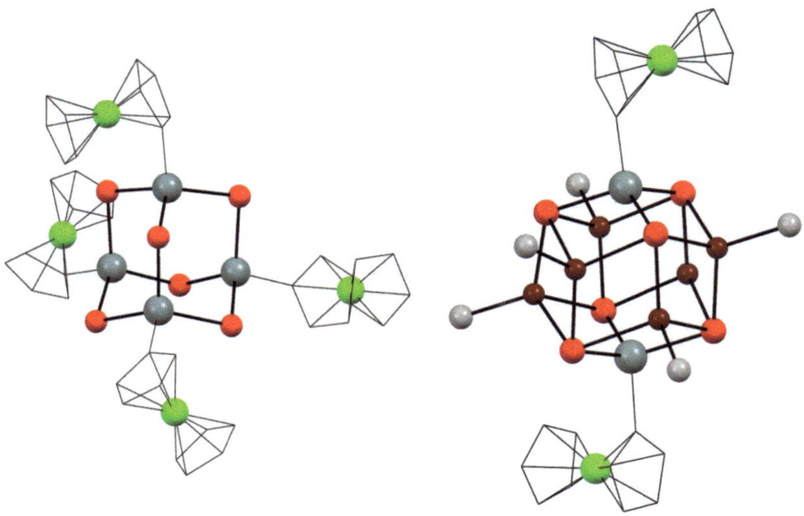

Figure 10.44: Molecular structure of $[(FcSn)_4Se_6]$ (*left*) and $[Cu_6Sn_2Se_{3.25}S_{2.75}Fc_2(PPh_3)_6]$ (*right*) [195]. The organic groups of phosphine ligands are omitted for clarity. Colour scheme: Cu, *brown*; Sn, *dark grey*; Fe, *green*; Se/S, *red*; P, *light grey*.

10.4 Summary

Group 14–16 reagents offer a multitude of opportunities in the preparation of metal chalcogen polynuclear compounds. Their reactive nature enables the use of mild reaction conditions and results in selective formation of crystalline products. In many cases, reaction workup is facile or even unnecessary. A wide range of binary and multinary chalcogen compounds of the d-block and main group metals with highly tunable chemical and physical properties have been prepared from 14–16 precursors. Metal chalcogen materials with advanced properties have been obtained by employing 14–16 reagents derivatized with organic functional groups. In such syntheses, the intrinsic semiconducting nature of the metal chalcogenide is integrated as a platform to an active organic surface. Alternatively, the metal chalcogenide phase may be incorporated into a porous solid through the self-assembly of 14–16 reagents in solution. The use of these reagents provides a promising avenue for continued discovery and development of metal chalcogenide compounds.

Notes

JFC thanks the Natural Sciences and Engineering Research Council (NSERC) of Canada for its continued support of his research program. AMP is most grateful to NSERC for a Graduate Scholarship.

References

[1] Dance, IG, and K Fisher. Metal chalcogenide cluster chemistry. Prog Inorg Chem. 1994;41:637–803.

[2] Dehnen, S, A Eichhöfer, and D Fenske. Chalcogen-bridged copper clusters. Eur J Inorg Chem. 2002;2002:279–317.

[3] Corrigan, JF, O Fuhr, and D Fenske. Metal chalcogenide clusters on the border between molecules and materials. Adv Mater. 2009;21:1867–71.

[4] Fuhr, O, S Dehnen, and D Fenske. Chalcogenide clusters of copper and silver from silylated chalcogenide sources. Chem Soc Rev. 2013;42:1871–906.

[5] Levchenko, TI, Y Huang, and JF Corrigan. Large metal chalcogenide clusters and their ordered superstructures via solvothermal and ionothermal syntheses. In S. Dehnen, ed. Clusters – Contemporary insights in structure and bonding. Structure and bonding, vol. 174. Basel Switzerland: Springer 2017.

[6] Xie, Y-P, J-L Jin, G-X Duan, X Lu, and TCW Mak. High-nuclearity silver(I) chalcogenide clusters: a novel class of supramolecular assembly. Coord Chem Rev. 2017;331:54–72.

[7] Kanatzidis, MG, and S-P Huang. Coordination chemistry of heavy polychalcogenide ligands. Coord Chem Rev. 1994;130:509–621.

[8] Kanatzidis, MG, and AC Sutorik. The application of polychalcogenide salts to the exploratory synthesis of solid state multinary chalcogenides at intermediate temperatures. In K.D. Karlin, ed. Progress in inorganic chemistry. Vol. 43. New York: John Wiley & Sons, Inc.; 1995. p. 151–265.

[9] Kanatzidis, MG. Discovery – Synthesis, design, and prediction of chalcogenide phases. Inorg Chem. 2017;56:3158–73.

[10] Herskovitz, T, BA Averill, RH Holm, JA Ibers, WD Phillips, and JF Weiher. Structure and properties of a synthetic analogue of bacterial iron-sulfur proteins. Proc Nat Acad Sci USA. 1972;69:2437–41.

[11] Howard, JB, and DC Rees. Structural basis of biological nitrogen fixation. Chem Rev. 1996;96:2965–82.

[12] J-M, M, and B Lamotte. Iron–Sulfur clusters and their electronic and magnetic properties. Coord Chem Rev. 1998;178-180:1573–614.

[13] Ohki, Y, Y Sunada, M Honda, M Katada, and K Tatsumi. Synthesis of the P-cluster inorganic core of nitrogenases. J Am Chem Soc. 2003;125:4052–3.

[14] Riaz, U, O Curnow, and MD Curtis. Desulfurization of thiophene and thiophenol by a sulfido-cobalt-molybdenum cluster: toward a homogeneous hydrodesulfurization catalyst. J Am Chem Soc. 1991;113:1416–7.

[15] Kuwata, S, and M Hidai. Hydrosulfido complexes of transition metals. Coord Chem Rev. 2001;213:211–305.

[16] Nirmal, M, and L Brus. Luminescence photophysics in semiconductor nanocrystals. Acc Chem Res. 1999;32:407–14.

[17] Owen, J, and L Brus. Chemical synthesis and luminescence applications of colloidal semiconductor quantum dots. J Am Chem Soc. 2017;139:10939–43.

[18] Lokhande AC, Chalapathy RBV, He M, Jo E, Gang M, Pawar SA, et al. Development of Cu_2SnS_3 (CTS) thin film solar cells by physical techniques: a status review. Sol Energy Mater Sol Cells. 2016;153:84–107.

[19] Hou, H-W, X X-Q, and S Shi. Mo(W,V)-Cu(Ag)-S(Se) cluster compounds. Coord Chem Rev. 1996;153:25–56.

[20] Zhang, Q-F, L W-H, and X Xin. Heteroselenometallic cluster compounds with tetraselenometalates. Coord Chem Rev. 2002;224:35–49.

[21] Chung, I, and MG Kanatzidis. Metal chalcogenides: a rich source of nonlinear optical materials. Chem Mater. 2014;26:849–69.

[22] Hsu KF, Loo S, Guo F, Chen W, Dyck JS, Uher C, et al. Cubic $AgPb_mSbTe_{2+m}$: bulk thermoelectric materials with high figure of merit. Science. 2004;303:818–21.

[23] Sootsman, JR, DY Chung, and MG Kanatzidis. New and old concepts in thermoelectric materials. Angew Chem Int Ed. 2009;48:8616–39.

[24] Kanatzidis, MG. Nanostructured thermoelectrics: the new paradigm? Chem Mater. 2010;22:648–59.

[25] Schulz, S. Covalently bonded compounds of heavy group 15/16 elements - synthesis, structure and potential application in material sciences. Coord Chem Rev. 2015;297-298:49–76.

[26] Beaulac, R, PI Archer, ST Ochsenbein, and DR Gamelin. Mn^{2+}-doped CdSe quantum dots: new inorganic materials for spin-electronics and spin-photonics. Adv Funct Mater. 2008;18:3873–91.

[27] Mulrooney, RC, N Singh, N Kaur, and JF Callan. An “off-on” sensor for fluoride using luminescent CdSe/ZnS quantum dots. Chem Commun. 2009;686–8.

[28] Bowes, CL, and GA Ozin. Self-assembling frameworks: beyond microporous oxides. Adv Mater. 1996;8:13–28.

[29] Bu, X, N Zheng, and P Feng. Tetrahedral chalcogenide clusters and open frameworks. Chem – Eur J. 2004;10:3356–62.

[30] Feng, P, X Bu, and N Zheng. The interface chemistry between chalcogenide clusters and open framework chalcogenides. Acc Chem Res. 2005;38:293–303.

[31] Kanatzidis, MG. Beyond silica: nonoxidic mesostructured materials. Adv Mater. 2007;19:1165–81.

[32] DeGroot, MW, and JF Corrigan. High nuclearity clusters: metal-chalcogenide polynuclear complexes. In M. Fujita, A. Powell and C.A. Creutz, ed. Comprehensive coordination chemistry II. Vol. 7. Oxford, UK: Elsevier; 2004. p. 57–123.

[33] DeGroot, MW, and JF Corrigan. Metal - chalcogenolate complexes with silyl functionalities: synthesis and reaction chemistry. Z Anorg Allg Chem. 2006;632:19–29.

[34] MacDonald, DG, and JF Corrigan. Metal chalcogenide nanoclusters with ‘tailored’ surfaces via ‘designer’ silylated chalcogen reagents. Phil Trans R Soc A. 2010;368:1455–72.

[35] Sheldrick, WS, and M Wachhold. Chalcogenidometalates of the heavier Group 14 and 15 elements. Coord Chem Rev. 1998;176:211–322.

[36] Dehnen, S, and M Melullis. A coordination chemistry approach towards ternary M/14/16 anions. Coord Chem Rev. 2007;251:1259–80.

[37] Heine, J, and S Dehnen. From simple chalcogenidotetrelate precursors to complex structures and functional compounds. Z Anorg Allg Chem. 2012;638:2425–40.

[38] Santner, S, J Heine, and S Dehnen. Synthesis of crystalline chalcogenides in ionic liquids. Angew Chem Int Ed. 2016;55:876–93.

[39] Abel, EW, and DB Brady. The preparation and properties of some alkylthio-compounds of tin. J Chem Soc. 1965;1192–7.

[40] Schumann, H, and M Schmidt. New products of reaction of organometallic compounds with sulfur, selenium, tellurium, and phosphorous. Angew Chem Int Ed Engl. 1965;4:1007–13.

[41] Abel, EW, DB Brady, and BC Crosse. The cleavage of organotin-sulphur compounds by halides. J Organomet Chem. 1966;5:260–2.

[42] Woodward, P, LF Dahl, EW Abel, and BC Crosse. A new type of cyclic transition metal complex, $[Ni(SC_2H_5)_2]_6$. J Am Chem Soc. 1965;87:5251–3.

[43] Do, Y, ED Simhon, and RH Holm. Improved syntheses of $[Fe_2S_2Cl_4]^{2-}$ and $[Fe_2OCl_6]^{2-}$ and oxo/sulfido ligand substitution by use of silylsulfide reagents. Inorg Chem. 1983;22:3809–12.

[44] Fenske, D, J Hachgenei, and J Ohmer. Novel cobalt- and nickel-clusters with S and PPh_3 as ligands; crystal structures of $[Co_7S_6(PPh_3)_5Cl_2]$, $[Co_6S_8(PPh_3)_6]^+[CoCl_3(THF)]^-$, $[Ni_8S_6Cl_2(PPh_3)_6]$, and $[Ni_8S_5(PPh_3)_7]$. Angew Chem Int Ed Engl. 1985;24:706–9.

[45] Fenske, D, J Ohmer, and J Hachgenei. New Co and Ni clusters with Se and PPh_3 as ligands: $[Co_4(\mu_3\text{-}Se)_4(PPh_3)_4]$, $[Co_6(\mu_3\text{-}Se)_8(PPh_3)_6]$, $[Co_9(\mu_4\text{-}Se)_3(\mu_3\text{-}Se)_8(PPh_3)_6]$, and $[Ni_{34}(\mu_5\text{-}Se)_2(\mu_4\text{-}Se)_{20}(PPh_3)_{10}]$. Angew Chem Int Ed Engl. 1985;24:993–5.

[46] Christou, V, and J Arnold. Synthesis of reactive homoleptic tellurolates of zirconium and hafnium and their conversion to terminal tellurides: a model for the first step in a molecule-to-solid transformation. J Am Chem Soc. 1992;114:6240–2.

[47] Christou, V, and J Arnold. Formation of monomeric terminal chalcogenides by template-induced disilylchalcogenide elimination; the crystal structures of $[ETa\{(Me_3SiNCH_2CH_2)_3N\}]$ (E = Se, Te). Angew Chem Int Ed Engl. 1993;32:1450–2.

[48] Liang, H-C, and PA Shapley. Synthesis and reactivity of $[PPh_4][Ru(N)Me_3(SSiMe_3)]$, a ruthenium (VI) trimethylsilanethiolate complex. Organometallics. 1996;15:1331–3.

[49] Tran, DTT, NJ Taylor, and JF Corrigan. Copper chalcogenolate complexes as precursors to ternary nanoclusters: synthesis and characterization of $[Hg_{15}Cu_{20}S_{25}(^nPr_3P)_{18}]$. Angew Chem Int Ed. 2000;39:935–7.

[50] Tran, DTT, LMC Beltran, CM Kowalchuk, NR Trefiak, NJ Taylor, and JF Corrigan. Ternary nanoclusters of CuHgS, CuHgSe, and CuInS. Inorg Chem. 2002;41:5693–8.

[51] Komuro, T, H Kawaguchi, and K Tatsumi. Synthesis and reactions of triphenylsilanethiolato complexes of manganese(II), iron(II), cobalt(II), and nickel(II). Inorg Chem. 2002;41:5083–90.

[52] Komuro, T, T Matsuo, H Kawaguchi, and K Tatsumi. Synthesis and structural characterization of silanethiolato complexes having tert-butyldimethylsilyl and trimethylsilyl groups. Dalton Trans. 2004;1618–25.

[53] Steigerwald ML, Alivisatos AP, Gibson JM, Harris TD, Kortan R, Muller AJ, et al. Surface derivatization and isolation of semiconductor cluster molecules. J Am Chem Soc. 1988;110:3046–50.

[54] García-Rodríguez, R, MP Hendricks, BM Cossairt, H Liu, and JS Owen. Conversion reactions of cadmium chalcogenide nanocrystal precursors. Chem Mater. 2013;25:1233–49.

[55] Eichhöfer, A, G Buth, S Lebedkin, M Kühn, and F Weigend. Luminescence in phosphine-stabilized copper chalcogenide cluster molecules - A comparative study. Inorg Chem. 2015;54:9413–22.

[56] Sevillano P, Fuhr O, Hampe O, Lebedkin S, Neiss C, Ahlrichs R, et al. Synthesis, characterization and quantum mechanical calculations of $[Au_{18}Se_8(dppthph)_6]Cl_2$. Eur J Inorg Chem. 2007;2007:5163–7.

[57] Sevillano, P, O Fuhr, and D Fenske. Synthesis and structure of $[Au_{10}Se_5(dppa)_4\{Co_2(CO)_5\}_4]$. Z Anorg Allg Chem. 2007;633:1783–6.

[58] Yu, W, O Fuhr, and D Fenske. Derivatives of bis(diphenylphosphino)maleic anhydride as ligands in polynuclear gold(I) complexes. J Clust Sci. 2012;23:753–66.

[59] Kumar, GA, RE Riman, and JG Brennan. NIR emission from molecules and clusters with lanthanide-chalcogen bonds. Coord Chem Rev. 2014;273-274:111–24.

[60] Bechlars, B, R Feuerhake, and D Fenske. Syntheses and crystal structures of novel chalcogenido-bridged niobium copper clusters. Z Anorg Allg Chem. 2007;633:2603–13.

[61] Bechlars, B, I Issac, R Feuerhake, R Clérac, O Fuhr, and D Fenske. Syntheses, structures and magnetic properties of new chalcogen-bridged heterodimetallic cluster compounds with heterocubane structure. Eur J Inorg Chem. 2008;2008:1632–44.

[62] Sola, J, Y Do, JM Berg, and RH Holm. Soluble sulfides of niobium(V) and tantalum(V): synthesis, structures, and properties of the fivefold symmetric cages $[M_6S_{17}]^{4-}$. Inorg Chem. 1985;24:1706–13.

[63] Lorenz, A, and D Fenske. Chalcogenoniobates as reagents for the synthesis of new heterobimetallic niobium coinage metal chalcogenide clusters. Z Anorg Allg Chem. 2001;627:2232–48.

[64] Robles, V, JF Trigo, C Guillén, and J Herrero. Copper tin sulfide (CTS) absorber thin films obtained by co-evaporation: influence of the ratio Cu/Sn. J Alloys Cmpd. 2015;642:40–4.

[65] Eichhöfer A, Jiang J, Lebedkin S, Fenske D, McDonald DG, Corrigan JF, et al. A ternary Cu–Sn–S cluster complex - $(NBu_4)[Cu_{19}S_{28}(SnPh)_{12}(PEt_2Ph)_3]$. Dalton Trans. 2012;41:3321–7.

[66] Kühn, M, S Lebedkin, F Weigend, and A Eichhöfer. Optical properties of trinuclear metal chalcogenolate complexes – Room temperature NIR fluorescence in $[Cu_2Ti(SPh)_6(PPh_3)_2]$. Dalton Trans. 2017;46:1502–9.

[67] Eichhöfer, A, M Kühn, S Lebedkin, M Kehry, MM Kappes, and F Weigend. Synthesis and optical properties of $[Cu_6E_6(SnPh)_2(PPh_2Et)_6]$ (E = S, Se, Te) cluster molecules. Inorg Chem. 2017;56:9330–6.

[68] Hauser, R, and K Merzweiler. $[(PhSnS_3)_2(CuPPhMe_2)_6]$, a hexanuclear copper(I) complex with $PhSnS_3$ ligands. Z Anorg Allg Chem. 2002;628:905–6.

[69] Turner, EA, Y Huang, and JF Corrigan. Me_3Si-SeS-$SiMe_3$: a reagent for the synthesis of the mixed sulfo-selenide cluster $[Cu_{84}Se_{42-x}S_x(PEt_2Ph)_{24}]$. Z Anorg Allg Chem. 2007;633:2135–7.

[70] Tolman, CA. Steric effects of phosphorus ligands in organometallic chemistry and homogeneous catalysis. Chem Rev. 1977;77:313–48.

[71] Choi B, Capozzi B, Ahn S, Turkiewicz A, Lovat G, Nuckolls C, et al. Solvent-dependent conductance decay constants in single cluster junctions. Chem Sci. 2016;7:2701–5.

[72] Lovat, G, B Choi, DW Paley, ML Steigerwald, L Venkataraman, and X Roy. Room-temperature current blockade in atomically defined single-cluster junctions. Nat Nanotechnol. 2017;12:1050–4.

[73] Khadka, CB, B Khalili Najafabadi, M Hesari, MS Workentin, and JF Corrigan. Copper chalcogenide clusters stabilized with ferrocene-based diphosphine ligands. Inorg Chem. 2013;52:6798–805.

[74] Dance, IG. The structural chemistry of metal thiolate complexes. Polyhedron. 1986;5:1037–104.

[75] Concepción Gimeno, M. Thiolates, selenolates, and tellurolates. In F.A. Devillanova, ed. Handbook of chalcogen chemistry: new perspectives in sulfur, selenium and tellurium. Cambridge, UK: Royal Society of Chemistry; 2007. p. 33–80.

[76] Gysling, HJ. The ligand chemistry of tellurium. Coord Chem Rev. 1982;42:133–244.

[77] Arnold, J. The chemistry of metal complexes with selenolate and tellurolate ligands. In K.D. Karlin, ed. Progress in inorganic chemistry. Vol. 43. New York: John Wiley & Sons Inc; 1995. p. 353–417.

[78] Veselska, O, and A Demessence. d^{10} coinage metal organic chalcogenolates: from oligomers to coordination polymers. Coord Chem Rev. 2018;325:240–70.

[79] Eichhöfer A, Jiang J-J, Sommer H, Weigend F, Fuhr O, Fenske D, et al. 1-D-tin(II) phenylchalcogenolato complexes $Sn(EPh)_2$ (E = S, Se, Te) - Synthesis, structures, quantum chemical studies and thermal behaviour. Eur J Inorg Chem. 2010;2010:410–8.

[80] Eichhöfer, A, PT Wood, R Viswanath, and RA Mole. Synthesis, structure and magnetic behaviour of manganese(II) selenolate complexes $Mn(SePh)_2]$, $[Mn(SePh)_2(bipy)_2]$ and $[Mn(SePh)_2(phen)_2]$ (bipy = bipyridyl, phen = phenanthroline). Eur J Inorg Chem. 2007;2007:4794–9.

[81] Sommer, H, A Eichhöfer, and D Fenske. Synthesis and crystal structures of bismuth chalcogenolato compounds $Bi(SC_6H_5)_3$, $Bi(SeC_6H_5)_3$, and $Bi(S\text{-}4\text{-}CH_3C_6H_4)_3$. Z Anorg Allg Chem. 2008;634:436–40.

[82] Eichhöfer, A, G Buth, F Dolci, K Fink, RA Mole, and PT Wood. Homoleptic 1-D iron selenolate complexes-synthesis, structure, magnetic and thermal behaviour of $Fe(SeR)_2]$ (R=Ph, Mes). Dalton Trans. 2011;40:7022–32.

[83] Kluge, O, K Grummt, R Biedermann, and H Krautscheid. Trialkylphosphine-stabilized copper(I) phenylchalcogenolate complexes - Crystal structures and copper–Chalcogenolate bonding. Inorg Chem. 2011;50:4742–52.

[84] Fu, M-L, D Fenske, B Weinert, and O Fuhr. One-dimensional coordination polymers containing polynuclear (selenolato)copper complexes linked by bipyridine ligands. Eur J Inorg Chem. 2010;2010:1098–102.

[85] Humenny, WJ, S Mitzinger, CB Khadka, B Khalili Najafabadi, I Vieira, and JF Corrigan. N-heterocyclic carbene stabilized copper- and silver-phenylchalcogenolate ring complexes. Dalton Trans. 2012;41:4413–22.

[86] Choi, B, DW Paley, T Siegrist, ML Steigerwald, and X Roy. Ligand control of manganese telluride molecular cluster core nuclearity. Inorg Chem. 2015;54:8348–55.

[87] Malik, MA, M Afzaal, and P O'Brien. Precursor chemistry for main group elements in semiconducting materials. Chem Rev. 2010;110:4417–46.

[88] Bendt, G, A Weber, S Heimann, W Assenmacher, O Prymak, and S Schulz. Wet-chemical synthesis of different bismuth telluride nanoparticles using metal organic precursors – Single source vs. dual source approach. Dalton Trans. 2015;44: 14272–80. and references therein.

[89] Knapp, CE, and CJ Carmalt. Solution based CVD of main group materials. Chem Soc Rev. 2016;45:1036–64.

[90] Schulz, S, S Heimann, J Friedrich, M Engenhorst, G Schierning, and W Assenmacher. Synthesis of hexagonal Sb_2Te_3 nanoplates by thermal decomposition of the single-source precursor $(Et_2Sb)_2Te$. Chem Mater. 2012;24:2228–34.

[91] Bendt, G, S Schulz, S Zastrow, and K Nielsch. Single-source precursor-based deposition of Sb_2Te_3 films by MOCVD. Chem Vap Deposition. 2013;19:235–41.

[92] Benjamin SL, De Groot CH, Gurnani C, Hector AL, Huang R, Koukharenko E, et al. Controlling the nanostructure of bismuth telluride by selective chemical vapour deposition from a single source precursor. J Mater Chem A. 2014;2:4865–69.

[93] Traut, S, AP Hähnel, and C Von Hänisch. Dichloro organosilicon bismuthanes as precursors for rare compounds with a bismuth–Pnictogen or bismuth–Tellurium bond. Dalton Trans. 2011;40:1365–71.

[94] Heimann, S, D Bläser, C Wölper, and S Schulz. Solid-state structures of bis(diethylbismuthanyl) sulfane, -selenane, and -tellurane. Organometallics. 2014;33:2295–300.

[95] Heimann S, Kuczkowski A, Bläser D, Wölper C, Haack R, Jansen G, et al. Syntheses and solid-state structures of $Et_2SbTeEt$ and $Et_2BiTeEt$. Eur J Inorg Chem. 2014;2014:4858–64.

[96] Traut, S, S Stahl, A Eichhöfer, and C Von Hänisch. Synthesis, structure and thermal decomposition of the bismuth silyltellurolate $Bi(TeSitBu_2Ph)_3$. Z Anorg Allg Chem. 2015;641:1200–2.

[97] Traut, S, C Von Hänisch, AP Hähnel, and S Stahl. Cyclic and polycyclic tellurium–Tin and tellurium–Lead compounds – Synthesis, structures and thermal decomposition. Chem Commun. 2012;48:6984–6.

[98] Seligson, AL, and J Arnold. Synthesis, structure, and reactivity of homoleptic tin(II) and lead(II) chalcogenolates and their conversion to metal chalcogenides. X-ray crystal structures of $\{Sn[TeSi(SiMe_3)_3]_2\}_2$ and $(PMe_3)Sn[TeSi(SiMe_3)_3]_2$. J Am Chem Soc. 1993;115:8214–20.

[99] Fuhrmann, D, T Severin, and H Krautscheid. Synthesis and crystal structures of copper zinc phenylthiolate and the first copper zinc selenolate and tellurolate complexes. Z Anorg Allg Chem. 2017;643:932–7.
[100] Sommer, H, A Eichhöfer, N Drebov, R Ahlrichs, and D Fenske. Preparation, geometric and electronic structures of $[Bi_2Cu_4(SPh)_8(PPh_3)_4]$ with a Bi_2 dumbbell, $[Bi_4Ag_3(SePh)_6Cl_3(PPh_3)_3]_2$ and $[Bi_4Ag_3(SePh)_6X_3(PPh^iPr_2)_3]_2$ (X = Cl, Br) with a Bi_4 unit. Eur J Inorg Chem. 2008;2008:5138–45.
[101] Hu, B, C-Y Su, D Fenske, and O Fuhr. Synthesis, characterization and optical properties of a series of binuclear copper chalcogenolato complexes. Inorg Chim Acta. 2014;419:118–23.
[102] Azizpoor Fard, M, B Khalili Najafabadi, M Hesari, MS Workentin, and JF Corrigan. New polydentate trimethylsilyl chalcogenide reagents for the assembly of polyferrocenyl architectures. Chem – Eur J. 2014;20:7037–47.
[103] Azizpoor Fard, M, MJ Willans, B Khalili Najafabadi, TI Levchenko, and JF Corrigan. Polydentate chalcogen reagents for the facile preparation of Pd_2 and Pd_4 complexes. Dalton Trans. 2015;44:8267–77.
[104] Labande, A, J Ruiz, and D Astruc. Supramolecular gold nanoparticles for the redox recognition of oxoanions: syntheses, titrations, stereoelectronic effects, and selectivity. J Am Chem Soc. 2002;124:1782–9.
[105] Stiles, RL, R Balasubramanian, SW Feldberg, and RW Murray. Anion-induced adsorption of ferrocenated nanoparticles. J Am Chem Soc. 2008;130:1856–65.
[106] Uosaki, K, T Kondo, M Okamura, and W Song. Electron and ion transfer through multilayers of gold nanoclusters covered by self-assembled monolayers of alkylthiols with various functional groups. Faraday Discuss. 2002;121:373–89.
[107] Nitschke, C, AI Wallbank, D Fenske, and JF Corrigan. Facile synthesis of high nuclearity silver-ferrocenyldiselenolate clusters. J Clust Sci. 2007;18:131–40.
[108] The only exception to this is for platinum(II) and palladium(II) complexes$[(fcSe_2)M(P^nBu_3)]_2$ (M = Pd, Pt), and $[(fcSe_2)Pd(P^nBu_2)_2]$ from Brown MJ, Corrigan JF. Synthesis, characterization and electrochemistry of ferrocenylselenolate bridged palladium(II) and platinum(II) complexes. J Organomet Chem. 2004;689:2872–9.
[109] Ahmar, S, C Nitschke, N Vijayaratnam, DG MacDonald, D Fenske, and JF Corrigan. A ferrocenylmethylselenolate complex of Ag(I): preparation of the polyferrocenyl cluster $[Ag_8(SeCH_2Fc)_8(PPh_3)_4]$ from the new silylated reagent $FcCH_2SeSiMe_3$. New J Chem. 2011;35:2013–7.
[110] Ahmar, S, DG MacDonald, N Vijayaratnam, TL Battista, MS Workentin, and JF Corrigan. A nanoscopic 3D polyferrocenyl assembly: the triacontakaihexa(ferrocenylmethylthiolate) $[Ag_{48}(\mu_4\text{-}S)_6(\mu_{2/3}\text{-}SCH_2Fc)_{36}]$. Angew Chem Int Ed. 2010;49:4422–4.
[111] MacDonald, DG, and JF Corrigan. New reagents for the synthesis of a series of ferrocenoyl functionalized copper and silver chalcogenolate complexes. Dalton Trans. 2008;5048–53.
[112] MacDonald, DG, A Eichhöfer, CF Campana, and JF Corrigan. Ferrocene-based trimethylsilyl chalcogenide reagents for the assembly of functionalized metal-chalcogen architectures. Chem – Eur J. 2011;17:5890–902.
[113] Anson CE, Eichhöfer A, Issac I, Fenske D, Fuhr O, Sevillano P, et al. Synthesis and crystal structures of the ligand-stabilized silver chalcogenide clusters $[AgSe_{77}(dppxy)_{18}]$, $[Ag_{320}(StBu)_{60}S_{130}(dppp)_{12}]$, $[Ag_{352}S_{128}(StC_5H_{11})_{96}]$, and $[Ag_{490}S_{188}(StC_5H_{11})_{114}]$. Angew Chem Int Ed. 2008;47:1326–31.
[114] Duan, T, X-Z Zhang, and Q-F Zhang. Synthesis and crystal structure of a polynuclear copper-selenide cluster $[Cu^I_{36}(Cu^{II}Cl)_2Se_{13}(SePh)_{12}(dppe)_6]\cdot 3EtOH$. Z Naturforsch B. 2008;63:941–4.
[115] Nayek, HP, W Massa, and S Dehnen. Presence or absence of a central Se atom in silver selenide/selenolate clusters with halite topology: syntheses and properties of $[(Ph_3PAg)_8Ag_6(\mu_6\text{-}Se)_{1-x/2}(SePh)]^{x+}$ (x = 0, 1). Inorg Chem. 2010;49:144–9.

[116] Corrigan, JF, and D Fenske. New copper telluride clusters by light-induced tellurolate-telluride conversions. Angew Chem Int Ed Engl. 1997;36:1981–3.

[117] Levchenko TI, Kübel C, Khalili Najafabadi B, Boyle PD, Cadogan C, Goncharova LV, et al. Luminescent CdSe superstructures: a nanocluster superlattice and a nanoporous crystal. J Am Chem Soc. 2017;139:1129–44.

[118] Fu, M-L, I Issac, D Fenske, and O Fuhr. Metal-rich copper chalcogenide clusters at the border between molecule and bulk phase: the structures of $[Cu_{93}Se_{42}(SeC_6H_4SMe)_9(PPh_3)_{18}]$, $[Cu_{96}Se_{45}(SeC_6H_4SMe)_6(PPh_3)_{18}]$, and $[Cu_{136}S_{56}(SCH_2C_4H_3O)_{24}(dpppt)_{10}]$. Angew Chem Int Ed. 2010;49:6899–903.

[119] Fenske, D, N Zhu, and T Langetepe. Synthesis and structure of new Ag–Se clusters: $[Ag_{30}Se_8(SetBu)_{14}(PnPr_3)_8]$, $[Ag_{90}Se_{38}(SetBu)_{14}(PEt_3)_{22}]$, $[Ag_{114}Se_{34}(SenBu)_{46}(PtBu_3)_{14}]$, $[Ag_{112}Se_{32}(SenBu)_{48}(PtBu_3)_{12}]$, and $[Ag_{172}Se_{40}(SenBu)_{92}(dppp)_4]$. Angew Chem Int Ed. 1998;37:2639–44.

[120] Eichhöfer, A, PT Wood, RN Viswanath, and RA Mole. Synthesis, structure and physical properties of the manganese(II) selenide/selenolate cluster complexes $[Mn_{32}Se_{14}(SePh)_{36}(P^nPr_3)_4]$ and $[Na(benzene\text{-}15\text{-}crown\text{-}5)(C_4H_8O)_2]_2[Mn_8Se(SePh)_{16}]$. Chem Commun. 2008;14:1596–8.

[121] Matheis, K, A Eichhöfer, F Weigend, OT Ehrler, O Hampe, and MM Kappes. Probing the influence of size and composition on the photoelectron spectra of cadmium chalcogenide cluster dianions. J Phys Chem C. 2012;116:13800–9.

[122] Eichhöfer A, Olkowska-Oetzel J, Fenske D, Fink K, Mereacre V, Powell AK, et al. Synthesis and structure of an "iron-doped" copper selenide cluster molecule: $[Cu_{30}Fe_2Se_6(SePh)_{24}(dppm)_4]$. Inorg Chem. 2009;48:8977–84.

[123] Eichhöfer, A, D Fenske, and J Olkowska-Oetzel. Synthesis and structure of $[nPr_3N(CH_2)_6NnPr_3][CuFe_3Br_3(SePh)_6]$, $[Cu_5Fe(SePh)_7(PPh_3)_4]$, and $[Cu_4Fe_3(SePh)_{10}(PPh_3)_4]$. Eur J Inorg Chem. 2007;2007:74–9.

[124] Zhu, N, and D Fenske. Novel Cu–Se clusters with Se–Layer structures: $[Cu_{32}Se_7(SenBu)_{18}(PiPr_3)_6]$, $[Cu_{50}Se_{20}(SetBu)_{10}(PiPr_3)_{10}]$, $[Cu_{73}Se_{35}(SePh)_3(PiPr_3)_{21}]$, $[Cu_{140}Se_{70}(PEt_3)_{34}]$ and $[Cu_{140}Se_{70}(PEt_3)_{36}]$. J Chem Soc, Dalton Trans. 1999;1067–75.

[125] Eichhöfer, A, O Hampe, S Lebedkin, and F Weigend. Bistrimethylsilylamide transition-metal complexes as starting reagents in the synthesis of ternary Cd–Mn–Se cluster complexes. Inorg Chem. 2010;49:7331–9.

[126] MacDonald, DG, C Kübel, and JF Corrigan. Ferrocenyl functionalized silver-chalcogenide nanoclusters. Inorg Chem. 2011;50:3252–61.

[127] Fuhr, O, L Fernandez-Recio, A Castiñeiras, and D Fenske. Synthesis and structure of the clusters $[Cu_{50}Se_{24}(S\text{-}thiaz)_2(dppm)_{10}]$ and $[Cu_{48}Se_{24}(S\text{-}thiazH)_2(dppm)_{10}]$. Z Anorg Allg Chem. 2007;633:700–4.

[128] Fernandez-Recio, L, D Fenske, and O Fuhr. Copper chalcogenide cluster compounds with nitro-functionalized ligand shell. Z Anorg Allg Chem. 2008;634:2853–7.

[129] Langer, R, B Breitung, L Wünsche, D Fenske, and O Fuhr. Functionalised silver chalcogenide clusters. Z Anorg Allg Chem. 2011;637:995–1006.

[130] Langer, R, W Yu, L Wünsche, G Buth, O Fuhr, and D Fenske. Syntheses and structure determination of trimethylsiloxy functionalized copper chalcogenide clusters. Z Anorg Allg Chem. 2011;637:1834–40.

[131] Langer, R, D Fenske, and O Fuhr. $[Cu_{40}Se_{16}(S\text{-}C_6H_4\text{-}CN)_8(dppm)_8]$: a disc-like copper cluster with a nitrile-functionalized ligand shell. Z Naturforsch B. 2013;68:575–80.

[132] Vahrenkamp, H. Metallorganische lewis-basen, II. Organometallische Schwefelkomplexe Des Molybdäns, Mangans, Rheniums Und Eisens. Chem Ber. 1970;103:3580–90.

[133] Vergamini, PJ, H Vahrenkamp, and LF Dahl. Organometallic chalcogen complexes. XXIII. Preparation and structural characterization of a mixed transition metal cluster complex, $[Re_2Mo(\eta^5\text{-}C_5H_5)(CO)_8](S)[SMo(\eta^5\text{-}C_5H_5)(CO)_3]$, containing triply and quadruply bridging sulfur

atoms. New synthetic route to highly clustered metal-sulfur systems. J Am Chem Soc. 1971;93:6326–7.
[134] S-B, Y. The Preparation and Structure of $Mo_2(OAc)_2(SSiMe_3)_2(PEt_3)_2$. Polyhedron. 1992;11:2115–7.
[135] Dehnen, S, and D Fenske. $[Cu_{24}S_{12}(PMeiPr_2)_{12}]$, $[Cu_{28}S_{14}(PtBu_2Me)_{12}]$, $[Cu_{50}S_{25}(PtBu_2Me)_{16}]$, $[Cu_{70}Se_{35}(PtBu_2Me)_{21}]$, $[Cu_{31}Se_{15}(SeSiMe_3)(PtBu_2Me)_{12}]$ and $[Cu_{48}Se_{24}(PMe_2Ph)_{20}]$: new sulfur- and selenium-bridged copper clusters. Chem – Eur J. 1996;2:1407–16.
[136] DeGroot, MW, NJ Taylor, and JF Corrigan. Zinc chalcogenolate complexes as capping agents in the synthesis of ternary II–II′–VI nanoclusters: structure and photophysical properties of $[(N, N'\text{-tmeda})_5Zn_5Cd_{11}Se_{13}(SePh)_6(thf)_2]$. J Am Chem Soc. 2003;125:864–65.
[137] DeGroot, MW, KM Atkins, A Borecki, H Rösner, and JF Corrigan. A molecular precursor approach for the synthesis of composition-controlled $Zn_xCd_{1-x}S$ and $Zn_xCd_{1-x}Se$ nanoparticles. J Mater Chem. 2008;18:1123–30.
[138] Biedermann, R, O Kluge, D Fuhrmann, and H Krautscheid. Synthesis and crystal structures of $[(iPr_3P)_2Cu(\mu\text{-}ESiMe_3)(InMe_3)]$ (E = S, Se): lewis acid-base adducts with chalcogen atoms in planar coordination. Eur J Inorg Chem. 2013;2013:4727–31.
[139] Kluge, O, M Puidokait, R Biedermann, and H Krautscheid. Synthesis and crystal structures of spirocyclic gallium and indium chalcogen heterocycles $[(Me_2Ga)_6S(SSiMe_3)_4]$, $[(Me_2Ga)_6Se(SeSiMe_3)_4]$ and $[(Me_2In)_6S(SSiMe_3)_4]$. Z Anorg Allg Chem. 2007;633:2138–40.
[140] Kluge, O, R Biedermann, J Holldorf, and H Krautscheid. Organo-gallium/indium chalcogenide complexes of copper(I): molecular structures and thermal decomposition to ternary semiconductors. Chem – Eur J. 2014;20:1318–31.
[141] Fuhrmann, D, S Dietrich, and H Krautscheid. Zinc tin chalcogenide complexes and their evaluation as molecular precursors for Cu_2ZnSnS_4 (CZTS) and $Cu_2ZnSnSe_4$ (CZTSe). Inorg Chem. 2017;56:13123–31.
[142] Khadka, CB, DG MacDonald, Y Lan, AK Powell, D Fenske, and JF Corrigan. Trimethylsilylchalcogenolates of Co(II) and Mn(II): from mononuclear coordination complexes to clusters containing $-ESiMe_3$ moieties (E = S, Se). Inorg Chem. 2010;49:7289–97.
[143] Khadka, CB, A Eichhöfer, F Weigend, and JF Corrigan. Zinc chalcogenolate complexes as precursors to ZnE and Mn/ZnE (E = S, Se) clusters. Inorg Chem. 2012;51:2747–56.
[144] Borecki, A, and JF Corrigan. New copper and silver trimethylsilylchalcogenolates. Inorg Chem. 2007;46:2478–84.
[145] Azizpoor Fard, M, F Weigend, and JF Corrigan. Simple but effective: thermally stable $Cu\text{-}ESiMe_3$ via NHC ligation. Chem Commun. 2015;51:8361–4.
[146] Azizpoor Fard, M, TI Levchenko, C Cadogan, WJ Humenny, and JF Corrigan. Stable $-ESiMe_3$ complexes of Cu^I and Ag^I (E=S, Se) with NHCs: synthons in ternary nanocluster assembly. Chem – Eur J. 2016;22:4543–50.
[147] Polgar, AM, F Weigend, A Zhang, MJ Stillman, and JF Corrigan. A N-heterocyclic carbene-stabilized coinage metal-chalcogenide framework with tunable optical properties. J Am Chem Soc. 2017;139:14045–8.
[148] Krebs, B. Thio- and seleno-compounds of main group elements - Novel inorganic oligomers and polymers. Angew Chem Int Ed Engl. 1983;22:113–34.
[149] Kromm, A, and WS Sheldrick. Synthesis and structures of dimanganese(II) complexes with spirotricyclic $[Mn(\mu\text{-}Ge_2Se_7)Mn]$ and $[Mn(\mu\text{-}Sn_2Se_6)Mn]$ cores. Z Anorg Allg Chem. 2008;634:1005–10.
[150] Zhang, G, P Li, and J Ding. Surfactant-thermal syntheses, structures, and magnetic properties of Mn–Ge–Sulfides/selenides. Inorg Chem. 2014;53:10248–56.
[151] Yaghi, OM, Z Sun, DA Richarson, and TL Groy. Directed transformation of molecules to solids: synthesis of a microporous sulfide from molecular germanium sulfide cages. J Am Chem Soc. 1994;116:807–8.

[152] Bag, S, PN Trikalitis, PJ Chupas, GS Armatas, and MG Kanatzidis. Porous semiconducting gels and aerogels from chalcogenide clusters. Science. 2007;317:490–3.

[153] Bag, S, IU Arachchige, and MG Kanatzidis. Aerogels from metal chalcogenides and their emerging unique properties. J Mater Chem. 2008;18:3628–32.

[154] Bag, S, and MG Kanatzidis. Chalcogels: porous metal-chalcogenide networks from main-group metal ions. Effect of Surface Polarizability on Selectivity in Gas Separation. J Am Chem Soc. 2010;132:14951–9.

[155] Ruzin, E, and S Dehnen. Influence of the counterions on the structures of ternary Zn/Sn/Se anions: synthesis and properties of $[Rb_{10}(H_2O)_{14.5}][Zn_4(\mu_4\text{-}Se)_2(SnSe_4)_4]$ and $[Ba_5(H_2O)_{32}][Zn_5Sn(\mu_3\text{-}Se)_4(SnSe_4)_4]$. Z Anorg Allg Chem. 2006;632:749–55.

[156] Melullis, M, R Clérac, and S Dehnen. Ternary Mn/Ge/Se anions from reactions of $[Ba_2(H_2O)_9][GeSe_4]$: synthesis and characterization of compounds containing discrete or polymeric $[Mn_6Ge_4Se_{17}]^{6-}$ Units. Chem Commun. 2005;6008–10.

[157] Zimmermann, C, M Melullis, and S Dehnen. Reactivity of chalcogenostannate salts: unusual synthesis and structure of a compound containing ternary cluster anions $[Co_4(\mu_4\text{-}Se)(SnSe_4)_4]^{10-}$. Angew Chem Int Ed. 2002;41:4269–72.

[158] Dehnen, S, and MK Brandmeyer. Reactivity of chalcogenostannate compounds: syntheses, crystal structures, and electronic properties of novel compounds containing discrete ternary anions $[M^{II}_4(\mu_4\text{-}Se)(SnSe_4)_4]^{10-}$ (M^{II} = Zn, Mn). J Am Chem Soc. 2003;125:6618–9.

[159] Brandmeyer, MK, R Clérac, F Weigend, and S Dehnen. Ortho-chalcogenostannates as ligands: syntheses, crystal structures, electronic properties, and magnetism of novel compounds containing ternary anionic substructures $[M_4(\mu_4\text{-}Se)(SnSe_4)_4]^{10-}$ (M=Mn, Zn, Cd, Hg), $[Hg_4(\mu_4\text{-}Se)(SnSe_4)_3]^{6-}\}$, or $[HgSnSe_4]^{2-}\}$. Chem – Eur J. 2004;10:5147–57.

[160] Ruzin, E, A Fuchs, and S Dehnen. Fine-tuning of optical properties with salts of discrete or polymeric, heterobimetallic telluride anions $[M_4(\mu_4\text{-}Te)(SnTe_4)_4]^{10-}$ (M = Mn, Zn, Cd, Hg) and $[Hg_4(\mu_4\text{-}Te)(SnTe_4)_3]^{6-}\}$. Chem Commun. 2006;4796–8.

[161] Ruzin, E, C Zimmermann, P Hillebrecht, and S Dehnen. Syntheses and structures of solvated tetrasodiumtetrachalcogenostannates $[Na_4(en)_4][SnE_4]$ (en = 1,2-diaminoethane; E = Te, Se), and their reactions toward further ternary or quaternary chalcogenometallates $[Mn(en)_3]_2[Sn_2Te_6]\cdot 4H_2O$ or $[Na_{10}(H_2O)_{34}][Mn_4(\mu_4\text{-}Se)(SnSe_4)_4]$. Z Anorg Allg Chem. 2007;633:820–9.

[162] Santner, S, and S Dehnen. $[M_4Sn_4Se_{17}]^{10-}$ Cluster anions (M = Mn, Zn, Cd) in a Cs+ environment and as ternary precursors for ionothermal treatment. Inorg Chem. 2015;54:1188–90.

[163] Dhingra, SS, and RC Haushalter. One dimensional inorganic polymers: synthesis and structural characterization of the main-group metal polymers $K_2HgSnTe_4$, $(Et_4N)_2HgSnTe_4$, $(Ph_4P)GeInTe_4$, and $RbInTe_2$. Chem Mater. 1994;6:2376–81.

[164] Zimmermann, C, and S Dehnen. $Cs_2[MnSnTe_4]$: uncommon synthesis of a quaternary phase based on one-dimensional, ternary anionic chains. Z Anorg Allg Chem. 2003;629:1553–6.

[165] Thiele G, Peter S, Schwarzer M, Ruzin E, Clérac R, Staesche H, et al. Heterobimetallic chalcogenidometallate strands: synthesis, structure, magnetism, and conductivity. Inorg Chem. 2012;51:3349–51.

[166] Haddadpour S, Melullis M, Staesche H, Mariappan CR, Roling B, Clérac R, et al. Inorganic frameworks from selenidotetrelate anions $[T_2Se_6]^{4-}$ (T = Ge, Sn): synthesis, structures, and ionic conductivity of $[K_2(H_2O)_3][MnGe_4Se_{10}]$ and $(NMe_4)_2[MSn_4Se_{10}]$ (M = Mn, Fe). Inorg Chem. 2009;48:1689–98.

[167] Tsamourtzi, K, J-H Song, T Bakas, AJ Freeman, PN Trikalitis, and MG Kanatzidis. Straightforward route to the adamantane clusters $[Sn_4Q_{10}]^{4-}$ (Q = S, Se, Te) and use in the assembly of open-framework chalcogenides $(Me_4N)_2M[Sn_4Se_{10}]$ (M = Mn^{II}, Fe^{II}0, Co^{II}, Zn^{II}) including the first telluride member $(Me_4N)_2Mn[Ge_4Te_{10}]$. Inorg Chem. 2008;47:11920–9.

[168] Lips, F, and S Dehnen. One-size adjustable-gap cluster anions: systematic approach to multinary, water-soluble chalcogenidometalate compounds. Inorg Chem. 2008;47:5561–3.
[169] Ding, N, C D-Y, and K Mg. $K_6Cd_4Sn_3Se_{13}$: a polar open-framework compound based on the partially destroyed supertetrahedral $[Cd_4Sn_4Se_{17}]^{10-}$ cluster. Chem Commun. 2004;1170–1.
[170] Ruzin, E, E Zent, E Matern, W Massa, and S Dehnen. Syntheses, structures, and comprehensive NMR spectroscopic investigations of hetero-chalcogenidometallates: the right mix toward multinary complexes. Chem – Eur J. 2009;15:5230–44.
[171] Finger, LH, B Scheibe, and J Sundermeyer. Synthesis of organic (trimethylsilyl)chalcogenolate salts Cat[TMS-E] (E = S, Se, Te): the methylcarbonate anion as a desilylating agent. Inorg Chem. 2015;54:9568–75.
[172] Finger, LH, and J Sundermeyer. Halide-free synthesis of hydrochalcogenide ionic liquids of the type [Cation][HE] (E=S, Se, Te). Chem – Eur J. 2016;22:4218–30.
[173] Donsbach, C, G Thiele, LH Finger, J Sundermeyer, and S Dehnen. Mercurates from a revised ionothermal synthesis route: the pseudo-flux approach. Inorg Chem. 2016;55:6725–30.
[174] Lin, Y, W Massa, and S Dehnen. “Zeoball” $[Sn_{36}Ge_{24}Se_{132}]^{24-}$: a molecular anion with zeolite-related composition and spherical shape. J Am Chem Soc. 2012;134:4497–500.
[175] Lin, Y, W Massa, and S Dehnen. Controlling the assembly of chalcogenide anions in ionic liquids: from binary Ge/Se through ternary Ge/Sn/Se to binary Sn/Se frameworks. Chem – Eur J. 2012;18:13427–34.
[176] Lin Y, Xie D, Massa W, Mayrhofer L, Lippert S, Ewers B, et al. Changes in the structural dimensionality of selenidostannates in ionic liquids: formation, structures, stability, and photoconductivity. Chem – Eur J. 2013;19:8806–13.
[177] Sheldrick, WS, and H-G Braunbeck. Preparation and crystal structure of $Cs_2Sn_3Se_7$, a cesium selenostannate(IV) with pentacoordinated tin. Z Naturforsch B. 1990;45:1643–6.
[178] Li, J-R, -W-W Xiong, Z-L Xie, C-F Du, G-D Zou, and X-Y Huang. From selenidostannates to silver-selenidostannate: structural variation of chalcogenidometallates synthesized in ionic liquids. Chem Commun. 2013;49:181–3.
[179] Lin, Y, and S Dehnen. $[BMIm]_4[Sn_9Se_{20}]$: ionothermal synthesis of a selenidostannate with a 3D open-framework structure. Inorg Chem. 2011;50:7913–5.
[180] Xu, G-H, C Wang, and P Guo. Poly[[(pentaethylenehexamine)manganese(II)] [hepta-μ-selenido-tritin(IV)]]: a tin–selenium net with remarkable flexibility. Acta Crystallogr. 2009;C65: m171–3.
[181] Dehnen, S, and C Zimmermann. A new route into single-crystalline partially oxidized cobalt compounds: reactions with Zintl-type hexaselenodistannate(III) $K_6Sn_2Se_6$ as mild oxidant. Chem – Eur J. 2000;6:2256–61.
[182] Zimmermann, C, CE Anson, and S Dehnen. Chalcogenido-bridged clusters by reactions of chalcogenostannate salts. J Clust Sci. 2007;18:618–29.
[183] Thiele, G, Z You, and DS Molecular. CHEVREL-like clusters $[(RhPPh_3)_6(\mu_3\text{-}Se)_8]$ and $[Pd_6(\mu_3\text{-}Te)_8]^{4-}$. Inorg Chem. 2015;54:2491–3.
[184] Thiele, G, Y Franzke, F Weigend, and S Dehnen. {μ-PbSe}: a heavy CO homologue as an unexpected ligand. Angew Chem Int Ed. 2015;54:11283–8.
[185] Eußner, JP, RO Kusche, and S Dehnen. Synthesis and thorough investigation of discrete organotin telluride clusters. Chem – Eur J. 2015;21:12376–88.
[186] Dorfelt, C, A Janeck, D Kobelt, EF Paulus, and H Scherer. Röntgenstrukturanalyse von alkylthiostannonsäuren (monoalkylzinnsesquisulfiden). J Organomet Chem. 1968;14:P22–4.
[187] Bouška M, Dostál L, Padělková Z, Lyčka A, Herres-Pawlis S, Jurkschat K, et al. Intramolecularly coordinated organotin tellurides: stable or unstable? Angew Chem Int Ed. 2012;51:3478–82.
[188] Heimann, S, M Hołynska, and S Dehnen. Synthesis, structures, and light-induced transformation of keto-functionalized selenidogermanate complexes. Z Anorg Allg Chem. 2012;638:1663–6.

[189] Barth, BEK, BA Tkachenko, JP Eußner, PR Schreiner, and S Dehnen. Diamondoid hydrazones and hydrazides: sterically demanding ligands for Sn/S cluster design. Organometallics. 2014;33:1678–88.
[190] Leusmann, E, M Wagner, NW Rosemann, S Chatterjee, and S Dehnen. Synthesis, crystal structure, and photoluminescence studies of a ruthenocenyl-decorated Sn/S cluster. Inorg Chem. 2014;53:4228–33.
[191] Leusmann, E, F Schneck, and S Dehnen. Functionalization of Sn/S clusters with hetero- and polyaromatics. Organometallics. 2015;34:3264–71.
[192] Rinn, N, K Hanau, L Guggolz, A Rinn, S Chatterjee, and S Dehnen. Trigonal bipyramidal metalselenide clusters with palladium and tin atoms in various positions. Z Anorg Allg Chem. 2017;643:1508–12.
[193] Rinn, N, L Guggolz, K Gries, K Volz, J Senker, and S Dehnen. Formation and structural diversity of organo-functionalized tin-silver selenide clusters. Chem – Eur J. 2017;23:15607–11.
[194] Eußner, JP, and S Dehnen. Bronze, silver and gold: functionalized group 11 organotin sulfide clusters. Chem Commun. 2014;50:11385–8.
[195] You, Z, J Bergunde, B Gerke, R Pöttgen, and S Dehnen. Ferrocenyl-functionalized Sn/Se and Sn/Te complexes: synthesis, reactivity, optical, and electronic properties. Inorg Chem. 2014;53:12512–8.
[196] Komuro, T, T Matsuo, H Kawaguchi, and K Tatsumi. Copper and silver complexes containing the $S(SiMe_2S)_2^{2-}$ ligand: efficient entries into heterometallic sulfido clusters. Angew Chem Int Ed. 2003;42:465–8.
[197] Fard, HZ, L Xiong, C Müller, M Hołynska, and S Dehnen. Synthesis and reactivity of functionalized binary and ternary thiometallate complexes $[(RT)_4S_6]$, $[(RSn)_3S_4]^{2-}$, $[(RT)_2(CuPPh_3)_6S_6]$, and $[(RSn)_6(OMe)_6Cu_2S_6]^{4-}$ (R = C_2H_4COOH, CMe_2CH_2COMe; T = Ge, Sn). Chem – Eur J. 2009;15:6595–604.

Vimal K. Jain and G. Kedarnath

11 Applications of metal selenium/tellurium compounds in materials science

Abstract: Metal chalcogenides are technologically important materials. Physical, chemical, electrical and mechanical properties of these materials can be fine-tuned by manipulating their shape, size and composition. Although several methods are employed for their synthesis, single-source molecular precursor route has emerged as a versatile strategy for their synthesis and in controlling shape, size and composition of the material under moderate conditions. This chapter gives a brief coverage on the design and development of single-source molecular precursors which have been employed for the preparation of metal selenide/telluride nanocrystals and for deposition of thin films. The discussion includes synthesis of transition-, main group and f-block metal chalcogenolate and/or chalcogenide clusters as precursors and their conversion into metal chalcogenides in the form of thin films and nanostructures. Precursors for ternary metal chalcogenides are also included.

Keywords: metal chalcogenolate, metal chalcogenide, chalcogenide clusters, selenolate, tellurolate, single-source molecular precursor, nanostructure, thin film

11.1 Introduction

Metal complexes of seleno- and telluro-ligands, such as $[PtCl_2(SeR_2)_2]$ (R = Me, Et, Pr^n, Ph) [1] and $[PtCl_2(TeBz_2)_2]$ [2], have been known for more than a century, yet the chemistry of complexes containing M-E (E = Se or Te) bond received scant attention until early 1990s [3]. There have been obvious reasons for this sluggish progress on metal complexes with heavier chalcogen ligands. These include non-availability of suitable synthetic methods for desirable ligands, malodorous reputation of their compounds, toxicity and also the belief that the chemistry of Se/Te complexes would be similar to those of sulfur compounds. However, such perceptions could be defied by the rapid growth of M-E derivatives during the last quarter century as several organoselenium/tellurium ligands show not only unusual reactivity but also display structural features which are markedly different from their sulfur counterparts. It is evident from Figure 11.1 that about 5,500 metal Se/Te-containing compounds have been structurally characterized till December 2016, whereas X-ray structures of only 130 selenolate and 5 tellurolate complexes were reported till July 1992 [3]. Further

This article has previously been published in the journal *Physical Sciences Reviews*. Please cite as: Jain, V.K., Kedarnath, G. Applications of Metal-Selenium/-Tellurium Compounds in Materials Science. *Physical Sciences Reviews* [Online] **2018** DOI: 10.1515/psr-2017-0127

https://doi.org/10.1515/9783110529340-011

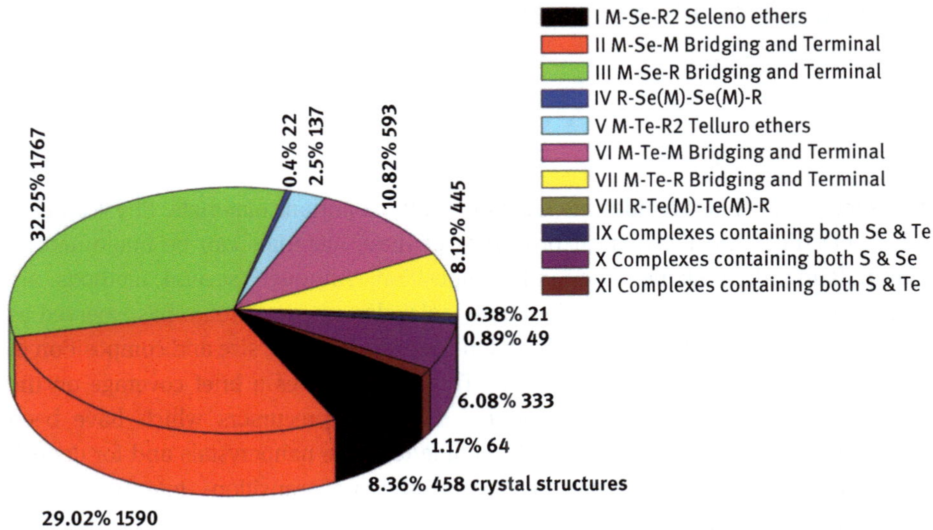

Figure 11.1: Pie chart of metal Se and Te compounds structurally characterized. Data based on CCDC till December 2016 [3].

impetus to the chemistry of the complexes of M-E linkage is received by the utility of these complexes in catalysis [4–6], biology [7] and materials science.

The last quarter century has also witnessed an exponential growth in the design and development of numerous inorganic materials. Several synthetic processes to tailor materials of desired properties have emerged. Metal chalcogenides represent an important family of inorganic materials. Various chemical, physical, optical and magnetic properties of metal chalcogenides are greatly influenced by their shape, size and composition. Among various preparative methods, single-source molecular precursor strategy has emerged as the most versatile route to prepare delicate structures of metal chalcogenides and even for isolation of metastable phases. These precursors have an advantage over the conventional precursors in terms of controlling nanostructure shape and stoichiometry. As the results of bulk decomposition of single-source molecular precursor quite often differ from film growth experiments and also from solvothermal preparation of nanocrystals, a single precursor can thus be employed for the preparation of materials of different compositions and phases. In fact, rapid development of metal chalcogenolate chemistry and ever-increasing interest in metal chalcogenide materials in the last two decades have contributed immensely to the progress of both the areas in a complimentary manner.

In addition to chalcogenolate ligands (RE^-), several other anionic organochalcogen ligands have been used in the design and development of single-source molecular precursors which are employed for the preparation of metal chalcogenide nanocrystals and for deposition of thin films. These include *N,N′*-dimethylaminoalkylchalcogenolate (**I, II**) [8], 2-pyridyl and 2-pyrimidyl chalcogenolate (**III**) [9], selenocrboxylate (**IV**) [10], diselenocarbamates (**V**) [11], diselenophosphate/phosphinates (**VI**) [12], phosphinochalcogenic amides (**VII**) [13] and dichalcogenoimidophosphinates (**VIII**) [14, 15] (Figure 11.2).

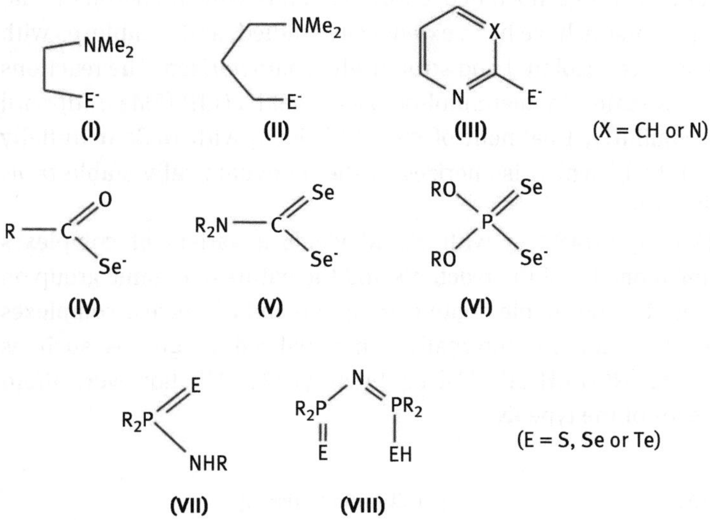

Figure 11.2: Some anionic organochalcogen ligands [8–15].

11.2 Synthesis of chalcogenolate/chalcogenide complexes

A number of reaction routes have been adopted for the synthesis of metal chalcogenolate/chalcogenide complexes, the metathesis reactions and oxidative addition of chalcogenide to low-valent metal precursors which are the most common for their synthesis. Various preparative methods are briefly described here which are representative rather than comprehensive.

11.2.1 Reactions of nucleophilic chalcogen reagents with metal salts

Both inorganic (NaEH, Na_2E, Na_2E_2, $(Me_3Si)_2E$) and organic (RE^-) nucleophiles have been extensively used for the synthesis of a variety of metal chalcogenide/chalcgenolate complexes [16]. Organochalcogenolates are readily obtained either by insertion of chalcogen into a M–C bond or by reductive cleavage of E–E bond of diorganodichalcogenides (eqs. (11.1)–(11.3)).

$$RLi + E \rightarrow RELi \tag{11.1}$$

$$RMgX + E \rightarrow REMgX \quad (X = Cl,\ Br\ or\ I) \tag{11.2}$$

$$\text{RE–ER} \xrightarrow{\text{Reducing agent}} 2\ \text{RE}^- \qquad \text{Reducing agent} = \text{NaH, NaBH}_4\text{, LiBEt}_3\text{H, etc.} \tag{11.3}$$

Reactions of sodium salt of organochalcogenolate ligands with numerous palladium and platinum precursors have been extensively studied, and complexes with diverse nuclearity have been isolated and structurally characterized. The reactions of $PtCl_2$(dppm) with NaEMes in methanol-benzene yield *cis*-[Pt$(EMes)_2$(dppm)] (E = Se or Te) [17]. Similarly, treatment of *cis*-$PtCl_2(PPh_3)_2$ with NaSeAr initially yields *cis*-[Pt$(SeAr)_2(PPh_3)_2$] which isomerizes to thermodynamically stable *trans* isomer in solution [18, 19].

Treatment of [$M_2Cl_2(\mu\text{-}Cl)_2(PR_3)_2$] with NaER′ yields a variety of complexes depending on the stoichiometry of the reactants and the nature of organic group on chalcogen atom. When R is the simple organic group, bi- and tri-nuclear complexes are formed (Figure 11.3) [20, 21]. Internally functionalized R′ groups such as $Me_2NCH_2CH_2$ [22–24], $Me_2NCH_2CH_2CH_2$ [25] or 2-pyridyl [26, 27], however, afford mononuclear complexes of the type **IX**.

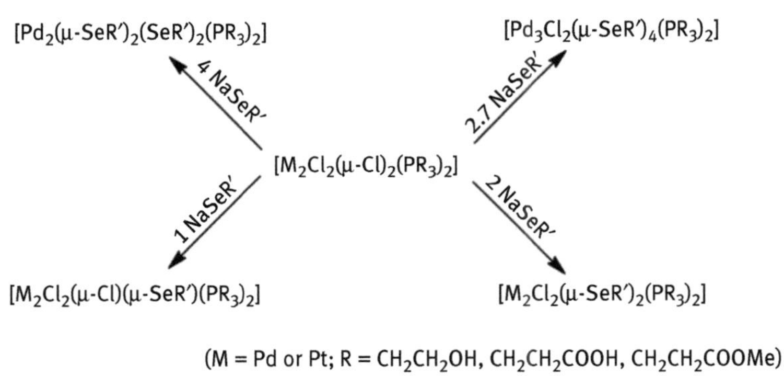

Figure 11.3: Treatment of [$M_2Cl_2(\mu\text{-}Cl)_2(PR_3)_2$] with NaER′ [20, 21].

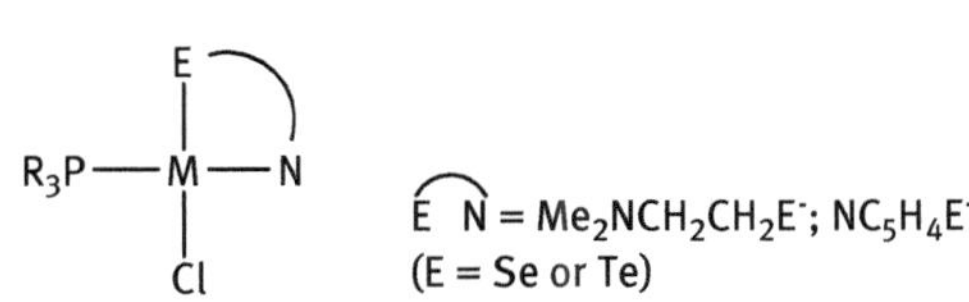

IX

11.2.2 Oxidative addition of dichalcogenides

Two-electron oxidative addition of diorganochalcogenides either on free metal or on low-valent metal precursors has emerged as a useful strategy to prepare metal chalcogenolate complexes [28]. Oxidative addition of diorganodichalcogenides on elemental mercury in toluene/methanol has been employed for the synthesis of $Hg(ER)_2$ complexes (eq. (11.4)) [29]. Pyridyl selenolate complexes of copper, silver and indium have been isolated either by the reaction of metal powder with dipyridyldiselenide in refluxing toluene [30] or by anodic oxidation of metal in acetonitrile solution of diselenide [31–33]. A series of lanthanide selenolates have been isolated by the reaction of dioraganodiselenide either with lanthanum amalgam (LnHg) or with elemental lanthanum in the presence of catalytic amount of Hg [34].

$$\begin{gathered} Hg + R_2E\ _2 \rightarrow Hg(ER)_2 \\ (E = Se \text{ or } Te) \end{gathered} \tag{11.4}$$

Low-valent metal precursors have been conveniently used in the oxidative addition reactions of dioganodichalcogenides, leading to the formation of metal chalcogenolate complexes (eq. (11.5)). The reactions involving ditellurides often result into several serendipitous products which are formed due to competitive cleavage of both Te–Te and Te–C bond because of their comparable bond energies [28].

$$L_nM + R_2E_2 \ \rightarrow \left[L_nM(ER)_2\right] \tag{11.5}$$

The reaction of Na_2PdCl_4 or other Pd(II) precursors with R_2E_2 (E = S, Se, Te) yields complexes of composition $[PdCl(ER)]_n$ retaining palladium in divalent state [28, 35, 36]. The reaction is intriguing in the sense that how the chalcogenide (E_2^{2-}) is oxidized to chalcogenolate ion. It has been proposed that initially R_2E_2 oxidatively adds to Pd^{2+} ion to generate a Pd(IV) species, "$Pd(ER)_2Cl_2$" which undergoes reductive elimination of RECl with concomitant formation of $[PdCl(ER)]_n$ [35].

Oxidative addition reactions of dioraganodiselenides with $[Pd(PR_3)_4]$ in general afford dinuclear complexes $[Pd_2(\mu\text{-}SeR)_2(SeR)_2(PR_3)_2]$ [37–39], whereas reactions with ditellurides are more complex giving several products. The reaction of $Pd(PPh_3)_4$ with Ph_2Te_2/Th_2Te_2 in dichloromethane yields hexanuclear complexes, $[Pd_6Cl_2Te_4(TeAr)_2(PPh_3)_6]$ (Ar = Ph or Th) [40, 41]. However, the reaction with Th_2Te_2 in toluene gives yet another hexanuclear complex, $[Pd_6Te_4(ThTe)_4(PPh_3)_6]$ [40].

Unlike reactions of Pd(0)/Pt(0) phosphine complexes with ditellurides containing simple aryl groups, oxidative addition reactions of *N*-heterocyclic ditellurides are more complex resulting into the formation of numerous serendipitous products (**X-XII**) [42–45]. For instance, reaction of $[Pt(PPh_3)_4]$ with $(NC_5H_3R\text{-}3)_2Te_2$ (R = H or Me) in toluene at room temperature, in addition to expected oxidative addition product, $[Pt\{TeC_5H_3(R\text{-}3)N\}_2(PPh_3)_2]$, a minor (~ 15 %) orange product, $[Pt(Te)\{TeC_5H_3(R\text{-}3)N\}_2(PPh_3)]$ (**X**) containing formally Te^0 as a ligand, is also isolated [42]. The reaction of $Pt(PPh_3)_4$ with Th_2Te_2 in dichloromethane, however, yields a tri-nuclear complex, $[Pt_3Te_2(Th)(PPh_3)_5]Cl$ and trans-$[Pt(Cl)(Th)(PPh_3)_2]$ [40].

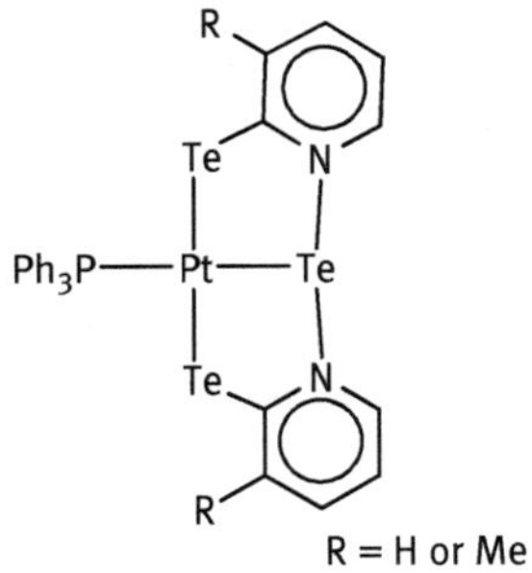

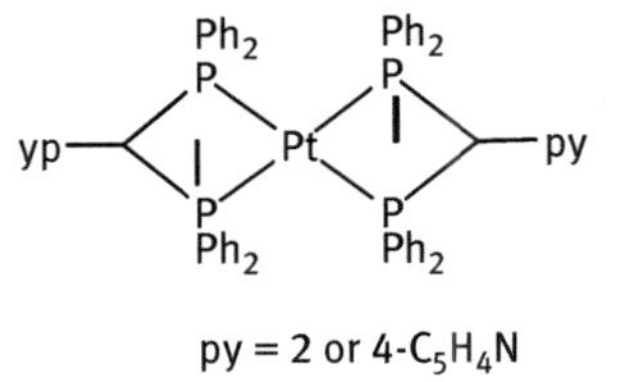

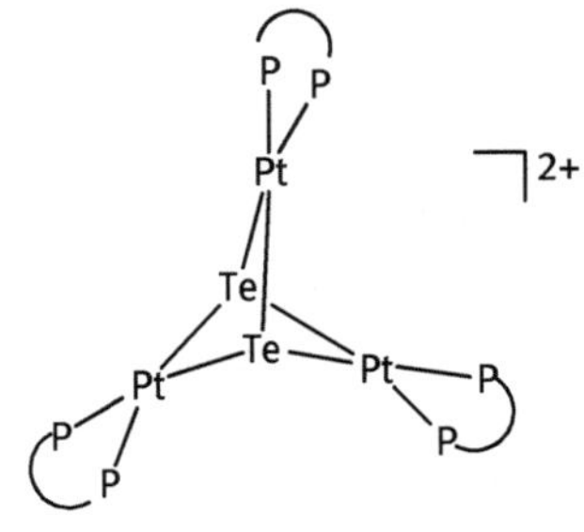

(X) (XI) (XII)

Reactions of $Pt(PPh_3)_4$ or $Pt(P^{\cap}P)_n$ with diorganodichalcogenides yield $[Pt(ER)_2(P^{\cap}P)]$ ($P^{\cap}P = 2PPh_3$, dppm, dppe, dppp) [46, 47]. Treatment of $[Pt(PPh_3)_2(CH_2 = CH_2)]$ with Ph_2Se_2 in benzene yields initially *cis*-$[Pt(SePh)_2(PPh_3)_2]$ which isomerizes to the trans product in solution [48, 49]. Similarly, reaction of Pt(dppn)(norbornene) with $Bu_2{}^nTe_2$ gives $[Pt(TeBu^n)_2(dppn)]$, whereas Ar_2Te_2 affords [Pt(TeAr)(Ar)(dppn)] (Ar = Ph or Th) [50]. In contrast to platinum(0) phosphine precursor, the platinum(0) olefin complexes containing chelating diimine ligands, [Pt(N,N-chelate)(olefin)], react with diorganodichalcogenides or phenylselenyl halides to give five coordinate Pt(II) chalcogenolate complexes (Figure 11.4) [51–53].

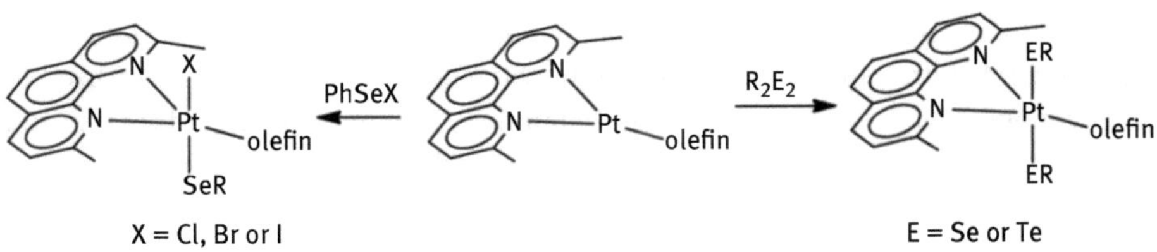

Figure 11.4: Five coordinate Pt(II) chalcogenolate complexes [51–53].

11.2.3 Ligand-exchange reactions

Ligand-exchange/redistribution reactions have also been employed to some extent for the synthesis of metal chalcogenolate complexes. Lead and mercury chalcogenolates react with allylpalladium chloride at room temperatures to give chalcogenolato-bridged palladium derivatives (eqs. (11.6)–(11.8)) [54, 55]. The tellurolate derivative decomposes in solution at room temperature to give a black powder of Pd_3Te_2 [54]. Organotin selenolate has been employed as a mild selenolate-donating species. Accordingly, the reaction between $[(PhSe)_3SnCH_2CH_2]_2$ and palladium acetate affords a tetranuclear selenolate complex, $[Pd(SePh)(OAc)]_4$ [56]. Redistribution reactions in main group chalcogenolate complexes are quite facile. For example, treatment of diorganotin dichlorides with diorganotin bis(selenolates) yields quantitatively [$R_2SnCl(SeAr)$] complexes (eq. (11.8)) [57].

$$[Pd_2Cl_2(\eta^3-\text{allyl})_2] + Pb(SeMes)_2 \rightarrow [Pd_2(\mu-SeMes)_2(\eta^3-\text{allyl})_2] + PbCl_2 \quad (11.6)$$

$$[Pd_2Cl_2(\eta^3-C_4H_7)_2] + Hg(TePh)_2 \rightarrow [Pd_2(\mu-TePh)_2(\eta^3-C_4H_7)_2] + HgCl_2 \quad (11.7)$$

$$R_2SnCl_2 + [R_2Sn\{SeC_4H(4,6-Me)_2N_2\}_2] \rightarrow 2[R_2SnCl\{SeC_4H(4,6-Me)_2N_2\}] \quad (11.8)$$

Trimethylsilyl chalcogenolates, Me_3SiER have been used as a mild ER group transfer reagent. These compounds react readily with metal halide or acetate precursors to give metal chalcgenolates and volatile Me_3SiX (X = Cl or OAc) (eq. (11.9)) [3].

$$L_nM-X + Me_3SiER \rightarrow L_nM-ER + Me_3SiX \quad (11.9)$$

11.2.4 Chalcogenolysis method

Although this route is very convenient for the preparation of metal chalcogenolates (eqs. (11.10)–(11.12)) [58–60], due to air sensitivity and unstability of free chalcogenol (REH), it is less frequently employed and is restricted to selenols/tellurols containing bulky R groups such as supermesityl (2,4,6-$Bu_3{}^tC_6H_2$), $C(SiMe_3)_3$, $Si(SiMe_3)_3$, etc. The chalcogenolysis reactions, often carried out in non-polar solvents, proceed under mild conditions giving pure products in high yields. The required selenol/tellurol (REH) is usually isolated by protonolysis of the corresponding alkali metal chalcogenolates [60–63].

$$[Pd(OAc)_2]_3 + 6\ PhSeH \rightarrow 3\ [Pd(SePh)_2]_n + 6\ AcOH \quad (11.10)$$

$$[Cd\{N(SiMe_3)_2\}_2 + 2\ MesTeH \rightarrow [Cd(TeMes)_2]_\infty + 2\ HN(SiMe_3)_2 \quad (11.11)$$

$$[Sn\{N(SiMe_3)_2\}_2] + 2\ HESi(SiMe_3)_3 \rightarrow [Sn\{ESi(SiMe_3)_3\}_2] + 2\ HN(SiMe_3)_2 \quad E = \text{Se or Te} \quad (11.12)$$

11.2.5 Insertion of chalcogen into M-C bond

Insertion of chalcogens into M–C bonds of alkali (e. g. organolithium) and alkaline earth metals (e. g. Grignard reagents) is a well-recognized route for the generation of organochalcogenolate ions [64, 65]. A number of these compounds have been isolated as adducts and have been structurally characterized. Insertion of chalcogens into main group, transition metal and f-block element M–C bonds is also reported.

For instance, reaction of tellurium with $AlBu^t_3$ yields binuclear tellurolato-bridged aluminium complex, $[Bu^t_2Al(\mu\text{-}TeBu^t)]_2$ [66]. Although metallic tellurium does not insert into Mn–C bond, R'_3PTe reacts readily with $MnR(CO)_5$ to give phosphine-substituted octahedral tellurolate complexes, $[Mn(TeR)(CO)_3(PR'_3)_2]$ (R′ = Me, Bz; R = Me or Et) [67]. Similarly, tellurium insertion into Sc–alkyl bond yields scandium tellurolate complex (eq. (11.13)) [68].

$$Cp^*_2ScR \xrightarrow{Te} [Cp^*_2Sc(TeR)]\,(R = Me,\ Bz,\ CH_2CH_2Bu^t,\ CH_2SiMe_3) \tag{11.13}$$

11.3 Transition metal chalcogenolate/chalcogenide clusters

Transition metals, particularly late transition metals, form numerous polynuclear chalcogenolate and/or chalcogenide clusters which have been isolated both as neutral and as ionic (cationic and anionic) derivatives. This family of compounds has been extensively explored by the research groups of Steigerwald in USA and Fenske in Germany. These compounds are considered as intermediary molecules with distorted (reconstructed) fragments in route to solid-state chalcogenide materials. This approach is referred as "bottom-up," "retro-synthetic" or "divide and protect" [69] route to material synthesis. Figure 11.5 depicts progression in particle size from small molecules through clusters, nanoparticles to colloid and finally bulk material.

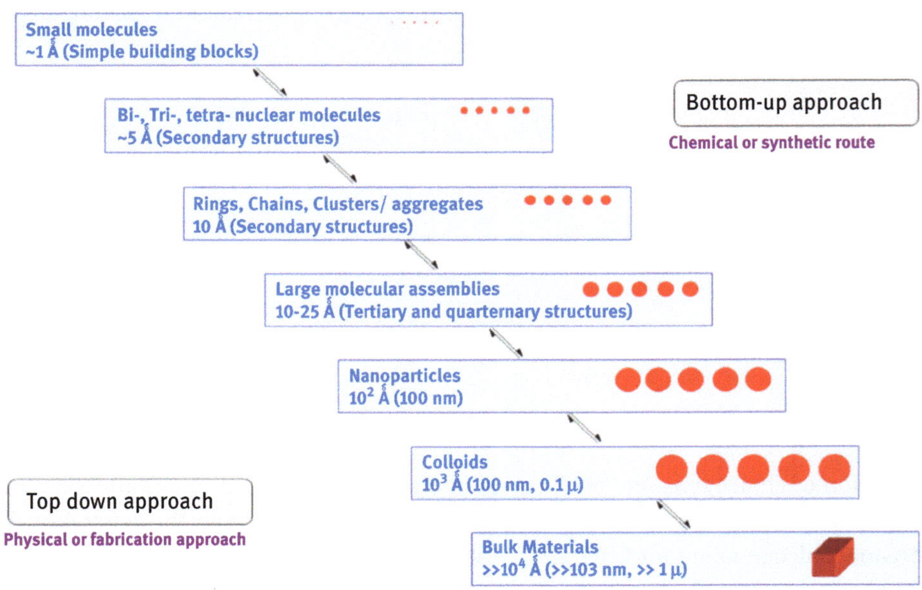

Figure 11.5: Progression of particle size from molecules to materials.

These compounds provide invaluable information about formation, structural aspects and various physicochemical properties of layered nanocrystals. The properties of these compounds are usually influenced by the nature of individual cluster compound, their spatial arrangement and the nature of auxiliary ligand. The metal centres show M–M bonding interaction, and the distance between them can be varied by changing the chalcogen and/or auxiliary ligand and charge on the cluster. For instance, Co–Co distance in $[Co_6(\mu_3\text{-}Te)_8(PH_3)_6]^{n+}$ is 3.05, 3.01 and 2.99 for $n = 0$, 1 and 2, respectively [70]. The chalcogenide ligand in general acts as a triply bridging ligand and also show E ... E interactions. These compounds are usually prepared by one of the following routes:

i. Oxidative addition of R_3PE on low-valent (often zero-valent) metal precursors.
ii. Reactions of metal halides/carboxylates with trimethylsilylated chalcogen compound, $(Me_3Si)_2E$ and Me_3SiER. The latter is employed to introduce surface chalcogenolate groups and/or chalcogenide via E–C bond cleavage [71].

Homoleptic neutral transition metal chalcogenolates of the general formula "$[M(ER)_2]$" show a wide structural diversity which is governed by the interplay of the steric demand of the organic group on chalcogen atom versus propensity of metal cluster to acquire coordination number greater than 2. The nature of chalcogen ligand, however, has a minor influence on the resulting structures. Less bulky *R* groups produce either ring-like oligomeric molecules (e. g. $[Pd(Se^\frown N)_2]_6$ ($Se^\frown N = SeCH_2CH_2NMe_2$ [72] or $SeCH_2CH_2COOMe$ [20]), $[Pd(\mu\text{-}TePh)_2]_{12}.2dmf$ [73]) or polymeric one-dimensional chains (e. g. ${}_\infty^1[Mn(SeMes)_2]$ [74], ${}_\infty^1[Fe(SeR)_2]$ (R = Ph or Mes) [75]). The chalcogenolate groups act as bridging ligands to form a four-membered "M_2E_2" rings.

A myriad of chalcogenide clusters differing in nuclearity have been isolated and structurally characterized. These compounds often preserve the primary lattice of the bulk material, though significant distortion in the structure due to surface passivation by phosphine ligands is observed. Cubane-like tetranuclear clusters, such as $[Fe_4(Te)_4(EPh)_4]^{3+}$ [76, 77], $[Fe_4(Te)_4(PR_3)_4]$ ($PR_3 = PEt_3$, PPr^i_3) (Figure 11.6) [78, 79], $[Co_4(\mu_3\text{-}Se)_4(PPh_3)_4]$ [80] and $[Mn_4(Te)_4(PR_3)_4]$ ($PR_3 = PEt_3$, PPr^i_3, c-Hx_3P) [81], are comprised of tetrahedral core of metal atoms where each metal centre is coordinated to E^{2-} and PR_3 ligand. Each face of the tetrahedron is capped with a single chalcogen atom. The neutral cluster, $[Fe_4(Te)_4(PEt_3)_4]$, releases PEt_3 on heating in evacuated Pyrex tube at 280 °C to give a mixture of β-FeTe and ε-FeTe (Fe and Te rich phases, respectively) [78].

The hexanuclear chalcogenide clusters built from "M_6E_8" units, $[M_6(\mu\text{-}E)_8(PR_3)_6]$ (M = Cr, Mo, Fe, Co; E = Se or Te; $PR_3 = PEt_3$, PBu^n_3 or PPh_3) (Figure 11.7), represent Chevrel phase [80, 82–85]. The molecular structure of these compounds is comprised of octahedrally arranged transition metal atoms with all the triangular faces symmetrically bonded by triply bridging chalcogen atoms. Each metal atom is coordinated with a phosphine ligand which passivates the surface of the cluster. The cobalt and chromium telluride clusters, $[M_6(\mu\text{-}Te)_8(PEt_3)_6]$ (M = Co or Cr), are

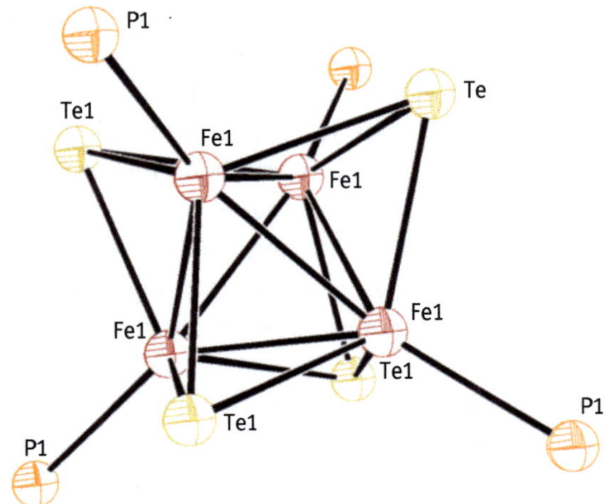

Figure 11.6: Molecular structure of $[Fe_4(Te)_4(PEt_3)_4]$ (adapted with permission from ref. [78], Copyright (1992) American Chemical Society).

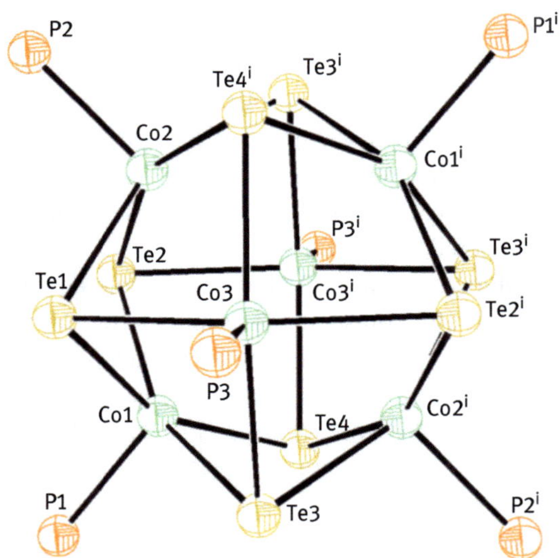

Figure 11.7: Molecular structure of $[Co_6(\mu\text{-}Te)_8(PEt_3)_6]$ (adapted with permission from ref. [83] Copyright (1991) American Chemical Society).

viewed as distorted fragment of β-cobalt telluride (β-CoTe) and Cr_3Te_4 which have NiAs-type lattice and on thermolysis release phosphine ligands to give β-CoTe [83] and Cr_3Te_4 [84].

The molecular structure of the cluster compound is markedly influenced by the size of the capping ligand. Accordingly compounds of different nuclearities can readily be isolated by varying the size of the phosphine ligands. For instance, reactions of Mn precursors with R_3PTe yield cubane-type compounds with bulky ligands, $[Mn_4(Te)_4(PR_3)_4]$ ($PR_3 = PEt_3$, PPr^i_3, c-Hx_3P), while cylindrical prismane type of molecules, $[Mn_6(Te)_6(PR_3)_6]$ (R = Me or Et), are isolated with smaller ligands ($PR_3 = PMe_3$ or PEt_3) [81]. Both cubane and prismane types of compounds could be isolated with intermediate-size ligand (e. g. PEt_3). The cubic $[Mn_4(Te)_4L_4]$ and hexagonal $[Mn_6(Te)_6L_6]$ compounds have been described as fragments of zinc blend and wurtzite structures, respectively [81].

Group 10 (Ni, Pd, Pt) chalcogenide compounds have been synthesized by the reactions of zero-valent metal precursors with R_3PTe. The reaction of $[Ni(COD)_2]$ with Et_3PTe yields $[Ni_9(Te)_6(PEt_3)_6]$ and $[Ni_{20}(Te)_{18}(PEt_3)_{12}]$ [86]. The structural relationship of these compounds with that of bulk NiTe structure has been demonstrated [87]. The hexanuclear palladium compound, $[Pd_6(Te)_6(PEt_3)_8]$, isolated in 10 % yield by the reaction of $[Pd(PPh_3)_4]$ with Et_3PTe, is converted readily to bulk PdTe (NiAs-type structure) by thermolysis (Figure 11.8) [88]. The compound is thought to be formed by dimerization and substitution of PEt_3 ligand by Te^{2-} of a trinuclear complex, $[Pd_3(\mu\text{-}Te)_2L_6]^{2+}$, a well-studied family of derivatives [89, 90].

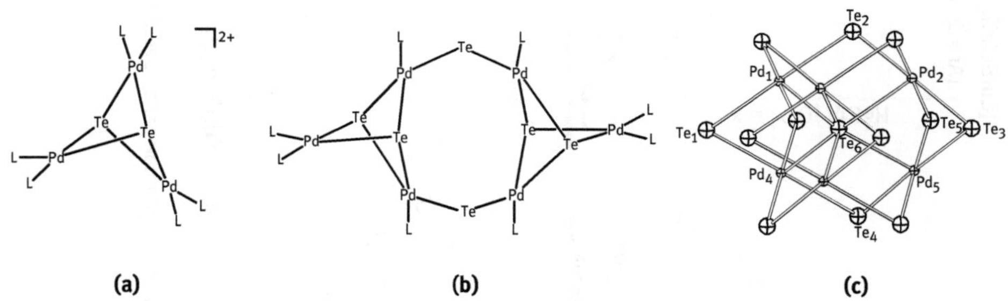

Figure 11.8: Conversion of $[Pd_6(Te)_6(PEt_3)_8]$ to bulk PdTe (Figure 8c is adapted with permission from ref. [88] Copyright (1990) American Chemical Society).

The chemistry of Groups 11 and 12 chalcogenide and chalcogenolate/chalcogenide cluster compounds has been extensively explored by Fenske and co-workers and has been reviewed [71, 91–94]. Reactions of copper(I) salts (CuCl, CuOAc or Cu_2O) with selenol or trimethylsilylated chalcogen compounds ($(Me_3Si)_2E$ or Me_3SiER) in the presence of tertiary phosphine result into the formation of copper chalcogenide or copper chalcogenolate/chalcogenide cluster compounds of different nuclearity varying from binuclear (e. g. $[Cu_2(\mu\text{-}SePh)_2(PPh_3)_3]$ [91]) to as large as Cu_{146} cluster (e. g. $[Cu_{146}(Se)_{73}(PPh_3)_{30}]$ [95]) (Figure 11.9 [95–102]). The formation of the resulting complex in these reactions is influenced by the reactants, ligands, reaction temperature as well as the reaction medium. Small- and medium-sized clusters adopt spherical shapes and possess hexagonal lattices of selenium atoms, while larger clusters represent the fragments of the bulk solid-state structures in which cubic lattice of

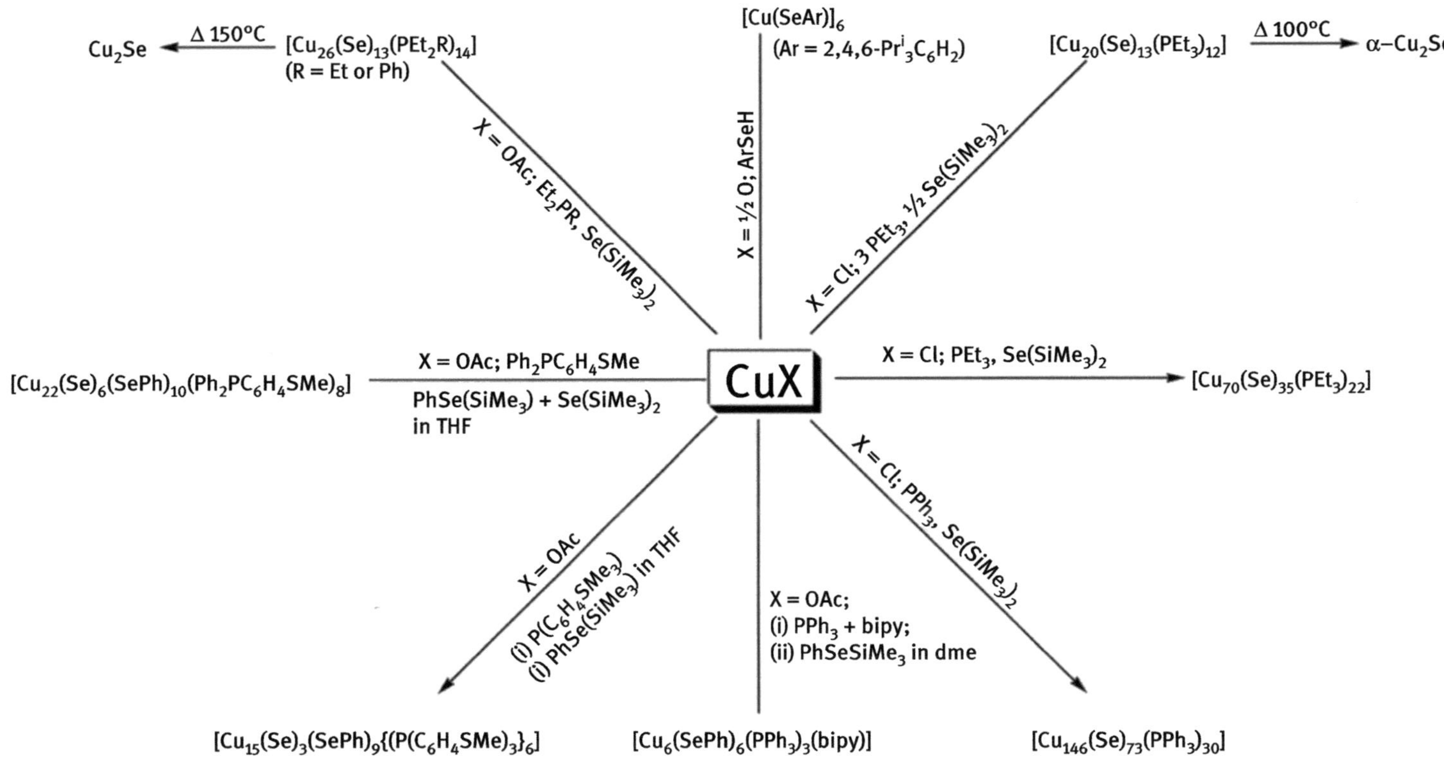

Figure 11.9: The formation of copper chalcogenide or copper chalcogenolate/chalcogenide clusters compounds of different nuclearity [95–102].

selenium is occupied by tetrahedral/trigonal copper atoms [71]. For example, the structure of $[Cu_{146}(Se)_{73}(PPh_3)_{30}]$ can be compared with bulk Cu_2Se where the surface of the fragment is passivated by PPh_3 ligands [95]. Thermal behaviours of several of these clusters have been studied. They cleanly eliminate the coordinated phosphine ligand to generate nanocrystals of 12–16 nm diameter of Cu_2Se [97, 99–102].

Like copper chalcogenide clusters, reactions of silver carboxylates with trimethylsilyl chalcogen compounds produce a myriad of silver chalcogenolate/chalcogenide cluster compounds and cluster as large as $[Ag_{172}(Se)_{40}(SeBu^n)_{92}(dppm)_4]$ is isolated [71, 92, 93]. Silver chalcogenide clusters as well as copper telluride systems tend to yield non-stoichiometric derivatives owing to similar electronegativities of metal atom (Ag or Cu) and chalcogen.

Gold forms numerous chalcogenolate protected clusters [94]. These clusters are comprised of core of metal atoms surrounded by a shell which is formed by a number of ligands (e. g. halide, SR^-, SeR, PR_3, etc.). Both core and shell have the fixed number of gold atoms. Recently, selenolate protected gold clusters, such as $[Au_{25}(SeR)_{18}]^-$ (R = Ph or $C_{12}H_{25}$) and $[Au_{38}(SeC_{12}H_{25})_{24}]$, have been described.

Numerous chalcogenolate/chalcogenide clusters of group 12 (Zn, Cd, Hg) have been reported. They are generally comprised from cubic (adamantoid) and hexagonal (barrelanoid) units and show clear relationship with the bulk chalcogenide materials [71]. Both ionic and neutral compounds have been isolated. Trimethylsilylated chalcogen compounds are usually employed for their synthesis. The reaction of $[CdCl_2(PPh_3)_2]$ with $PhSeSiMe_3$ in acetone or diethylether yields $[Cd_{10}(Se)_4(SePh)_{12}(PPh_3)_4]$ which is composed of four fused adamantoid cages (Figure 11.10) [103, 104]. When this reaction is carried out in diglyme, another cluster $[Cd_{16}(SePh)_{32}(PPh_3)_2]$ is formed in which four adamantoid Cd_4Se_6 cages are connected through $PhSe^-$ ligand [103]. Similarly, reaction of $HgCl_2$ with $PhSeSiMe_3$ in toluene–heptane gives $[Hg_{32}(Se)_{14}(SePh)_{36}]$ which comprises of 13 fused HgSe adamantane cages, while the barrelanoid HgSe cages fill the corners of the cluster [104]. The telluro complexes have also been synthesized. Thus, the reaction between $HgCl_2$ and $Te(SiBu^n{}_3)_2$ and $PhESiMe_3$ (E = Se or Te) in the

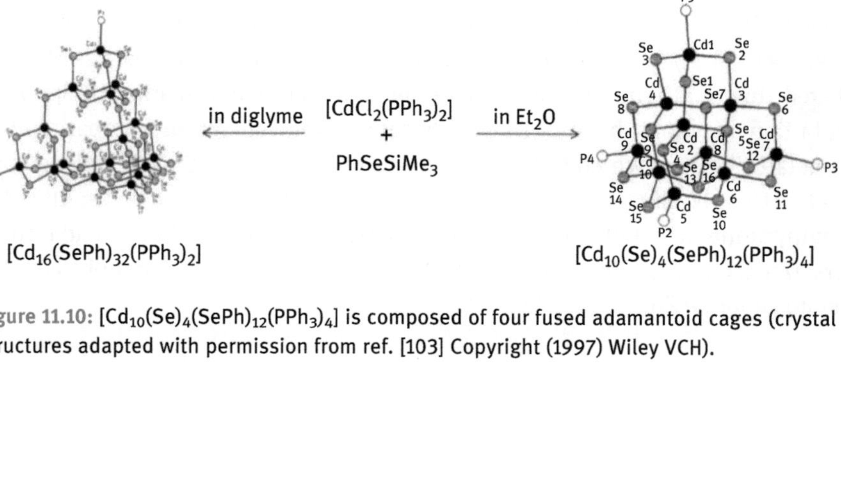

Figure 11.10: $[Cd_{10}(Se)_4(SePh)_{12}(PPh_3)_4]$ is composed of four fused adamantoid cages (crystal structures adapted with permission from ref. [103] Copyright (1997) Wiley VCH).

presence of $PhPPr^n_2$ in dme (dimethyleneglycol) yields metastable compounds, $[Hg_{10}(Te)_4(EPh)_{12}(PhPPr^n_2)_4]$ (E = Se or Te) [105]. When the selenolate compound is dissolved in benzene/toluene at room temperature, orange crystals of a new cluster $[Hg_{34}(Te)_{16}(SePh)_{36}(PhPPr^n_2)_4]$ are isolated (Figure 11.11) [105].

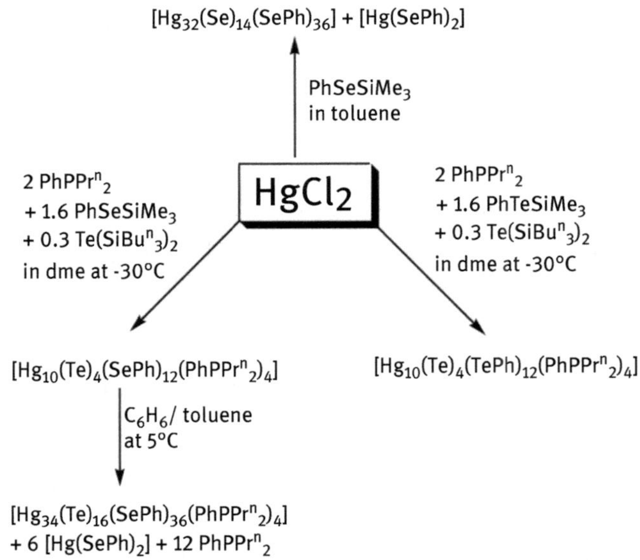

Figure 11.11: Metastable $[Hg_{10}(Te)_4(EPh)_{12}(PhPPr^n_2)_4]$ (E = Se or Te) [105].

Compounds of composition $[M(EPh)_2]$ (M = Cd or Hg; E = Se or Te) have been conveniently employed for the preparation of nanoclusters using redistribution reactions [106–108]. Thus, treatment of $[Cd(SePh)_2]$ with CdX_2 (X = Cl or Br) in the presence of PPh_3 in methanol affords tetranuclear adamantoid cluster $[Cd_4(SePh)_7(X)(PPh_3)]$ [108] which are isostructural to mercury analogue, $[Hg_4(SePh)_7(Br)(py)]$ [109]. Reaction of $[Hg(SeR)_2]$ with equimolar $HgCl_2$ in pyridine yields tetranuclear complexes, $[Hg_4(\mu\text{-}SeR)_4Cl_2(py)_4]$ (R = Et or Bu^t) [110]. However, a similar reaction of $[Hg(EPh)_2]$ with HgX_2 in dmso affords trinuclear complexes, $[Hg_3(\mu\text{-}EPh)_3X_3]$.2dmso (E = Se or Te; X = Cl, Br, I). The molecular structure of these complexes is comprised of six-membered ring of alternating Hg and E atoms in a chair conformation [107]. These redistribution reactions have been successfully employed for the preparation of heterometallic clusters which have been used as precursors for ternary chalcogenide materials [111–113]. Thus, the reactions of $[(lut)_2Zn(ESiMe_3)_2]$ (E = Se or Te; lut = 3,5-dimethylpyridine) with $Cd(OAc)_2$ in the presence of $PhESiMe_3$ and PPr^n_3 yield ternary clusters, $[Zn_xCd_{10\text{-}x}(E)_4(EPh)_{12}(PPr^n_3)_4]$ [113]. These compounds on thermolysis in the solid state generate $Zn_xCd_{1-x}E$ [113].

11.4 Transition metal chalcogenolate precursors for metal chalcogenides

With the isolation of graphene by Geim and Novoselov, interest in two-dimensional inorganic materials has grown exponentially. Many transition metal dichalcogenides (TMDs, ME_2) crystallize in layered structures [114]. Each layer of TMD comprises hexagonally packed layer of metal atoms which are sandwiched between two layers of chalcogen atoms. Each layer has a thickness of 6–7 Å and is interconnected by weak van der Waals interactions [115]. The interlayer interaction energy increases on moving down the group [116]. The M–E bond in each layer is primarily covalent in nature, with the metal atom acquiring either trigonal prismatic or octahedral configuration. The M–M distance varies in the range 3.15–4.03 Å depending on the size of M and E atoms.

The layered structures of TMDs give rise strong anisotropy in their chemical and physical (e. g. electrical, thermal and mechanical) properties, which may range from semiconducting (e. g. WS_2), semimetals (e. g. $TiSe_2$), true metals (e. g. VSe_2) through superconductors (e. g. $NbSe_2$) [117]. The coordination environment of the metal atom and its "*d*" electron count give rise to an array of electronic and magnetic properties, whereas such properties are little influenced by the nature of chalcogen atom [118]. For instance, the band gap of 2H-MoS_2, 2H-$MoSe_2$ and 2H-$MoTe_2$ decreases gradually from 1.3 to 1.0 eV [119]. Among chalcogenide materials, molecular structures and electronic and physical properties of metal tellurides often differ from those of metal sulfide and selenides [120]. Several methods have evolved for the synthesis of TMD thin layer, and chemical vapour deposition (CVD) route using different precursors has gained momentum during the last decade [118, 121].

11.4.1 Group 4 (Ti, Zr, Hf) chalcogenides

Both homoleptic, $M(ER)_4$, and metallocene bis(chalcogenolates), $Cp_2M(ER)_2$, have been isolated and characterized by numerous methods. The metallocene derivatives are generally prepared by the reactions of Cp_2MCl_2 with lithium chalcogenolates [122–125], selenol in the presence of Et_3N (eq. (11.14)) [126] or by treatment of Cp_2MMe_2 with selenium in boiling heptane [127]. The homoleptic derivatives are obtained either by chalcogenolysis reaction between MR_4 (R = Bz, $Si(SiMe_3)_3$) and $HESi(SiMe_3)_3$ [128, 129]) or by a metathesis reaction [129]. These complexes are brightly coloured crystalline solids, sensitive to air and moisture. The metallocene tellurolates react with neutral donor ligands such as phosphines, pyridine and isocyanides, resulting into either reduction of the metal centre from M(IV) to M(III) (M = Ti) or elimination of telluroethers [128, 129].

$$
\begin{gathered}
[Cp_2Ti(SePh)_2] \xleftarrow{PhSeH,Et_3N} Cp_2MCl_2 \xrightarrow{2LiER} [Cp_2M(ER)_2] \\
M = Ti,\ Zr,\ Hf;\ E = Se,\ Te; \\
Cp = C_5H_5,\ MeC_5H_4,\ Bu^tC_5H_4; \\
R = Me,\ Bu^t,\ Mes,\ Ph,\ Si(SiMe_3)_3,\ SiPh_3
\end{gathered}
\tag{11.14}
$$

Metallocene selenolates, $[Cp_2M(SeBu^t)_2]$ (M = Ti, Zr, Hf), have been used as single-source precursors for deposition of MSe_2 thin films on a silica substrate by low-pressure CVD (LPCVD) method (Figure 11.12) [125]. Good-quality films are deposited at 450 °C for $TiSe_2$ and at 600 °C for $ZrSe_2$ (dark green) and $HfSe_2$ (dark blue). The films grow as hexagonal platelets perpendicular to the surface of the substrate and undergo surface oxidation on exposure to air. The tellurolates, $[Cp_2M(TeBu^t)_2]$, however yield grey elemental tellurium in the temperature range 450–600 °C rather than MTe_2 [125].

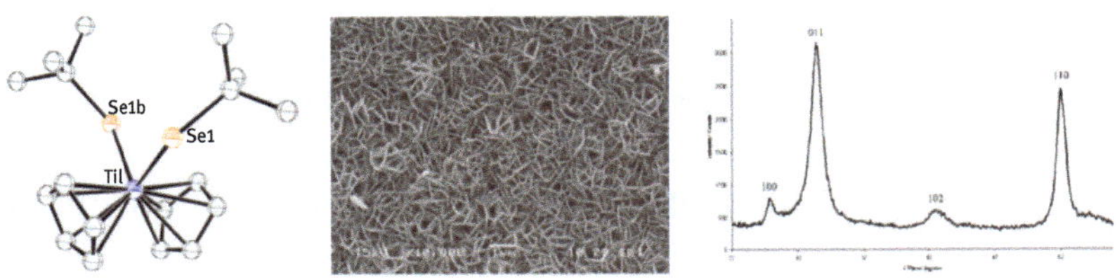

(a) (b) (c)

Figure 11.12: (a) Molecular structure of $[Cp_2Ti(SeBu_t)_2]$; (b) SEM micrographs of $TiSe_2$ deposited at 450 °C; (c) PXD pattern of $TiSe_2$ obtained at 600 °C (adapted with permission from ref. [125] Copyright (2008) American Chemical Society).

The selenoether adducts of titanium, $[TiCl_4(SeR_2)_2]$ (SeR_2 = $SeEt_2$ [130], $SeBu^t_2$ [130, 131], ½ o-$C_6H_4(CH_2SeMe)_2$ [132]), obtained by the reaction of $TiCl_4$ with a selenoether, have been used for deposition of $TiSe_2$ films on glass or silica substrate. Intensely purple films are deposited in the temperature range 500–600 °C with a regular morphology comprising of hexagonal platelets which grow perpendicular to the plane of the substrate. The films are sensitive to areal oxidation.

11.4.2 Group 5 (V, Nb, Ta) chalcogenides

Although metallocene chalcogenolates, such as $[CpM(CO)_2(\mu\text{-}ER)]_2$ (Cp = C_5H_5 or C_5Me_5; M = V or Ta), $[Cp_2Nb(Te_2)H]$ (Cp = C_5H_5, $Bu^tC_5H_9$, C_5Me_5), *etc.*, have been synthesized and investigated [133–136]. They have hardly been employed as precursors for the synthesis of metal chalcogenides. Chalcogenoether complexes have, however, been used to a limited extent. $[VCl_3(SeMe_2)_2]$, obtained by the reaction of VCl_4 with an excess of $SeMe_2$ in refluxing CH_2Cl_2, has been used to deposit thin black films of VSe_2 on silica substrate at 600 °C by LPCVD method [137]. Similarly, $[NbCl_5$

$(SeBu^n{}_2)$] has been used to deposit brown-black films of $NbSe_2$ at 650 °C. The films are formed of hexagonal plate-like microcrystals of 3R phase [138]. The tantalum pentachloride chalcogenoether complexes do not show any deposition under similar conditions, except [$TaCl_5(TeBu^n{}_2)$] which deposited tellurium [138].

11.4.3 Group 6 (Cr, Mo, W) chalcogenide materials

Homoleptic and organometallic compounds of group 6 with chalcogen ligands have been synthesized by chalcogenolysis, oxidative addition of diorganodichalcogenides or metathesis reactions. Examples of such derivatives are [$Cr(SePh)_3$] [139], [$M(SeMes)_3]_2$ (M = Mo or W with M-M bond) [140], [$M_2(OR)_2(SeAr)_4$] [140], [$Cp_2M(ER)_2$] (E = S, Se, Te; R = Me or Ph) [133], [$CpCr(SePh)(CO)_3$] [141], [$CpCr(SePh)(CO)_2]_2$ [141], [η^7-$C_7H_7)M(CO)_2$ (EPh)] (M = Mo or W; E = Se, Te) [142], [$Mo(\mu$-$TePh)(CO)_4]_2$ [143], etc. Only a few complexes have been utilized as precursors. Chromium(II) complexes of di-*tert*-butylphosphino(chalcogeno) amides, [$Cr\{EPBu^t{}_2(NR)\}_2$] (E = Se or Te; R = Pr^i or *c*-Hx), have been prepared by treatment of [$Cr\{N(SiMe_3)_2\}_2(thf)_2$] with $Bu^t{}_2P(E)NHR$ in light petroleum at room temperature [13]. The selenolate complexes have been used to deposit thin films of $Cr_{0.68}Se$ at ~400 °C [13]. The chromium cluster complex, [$Cr_6Te_8(PEt_3)_6$], prepared by the reaction between [$Cr(C_3H_5)_2(PEt_3)_2$] with Et_3PTe or bis(2,4-dimethylpentadienyl)chromium with Et_3PTe in the presence of excess PEt_3, on heating in an evacuated Pyrex tube at 315 °C, yields ferromagnetic black Cr_3Te_4 [84]. The cluster is comprised of octahedron of six chromium atoms concentric with a cube of eight Te atoms. Each chromium is also coordinated with a PEt_3 ligand. The Cr-Cr and Te-Te distances are 2.94 and 3.74 Å, respectively [84].

11.4.4 Group 7 (Mn, Tc, Re) chalcogenides

Chalcogenolysis and oxidative addition reactions, in general, have been used to prepare group 7 chalcogenolate complexes. Chalcogenolysis of [$Mn\{N(SiMe_3)_2\}_2$]. THF with HSe(2,4,6-$Pr^i{}_3C_6H_2$) and $HTeSi(SiMe_3)_3$ in the presence of a neutral ligand yields [$Mn\{Se(2,4,6$-$Pr^i{}_3C_6H_2)\}_2(THF)$] [144] and [$Mn\{TeSi(SiMe_3)_3\}_2$. L] (L = dmpe or 4-$Bu^tC_5H_4N$) [145], respectively. Treatment of $RM(CO)_5$ with R'_3PTe yields [Mn(TeR) $(CO)_3(PR'_3)_2$] (R = Me or Bz; R′ = Me or Et) by insertion of Te into Mn–C bond with concomitant elimination of CO [67, 146]. The latter complexes can also be obtained by the reaction of [$Mn(\mu$-$TeR)(CO)_4]_2$ (R = Me, Et, Pr^i, Ph, CH_2SiMe_3, $(SiMe_3)_3$) with a phosphine ligand [67]. Cyclopentadienyl manganese complexes, [$CpMn(\mu$-TePh) $(CO)_2]_2$ [147] and [$(\eta^5$-$MeC_5H_4)Mn(EAr)(CO)_2$] (E = Se or Te; Ar = Pr^i, tol, Mes) [148], have also been isolated and characterized. The Mn(II) complexes of chalcogenophosphane ligands, such as [$Mn\{Bu^t{}_2P(E)NR\}_2$] (E = Se or T; R = Pr^i, *c*-Hx) [13] and [Mn $\{Ph_2P(Se)NP(Se)Ph_2\}_2$] [149], have been obtained by the reactions of $MnCl_2$ with the alkali metal salt of the ligands. The latter adopts a slightly distorted octahedral configuration with Mn–Se distance of av. 2.55 Å [149]. The oxidative addition of Ph_2Se_2 to $[Re_2Cl_4(\mu\text{-dppm})_2]$ yields Re–Re bonded (2.656(1) Å) binuclear selenolate, $[Re_2Cl_4(\mu\text{-SePh})_2(\mu\text{-dppm})_2]$ [150].

Pyrolysis of $[Mn(TeBz)(CO)_3(PEt_3)_2]$ and $[Mn(\mu\text{-TeR})(CO)_4]_2$ (R = Me or Et) at 300 °C under a hydrogen stream yields MnTe [67]. In contrast, the complexes $[Mn\{Bu^t{}_2P(Te)NR\}_2]$ (R = Pr^i; *c*-Hx) at 350–380 °C/10^{-6} torr deposit polycrystalline films of $MnTe_2$, whereas the corresponding seleno derivatives produce films of cubic MnSe [13].

11.4.5 Group 8 (Fe, Ru, Os) chalcogenides

Although a wide variety of chalcogenolate complexes, both classical and organometallic, of group 8 metals have been reported, only a few have been used for the preparation of metal chalcogenide nanocrystals and for deposition of thin films. The iron chalcogenides, predominantly sulfides, have been investigated due to their remarkable magnetic properties. Iron selenide (FeSe (α- and β-phase), $FeSe_2$, Fe_3Se_4 and Fe_7Se_8) and telluride (FeTe, $Fe_{1.25}Te$ and $FeTe_2$) are known in different phases [151], while ruthenium and osmium chalcognides exist in cubic ME_2 (M = Ru or Os; E = S, Se, Te) phase adopting a chalcopyrite structure [152].

Homoleptic neutral selenolates of iron, both ring-like oligomeric forms $[Fe(SePh)_2]_{12}$ [153] and polymeric ${}_\infty^1[Fe(SeAr)_2]$ (Ar = Ph or Mes) [75], have been synthesized. The latter, which comprises of infinite chains of selenolate-bridged distorted tetrahedral iron atoms, are obtained in several ways (Figure 11.13) [75]. The ${}_\infty^1[Fe(SePh)_2]$ undergoes a two-step decomposition under vacuum below 175 °C to give iron selenide with the elimination of $SePh_2$. The sample heated up to 550 °C in vacuum consists of a mixture of tetragonal $Fe_{1-x}Se$ and hexagonal Fe_7Se_8 phases [75].

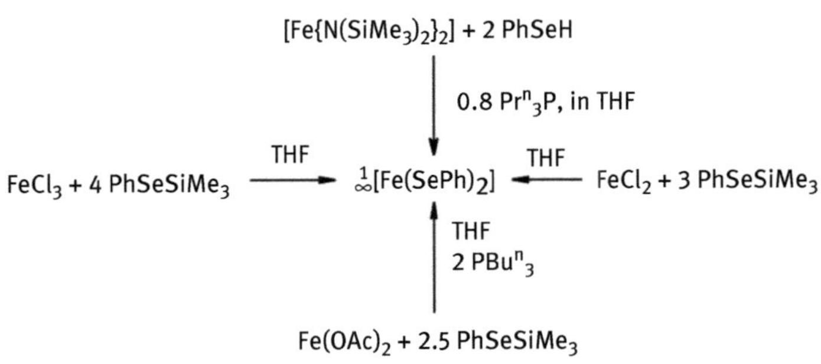

Figure 11.13: Preparation of ${}_\infty^1[Fe(SeAr)_2]$ (Ar = Ph or Mes) [75].

Thermolysis of *N,N*-diethyl-*N′*-naphthylselenourea complex of iron(III), $[Fe\{napC(O)NC(Se)NEt_2\}_3]$ (nap = naphthyl) in oleylamine (OA) at different temperatures (190, 240, 290 °C) yields nanocrystals of a mixture of iron selenides [154]. However, nanocrystals of pure orthorhombic ferroselite $FeSe_2$ phase could be isolated by thermolysis of tetraorganodiseleno imidophosphinate complex of iron(II), [Fe{$(SePR_2)_2N$}$_2$] (R = Pr^i or Ph) in OA [155]. Different shapes of nanocrystals, rod like (at 190 °C), flower like (at 240 °C) and plate like (at 290 °C), have been prepared by varying the pyrolysis temperature [155]. The complex [Fe{$(SePPr^i_2)_2N$}$_2$] has also been used for deposition of iron selenide thin films on silicon substrate by aerosol assisted chemical vapour deposition (AACVD) process in the temperature range 500–600 °C. The films, in contrast to nanocrystals, are comprised of a mixture of Fe_7Se_8 and $FeSe_2$ [155]. The phosphinochalcogenoic amidate complexes of iron(II), [Fe{$Bu^t_2P(E)NR$}$_2$] (E = Se or Te; R = Pr^i or *c*-Hx), have been used for deposition of thin films of FeSe and $FeTe_2$ on a glass substrate at 400 °C [13]. Cyclopentadienyliron telluride complexes, [{CpFe(CO)(PEt_3)}$_2$(μ-Te)] and [{CpFe(CO)(PEt_3)}$_2$(μ-Te_2)], prepared by oxidative addition of Et_3PTe on [CpFe(CO)$_2$]$_2$ (Figure 11.14), on pyrolysis in a Pyrex tube in a furnace (260–290 °C) yield FeTe and $FeTe_2$, respectively [156].

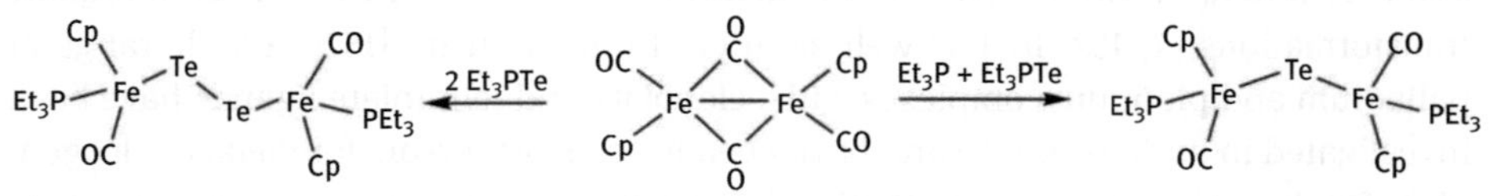

Figure 11.14: Cyclopentadienyliron telluride complexes prepared by oxidative addition of Et_3PTe on [CpFe(CO)$_2$]$_2$ [156].

Homoleptic ruthenium(III) selenolates, [Ru(SeC_6H_4R-4)$_3$]$_n$ (R = H or Bu^t) on pyrolysis at 450 °C, yield nanocrystals of $RuSe_2$ of 3–15 nm size [157]. The nano-$RuSe_2$ supported on graphite has been shown to be an efficient catalyst for chemoselective reduction of aromatic nitro compounds by hydrazine hydrate in the presence of other functionalities [157].

Selenolates and tellurolates of organoruthenium and osmium have been described. The reactions of [Cp*$RuCl_2$]$_2$ with Me_3SiEAr (E = Se or Te; Ar = Ph or tol) yield [(Cp*Ru)$_2$(μ-SeAr)$_3$][Cl] and [(Cp*Ru)$_2$(μ-TeAr)$_2$(μ-Te_2Ar_2)] [158]. Diorganodiselenide oxidatively adds to triosmium carbonyl clusters giving two isomers of [$Os_3(CO)_{10}$(μ-SeR)$_2$] (R = Me or Ph) [159].

11.4.6 Group 9 (Co, Rh, Ir) chalcogenides

Group 9 selenolates and tellurolates are scantly investigated. The cobalt complexes with phosphorus-based selenium ligands, such as [Co{$(SePPr^i_2)_2N$}$_2$] [160], [Co($Se_2PPr^i_2$)$_2$] [161] and [Co{$Bu^t_2P(Se)NR$}$_2$] (R = Pr^i or *c*-Hx) [13], show complex decomposition behaviour and invariably yield a mixture of cobalt selenides (e. g. Co_3Se_4, $CoSe_2$, CoSe, etc.) on attempting to deposit thin films of cobalt selenide above 400 °C [13, 161]. The diselenocarbamate complex, [Co(Se_2CNEt_2)$_2$], however, proved a promising precursor as it deposited cubic $CoSe_2$ films at 450 °C in the form of bundled fibres [161]. Thermolysis of [Co{$Bu^t_2P(Te)NR$}$_2$] at 350 °C yields films of $CoTe_2$ [13].

11.4.7 Group 10 (Ni, Pd, Pt) chalcogenides

Several neutral and anionic selenolates of nickel with the general formula [Ni(SeAr)$_2$L$_2$] and [Ni(SeAr)$_4$]$^{2-}$ have been isolated and structurally characterized [162, 163]. The bidentate seleno ligands yield discrete monomeric complexes, [Ni{(SePPr$^i_2$)$_2$N}$_2$] [161], [Ni(Se$_2$PR$_2$)$_2$] (R = Pr^i, Bu^t, Ph) [161, 164] and [Ni(Se$_2$CNEt$_2$)$_2$] [165], in which square planar nickel is surrounded by four Se atoms from two chelating ligands. The former two series of complexes, [Ni{(SePPr$^i_2$)$_2$N}$_2$] and [Ni(Se$_2$PPr$^i_2$)$_2$], have been used for deposition of thin films of nickel selenide (Ni$_{0.85}$Se) which grow along the (101) direction [161, 164, 166], but at higher thermolysis temperatures (>400 °C) nickel phosphide (Ni$_2$P) is deposited [164, 166]. The diselenocarbamate complex, [Ni (Se$_2$CNEt$_2$)$_2$], deposits cleanly cubic NiSe$_2$ films at 450 °C [161].

There are several phases of binary palladium and platinum chalcogenides [152] and they find applications, particularly tellurides, as catalysts in a number of organic transformations [4, 152, 167] as well as in electronic industry [152]. A wide range of palladium and platinum complexes with selenolate and tellurolate ligands have been investigated in authors' group and have been used as precursors for metal chalcogenides. The homoleptic complex, [Pd(SeBz)$_2$]$_n$, undergoes a two-step decomposition. In the first step of decomposition (~210 °C), a mixture of PdSe$_2$ and Pd$_{17}$Se$_{15}$ is formed, whereas decomposition in the second step yields exclusively Pd$_{17}$Se$_{15}$ [18]. The heteroleptic mono- (e. g. [PdCl(SeCH$_2$CH$_2$NMe$_2$)(PPh$_3$)] [22], [PdCl{SeCH$_2$CH(Me)NMe$_2$}(PMePh$_2$)] [168], [Pd(SeCOPh)$_2$(PPr$_3$)$_2$] [169], [Pd(SeCb°Ph)$_2$(PEt$_3$)$_2$] (Cb°Ph = 2-phenyl-*o*-carborane) [170], bi- (e. g. [Pd(OAc)(SeCH$_2$CH$_2$CH$_2$NMe$_2$)]$_2$ [25], [Pd$_2$Cl$_2$(μ-SeCH$_2$CH$_2$X)$_2$(PPr$^n_3$)$_2$] (X = OH, COOMe, COOEt) [20, 21] and tri-nuclear (e. g. [PdCl(SeCH$_2$CH$_2$NMe$_2$)]$_3$ [22]) palladium complexes with internally functionalized selenolate ligands usually undergo a two-step decomposition in the temperature range 200–350 °C to yield single-phasic Pd$_{17}$Se$_{15}$ irrespective of the substituent on selenium. However, the substituent on selenolate ligand influenced the decomposition temperature. Thermolysis in TBP or hexadecyl amine (HDA) yields nanocrystals of Pd$_{17}$Se$_{15}$ as their X-Ray Diffraction (XRD) peaks are considerably broadened than those samples obtained by decomposition in a furnace under an inert atmosphere. The binuclear platinum complex, [Pt$_2$Cl$_2$(μ-SeCH$_2$CH$_2$COOMe)$_2$(PPr$^n_3$)$_2$], on pyrolysis yields Pt$_5$Se$_4$ [20].

The allylpalladium and cyclopalladated selenolate complexes on thermolysis yield different phases of palladium selenides. The allylpalladium complex, [Pd(μ-SeCH$_2$CH$_2$COOEt)(η^3-C$_3$H$_5$)]$_n$ (n = 2 or 3), which exists in dimeric (*syn* and *anti* forms) and trimeric species in solution (Figure 11.15), on thermolysis at 180° C in diphenyl ether and hexadecylamine yields nanocrystals of Pd$_7$Se$_4$, but at higher temperature (220°C) Pd$_{17}$Se$_{15}$ is formed [171]. In contrast to this, the methylallylpalladium selenolate complexes, [Pd(μ-SeAr)$_2$(η^3-C$_4$H$_7$)$_2$] (Ar = Ph or Mes), on thermolysis either in a furnace (at 160 °C) or in refluxing xylene give polycrystalline Pd$_4$Se [54, 55]. Selenolato-bridged cyclopalladated complexes, [Pd(μ-SeAr)$_2$(C$^\frown$N)$_2$], on thermolysis yield a mixture of palladium selenides. The thermogravimetric (TG) analysis (TGA) of

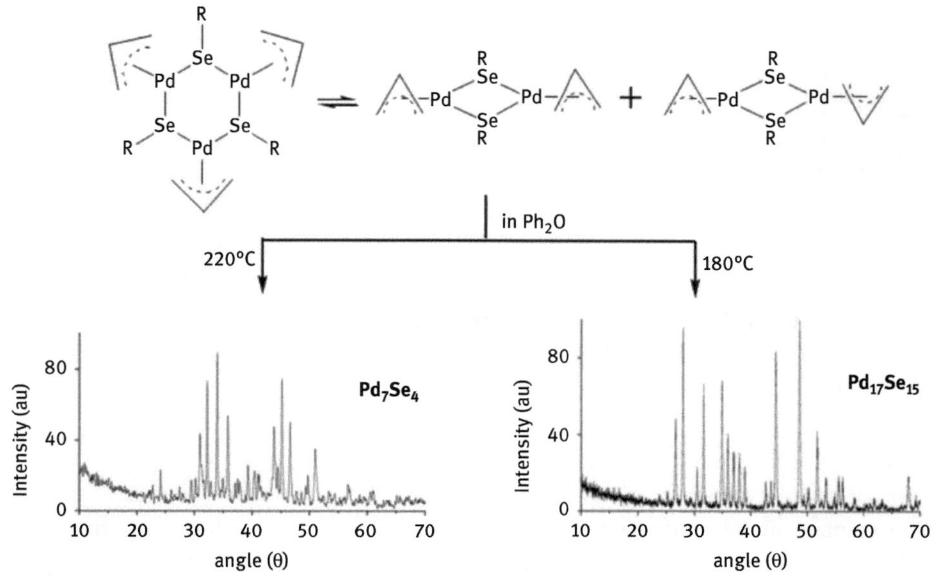

Figure 11.15: XRD pattern of the palladium selenides obtained by decomposition of [Pd(μ-SeCH$_2$CH$_2$COOEt)(η^3-C$_3$H$_5$)]$_n$ in diphenyl ether at 220 °C ($Pd_{17}Se_{15}$), in diphenyl ether at 180 °C (Pd_7Se_4) (XRD patterns of Pd_7Se_4 and $Pd_{17}Se_{15}$ are adapted with permission from ref. [171] Copyright (2011) Royal Society of Chemistry).

[Pd_2(μ-SePh)$_2$(Me$_2$NCH$_2$C$_6$H$_4$-C,N)$_2$] revealed that it undergoes a single-step decomposition at ~290 °C into a single phasic Pd_7Se_4. However, decomposition in a furnace at 200 °C under flowing argon, air or H_2 results into Pd_4Se as a major product together with Pd_7Se_4 [172].

Palladium tellurides have also been synthesized using molecular precursors. The mononuclear tellurolate complexes, [PdCl{TeC$_5$H$_3$(Me-3)N}(PR$_3$)] (PR$_3$ = PEt$_3$ or PPh$_3$), undergo a single-step decomposition ~290 °C under an argon atmosphere yielding aggregates of microcrystalline PdTe (Figure 11.16) [27]. The cyclopalladated complex, [Pd$_2$(μ-TeMes)$_2$(C$_{10}$H$_6$N-C,N)$_2$], decomposes in two steps to give Pd_7Te_3 at 450 °C; however at lower temperatures (265 °C), a mixture of tellurides (Pd_9Te_4 and Pd_2Te) is formed. Thermolysis in refluxing *n*-trioctylphoshine oxide (TOPO) (202 °C) yields $PdTe_2$ [173]. Cyclometallated telluroether complex, [PdCl{4-MeOC$_6$H$_4$TeCH$_2$CH$_2$CH$_2$NHCH (C$_6$H$_4$OH-2)(C$_6$H$_5$)], on thermolysis at 450 °C affords nano-rods of Pd_3Te_2 [4]. The telluro ether complex, [PdCl{4-MeOC$_6$H$_4$TeCH$_2$CH$_2$CH$_2$(pzBr-4)}], in contrast, in TOP yields rod-shaped nanocrystals (217 nm diameter and 2,175 nm length by TEM) of PdTe [174]. The oxidative addition of Et$_3$PTe to Pd(PPh$_3$)$_4$ in refluxing toluene yields a black powder of PdTe [88]. The reaction proceeds via the formation of telluride-bridged intermediate complexes, [Pd$_2$(μ-Te)$_2$(PEt$_3$)$_4$] and [Pd$_6$(μ-Te)$_6$(PEt$_3$)$_8$], which on decomposition results into polycrystalline PdTe. These intermediates are comprised of four-membered Pd_2Te_2 ring which can be represented as a fragment of the PdTe solid lattice [88].

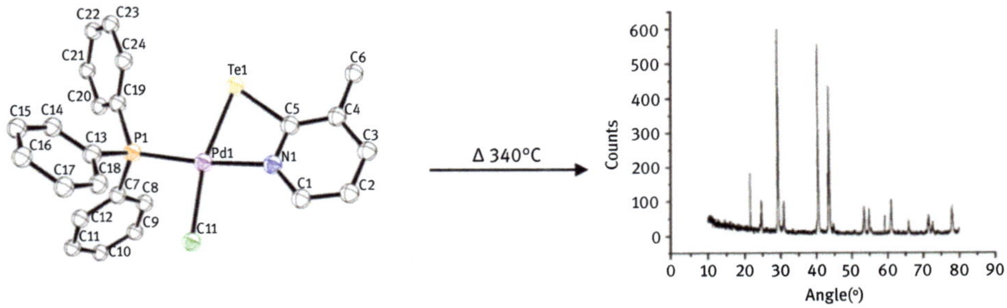

Figure 11.16: Molecular structure of [PdCl{Te(3-MeC$_5$H$_3$N)}(PPh$_3$)], and the XRD pattern of PdTe obtained from [PdCl{Te(3-MeC$_5$H$_3$N)}(PPh$_3$)] (adapted with permission from ref. [27] Copyright (2003) Wiley VCH).

11.4.8 Group 11 (Cu, Ag, Au) chalcogenides

The coinage metal chalcogenides are an interesting family of inorganic materials exhibiting a wide compositional and stoichiometric versatility. Temperature-dependent phase transitions as well as composition variations are frequently encountered. For instance, CuSe undergoes phase transition from hexagonal phase at room temperature to orthorhombic phase at 48 °C and back to hexagonal phase at 120 °C. It disproportionates at higher temperature to $Cu_{2-x}Se$ and Se. Similarly, Cu_3Se_2 disproportionates into $Cu_{1.8}Se$ and CuSe in the temperature range 120–135 °C. Silver selenide (Ag_2Se) exists in two phases. The high-temperature phase (β-Ag_2Se, 133 °C) is a superionic conductor, whereas the low-temperature phase (α-Ag_2Se) is a narrow-band gap material. The former finds applications as a solid electrolyte in photochargeable batteries, while the latter is widely used as a photosensitizer in photographic films. The coinage metal chalcogenides find numerous applications such as photovoltaic, thermoelectric, batteries, photocatalytic, etc. [175].

The chemistry of copper [91, 92, 98, 176], silver [92, 177] and gold [88, 94, 178–180] chalcogenolate/chalcogenide clusters is fairly well developed and reviewed recently. Metallophilic interactions are generally encountered in these compounds which influence a variety of physical and chemical properties such as photoluminescence, redox behaviour and catalytic activity. Structural diversity in these compounds is rampant.

Copper(I), silver(I) and gold(I) complexes with pyridyl and pyrimidyl selenolates have been synthesized and used as precursors for the preparation of metal selenides [30, 181–184]. Copper pyridylselenolates, [Cu{SeC$_5$H$_3$(3-R)N}]$_4$ (R = H [30], Me [181], SiMe$_3$ [30]), have been prepared either by the reaction of copper powder with dipyridyldiselenides in refluxing toluene [30] or by treatment of CuCl with sodium 2-pyridylselenolates in benzene–methanol mixture at room temperature [181]. Pyrolysis of [Cu(Sepy)]$_4$ at 210 °C yields a mixture of α-CuSe and $Cu_{2-x}Se$ [30], whereas [Cu{SeC$_5$H$_3$(3-R)N}]$_4$ (R = Me at 400 °C and SiMe$_3$ at 220 °C) affords cubic $Cu_{2-x}Se$ [30,

181]. Thermolysis of $[Cu\{SeC_5H_3(3\text{-}Me)N\}]_4$ [181] and $[Cu\{SeC_4H(4,6\text{-}Me_2)N_2\}]_6$ [182] in coordinating solvents like HDA/TOPO and dodecanethiol (DDT), respectively, yield cubic phase of Cu_7Se_4 (Figure 11.17). These precursors have been successfully used for deposition of thin films of copper selenide by AACVD. The pyridylselenolate complex has been utilized to deposit films of orthorhombic Cu_5Se_4 in the temperature range 360–400 °C [181]. In contrast, $[Cu\{SeC_4H(4,6\text{-}Me_2)N_2\}]_6$ produced films of cubic Cu_7Se_4 at 400 °C [182]. The films grown on glass substrate had wool-like appearance, while on silicon substrate they were comprised of high-quality vertically oriented hexagonal sheets. Silver pyridyl/pyrimidylselenolates either in the solid state or in DDT solution on thermolysis give orthorhombic Ag_2Se (Figure 11.18) [182, 183].

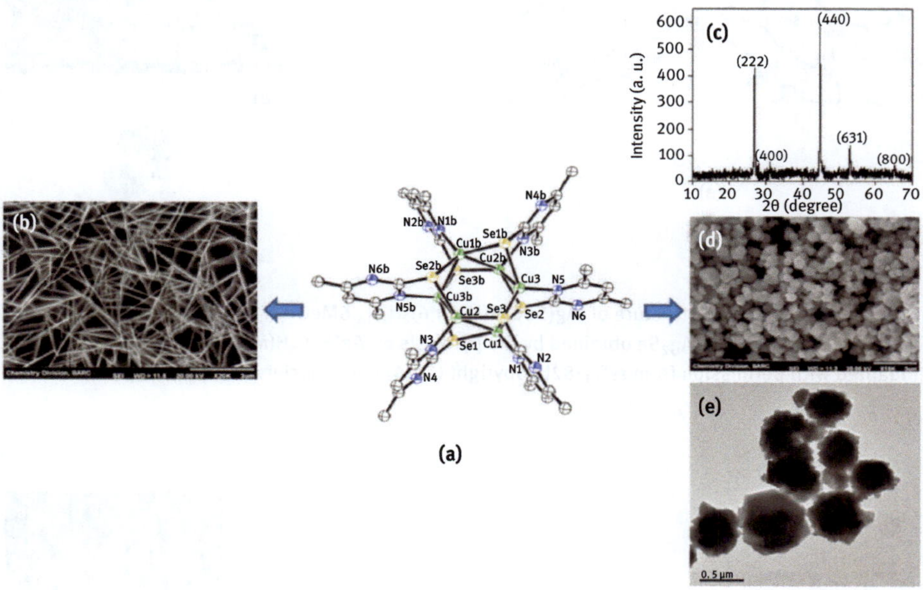

Figure 11.17: (a) Crystal structure of $[Cu\{SeC_4H(Me\text{-}4,6)_2N_2\}]_6.2H_2O$; (b) SEM image of Cu_7Se_4 thin film obtained by AACVD of $[Cu\{SeC_4H(Me\text{-}4,6)_2N_2\}]_6$; (c) XRD pattern, (d) SEM image and (e) TEM image of Cu_7Se_4 nanocrystals obtained by thermolysis of $[Cu\{SeC_4H(Me\text{-}4,6)_2N_2\}]_6$ in DDT at 150ºC (adapted with permission from ref. [182] Copyright (2011) Royal Society of Chemistry).

Tetranuclear copper pyridyltellurolate, $[Cu\{TeC_5H_3(3\text{-}R)N\}]_4$ (R = H or Me) in TOPO at 150 °C, yields spherical nanoparticles of hexagonal $Cu_{2-x}Te$ and tetragonal $Cu_{2.8}Te_2$, respectively [181]. High-quality films of $Cu_{1.85}Te$ could also be deposited on a glass substrate at 350 °C using $[Cu\{TeC_5H_3(3\text{-}Me)N\}]_4$ (Figure 11.19) [181]. Cu_2Te nanoparticles have been produced in the channels of MCM-41 when the complex $[Cu_6(TePh)_6(PEtPh_2)_5]$ loaded into the pores of MCM-41 is heated above 550 °C [185].

Unsymmetrical binary selenocarboxylates of copper(I), $[Cu_2\{\mu\text{-}SeC(O)R\}_2(PPh_3)_3]$ (R = Me, $4\text{-}XC_6H_4$ (X = H, F, Cl, Me, OMe), containing a tetrahedral and a trigonal

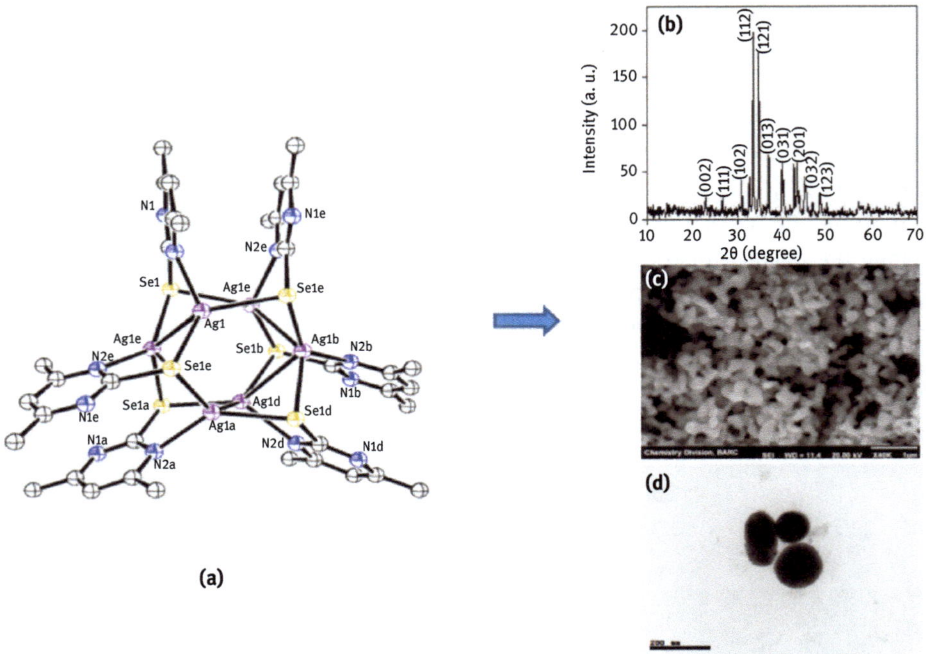

Figure 11.18: (a) Crystal structure of $[Ag\{SeC_4H(Me\text{-}4,6)_2N_2\}]_6.6MeOH.H_2O$, (b) XRD pattern, (c) SEM image, (d) TEM image of Ag_2Se obtained by the pyrolysis of $[Ag\{SeC_4H(Me\text{-}4,6)_2N_2\}]_6$ in DDT at 150ºC (adapted with permission from ref. [182] Copyright (2014) Royal Society of Chemistry).

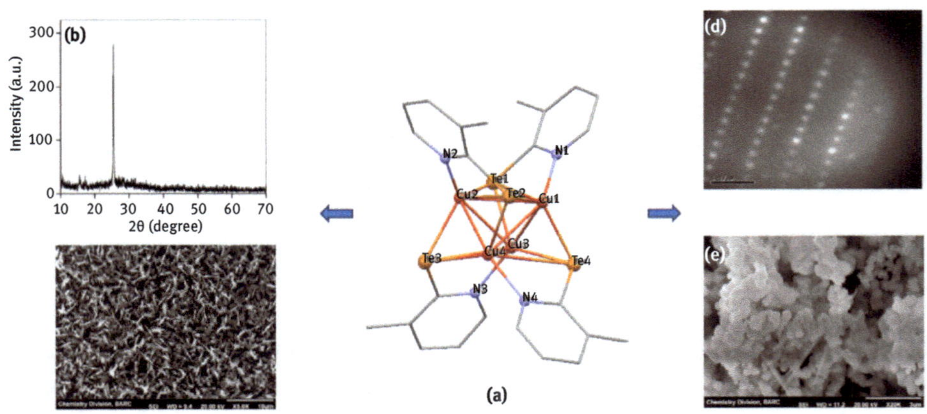

Figure 11.19: (a) Crystal structure of $[Cu\{TeC_5H_3(Me\text{-}3)N\}]_4$; (b) XRD pattern, (c) SEM image of $Cu_{1.85}Te$ thin film obtained by AACVD of $[Cu\{TeC_5H_3(Me\text{-}3)N\}]_4$ at 350°C, (d) SAED pattern, (e) SEM of $Cu_{2-x}Te$ obtained by the pyrolysis of $[Cu\{TeC_5H_3(Me\text{-}3)N\}]_4$ in HDA/TOPO at 150°C (adapted with permission from ref. [181] Copyright (2011) Royal Society of Chemistry).

planar copper centres, have been prepared by the reactions of $[Cu(NO_3)(PPh_3)_2]$ with NaSeCOR in acetonitrile [186]. These compounds on heating in vacuum at 300 °C give a mixture of Cu_2Se and $Cu_{2-x}Se$. However, thermolysis in TOP and TOPO mixture yields nanoparticles of $Cu_{2-x}Se$, with the size ranging from 40 to 100 nm (from TEM) [186]. The analogous silver complex, $[Ag_2\{SeC(O)Ph\}_2(PPh_3)_3]$, in TOP–HDA mixture yields monodispersed nanocubes of Ag_2Se. The size (20–74 nm) and morphology of the nanocubes can be controlled by varying the HDA to precursor ratio, heating time and thermolysis temperature [187].

Both copper(I)/(II) diselenocarbamate complexes have been utilized for the preparation of copper selenides. The copper(II) complexes, $[Cu(Se_2CNRR')_2]$ (R, R′ = alkyl group) are monomeric derivatives containing chelating diselenocarbamate ligands [188, 189]. Thermolysis of $[Cu(Se_2CNEt_2)_2]$ in TOPO yields monodispersed nanoparticles of CuSe [190]. The complexes, $[Cu(Se_2CNMeR)_2]$ (R = Bu^n, Hex^n), have been used for deposition of CuSe/$Cu_{2-x}Se$ films on glass substrate in the temperature range 450–500 °C by LPCVD [11, 189].

Copper(I) diselenocarbamates, $[Cu(Se_2CNPr^n{}_2)_2]$, are tetrameric air-sensitive derivatives [191]. The complex, comprising of tetrahedral framework of copper atoms capped by tri-connective diselenocarbamate ligands, is obtained by the reaction of $[Cu(MeCN)_4][PF_6]$ with $[Pr^nNH_2][Se_2CNPr^n{}_2]$. If $LiBH_4$ is added in the reaction mixture, octanuclear hydrido-bridged complexes, $[Cu_4(\mu_4\text{-}H)(\mu_3\text{-}Cu)_4(Se_2CNR_2)_6][PF_6]$ (R = Et or Pr^n), are isolated. The latter (R = Pr^n) on heating in a furnace at 200 °C yields hexagonal CuSe as revealed from XRD pattern [191]. The diselenophosphate, $[Cu\{Se_2P(OPr^i)_2\}]_4$, analogous to diselenocarbamate, has been employed for deposition of $Cu_{2-x}Se$ nanowires on silicon substrate at 360 °C [192].

Ditelluroimidodiphosphinate complexes of coinage metals, such as $[Cu\{TePPr^i{}_2)_2N\}]_3$, $[Ag\{TePR_2)_2N\}]_6$ (R = Pr^i or Ph) and $[Au\{TePPr^i{}_2)_2N\}(PPh_3)]$, have been used for deposition of metal telluride thin films on glass substrate by AACVD in the temperature range 300–500 °C [193]. Films showed the formation of mixture of hexagonal tellurium and metal tellurides (CuTe and Ag_7Te_4) from copper and silver precursors. However, the gold complex produced a mixture of cubic Au and monoclinic $AuTe_2$ [193].

11.4.9 Group 12 (Zn, Cd, Hg) chalcogenides

Group 12 chalcogenides are technologically important family of semiconductor materials, with CdTe being one of the most studied materials. The shape- and size-dependent optical properties of these materials make them useful candidates for a wide range of optoelectronic devices such as light-emitting displays, photovoltaic and biomedical tagging.

Zinc and cadmium chalcogenolates are polymeric, high melting solids showing a wide structural diversity. The structures of these complexes are strongly influenced by the nature of organic group on chalcogen atom. Thus, admantoid cages to one- (e. g. $[M(SeMes)_2]_\infty$ [194], $[M(TeMes)_2]_\infty$ [60, 195] (M = Zn, Cd), two- and three-dimensional lattice frameworks are formed. However, with very bulky R group, low nuclearity molecules can be isolated. For example, $[Cd(SeC_6H_2Bu^t{}_3\text{-}2,4,6)_2]$ adopts a dimeric structure with trigonal planar coordination around Cd in the solid state, but in solution a monomeric structure persists [61, 194]. Oligomeric compounds dissolve in coordinating solvents, such as pyridine, tmeda, to give monomeric adducts, e. g. $[Zn(TeMes)_2(py)_2]$ [195], $[Zn(SeSiMe_3)_2(tmeda)]$ [112]. Mercury chalcogenolates tend to adopt a two-coordinate linear E-Hg-E arrangement, although polymeric compounds are also reported, e. g. $[Hg(SeMe)_2]$ which forms infinite one-dimensional chain [110].

Several zinc, cadmium and mercury chalcogenolates, such as $[Cd(SePh)_2]_n$ [196, 197], $[Cd(SeC_6H_2Bu^t{}_3\text{-}2,4,6)_2]_2$ [61], $[M(Sepy)_2]$ (M = Cd or Hg) [198], $[M(TeAr)_2]$ (M = Cd or Hg; Ar = Ph or tol) [199], $[Zn(TeMes)_2(py)_2]$ [195], $[M\{TeSi(SiMe_3)_2\}_2]_2$ (M = Zn or Cd) [200], on pyrolysis in the solid state yield ME (M = Zn, Cd or Hg; E = Se or Te). When a trioctylphosphine (TOP) solution of monomeric zinc complexes, $[Zn(EPh)_2(tmeda)]$, obtained by dissolution of oligomeric $[Zn(EPh)_2]_\infty$ in tmeda [201, 202], is injected in hot TOPO (when E = Se) or dodecylamine/trioctylamine (when E = Te) monodispersed quantum dots (QDs) of ZnE are isolated. The size of QDs depends on the growth temperature [201, 202]. Solution phase photolysis of $Hg(TeBu^n)_2$ gives nanocrystals of HgTe [197].

Group 12 metal complexes with internally functionalized chalcogen ligands such as *N,N'*-dimethylaminoalkylchalcogenolate and 2-pyridyl/2-pyrimidyl chalcogenolate have been developed in authors' group as molecular precursors for metal selenides/tellurides [203–210]. The monomeric complexes, $[M(Se(CH_2)_nNMe_2)_2]$ (M = Zn, Cd; n = 2 or 3), obtained by the reactions of metal salts with $NaSe(CH_2)_nNMe_2$, undergo a one- or two-(overlapping) step decomposition to give exclusively MSe [203–205]. Both hexagonal and cubic CdSe could be isolated by pyrolysis of $[Cd(SeCH_2CH_2NMe_2)_2]$ in vacuum and flowing inert gas, respectively [204]. Thermolysis in HDA–TOPO mixture affords HAD-capped cubic CdSe nanoparticles which display typical band edge absorption [204]. Attempts to prepare analogous mercury complex, $[Hg(SeCH_2CH_2NMe_2)_2]$ result into the formation of a black powder identified as cubic HgSe [204]. In contrast, the complexes $[Hg(TeCH_2CH_2NMe_2)_2]$ [206] and $[Hg(SeCH_2CH_2CH_2NMe_2)_2]$ [205] could be isolated in pure form which on pyrolysis under flowing argon in a furnace gives polycrystalline cubic HgE (E = Se or Te).

Monomeric cadmium $[Cd(SeAr)_2(tmeda)]$ and mercury $[Hg(SeAr)_2]/[Hg(SeAr)_2(tmeda)]$ (Ar = 2-py, 4,6-$Me_2C_4HN_2$) have been used for the synthesis of metal selenide nanocrystals [209, 210]. Thermolysis of cadmium complexes in coordinating solvents (TOPO, OA or HDA) gives red-coloured hexagonal CdSe nanocrystals (Figure 11.20), whereas mercury complexes in HDA or OA at 100–145 °C yield cubic HgSe nanoparticles (Figure 11.21) [209, 210].

2-Pyridyl tellurolate complexes $[M\{TeC_5H_3(R\text{-}3)N\}_2]$ (M = Cd or Hg; R = H or Me) on pyrolysis either in the solid state or in coordinating solvents yield metal tellurides

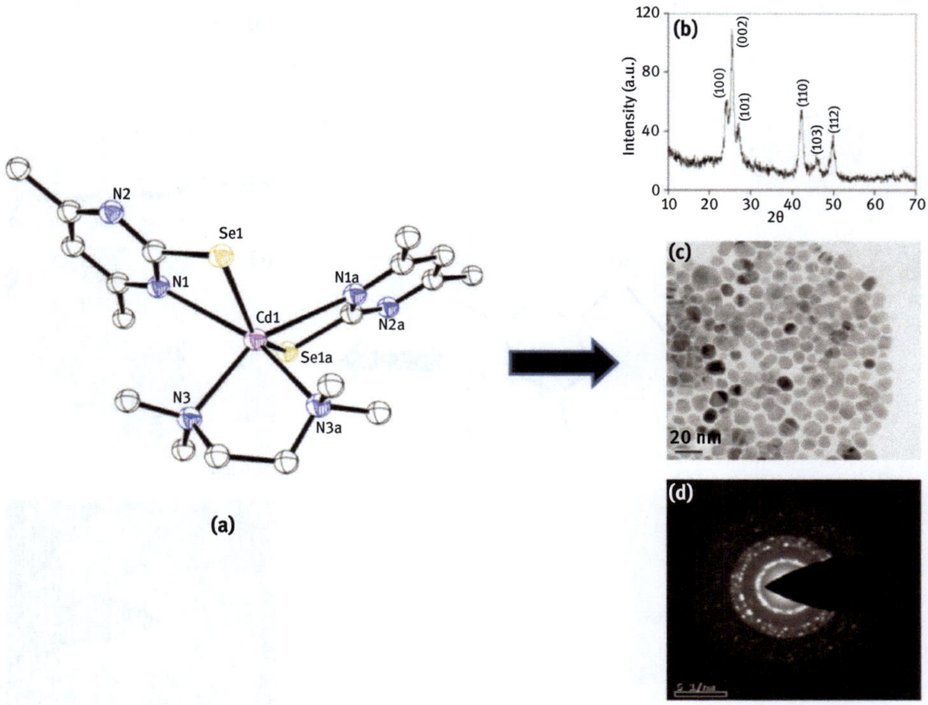

Figure 11.20: (a) Molecular structure of [Cd{$SeC_4H(Me\text{-}4,6)_2N_2$}$_2$(tmeda)], (b) XRD pattern, (c) TEM image and (d) SAED pattern of CdSe obtained by pyrolysis of [Cd{$SeC_4H(Me\text{-}4,6)_2N_2$}$_2$(tmeda)] in OA/TOPO at 230 °C (adapted with permission from ref. [210] Copyright (2014) Elsevier Publishers).

[208]. The cadmium complex under an argon atmosphere at 350 °C yields cubic CdTe, whereas in HDA or TOPO at 160 °C affords red or yellow spherical nanocrystals of hexagonal CdTe. In contrast, mercury complexes do not show such anomaly and invariably give cubic HgTe (Figure 11.22) [208].

Besides chalcogenolate derivatives discussed above, several other complexes containing bidentate chalcogen ligands have been developed as precursors for the synthesis of metal chalcogenides. The facile cleavage of the C–Se linkage in selenocarboxylates offers an attractive potential for MSe synthesis. Accordingly, zinc and cadmium complexes, [M(SeCOAr)$_2$(tmeda)], have been synthesized by the reactions of MCl_2(tmeda) with KSeCOAr (Ar = Ph or tol). Similar reaction with mercury, however, gives a black powder identified as cubic HgSe. The zinc and cadmium complexes undergo a three-step decomposition (by TGA), leading to the formation of polycrystalline metal selenide nanocrystals. Pyrolysis of [Cd(SeCOtol)$_2$(tmeda)] in HDA at 230 °C yields red-coloured cubic CdSe nanoparticles of average diameter ~15 nm [211].

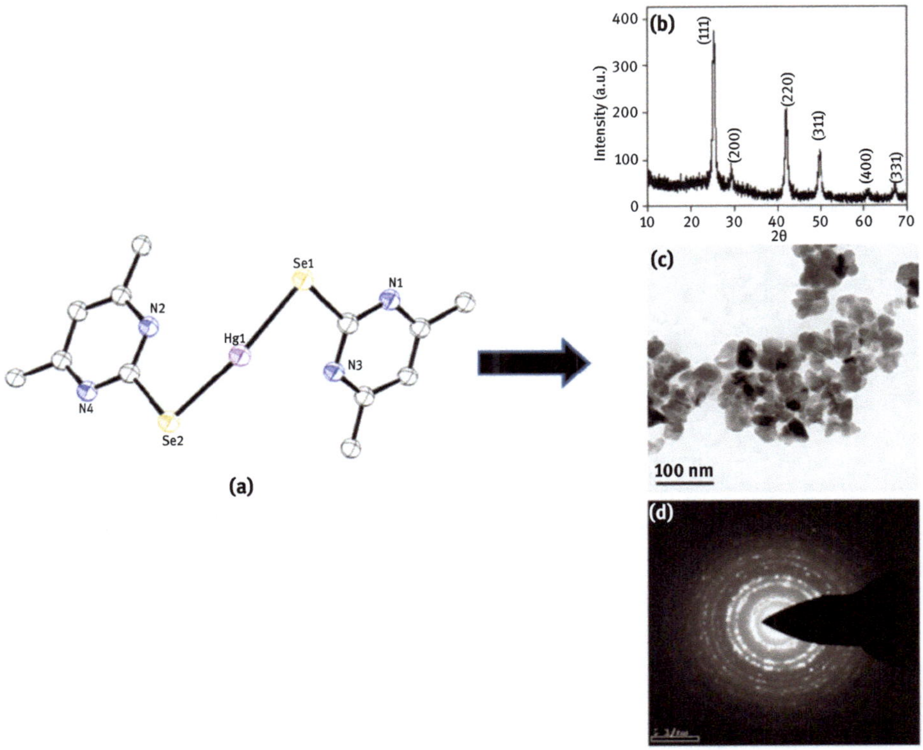

Figure 11.21: (a) Crystal structure of $[Hg\{SeC_4H(Me\text{-}4,6)_2N_2\}_2]$, (b) XRD pattern, (c) TEM image, (d) SAED pattern of HgSe obtained by pyrolysis of $[Hg\{SeC_4H(Me\text{-}4,6)_2N_2\}_2]$ in OA/OA at 100 °C (adapted with permission from ref. [210] Copyright (2014) Elsevier Publishers).

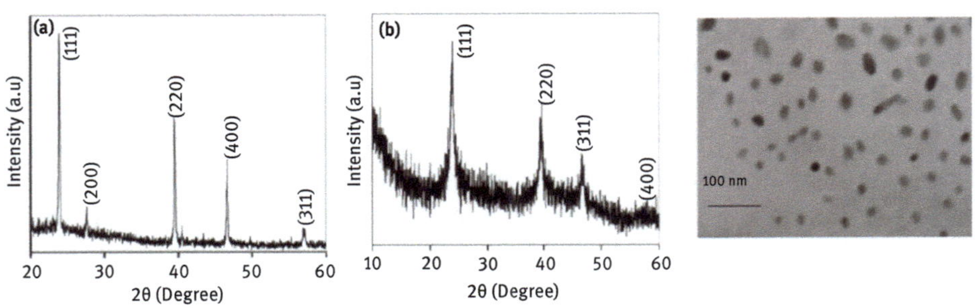

Figure 11.22: XRD pattern of HgTe obtained by the pyrolysis of (a) $[Hg(TepyMe)_2]$ in a furnace at 250 °C for 1 h, (b) XRD pattern and (c) TEM image of HgTe nanoparticles obtained by the pyrolysis of $[Hg(Tepy)_2]$ in TOPO at 145 °C for 20 min (adapted with permission from ref. [208] Copyright (2009) Royal Society of Chemistry).

Organo-zinc and -cadmium diselenocarbamates, $[RM(Se_2CNR'_2)]_2$ (R = Me, Et, Bu^tCH_2; M = Zn or Cd) [212–215] and bis(diselenocarbamates), $[M(Se_2CNR'_2)_2]$ (M = Zn or Cd) [216–220], have been used for the preparation of metal selenide nanocrystals and for deposition of MSe films by LP–MOCVD method. The complexes $[M(Se_2CNMeHex^n)_2]$ in TOPO at 200 °C yield TOPO-capped monodispersed MSe nanocrystals [219, 220]. These complexes have also been used for deposition of thin films of MSe on a glass substrate in the temperature range 400–450 °C [218]. In contrast, $[Cd(Se_2CNEt_2)_2]$ under similar conditions gives only elemental selenium, although bulk pyrolysis results into hexagonal CdSe [216]. Unlike $[Cd(Se_2CNEt_2)_2]$, bulk decomposition and MOCVD film growth experiments employing organometallic derivatives, $[RM(Se_2CNEt_2)]_2$ (R = Me or Et), prepared by comproportionation reactions between R_2M and $[M(Se_2CNEt_2)_2]$, results into MSe [212, 214]. Formation of different phases can, however be encountered. For instance, pyrolysis of $[RZn(Se_2CNEt_2)]_2$ (R = Me or Et) in a sealed tube at 285 °C gives hexagonal ZnSe, whereas films grown at 450 °C/10^{-2} torr consist of cubic phase of ZnSe [212].

Phosphinochalcogenoic amidated are yet another family of bidentate ligands used for the synthesis of group 12 complexes, $[M\{(EPR_2)_2N\}_2]$ (M = Zn, Cd or Hg; E = Se or Te; R = Pr^i, c-Hx) [221–226]. These complexes are monomeric containing chelating chalcogen ligand, with the metal centre acquiring a distorted tetrahedral configuration [221, 227]. Thermolysis of $[Cd\{(SePPr^i_2)_2N\}_2]$ in TOP–TOPO mixture yields monodispersed TOPO-capped CdSe nanocrystals, size (5.4–6.4 nm) of which depends on the growth time [223]. Zinc and cadmium complexes, $[M\{(EPR_2)_2N\}_2]$ (M = Zn or Cd; R = Pr^i or Ph), have been employed as precursors for deposition of polycrystalline films of hexagonal MSe by LP–MOCVD method [222, 224]. The tellurium analogue, $[Cd\{(TePPr^i_2)_2N\}_2]$, has also been successfully used for the deposition of cubic CdTe films at 425–475 °C [225, 226]. In contrast, the mercury derivative $[Hg\{(TePPr^i_2)_2N\}_2]$ deposits hexagonal tellurium films under similar conditions, which is attributed to reductive elimination of mercury from the complex at higher temperature [225].

11.5 Main group chalcogenolates as precursors for metal chalcogenides

11.5.1 Group 13 chalcogenides

Group 13 elements, particularly gallium and indium, form several chalcogenides of varying compositions and exist in a number of phases [228]. These chalcogenides (ME and M_2E_3) find applications in optoelectronic and photovoltaic devices as well as serve as passivating and dielectric material for III–V (e. g. GaAs) devices. Most of the precursors have been employed for deposition of metal chalcogenide thin films on various substrates (e. g. glass, Si(100), GaAs(100)), and only few reports deal with the preparation of nanocrystals [229].

The tris-selenolate of indium, $[In(SeC_5H_3(Me\text{-}3)N\}_3]$, prepared by the reaction of $InCl_3$ with $NaSeC_5H_3(Me\text{-}3)N$ in benzene–methanol mixture, on thermolysis in HDA, TOPO or OA at 330 °C yield agglomerated spheres (in HDA), spherical morphology (in TOPO) and sheet-like structures (in OA) of In_2Se_3. The average diameter and length of the rods obtained from HDA are 57–76 nm and 400 nm, respectively (Figure 11.23) [230]. Gallium and indium selenide (MSe) nanoparticles have been prepared by vapour phase thermolysis of cubane precursors, $[Bu^tGaSe]_4$ and $[(EtMe_2C)InSe]_4$ [231]. The average diameters of gallium and indium selenide nanoparticles are 42 and 88 nm, respectively.

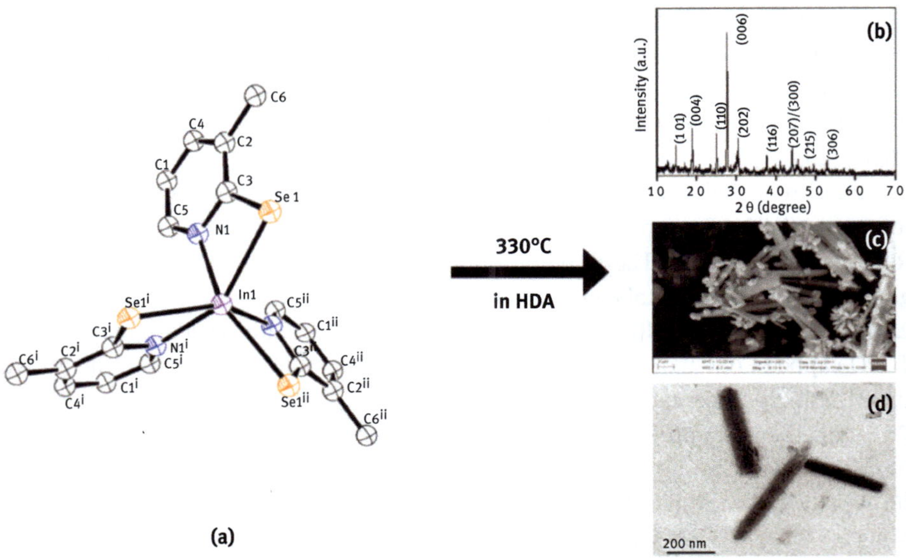

Figure 11.23: (a) Molecular structure of $[In\{SeC_5H_3(Me\text{-}3)N\}_3]$, (b) XRD pattern, (c) SEM, (d) TEM images of In_2Se_3 obtained by thermolysis of $[In\{SeC_5H_3(Me\text{-}3)N\}_3]$ at 330 °C for 10 min in HDA (adapted with permission from ref. [230] Copyright (2013) Elsevier Publishers).

In contrast to the extensively investigated MOCVD of III–V materials, there are only a few MOCVD approaches for Group 13 chalcogenide materials. Both homoleptic and organometallic gallium and indium chalcogenolates/chalcogenides have been used as molecular precursors for deposition of thin films. Homoleptic indium selenolates, $[In(SePh)_3]$ [232], $[In\{SeC(SiMe_3)_3\}_3]$ [233], have been used for deposition of In_2Se_3 films. Deposition on GaAs(100) substrate at 470–530 °C from the former precursor produced highly oriented In_2Se_3 films, but on Si(100) substrate at 400 °C gave mainly hexagonal In_2Se_3 with minor contamination with the cubic phase of InSe [232]. Similarly diselenocarbamate complexes, $[In\{Se_2CNMeR\}_3]$ (R = Bu^n, Hx^n), have been used to grow films of cubic In_2Se_3 on glass or InP(111) substrate in the temperature range 450–500 °C [11, 234].

Monomeric diorgano-gallium and -indium complexes of selenophosphonylamide, $[Me_2Ga\{(SePPr^i_2)_2N\}]$ and $[R_2In\{(SePPr^i_2)_2N\}]$ (R = Me or Et) in a MOCVD process deposit cubic Ga_2Se_3 and hexagonal γ-In_2Se_3 films on a glass substrate in the temperature range 450–500 °C. The band gap values of these films, determined using UV-vis spectroscopy, are 2.74 and 1.78 eV for Ga_2Se_3 and In_2Se_3, respectively [235]. In contrast, the complex $[Me_2In(SePh)]_2$, prepared by redistribution reaction between $InMe_3$ and $[In(SePh)_3]$, yields films of InSe by spray MOCVD process [232]. The films grown at 365 °C on GaAs showed cubic phase of InSe and grew in perpendicular direction to the substrate surface. However, deposition in the temperature range 400–476 °C primarily resulted into hexagonal phase of β-InSe [232].

Barron and co-workers employed cubane clusters of gallium, $[RGa(\mu_3\text{-}E)]_4$ (**X**), for deposition of hexagonal GaE films [231, 236], which are polycrystalline in nature and grew as clusters. The tellurium precursors decompose at lower temperatures (285–310 °C) than the corresponding selenium compound (325–370 °C). The cyclopentadienyl cluster, $[Cp^*Ga(\mu_3\text{-}E)]_4$, yields Ga_2Se_3 films rather than GaSe [237].

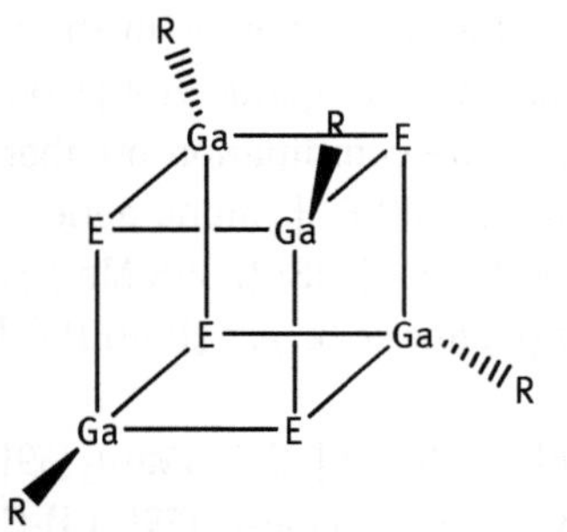

R = Bu^t, $CEtMe_2$, CEt_2Me; Cp*; E = Se or Te

X

Tellurophosphorylamide complexes, $[M(\mu\text{-}Te)\{(TePPr^i)_2N\}]_3$ (M = Ga or In), have been used as molecular precursors for deposition of metal telluride films on glass and silicon(100) substrates at deposition temperature 325–475 °C [238]. The indium complex deposited pure cubic In_2Te_3 films, whereas the gallium complex gave a mixture comprising of cubic Ga_2Te_3, monoclinic GaTe and hexagonal Te. The grain size of In_2Te_3 increases with an increase in deposition temperature.

11.5.2 Group 14 chalcogenides

Group 14 chalcogenide materials have potential applications in infrared detectors, photovoltaic solar cells and thermoelectric devices, PbTe being an important mid-range temperature (325–525 °C) thermoelectric material [239]. The ME (M = Ge, Sn or Pb; E = S, Se or Te) crystallizes in the rock salt (NaCl) structure, with M occupying the cation sites and chalcogen forming the anionic lattice. Among group 14 chalcogenide nanomaterials, lead chalcogenides were the first candidates explored since mid-1980s [229, 240]. They show strong quantum confinement effect due to large Bohr exciton radii (20 nm for PbS and 46 nm for PbSe) which is smaller for other classes of semiconductors such as II–VI and III–V (e. g. CdTe and GaAs) [240]. The large Bohr exciton radii of PbE allow tailoring band gap energies of nanocrystals. This ease of size tumeability of PbE nanocrystals helps in device design. However, the toxicity issues associated with lead has prompted exploration of alternative materials with less toxicity and similar properties, such as germanium and tin chalcogenides [240]. Germanium chalcogenide nanocrystals are still under explored, and only recently synthesis of photoresponsive orthorhombic GeS nanocrystals using $[Ge(S_2COR)_4]$ as precursors has been reported [241].

A wide variety of germanium, tin and lead chalcogenolate complexes have been synthesized. Tin compounds have been isolated both in +2 and +4 oxidation states. The lead derivatives have been obtained in +2 oxidation state in consonance with the prevalence of the relativistic effects in the sixth period. The chalcogenolate complexes in +4 oxidation state in general have discrete monomeric structures. In divalent state, several coordination geometries are possible [242] which are governed by the effect of a lone pair of electron and the interplay of steric demands of organic groups on chalcogen versus the tendency of the metal centre to acquire coordination number greater than two. Thus, for compounds of general formula $[M(ER)_2]$, monomeric (e. g. $[Pb(SeCH_2CH_2NMe_2)_2]$ [243]), dimeric (e.g. $[Sn(Sepy)_2]_2$ [244], $[Sn\{ESi(SiMe_3)_3\}_2]_2$ [59]), trimeric (e. g. $[Sn\{SC_6H_3Pr^i_2\text{-}2,6\}_2]_3$ [245]) and 1-D polymeric (e. g. $^1_\infty[Sn(EPh)_2]$ (E = S, Se, Te) [246]), have been isolated.

Tin(II) compounds, $[Sn(ER)_2]$, (ER = SePh [246], TePh [246], $ESi(SiMe_3)_3$ [59], 2-Sepy [244]) and lead selenolates $[Pb(SeR)_2]$ (R = 2-py [244], $Si(SiMe_3)_3$ [59], $CH_2CH_2NMe_2$ [243], CH_2CH_2COOMe [243]) undergo a single-step decomposition to yield a black powder identified as ME (M = Sn or Pb; E = Se or Te) by XRD. The tellurolate complexes decompose at much lower temperatures than the corresponding selenolates. The decomposition is accompanied by elimination of chalcogeno ethers (R_2E) [59, 246]. Thermolysis of $[Pb(SeCH_2CH_2NMe_2)_2]$ in HDA at 120 °C results into quantum confined spherical nanocrystals (av. 15 nm diameter from TEM) of cubic PbSe (Figure 11.24) [243].

Different shapes of cubic PbSe have been prepared via a solvothermal process using lead diselenophosphate, $[Pb\{Se_2P(OPr^i)_2\}_2]$ [247]. Thermolysis in H_2O/polyvinylpyrolidone (10:0.4 ml, at 150 °C), H_2O/ethylenediamine (2:8 ml, at 220 °C) and ethylenediamine (220 °C) in an autoclave gives cubic like, truncated octahedral and eight-horned rod dendrites, respectively (Figure 11.25) [247]. Dichalcogenido imidodiphophates, $[M\{(EPR_2)_2N\}_2]$ (M = Sn or Pb; E = Se or Te; R = Pr^i or Ph) isolated by salt elimination reactions [248–250], containing chelating dichalcogenolate ligands, are yet another promising precursor. The lead tellurolate complex, $[Pb\{(TePPr^i_2)_2N\}_2]$, has been employed as a suspension in TOP for deposition of thin films of cubic PbTe on a preheated Si/SiO_2 (100) substrate at 200–220 °C. The films comprise of wires of 0.2–1.1 µm width and several tens of micrometre length [250]. Poor-quality films are

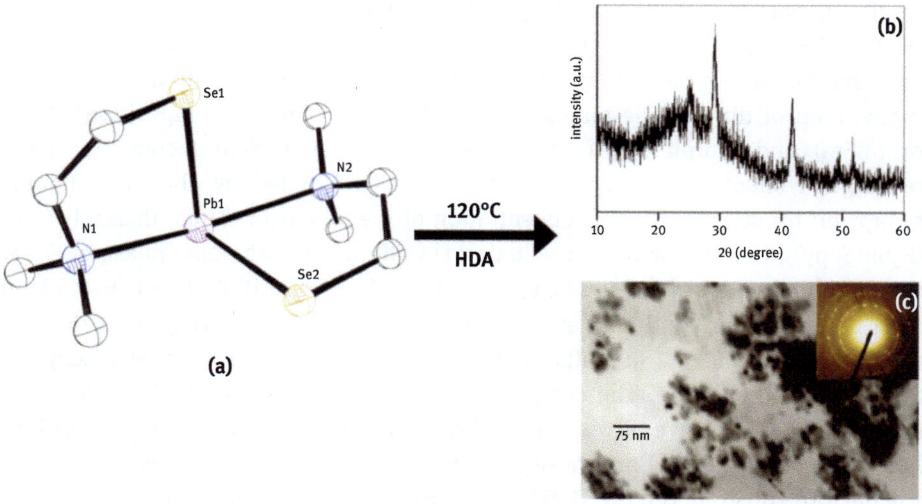

Figure 11.24: (a) Molecular structure of X-ray of $[Pb(SeCH_2CH_2NMe_2)_2]$, (b) XRD pattern and (c) TEM image of PbSe nanocrystals obtained by thermolysis of $[Pb(SeCH_2CH_2NMe_2)_2]$ in HDA at 120 °C (adapted with permission from ref. [243] Copyright (2009) Elsevier Publishers).

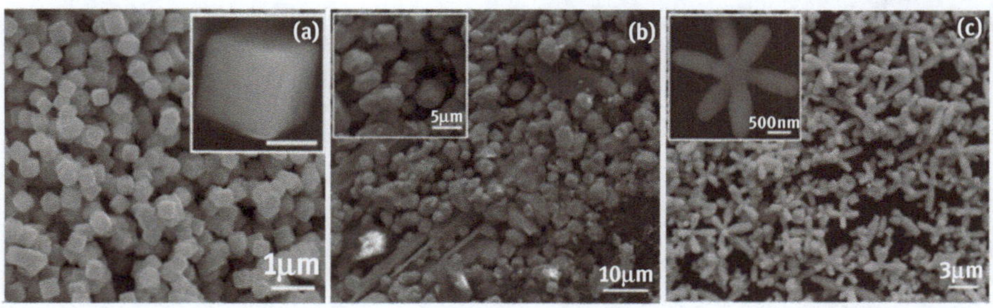

Figure 11.25: SEM images of PbSe (a) cubes fabricated at 150 °C in H_2O/PVP, (b) truncated octahedra at 220 °C using ethylenediamine/H_2O (8:2), (c) eight-horned rod dendrite at 220 °C using ethylenediamine, by solvothermolysis of $[Pb\{Se_2P(O^iPr)_2\}_2]_n$ (adapted with permission from ref. [247] Copyright (2010) Royal Society of Chemistry).

produced on a glass substrate at 375–475 °C when a THF/CH_2Cl_2 solution of the precursor was used [250]. The X-ray photoelectron spectroscopy (XPS) analysis revealed highly oxidized surface of PbTe with oxygen contents of 20 wt.%.

Tin(IV) precursors yield tin mono- or di-chalcogenides (SnE and SnE_2) on pyrolysis. The homoleptic tin selenolate, $[Sn(SePh)_4]$, on pyrolysis at 300 °C yields nanocrystals of SnSe [251], whereas analogous pyridine selenolate, $[Sn(Sepy)_4]$, decomposes at this temperature to $SnSe_2$ [244]. The organotin chalcogenides, such as $(Ph_3Sn)_2E$ (E = S, Se, Te) [252, 253] and $[\{(Me_2Si)_2CH\}_2Sn(\mu\text{-}Se)]_2$ [254], on decomposition give SnE. The latter has also been employed for deposition of SnSe thin films by LPCVD [254].

Diorganotin(IV) 2-pyridyl and 2-pyrimidyl selenolates, $[R_2Sn(SeAr)_2]$ and $[R_2SnCl(SeAr)]$ (R = Me, Et, Bu^n, Bu^t; Ar = 2-C_5H_4N, 2-C_5H_3(3-Me)N, 2-C_4H(4,6-Me_2)N_2), have been synthesized in the authors' group and have been successfully utilized for the preparation of tin selenide nanostructures [57, 255]. Different compositions (SnSe or $SnSe_2$) and morphologies of tin selenides ranging from rectangular bars, rectangular sheets to hexagonal sheets could be isolated by changing organic groups on tin/selenium or by concentration of the precursor. Thus, thermolysis of $[R_2Sn(Sepy)_2]$ (R = Me or Bu^t) in refluxing OA affords orthorhombic phase of SnSe, whereas $[Et_2Sn(SeAr)_2]$ (Ar = 2-C_5H_4N, 2-C_5H_3(3-Me)N) and $[Bu^t_2Sn\{SeC_4H(4,6\text{-}Me_2)N_2\}_2]$ under similar conditions give hexagonal phase of $SnSe_2$ (Figure 11.26). The dilute OA solution of $[Bu^t_2Sn(Sepy)_2]$ at 210 °C yield nanocrystals of hexagonal $SnSe_2$ rather than SnSe. The complexes, $[Bu^t_2Sn(Sepy)_2]$ [255] and $[Bu^t_2Sn\{SeC_4H(4,6\text{-}Me_2)N_2\}_2]$ [57], have also been employed for deposition of phase pure orthorhombic SnSe (at 490 °C) and hexagonal $SnSe_2$ (at 375 °C), respectively, on *p*-type silicon(100) substrate (Figure 11.27). The SnSe films show continuous grain boundaries that overlap on the substrate. These films showed photo response on a Si/SnSe/Au configuration with 20 % increase in the current [255].

Figure 11.26: (a) SEM and (b) TEM images of SnSe obtained by the thermolysis of $[Et_2Sn\{2\text{-}SeC_5H_3(Me\text{-}3)N\}_2]$ in 6 ml of OLA at 215 °C (adapted with permission from ref. [255] Copyright (2012) Royal Society of Chemistry).

11.5.3 Group 15 chalcogenides

Antimony and bismuth chalcogenides of composition M_2E_3 (M = Sb or Bi; E = Se or Te) are layered semiconductor materials that are stacked along the *c*-axis and find several applications in room-temperature thermoelectric devices [239]. Schulz has reviewed the synthesis and structural characterization of antimony and bismuth compounds (Figure 11.28) with selenium and tellurium ligands and their use as single-source molecular precursors for the preparation of M_2E_3 materials [256]. These compounds generally adopt distorted structures due to the presence of stereochemically active lone pair of electrons at Sb/Bi [256].

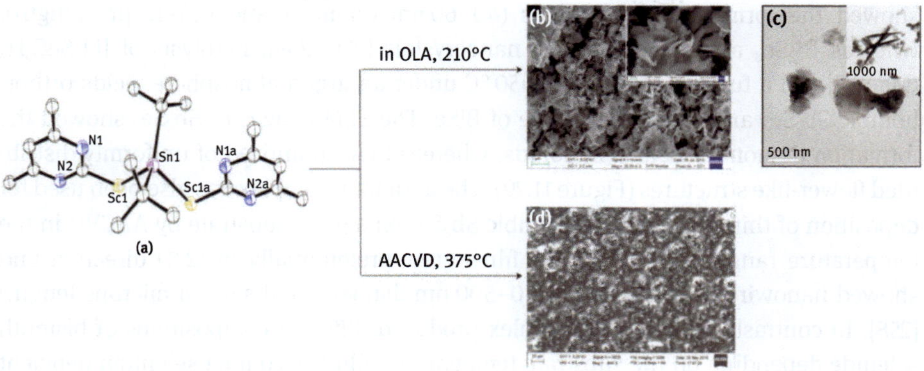

Figure 11.27: (a) Molecular structure of $[Bu^t{}_2Sn\{SeC_4H(Me\text{-}4,6)_2N_2\}_2]$, (b) and (c) SEM and TEM images of $SnSe_2$ nanosheets, (d) SEM image of $SnSe_2$ thin film (adapted with permission from ref. [57] Copyright (2016) Royal Society of Chemistry).

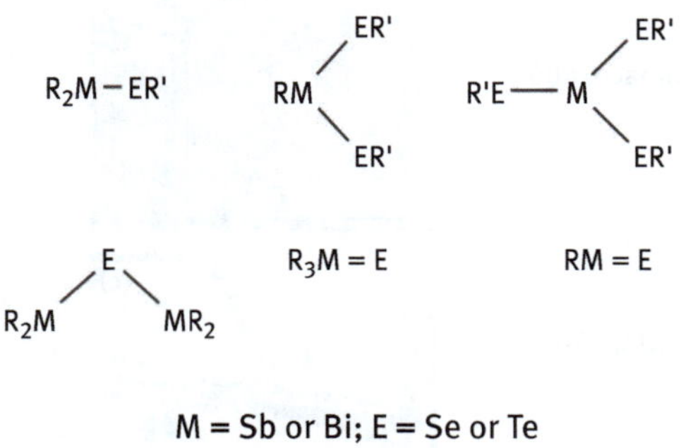

Figure 11.28: Antimony and bismuth compounds with selenium and tellurium ligands [256].

Tris selenolate complexes of antimony and bismuth have been employed for the synthesis of antimony and bismuth selenides. The antimony derivatives generally yield orthorhombic Sb_2Se_3, while the bismuth compounds give different products including Bi_2Se_3 depending on the nature of precursor and thermolysis conditions. The complexes $[M(SeC_6H_2Bu^t{}_3\text{-}2,4,6)_3]$ (M = Sb or Bi) on pyrolysis at 200 °C yield Sb_2Se_3 (when M = Sb) elemental bismuth and Bi_2Se_3 (when M = Bi) with the elimination of the selenoether, $(2,4,6\text{-}Bu^t{}_3C_6H_2)_2Se$ [257]. In contrast, the analogous compounds with less bulky selenolate group, $[M(SeC_6H_2Me_3\text{-}2,4,6)_3]$, decomposed to elemental Sb/Bi and diselenide Mes_2Se_2 [257]. The 2-pyridyl and 2-pyrimidyl selenolates, $[M(SeAr)_3]$ (M = Sb or Bi; Ar = $2\text{-}C_5H_4N$, $2\text{-}C_5H_3(Me\text{-}3)N$, $2\text{-}C_4H\,(Me_2\text{-}4,6)N_2$), developed in the authors' group [258, 259], have been successfully employed for the synthesis of M_2Se_3 nanomaterials. Thermolysis of $[M\{SeC_5H_3(Me\text{-}3)N\}_3]$ (in HDA) [258] and $[M\{SeC_4H(Me_2\text{-}4,6)_2N_2\}_3]$ (in OA) [259] at 200 °C under an argon atmosphere affords orthorhombic and rhombohedral phases of Sb_2Se_3 and Bi_2Se_3, respectively. The TEM images of Sb_2Se_3

showed the formation of nanorods (40–60 nm diameter and 0.6–1.4 μm length), whereas Bi_2Se_3 appeared flake-like nanocrystals [258, 259]. Pyrolysis of [M{SeC_5H_3(Me-3)N}$_3$] in a furnace at 400 and 450 °C under an argon atmosphere yields orthorhombic Sb_2Se_3 and hexagonal phase of BiSe. The SEM images of Sb_2Se_3 showed the formation of monodispersed nanorods, whereas BiSe comprised of uniformly distributed flower-like structures (Figure 11.29). The antimony complex has also been used for deposition of thin films of orthorhombic Sb_2Se_3 on a glass substrate by AACVD in the temperature range 375–500 °C. The films grew preferentially in (221) direction and showed nanowire-like structures (240–500 nm diameter and several microns length) [258]. In contrast, the bismuth complex produced different compositions of bismuth selenide depending on the substrate temperature, films becoming selenium deficient with increasing temperature. Thus, films of rhombohedral Bi_3Se_4 (at 350 °C), hexagonal BiSe (at 375–425 °C) and hexagonal Bi_3Se_2 (at 475 °C) could be deposited [258].

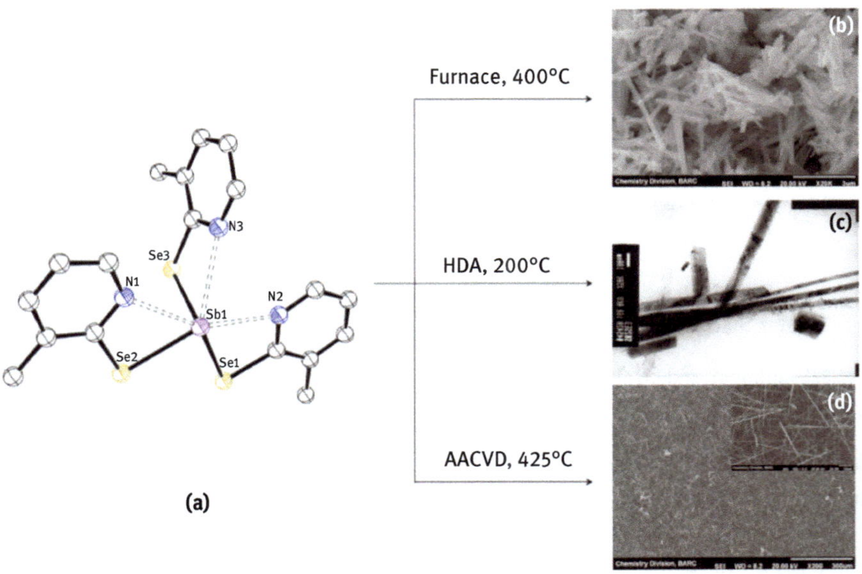

Figure 11.29: (a) Molecular structure of [Sb{Se-C_5H_3(Me-3)N}$_3$].1.5H_2O, (b) SEM image of Sb_2Se_3 obtained by pyrolysis of [Sb{Se-C_5H_3(Me-3)N}$_3$].1.5H_2O in a furnace (400 °C), (c) TEM image of Sb_2Se_3 nanorods obtained in HDA (200 °C) and (d) SEM image of Sb_2Se_3 film obtained by AACVD of [Sb{Se-C_5H_3(Me-3)N}$_3$] at 425 °C (adapted with permission from ref. [258] Copyright (2010) Royal Society of Chemistry).

The tris chelates of diseleno ligands, such as diselenocarbamate [260], imino bis (diorganophophine selenide) [261, 262] and diselenophosphate [263, 264], have been employed for the preparation of M_2Se_3. Bismuth tris(diorganoselenocarbamate), [Bi(Se_2CNRR′)$_3$] (R/R′ = Et/Et; Bu^n/Bu^n; Me/Bu^n; Me/*c*-Hx), obtained from a reaction mixture containing primary amine, CSe_2 and Bi_2O_3 in methanol at RT, have been used for deposition of thin films of rhombohedral Bi_2Se_3 at 475 °C on a glass substrate [260]. Different compositions of bismuth selenide films have been deposited on glass substrate by subtle variation of substituents in $(R_2PSe)_2N$ ligand [262]. Accordingly, monophasic Bi_2Se_3 and BiSe films have been prepared from [Bi{(SePPr$^i_2$)$_2$N}$_3$] and [Bi{(SePPh$_2$)$_2$N}$_3$], respectively (Figure 11.30) [262]. Solvothermal decomposition of a tris chelate, [Sb{Se$_2$P(OPr^i)$_2$}$_3$] in methanol at 100 and 150 °C in an autoclave affords nanowires of orthorhombic Sb_2Se_3 which show direct band gap absorption of 1.49 eV [263]. The nanowires formed at 150 °C are longer (70–80 nm diameter and 3–5 µm length) than those formed at 100 °C (30–50 nm diameter and 2–3 µm length) [263]. In contrast, the analogous bismuth complex, [Bi{Se$_2$P(OPr^i)$_2$}$_3$], yields nanowires of $BiPO_4$ and Bi_2Se_3 nanoplates simultaneously [264].

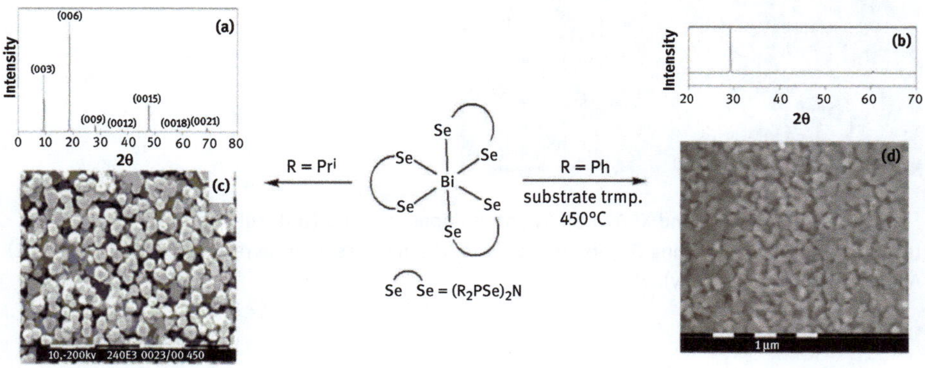

Figure 11.30: (a, c) XRD pattern and SEM image of Bi_2Se_3 film and (b, d) XRD pattern and SEM image of thin film of BiSe grown via LP–MOCVD of Bi[(SePR$_2$)$_2$N]$_3$ (R = ^{i}Pr or Ph) at T_{subs} = 475 and T_{subs} = 450 °C, respectively (adapted with permission from ref. [262] Copyright (2004) American Chemical Society).

Organoantimony tellurolates and tellurides have been used for the preparation of Sb_2Te_3 nanocrystals and thin films. Gupta and Kim [265] synthesized diphenylantimony tellurolates, Ph_2SbTeR (R = Et or Ph), as brown liquids by the reaction between Ph_2SbCl and LiTeR in THF. Both the compounds on decomposition in OA at 250 °C (R = Ph) and 300 °C (R = Et) yield nanoplates of Sb_2Te_3. The hexagonal shaped Sb_2Te_3 nanoparticles have an edge length of 0.4–2.0 µm and thickness of 20–50 nm [265]. Thermal decomposition of $Ph_2SbTePh$ in a closed glass ampule in the absence of a solvent, however, yields black particles of Sb [265].

Schulz and co-workers isolated hexagonal nanoplates of Sb_2Te_3 (400 nm diameter and 35 nm thickness) by thermolysis of $(Et_2Sb)_2Te$ in 1,3-diisopropylbenzene at 180 °C [266]. Cold-pressed nanoplates of Sb_2Te_3 showed favourable thermoelectric properties with Seebeck coefficients varying between 145 and 170 µV/K at room temperature [266]. $(Et_2Sb)_2Te$ has also been employed for deposition of rhombohedral

Sb_2Te_3 films on silicon(100) substrate using argon as a carrier gas in a cold-walled MOCVD reactor at 200–220 °C (Figure 11.31) [267]. Higher deposition temperature resulted in Sb-rich films. Earlier attempts to prepare Sb_2Te_3 using H_2 as a carrier gas resulted in hydrogenolysis of the precursor [268]. Phase pure Sb_2Te_3 films comprising of hexagonal nanoplates of 100–200 nm diameters have been deposited using $[Sb\{(TePPr^i)_2N\}_3]$ in the temperature range 375–475 °C by AACVD process [269]. The nanoplates are randomly oriented with a preferred orientation along the (003) plane.

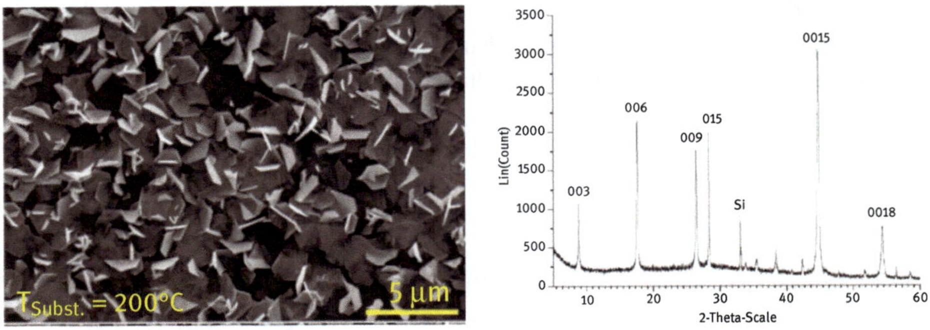

Figure 11.31: SEM image and XRD of Sb_2Te_3 films deposited on Si(100) substrates at substrate temperatures of 200 °C using $(Et_2Sb)_2Te$ (adapted with permission from ref. [266] Copyright (2012) American Chemical Society).

11.6 f-block metal chalcogenolates as precursors for metal chalcogenides

The chemistry of lanthanide [34, 270, 271] and actinide [270, 272] complexes with chalcogen ligands remained under developed for a long time because of the general belief that the linkage between the hard metal centre and soft chalcogen atom would be disfavoured. However, in the last two decades or so, the chemistry of f-block elements with chalcogen ligands is blossomed and reviewed [34, 270–272]. Lanthanide chalcogenides, particularly monochalcogenides (LnE), are of interest due to their dilute magnetic semiconducting properties.

Numerous di- and trivalent lanthanide chalcogenolates have been synthesized and several of them characterized structurally. In order to isolate stable complexes, bulkier organic groups on the metal centre (e. g. $[Sm(TpMe_2)_2ER]$ ($TpMe_2$ = tris(3,5-dimethyl-pyrazolyl)borate; E = O, S, Se, Te [273])/chalcogen atom (e. g. $[Ln(EMes)_2(THF)_n]$; Ln = Sm, Eu, Yb; E = Se or Te [274]) or on both (e. g. $[Yb\{TeSi(SiMe_3)_3\}_2(TMEDA)_2]$; TMEDA = $Me_2NCH_2CH_2NMe_2$ [275]) have been employed. Metathesis of halide ligand on metal centre, chalcogenolysis or oxidative addition reactions of chalcogen/dichalcogenide have generally been used for the synthesis of chalcogenolate complexes.

Metathesis of halide ligand of divalent lanthanide complexes with KEAr in THF yields lanthanide chalcogenolates (eq. (11.15)) [274]. Arnold and co-workers used chalcogenolysis reactions for the synthesis of both lanthanide(II)/(III) chalcogenolates [275, 276]. Thus, the reactions between [Ln{N(SiMe$_3$)$_2$}$_2$.2THF] (Ln = Sm, Eu, Yb) and HESi(SiMe$_3$)$_3$ (E = Se or Te) in diethylether in the presence of TMEDA or DMPE give Ln(II) complexes, [Ln{ESi(SiMe$_3$)$_3$}$_2$(TMEDA)$_2$] and [Ln{ESi(SiMe$_3$)$_3$}$_2$(DMPE)$_2$(μ-DMPE)] (TMEDA = Me$_2$NCH$_2$CH$_2$NMe$_2$; DMPE = Me$_2$PCH$_2$CH$_2$PMe$_2$) (eq. (11.16)) [275], whereas the reactions between [Ln{N(SiMe$_3$)$_2$}$_3$] and HTeSi(SiMe$_3$)$_3$ in hexane yield an unstable tellurolate, [Ln{TeSi(SiMe$_3$)$_3$}$_3$], which converts to a pentanuclear cluster [M$_5$Te$_3${TeSi(SiMe$_3$)$_3$}$_9$] at 20 °C but in the presence of DMPE gives a stable adduct [Ln{TeSi(SiMe$_3$)$_3$}$_3$(DMPE)$_2$] (Ln = La and Ce) [276].

$$\mathrm{LnX_2(THF)}_n + 2\ \mathrm{KEMes} \xrightarrow{\mathrm{THFDME}} [\mathrm{Ln(EMes)_2(THF)}_n] \quad (\mathrm{Ln = Sm,\ Eu,\ Yb;\ E = Se\ or\ Te}) \tag{11.15}$$

$$\mathrm{Ln\{N(SiMe_3)_2\}_2.2THF} + 2\ \mathrm{HESi(SiMe_3)_3} \xrightarrow{\mathrm{TMEDA,\ Et_2O}} [\mathrm{Ln\{ESi(SiMe_3)_3\}_2(TMEDA)_2}] \tag{11.16}$$

Oxidative addition of chalcogen (such as Ph$_3$PSe, Bu$^n{}_3$PTe or Te) on divalent pentamethylcyclopentadienyl lanthanide, [Cp*$_2$Yb(OEt$_2$)], yields chalcogen-bridged paramagnetic binuclear complexes, [(Cp*$_2$Yb)$_2$(μ-E)] (E = Se or Te) [277]. Oxidative addition of diaryldiselenides/ditellurides on Yb(II) benzamidinates, [{RC$_6$H$_4$C(NSiMe$_3$)$_2$}$_2$Yb.2THF] (R = H or OMe) affords Yb(III) chalcogenolates, [{RC$_6$H$_4$C(NSiMe$_3$)$_2$}$_2$Yb(ER′)THF] (E = Se or Te; R′ = Ph or Mes) [278].

The divalent tellurolate complexes, [Yb(TeMes)$_2$(THF)$_n$], either in refluxing toluene or heating (250–300 °C) in vacuum yield YbTe with the elimination of telluroether, TeMes$_2$ [274]. Similarly pyrolysis (150–200 °C) of [Yb{ESi(SiMe$_3$)$_3$}$_2$(TMEDA)$_2$] in vacuum gives YbE (E = Se or Te) [275]. The trivalent tellurolate complexes, [M{TeSi(SiMe$_3$)$_3$}$_3$(DMPE)$_2$] and [M$_5$Te$_3${TeSi(SiMe$_3$)$_3$}$_9$], on pyrolysis (500–600 °C) under vacuum yield M$_2$Te$_3$ (M = La or Ce) with the elimination of telluroether [276].

11.7 Ternary metal chalcogenides

Ternary chalcogenides can be constructed from binary systems either through substitution of cation (e. g. Mo$_{1-x}$W$_x$Se$_2$ or Cu$_2$SnSe$_3$) or anion (e. g. Cu$_2$Se$_x$S$_{1-x}$, MoSe$_{2(1-x)}$S$_{2x}$) retaining tetrahedral coordination around the chalcogen atom as well as charge neutrality around both the cation and anion [279, 280]. Among ternary chalcogenides, I–III–VI$_2$ nanocrystals have attracted considerable attention due to their remarkable photovoltaic and thermoelectric properties [281]. The optical band gaps of their QDs cover a wide wavelength ranging from ultraviolet to near infrared [281, 282]. Although several methods for the synthesis of ternary chalcogenides have been reported [283], the molecular precursor route for their preparation is still underexplored and only a few reports are available. Metal selenolates and selenocarboxylates have been employed for the synthesis of ternary metal chalcogenides.

Selenolate complexes, $[(PPh_3)_2CuIn(SeR)_4]$ (R = Et [284], Ph [285, 286]), have been synthesized by the reactions of $[Cu(PPh_3)_2(MeCN)]^+$ with $[In(SeR)_4]^-$ (eq. (11.17)). A ligand-exchange reaction has also been employed for their synthesis. Thus, the reaction of $[(PPh_3)_2CuIn(SEt)_4]$ with four equivalents of PhSeH in benzene at room temperature affords the corresponding phenylselenolate derivative in 97 % yield [287]. The X-ray structural analysis of $[(PPh_3)_2CuIn(SeEt)_4]$ revealed that both copper and indium are tetrahedrally coordinated with a planar "$CuIn(SeEt)_4$" core [284]. The TGA shows that these complexes undergo a single-step decomposition to $CuInSe_2$ [284]. Thermolysis of $[(PPh_3)_2CuIn(SePh)_4]$ in dioctylphthalate at 300°C yields nanocrystalline tetragonal $CuInSe_2$ [285], whereas coordinating solvents like TOP and oleic acid mixture in the presence of a catalyst give high yield of $CuInSe_2$ nanowires (33 ± 10 nm diameter and several micrometre length) [286].

$$[Cu(PPh_3)_2(MeCN)][PF_6] + M[In(SeR)_4] \rightarrow [(PPh_3)_2Cu(\mu - SeR)_2In(SeR)_2] + MPF_6 (R = Et \text{ or } Ph;\ M = Na \text{ or } K) \quad (11.17)$$

Organogallium/indium chacogenolate/selenide complexes of copper have also been synthesized [288, 289]. Treatment of trialkylphosphine-stabilized copper chalcogenolate with $[Me_2Ga(EPh)]_n$ results in the formation of ternary copper-dimethylgallium chalocgenolate complexes (**XI**) which depending on the nature of the phosphine ligands are isolated as ionic and neutral derivatives [288]. The **XI** reacts with free PhEH to give monomethylgallium complexes (**XII**) (Figure 11.32) [288]. Several of these complexes have been structurally characterized, and metal atoms (both Cu and Ga) in all of them acquire a tetrahedral configuration [288]. Trimethylsilylchalcogenolate

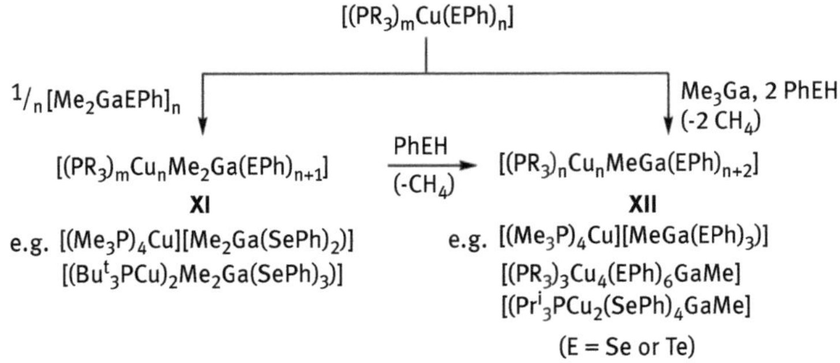

Figure 11.32: Treatment of trialkylphosphine-stabilized copper chalcogenolate with $[Me_2Ga(EPh)]_n$ and Me_3Ga [288].

complexes of copper and dimethylgallium/indium have been used for construction of ternary organometallic derivatives [289]. Thus, the reactions of either $[(Me_2M)_6Se(SeSiMe_3)_4]$ with cuprous acetate in the presence of tertiary phosphine in methanol or *in situ* generated $[(PEt_3)_3CuSeSiMe_3]$ with Me_3M in methanol afford $[(R_3PCu)_4(\mu\text{-}E)_6(MMe)_4]$ (M = Ga or In; E = Se or Te; $PR_3 = PEt_3$, PPr^i_3, PEt_2Pr^i). The structures of these complexes are comprised of a cube of metal atoms (Cu and Ga/In) with all the faces capped by selenide ions (Figure 11.33) [289]. These organometallic derivatives turned out to be poor precursors in the sense they give impure $CuME_2$ on decomposition [288, 289]. For instance, ionic complexes $[(PMe_3)_4Cu][Me_2Ga(SePh)_2]$ and $[(Me_3P)_4Cu][MeGa(EPh)_3]$ (E = Se or Te) undergo a single-step decomposition to give a mixture of $CuGaE_2$ and $Cu_{2-x}E$ [288]. The complexes, $[(R_3PCu)_4(\mu\text{-}Se)_6(MMe)_4]$, decompose below 300°C yielding tetragonal phase of $CuMSe_2$ in addition to small amounts of binary copper selenide or CuM (M = Ga/In) alloys [289].

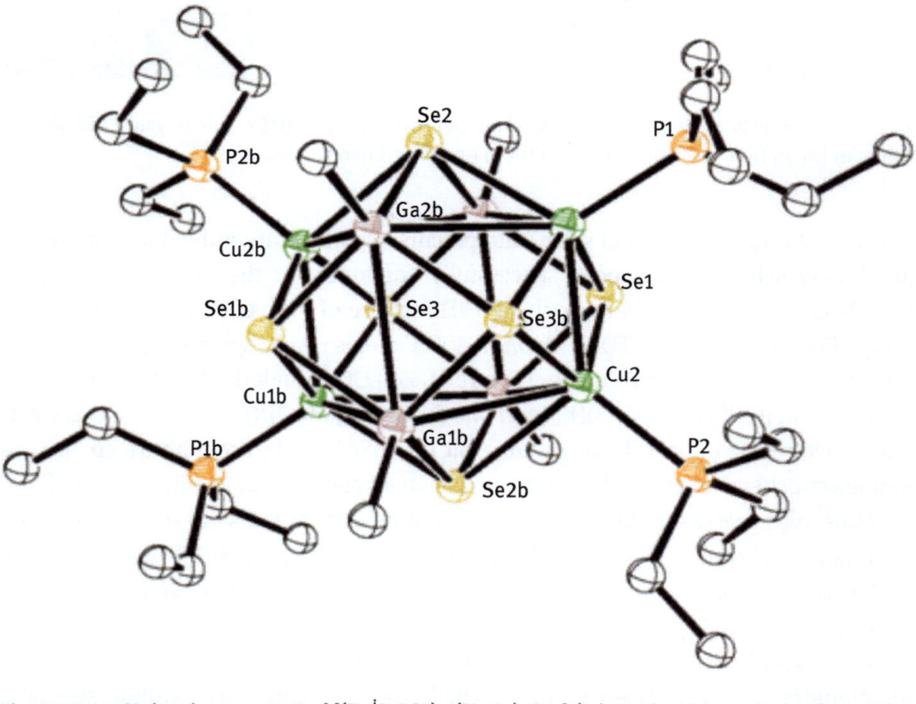

Figure 11.33: Molecular structure of $[(Et_2{}^iPrPCu)_4\text{-}(MeGa)_4Se_6]$ (adapted with permission from ref. [289] Copyright (2014) Wiley VCH).

Selenocarboxylates of the type $[(PPh_3)_2MIn(SeCOAr)_4]$ (M = Cu or Ag; Ar = Ph or *p*-tol) have been used as single-source molecular precursors for $MInSe_2$ [290–292]. The structures of Cu and Ag selenocarboxylate complexes $[(PPh_3)_2MIn(SeCOPh)_4]$ [291] are isoelectronic containing two bridging selenocarboxylate ligands binding through Se atom and forming a four-membered "$MInSe_2$" core, while the indium metal centre is six coordinated (Figure 11.23) [291]. Pyrolysis of these precursors at ~300°C yields tetragonal phase of bulk $MInSe_2$ [290–292]. Pyrolysis of copper complexes in refluxing ethylene glycol (196° C) gives nanorods (17–29 × 10 nm) of $CuInSe_2$ [230], whereas the silver complex (Ar = Ph) in 1:1 mixture of OA and dodecanethiol at 185°C affords nanorods (50.3 ± 5.0 × 14.5 ± 1.8 nm) of metastable orthorhombic phase of $AgInSe_2$ (Figure 11.34) [290].

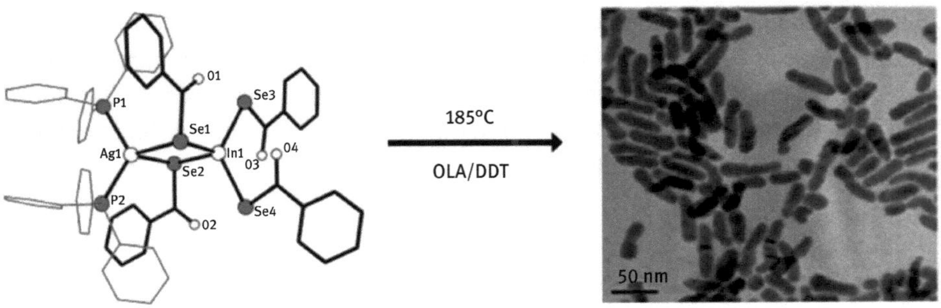

Figure 11.34: Molecular structure of $[(PPh_3)_2AgIn(SeC\{O\}Ph)_4]$ and $AgInSe_2$ nanocrystals (adapted with permission from ref. [10] Copyright (2006) American Chemical Society).

Pyrolysis of a mixture of dual selenolate precursors, which are individually known to give binary selenides, has been successfully employed for the synthesis of ternary derivatives. Accordingly, thermolysis of a 1:1 mixture of $[Cu\{SeC_5H_3(Me\text{-}3)N\}]_4$ and $[In\{SeC_5H_3(Me\text{-}3)N\}_3]$ in HDA, TOPO or OA at 300°C yields nearly spherical particles (400 nm diameter) of tetragonal $CuInSe_2$ (Figure 11.35) [230]. Similarly, thermolysis of a 1:2 molar ratio of $[Cu\{SeC_5H_3(Me\text{-}3)N\}]_4$ and $[Me_2Sn\{SeC_5H_3(Me\text{-}3)N\}_2]$ in OA at 300°C yields nanostructures (11 ± 1 nm) of phase pure cubic Cu_2SnSe_3 which show photo response in a photoelectrochemical cell with power conversion efficiency of 1.7 % (Figure 11.36) [293].

Molecular precursors for anion-substituted ternary systems are scanty. Pyrolysis of a mixture of trinuclear dibenzyltin sulfide and selenide, $[Bz_2SnE]_3$ (E = S or Se), in 3:1, 1:1 and 1:3 molar ratio under nitrogen in a tube furnace at 450°C, yields ternary tin chalcogenides, $SnS_{0.75}Se_{0.25}$, $SnS_{0.5}Se_{0.5}$ and $SnS_{0.25}Se_{0.75}$, respectively, as plate-shaped (~4 μm) nanocrystals [294]. Single-source molecular precursors for ternary tin selenides have also been reported. Oxidation of organotin(II) sulfide/selenide, $(RSn)_2S$ with chalcogen affords $(RSnE)_2S$ (R = 2,6-$(Me_2NCH_2)_2C_6H_3^-$; E = S, Se, Te) (Figure 11.37) [295]. Similarly, oxidation of $(RSn)_2Se$ with tellurium yields $(RSnE)_2Se$ (E = Se, Te) [295, 296]. The structural analyses of these compounds with the general formula, $R_2Sn_2E_2X$ (E = S, Se, Te; X = S or Se), revealed that they contain two terminal Sn = E bonds and are considered as "formal" diseleno-/ditelluro-thio(seleno)stannoic acid anhydrides [295]. These compounds undergo a multi-step thermal decomposition, leading to the formation of tin chalcogenides. The compound, $(RSnSe)_2S$, on

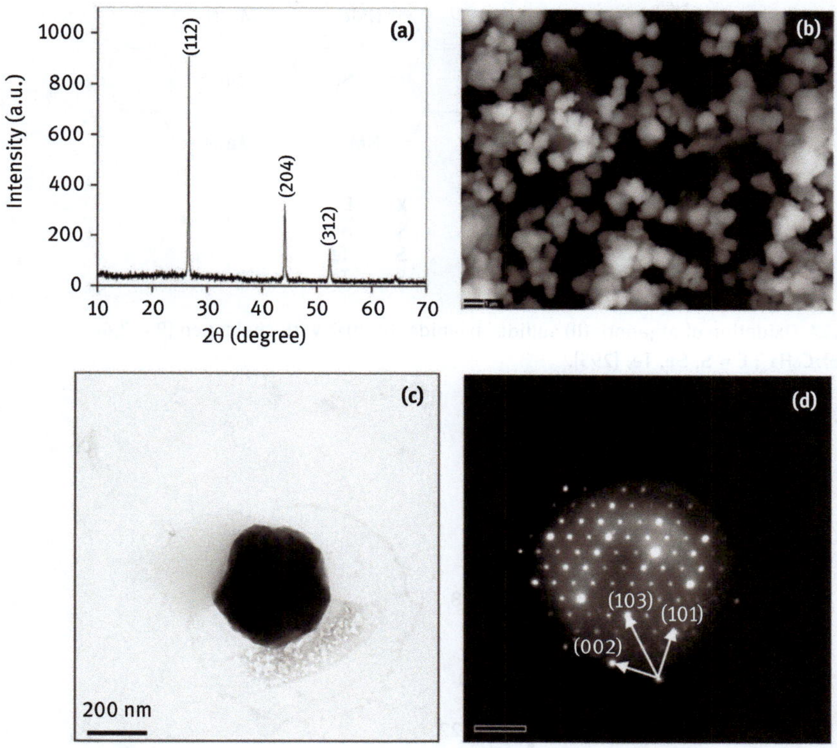

Figure 11.35: (a) XRD pattern, (b) SEM image, (c) TEM image and (d) SAED pattern of $CuInSe_2$ obtained by thermolysis of $[Cu\{SeC_5H_3(Me\text{-}3)N\}]_4$ and $[In\{SeC_5H_3(Me\text{-}3)N\}_3]$ in TOPO at 330°C for 10 min (adapted with permission from ref. [230] Copyright (2013) Elsevier Publishers).

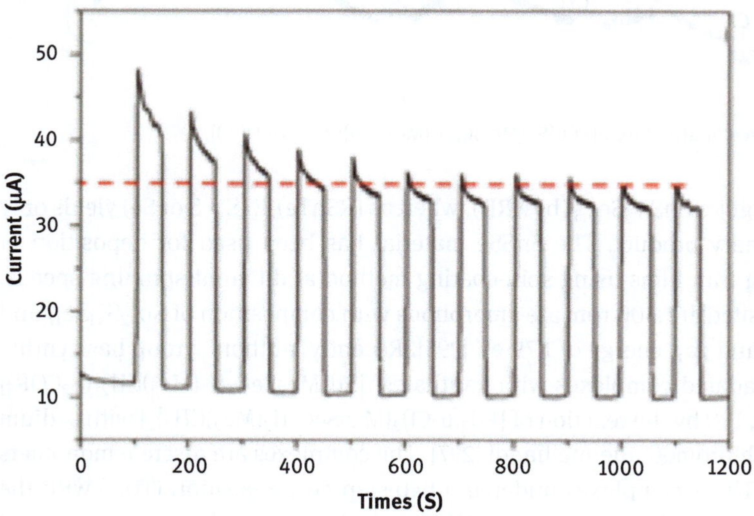

Figure 11.36: Current versus time plot of Cu_2SnSe_3 nanostructures obtained by thermolysis of $[Me_2Sn(2\text{-}SeC_5H_3(Me\text{-}3)N)_2]$ and $[Cu\{SeC_5H_3(Me\text{-}3)N\}]_4$ [293].

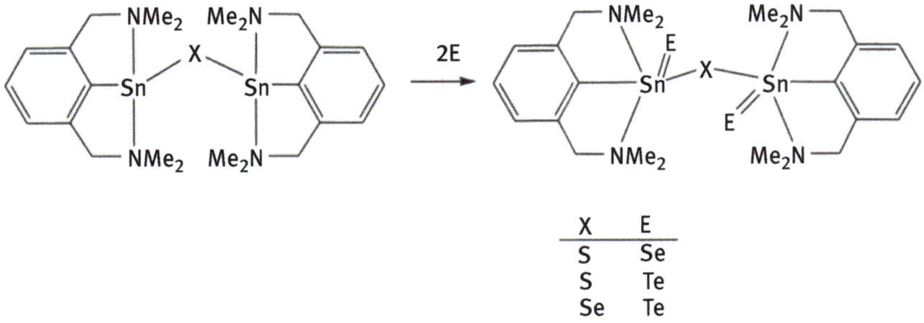

Figure 11.37: Oxidation of organotin(II) sulfide/selenide, $(RSn)_2X$ with chalcogen (R = 2,6-$(Me_2NCH_2)_2C_6H_3^-$; E = S, Se, Te) [295].

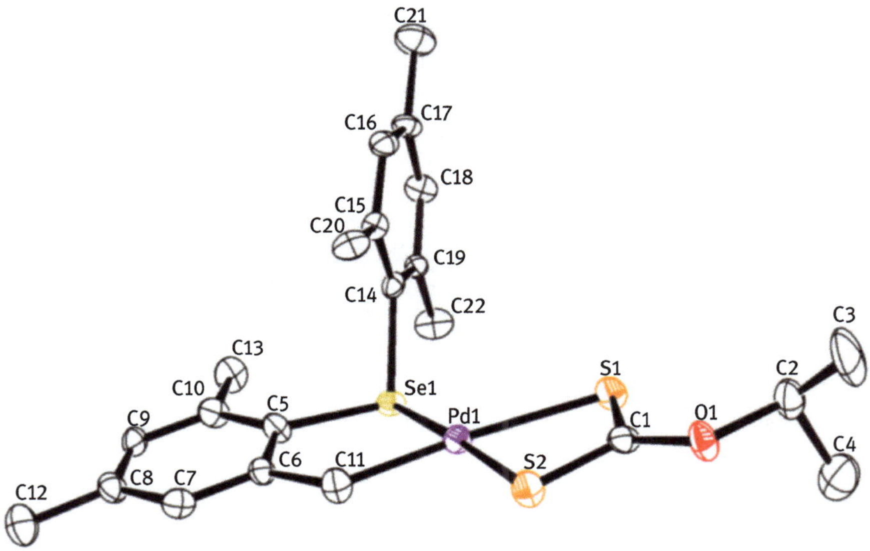

Figure 11.38: Molecular structure of [Pd{Se(Mes)$C_6H_2(Me_2$-4,6)CH_2}(S_2COPr^i)].

decomposition gives $SnS_{0.5}Se_{0.5}$ (by XRD), whereas $(RSnTe)_2E$ (E = S or Se) yields only SnTe as a primary product. The SnSSe material has been used for deposition of semiconducting thin films using spin-coating method at different spinning speeds. The films deposited at 1,500 rpm are amorphous with composition of $Sn_{42}S_{41}Se_{17}$ and show optical band gap energy of 1.79 eV [295]. Recently, authors' group has synthesized cyclopalladated complexes with xanthates, [Pd{MesSe$C_6H_2(Me_2)CH_2$}(S_2COR)] (R = Me, Et, Pr^n, Pr^i) by the reaction of [$Pd_2(\mu$-$Cl)_2${MesSe$C_6H_2(Me_2)CH_2$}$_2$] with sodium xanthate in dichloromethane-methanol [297]. The complexes are discrete monomers (Figure 11.38). These complexes undergo a two-step decomposition (TGA) with the weight loss corresponding to PdSeS. The XRD analysis revealed that there is a partial substitution by S at Se sites in a cubic $Pd_{17}Se_{15}$ structure [297].

Abbreviations

CVD	Chemical vapour deposition
LPCVD	Low-pressure chemical vapour deposition
AACVD	Aerosol assisted chemical vapour deposition
HDA	Hexadecyl amine
TOPO	*n*-Trioctylphoshine oxide
OA	Oleylamine
dppm	bis(diphenylphosphino)methane
dppe	1,2-bis(diphenylphosphino)ethane
dppp	1,3-bis(diphenylphosphino)propane
dppn	1,2-bis(diphenylphosphino)naphthalene
Th	2-thienyl

References

[1] Fritzmann E. Über komplexverbindungen des platins mit organischen seleniden I. Z Anorg Allg Chem. 1912;73:239–55.

[2] Fritzmann E. Complex compounds of platinum with tellurium ethers. J Russ Phys Chem Soc. 1915;47:588–90.

[3] Arnold J. The chemistry of metal complexes with selenolate and tellurolate ligands. Prog Inorg Chem. 1995;43:353–417.

[4] Rao GK, Kumar A, Singh MP, Singh AK. Palladium(II) complex of an organotellurium ligand as a catalyst for Suzuki Miyaura coupling: generation and role of nano-sized Pd_3Te_2. J Organomet Chem. 2014;749:1–6.

[5] Vivekanada KV, Dey S, Wadawale A, Bhuvanesh N, Jain VK. Syntheses of Pd(II)/Pt(II) complexes with non-chelating 4-pyridylselenolate ligand ranging from mononuclear to macrocyclic structures and their utility as catalysts in Suzuki C–C coupling reaction. Dalton Trans. 2013;42:14158–67.

[6] Paluru DK, Dey S, Wadawale A, Maity DK, Bhuvanesh N, Jain VK. Structural variation in [PdX_2 {$RE(CH_2)_nNMe_2$}] (E = Se, Te; X = Cl, OAc) complexes: experimental results, computational analysis, and catalytic activity in suzuki coupling reactions. Eur J Inorg Chem. 2015;2015:397–407.

[7] Jain VK, Priyadarsini KI, editors. Organoselenium compounds in biology and medicine. UK: RSC; 2018.

[8] Jain VK. Synthesis and characterization of single source molecular precursors for the preparation of metal chalcogenides. J Chem Sci. 2006;118:547–52.

[9] Kedarnath G, Jain VK. Pyridyl and pyrimidyl chalcogen (Se and Te) compounds: a family of multi utility molecules. Coord Chem Rev. 2013;257:1409–35.

[10] Vittal JJ, Ng MT. Chemistry of metal thio- and selno-carboxylates: precursors for metal sulfides/selenides materials, thin films and nanocrystals. Acc Chem Res. 2006;39:869–77.

[11] Malik MA, O'Brien P, Otway DJ. Novel approaches to the deposition of selenium containing materials. Phosphorus Sulfur Silicon. 1998;136-138:431–46.

[12] Lobana TS, Wang JC, Liu CW. Recent advances in the coordination chemistry of diselenophosphates and allied ligands. Coord Chem Rev. 2007;251:91–110.

[13] Song X, Bochmann M. Synthesis of phosphinochalcogenoic amidato complexes of divalent transition metals and their thermolysis to metal selenide and telluride phases. J Chem Soc Dalton Trans. 1997;2689–92.

[14] Woollins JD. P-N-S/Se- containing metallacycles. J Chem Soc Dalton Trans. 1996;2893–901.

[15] Ritch JS, Chivers T, Afzaal M, O'Brien P. The single molecular precursor approach to metal telluride thin films: imino-bis(diisopropylphosphine tellurides) as examples. Chem Soc Rev. 2007;36:1622–31.

[16] Iwaoka M. Organoselenium chemistry: synthesis and reactions. Wirth T, editor. GmbH & Co. KGoA: Wiley-VCH Verlag; 2012.

[17] Chauhan RS, Kedarnath G, Wadawale A, Maity DK, Golen JA, Rheingold AL, et al. Cis configured bis phosphine platinum(II) chalcogenolate complexes: structures, NMR and computational studies. J Organomet Chem. 2013;737:40–6.

[18] Dey S, Jain VK, Varghese B. Synthesis and characterization of benzylselenolate complexes of palladium(II) and platinum(II). J Organomet Chem. 2001;623:48–55.

[19] Hannu-Kuure MS, Komulainen J, Oilunkaniemi R, Laitinen RS, Suontamo R, Ahlgrén M. An experimental and theoretical study of the isomerization of mononuclear bis(arylselenolato)bis(triphenylphosphine)platinum complexes $[Pt(SeR)_2(PPh_3)_2]$. J Organomet Chem. 2003;666:111–20.

[20] Kumbhare LB, Jain VK, Phadnis PP, Nethaji M. Palladium(II) and platinum(II) 2-(methoxycarbonyl)ethylselenolates: synthesis, spectroscopy, structures and their conversion into metal selenide. J Organomet Chem. 2007;692:1546–56.

[21] Kumbhare LB, Wadawale AP, Jain VK, Kolay S, Nethaji M. Palladium(II) and platinum(II) complexes of β-functionalized ethyl selenolates: effect of substitution on synthesis, reactivity, spectroscopy, structures and thermal behaviour. J Organomet Chem. 2009;694:3892–901.

[22] Dey S, Jain VK, Chaudhury S, Knoedler A, Lissner F, Kaim W. 2-(Dimethylamino)ethaneselenolates of palladium(II): synthesis, structure, spectroscopy and transformation into palladium selenide. J Chem Soc Dalton Trans. 2001;723–8.

[23] Dey S, Jain VK, Knoedler A, Kaim W, Zalis S. Platinum(II) complexes of 2-(Dimethylamino) ethylselenolate – donor–acceptor inter-ligand interactions as evident from experimental and TD-DFT computational analysis. Eur J Inorg Chem. 2001;2001:2965–73.

[24] Dey S, Jain VK, Knödler A, Klein A, Kaim W, Záliš S. Structural basis for unusually long wavelength charge transfer transitions in complexes $[MCl(ECH_2CH_2NMe_2)(PR_3)]$ (E = Te, Se; M = Pt, Pd): experimental results and TD-DFT calculations. Inorg Chem. 2002;41:2864–70.

[25] Dey S, Jain VK, Varghese B, Schurr T, Niemeyer M, Kaim W, et al. 3-Dimethylaminopropyl chalcogenolate complexes of palladium(II): syntheses and characterization, including the crystal structures of $[Pd(OAc)(ECH_2CH_2CH_2NMe_2)]_2 \cdot H_2O$ (E=S, Se) and $[PdCl(TeCH_2CH_2CH_2NMe_2)]_2$. Inorg Chim Acta. 2006;359:1449–57.

[26] Narayan S, Jain VK, Varghese B. Pyridine-2-selenolate complexes of palladium(II) and platinum(II): crystal structure of $[(Pr^n{}_3P)Cl_2Pd(NC_5H_4Se)PdCl(PPr^n{}_3)]$. J Chem Soc Dalton Trans. 1998;2359–66.

[27] Dey S, Jain VK, Singh J, Trehan V, Bhasin KK, Varghese B. Pyridine- and 3-/6-Methylpyridine-2-tellurolate complexes of Palladium(II) and Platinum(II). Eur J Inorg Chem. 2003;2003:744–50.

[28] Jain VK, Chauhan RS. New vistas in the chemistry of platinum group metals with tellurium ligands. Coord Chem Rev. 2016;306:270–301.

[29] Bochmann M, Webb KJ. Sterically hindered thiolato, selenolato and tellurolato complexes of mercury(II). J Chem Soc Dalton Trans. 1991;2325–9.

[30] Cheng Y, Emge TJ, Brennan JG. Pyridineselenolate complexes of copper and indium: precursors to $CuSe_x$ and In_2Se_3. Inorg Chem. 1996;35:7339–44.

[31] Romero J, Durán ML, Vázquez JAG, Castiñeiras A, Sousa A, Christiaens L, et al. Direct electrochemical synthesis and crystal structure of tris(pyridine-2-selenolato)indium(III). Inorg Chim Acta. 1997;255:307–11.

[32] Rodriguez A, Romero J, Vázquez JA, Durán ML, Pedrares AS, Sousa A, et al. Electrochemical synthesis of copper and silver pyridine-2-selenolato complexes: crystal structure of (6-trimethylsilylpyridine-2-selenolato)silver(I), $[Ag_6(6\text{-}Me_3SipySe)_6]$. Inorg Chim Acta. 1999;284:133–8.

[33] Pedrares AS, Durán-Carril ML, Romero J, Vázquez JA, Sousa A. Synthesis and characterization of copper(I) and silver(I) complexes with heterocyclic bidentate ligands (N, X), X=S, Se. Inorg Chim Acta. 2010;363:1212–21.

[34] Li HX, Zhu YJ, Cheng ML, Ren ZG, Lang JP, Shen Q. Lanthanide chalcogenolate complexes: syntheses, structures and applications in organic chemistry. Coord Chem Rev. 2006;250:2059–92.

[35] Ghavale N, Wadawale A, Dey S, Jain VK. Palladium assisted formation of arylselenoacetone from diaryldiselenides in methanol-acetone. Indian J Chem A. 2009;48:1510–4.

[36] Chakraborty T, Srivastava K, Singh HB, Butcher RJ. Palladium selenolates via oxidative addition of organylselenenyl halides to palladium(0) precursor and via cleavage reaction of diselenides: synthesis, structure and spectroscopic investigation. J Organomet Chem. 2011;696:2782–8.

[37] Oilunkaniemi R, Laitinen RS, Ahlgrén M. The NMR spectroscopic and X-ray crystallographic study of the oxidative addition of bis(2-thienyl) diselenide to zerovalent palladium and platinum centers. J Organomet Chem. 1999;587:200–6.

[38] Ananikov VP, Beletskaya IP, Aleksandrov GG, Eremenko IL. Mechanistic investigation and new catalyst design in palladium- and platinum-catalyzed Se–Se bond addition to alkynes. Organometallics. 2003;22:1414–21.

[39] Ananikov VP, Kabeshov MA, Beletskaya IP, Aleksandrov GG, Eremenko IL. Mechanistic study of palladium catalyzed S–S and Se–Se bonds addition to alkynes. J Organomet Chem. 2003;687:451–61.

[40] Oilunkaniemi R, Laitinen RS, Ahlgrén M. The X-ray crystallographic study of the reaction of bis (2-thienyl)ditelluride with tetrakis(triphenylphosphine)platinum or –palladium. J Organomet Chem. 2000;595:232–40.

[41] Oilunkaniemi R, Laitinen RS, Ahlgrén M. The oxidative addition of diphenyl diselenide and ditelluride to tetrakis(triphenylphosphine)palladium. J Organomet Chem. 2001;623:168–75.

[42] Chauhan RS, Kedarnath G, Wadawale A, Castro AM, Perez RA, Jain VK, et al. Tellurium(0) as a ligand: synthesis and characterization of 2-pyridyltellurolates of platinum(ii) and structures of $[Pt\{2\text{-}Te\text{-}3\text{-}(R)C_5H_3N\}_2Te(PR'_3)]$ (R = H or Me). Inorg Chem. 2010;49:4179–85.

[43] Chauhan RS, Kedarnath G, Wadawale A, Rheingold AL, Castro AM, Perez RA, et al. Reactivity of dipyridyl ditellurides with (Diphosphine)Pt^0 and 2-pyridyltellurolates with (Diphosphine)$PtCl_2$ and isolation of different structural motifs of platinum(II) complexes. Organometallics. 2012;31:1743–50.

[44] Chauhan RS, Kedarnath G, Wadawale A, Slawin AM, Jain VK. Reactivity of 2-chalcogenopyridines with palladium–phosphine complexes: isolation of different complexes depending on the nature of chalcogen atom and phosphine ligand. Dalton Trans. 2013;42:259–69.

[45] Dey S, Vivekananda KV, Wadawale AP, Jain VK, Bhuvanesh N. Reactivity of 4-pyridyltellurolate with Pd(II)/Pt(II) complexes. Chem Select. 2017;2:5073–9.

[46] Beletskaya IP, Ananikov VP. Addition reactions of E-E and E-H bonds to triple bond of alkynes catalyzed by Pd, Pt, and Ni complexes (E=S, Se). Pure Appl Chem. 2007;79:1041–56.

[47] Chauhan RS, Sharma RK, Kedarnath G, Cordes DB, Slawin AM, Jain VK. Reactivity of dipyrimidyldiselenides with $[M(PPh_3)_4]$ and 2-pyrimidylchalcogenolates with $[MCl_2(\text{diphosphine})]$ (M = Pd or Pt). J Organomet Chem. 2012;717:180–6.

[48] Jain VK, Kannan S, Tiekink ER. Mononuclear platinum(II) complexes with organochalcogenides. J Chem Res (M). 1994;501-32.

[49] Morley CP, Webster CA, Di Vaira M. Oxidative addition of $(PhSe)_2$ and $(FcSe)_2$ to zerovalent palladium and platinum trialkylphosphine complexes (Fc = ferrocenyl, $[Fe(\eta^5\text{-}C_5H_5)(\eta^5\text{-}C_5H_4)]$). J Organomet Chem. 2006;691:4244–9.

[50] Karjalainen MM, Wiegand T, Rautiainen JM, Wagner A, Görls H, Weigand W, et al. Competitive Te-Te and C-Te bond cleavage in the oxidative addition of diaryl and dialkyl ditellurides to Pt(0) centers. J Organomet Chem. 2017;836-837:17–25.

[51] Albano VG, Monari M, Orabona I, Panunzi A, Ruffo F. Oxidative additions of E–E bonds (E = Chalcogen) to group 10 metals: "Tunable" cleavage of Se–Se bonds by Pt(0) complexes. J Am Chem Soc. 2001;123:4352–3.

[52] Albano VG, Monari M, Orabona I, Panunzi A, Roviello G, Ruffo F. Synthesis and characterization of trigonal-bipyramidal platinum(II) olefin complexes with chalcogenide ligands in axial positions. X-ray molecular structures of $[Pt(SMe)_2(dmphen)(diphenyl\ fumarate)]$, its cationic dipositive derivative $[Pt(SMe_2)_2(dmphen)(diphenyl\ fumarate)][BF_4]_2$, and free diphenyl fumarate. Organometallics. 2003;22:1223–30.

[53] Albano VG, De Felice V, Monari M, Roviello G, Ruffo F. Oxidative addition of phenylselenyl halides to platinum(0) complexes: characterisation and reactivity of the products [PtX(SePh)(N,N-chelate)(olefin)] (X = Cl, Br, I). Eur J Inorg Chem. 2005;2005:416–22.

[54] Singhal A, Jain VK, Mishra R, Varghese B. Organochalcogenido-bridged dimeric 2-methylallyl-palladium complexes: synthesis, structure and their transformation into palladium chalcogenides. J Mater Chem. 2000;10:1121–4.

[55] Ghavale N, Dey S, Wadawale A, Jain VK. Synthesis and characterization of chalcogenolato-bridged allyl palladium complexes: versatile precursors for palladium chalcogenides. Organometallics. 2008;27:3297–302.

[56] Nayek HP, Niedermeyer H, Dehnen S. Synthesis and properties of the tetranuclear palladium chalcogenolato-acetato complex $[Pd(SePh)(OOCCH_3)]_4$ with Pd–Pd bonding: an experimental and theoretical study. Dalton Trans. 2009;4208–12.

[57] Tyagi A, Kedarnath G, Wadawale A, Shah AY, Jain VK, Vishwanadh B. Diorganotin(IV) 4,6-dimethyl-2-pyrimidyl selenolates: synthesis, structures and their utility as molecular precursors for the preparation of $SnSe_2$ nano-sheets and thin films. RSC Adv. 2016;6:8367–76.

[58] Ananikov VP, Orlov NV, Beletskaya IP. Highly efficient nickel-based heterogeneous catalytic system with nanosized structural organization for selective Se–H bond addition to terminal and internal alkynes. Organometallics. 2007;26:740–50.

[59] Seligson AL, Arnold J. Synthesis, structure, and reactivity of homoleptic tin(II) and lead(II) chalcogenolates and conversion to metal chalcogenides. X-ray crystal structures of $\{Sn[TeSi(SiMe_3)_3]_2\}_2$ and $(PMe_3)Sn[TeSi(SiMe_3)_3]_2$. J Am Chem Soc. 1993;115:8214–20.

[60] Bochmann M, Coleman AP, Webb KJ, Hursthouse MB, Mazid M. Synthesis of sterically hindered tellurophenols and the structure of $[Cd(\mu\text{-}TeC_6H_2M_3)_2]_\infty$. Angew Chem Int Ed Engl. 1991;30:973–5.

[61] Bochmann M, Webb K, Harman M, Hursthouse MB. Synthesis, structure, and gas-phase decomposition of $[Cd(EC_6H_2{}^tBu_3)_2]_2$: first examples of low-coordinate volatile cadmium chalcogenolato complexes. Angew Chem Int Ed Engl. 1990;29:638–9.

[62] Bonasia PJ, Christou V, Arnold J. Alkyl-, silyl-, and germyl-substituted thiolate, selenolate, and tellurolate derivatives and interconversion of silyl species by chalcogen metathesis. J Am Chem Soc. 1993;115:6777–81.

[63] Bonasia PJ, Gindelberger DE, Dabbousi BO, Arnold J. New reagents for the synthesis of compounds containing metal-tellurium bonds: sterically hindered silyltellurolate derivatives and the x-ray crystal structures of $[(THF)_2LiTeSi(SiMe_3)_3]_2$ and $[(12\text{-}crown\text{-}4)_2Li][TeSi(SiMe_3)_3]$. J Am Chem Soc. 1992;114:5209–14.

[64] Mugesh G, Singh HB. Heteroatom-directed aromatic lithiation: a versatile route to the synthesis of organochalcogen (Se, Te) compounds. Acc Chem Res. 2002;35:226–36.

[65] Englich U, Ruhlandt-Senge K. Thiolates, selenolates, and tellurolates of the s-block elements. Coord Chem Rev. 2000;210:135–79.

[66] Cowley AH, Jones RA, Harris PR, Atwood DA, Contreras L, Burek CJ. Tetrameric gallium and aluminum chalcogenides, $[^{t}BuME]_4$ (M = Al, Ga; E = S, Se, Te): a new class of heterocubanes. Angew Chem Int Ed Engl. 1991;30:1143–5.

[67] Coleman AP, Dickson RS, Deacon GB, Fallon GD, Ke M, McGregor K, et al. The synthesis of some alkyltelluride-manganese(I) complexes, and an assessment of their suitability for MOCVD applications. Polyhedron. 1994;13:1277–90.

[68] Piers WE, MacGillivray LR, Zaworotko M. Reversible interconversion of permethylscandocene tellurolates and tellurides. X-ray structure of $[(C_5Me_5)_2Sc]_2(\mu\text{-Te})$. Organometallics. 1993;12:4723–5.

[69] Häkkinen H, Walter M, Grönbeck H. Divide and protect: capping gold nanoclusters with molecular gold–thiolate rings. J Phys Chem B. 2006;110:9927–31.

[70] Cecconi F, Ghilardi CA, Midollini S, Orlandini A, Bencini A. Synthesis of the cobalt–tellurium 'stellated' octahedral clusters $[Co_6(\mu_3\text{-Te})_8(PEt_3)_6]^{n+}$ (n = 1 or 2). Molecular structure of $[Co_6(\mu_3\text{-Te})_8(PEt_3)_6][PF_6]_2$. J Chem Soc Dalton Trans. 1996;3991–4.

[71] Corrigan JF, Fuhr O, Fenske D. Metal chalcogenide clusters on the border between molecules and materials. Adv Mater. 2009;21:1867–71.

[72] Dey S, Jain VK, Klein A, Kaim W. Interconversion of structurally characterised complexes $[PdCl(Se^{\cap}N)]_2$ and $[Pd(Se^{\cap}N)_2]_6$ with the hemi-labile ligand $Se^{\cap}N{=}Me_2NCH_2CH_2CH_2Se^-$. Inorg Chem Commun. 2004;7:601–3.

[73] Tirloni B, Lang ES, de Oliveira GM, Piquini P, Hörner M. Synthesis, crystal structure, and optical characteristics of $[Pd_2Hg_4Cl_6\{Te(DMB)\}_6]\cdot 2DMF$, $[HgClTe(DMB)]_4$, and the ring-forming cluster $[Pd_{12}(TePh)_{24}]\cdot 2DMF$. New J Chem. 2014;38:2394–9.

[74] Bochmann M, Powell AK, Song X. Synthesis and characterisation of manganese(II) chalcogenolato complexes. Crystal and molecular structure of $[\{Mn(\mu\text{-SeC}_6H_2Me_3\text{-}2,4,6)_2\}_\infty]$. J Chem Soc Dalton Trans. 1995;1645–8.

[75] Eichhöfer A, Buth G, Dolci F, Fink K, Mole RA, Wood PT. Homoleptic 1-D iron selenolate complexes-synthesis, structure, magnetic and thermal behaviour of ${}^1_\infty[Fe(SeR)_2]$ (R = Ph, Mes). Dalton Trans. 2011;40:7022–32.

[76] Simon W, Wilk A, Kerbs B, Henkel G. $[Fe_4Te_4(TePh)_4]^{3-}$, the first telluride–tellurolate complex. Angew Chem Int Ed. 1987;26:1009–10.

[77] Barbaro P, Bencini A, Bertini I, Briganti F, Midollini S. The tetranuclear trianion $[Fe_4Te_4(SC_6H_5)_4]^{3-}$ crystal and molecular structure and magnetic properties. J Am Chem Soc. 1990;112:7238–46.

[78] Steigerwald ML, Siegrist T, Stuczynski SM, Kwon YU. Iron telluride $(Et_3P)_4Fe_4Te_4$: an intermediate between molecular reagents and solid state products. J Am Chem Soc. 1992;114:3155–6.

[79] Steigerwald ML, Siegrist T, Gyorgy EM, Hessen B, Kwon YU, Tanzler SM. Effect of diverse ligands on the course of a molecules-to-solids process and properties of its intermediates. Inorg Chem. 1994;33:3389–95.

[80] Fenske D, Ohmer J, Hachgenei J. New Co and Ni clusters with Se and PPh_3 as ligands: $[Co_4(\mu_3\text{-Se})_4(PPh_3)_4]$, $[Co_6(\mu_3\text{-Se})_8(PPh_3)_6]$, $[Co_9(\mu_4\text{-Se})_3(\mu_3\text{-Se})_8(PPh_3)_6]$, and $[Ni_{34}(\mu_5\text{-Se})_2(\mu_4\text{-Se})_{20}(PPh_3)_{10}]$. Angew Chem Int Ed. 1985;24:993–5.

[81] Choi B, Paley DW, Siegrist T, Steigerwald ML, Roy X. Ligand control of manganese telluride molecular cluster core nuclearity. Inorg Chem. 2015;54:8348–55.

[82] Saito T, Yamamoto N, Nagase T, Tsuboi T, Kobayashi K, Yamagata T, et al. Molecular models of the superconducting chevrel phases: syntheses and structures of $[Mo_6X_8(PEt_3)_6]$ and $[PPN][Mo_6X_8(PEt_3)_6]$ (X = S, Se; PPN = $(Ph_3P)_2N$). Inorg Chem. 1990;29:764–70.

[83] Steigerwald ML, Siegrist T, Stuczynski SM. Octatelluridohexakis(triethylphosphine)hexacobalt and a connection between Chevrel clusters and the NiAs structure. Inorg Chem. 1991;30:2256–7.

[84] Hessen B, Siegrist T, Palstra T, Tanzler SM, Steigerwald ML. Hexakis(triethylphosphine)octatelluridohexachromium and a molecule-based synthesis of chromium telluride, Cr_3Te_4. Inorg Chem. 1993;32:5165–9.

[85] Stuczynski SM, Kwon YU, Steigerwald ML. The use of phosphine chalcogenides in the preparation of cobalt chalcogenides. J Organomet Chem. 1993;449:167–72.

[86] Brennan JG, Siegrist T, Stuczynski SM, Steigerwald ML. The transition from molecules to solids: molecular syntheses of $Ni_9Te_6(PEt_3)_8$, $Ni_{20}Te_{18}(PEt_3)_{12}$ and NiTe. J Am Chem Soc. 1989;111:9240–1.

[87] Nomikou Z, Schubert B, Hoffmann R, Steigerwald ML. Relationships between extended structures and molecular clusters of nickel and tellurium. Inorg Chem. 1992;31:2201–9.

[88] Brennan JG, Siegrist T, Stuczynski SM, Steigerwald ML. Cluster intermediates in an organometallic synthesis of palladium telluride PdTe. J Am Chem Soc. 1990;112:9233–6.

[89] Ma AL, Thoden JB, Dahl LF. Synthesis and structural, electrochemical and NMR analysis of the $[\{Pt(PEt_3)_2\}_2Te_2]^n$ series (n = 0, 2+) and the bicapped triplatinum $[\{Pt(PEt_3)_2\}_3(\mu_3\text{-}Te)_2]^{2+}$ dication: Te–Te bond formation in a cyclo-Pt_2Te_2 core upon a chemically reversible two-electron oxidation. J Chem Soc Chem Commun. 1992;1516–8.

[90] Jain VK, Jain L. The chemistry of tri- and high-nuclearity palladium(II) and platinum(II) complexes. Coord Chem Rev. 2010;254:2848–903.

[91] Dehnen S, Eichhöfer A, Fenske D. Chalcogen-bridged copper clusters. Eur J Inorg Chem. 2002;2002:279–317.

[92] Fuhr O, Stefanie Dehnen S, Fenske D. Chalcogenide clusters of copper and silver from silylated chalcogenide sources. Chem Soc Rev. 2013;42:1871–906.

[93] Xie YP, Jin JL, Duan GX, Lu X, Mak TC. High-nuclearity silver(I) chalcogenide clusters: A novel class of supramolecular assembly. Coord Chem Rev. 2017;331:54–72.

[94] Pichugina DA, Kuz'menko NE, Shestakov AF. Ligand-protected gold clusters: the structure, synthesis and applications. Russian Chem Rev. 2015;84:1114–44.

[95] Krautscheid H, Fenske D, Baum G, Semmelmann M. A new copper selenide cluster with PPh_3 ligands: $[Cu_{146}Se_{73}(PPh_3)_{30}]$. Angew Chem Int Ed. 1993;32:1303–5.

[96] Deveson A, Dehnen S, Fenske D. Syntheses and structures of four new copper(I)–selenium clusters: size dependence of the cluster on the reaction conditions. J Chem Soc Dalton Trans. 1997;4491.

[97] Fenske D, Krautscheid H. New copper clusters containing Se and PEt_3 as ligands: $[Cu_{70}Se_{35}(PEt_3)_{22}]$ and $[Cu_{20}Se_{13}(PEt_3)_{12}]$. Angew Chem Int Ed. 1990;29:1452–4.

[98] Fuhr O, Meredith A, Fenske D. New copper selenium clusters with a sulfur functionalised ligand shell. J Chem Soc Dalton Trans. 2002;4091–4.

[99] Eichhöfer A, Fenske D, Scheer P. Investigation of the thermal properties of $[Cu_{20}Se_{13}(PEt_3)_{12}]$. Eur J Inorg Chem. 2004;2004:93–7.

[100] Fu ML, Fenske D, Weinert B, Fuhr O. One-dimensional coordination polymers containing polynuclear (Selenolato)copper complexes linked by bipyridine ligands. Eur J Inorg Chem. 2010;2010:1098–102.

[101] Ohlmann D, Pritzkow H, Grützmacher H, Anthamatten M, Glaser R. A hexanuclear copper arylselenolate: synthesis, structure and proposal for its rearrangement. J Chem Soc Chem Commun. 1995;1011–2.

[102] Cave D, Corrigan JF, Eichhöfer A, Fenske D, Kowalchuk CM, Rösner H, et al. Investigation of the thermal properties of a series of copper selenide cluster molecules. J Cluster Science. 2007;18:157–72.

[103] Behrens S, Bettenhausen M, Eichhöfer A, Fenske D. Synthesis and crystal structure of $[Cd_{10}Se_4(SePh)_{12}(PPh_3)_4]$ and $[Cd_{16}(SePh)_{32}(PPh_3)_2]$. Angew Chem Int Ed. 1997;36:2797–9.

[104] Behrens S, Bettenhausen M, Deveson AC, Eichhöfer A, Fenske D, Lohde A, et al. Synthesis and structure of the nanoclusters $[Hg_{32}Se_{14}(SePh)_{36}]$, $[Cd_{32}Se_{14}(SePh)_{36}(PPh_3)_4]$, $[P(Et)_2(Ph)C_4H_8OSiMe_3]_5$, $[Cd_{18}I_{17}(PSiMe_3)_{12}]$, and $[N(Et)_3C_4H_8OSiMe_3]_5$, $[Cd_{18}I_{17}(PSiMe_3)_{12}]$. Angew Chem Int Ed. 1996;35:2215–8.

[105] Eichhöfer A, Deglmann P. Mercury-chalcogenide clusters: synthesis and structure of $[Hg_{10}Te_4(SePh)_{12}(PPh\textit{n}Pr_2)_4]$, $[Hg_{10}Te_4(TePh)_{12}(PPh\textit{n}Pr_2)_4]$ and $[Hg_{34}Te_{16}(SePh)_{36}(PPh\textit{n}Pr_2)_4]$. Eur J Inorg Chem. 2004;2004:349–55.

[106] Fenske D, Persau C, Dehnen S, Anson CE. Syntheses and crystal structures of the Ag-S cluster compounds $[Ag_{70}S_{20}(SPh)_{28}(dppm)_{10}]$ $(CF_3CO_2)_2$ and $[Ag_{262}S_{100}(S^tBu)_{62}(dppb)_6]$. Angew Chem Int Ed. 2004;43:305–9.

[107] Back DF, de Oliveira GN, Burrow RA, Castellano EE, Abram U, Lang ES. Mercury bis(phenyltellurolate) as a precursor for the synthesis of binary and ternary nanoclusters. Inorg Chem. 2007;46:2356–8.

[108] Lang ES, Burrow RA, Stieler R, Villetti MA. Cadmium bis(phenylselenolate) as a precursor for the synthesis of polymeric Cd(μ-Se)clusters: crystal and molecular structures of $[Cd_4(SePh)_7(PPh_3)X]_n$ (X = Cl, Br). J Organomet Chem. 2009;694:3039–42.

[109] Casagrande GA, Lang ES, de Oliveira GM, Hörner M, Broch F. Dealing with 1,3-bis(4-nitrophenyl) triazene as intermediary ligand in the synthesis of polymeric (μ-Se)Hg-clusters: one-pot synthetic procedures and X-ray structural characterization of $[(PhSe)_7Hg_4BrPy]_n$. Inorg. Chim. Acta. 2007;360:1776–9.

[110] Arnold AP, Canty AJ, Skelton BW, White AH. Mercury(II) selenolates. Crystal structures of polymeric $Hg(SeMe)_2$ and the tetrameric pyridinates $[\{HgCl(py)(SeEt)\}_4]$ and $[\{HgCl(py)_{0.5}(SeBu^t)\}_4]$. J Chem Soc Dalton Trans. 1982;607–13.

[111] DeGroot MW, Rösner H, Corrigan JF. Control of metal-ion composition in the synthesis of ternary II-II′-VI nanoparticles by using a mixed-metal cluster precursor approach. Chem Eur J. 2006;12:1547–54.

[112] DeGroot MW, Taylor NJ, Corrigan JF. Zinc chalcogenolate complexes as capping agents in the synthesis of ternary II–II'–VI nanoclusters: structure and photophysical properties of $[(N,N'\text{-}tmeda)_5Zn_5Cd_{11}Se_{13}(SePh)_6(thf)_2]$. J Am Chem Soc. 2003;125:864–5.

[113] DeGroot MW, Taylor NJ, Corrigan JF. Controlled synthesis of ternary II–II'–VI nanoclusters and the effects of metal ion distribution on their spectral properties. Inorg Chem. 2005;44:5447–58.

[114] Chhowalla M, Liu Z, Zhang H. Two-dimensional transition metal dichalcogenide (TMD) nanosheets. Chem Soc Rev. 2015;44:2584–6.

[115] Voiry D, Mohite A, Chhowalla M. Phase engineering of transition metal dichalcogenides. Chem Soc Rev. 2015;44:2702–12.

[116] Miró P, Ghorbani-Asl M, Heine T. Two dimensional materials beyond MoS_2: noble-transition-metal dichalcogenides. Angew Chem Int Ed. 2014;53:3015–8.

[117] Lv R, Robinso JA, Schaak RE, Sun D, Sun Y, Mallouk TE, et al. Transition metal dichalcogenides and beyond: synthesis, properties, and applications of single- and few-layer nanosheets. Acc Chem Res. 2015;48:56–64.

[118] Duan X, Wang C, Pan A, Yu R, Duan X. Two-dimensional transition metal dichalcogenides as atomically thin semiconductors: opportunities and challenges. Chem Soc Rev. 2015;44:8859–76.

[119] Wang H, Yuan H, Hong SS, Li Y, Cui Y. Physical and chemical tuning of two-dimensional transition metal dichalcogenides. Chem Soc Rev. 2015;44:2664–80.

[120] Jiao WH, Tang ZT, Sun YL, Liu Y, Tao Q, Feng CM, et al. Superconductivity in a Layered $Ta_4Pd_3Te_{16}$ with $PdTe_2$ Chains. J Am Chem Soc. 2014;136:1284–7.

[121] Chhowalla M, Shin HS, Eda G, Li LJ, Loh KP, Zhang H. The chemistry of two-dimensional layered transition metal dichalcogenide nanosheets. Nature Chemistry. 2013;5:263–75.

[122] (a) Dabbousi BO, Bonasia PJ, Arnold J. Tris(trimethylsilyl)silanetellurol: preparation, characterization, and synthetic utility of a remarkably stable tellurol. J Am Chem Soc. 1991;113:3186. (b) Gindelberger DE, Arnold J. Preparation, Characterization, and Kinetic Studies of Group 4 Metallocene Complexes with Triphenylsilanetellurolate Ligands. Organometallics 1994, 13, 4462.

[123] Chirstou V, Wuller SP, Arnold J. Synthesis, structure, and reactivity of Group 4 metallocene-tellurolates. X-ray crystal structures of $Cp_2Zr[TeSi(SiMe_3)_3]_2$, $Cp'_2Ti[TeSi(SiMe_3)_3]_2$, $Cp_2Zr(\eta^2\text{-}COMe)[TeSi(SiMe_3)_3]$, and $Cp_2Ti[TeSi(SiMe_3)_3]PMe_3$. J Am Chem Soc. 1993;115:10545.

[124] Hector AL, Levason W, Reid G, Reid SD, Webster M. Bis(η^5–cyclo–penta–dien–yl)bis–(2,4,6–tri–methyl–phenyl–tellurolato)zirconium(IV). Acta Cryst. 2008;64:667.

[125] Hector AL, Levason W, Reid G, Reid SD, Webster M. Evaluation of group 4 metal bis-cyclopentadienyl complexes with selenolate and tellurolate ligands for CVD of ME_2 films (E = Se or Te). Chem Mater. 2008;20:5100–6.

[126] Kopf H, Block B, Schmidt M. Zur Bildung gemischter Pinakole bei der Photoreduktion des Benzophenons. Z Naturforsch. 1967;22b:1077.

[127] Gautheron B, Tainturier G, Pouly S, Theobald F, Vivier H, Laarif A. Trapping of (aryne)metallocene complexes by elemental selenium. Crystal structure of $(\eta^{5\text{-}t}BuC_5H_4)_2ZrSe_2C_6H_4\text{-o}$. Organometallics. 1984;3:1495–9.

[128] Chirstou V, Arnold J. Synthesis of reactive homoleptictellurolates of zirconium and hafnium and their conversion to terminal tellurides: a model for the first step in a molecule-to-solid transformation. J Am Chem Soc. 1992;114:6240–2.

[129] Gerlach CP, Chirstou V, Arnold J. Synthesis and reactivity of group 4 homolepticselenolates and tellurolates: lewis base induced conversion to terminal and bridging chalcogenides. Inorg Chem. 1996;35:2758–66.

[130] McKarns PJ, Lewkebandara TS, Yap GP, Liable-Sands LM, Rheingold AL, Winter CH. Adducts of titanium tetrachloride with alkylselenium compounds: molecular precursors to titanium diselenide films. Inorg Chem. 1998;37:418–24.

[131] Boscher ND, Carmalt CJ, Parkin IP. Atmospheric pressure CVD of $TiSe_2$ thin films on glass. Chem Vap Deposition. 2006;12:54–8.

[132] Reid SD, Hector AL, Levason W, Reid G, Waller BJ, Webster M. Thio- and seleno-ether complexes with Group 4 tetrahalides and tin tetrachloride: preparation and use in CVD for metal chalcogenide films. Dalton Trans. 2007;4769–77.

[133] Sato M, Yoshida T. Selenium and tellurium derivatives of di-π-cyclopentadienyl-vanadium(IV), -niobium(IV), -molybdenum(IV) and -tungsten(IV). J Organomet Chem. 1975;87:217–22.

[134] Herberhold M, Peukert J, Krüger M, Daschner D, Milius W. Binuclear CpV, Cp*V, and Cp*Ta complexes containing organochalcogenolato bridges, μ-ER (E = sulfur, selenium, tellurium; R = methyl, phenyl, and ferrocenyl). Z Anorg Allg Chem. 2000;626:1289–95.

[135] Galindo A, Gómez M, del Río D, Sánchez F. Synthesis and reactivity of ene-diamido and ene-diolato [(Trimethylsilyl)cyclopentadienyl]niobium(V) complexes and a comparative DFT study of the bonding capabilities of diazabutadiene and butadiene ligands. Eur J Inorg Chem. 2002;2002:1315–25.

[136] Brandl M, Ebner A, Kubicki MM, Mugnier Y, Wachter J, Vigier-Juteau E, et al. Syntheses, properties and structures of $[\{(C_5Me_5)_2Nb\}_2NiTe_4]$ and $[(^tBuC_5H_4Nb)_2Ni_5Te_7(Ph_2PCH_2PPh_2)_2]$: the quest for tetratelluronickelate clusters. Eur J Inorg Chem. 2007;2007:994–1003.

[137] Hector AL, Jura M, Levason W, Reid SD, Reid G. Vanadium selenoether and selenolate complexes, potential single-source precursors for CVD of VSe_2 thin films. New J Chem. 2009;33:641–5.

[138] Benjamin SL, Chang YP, Gurnani C, Hector AL, Huggon M, Levason W, et al. Niobium(V) and tantalum(V) halide chalcogenoether complexes – towards single source CVD precursors for ME_2 thin films. Dalton Trans. 2014;43:16640–8.

[139] Andrä K. Spezielle Organoschwefel- und -selenverbindungen einiger Übergangsmetalle. Z Anorg Allg Chem. 1970;373:209–16.

[140] Chisholm MH, Huffman JC, Parkin IP, Streib WE. Preparation and characterization of $M_2(SeAr')_6$ and mixed ligand $M_2(OR)_2(SeAr')_4$ species (M = Mo, W). Polyhedron. 1990;9:2941–52.

[141] Goh LY, Lim YY, Tay MS, Mak TC, Zhou ZY. Selenolato chromium complexes. Synthesis, thermolytic degradation and crystal structure of $[Cr(\eta\text{-}C_5H_5)(CO)_3(SePh)]$. J Chem Soc Dalton Trans. 1992;1239–42.

[142] Rettenmeier A, Weidenhammer K, Ziegler ML. Strukturuntersuchungen an metallorganischen Chalkogenatokomplexen von Molybdän und Wolfram. Die Molekül- und Kristallstrukturen von Tricarbonyl-1-7-η-cycloheptatrienyl-tris-μ-phenylselenodimolybdän, $\eta^7\text{-}C_7H_7Mo(\mu\text{-}Se(C_6H_5))_3Mo(CO)_3$; Dicarbonyl-1-7-ηcycloheptatrienylphenyltelluromolybdän, $\eta^7\text{-}C_7H_7Mo(CO)_2Te(C_6H_5)$, und Dicarbonyl-1-7-η-cycloheptatrienylphenylselenowolfram, $\eta^7\text{-}C_7H_7W(CO)_2Se(C_6H_5)$. Z Anorg Allg Chem. 1981;473:91–100.

[143] Vogt T, Strähle J. Synthese und struktur von $[Mo(CO)_4(TeC_6H_5)]_2$. Z Naturforsch. 1985;40b:1599.

[144] Bochmann M, Powell AK, Song X. New sterically hindered manganese selenolato complexes. Isolation and structural characterization of a reaction intermediate, $[Mn\{N(SiMe_3)_2\}(\mu SeC_6H_2\text{-}Pr^i_3\text{-}2,4,6)(THF)]_2$. Inorg Chem. 1994;33:400–01.

[145] Tremel W, Krebs W, Greiwe K, Simon W, Stephan HO, Henkel G. $[Mn(SPh)_3Cl]^{2-}$, $[Mn(SPh)_3Br)]^{2-}$, $[Mn(SePh)_4]^{2-}$, $[Mn(TePh)_4]^{2-}$, and $[Co_4(SPh)_6Cl_4]^{2-}$: new mixed halide/thiolate and chalcogenolate complexes of manganese and cobalt. Z Naturforsch. 1992;47b:1580.

[146] McGregor K, Deacon G, Dickson RS, Fallon GD, Rowe RS, West BO. Insertion of tellurium into an alkylpentacarbonylmanganese complex and a convenient new route to manganese telluride. J Chem Soc Chem Commun. 1990;1293–4.

[147] Jaitner P, Wohlgenannt W, Gieren A, Betz H, Hübner T. Die struktur von $[(\eta^5\text{-}Cp)Mo(CO)_2(\mu\text{-}TePh)]_2$, einem doppelt tellurolato-überbrücktendimolybdän complex. J Organomet Chem. 1985;297:1277–90.

[148] Lau P, Hüttner G, Zsolnai L. Reversible Öffnung von Metall-Metall-Bindungen: Quantitative Untersuchung des Ringöffnungsgleichgewichts von $[Cp'(CO)_2Mn]_2SeAryl^+$. Z Naturforsch. 1991;46b:719–28.

[149] Maganas D, Staniland SS, Grigoropoulos A, White F, Parsons S, Robertson N, et al. S tructural, spectroscopic and magnetic properties of $M[R_2P(E)NP(E)R'_2]_2$ complexes, M = Co, Mn, E = S, Se and R, R′ = Ph or ^iPr. Covalency of M–S bonds from experimental data and theoretical calculations. Dalton Trans. 2006;2301–15.

[150] Cotton FA, Dunbar KR. Conversion of an electron-rich triple bond to a double bond by oxidative addition of diphenyldiselenide to $Re_2Cl_4(\mu\text{-}dppm)_2$. Preparation and characterization of $Re_2Cl_4(\mu\text{-}SePh)_2(\mu dpmo)_2$ (dpmo = bis(diphenylphosphino)methane). Inorg Chem. 1987;26:1305–9.

[151] Matthews PD, Akhtar M, Malik MA, Revaprasadu N, O'Brien P. Synthetic routes to iron chalcogenide nanoparticles and thin films. Dalton Trans. 2016;45:18803–12.

[152] Dey S, Jain VK. Platinum group metal chalcogenides. Platinum Metals Rev. 2004;48:16–29.

[153] Fenske D, Fischer A. New selenolato-bridged clusters of iron and nickel; the structures of $[Fe_{12}(SePh)_{24}]$ and $[Na_2(POPh_3)_6][Ni_{20}Se_{12}(SeMe)_{10}]$. Angew Chem Int Ed Engl. 1995;34:307–9.

[154] Akhtar M, Akhtar J, Malik MA, Tuna F, Helliwell M, O'Brien P. Deposition of iron selenide nanocrystals and thin films from tris(*N, N*-diethyl-*N'*-naphthoylselenoureato)iron(III). J Mater Chem. 2012;22:14970–5.

[155] Akhtar M, Malik MA, Raftery J, O'Brien P. Synthesis of iron selenide nanocrystals and thin films from bis(tetraisopropyldiselenoimidodiphosphinato)iron(II) and bis(tetraphenyldiselenoimidodiphosphinato)iron(II) complexes. J Mater Chem A. 2014;2:20612–20.

[156] Steigerwald ML. Selective syntheses of iron monotelluride and iron ditelluride from organometallic precursors. Synthesis and pyrolysis of $[Cp(Et_3P)(CO)Fe]_2(Te)_n$. Chem Mater. 1989;1:52–7.

[157] Chan SLF, Low KH, So GK, Chui SSY, Che CM. Homoleptic ruthenium(III) chalcogenolates: a single precursor to metal chalcogenide nanoparticles catalyst. Chem Commun. 2011;47:8808–10.
[158] Matsuzaka H, Ogino T, Nishio M, Hidai M, Nishibayashi YA, Uemura S. Dinuclear (η^5-C_5Me_5)Ru complexes triply bridged by tellurium or selenium ligands–syntheses and characterisation of (η^5-C_5Me_5)Ru(π_2-RTeTeR)(π_2-TeR)$_2$Ru(η^5-C_5Me_5) and [(η^5-C_5Me_5)Ru(π_2-SeR)$_3$Ru(η^5-C_5Me_5)]Cl (R = Tol, Ph). J Chem Soc, Chem Commun. 1994;223.
[159] Arce AJ, Arrojo P, Sanctis YD, Deeming AJ, West DJ. Addition of diselenides RSe_2R to [$Os_3(CO)_{10}$ $(MeCN)_2$] to give isomers of [Os_3(μ-SeR)$_2$$(CO)_{10}$]. Polyhedron. 1992;11:1013–21.
[160] Gilby LM, Piggott B. The synthesis and X-ray structure of cobalt(II) complexes of iminobis (phosphinechalcogenides), [Co{N(XPR_2)$_2$–X,X′}$_2$] (X = S or Se; R = Ph or Pr^i). Polyhedron. 1999;18:1077–82.
[161] Panneerselvam A, Nguyen CQ, Waters J, Malik MA, O'Brien P, Raftery J, et al. Ligand influence on the formation of P/Se semiconductor materials from metal–organic complexes. Dalton Trans. 2008;4499–506.
[162] Baidya N, Noll BC, Olmstead MM, Mascharak PK. Nickel(II) complexes with the [NiN_xSe_y] chromophore in different coordination geometries: search for a model of the active site of [FeNiSe] hydrogenases. Inorg Chem. 1992;31:2999–3000.
[163] Goldman CM, Olmstead MM, Mascharak PK. Discrete mononuclear and dinuclear nickel(II) complexes of alkane- and areneselenolates: syntheses, structures, and properties of $(Et_4N)_2$ [Ni_2(Se$(CH_2)_3$Se)$_3$], $(Ph_4P)_2$[Ni(SePh)$_4$], and $(Ph_4P)_2$[Ni_2(μ-2,4,6-(Me)$_3C_6H_2$Se)$_2$(2,4,6-(Me)$_3C_6H_2$Se)$_4$]·8CH_3CN. Inorg Chem. 1996;35:2752–7.
[164] Maneeprakorn W, Nguyen CQ, Malik MA, O'Brien P, Raftery J. Synthesis of the nickel selenophosphinates [Ni(Se_2PR_2)$_2$] (R = ^{i}Pr, ^{t}Bu and Ph) and their use as single source precursors for the deposition of nickel phosphide or nickel selenide nanoparticles. Dalton Trans. 2009;2103–8.
[165] Furlani C, Cervone E, Camassei FD. Transition metal N,N-diethyldiselenocarbamates. Inorg Chem. 1968;7:265–8.
[166] Panneerselvam A, Malik MA, Afzaal M, O'Brien P, Helliwell M. The chemical vapor deposition of nickel phosphide or selenide thin films from a single precursor. J Am Chem Soc. 2008;130:2420–1.
[167] Li HH, Zhao S, Gong M, Cui CH, He D, Liang HW, et al. Ultrathin PtPdTe nanowires as superior catalysts for methanol electrooxidation. Angew Chem Int Ed. 2013;52:7472–6.
[168] Dey S, Kumbhare LB, Jain VK, Schurr T, Kaim W, Klein A, et al. Structural variety and multiple isomerism in 1-(Dimethylamino)propyl-2-chalcogenolate and 2-(Dimethylamino)propyl-1-chalcogenolate complexes of palladium(II) and platinum(II): synthesis, spectroscopy and structures. Eur J Inorg Chem. 2004;2004:4510–20.
[169] Kumbhare LB, Jain VK, Varghese B. Palladium(II) and platinum(II) monoselenocarboxylates: synthesis, spectroscopy, structure and their use as molecular precursors for metal selenides. Inorg Chim Acta. 2006;359:409–16.
[170] Pal MK, Jain VK, Kushwah NK, Wadawale A, Glazun SA, Starikova ZA. Bregadze VI.1-Selenolato-2-phenyl-o-carborane complexes of palladium(II) and platinum(II): synthesis, spectroscopy and structures. J Organomet Chem. 2010;695:2629–34.
[171] Kumbhare LB, Wadawale A, Zade SS, Jain VK. Oligomeric allyl-palladium(II) complexes of β-substituted ethylselenolates: syntheses, structures and thermal decomposition. Dalton Trans. 2011;40:7957–66.
[172] Kolay S, Ghavale N, Wadawale A, Das D, Jain VK. Binuclear orthometalated N,N-dimethylbenzylamine complexes of palladium(II): synthesis, structures and thermal behavior. Phosphorus Sulfur Silicon. 2013;188:1449–61.
[173] Ghavale ND, Dey S, Jain VK, Nethaji M. Chalcogenolato-bridged cyclometallated binuclear palladium complexes: synthesis, spectroscopy, structures of [Pd_2(μ-Cl)(μ-SMes)($C_{10}H_6NMe_2$–C,N)$_2$] and [Pd_2(μ-SePh)$_2$($C_{10}H_6NMe_2$–C,N)$_2$]. Inorg Chim Acta. 2008;361:2462–70.

[174] Sharma KN, Joshi H, Sharma AK, Prakash O, Singh AK. Single source precursor routes for synthesis of PdTe nanorods and particles: solvent dependent control of shapes. Chem Commun. 2013;49:9344–6.
[175] Coughlan C, Ibáñez M, Dobrozhan O, Singh A, Cabot A, Ryan KM. Compound copper chalcogenide nanocrystals. Chem Rev. 2017;117:5865–6109.
[176] Khadka CB, Najafabadi BK, Hesari M, Workentin MS, Corrigan JF. Copper chalcogenide clusters stabilized with ferrocene-based diphosphine ligands. Inorg Chem. 2013;52:6798–805.
[177] Xie YP, Jin JL, Duan GX, Lu X, Mak TC. High-nuclearity silver(I) chalcogenide clusters: a novel class of supramolecular assembly. Coord Chem Rev. 2017;331:54–72.
[178] Molter A, Mohr M. Gold complexes containing organoselenium and organotellurium ligands. Coord Chem Rev. 2010;254:19–45.
[179] Kurashige W, Yamaguchi M, Nobusada K, Negishi Y. J Phys Chem Lett. 2012;3:2649–52.
[180] Hossain S, Kurashige W, Wakayama S, Kumar B, Nair LV, YNiihori Y, et al. Ligand exchange reactions in thiolate-protected Au_{25} nanoclusters with selenolates or tellurolates: preferential exchange sites and effects on electronic structure. J Phys Chem C. 2016;120:25861–9.
[181] Sharma RK, Kedarnath G, Jain VK, Wadawale A, Pillai CG, Nalliath M, et al. Copper(I) 2-pyridyl selenolates and tellurolates: synthesis, structures and their utility as molecular precursors for the preparation of copper chalcogenide nanocrystals and thin films. Dalton Trans. 2011;40:9194–202.
[182] Sharma RK, Wadawale A, Kedarnath G, Manna D, Ghanty TK, Vishwanadh B, et al. Synthesis, structures and DFT calculations of 2-(4,6-dimethyl pyrimidyl)selenolate complexes of Cu(I), Ag(I) and Au(I) and their conversion into metal selenide nanocrystals. Dalton Trans. 2014;43:6525–35.
[183] Kole GK, Vivekananda KV, Kumar M, Ganguly R, Dey S, Jain VK. Hemilabile silver(I) complexes containing pyridyl chalcogenolate (S, Se) ligands and their utility as molecular precursors for silver chalcogenides. CrystEngComm. 2015;17:4367–76.
[184] Kole GK, Wadawale AP, Nigam S, Majumder C, Jain VK. Intermolecular Aurophilic versus Intramolecular Au···N secondary interactions in two-coordinate gold(I) selenolate complexes. Chemistry Select. 2016;1:4131–6.
[185] Kowalchuk CM, Schmid G, Meyer-Zaika W, Huang Y, Corrigan JF. Preparation, characterization, and condensation of copper tellurolate clusters in the pores of periodic mesoporous silica MCM-41. Inorg Chem. 2004;43:173–80.
[186] Lu Z, Huang W, Vittal JJ. Synthesis, structural characterization, photoluminescence and thermal properties of $[(Ph_3P)_2Cu(\mu\text{-}SeC\{O\}R)_2Cu(PPh_3)]$. New J Chem. 2002;26:1122–9.
[187] Ng MT, Boothroyd C, Vittal JJ. Shape and size control of Ag_2Se nanocrystals from a single precursor $[(Ph_3P)_3Ag_2(SeC\{O\}Ph)_2]$. Chem Commun. 2005;3820–2.
[188] Bonamico M, Dessy G. Structural studies of metal diselenocarbamates. Crystal and molecular structures of nickel(II), copper(II), and zinc(II) diethyldiselenocarbamates. J Chem Soc A. 1971;264–9.
[189] Kemmler M, Lazell M, O'Brien P, Otway DJ, Park JH, Walsh JR. The growth of thin films of copper chalcogenide films by MOCVD and AACVD using novel single-molecule precursors. J Mater Sci Mater Electron. 2002;13:531–5.
[190] Malik MA, O'Brien P, Revaprasadu N, Novel A. Route for the preparation of CuSe and $CuInSe_2$ nanoparticles. Adv Mater. 1999;11:1441–4.
[191] Dhayal RS, Liao JH, Hou HN, Ervilita R, Liao PK, Liu CW. Copper(I) diselenocarbamate clusters: synthesis, structures and single-source precursors for Cu and Se composite materials. Dalton Trans. 2015;44:5898–908.
[192] Hsu YJ, Hung CM, Lin YF, Liaw BJ, Lobana TS, Lu SY, et al. $[Cu_4\{Se_2P(O^iPr)_2\}_4]$: A novel precursor enabling preparation of nonstoichiometric copper selenide ($Cu_{2-x}Se$) nanowires. Chem Mater. 2006;18:3323–9.

[193] Copsey MC, Panneerselvam A, Afzaal M, Chivers T, O'Brien P. Syntheses, X-ray structures and AACVD studies of group 11 ditelluroimidodiphosphinate complexes. Dalton Trans. 2007;1528–38.
[194] Bochmann M, Webb KJ, Hursthouse MB, Mazid M. Sterically hindered chalcogenolato complexes. Mono- and di-meric thiolates and selenolates of zinc and cadmium; structure of [{Cd $(SeC_6H_2Bu^t{}_3\text{-}2,4,6)_2\}_2$], the first three-co-ordinate cadmium–selenium complex. J Chem Soc Dalton Trans. 1991;2317–2323.
[195] Bochmann M, Bwembya GC, Powell AK, Song X. Zinc(II) arene tellurolato complexes as precursors to zinc telluride. The crystal and molecular structure of $[Zn(TeC_6H_2Me_3\text{-}2,4,6)_2(pyridine)_2]$. Polyhedron 14. 1995;14:3495–500.
[196] Brennan JG, Siegrist T, Carroll PJ, Stuczynski SM, Brus LE, Steigerwald ML. The preparation of large semiconductor clusters via the pyrolysis of a molecular precursor. J Am Chem Soc. 1989;111:4141–3.
[197] Brennan JG, Siegrist T, Carroll PJ, Stuczynski SM, Reynders P, Brus LE, et al. Bulk and nanostructure Group II-VI compounds from molecular organometallic precursors. Chem Mater. 1990;2:403–9.
[198] Cheng Y, Emge TJ, Brennan JG. polymeric $Cd(Se\text{-}2\text{-}NC_5H_4)_2$ and Square Planar $Hg(Se\text{-}2\text{-}NC_5H_4)_2$: volatile CVD Precursors to II-VI Semiconductors. Inorg Chem. 1994;33:3711–4.
[199] Steigerwald ML, Sprinkle CR. Organometallic synthesis of II-VI semiconductors. Formation and decomposition of bis(organotelluro)mercury and bis(organotelluro)cadmium compounds. J Am Chem Soc. 1987;109:7200–01.
[200] Bonasia PJ, Arnold J. Zinc, cadmium, and mercury tellurolates: hydrocarbon solubility and low coordination numbers enforced by sterically encumbered silyltellurolate ligands. Inorg Chem. 1992;31:2508–14.
[201] Jun Y, Koo JE, Cheon J. One-step synthesis of size tuned zinc selenide quantum dots via a temperature controlled molecular precursor approach. Chem Commun. 2000;1243–4.
[202] Jun J, Choi CS, Cheon J. Size and shape controlled ZnTe nanocrystals with quantum confinement effect. Chem Commun. 2001;101–2.
[203] Dey S, Jain VK, Chaudhury S, Knoedler A, Kaim W. Synthesis and structure of bis(2-N,N-dimethylamioethyl-selenolato)zinc and its transformation to ZnSe. Polyhedron. 2003;22:489–91.
[204] Kedarnath G, Dey S, Jain VK, Dey GK, Varghese B. 2-(N,N-Dimethylamino)ethylselenolates of cadmium(II): syntheses, structure of $[Cd_3(OAc)_2(SeCH_2CH_2NMe_2)_4]$ and their use as single source precursors for the preparation of CdSe nanoparticles. Polyhedron. 2006;25:2383–91.
[205] Kedarnath G, Dey S, Jain VK, Dey GK. Synthesis and characterization of metal selenide (ZnSe, CdSe, HgSe) nanoparticles. J Nanosci Nanotechnol. 2006;6:1031–7.
[206] Kedarnath G, Dey S, Jain VK, Dey GK, Kadam RM. Synthesis of undoped and manganese-doped HgTe nanoparticles using $[Hg(TeCH_2CH_2NMe_2)_2]$ as a single source precursor. J Nanosci Nanotechnol. 2008;8:4500–5.
[207] Kedarnath G, Kumbhare LB, Jain VK, Wadawale A, Dey GK, Thinaharan C, et al. Bis(1-methylimidazolyl)diselenide and 1-methylimidazole-2-selenolate complexes of zinc, cadmium, and mercury: synthesis, characterization, and their conversion to metal selenide quantum dot. Bull Chem Soc Jpn. 2008;81:489–94.
[208] Kedarnath G, Jain VK, Wadawale A, Dey GK. Bis(3-methyl-2-pyridyl)ditelluride and pyridyl tellurolate complexes of zinc, cadmium, mercury: synthesis, characterization and their conversion to metal telluride nanoparticles. Dalton Trans. 2009;8378–85.
[209] Sharma RK, Kedarnath G, Wadawale A, Jain VK, Vishwanadh B. Monomeric pyridyl-2-selenolate complexes of cadmium and mercury: synthesis, characterization and their conversion to metal selenide nanoparticles. Inorg Chim Acta. 2011;365:333–9.

[210] Sharma RK, Wadawale A, Kedarnath G, Vishwanadh B, Jain VK. Pyrimidyl-2-selenolates of cadmium and mercury: synthesis, characterization, structures and their conversion to metal selenide nano-particles. Inorg Chim Acta. 2014;411:90–6.

[211] Kedarnath G, Kumbhare LB, Jain VK, Phadnis PP, Nethaji M. Group 12 metal monoselenocarboxylates: synthesis, characterization, structure and their transformation to metal selenide (MSe; M = Zn, Cd, Hg) nanoparticles. Dalton Trans. 2006;2714–8.

[212] Malik MA, O'Brien P. Mixed methyl and ethylzinc complexes with diethylselenocarbamate: novel precursors for zinc selenide. Chem Mater. 1991;3:999–1000.

[213] Malik MA, Motevalli M, Walsh JR, O'Brien P. Neopentyl- or tert-butylzinc complexes with diethylthio- or diethylselenocarbamates: precursors for zinc chalcogens. Organometallics. 1992;11:3136–9.

[214] Hursthouse MB, Malik MA, Motevalli M, O'Brien P. Synthesis and characterization of some mixed alkyl selenocarbamates of zinc and cadmium: novel precursors for II/VI materials. J Mater Chem. 1992;2:949–55.

[215] Abrahams I, Malik M, Motevalli M, O'Brien P. Some complexes of neopentylcadmium species with dithio-and di-selenocarbamates: the synthesis, characterization and single crystal X-ray structure of a mixed neopentyl/diethyldiselenocarbamate of cadmium: $[(CH_3)_3CCH_2CdSe_2CNEt_2]_2$. J Organomet Chem. 1994;465:73–7.

[216] Hursthouse MB, Malik MA, Motevalli M, O'Brien P. The crystal and molecular structure of N,N-diethyldiselenocarbamatocadmium(II): cadmium and zinc diethyldiselenocarbamates as precursors for selenides. Polyhedron. 1992;11:45–8.

[217] Trindade T, O'Brien P, Zhang X. Synthesis of CdS and CdSe nanocrystallites using a novel single-molecule precursors approach. Chem Mater. 1997;9:523–30.

[218] Chunggaze M, McAleese J, O'Brien P, Otway DJ. Deposition of thin films of CdSe or ZnSe by MOCVD using simple air stable precursors. Chem Commun. 1998;833–4.

[219] Ludolph B, Malik MA. Novel single molecule precursor routes for the direct synthesis of highly monodispersed quantum dots of cadmium or zinc sulfide or selenide. Chem Commun. 1849-1850;1998.

[220] Malik MA, Revaprasadu N, O'Brien P. Air-stable single-source precursors for the synthesis of chalcogenide semiconductor nanoparticles. Chem Mater. 2001;13:913–20.

[221] Cupertino D, Birdsall DJ, Slawin AM, Woollins JD. The preparation and coordination chemistry of $^{i}Pr_2P(E)NHP(E')^{i}Pr_2$ (E, E'= Se; E = Se, E' = S; E = S, E' = O; E, E' = O). Inorg Chim Acta. 1999;290:1–7.

[222] Afzaal M, Aucott SM, Crouch D, O'Brien P, Woollins JD, Park JH. Deposition of MSe (M = Cd, Zn) films by LP-MOCVD from novel single-source precursors $M[(SePPh_2)_2N]_2$. Chem Vap Deposit. 2002;8:187–9.

[223] Crouch DJ, O'Brien P, Malik MA, Skabara PJ, Wright SP. A one-step synthesis of cadmium selenide quantum dots from a novel single source precursor. Chem Commun. 2003;1454–5.

[224] Afzaal M, Crouch D, Malik MA, Motevalli M, O'Brien P, Park JH, et al. Deposition of II-VI thin films by LP-MOCVD using novel single-source precursors. Eur J Inorg Chem. 2004;2004:171–7.

[225] Garje SS, Ritch JS, Eisler DJ, Afzaal M, O'Brien P, Chivers T. Chemical vapour deposition of II–VI semiconductor thin films using $M[(TeP^{i}Pr_2)_2N]_2$ (M = Cd, Hg) as single-source precursors. J Mater Chem. 2006;16:966–9.

[226] Ahmad K, Afzaal M, Ritch JS, Chivers T, PO'Brien P. Epitaxial CdTe rods on Au/Si Islands from a molecular compound. J Am Chem Soc. 2010;132:5964–5.

[227] Bochmann M, Bwembya GC, Hursthouse MB, Coles SJ. Synthesis of phosphinochalcogenoic amidato complexes of zinc and cadmium. The crystal and molecular structure of $[Zn\{Bu^t_2P(Se)NPr^i\}_2]$. J Chem Soc Dalton Trans. 1995;2813–7.

[228] Barron AR. MOCVD of group III chalcogenides. Adv Mater Opt Electron. 1995;5:245–58.

[229] Malik MA, Afzaal M, O'Brien P. Precursor chemistry for main group elements in semiconducting materials. Chem Rev. 2012;110:4417–46.

[230] Sharma RK, Kedarnath G, Kushwah N, Pal MK, Wadawale A, Vishwanadh B, et al. Indium(III) (3-methyl-2-pyridyl)selenolate: synthesis, structure and its utility as a single source precursor for the preparation of In_2Se_3 nanocrystals and a dual source precursor with $[Cu\{SeC_5H_3(Me\text{-}3)N\}]_4$ for the preparation of $CuInSe_2$. J Organomet Chem. 2013;747:113–8.

[231] Stoll SL, Bott SG, Barron AR. Selenide and selenolate compounds of indium: a comparative study of In–Se bond-forming reactions. J Chem Soc Dalton Trans. 1997;1315–22.

[232] Gysling HJ, Wernberg AA, Blanton TN. Molecular design of single-source precursors for 3-6 semiconductor films: control of phase and stoichiometry in indium selenide (In_xSe_y) films deposited by a spray MOCVD process using single-source reagents. Chem Mater. 1992;4:900–5.

[233] Cheon J, Arnold J, Yu KM, Bourret ED. Metalorganic chemical vapor deposition of semiconducting III/VI In_2Se_3 thin films from the single-source precursor: $In[SeC(SiMe_3)_3]_3$. Chem Mater. 1995;7:2273–6.

[234] O'Brien P, Otway DJ, Walsh JR. The growth of indium selenide thin films from a novel asymmetric dialkyldiselenocarbamate of indium. Chem Vap Deposition. 1997;3:227–9.

[235] Park JH, Afzaal M, Helliwell M, Malik MA, O'Brien P, Raftery J. Chemical vapor deposition of indium selenide and gallium selenide thin films from mixed alkyl/dialkylselenophosphorylamides. Chem Mater. 2003;15:4205–10.

[236] Gillan EG, Barron AR. Chemical vapor deposition of hexagonal gallium selenide and telluride films from cubane precursors: understanding the envelope of molecular control. Chem Mater. 1997;9:3037–48.

[237] Schulz S, Gillan EG, Ross JL, Rogers LM, Rogers RD, Barron AR. Synthesis of gallium chalcogenide cubanes and their use as CVD precursors for Ga_2E_3 (E = S, Se). Organometallics. 1996;15:4880–3.

[238] Garje SS, Copsey MC, Afzaal M, O'Brien P, Chivers T. Aerosol-assisted chemical vapour deposition of indium telluride thin films from $\{In(\mu\text{-}Te)[N(^iPr_2PTe)_2]\}_3$. J Mater Chem. 2006;16:4542–7.

[239] Sootsman JR, Chung DY, Kanatzidis MG. New and old concepts in thermoelectric materials. Angew Chem Int Ed. 2009;48:8616–39.

[240] Antunez PD, Buckley JJ, Brutchey RL. Tin and germanium monochalcogenide IV–VI semiconductor nanocrystals for use in solar cells. Nanoscale. 2011;3:2399–411.

[241] Shah AY, Kedarnath G, Tyagi A, Betty CA, Jain VK, Vishwanadh B. Germanium xanthates: versatile precursors for photo responsive germanium sulfide nanostructures. Chemistry Select. 2017;2:4598–604.

[242] Davidovich RL, Stavila V, Whitmire KH. Stereochemistry of lead(II) complexes containing sulfur and selenium donor atom ligands. Coord Chem Rev. 2010;254:2193–226.

[243] Kedarnath G, Kumbhare LB, Dey S, Wadawale AP, Jain VK, Dey GK. β-Functionalized ethylchalcogenolate complexes of lead(II): synthesis, structures and their conversion into lead chalcogenide nanoparticles. Polyhedron. 2009;28:2749–53.

[244] Cheng Y, Emge TJ, Brennan JG. Pyridineselenolate complexes of tin and lead: $Sn(2\text{-}SeNC_5H_4)_2$, $Sn(2\text{-}SeNC_5H_4)_4$, $Pb(2\text{-}SeNC_5H_4)_2$, and $Pb(3\text{-}Me_3Si\text{-}2\text{-}SeNC_5H_3)_2$. volatile CVD precursors to group IV–group VI semiconductors. Inorg Chem. 1996;35:342–6.

[245] Hitchcock PB, Lappert MF, Samways BJ, Weinberg EL. Metal (Li, GeII, GeIII, SnII, and PbII) 2,6-dialkylbenzenethiolates; X-ray crystal structures of $Sn(SAr)_2$($Ar = C_6H_2Bu^t_3$-2,4,6) and $[M(SAr)_2]_3$ (M = Sn or Pb, $Ar = C_6H_3Pr^i_2$-2,6). J Chem Soc Chem Commun. 1983;1492–4.

[246] Eichhöfer A, Jiang JJ, Sommer H, Weigend F, Fuhr O, Fenske D, et al. 1-D-Tin(II) phenylchalcogenolato complexes $^{\infty}{}_1[Sn(EPh)_2]$ (E = S, Se, Te) – synthesis, structures, quantum chemical studies and thermal behaviour. Eur J Inorg Chem. 2010;2010:410–8.

[247] Chang WS, Lin YF, Sarkar B, Chang YM, Liu LK, Liu CW. Synthesis and characterization of $[Pb\{Se_2P(O^iPr)_2\}_2]_n$ pseudo polymorphs: polymeric, single source precursor enabling preparation of shape-controlled lead selenide structures. Dalton Trans. 2010;39:2821–30.

[248] García-Montalvo V, Novosad J, Kilian P, Woollins JD, Slawin AM, García PG, et al. Structural diversity of four-co-ordinatemetal(II) compounds with the bidentatebis(diphenylselenophosphoryl)amide ligand,$[M\{N(PPh_2Se)_2\text{-}Se,Se'\}_2]$ (M = Sn, Pb, Zn, Cd or Hg). J Chem Soc Dalton Trans. 1997;1025–30.

[249] Cea-Olivares R, Moya-Cabrera M, García-Montalvo V, Castro-Blanco R, Toscano RA, Hernández-Ortega S. True square planar $[M\{N(SeP^iPr_2)_2\text{-}Se,Se'\}_2]$ [M = Sn, Se] complexes. an extraordinary geometrical arrangement for well known centers [Sn(II), Se(II)]. Dalton Trans. 2005;1017–8.

[250] Ritch JS, Chivers T, Ahmad K, Afzaal M, O'Brien P. Synthesis, structures, and multinuclear NMR spectra of tin(II) and lead(II) complexes of tellurium-containing imidodiphosphinate ligands: preparation of two morphologies of phase-pure PbTe from a single-source precursor. Inorg Chem. 2010;49:1198–205.

[251] Schlecht S, Budde M, Kienle L. Nanocrystalline tin as a preparative tool: synthesis of unprotected nanoparticles of SnTe and SnSe and a new route to $(PhSe)_4Sn$. Inorg Chem. 2002;41:6001–5.

[252] Bahr SR, Boudjouk P, McCarthy GJ. Tin-sulfur and tin-selenium phenylated ring systems as organometallic precursors to tin sulfide and tin selenide. Chem Mater. 1992;4:383–8.

[253] Boudjouk P, Seidler DJ, Bahr SR, McCarthy GJ. Bis(triphenyltin) chalcogenides as convenient precursors to phase-pure binary semiconductors. Chem Mater. 1994;6:2108–12.

[254] Chuprakov IS, Dahmen KH, Schneider IJ, Hagon J. Bis(bis(trimethylsilyl)methyl)tin(IV) chalcogenides as possible precursors for the metal organic chemical vapor deposition of tin(II) selenide and tin(II) telluride films. Chem Mater. 1998;10:3467.

[255] Sharma RK, Kedarnath G, Wadawale A, Betty CA, Vishwanadh B, Jain VK. Diorganotin(IV) 2-pyridyl selenolates: synthesis, structures and their utility as molecular precursors for the preparation of tin selenide nanocrystals and thin films. Dalton Trans. 2012;41:12129–38.

[256] Schulz S. Covalently bonded compounds of heavy group 15/16 elements–synthesis, structure and potential application in material sciences. Coord Rev. 2015;297-298:49–76.

[257] Bochmann M, Song X, Hursthouse MB, Karaulov A. Chalcogenolato complexes of bismuth and antimony. Syntheses, thermolysis reactions, and crystal structure of $Sb(SC_6H_2Pr^i{}_3\text{-}2,4,6)_3$. J Chem Soc Dalton Trans. 1995;1649–52.

[258] Sharma RK, Kedarnath G, Jain VK, Wadawale A, Nalliath M, Pillai CG, et al. 2-pyridyl selenolates of antimony and bismuth: synthesis, characterization, structures and their use as single source molecular precursor for the preparation of metal selenide nanostructures and thin films. Dalton Trans. 2010;39:8779–87.

[259] Sharma RK, Kedarnath G, Vishwanadh B, Jain VK. Antimony and bismuth complexes containing 2-(4,6-dimethylpyrimidyl)selenolate: synthesis, characterization and their conversion to metal selenide nanocrystals. Int J Chem. 2013;2:423–6.

[260] Monteiro OC, Trindade T, Paz FA, Klinowski J, Waters J, O'Brien P. Aerosol-assisted metallo-organic chemical vapour deposition of Bi_2Se_3 films using single-molecule precursors. The crystal structure of bismuth(III) dibutyldiselenocarbamate. J Mater Chem. 2003;13:3006–10.

[261] Crouch DJ, Helliwell M, O'Brien P, Park JH, Waters J, Williams DJ. Imino-bis(diisopropylphosphine chalcogenide) complexes of arsenic, antimony and bismuth. Synthesis, CVD studies and X-ray structure of $M[N(EP^iPr_2)_2]_n$ (E = Se, S; M = As, Sb, Bi). Dalton Trans. 2003;1500–4.

[262] Waters J, Crouch D, Raftery J, O'Brien P. Deposition of bismuth chalcogenide thin films using novel single-source precursors by metal-organic chemical vapor deposition. Chem Mater. 2004;16:3289–98.

[263] Chang HW, Sarkar B, Liu CW. Synthesis of Sb_2Se_3 nanowires via a solvothermal route from the single source precursor $Sb[Se_2P(O^iPr)_2]_3$. Cryst Growth Des. 2007;7:2691–5.

[264] Lin YF, Chang HW, Lu SY, Liu CW. Preparation, characterization, and electrophysical properties of nanostructured $BiPO_4$ and Bi_2Se_3 derived from a structurally characterized, single-source precursor $Bi[Se_2P(O^iPr)_2]_3$. J Phys Chem. 2007;111:18538–44.

[265] Gupta G, Kim J. Facile synthesis of hexagonal Sb_2Te_3 nanoplates using Ph_2SbTeR (R = Et, Ph) single source precursors. Dalton Trans. 2013;42:8209–11.

[266] Schulz S, Heimann S, Friedrich J, Engenhorst M, Schierning G, Assenmacher W. Synthesis of hexagonal Sb_2Te_3 nanoplates by thermal decomposition of the single-source precursor $(Et_2Sb)_2Te$. Chem Mater. 2012;24:2228–34.

[267] Bendt G, Schulz S, Zastrow S, Nielsch K. Single-source precursor-based deposition of Sb_2Te_3 films by MOCVD. Chem Vap Deposition. 2013;19:235–41.

[268] Dickson RS, Heazle KD. The assessment of some Sb-Te single-source compounds for MOCVD applications. J Organomet Chem. 1995;493:189–97.

[269] Garje SS, Eisler DJ, Ritch JS, Afzaal M, O'Brien P, Chivers T. A new route to antimony telluride nanoplates from a single-source precursor. J Am Chem Soc. 2006;128:3120–1.

[270] Nief F. Complexes containing bonds between group 3, lanthanide or actinide metals and non-first-row main group elements (excluding halogens). Coord Chem Rev. 1998;178-180:13–81.

[271] Kumar GA, Riman RE, Brennan JG. NIR emission from molecules and clusters with lanthanide–chalcogen bonds. Coord Chem Rev. 2014;273-274:111–24.

[272] Ephritikhine M. Molecular actinide compounds with soft chalcogen ligands. Coord Chem Rev. 2016;319:35–62.

[273] Hillier AC, Liu SY, Sella A, Elsegood MR. Lanthanide chalcogenolate complexes: synthesis and crystal structures of the isoleptic series $[Sm(Tp^{Me,Me})_2ER]$ (E = O, S, Se, Te; $Tp^{Me,Me}$ = *tris*-3,5-Dimethylpyrazolylborate). Inorg Chem. 2000;39:2635–44.

[274] Strzelecki AR, Timinski PA, Helsel BA, Bianconi PA. Synthesis of lanthanide(II) complexes of aryl chalcogenolate ligands: potential precursors to magnetic semiconductors. J Am Chem Soc. 1992;114:3159–60.

[275] Cary DR, Arnold J. Synthesis and characterization of divalent lanthanide selenolates and tellurolates. X-ray crystal structures of $Yb[SeSi(SiMe_3)_3]_2(TMEDA)_2$ and $\{Eu[TeSi(SiMe_3)_3]_2(DMPE)_2\}_2(\mu\text{-DMPE})$. Inorg Chem. 1994;33:1791–6.

[276] Cary DR, Arnold J. Preparation of lanthanide tellurolates and evidence for the formation of cluster intermediates in their thermal decomposition to bulk metal tellurides. J Am Chem Soc. 1993;115:2520–1.

[277] Berg DJ, Burns CJ, Andersen RA, Zalkin Z. Electron-transfer reactions of divalent ytterbium metallocenes. Synthesis of the series $[(Me_5C_5)_2Yb]_2[\mu\text{-E}]$ (E = oxygen, sulfur, selenium, or tellurium) and crystal structure of $[(Me_5C_5)_2Yb]_2[\mu\text{-Se}]$. Organometallics. 1989;8:1865–70.

[278] Welder M, Recknagel A, Gilje JW, Nottemeyer M, Edelmann FT. Struktur und reaktivität von ytterbiumbenzamidinaten. J Organomet Chem. 1992;426:295–306.

[279] Chen CS, Gong XG, Walsh A, Wei SH. Electronic structure and stability of quaternary chalcogenide semiconductors derived from cation cross-substitution of II-VI and I-III-VI_2 compounds. Phys Rev B. 2009;79:165211–10.

[280] Aldakov DD, Lefrancois A, Reiss P. Ternary and quaternary metal chalcogenide nanocrystals: synthesis, properties and applications. J Mater Chem C. 2013;1:3756–76.

[281] Fan FJ, Klu L, Yu SH. Energetic I–III–VI_2 and I_2–II–IV–VI_4 nanocrystals: synthesis, photovoltaic and thermoelectric applications. Energy Environ Sci. 2014;7:190–208.

[282] Omata T, Nose K, Otsuka-Yao-Matsuo S. Size dependent optical band gap of ternary I-III-VI_2 semiconductor nanocrystals. J Appl Phys. 2009;105:073106.

[283] Coughlan C, Ibanez M, Dobrozhan O, Singh A, Cabot A, Ryan KM. Compound copper chalcogenide nanocrystals. Chem Rev. 2017;117:5865–6109.

[284] Hirpo W, Dhingra S, Sutorik AC, Kanatzidis MG. Synthesis of mixed copper-indium chalcogenolates. Single-source precursors for the photovoltaic materials $CuInQ_2$ (Q = S, Se). J Am Chem Soc. 1993;115:1597–9.

[285] Castro SL, Bailey SG, Raffaelle RP, Banger KK, Hepp AF. Nanocrystalline chalcopyrite materials ($CuInS_2$ and $CuInSe_2$) via low-temperature pyrolysis of molecular single-source precursors. Chem Mater. 2003;15:3142–7.

[286] Wooten AJ, Werder DJ, Williams DJ, Casson JL, Hollingsworth JA. Solution–liquid–solid growth of ternary Cu–In–Se semiconductor nanowires from multiple- and single-source precursors. J Am Chem Soc. 2009;131:16177–88.

[287] Sun C, Westover RD, Margulieux KR, Zakharov LN, Holland AW, Pak JJ. Divergent syntheses of copper–indium bimetallic single-source precursors via thiolate ligand exchange. Inorg Chem. 2010;49:4756–8.

[288] Kluge O, Krautscheid H. Trialkylphosphine-stabilized copper(I) gallium(III) phenylchalcogenolate complexes: crystal structures and generation of ternary semiconductors by thermolysis. Inorg Chem. 2012;51:6655–66.

[289] Kluge O, Biedermann R, Holldorf J, Krautscheid H. Organo-gallium/indium chalcogenide complexes of copper(I): molecular structures and thermal decomposition to ternary semiconductors. Chem Eur J. 2014;20:1318–31.

[290] Ng MT, Boothroyd CB, Vittal JJ. One-pot synthesis of new-phase $AgInSe_2$ nanorods. J Am Chem Soc. 2006;128:7118–9.

[291] Ng MT, Vittal JJ. New Heterobimetallic and polymeric selenocarboxylates derived from $[M(SeC\{O\}Ph)_4]^-$ (M = Ga and In) as molecular precursors for ternary selenides. Inorg Chem. 2006;45:10147–54.

[292] Ghoshal S, Kumbhare LB, Jain VK, Dey GK. A facile synthesis of $MInSe_2$ (M = Cu, Ag) via low temperature pyrolysis of single source molecular precursors, $[(R_3P)_2MIn(SeCOAr)_4]$. Bull Mater Sci. 2007;30:173–8.

[293] Tyagi A, Shah AY, Kedarnath G, Wadawale A, Singh V, Tyagi D, et al. Synthesis, characterization and photovoltaic properties of colloidal phase pure Cu_2SnSe_3 nanostructures using molecular precursors. Mater. Sci.: Mater. Electron. 2018;29:8937–46.

[294] Boudjouk P, Seidler DJ, Grier D, McCarthy GJ. Benzyl-substituted tin chalcogenides. efficient single-source precursors for tin sulfide, tin selenide, and $Sn(S_xSe_{1-x})$ solid solutions. Chem Mater. 1196;8:1189–96.

[295] Bouska M, Strizik L, Dostal L, Bouška M, Střižík L, Dostál L, et al. Mixed organotin(IV) chalcogenides: from molecules to Sn-S-Se semiconducting thin films deposited by spin-coating. Chem Eur J. 2013;19:1877–81.

[296] Bouška M, Dostál L, de Proft F, Růžička A, Lyčka A, Jambor A. Intramolecularly coordinated tin(II) selenide and triseleneoxostannonic acid anhydride. Chem Eur j. 2011;17:455–9.

[297] Kolay S, Wadawale A, Jain VK Unpublished Results.

Formula Index of Selenium and Tellurium Reactants

$(1.5,9\text{-}TeCl_2)(CH_2)_9$ 47
$(2\text{-}Me_2NCH_2C_6H_4)_2Te$ 48
$(2\text{-}Me_2NCH_2C_6H_4)_2TeF_2$ 48
$(2\text{-}Me_2NCH_2C_6H_4)TeF_3$ 44
$(2\text{-}O_2NC_6H_4)SeOH$ 77
$(2,4,6\text{-}^iPr_3C_6H_2)_2TeO$ 107
$(2,4,6\text{-}^iPr_3C_6H_2)_2TeO_2$ 107
$(3\text{-}FC_6H_4)_2Se_2$ 64
$(3\text{-}Me\text{-}4\text{-}HOC_6H_4)_2TeO$ 105
$(4\text{-}HOC_6H_4)_2TeO$ 105
$(4\text{-}MeOC_6H_4)(4\text{-}Me_2NC_6H_4)TeO$ 105
$(4\text{-}MeOC_6H_4)_2Se$ 78
$(4\text{-}MeOC_6H_4)_2Te$ 103, 105, 108
$(4\text{-}MeOC_6H_4)_2Te(OAc)_2$ 103
$(4\text{-}MeOC_6H_4)_2Te(OTs)_2$ 103
$(4\text{-}MeOC_6H_4)_2TeO$ 99, 100, 103, 107, 108
$(4\text{-}MeOC_6H_4)_2TeO_2$ 107, 108
$(AdN)_2Se$ 136, 138
$(ArNCH)_2CSe$ 225
$(BzSnSe)_3$ 424
$(C_{10}H_6O)_2P(Se)Cl$ 161, 163
$(C_3H_6O)_2Te$ 98
$(C_6H_5)_2SeF_2$ 28
$(CH_2)_3Se(O)O$ 73
$(DippNCH)_2CSe$ 230
$(Et_2NH_2)[R_2PSe_2]$ 169
$(Et_2Sb)_2Te$ 419
$(Et_3HN)[(Me)_2C(CH_2O)_2P(S)Se]$ 164
$(EtNCH_2)_2CTe$ 236
$(HO)_2C_4H_6Se(O)$ 83
$(Me_2N)_2SeO$ 139
$(Me_2N)_3PTe$ 165
$(Me_3N)_2[Cd(SePh)_4]$ 346
$(Me_3Si)_2Se$ 333, 334, 335, 337, 345, 346, 347, 348, 349, 350, 351, 352, 368, 385, 391, 393
$(Me_3Si)_2SeS$ 335
$(Me_3Si)_2Te$ 155, 333, 334, 337, 345, 347, 348, 349, 352, 367, 385, 391, 393
$(Me_3Si)_3CTeSiMe_3$ 341
$(Me_3Si)_3SiTeSiMe_3$ 341
$(Me_3SiN)_2Se$ 30, 32, 126, 132, 133, 135, 136
$(Me_3SiNSN)_2Se$ 134
$(Me_3Sn)_2Se$ 154
$(Me_3Sn)_2Te$ 154
$(Me_4N)_4[Ge_4Se_{10}]$ 356
$(MeCNiPr)_2CSe$ 221
$(MeN)_2Te$ 140
$(MeNCH)_2CSe$ 222, 223, 226, 227, 228, 229, 232, 233, 235
$(Mes^*)_2Te_2$ 44
$(Mes^*)TeCl_3$ 46
$(NC_5H_3R\text{-}3)_2Te_2$ 387
$(NsN)_2Se$ 135
(**P**)$C_6H_4Te(O)(4\text{-}MeOC_6H_4)$ 101
(**P**)$C_6H_4Te(O)OH$ 97
$(Ph_3Sn)_2Se$ 415
$(Ph_3Sn)_2Te$ 415
$(PhP)_2Se_4$ (*Woollins' reagent*) 173, 174, 175, 176, 177, 178, 179, 180, 181, 182, 183, 184, 185, 307, 308
$(PhP)_4(MeN)_2Se_2$ 155
$(PhSe)[SbCl_6]$ 49
$(RN)_2Se$ 126, 136, 137
$(RNH)_2CSe$ 215
$(RSe)_2InX$ 319
$(RSn)_2Se$ 424
$(RSn)_4Se_6$ 368
$(RSnCl_2)_2Te$ 367
$(RSnCl_2)_2Te_2$ 367
$(RSnSe)_2S$ 424
$(RSnSe)_2Se$ 424
$(RSnTe)_2Se$ 424, 426
$(Se_3N_2)^{2-}$ 28
$(^tBuN)_2Se$ 134, 135, 136, 137
$(^tBuN)_2Te$ 136
$(tmtu)TeCl_2$ 43
$(TsN)_2Se$ 135

$[(2\text{-naphthyl})Te(O)]_2O$ 93
$[(3\text{-}Me\text{-}4\text{-}HOC_6H_4)Te(O)]_2$ 93
$[(4\text{-}MeOC_6H_4)Te(O)]_2O$ 93, 94, 98
$[(4\text{-}ROC_6H_4)Te(O)]_2O$ 93
$[(CH_2CH_2Ph)_2PSe_2]^-$ 168
$[(CuPR_3)_4(\mu\text{-}Se)_6(MMe)_4]$ 423
$[(Dipp_2BIAN)TeI_2]$ 43
$[(EtMe_2C)InSe]_4$ 412
$[(^iPr_3P)_3(CuSePh)_4]$ 342
$[(^iPr_3P)_4(CuSePh)_6]$ 342
$[(^iPr_3PCu)_2SeC_2H_4Se)]_2$ 354
$[(Me_2M)_6Se(SeSiMe_3)_4]$ 352
$[(Me_3Si)_2N]_2Se$ 132, 133
$[(Me_3Si)_2N]_2Te$ 140
$[(Me_3Si)^tBuN]_2Se$ 133
$[(PhSe)_3SnCH_2CH_2]_2$ 388

https://doi.org/10.1515/9783110529340-012

$[(SeCl))_2N][SbCl_6]$ 128
$[(SeCl_2)_2N][AsF_6]$ 128
$[(^tBu_2PhSiTe)_2Pb]_2$ 341
$[(^tBu_2PhSiTe)_2Sn]_2$ 341
$[(TiCp^*{}_2)_2Te$ 165
$[\{(ArTeO)_2O\}_2(Ar_2TeO_2)_2]$ 108
$[\{(Me_2Si)_2CH\}_2Sn(\mu\text{-}Se)]_2$ 415
$[\{FeCp(CO)(PEt_3)\}_2(\mu\text{-}Te_2)]$ 401
$[1\text{-}(8\text{-}Me_2NC_{10}H_6)]_2TeO$ 101, 105, 107
$[1\text{-}(8\text{-}Me_2NC_{10}H_6)]_2TeO_2$ 108
$[2\text{-}((Me_2NCH_2)C_6H_4](PhCH_2)SeO$ 81
$[2\text{-}(TfO)C_6H_4]Se(O)OH$ 71
$[2\text{-}\{(S)\text{-}Me(MeS)CH\}C_6H_4]_2Se_2$ 67
$[2\text{-}\{(S)\text{-}Me(MeS)CH\}C_6H_4]SeOSO_3H$ 67
$[2\text{-}\{(S)\text{-}Me(MeS)CH\}C_6H_4]SeOTf$ 66
$[Ag\{(TePR_2)_2N\}]_6$ 407
$[Ag_{180}Se_{54}(SeCH_2CH_2O\{O\}CFe(C_5H_5)(C_5H_4)_{72}$ $(PPh_3)_{14}]$ 350
$[Ag_2\{SeC(O)Ph\}_2(PPh_3)_3]$ 407
$[ArTe(O)]_2O$ 92, 93, 94
$[Au\{TeP^iPr_2)_2N\}(PPh_3)]$ 407
$[Bi(Se_2CNRR')_3]$ 418
$[Bi\{(SeP^iPr_2)_2N\}_3]$ 419
$[Bi\{(SePPh_2)_2N\}_3]$ 419
$[Bi\{Se_2P(O^iPr)_2\}_3]$ 419
$[C_6H_3Me_2\text{-}2,6]_2Te_2$ 221
$[Cd(S_2CNMe_nHex)_2]$ 171
$[Cd(Se_2CNEt_2)_2]$ 411
$[Cd(Se_2P^iPr_2)_2]$ 171
$[Cd(SeC_6H_2{}^tBu_3\text{-}2,4,6)_2]_2$ 408
$[Cd(SeCH_2CH_2NMe_2)_2]$ 408
$[Cd(SeCOtol)_2(tmeda)]$ 409
$[Cd(SePh)_2]_n$ 408
$[Cd\{(SeP^iPr_2)_2N\}_2]$ 411
$[Cl_2Te(\mu\text{-}N^tBu)_2TeCl_2]_2$ 130
$[ClSe_2SN_2]^+$ 131
$[Co(Se_2CNEt_2)_2]$ 401
$[Co(Se_2P^iPr_2)_2]$ 401
$[Co(Se_2PR_2)_2]$ 171
$[Co\{(SeP^iPr_2)_2N\}_2]$ 401
$[Co\{^tBu_2P(Se)NR\}_2]$ 401
$[Co\{^tBu_2P(Te)NR\}_2]$ 401
$[Cp^*Ga(\mu_3\text{-}Se)]_4$ 413
$[Cr_6Te_8(PEt_3)_6]$ 399
$[Cu(PEt_3)_3SeSiMe_3]$ 423
$[Cu(P^iPr)_2TeSiMe_3]$ 352
$[Cu(P^iPr_3)_2SeSiMe_3]$ 352
$[Cu(PMe)_4][GaMe(SePh)_3]$ 423
$[Cu(PMe_3)_4][GaMe(TePh)_3]$ 423
$[Cu(PMe_3)_4][GaMe_2(SePh)_2]$ 423
$[Cu(PPh_3)_2In(SePh)_4]$ 422
$[Cu(Se_2CNEt_2)_2]$ 407
$[Cu(Se_2CNMeR)_2]$ 407
$[Cu(SeC_5H_3(R\text{-}3)N]_4$ 404
$[Cu(Sepy)]_4$ 404
$[Cu\{Se(SiMe_3)_2\}]$ 394
$[Cu\{Se_2P(O^iPr)_2\}]_4$ 407
$[Cu\{SeC_4H(Me_2\text{-}4,6)N_2\}]_6$ 405
$[Cu\{SeC_5H_3(Me\text{-}3)N\}]_4$ 405, 424
$[Cu\{TeP^iPr_2)_2N\}]_3$ 407
$[Cu_4(\mu_4\text{-}H)(\mu_3\text{-}Cu)_4(Se_2CNR_2)_6][PF_6]$ 407
$[Cu_6(TePh)_6(PEtPh_2)_5]$ 405
$[Et_2NH_2][(CH_2CH_2Ph)_2PSe_2]$ 168
$[Et_2Sn(SeAr)_2]$ 416
$[Eu(Se_2PPh_2)_2]$ 171
$[Fe(C_5H_5)(C_5H_4)C(O)CH_2CH_2SeSiMe_3]$ 350
$[Fe(C_5H_5)(C_5H_4)C(O)SeSiMe_3]$ 344
$[Fe(C_5H_5)(C_5H_4)SnSe_6]$ 370
$[Fe\{(SeP^iPr_2)_2N\}_2]$ 401
$[Fe\{(SePR_2)_2N\}_2]$ 401
$[Fe\{napC(O)NC(Se)NEt_2\}_3]$ 400
$[Fe\{^tBu_2P(Se)NR\}_2]$ 401
$[Fe\{^tBu_2P(TeNR\}_2]$ 401
$[Fe_4(Te)_4(PEt_3)_4]$ 391
$[FeCp(CO)(PEt_3)\}_2(\mu\text{-}Te)]$ 401
$[Ge_4Te_{10}]^{4-}$ 357
$[Hg(SeCH_2CH_2CH_2NMe_2)_2]$ 408
$[Hg(SePh)_2]$ 396
$[Hg(SeR)_2]$ 396
$[Hg(TeCH_2CH_2NMe_2)_2]$ 408
$[Hg(Te^nBu)_2]$ 408
$[Hg(TePh)_2]$ 396
$[Hg\{(TeP^iPr_2)_2N\}_2]$ 411
$[In\{SeC_5H_3(Me\text{-}3)N\}_3]$ 412
$[In(SePh)_3]$ 412, 413
$[In\{Se_2CNMeR\}_3]$ 412
$[In\{SeC(SiMe_3)_3\}_3]$ 412
$[In\{SeC_5H_3(Me\text{-}3)N\}_3]$ 424
$[^iPr_2P(Te)NHP(Te)^iPr_2]$ 188
$[K([2.2.2]crypt)]_2[Pb_2Se_3]$ 365
$[K(18\text{-}crown\text{-}6)]_2[Pb_2Se_3]$ 364
$[K_4(OH_2)_3][Ge_4Se_{10}]$ 360
$[KSe_2PPh_2)]_2$ 168, 169
$[Li(en)]_2[Pb_2Te_3]$ 365
$[Li(TMEDA)]_2[^tBuN(Te)P(\mu\text{-}N^tBu)_2P(Te)N^tBu]$ 193, 195
$[M(L)_n(TeR)_m]$ 341
$[M(L_n)(SeSiMe_3)]_m$ 351, 352

$[M(L_n)(TeSiMe_3)]_m$ 351, 352
$[M(PPh_3)_2In(SeCOAr)_4]$ 423
$[M(PPh_3)_2In(SeCOPh)_4]$ 423
$[M(Se(CH_2)_nNMe_2)_2]$ 408
$[M(Se_2CNMe^nHex)_2]$ 411
$[M(Se_2CNR_2)_2]$ 411
$[M(SeAr)_3]$ 417
$[M(SeC_6H_2{}^tBu_3\text{-}2,4,6)_3]$ 417
$[M(Sepy)_2]$ 408
$[M(SeSiMe_3)(PR_3)_n]$ 355
$[M(TeAr)_2]$ 408
$[M(TeSiMe_3)(PR_3)_n]$ 355
$[M(\mu\text{-}Te)\{(TeP^iPr)_2N\}]_3$ 413
$[M\{(SePR_2)_2N\}_2]$ 414
$[M\{(TePR_2)_2N\}_2]$ 414
$[M\{SeC_4H(Me_2\text{-}4,6)_2N_2\}_2]$ 417
$[M\{SeC_5H_3(Me\text{-}3)N\}_3]$ 417, 418
$[M\{TeC_5H_3(R\text{-}3)N\}_2]$ 408
$[M\{TeSi(SiMe_3)_2\}]_2$ 408
$[M\{TeSi(SiMe_3)_3\}_3(DMPE)_2]$ 421
$[M_5Te_3\{TeSi(SiMe_3)_3\}_9]$ 421
$[MCp_2(Te^tBu)_2]$ 398
$[Me_2Ga(SePh)]_n$ 422
$[Me_2Ga(TePh)]_n$ 422
$[Me_2Ga\{(SeP^iPr_2)_2N\}]$ 413
$[Me_2NC(O)]_2Te_2$ 309
$[Me_2Sn\{SeC_5H_3(Me\text{-}3)N\}_2]$ 424
$[MeN_2C_2H_2C(Se)]_2(CH_2)$ 227, 230, 231, 234
$[MeN_2C_2H_2C(Se)]_2(CH_2)_n$ 219, 231, 234
$[Mn(SePh)_2]_n$ 346, 347
$[Mn(TeBz)(CO)_3(PEt_3)_2]$ 400
$[Mn(\mu\text{-}TeR)(CO)_4]_2$ 399, 400
$[Mn\{^tBu_2P(Te)NR\}_2]$ 400
$[N(P^tBu_2Se)_2]^-$ 189
$[N(P^tBu_2Te)_2]^-$ 189
$[Na(TMEDA)_2]_2[Te_2(\mu\text{-}N^tBuP)_2(N^tBu)_2]$ 193, 194
$[NbCl_5(Se^nBu_2)]$ 398
$[^nBuN(CH)_2NMe]CSe$ 233
$[Ni(Se_2CNEt_2)_2]$ 402
$[Ni(Se_2PR_2)_2]$ 171, 402
$[Ni\{(SeP^iPr_2)_2N\}_2]$ 402
$[^nPrNH_2][Se_2CN^nPr_2]$ 407
$[Pb(SeCH_2CH_2NMe_2)_2]$ 414
$[Pb(SeR)_2]$ 414
$[Pb\{(TeP^iPr_2)_2N\}_2]$ 414
$[Pb\{Se_2P(O^iPr)_2\}_2]$ 414
$[Pd(OAc)(SeCH_2CH_2CH_2NMe_2)]_2$ 402
$[Pd(SeBz)_2]_n$ 402
$[Pd(SeCb^oPh)_2(PEt_3)_2]$ 402
$[Pd(SeCOPh)_2(P^nPr_3)_2]$ 402
$[Pd(\mu\text{-}SeAr)_2(C\cap N)_2]$ 402
$[Pd(\mu\text{-}SeAr)_2(\eta^3\text{-}C_4H_7)_2]$ 402
$[Pd(\mu\text{-}SeCH_2CH_2COOEt)(\eta^3\text{-}C_3H_5)]_n$ 402
$[Pd_2(\mu\text{-}Cl)_2\{MesSeC_6H_2(Me_2)CH_2\}_2]$ 426
$[Pd_2(\mu\text{-}SePh)_2(Me_2NCH_2C_6H_4\text{-}C,N)_2]$ 403
$[Pd_2(\mu\text{-}TeMes)_2(C_{10}H_6N\text{-}C,N)_2]$ 403
$[Pd_2Cl_2(\mu\text{-}SeCH_2CH_2X)_2(P^nPr_3)_2]$ 402
$[PdCl(SeCH_2CH_2NMe_2)(PPh_3)]$ 402
$[PdCl(SeCH_2CH_2NMe_2)]_3$ 402
$[PdCl\{4\text{-}MeOC_6H_4TeCH_2CH_2CH_2(pzBr\text{-}4)\}]$ 403
$[PdCl\{4\text{-}MeOC_6H_4TeCH_2CH_2CH_2NHCH(C_6H_4OH\text{-}2)(C_6H_4)]$ 403
$[PdCl\{SeCH_2CH(Me)NMe_2\}(PMePh_2)]$ 402
$[PdCl\{TeC_5H_3(Me\text{-}3)N\}(PR_3)]$ 403
$[Ph_2P(Se)NHP(Se)Ph_2]$ 188
$[Ph_4As][TeCN]$ 164
$[PhTe(O)]_2O$ 91, 92, 94
$[PhTe(O)X]$ 94, 95
$[Pt(Se_2N_2)(PR_3)_2]$ 127
$[Pt_2Cl_2(\mu\text{-}SeCH_2CH_2COOMe)_2(P^nPr_3)_2]$ 402
$[pyrenyl(CH_2)_n](2\text{-}NO_2C_6H_4)Se$ 77
$[pyrenyl(CH_2)_n](2\text{-}NO_2C_6H_4)SeO$ 77
$[R_2In\{(SeP^iPr_2)_2N\}]$ 413
$[R_2Sn(SeAr)_2]$ 416
$[R_2Sn(Sepy)_2]$ 416
$[R_2SnCl(SeAr)]$ 416
$[RGa(\mu_3\text{-}Se)]_4$ 413
$[RGa(\mu_3\text{-}Te)]_4$ 413
$[RhCl(C_8H_{12})(OSePh_2)]$ 80
$[RM(Se_2CNR_2)]_2$ 411
$[RN(Se)P(\mu\text{-}NR)_2P(Se)NR]^{2-}$ 190
$[RN(Te)P(\mu\text{-}NR)_2P(Te)NR]^{2-}$ 190
$[RR'SeOH]^+$ 81
$[RSn(Te)_2]_2Te$ 367
$[Ru(SeC_6H_4R\text{-}4)_3]_n$ 401
$[RZn(Se_2CNEt_2)]_2$ 411
$[S(Se)P(\mu\text{-}N^tBu)]_2{}^{2-}$ 192
$[Sb\{(TeP^iPr)_2N\}_3]$ 420
$[Sb\{Se_2P(O^iPr)_2\}_3]$ 419
$[Se_4][AsF_6]$ 124
$[SeCl_3][AsF_6]$ 128
$[Sn(SePh)_4]$ 415
$[Sn(Sepy)_4]$ 415
$[Sn(SeR)_2]$ 414
$[^tBu_2Sn(Sepy)_2]$ 416
$[^tBu_2Sn\{SeC_4H(Me_2\text{-}4,6)N_2\}_2]$ 416
$[^tBuGaSe]_4$ 412
$[^tBuNTe(\mu\text{-}N^tBu)_2TeN^tBu]]_2$ 39

$[Te(NMe_2)_2]$ 140
$[TiCl_4(SeR_2)_2]$ 398
$[TiCp_2(\mu\text{-}Se_2)_2TiCp_2]$ 4, 5
$[TiCp_2Se_5]$ 4, 5, 8
$[TiCp_2Se_xS_{5-x}]$ 5
$[Yb(TeMes)_2(thf)_n]$ 421
$[Yb\{SeSi(SiMe_3)_3\}_2(TMEDA)_2]$ 421
$[Yb\{TeSi(SiMe_3)_3\}_2(TMEDA)_2]$ 421
$[Zn(lut)_2(SeSiMe_3)_2]$ 396
$[Zn(lut)_2(TeSiMe_3)_2]$ 396
$[Zn(N,N'\text{-tmeda})(Sn^tBu_2)_2Se_3]$ 354
$[Zn(N,N'\text{-tmeda})(SSeSiMe_3)_2]$ 354
$[Zn(SePh)_2(tmeda)]$ 408
$[Zn(SePh)_2]_\infty$ 408
$[Zn(TeMes)_2(py)_2]$ 408
$[Zn(TePh)_2(tmeda)]$ 408
$[Zn(TePh)_2]_\infty$ 408

$\{(Me_3Si)_3C\}_2Te_2$ 50
$\{[(MeN)(CH)_2C(Se)N]_2\text{-}1,3\text{-}C_6H_4\}$ 232
$\{[(MesN)(CH)_2C(Se)NCH_2]_2\text{-}2,6\}_2C_6H_3Br$ 233
$\{[BMIm]_2[Ge_4Se_9]\}_n$ 361
$\{A_2[MSn_4Se_{10}]\}_n$ 358

${}_\infty^1[Fe(SeAr)_2]$ 400
${}_\infty^1[Fe(SePh)_2]$ 400

$1,5\text{-}Ph_4P_2N_4Se_2$ 141
$1,5\text{-}Ph_4P_2N_4Se_2R_2$ 142
$1,5\text{-}Te_2(CH_2)_6$ 47
$2\text{-}CH_2COOH\text{-}4\text{-}Ar[2,4\text{-}N_2(CH)_2]CSe$ 222, 223
$2\text{-}Me\text{-}4\text{-}R\text{-}(2,4\text{-}N_2C_2H_2)SeR$ 214
$2\text{-}Me_2NCH_2C_6H_4SeCl$ 50
$2,4,6\text{-}{}^iPr_3C_6H_2TeI$ 50
$4\text{-}FC_6H_4Se(O)OH$ 69
$4\text{-}MeC_6H_4Se(O)OH$ 69

$A(SeSiMe_3)$ 352
$A(TeSiMe_3)$ 352
$A_2[Pb_2Se_3]$ 364, 365
$A_2[Pb_2Te_3]$ 364
$A_4[SnSe_4]$ 357, 363
$A_4[SnTe_4]$ 357
$A_6[Sn_2Se_6]$ 364
$A_6[Sn_2Te_6]$ 364
Al_2Se_3 302
Al_2Te_3 301
$Ar(CH_2SeSiMe_3)_n$ 343
$Ar(OH)CH(CH_2)_4SePh$ 91
Ar_2Se 79
Ar_2Se_2 49, 64, 65, 69, 259, 260, 264, 270, 272, 318
Ar_2SeO 79
Ar_2Te 108, 109, 288
Ar_2Te_2 44, 48, 49, 97, 98
Ar_2TeCl_2 48
Ar_2TeO 99, 100, 101, 103, 105, 108
ArNSe 134
ArSe(O)OOH 69
ArSe(O)OSe(O)Ar 69
ArSeCC(H)SeAr 78
ArSeCl 49
$ArSeN_3$ 49
ArSeOC(O)R 67
ArSeOH 64
ArSeOTf 65
ArSeX 77, 283
ArSeZnSeAr 318, 320
$ArSiMe_3$ 49
ArTe(O)OH 92, 97
ArTe(O)OOH 97
Arte(O)X 92, 93
$ArTeCl_3$ 46, 47
$ArTe^nBu$ 285, 286, 287
$ArTeO_2$ 105
ArTeZnTeAr 320
ArX_2TeAr 51

Bu_2Te_2 387
$BuN(CH)_2N(C_5H_4N)CSe$ 224, 225

$C_2H_3Se(O)_2Ph$ 85, 86, 87, 89, 90
$C_6H_4(BnN)(NCH_2SePh)C(Se)$ 234, 235
$C_6H_4(N^nBu)_2CSeI_2$ 219, 220
$C_6H_4(NR)_2CSe$ 224
$C_6H_4(N^tBu)_2CSeX_2$ 221
$C_6H_4[SeN(Ph)C(O)\text{-}1,2]$ 141
$C_6H_7(O)SeAr$ 76
CdSe 171
$CH_2(TeCl_3)_2$ 44
cis-$[Pt(SeAr)_2(PPh_3)_3]$ 386
$Cl_2Te(NSO)_2$ 131, 132
$Cl_2Te(\mu\text{-}Cl)(\mu\text{-}Cl)(\mu\text{-}NSN)TeCl_2$ 131
$ClSe[N(^tBu)Se]_nCl$ 31, 129, 130
CSe_2 419
Cu_3Se_2 404

$Et_2P(Se)Cl$ 162, 168
Et_2Te_2 44

Et_3PTe 165, 393, 399, 403
Et_3PTeCl_2 165

GaSeX 36

H_2Se 301, 302, 303, 366
H_2Te 301, 302, 366
$HSe(2,4,6\text{-}{}^iPr_3C_6H_2)$ 399
$HSeSi(SiMe_3)_3$ 421
$HTeSi(SiMe_3)_3$ 399, 421

$In(SeR)_4^-$ 422
${}^iPr_3PTe$ 165, 399

$K_2[Te(NH)_3]$ 127
K_2Se 13, 369
K_2Te 13, 369
$K_4(SnSe_4)$ 363
$K_4[Ge_2Se_6]$ 358
$K_4[Ge_4Se_{10}]$ 361, 362
KSeAr 421
KSeCN 156, 157, 163
KSeCOAR 409
KTeAr 421

Li(TeR) 385
$Li[HC(PPh_2Se)_2]$ 43
$Li[SeSIMe_3]$ 344, 352
$Li[SSeSiMe_3]$ 335
Li_2Se_2 173
LIALHSeH 305
LiTeR 419
$LTeX_2$ 43

$M(Se_2PR_2)$ 170
$M(SePh)_2$ 396
$M(TePh)_2$ 396
$M[Se_2P(OR)_2]$ 170
$M[{}^tBu_2P(Te)NR]_2$ 186
$Me_2NC(O)TeH$ 309
Me_2Se 398
Me_2Se_2 44
Me_3PTe 165
$Me_3SiSe(C_6H_4)SMe$ 346
$Me_3SiSeAr$ 401
Me_3SiSeR 388, 391, 393
$Me_3SiTeAr$ 401
Me_3SiTeR 388, 391, 393
MePhTeO 105
Mes_2Te 50, 103
Mes_2TeO 100
MesTeI 50
MeTeC(O)Ph 92
MeTeCHMePh 92

$Na(Se_2{}^tBuPh)$ 172
Na_2Se 154, 168, 170, 171, 174, 212, 367
Na_2Se_2 210, 212, 213, 385
Na_2Te 14, 156, 304
Na_2Te_2 153, 168, 210, 212, 385
$NaSe(CH_2)_nNMe_2$ 408
$NaSeC_5H_3(Me\text{-}3)N$ 412
NaSeCOR 407
NaSeH 162, 168, 215, 303, 304, 308, 385
NaSeMes 386
NaSeR 386
NaTeH 14, 303, 304, 308, 309, 385
NaTeMes 386
NaTeR 386
Na_xSe_y 168
${}^nBu_2Se_2$ 265
${}^nBu_2Te_2$ 388
${}^nBu_3PTe$ 165, 421
$NH_4[Se_2PPh_2]$ 168
${}^nPrSeI$ 220

P_2Se_5 152, 153
P_4Se_3 152
$Ph_2P(Se)Cl$ 159, 160
$Ph_2SbTePh$ 419
Ph_2SbTeR 419
Ph_2Se_2 44, 48, 64, 67, 68, 244, 246, 247, 252, 253, 258, 259, 261, 264, 271, 273, 275, 276, 277, 279, 280, 281, 317, 319, 320, 322, 388
Ph_2SeF_2 47
Ph_2SeO 80
Ph_2SeO_2 80
Ph_2Te 103
Ph_2Te_2 48, 51, 91, 97, 320, 322, 387
Ph_3PSe 421
Ph_3PTe 164
Ph_4Se 47
Ph_5TeLi 38
$PhPSe_2$ 173
PhSe(O)Cl 71
PhSe(O)OH 69, 70
PhSe(O)OOH 70, 75
PhSeBr 48, 49, 246, 324

$PhSeCl$ 48, 49, 141, 142, 246, 268, 269, 274, 324
$PhSeH$ 305, 306, 307, 316, 322, 389, 422
$PhSeI$ 246, 247, 249
$PhSeOC(O)Ph$ 67
$PhSeOH$ 64, 69, 75
$PhSeOSO_3H$ 67, 68
$PhSeOTf$ 65, 67, 92
$PhSeR$ 64
$PhSeSiMe_3$ 337, 338, 339, 342, 345, 349, 395, 396
$PhSeZnBr$ 324, 325
$PhSeZnCl$ (*Santi's reagent*) 324, 325, 326
$PhSO_2SeCl$ 31, 134
$PhTe(O)O(O)CCF_3$ 95, 96
$PhTe(O)OAc$ 94
$PhTe(O)OH$ 97
$PhTe(O)OTf$ 95
$PhTeH$ 306
$PhTeN^iPr_2$ 140
$PhTeOAc$ 91
$PhTeOTf$ 91
$PhTeSiMe_3$ 306, 338, 342, 395, 396
$PhTeTe(Ph)Br_2$ 92
$POCH_2C_4H_5Se[O_2C(Me)_2]$ 82
Py_2Se_2 74
$PySeOH$ 74

$R_2P(Se)Cl$ 168
$R_2P(Se)H$ 167
R_2PTe^- 172
$R_2PTe_2^-$ 172
R_2Se_2 68, 248, 249, 250, 254, 255, 256, 262, 263, 266, 267, 268, 269, 271, 316, 317, 318, 321, 387, 400
R_2SeO 98
R_2Te 47
R_2Te_2 51, 155, 387
R_2TeI_2 48
R_2TeO 48, 98
R_3PTe 390, 393, 399
$RC(O)SeH$ 308
$RC_2H_2Se(O)_2Ph$ 88, 89
R_fNSeCl_2 130
$RHN(Se)P(\mu\text{-}N^tBu)_2P(Se)NHR$ 190, 192
$RMeC_3HTeN$ 309
$RNCSe$ 215
$RR'(C_3HNTe(O)$ 102
$RR'CHCH_2SePy$ 74
$RR'GeSe_4$ 9
$RR'Se(OH)(OOH)$ 80
$RR'SeO$ 81
$RR'Te(OH)_2$ 105
$RR'Te(X)OH$ 107
RSe^- 316, 385
$RSeCl$ 273
$RSeH$ 347, 389
$RSeSiMe_3$ 345, 348, 389, 391, 393
$RSeX$ 317
$RSeX_3$ 44
$RSeZnSeR$ 317, 318, 319
$R^tBuP(Se)Cl$ 159
$RTeCl_2Bu$ 284, 285
$RTeH$ 306, 347, 389
$RTeNa$ 306
$RTeSiMe_3$ 345, 348, 389, 391, 393
$RTeX_3$ 44

Se 2, 3, 4, 5, 6, 8, 9, 10, 11, 13, 16, 18, 19, 31, 140, 154, 158, 159, 161, 163, 166, 167, 168, 173, 174, 185, 186, 187, 209, 210, 211, 212, 213, 214, 316, 343, 397
$Se(NSO)_2$ 131, 132
$Se(N^tBu)_3^{2-}$ 138
Se_2Br_2 34, 35, 36
Se_2Cl_2 4, 5, 34, 35, 36, 37, 125, 126, 132, 133, 193
Se_2NCl_3 129
Se_2NCl_5 129
Se_4N_4 124, 125, 127
$SeBr_2$ 31, 32, 33, 34
$SeBr_4$ 29, 31, 124, 129
$SeCl_2$ 4, 31, 32, 33, 34, 36, 37, 131, 134, 136, 265
$SeCl_4$ 28, 29, 31, 35, 36, 37, 124, 127, 131, 133, 135
Se_n^{2-} 4, 13
$SeOCl_2$ 29, 30, 36, 132, 139
$SePPh_2(CH_2)_nPh_2PSe$ 50

$TbtB(SeMe)_2$ 10, 11
$^tBu_2PhSiTeSiMe_3$ 341
$^tBu_2Se$ 398
$^tBu_3PTe$ 165
$^tBuHN(Se)P(\mu\text{-}N^tBu)_2P(Se)NH^tBu]$ 190, 191
$^tBuNTe(\mu\text{-}N^tBu)_2TeN^tBu$ 137, 138
$^tBuPhP(Se)Cl$ 171
$^tBuPhP(Se)SeR$ 162
$^tBuPhP(Se)TeR$ 162
Te 6, 9, 12, 13, 14, 15, 16, 17, 18, 19, 43, 47, 153, 165, 172, 173, 185, 186, 187, 209, 210, 343, 421

$Te(bipy)Cl_2$ 42
$Te(NSO)_2$ 131
$Te(N^tBu)_3^{2-}$ 138, 139
$Te(PEt_3)I_2$ 43
$Te(PEt_3)X_2$ 43
$Te(S_2CNEt_2)_2$ 156
$Te(SCF_3)_2$ 131
$Te(Si^nBu_3)$ 395, 396
Te_2X_2 43
$TeBr_2$ 41, 42, 131
$TeBr_4$ 14, 36, 38, 40, 41, 43, 131
$TeCl_2$ 41, 42, 131
$TeCl_2$.TMTU 193
$TeCl_4$ 14, 15, 36, 37, 38, 39, 40, 41, 43, 45, 131, 135
TeF_4 40
TeF_6 37
TeI_2 42
TeI_2.($SePPh_2CH_2Ph_2PSe$) 39
TeI_4 39
Te_n^{2-} 13
Th_2Te_2 387
TPPSe 157

$Zn(SePh)_2$ 322, 323
$Zn(SePh)_2$-(TMDEA) 322, 323

Bei Fragen zur Produktsicherheit wenden Sie sich bitte an:
If you have any questions regarding product safety,
please contact:

Walter de Gruyter GmbH
Genthiner Straße 13
10785 Berlin
productsafety@degruyterbrill.com